# DIOPHANTINE APPROXIMATION

Festschrift for Wolfgang Schmidt

# Developments in Mathematics

VOLUME 16

*Series Editor:*

Krishnaswami Alladi, University of Florida, U.S.A.

*Aims and Scope*

*Developments in Mathematics* is a book series publishing

(i) Proceedings of conferences dealing with the latest research advances,

(ii) Research monographs, and

(iii) Contributed volumes focusing on certain areas of special interest.

Editors of conference proceedings are urged to include a few survey papers for wider appeal. Research monographs, which could be used as texts or references for graduate level courses, would also be suitable for the series. Contributed volumes are those where various authors either write papers or chapters in an organized volume devoted to a topic of special/current interest or importance. A contributed volume could deal with a classical topic that is once again in the limelight owing to new developments.

# DIOPHANTINE APPROXIMATION

## Festschrift for Wolfgang Schmidt

Edited by

HANS PETER SCHLICKEWEI
Philipps-Universität Marburg, Marburg, Germany

KLAUS SCHMIDT
Universität Wien, Vienna, Austria

ROBERT F. TICHY
Technische Universität Graz, Graz, Austria

SpringerWienNewYork

2000 Mathematics Subject Classification: 11D, 11J, 11K


Printed in Germany

SpringerWienNewYork is part of Springer Science+Business Media
springer.at

Printed on acid-free and chlorine-free bleached paper

With 10 figures

ISBN 978-3-211-99909-7
e-ISBN 978-3-211-74280-8

# CONTENTS


Preface vii

The mathematical work of Wolfgang Schmidt 1
*Hans Peter Schlickewei*

Schäffer's determinant argument 21
*Roger C. Baker*

Arithmetic progressions and Tic-Tac-Toe games 41
*József Beck*

Metric discrepancy results for sequences $\{n_k x\}$ and Diophantine equations 95
*István Berkes, Walter Philipp, and Robert F. Tichy*

Mahler's classification of numbers compared with Koksma's, II 107
*Yann Bugeaud*

Rational approximations to a $q$-analogue of $\pi$ and some other $q$-series 123
*Peter Bundschuh and Wadim Zudilin*

Orthogonality and digit shifts in the classical mean squares problem in irregularities of point distribution 141
*William W. L. Chen and Maxim M. Skriganov*

Applications of the Subspace Theorem to certain Diophantine Problems: A survey of some recent results 161
*Pietro Corvaja and Umberto Zannier*

A generalization of the Subspace Theorem with polynomials of higher degree 175
*Jan-Hendrik Evertse and Roberto G. Ferretti*

On the Diophantine equation $G_n(x) = G_m(y)$ with $Q(x, y) = 0$ 199
*Clemens Fuchs, Attila Pethő, and Robert F. Tichy*

A criterion for polynomials to divide infinitely many $k$-nomials 211
*Lajos Hajdu and Robert Tijdeman*

Approximants de Padé des $q$-polylogarithmes 221
*Christian Krattenthaler et Tanguy Rivoal*

The set of solutions of some equation for linear recurrence sequences 231
*Viktor Losert*

Counting algebraic numbers with large height I 237
*David Masser and Jeffrey D. Vaaler*

Class number conditions for the diagonal case of the equation of Nagell and Ljunggren 245
*Preda Mihălescu*

Construction of approximations to zeta-values 275
*Yuri V. Nesterenko*

Quelques aspects diophantiens des variétés toriques projectives 295
*Patrice Philippon et Martín Sombra*

Une inégalité de Łojasiewicz arithmétique 339
*Gaël Rémond*

On the continued fraction expansion of a class of numbers 347
*Damien Roy*

The number of solutions of a linear homogeneous congruence 363
*Andrzej Schinzel*

A note on Lyapunov theory for Brun algorithm 371
*Fritz Schweiger*

Orbit sums and modular vector invariants 381
*Serguei A. Stepanov*

New irrationality results for dilogarithms of rational numbers 413
*Carlo Viola*

# PREFACE

This volume contains 22 research and survey papers on recent developments in the field of diophantine approximation. The first article by Hans Peter Schlickewei is devoted to the scientific work of Wolfgang Schmidt. Further contributions deal with the subspace theorem and its applications to diophantine equations and to the study of linear recurring sequences. The articles are either in the spirit of more classical diophantine analysis or of geometric or combinatorial flavor. In particular, estimates for the number of solutions of diophantine equations as well as results concerning congruences and polynomials are established. Furthermore, the volume contains transcendence results for special functions and contributions to metric diophantine approximation and to discrepancy theory. The articles are based on lectures given at a conference at the Erwin Schrödinger Institute in Vienna in 2003, in which many leading experts in the field of diophantine approximation participated. The editors are very grateful to the Erwin Schrödinger Institute and to the FWF (Austrian Science Fund) for the financial support and they express their particular thanks to Springer-Verlag for the excellent cooperation.

Robert F. Tichy

*Diophantine Approximation*
H. P. Schlickewei et al., Editors


# THE MATHEMATICAL WORK OF WOLFGANG SCHMIDT

Hans Peter Schlickewei
*Fachbereich Mathematik und Informatik, Philipps-Universität Marburg, Hans-Meerwein-Strasse, 35032 Marburg, Germany*
hps@mathematik.uni-marburg.de

## Introduction

Wolfgang Schmidt's mathematical activities started more than fifty years ago in 1955. In the meantime he has written more than 180 papers – many of them containing spectacular results and breakthroughs in different areas of number theory.

Studying the list of his publications we may classify Wolfgang Schmidt's scientific papers under the following headings: (1) geometry of numbers, (2) uniform distribution, (3) approximation of real numbers, (4) heights, (5) approximation of algebraic numbers (qualitative results), (6) norm form equations (qualitative results), (7) transcendental numbers, (8) elementary proof of the Riemann hypothesis for curves, (9) nonlinear approximation of real numbers, (10) zeros and small values of forms, (11) quadratic geometry of numbers, (12) approximation of algebraic numbers – quantitative results, (13) norm form equations – quantitative results, (14) linear recurrence sequences.

The ordering of this list is chronological according to the date when Schmidt has written his first paper on the respective subject. In the sequel we will discuss for each of these subjects one of the important results obtained by Schmidt in the respective area. In view of the large number of outstanding papers written by Wolfgang Schmidt, the choice was rather difficult. It certainly depends also upon personal taste and no doubt for most subjects also a different choice would have been well justified.

## 1 Geometry of numbers

(Schmidt's papers [1–3,5–7,9,10,12,14,21,24,32,40,65,91] deal with this subject.)

Schmidt's first mathematical paper [1] appeared in 1955, when he was not yet 22 years old. It is his doctoral dissertation, which was written under the guidance of his supervisor Edmund Hlawka in Vienna. Recall Minkowski's famous lattice point theorem:

*Let $S \subset \mathbb{R}^n$ be a symmetric convex body of volume $V(S)$. Let $m$ be a natural number. Then for any lattice $\Lambda$ in $\mathbb{R}^n$ with determinant $d(\Lambda) = d$ satisfying*

$$d < \frac{1}{m2^n} V(S) \tag{1.1}$$

*$S$ contains $m$ distinct pairs of points $\pm \boldsymbol{u}_1, \ldots, \pm \boldsymbol{u}_m \in \Lambda \backslash \{\boldsymbol{0}\}$.*

---

*2000 Mathematics subject classification.* 11-02.

In the opposite direction Hlawka (1943) had shown the following:

*Let $S \subset \mathbb{R}^n$ be a bounded symmetric star body of volume $V(S)$.*
*Then for any $d$ satisfying*

$$d < \frac{1}{2}\zeta(n)^{-1}V(S) \tag{1.2}$$

*there exists a lattice $\Lambda$ in $\mathbb{R}^n$ of determinant $d(\Lambda) = d$ such that $S$ contains no point $\boldsymbol{u} \in \Lambda \backslash \{\mathbf{0}\}$.*

In his thesis [1] Schmidt was able to improve upon Hlawka's result. Moreover, he got the following extension:

*For any natural number $m \geq m_0$ and for any $d$ satisfying*

$$d < \frac{1}{2m}\zeta(n)^{-1}V(S) \tag{1.3}$$

*there exists a lattice $\Lambda$ of determinant $d(\Lambda) = d$ such that $S$ does not contain $m$ distinct pairs of points $\pm\boldsymbol{u}_1, \ldots, \pm\boldsymbol{u}_m \in \Lambda\backslash\{\mathbf{0}\}$.*

We mention that Cassels in his monograph [C2] dedicated several pages to Schmidt's thesis.

## 2 Uniform distribution

In view of his origins from Edmund Hlawka's Vienna school of mathematics it seems to be quite natural that a topic of recurrent interest in Schmidt's scientific activities have been problems from the theory of uniform distribution. His series of papers [39,43–45,48,63,66,73,74,82] as well as his lecture notes [86] certainly should be mentioned in this context.

Here we quote two spectacular theorems on uniform distribution proved by Schmidt:

For an infinite sequence of points $\boldsymbol{x}_1, \boldsymbol{x}_2, \ldots$ in the $k$-dimensional unit-cube $[0,1)^k$, for $N \in \mathbb{N}$ and for a point $\boldsymbol{\alpha} = (\alpha_1, \ldots, \alpha_k)$ with $0 < \alpha_\kappa \leq 1$ $(\kappa = 1, \ldots, k)$ we write $A(\boldsymbol{\alpha}, N)$ for the number of points $\boldsymbol{x}_n$ with $1 \leq n \leq N$ satisfying $\boldsymbol{x}_n \in [0,\alpha_1) \times \ldots \times [0,\alpha_k)$. The discrepancy $D_N$ of the sequence $\{\boldsymbol{x}_n\}_{n\in\mathbb{N}}$ then is defined as

$$D_N = \sup_{\boldsymbol{\alpha}\in(0,1]^k} \left| \frac{A(\boldsymbol{\alpha}, N)}{N} - \alpha_1 \cdot \ldots \cdot \alpha_k \right|. \tag{2.1}$$

K. F. Roth [R1] proved in 1954:

*Suppose $k \geq 1$. There exists a positive constant $c_k$ depending only upon $k$ with the following property:*

*For any infinite sequence $\{\boldsymbol{x}_n\}_{n\in\mathbb{N}}$ as above there are infinitely many natural numbers $N$ such that we have*

$$D_N \geq c_k N^{-1}(\log N)^{k/2}. \tag{2.2}$$

In 1972 Schmidt [66] succeeded to obtain in the case $k = 1$ the following improvement of (2.2):

*There exists a positive constant c with the following property:*

*For any infinite sequence* $\{x_n\}_{n\in\mathbb{N}}$ *in the unit interval* $[0, 1)$ *there exist infinitely many natural numbers N such that we have*

$$D_N > cN^{-1}\log N. \tag{2.3}$$

As for an inequality in the opposite direction, it is well known that for any irrational real number $\alpha$ which has bounded partial quotients in its continued fraction expansion the sequence $\{n\alpha\}$ (where $\{x\}$ denotes the fractional part of $x$) satisfies

$$D_N \le c(\alpha)\frac{\log N}{N}.$$

Here $c(\alpha)$ is a positive constant depending only upon $\alpha$. Therefore (2.3) is best possible.

In a later paper Schmidt shows in a most spectacular way that results of type (2.2) or (2.3) depend very much upon the sets used in the definition of discrepancy. In (2.1) we use subcubes of the unit cube whose axes are parallel to the coordinate axes to define the discrepancy. If we allow rotations, the result changes dramatically. We quote in this context Schmidt's result from 1969 [45]:

For $k \in \mathbb{N}$ we let

$$\mathfrak{S}^k = \{\boldsymbol{x} \in \mathbb{R}^{k+1} \mid x_0^2 + \ldots + x_k^2 = 1\}$$

be the $k$-dimensional the unit-sphere.

On $\mathfrak{S}^k$ we introduce the normalized Lebesgue-measure $\sigma$ with $\sigma(\mathfrak{S}^k) = 1$. For a sequence of points $\boldsymbol{x}_1, \ldots, \boldsymbol{x}_N$ in $\mathfrak{S}^k$ and for a subset $M$ of $\mathfrak{S}^k$ we write $A(M)$ for the number of points $\boldsymbol{x}_i \in M$ $(i = 1, \ldots, N)$. Now Schmidt's result reads as follows:

*Let* $\boldsymbol{x}_1, \ldots, \boldsymbol{x}_N$ *be any N points in* $\mathfrak{S}^k$. *Then for any* $\varepsilon > 0$ *there exists a spherical cap K in* $\mathfrak{S}^k$ *such that*

$$\left|\frac{A(K)}{N} - \sigma(K)\right| > c(n, \varepsilon)N^{-1/2-1/2k-\varepsilon}. \tag{2.4}$$

Beck [Be] has improved the bound (2.4). Indeed he got rid of the $-\varepsilon$ in the exponent. Moreover, he showed that apart from the $\varepsilon$, the exponent in (2.4) is best possible.

## 3 Approximation of real numbers

Dirichlet's Theorem on rational approximation of real numbers says the following:

*Suppose* $\alpha$ *is an irrational real number. Then there are infinitely many rational numbers* $p/q$ *such that we have*

$$\left|\alpha - \frac{p}{q}\right| < q^{-2}. \tag{3.1}$$

For an algebraic number $\beta$ of degree $d$ let $P(X)$ be the defining polynomial of $\beta$ over $\mathbb{Q}$, i.e., the unique irreducible polynomial with coprime integral coefficients and positive leading coefficient having $P(\beta) = 0$. Write $H(\beta)$ for the maximum absolute value of the coefficients of $P$. The following conjecture would generalize Dirichlet's

Theorem to the case of approximation by algebraic numbers $\beta$ of degree $\leq d$ instead of approximation by rationals (i.e., by algebraic numbers of degree 1):

*Conjecture:* Suppose $\alpha$ is real but is not algebraic of degree $\leq d$. Suppose $\varepsilon > 0$. Then there are infinitely many real algebraic numbers $\beta$ of degree $\leq d$ satisfying

$$|\alpha - \beta| \leq \frac{c(\alpha, \varepsilon)}{H(\beta)^{d+1-\varepsilon}}. \tag{3.2}$$

Davenport and Schmidt [36] have proved the conjecture for $d = 2$. For *algebraic* $\alpha$, Wirsing [W1] has verified (3.2). For general real $\alpha$ and general $d$, the conjecture still remains open. Other papers belonging into this section are [29,30,37,41,46,47,180].

## 4 Heights

A central tool in diophantine approximation is the notion of height. In the simplest case of integer points we may define it in the following naive way: Given $n \geq 2$ and a point $\boldsymbol{x} \in \mathbb{Z}^n$, the height $H(\boldsymbol{x})$ of $\boldsymbol{x}$ is defined as the maximum of the absolute values of the components of the primitive point $\boldsymbol{x}' \in \mathbb{Z}^n$ which is proportional to $\boldsymbol{x}$. In his fundamental paper [33] Wolfgang Schmidt developed important generalizations for the notion of height. In particular he introduced the height of a subspace in terms of its Plücker coordinates. It appears that the paper [33] plays a pioneering rôle in the literature for the notion of height. It might be surprising, but it seems that in *Reviews in Number Theory* (edited by W. J. LeVeque, American Mathematical Society, 1974), [33] is the only paper in which heights are dealt with explicitly.

Other papers which have been crucial for applications are [155] and [156]. Here we quote the following result from [156]:

*Let $\Gamma$ be a finitely generated subgroup of $\overline{\mathbb{Q}}^{*n}$ of rank $r$. Suppose $c > 1$. Then the set of nondegenerate solutions $\mathbf{x} \in \Gamma$ of*

$$a_1x_1 + \ldots + a_nx_n = 1 \tag{4.1}$$

*with*

$$h(\mathbf{x}) \leq b \tag{4.2}$$

*has cardinality $\leq (c\,b)^r$. Here $c$ is an explicit constant which depends only upon $n$.*

The bound for the cardinality is completely uniform. It depends only upon the rank $r$ of the group, upon the dimension $n$ and upon the bound $b$ for the height.

This result should be seen in the context of the Bogomolov conjectures on lower bounds for the heights of points on varieties. (Cf. the papers by S. Zhang [Zh], by E. Bombieri and U. Zannier [BZ], and by S. David and P. Philippon [DP].)

Other outstanding contributions on heights may be found in the papers [145,150, 159,161].

## 5 Approximation of algebraic numbers by rationals

In 1955, K. F. Roth [R2] obtained the following striking result on rational approximations of algebraic numbers.

*Let $\alpha$ be an algebraic number. Then for any $\varepsilon > 0$, there are only finitely many rational numbers $p/q$ satisfying*

$$0 < \left| \alpha - \frac{p}{q} \right| < q^{-2-\varepsilon}. \tag{5.1}$$

In view of Dirichlet's theorem quoted in (3.1), Roth' theorem is best possible.

Wolfgang Schmidt was able to extend Roth' theorem to the case of simultaneous approximations of algebraic numbers by rationals. No doubt, this is his most famous result. In its simplest version from 1970 it reads as follows [49]:

*Let $n$ be a natural number. Let $\alpha_1, \ldots, \alpha_n$ be algebraic numbers such that $1, \alpha_1, \ldots, \alpha_n$ are linearly independent over the rationals. Then for any $\varepsilon > 0$ there are only finitely many integers $p_1, \ldots, p_n$, $q$ with $q > 0$ such that*

$$\left| \alpha_1 - \frac{p_1}{q} \right| < q^{-1-1/n-\varepsilon}, \ldots, \left| \alpha_n - \frac{p_n}{q} \right| < q^{-1-1/n-\varepsilon}. \tag{5.2}$$

For $n = 1$ inequality (5.2) becomes (5.1). Still more general is Schmidt's celebrated **Subspace Theorem** [64]: *Let $L_1(\boldsymbol{X}), \ldots, L_n(\boldsymbol{X})$ be linearly independent linear forms in $\boldsymbol{X} = (X_1, \ldots, X_n)$ with algebraic coefficients.*

*Suppose that $\varepsilon > 0$. Let $M$ be the set of solutions $\boldsymbol{x} \in \mathbb{Z}^n$ of the inequality*

$$|L_1(\boldsymbol{x}) \cdot \ldots \cdot L_n(\boldsymbol{x})| < |\boldsymbol{x}|^{-\varepsilon}, \tag{5.3}$$

*where $|\boldsymbol{x}| = \max\{|x_1|, \ldots, |x_n|\}$.*

*Then there are finitely many proper linear subspaces $T_1, \ldots, T_t$ of $\mathbb{Q}^n$ such that*

$$M \subset T_1 \cup \ldots \cup T_t. \tag{5.4}$$

The subspace theorem is a milestone in diophantine approximation. It has had numerous applications to other problems in number theory, and even in completely different fields such as group theory and ergodic theory. It is very likely that it will also serve for future developments.

It is easily seen that Schmidt's result (5.2) is a direct consequence of his subspace theorem.

In particular the special case $n = 2$ of the subspace theorem implies Roth' theorem. Indeed consider

$$L_1(X_1, X_2) = X_1\,\alpha - X_2$$
$$L_2(X_1, X_2) = X_1\,.$$

Notice that (5.1) is almost the same as

$$|q||q\alpha - p| < \max\{|p|, |q|\}^{-\varepsilon}. \tag{5.5}$$

Applying the subspace theorem with $n = 2$ and with the forms $L_1$ and $L_2$ from above we see that the set of solutions $(p, q) \in \mathbb{Z}^2$ of (5.5) is contained in finitely many proper linear subspaces $T_1, \ldots, T_t$ of $\mathbb{Q}^2$. A typical subspace $T$ may be defined by an equation

$$ap + bq = 0$$

with coprime integers $a$ and $b$. But this implies that each subspace $T$ gives rise to just *one* rational solution $p/q$ in (5.1). Since the number of subspaces is bounded, Roth' theorem follows.

For more recent developments of the Subspace Theorem see the paper by J.-H. Evertse and H. P. Schlickewei [ES] as well as the paper by J.-H. Evertse and R. Ferretti in this volume.

Further contributions by Wolfgang Schmidt to be mentioned in this section are [28,34,55,57–59,68,69,76,77] as well as his monograph [96].

## 6 Norm form equations

Let $F(X, Y)$ be a binary form with rational coefficients, and with at least 3 distinct linear factors.

Thue [Th] had shown that under these assumptions *given a nonzero rational number $m$ the diophantine equation*

$$F(x, y) = m \tag{6.1}$$

*has only finitely many solutions in integers $x, y$.*

Notice that the form $F$ may be written as

$$F(X, Y) = a(\beta_1 X + \alpha_1 Y)^{d_1} \cdot \ldots \cdot (\beta_s X + \alpha_s Y)^{d_s}, \tag{6.2}$$

where for the linear factors $\alpha_i X + \beta_i Y$ we may suppose that $\alpha_i$ and $\beta_i$ are algebraic numbers such that moreover $\beta_i$ equals 0 or 1. It turns out that then any integral solution $(x, y)$ of (6.1) yields a good rational approximation $x/y$ to one at least of the elements $\alpha_i$ $(1 \leq i \leq s)$.

Roth [R2] proved on the basis of his approximation result:

*Let $F(X, Y)$ be a binary form of degree $d \geq 3$ with rational coefficients and without multiple factors.*

*Let $G(X, Y)$ be a polynomial of degree $< d - 2$. Then the diophantine equation*

$$F(x, y) = G(x, y) \tag{6.3}$$

*has only finitely many solutions $(x, y) \in \mathbb{Z}^2$ with $F(x, y) \neq 0$.*

Using his subspace theorem on inequality (5.3), in 1972 Schmidt [64] succeeded to obtain the generalization of (6.3) to $n$ dimensions.

It turns out that both in Thue's and in Roth' result an essential feature is the fact that any binary form $F(X, Y)$ splits over $\mathbb{C}$ into a product of linear forms. On the other hand, forms $F(X_1, \ldots, X_n)$ with rational coefficients in $n$ variables with $n \geq 3$ in general do not split into a product of linear forms in analogy with (6.2).

Homogeneous polynomials $F(X_1, \ldots, X_n)$ that do split into a product of linear forms are the so-called norm forms or more generally decomposable forms:

Let $K$ be a number field of degree $d$. There are $d$ isomorphic embeddings say $\sigma_1, \ldots, \sigma_d$ of $K$ into the field of complex numbers. Given an element $\alpha \in K$ we write $\alpha^{(i)}$ for the image of $\alpha$ under $\sigma_i$. Let

$$L(X) = \alpha_1 X_1 + \ldots + \alpha_n X_n$$

be a linear form with coefficients $\alpha_1, \ldots, \alpha_n$ in $K$. The norm form $\mathfrak{N}(L(\boldsymbol{X}))$ then is defined as

$$\mathfrak{N}(L(\boldsymbol{X})) = \prod_{i=1}^{d} \left(\alpha_1^{(i)} X_1 + \ldots + \alpha_n^{(i)} X_n\right).$$

It is clear that $\mathfrak{N}(L(\boldsymbol{X}))$ will be a form of degree $d$ in $X_1, \ldots, X_n$ with rational coefficients.

Wolfgang Schmidt has obtained several results on diophantine equations involving norm forms or decomposable forms. Here we quote what seems to be nearest to (6.3).

Let $K$ and $L(\boldsymbol{X})$ be as above. Write $H$ for the normal hull of $K$, i.e., $H$ is the smallest normal extension of $\mathbb{Q}$ containing $K$. Let $G$ be the Galois group of $H$ over $\mathbb{Q}$. In [64] Schmidt has shown:

*Suppose $n < d$. Let $L(\boldsymbol{X}) = \alpha_1 X_1 + \ldots + \alpha_n X_n$ be a linear form with coefficients $\alpha_1, \ldots, \alpha_n$ in a number field $K$ of degree $d$. Suppose that the Galois group $G$ of the normal hull $H$ of $K$ over $\mathbb{Q}$ is $(n-1)$-times transitive. Moreover let $P(\boldsymbol{X}) = P(X_1, \ldots, X_n)$ be a polynomial of total degree $< d - n$.*

*Assume that any $n$ conjugates $L^{(i_1)} = \alpha_1^{(i_1)} X_1 + \ldots + \alpha_n^{(i_1)} X_n, \ldots, L^{(i_n)} = \alpha_1^{(i_n)} X_1 + \ldots + \alpha_n^{(i_n)} X_n$ are linearly independent.*

*Then the diophantine equation*

$$\mathfrak{N}(L(\boldsymbol{x})) = P(\boldsymbol{x})$$

*has only finitely many solutions $\mathbf{x} \in \mathbb{Z}^n$.*

Norm form equations as well as decomposable form equations are also treated in [35,58,61,70].

## 7 Transcendental numbers

Mahler [Ma] introduced a classification of the real numbers: Given a real number $\alpha$ and $k \in \mathbb{N}$ he defines the quantity $\omega_k(\alpha)$ as follows:

For a polynomial $P \in \mathbb{Z}[x]$ we write $H(P)$ for the maximum of the absolute values of its coefficients.

Now $\omega_k(\alpha)$ is the least upper bound of the numbers $\omega$ such that

$$0 < |P(\alpha)| < H(P)^{-\omega}$$

has infinitely many solutions in polynomials $P$ of degree $\leq k$.

Obviously, $\omega_1 \leq \omega_2 \leq \ldots$. It is easily seen that $\alpha$ is algebraic if and only if the sequence $\omega_1, \omega_2, \ldots$ is bounded.

Mahler calls such numbers $A$-numbers. The remaining numbers (i.e., the set of transcendental real numbers) are divided by Mahler into three classes as follows:

$\alpha$ will be an *$S$-number* if the sequence $(\omega_k(\alpha))_{k \in \mathbb{N}}$ is not bounded, but the sequence $\omega_k(\alpha)/k$ is bounded.

$\alpha$ will be a *$T$-number* if the sequence $(\omega_k(\alpha)/k)_{k \in \mathbb{N}}$ is not bounded but if for each $k$ $\omega_k(\alpha) < \infty$.

$\alpha$ will be called a *$U$-number* if there exists some $k$ such that $\omega_k(\alpha) = \infty$.

It is well known that almost all real numbers are $S$-numbers. Moreover, it is not difficult to construct explicit examples of $U$-numbers.

The problem, whether $T$-numbers do exist had been open for over 30 years.

In 1968 Wolfgang Schmidt [53,60], using a generalization of Roth' theorem due to Wirsing [W2] on the approximation of algebraic numbers by algebraic numbers of bounded degree, was able to close the gap and to prove the existence of $T$-numbers. His proof consists in a formidable inductive construction of a sequence that eventually gives the desired $T$-number.

## 8 Riemann hypothesis for curves

Let $K$ be a finite field with $q$ elements. Let $C$ be a curve defined over $K$, i.e., $C$ is given by a polynomial equation

$$f(x, y) = 0, \tag{8.1}$$

where $f$ is a polynomial with coefficients in $K$.

For any $r \in \mathbb{N}$ we denote by $K_r$ the unique extension of $K$ of degree $r$.

Let $A_r$ be the number of solutions $(x, y) \in K_r^2$ of equation (8.1). For the case of hyperelliptic curves $f = 0$ (and some generalizations thereof) Stepanov [St] has developed an elementary method to prove the estimate

$$|A_r - q^r| \leq cq^{r/2}. \tag{8.2}$$

Here $c$ is an explicit constant depending only upon $f$.

As is well known, estimate (8.2) is equivalent to the Riemann hypothesis for the curve $f = 0$.

We quote from Cassels' review of Stepanov's paper [St]: "In the case of genus 1, The 'Riemann Hypothesis' was first proved by Hasse, as he records, specifically as a demonstration of the superiority of abstract over elementary methods,.. The 'hypothesis' in its full generality for curves of any genus was the first fruit of A. Weil's new algebraic geometry in arbitrary characteristic.... It was generally felt to be a reproach to number-theoreticians that they could not prove a key result of their own without invoking alien resources." (The reader might also like the comments made by M. Fried in *his* review of Stepanov's paper [St]: "For years certain number theorists have railed at Weil's proof.... Those not at home with algebraic geometry (most mathematicians seem quite comfortable about not being at home with algebraic geometry) will be especially pleased with Stepanov's achievement, until they try to read the proof.") (In *Reviews in Number Theory* [edited by W. J. LeVeque, American Mathematical Society, 1974], Stepanov's paper [St] is reviewed twice, under G20-203 and under G20-213.)

Stepanov's method shows some elements similar to the Thue–Siegel method in diophantine approximation. It is the first elementary access to this deep problem.

In 1973, Schmidt [71] succeeded in generalizing Stepanov's elementary approach. He proved (8.2) for general curves with this elementary method. In [72] also some special cases of the analog of (8.2) for dimension $n$ are treated. However, up to now no elementary proof of Deligne's results has been given. The reader is also referred to Schmidt's monographs [81] and [176]. Finally, we mention that Bombieri [Bo] has extended in a different way Stepanov's elementary proof to get the Riemann hypothesis for general curves.

## 9 Nonlinear approximation of real numbers

Now we shall discuss a topic that is located rather in the complement of algebraic geometry. The proofs in this section are based on analytical methods.

For a real number $x$ we write $\|x\|$ for the distance of $x$ to the nearest integer.

In 1948, Heilbronn [He] proved:

*For any $\varepsilon > 0$ there exists a positive constant $c(\varepsilon)$ with the following property: For any real number $\alpha$ and for any $N > c(\varepsilon)$ there is a natural number $n$ satisfying*

$$\| \alpha n^2 \| < N^{-1/2+\varepsilon} \text{ and } n \leq N. \tag{9.1}$$

The significant feature in (9.1) is the uniformity, as the constant $c(\varepsilon)$ does not depend upon $\alpha$ at all.

For almost 30 years essentially no progress had been made in the area of Heilbronn's theorem when Wolfgang Schmidt took up the question in 1977.

Developing a very nice extension of Weyl's estimate for exponential sums involving quadratic polynomials he proved [84]:

*For any pair of real numbers $\alpha$, $\beta$ and for any $N$ there exists a natural number $n$ satisfying*

$$n \leq N \text{ and } \|\alpha n^2 + \beta n\| < N^{-1/2+\varepsilon} \tag{9.2}$$

*provided that $N > c(\varepsilon)$.*

In 1958, Danicic [Da] had extended Heilbronn's result to monomials $\alpha n^k$ $(k \geq 2)$. He proved:

*For any real number $\alpha$ and for any $N$ there exists $n \in \mathbb{N}$ with*

$$n \leq N \text{ and } \|\alpha n^k\| < N^{-1/K+\varepsilon} \tag{9.3}$$

*provided that $N > c(k, \varepsilon)$. Here $K = 2^{k-1}$.*

Schmidt's method inspired later on R. C. Baker [Ba] to generalize Danicic' result (9.3) to arbitrary polynomials

$$\alpha_k x^k + \alpha_{k-1} x^{k-1} + \ldots + \alpha_1 x \text{ of degree } k \tag{9.4}$$

(with constant term equal to zero). To derive the theorem on (9.2), Schmidt modifies Weyl's estimate of exponential sums $\sum_{x=1}^{N} e^{2\pi i f(x)}$ such as to bring in approximation properties of the leading coefficient as well as the second highest coefficient of the polynomial $f$. (The classical Weyl estimate depends only upon properties of the leading coefficient.) R. C. Baker generalized this to an estimate for exponential sums as above depending on the approximation properties of all coefficients of the polynomial $f$.

In 1977, Schmidt [83] extended Heilbronn's result towards simultaneous approximations:

*Let $\alpha_1, \ldots, \alpha_h$ be real numbers. Suppose that $N > c(h, \varepsilon)$. Then there is a natural number $n$ satisfying*

$$n \leq N \text{ and } \|\alpha_i n^2\| < N^{-1/h^2+h+\varepsilon} \; (i = 1, \ldots, h). \tag{9.5}$$

We observe that in all previous papers on simultaneous approximation in the literature the denominator of the exponent on the left-hand side in (9.5) was increasing exponentially in $h$. So it is not astonishing that the proof of (9.5) required a completely new method which nowadays in the literature is called Schmidt's lattice method. In the proof, either the analytical method based on Weyl's estimate for exponential sums leads to a good end or alternatively Schmidt is able to bring in strong tools from the geometry of numbers using Minkowski's theory of successive minima.

## 10 Zeros and small values of forms

A classical problem in number theory deals with small values of forms: *Given a form $F(X_1, \ldots, X_n)$ with real coefficients and given $\varepsilon > 0$, does there exist an integral point $\boldsymbol{x} \neq \mathbf{0}$ such that*

$$|F(\boldsymbol{x})| < \varepsilon\ ? \tag{10.1}$$

It is clear that in this generality the answer to the question is "no". Indeed, it suffices to consider forms of type

$$\alpha_1 X_1^{2d} + \ldots + \alpha_n x_n^{2d}$$

of even degree with $\alpha_1, \ldots, \alpha_n$ all of the same sign.

Things become quite different if we restrict ourselves to forms $F$ of odd degree. There had been the following

*Conjecture*: If F is a form of odd degree $k$ in $n \geq n_0(k)$ variables, inequality (10.1) has a nontrivial integral solution $\boldsymbol{x}$.

For $k = 1$ it is clear that the conjecture is true with $n_0(1) = 2$. If $F$ is a form with *integral* coefficients, then the problem of solving (10.1) is the same as finding a nontrivial solution $\boldsymbol{x} \in \mathbb{Z}^n$ of

$$F(\boldsymbol{x}) = 0. \tag{10.2}$$

A well-known theorem of Birch [Bi] says that *any form of odd degree with integral coefficients in "enough" variables admits a nontrivial zero*, in other words Birch proved *the conjecture to be true for forms with integral coefficients.*

In the more general situation when $F$ has real coefficients, Pitman [Pi] could show that *the conjecture is true for cubic forms.*

When it comes to forms $F$ of higher degree, Davenport on the last page of his lecture notes on "analytical methods for diophantine equations and diophantine inequalities" remarks about the general conjecture: *There seems to be a difficulty of principle in proving any analogous result for a form of degree* 5.

In 1980, Wolfgang Schmidt inventing an ingenious variant of the analytical Hardy–Littlewood circle method was able to overcome this difficulty of principle. He first proved the following result on diagonal equations [90]:

*Let $k$ be an odd integer. Suppose $\varepsilon > 0$. Then, given $n \geq c(k, \varepsilon)$ and integers $a_1, \ldots, a_n$ not all equal to* 0, *the equation*

$$a_1 x_1^k + \cdots + a_n x_n^k = 0 \tag{10.3}$$

*has a solution in integers $x_1, \ldots, x_n$ satisfying*

$$0 < \max_{1 \leq i \leq n} |x_i| \leq A^{\varepsilon}, \tag{10.4}$$

*where* $A = \max\{|a_1|, \ldots, |a_n|\}$.

It had been known since a long time that the results (10.3), (10.4) would imply the conjecture on (10.1) for forms $F$ of odd degree with *real* coefficients. Indeed, combining his theorem from [90] with the method of proof in Birch [Bi] Schmidt [93] finally was able to overcome "the problem of principle" discussed by Davenport and thus to prove the conjecture in full generality.

In Schmidt's list of papers we find many other results that would deserve to be discussed in this section. Here are examples: [88,94,99–103,106,109,111,112,116, 119].

## 11 Quadratic geometry of numbers

Cassels [C1] had proved the following result:

*Let $Q(x_1, \ldots, x_n)$ be a quadratic form with integral coefficients. Write $|Q|$ for the maximum absolute value of its coefficients. Then, if the equation*

$$Q(\boldsymbol{x}) = 0 \tag{11.1}$$

*has a solution $\boldsymbol{x} \in \mathbb{Z}^n \backslash \{0\}$, there exists such a solution of* (11.1) *having*

$$0 < \max |x_i| \leq c(n) Q^{(n-1)/2}. \tag{11.2}$$

In a series of joint papers Schmidt and I succeeded to extend (11.2) towards a theory we might call quadratic geometry of numbers.

A particular instance of the results obtained in [115,124,126,129,136] is as follows:

*Let $Q(x_1, \ldots, x_n)$ be a nonzero quadratic form with integral coefficients. Assume that $Q$ vanishes on a linear subspace $S_d$ of $\mathbb{Q}^n$ of dimension $d$. Then there exists a subspace $S_d$ on which $Q$ vanishes and which has an integral basis $\boldsymbol{x}_1, \ldots, \boldsymbol{x}_d$ satisfying*

$$|\boldsymbol{x}_1| \cdot \cdots \cdot |\boldsymbol{x}_d| \leq c(n) |Q|^{(n-d)/2}. \tag{11.3}$$

*Here $|\boldsymbol{x}|$ denotes the maximum norm.*

## 12 Approximation of algebraic numbers – quantitative results

In the second half of the 1980s Wolfgang Schmidt returned to his subspace theorem.

The most ambitious goal in this context no doubt would be to prove an effective version of (5.3), (5.4). This would mean to give an algorithm which allows it, given linear forms $L_1, \ldots, L_n$ to determine explicitly the subspaces $T_1, \ldots, T_t$ containing the set of solutions of inequality (5.3). As a consequence one would have for the class of diophantine equations that can be treated with the subspace theorem an algorithm which would yield the set of solutions of such equations. In the light of Matiasevich's celebrated theorem (according to which there is no universal algorithm with which any diophantine equation can be dealt with) this would be a most spectacular result. However, until today an effective subspace theorem seems to be far out of reach. Indeed, the method of Thue, Siegel, Roth, and Schmidt is highly noneffective.

However, the method allows to give an upper bound for the *number* of subspaces needed in the subspace theorem.

In 1989, Schmidt [131] got the following result:

*Let $L_1(\boldsymbol{X}), \ldots, L_n(\boldsymbol{X})$ be linearly independent linear forms with coefficients from a number field $K$ of degree $d$. Suppose $\varepsilon > 0$. Let $M$ be the set of points $\boldsymbol{x} \in \mathbb{Z}^n$ satisfying*

$$|L_1(\boldsymbol{x}) \ldots L_n(\boldsymbol{x})| < |\boldsymbol{x}|^{-\varepsilon} |\det(L_1, \ldots, L_n)|, \tag{12.1}$$

$$|\boldsymbol{x}| \geq c(n, \varepsilon, L_1, \ldots, L_n) \tag{12.2}$$

*(here the function $c$ is given explicitly). Then there are proper linear subspaces $T_1, \ldots, T_t$ of $\mathbb{Q}^n$ with*

$$t \leq (2d)^{2^{26n}\varepsilon^{-2}} \tag{12.3}$$

*such that $M$ is contained in the union $T_1 \cup \ldots \cup T_t$.*

An essential ingredient in the proof in [131] (as compared with the qualitative result discussed in section 5) are so called gap principles. We give a typical instance of such a gap principle:

The set of points $\boldsymbol{x} \in \mathbb{Z}^n$ satisfying inequalities of type

$$\begin{aligned} |L_1(\boldsymbol{x})| &\leq Q^{c_1} \\ &\vdots \\ |L_n(\boldsymbol{x})| &\leq Q^{c_n}, \end{aligned} \tag{12.4}$$

where $c_1 + \ldots + c_n \leq -\varepsilon$ and where $Q$ is a parameter that runs through an interval of type $[Q_1, Q_1^{1+\varepsilon/n})$ are contained in a single linear subspace of $\mathbb{Q}^n$ provided $Q_1$ is large enough.

The significant feature in (12.3) is the uniformity of the bound. It involves only the degree $d$ of the field $K$, the dimension $n$ and the parameter $\varepsilon$. Otherwise it is completely uniform in terms of $L_1, \ldots, L_n$. This is crucial in applications. But notice that this is a bound for the number of subspaces containing the *large* solutions. *Large* is defined in (12.2) and here the lower bound involves the linear forms and therefore is not uniform.

There is a series of papers by Wolfgang Schmidt that deal with related questions – also concerning Roth' theorem (cf. the articles [128, 138, 143, 154]). For more general versions of the quantitative Subspace Theorem the reader is referred to the paper by Evertse and Schlickewei [ES].

## 13 Norm form equations – quantitative results

In 1990, Schmidt using his quantitative version of the subspace theorem (12.3) was able to derive *uniform* upper bounds for the *number* of solutions of norm form equations [132]:

As in section 6, let $K$ be a number field of degree $d$. Let

$$L(\boldsymbol{X}) = \alpha_1 X_1 + \cdots + \alpha_n X_n$$

be a linear form with coefficients $\alpha_1, \ldots, \alpha_n$ in $K$ and

$$\mathfrak{N}(L(\boldsymbol{X})) = \prod_{i=1}^{d} \left(\alpha_1^{(i)} X_1 + \cdots + \alpha_n^{(i)} X_n\right).$$

the corresponding norm form. In [132] Schmidt studies the norm from equation

$$\mathfrak{N}(L(\boldsymbol{x})) = 1. \tag{13.1}$$

For "nondegenerate" forms $\mathfrak{N}(L(\boldsymbol{x}))$ he shows that the number of solutions $\boldsymbol{x} \in \mathbb{Z}^n$ of equation (13.1) is bounded by a function $c(n, d)$ which depends only upon the degree $d$ of the field $K$ and upon the dimension $n$. It is remarkable that he obtains a bound which apart from the degree $d$ is independent of the particular linear form $L$. In the proof, the *large* solutions $\boldsymbol{x}$ are treated with the quantitative subspace theorem as quoted in (12.1), (12.2), (12.3). Obviously, in view of (12.2), the small solutions cause some extra trouble, as with a direct application of (12.2) we would get a dependence on $L$. To avoid such a dependence, Schmidt uses a device "jacking" up the height. Such a device is also used in the joint paper with Bombieri [121] on upper bounds for the number of solutions of Thue equations (essentially norm form equations in dimension 2). We mention that the bound obtained in [121] for the number of solutions of Thue equations is best possible: In the particular case discussed in (13.1) they get for Thue equations the bound $c_0 d$, where $c_0$ is an absolute constant. Further papers related with the topic discussed in this section are, e.g., [120,123,125,127,134].

## 14 Linear recurrence sequences

A linear recurrence sequence of order $t$ is a sequence $(u_k)_{k\in\mathbb{Z}}$ of complex numbers where each term is a linear combination of the $t$ preceding terms with fixed coefficients $c_0, \ldots, c_{t-1}$ not all equal to zero:

$$u_{k+t} = c_{t-1}u_{k+t-1} + \cdots + c_0 u_k \ (k \in \mathbb{Z}). \tag{14.1}$$

In what follows we will assume that for a given sequence $(u_k)_{k\in\mathbb{Z}}$ we have fixed initial values $u_0, \ldots, u_{t-1}$ not all equal to zero. Moreover we will assume that $t$ is minimal, i.e., that $(u_k)_{k\in\mathbb{Z}}$ does not satisfy a nontrivial recurrence relation of type (14.1) for some $t' < t$.

The companion polynomial of relation (14.1) is given by

$$P(X) = X^t - c_{t-1}X^{t-1} - \cdots - c_0 = \prod_{i=1}^{r}(X - \alpha_i)^{\rho_i} \tag{14.2}$$

with distinct zeros $\alpha_1, \ldots, \alpha_r$ of respective multiplicities $\rho_1, \ldots, \rho_r$.

It is well known that given (14.1), (14.2) and initial values $u_0, \ldots, u_{t-1}$, the sequence $(u_k)_{k\in\mathbb{Z}}$ has a representation of the shape

$$u_k = \sum_{i=1}^{r} p_i(k)\alpha_i^k, \tag{14.3}$$

where $p_i$ is a polynomial of degree $\rho_i - 1$ $(i = 1, \ldots, r)$.

A famous problem in the literature deals with the equation

$$u_k = 0 \quad (k \in \mathbb{Z}). \tag{14.4}$$

Similarly as in the case of the norm form equation, the problem of giving an algorithm to determine explicitly the solutions $k$ of (14.4) seems to be currently out of reach.

On the other hand there is the well-known theorem of Skolem, Mahler, and Lech. It says the following:

*The set of solutions k of equation* (14.4) *is the union of finitely many arithmetic progressions and a finite set.*

If we assume moreover that the sequence $(u_k)_{k\in\mathbb{Z}}$ is nondegenerate, i.e., if we assume that for the roots $\alpha_i$ of the companion polynomial $P(X)$ in (14.2) none of the quotients $\alpha_i/\alpha_j$ $(i \neq j)$ is a root of unity, then it is an easy consequence of the Skolem–Mahler–Lech theorem that *equation* (14.4) *has only finitely many solutions.* Indeed by nondegeneracy the set of solutions of (14.4) cannot contain an arithmetic progression.

It has been a classical conjecture that for a nondegenerate sequence $(u_k)_{k\in\mathbb{Z}}$ the *number* of solutions of (14.4) *is bounded in terms of the order t only.*

This conjecture has been proved by Wolfgang Schmidt [162,165]. His theorem in fact makes the theorem of Skolem–Mahler–Lech quantitative – independently of any assumption on nondegeneracy. He showed:

*Let* $(u_k)_{k\in\mathbb{Z}}$ *be a linear recurrence sequence of order t (which also might be degenerate). Then the set M of solutions* $k \in \mathbb{Z}$ *of equation* (14.4), *i.e., of* $u_k = 0$ *consists of finitely many arithmetic progressions* $P_1, \dots, P_{t_1}$ *and of a finite set* $M_1$ *such that*

$$c(t) = t_1 + \mathrm{card} M_1 \leq \exp\exp\exp(20t).$$

*In particular for any nondegenerate sequence* $(u_k)_{k\in\mathbb{Z}}$ *equation* (14.4) *does not have more than* $c(t)$ *solutions.*

The remarkable feature in this theorem is the fact that the bound $c(t)$ apart from its dependence upon the order $t$ of the sequence is completely uniform. It does not depend upon the coefficients $c_0, \dots, c_{t-1}$ of the recurrence relation (14.1) nor does it depend upon the initial values $u_0, \dots, u_{t-1}$ of the sequence.

For *simple* recurrence sequences, i.e., for sequences where the companion polynomial $P$ in (14.2) has only simple zeros, the conjecture had been proved already in [172].

The proof of the conjecture in principle depends upon the quantitative version of the subspace theorem by Evertse and Schlickewei [ES] as well as on the lower bound for heights of points on varieties from [156] (cf. section 4). However, the step from simple sequences [172] to the general case, when in (14.3) we have nonconstant polynomial coefficients $p_i$ is everything else but trivial. Schmidt develops a very ingenious device to "jack" up the heights of the roots $\alpha_i$ to obtain uniform bounds also in the case of nonconstant polynomial coefficients.

Topics related to the subject discussed in this section are treated in [142,144,146, 153,160,168,173,175].

## Publications by W. Schmidt

1. Über höhere kritische Determinanten von Sternkörpern. *Monatsh. Math.* **59**, 274–304 (1955)
2. Eine neue Abschätzung der kritischen Determinante von Sternkörpern. *Monatsh. Math.* **60**, 1–10 (1956)
3. Eine Verschärfung des Satzes von Minkowski–Hlawka. *Monatsh. Math.* **60**, 110–113 (1956)
4. with K. Baumann: Quantentheorie der Felder als Distributionstheorie. *Nuovo Cimento X. Ser.* **4**, 860–886 (1956)

5. Mittelwerte über Gitter. *Monatsh. Math.* **61**, 269–276 (1957)
6. The measure of the set of admissible lattices. *Proc. Am. Math. Soc.* **9**, 390–403 (1958)
7. Mittelwerte über Gitter. II. *Monatsh. Math.* **62**, 250–258 (1958)
8. Flächenapproximation beim Jacobialgorithmus. *Math. Ann.* **136**, 365–374 (1958)
9. On the convergence of mean values over lattices. *Can. J. Math.* **10**, 103–110 (1958)
10. Maßtheorie in der Geometrie der Zahlen. *Acta Math.* **102**, 159–224 (1959)
11. On normal numbers. *Pac. J. Math.* **10**, 661–672 (1960)
12. A metrical theorem in geometry of numbers. *Trans. Am. Math. Soc.* **95**, 516–529 (1960)
13. A metrical theorem in diophantine approximation. *Can. J. Math.* **12**, 516–529 (1960)
14. Zur Lagerung kongruenter Körper im Raum. *Monatsh. Math.* **65**, 154–158 (1961)
15. Stetige Funktionen auf dem Torus. *J. Reine Angew. Math.* **207**, 86–95 (1961)
16. Bounds for certain sums; a remark on a conjecture of Mahler. *Trans. Am. Math. Soc.* **101**, 200–210 (1961)
17. Continuous functions defined on product-spaces. *Proc. Am. Math. Soc.* **12**, 918–920 (1961)
18. Über die Normalität von Zahlen zu verschiedenen Basen. *Acta Arith.* **7**, 299–309 (1962)
19. Two combinatorial theorems on arithmetic progressions. *Duke Math. J.* **29**, 129–140 (1962)
20. Simultaneous approximation and algebraic independence of numbers. *Bull. Am. Math. Soc.* **68**, 475–478 (1962)
21. On the Minkowski–Hlawka theorem. *Ill. J. Math.* **7**, 18–23 (1963)
22. Correction to my paper "On the Minkowski-Hlawka theorem". *Ill. J. Math.* **7**, 714 (1963)
23. Metrical theorems on fractional parts of sequences. *Trans. Am. Math. Soc.* **110**, 493–518 (1964)
24. Über Gitterpunkte auf gewissen Flächen. *Monatsh. Math.* **68**, 59–74 (1964)
25. Normalität bezüglich Matrizen. *J. Reine Angew. Math.* **214/215**, 227–260 (1964)
26. Ein kombinatorisches Problem von P. Erdős und A. Hajnal. *Acta Math. Acad. Sci. Hung.* **15**, 373–374 (1964)
27. Metrische Sätze über simultane Approximation abhängiger Größen. *Monatsh. Math.* **68**, 154–166 (1964)
28. Über simultane Approximation algebraischer Zahlen durch Rationale. *Acta Math.* **114**, 159–206 (1965)
29. On badly approximable numbers. *Mathematika* **12**, 10–20 (1965)
30. On badly approximable numbers and certain games. *Trans. Am. Math. Soc.* **123**, 178–199 (1966)
31. Simultaneous approximation to a basis of a real number-field. *Am. J. Math.* **88**, 517–527 (1966)
32. Maßtheorie in der Geometrie der Zahlen. *Colloq. Int. Centre Nat. Rech. Sci.* **143**, 225–229 (1966)
33. On heights of algebraic subspaces and diophantine approximations. *Ann. Math. (2)* **85**, 430–472 (1967)
34. On simultaneous approximations of two algebraic numbers by rationals. *Acta Math.* **119**, 27–50 (1967)
35. Some diophantine equations in three variables with only finitely many solutions. *Mathematika* **14**, 113–120 (1967)
36. with H. Davenport: Approximation to real numbers by quadratic irrationals. *Acta Arith.* **13**, 169–176 (1967)
37. with H. Davenport: A theorem on linear forms. *Acta Arith.* **14**, 209–223 (1968)
38. Asymptotic formulae for point lattices of bounded determinant and subspaces of bounded height. *Duke Math. J.* **35**, 327–339 (1968)
39. Irregularities of distribution. *Q. J. Math. Oxf. II. Ser.* **19**, 181–191 (1968)
40. A problem of Schinzel on lattice points. *Acta Arith.* **15**, 199–203 (1969)
41. with H. Davenport: Approximation to real numbers by algebraic integers. *Acta Arith.* **15**, 393–416 (1969)
42. Disproof of some conjectures on Diophantine approximations. *Stud. Sci. Math. Hung.* **4**, 137–144 (1969)
43. Irregularities of distribution. II. *Trans. Am. Math. Soc.* **136**, 347–360 (1969)
44. Irregularities of distribution. III. *Pac J. Math.* **29**, 225–234 (1969)
45. Irregularities of distribution. IV. *Invent. Math.* **7**, 55–82 (1969)
46. Badly approximable systems of linear forms. *J. Number Theory* **1**, 139–154 (1969)
47. with H. Davenport: Dirichlet's theorem on diophantine approximation. II. *Acta Arith.* **16**, 413–424 (1970)
48. Irregularities of distribution. V. *Proc. Am. Math. Soc.* **25**, 608–614 (1970)
49. Simultaneous approximation to algebraic numbers by rationals. *Acta Math.* **125**, 189–201 (1970)

50. with A. Baker: Diophantine approximation and Hausdorff dimension. *Proc. Lond. Math. Soc. III. Ser.* **21**, 1–11 (1970)
51. with H. Davenport: Supplement to a theorem on linear forms. In: Turán, P. (ed.) *Number Theory.* Colloq. Math. Soc. János Bolyai, vol. 2, pp. 15–25. North-Holland, Amsterdam (1970)
52. with H. Davenport: Dirichlet's theorem on diophantine approximation. In: *Teoria dei Numeri, Istituto Nazionale di Alta Matematica, Rome, 1968.* Symp. Math., 4, pp. 113–132. Academic Press, London (1970)
53. $T$-numbers do exist. In: *Teoria dei Numeri, Istituto Nazionale di Alta Matematica, Rome, 1968.* Symp. Math., 4, pp. 3–26. Academic Press, London (1970)
54. Remark on my paper: "Disproof of some conjectures on diophantine approximations". *Stud. Sci. Math. Hung.* **5**, 479 (1970)
55. Some recent progress in diophantine approximations. In: *Actes du Congrès International des Mathématiciens (Nice, 1970)*, tome 1, pp. 497–503. Gauthier-Villars, Paris (1971)
56. Diophantine approximation and certain sequences of lattices. *Acta Arith.* **18**, 195–178 (1971)
57. Linear forms with algebraic coefficients. I. *J. Number Theory* **3**, 253–277 (1971)
58. Linearformen mit algebraischen Koeffizienten. II. *Math. Ann.* **191**, 1–20 (1971)
59. Approximation to algebraic numbers. *Enseign. Math. II. Ser.* **17**, 187–253 (1971)
60. Mahler's $T$-numbers. In: *1969 Number Theory Institute – Proceedings of the 1969 Summer Institute on Number Theory – American Mathematical Society.* Proc. Symp. Pure Math., vol. 20, pp. 275–286. American Mathematical Society, Providence, R.I. (1971)
61. Diophantine equations involving norm forms. *Sem. Mod. Methods Number Theory, Inst. Stat. Math. Tokyo* **5** (1971)
62. On a problem of Heilbronn. *J. Lond. Math. Soc. II. Ser* **4**, 545–550 (1972)
63. Irregularities of distribution. VI. *Compos. Math.* **24**, 63–74 (1972)
64. Norm form equations. *Ann. Math. (2)* **96**, 526–551 (1972)
65. Volume, surface area and the number of integer points covered by a convex set. *Arch. Math. (Basel)* **23**, 537–543 (1972)
66. Irregularities of distribution. VII. *Acta Arith.* **21**, 45–50 (1972)
67. with G.H. Meisters: Translation-invariant linear forms on $L^2(G)$ for compact Abelian groups $G$. *J. Funct. Anal.* **11**, 407–424 (1972)
68. *Approximation to Algebraic Numbers.* Série des Conférences de l'Union Mathématique Internationale, 2; Monographies de l'Enseignement Mathématique, 19. L'Enseignement Mathématique, Université de Genève, Geneva (1972)
69. Simultaneous approximation to algebraic numbers by algebraic numbers in a given number field. In: *Proceedings of the 1972 Number Theory Conference, Boulder, Colo.*, pp. 189–193. University of Colorado, Boulder, Colo. (1972)
70. Inequalities for resultants and for decomposable forms. In: Osgood, C.F. (ed.) *Diophantine Approximation and Its Applications: Proceedings*, pp. 235–253. Academic Press, New York (1973)
71. Zur Methode von Stepanov. In: Collection of articles dedicated to Carl Ludwig Siegel on the occasion of his seventy-fifth birthday, IV. *Acta Arith.* **24**, 347–367 (1973)
72. A lower bound for the number of solutions of equations over finite fields. In: Collection of articles dedicated to K. Mahler on the occasion of his seventieth birthday. *J. Number Theory* **6**, 448–480 (1974)
73. Irregularities of distribution. VIII. *Trans. Am. Math. Soc.* **198**, 1–22 (1974)
74. Irregularities of distribution. IX. In: Collection of articles in memory of Juriĭ Vladimirovič Linnik. *Acta Arith.* **27**, 385–396 (1975)
75. The measure of the intersection of rotates of a set on the circle. *Proc. Am. Math. Soc.* **48**, 18–20 (1975)
76. Simultaneous approximation to algebraic numbers by elements of a number field. *Monatsh. Math.* **79**, 55–66 (1975)
77. Rational approximation to solutions of linear differential equations with algebraic coefficients. *Proc. Am. Math. Soc.* **53**, 285–289 (1975)
78. Applications of Thue's method in various branches of number theory. In: *Proceedings of the International Congress of Mathematicians – Canadian Mathematical Congress – International Mathematical Union, Vancouver, B.C., 1974*, vol. 1, pp. 177–185 (1975)
79. On Osgood's effective Thue theorem for algebraic functions. *Commun. Pure Appl. Math.* **29**, 749–763 (1976)
80. Two questions in Diophantine approximation. *Monatsh. Math.* **82**, 237–245 (1976)
81. *Equations over Finite Fields: an Elementary Approach.* Lect. Notes Math., vol. 536. Springer, Berlin (1976)

82. Irregularities of distribution. X. In: Zassenhaus, H. (ed.) *Number Theory and Algebra: Collected Papers Dedicated to Henry B. Mann, Arnold E. Ross, and Olga Taussky-Todd*, pp. 311–329. Academic Press, New York (1977)
83. *Small Fractional Parts of Polynomials*. Regional Conference Series in Mathematics, 32. American Mathematical Society, Providence, R.I. (1977)
84. On the distribution modulo 1 of the sequence $\alpha n^2 + \beta n$. *Can. J. Math.* **29**, 819–826 (1977)
85. Diophantine approximation in power series fields. *Acta Arith.* **32**, 275–296 (1977)
86. *Lectures on Irregularities of Distribution: Notes Taken by T. N. Shorey*. Tata Institute of Fundamental Research Lectures on Mathematics and Physics, 56. Tata Institute of Fundamental Research, Bombay (1977)
87. Thue's equation over function fields. *J. Aust. Math. Soc. Ser. A* **25**, 385–422 (1978)
88. Small zeros of additive forms in many variables. *Trans. Am. Math. Soc.* **248**, 121–133 (1979)
89. Contributions to Diophantine approximation in fields of series. *Monatsh. Math.* **87**, 145–165 (1979)
90. Small zeros of additive forms in many variables. II. *Acta Math.* **143**, 219–232 (1979)
91. with Y. Wang: A note on a transference theorem of linear forms. *Sci. Sin.* **22**, 276–280 (1979)
92. Polynomial solutions of $F(x, y) = z^n$. In: Ribenboim, P. (ed.) *Proceedings of the Queen's Number Theory Conference, 1979*. Queens Pap. Pure Appl. Math., 54, pp. 33–65. Queen's University, Kingston, Ont. (1980)
93. Diophantine inequalities for forms of odd degree. *Adv. Math.* **38** (1980), 128–151
94. Simultaneous $p$-adic zeros of quadratic forms. *Monatsh. Math.* **90**, 45–65 (1980)
95. with A. Schinzel and H.P. Schlickewei: Small solutions of quadratic congruences and small fractional parts of quadratic forms. *Acta Arith.* **37**, 241–248 (1980)
96. *Diophantine Approximation*. Lect. Notes Math., vol. 785. Springer, Berlin (1980)
97. with R.C. Baker: Diophantine problems in variables restricted to the values 0 and 1. *J. Number Theory* **12**, 460–486 (1980)
98. with R.C. Baker: Addendum: "Diophantine problems in variables restricted to the values 0 and 1". *J. Number Theory* **13**, 270 (1981)
99. Simultaneous rational zeros of quadratic forms. In: *Seminar on Number Theory, Paris 1980–81*. Prog. Math., vol. 22, pp. 281–307. Birkhäuser, Boston (1982)
100. On cubic polynomials. I. Hua's estimate of exponential sums. *Monatsh. Math.* **93**, 63–74 (1982)
101. On cubic polynomials. II. Multiple exponential sums. *Monatsh. Math.* **93**, 141–168 (1982)
102. On cubic polynomials. III. Systems of $p$-adic equations. *Monatsh. Math.* **93**, 211–223 (1982)
103. On cubic polynomials IV. Systems of rational equations. *Monatsh. Math.* **93**, 329–348 (1982)
104. The joint distribution of the digits of certain integer $s$-tuples. In: Erdős, P., Alpár, L., Halász, G., Sarözy, A. (eds.) *Studies in Pure Mathematics: to the Memory of Paul Turán*, pp. 605–622. Birkhäuser, Basel (1983)
105. with D.B. Leep: Systems of homogeneous equations. *Invent. Math.* **71**, 539–549 (1983)
106. The density of integer points on homogeneous varieties. In: *Seminar on Number Theory, Paris 1981–82*. Prog. Math., vol. 38, pp. 283–286. Birkhäuser, Boston (1983)
107. Open problems in Diophantine approximation. In: Bertrand, D., Waldschmidt, M. (eds.) *Approximations diophantiennes et nombres transcendants: colloque de Luminy, 1982*. Prog. Math., vol. 31, pp. 271–287. Birkhäuser, Boston (1983)
108. Diophantine approximation properties of certain infinite sets. *Trans. Am. Math. Soc.* **278**, 635–645 (1983)
109. Analytic methods for congruences, diophantine equations and approximations. In: *Proceedings of the International Congress of Mathematicians, Warsaw, 1983*, vol. 1, pp. 515–524. PWN, Warsaw (1984)
110. The solubility of certain $p$-adic equations. *J. Number Theory* **19**, 63–80 (1984)
111. Bounds for exponential sums. *Acta Arith.* **44**, 281–297 (1984)
112. *Analytische Methoden für Diophantische Gleichungen: einführende Vorlesungen*. DMV Seminar, vol. 5. Birkhäuser, Basel (1984)
113. Integer points on curves and surfaces. *Monatsh. Math.* **99**, 45–72 (1985)
114. Small solutions of congruences in a large number of variables. *Can. Math. Bull.* **28**, 295–305 (1985)
115. Small zeros of quadratic forms. *Trans. Am. Math. Soc.* **291**, 87–102 (1985)
116. The density of integer points on homogeneous varieties. *Acta Math.* **154**, 243–296 (1985)
117. Small solutions of congruences with prime modulus. In: Loxton, J.H., Van der Poorten, A.J. (eds.) *Diophantine Analysis: Proceedings of the Number Theory Section of the 1985 Australian Mathematical Society Convention (Kensington, N.S.W.)*. Lond. Math. Soc. Lect. Note Ser., 109, pp. 37–66. Cambridge University Press, Cambridge (1986)

118. Partitions of triangles into convex sets. *Österr. Akad. Wiss. Math. Naturwiss. Kl. Sitzungsber. II* **195**, 167–169 (1986)
119. Integer points on hypersurfaces. *Monatsh. Math.* **102**, 27–58 (1986)
120. Thue equations with few coefficients. *Trans. Am. Math. Soc.* **303**, 241–255 (1987)
121. with E. Bombieri: On Thue's equation. *Invent. Math.* **88**, 69–81 (1987)
122. with A. Florian: Zerlegung von Dreiecken in Dreiecke mit Nebenbedingung. *Geom. Dedicata* **24**, 363–368 (1987)
123. with J. Mueller: Trinomial Thue equations and inequalities. *J. Reine Angew. Math.* **379**, 76–99 (1987)
124. with H.P. Schlickewei: Quadratic geometry of numbers. *Trans. Am. Math. Soc.* **301**, 679–690 (1987)
125. The number of solutions of Thue equations. In: Baker, A. (ed.) *New Advances in Transcendence Theory*, pp. 337–346. Cambridge University Press, Cambridge (1988)
126. with H.P. Schlickewei: Quadratic forms which have only large zeros. *Monatsh. Math.* **105**, 295–311 (1988)
127. with J. Mueller: Thue's equation and a conjecture of Siegel. *Acta Math.* **160**, 207–247 (1988)
128. with J. Mueller: On the number of good rational approximations to algebraic numbers. *Proc. Am. Math. Soc.* **106**, 859–866 (1989)
129. with H.P. Schlickewei: Isotrope Unterräume rationaler quadratischer Formen. *Math. Z.* **201**, 191–208 (1989)
130. with E. Bombieri: Correction to: "On Thue's equation" [Invent. Math. **88** (1987), no. 1, 69–81, MR 88d:11026]. *Invent. Math.* **97**, 445 (1989)
131. The subspace theorem in Diophantine approximations. *Compos. Math.* **69**, 121–173 (1989)
132. The number of solutions of norm form equations. *Trans. Am. Math. Soc.* **317**, 197–227 (1990)
133. A remark on the heights of subspaces. In: Baker, A., Bollobaś, B., Hajnal, A. (eds.) *A Tribute to Paul Erdős*, pp. 359–360. Cambridge University Press, Cambridge (1990)
134. The number of solutions of norm form equations. In: Győry, K., Halaśz (eds.) *Number Theory*, vol. 2. Colloq. Math. Soc. János Bolyai, vol. 51, pp. 965–979. North-Holland, Amsterdam (1990)
135. Eisenstein's theorem on power series expansions of algebraic functions. *Acta Arith.* **56**, 161–179 (1990)
136. with H.P. Schlickewei: Bounds for zeros of quadratic forms. In: Győry, K., Halaśz (eds.) *Number Theory*, vol. 2. Colloq. Math. Soc. János Bolyai, vol. 51, pp. 951–964. North-Holland, Amsterdam (1990)
137. *Diophantine Approximations and Diophantine Equations*. Lect. Notes Math., vol. 1467. Springer, Berlin (1991)
138. On the number of good simultaneous approximations to algebraic numbers. In: Gong, S. (ed.) *International Symposium in Memory of Hua Loo Keng*, vol. I, pp. 249–264. Springer, Berlin (1991)
139. Construction and estimation of bases in function fields. *J. Number Theory* **39**, 181–224 (1991)
140. Integer points on curves of genus 1. *Compos. Math.* **81**, 33–59 (1992)
141. with J. Mueller: On the Newton polygon. *Monatsh. Math.* **113**, 33–50 (1992)
142. with H.P. Schlickewei: On polynomial-exponential equations. *Math. Ann.* **296**, 339–361 (1993)
143. Vojta's refinement of the subspace theorem. *Trans. Am. Math. Soc.* **340**, 705–731 (1993)
144. with H.P. Schlickewei: Equations $au_n^l = bu_m^k$ satisfied by members of recurrence sequences. *Proc. Am. Math. Soc.* **118**, 1043–1051 (1993)
145. Northcott's theorem on heights. I. A general estimate. *Monatsh. Math.* **115**, 169–181 (1993)
146. with H.P. Schlickewei: Linear equations in members of recurrence sequences. *Ann. Sc. Norm. Super. Pisa Cl. Sci. IV. Ser.* **20**, 219–246 (1993)
147. Bemerkungen zur Polynomdiskrepanz. *Österr. Akad. Wiss. Math. Naturwiss. Kl. Sitzungsber. II* **202**, 173–177 (1993)
148. Equations $\alpha^x = R(x, y)$. *J. Number Theory* **47**, 348–358 (1994)
149. Approximation to orthogonal bases in $\mathbb{R}^n$ by orthogonal bases with integer coordinates. Appendix to: "Approximation of viscosity solutions of elliptic partial differential equations on minimal grids" [*Numer. Math.* **72** (1995), 73–92] by M. Kocan. *Numer. Math.* **72**, 117–122 (1995)
150. Northcott's theorem on heights. II. The quadratic case. *Acta Arith.* **70**, 343–375 (1995)
151. with J. Mueller: The generalized Thue inequality. *Compos. Math.* **96**, 331–344 (1995)
152. Number fields of given degree and bounded discriminant. *Astérisque* **228**, 189–195 (1995)
153. with H.P. Schlickewei: The intersection of recurrence sequences. *Acta Arith.* **72**, 1–44 (1995)
154. The number of exceptional approximations in Roth's theorem. *J. Aust. Math. Soc. Ser. A* **59**, 375–383 (1995)
155. Heights of algebraic points lying on curves or hypersurfaces. *Proc. Am. Math. Soc.* **124**, 3003–3013 (1996)

156. Heights of points on subvarieties of $\mathbb{G}_m^n$. In: *Number Theory: Séminaire de theórie des nombres de Paris 1993–1994*. Lond. Math. Soc. Lect. Note Ser., 235, pp. 157–187. Cambridge University Press, Cambridge (1996)
157. with C.L. Stewart: Congruences, trees, and $p$-adic integers. *Trans. Am. Math. Soc.* **349**, 605–639 (1997)
158. The distribution of sublattices of $\mathbb{Z}^m$. *Monatsh. Math.* **125**, 37–81 (1998)
159. Heights of points on subvarieties of $G_m^n$. In: Murty, V.K., Waldschmidt. M. (eds.) *Number Theory: Ramanujan Mathematical Society, January 3–6, 1996, Tiruchirapalli, India.* Contemp. Math., 210, pp. 97–99. American Mathematical Society, Providence, R.I. (1998)
160. with H.P. Schlickewei and M. Waldschmidt: Zeros of linear recurrence sequences. *Manuscr. Math.* **98**, 225–241 (1999)
161. Heights of algebraic points. In: Yıldırım, C.Y., Stepanov, S.A. (eds.) *Number Theory and Its Applications: Proceedings of a Summer School at Bilkent University.* Lect. Notes Pure Appl. Math., vol. 204, pp. 185–225. Dekker, New York (1999)
162. The zero multiplicity of linear recurrence sequences. *Acta Math.* **182**, 243–282 (1999)
163. Solution trees of polynomial congruences modulo prime powers. In: Győry, K., Iwaniec, H., Urbanowicz, J. (eds.) *Number Theory in Progress: Proceedings of the International Conference on Number Theory, in Honor of the 60th Birthday of Andrzej Schinzel, Zakopane, Poland, June 30–July 9, 1997*, vol. 1, pp. 451–471. de Gruyter, Berlin (1999)
164. with H.P. Schlickewei: The number of solutions of polynomial-exponential equations. *Compos. Math.* **120**, 193–225 (2000)
165. Zeros of linear recurrence sequences. *Publ. Math. (Debrecen)* **56**, 609–630 (2000)
166. with S. Caulk: Simultaneous approximation in positive characteristic. *Monatsh. Math.* **131**, 15–28 (2000)
167. On continued fractions and Diophantine approximation in power series fields. *Acta Arith.* **95**, 139–166 (2000)
168. with B. Brindza and Á. Pintér: Multiplicities of binary recurrences. *Can. Math. Bull.* **44**, 19–21 (2001)
169. with A. Schinzel: Comparison of $L^1$- and $L^\infty$-norms of squares of polynomials. *Acta Arith.* **104**, 283–296 (2002)
170. with J.-H. Evertse and H.P. Schlickewei: Linear equations in variables which lie in a multiplicative group. *Ann. Math. (2)* **155**, 807–836 (2002)
171. with D. Evans and A.J. Jones: Asymptotic moments of near-neighbour distance distributions. *R. Soc. Lond. Proc. Ser. A Math. Phys. Eng. Sci* **458**, 2839–2849 (2002)
172. Complementary sets of finite sets. *Monatsh. Math.* **138**, 61–71 (2003)
173. Linear recurrence sequences. In: Amoroso, F., Zannier, U. (eds.) *Diophantine Approximation: Lectures Given at the C.I.M.E. Summer School Held in Cetraro, Italy, June 28–July 6, 2000.* Lect. Notes Math., vol. 1819, pp. 171–247. Springer, Berlin (2003)
174. *Equations over Finite Fields: an Elementary Approach*, 2nd edn. Kendrick Press, Heber City, Utah (2004)
175. Diophantine equations $E(x) = P(x)$ with $E$ exponential, $P$ polynomial. *Acta Arith.* **113**, 351–362 (2004)
176. Rationality of exponential functions at integer arguments. *J. Number Theory* **106**, 285–298 (2004)
177. with I. Aliev and A. Schinzel: On vectors whose span contains a given linear subspace. *Monatsh. Math.* **144**, 177–191 (2005)
178. with A. Schinzel: The mathematical work of Eduard Wirsing. *Funct. Approximatio Comment. Math.* **35**, 7–18 (2006)
179. Mahler and Koksma classification of points in $\mathbb{R}^n$ and $\mathbb{C}^n$. *Funct. Approximatio Comment. Math.* **35**, 307–319 (2006)
180. Diophantine approximation by algebraic hypersurfaces and varieties. *Trans. Am. Math. Soc.* **359**, 2221–2241 (2007)
181. Covering and packing I. *Monatsh. Math.* (to appear)
182. The diophantine equation $\alpha_1^{(x_1)} \dots \alpha_n^{(x_n)} = f(x_1, \dots, x_n)$. In: Chen, W., Gowers, T., Halberstam, H., Schmidt, W., Vaughan, R.C. (eds.) *Analytical Number Theory: Essays in Honour of Klaus F. Roth.* Cambridge University Press, Cambridge (to appear)


## Additional cited references


[Ba] Baker, R.C.: Weyl sums and Diophantine approximation. *J. Lond. Math. Soc. II. Ser.* **25**, 25–34 (1982)

[Be] Beck, J.: Sums of distances between points on a sphere – an application of the theory of irregularities of distributions to discrete geometry. *Mathematika* **31**, 33–41 (1984)

[Bi] Birch, B.J.: Homogeneous forms of odd degree in a large number of variables. *Mathematika* **4**, 102–105 (1957)

[Bo] Bombieri, E.: Counting points on curves over finite fields (d'après S.A. Stepanov). In: *Séminaire Bourbaki, 1972/73, Exposes 418–435*. Lect. Notes Math., vol. 383, pp. 234–241. Springer, Berlin (1974)

[BZ] Bombieri, E., Zannier, U.: Algebraic points on subvarieties of $G_m^n$. *Int. Math. Res. Not.* **7**, 333–347 (1995)

[C1] Cassels, J.W.S.: Bounds for the least solutions of homogeneous quadratic equations. *Proc. Camb. Philos. Soc.* **51**, 262–264 (1955)

[C2] Cassels, J.W.S.: *An Introduction to the Geometry of Numbers*. Die Grundlehren der mathematischen Wissenschaften in Einzeldarstellungen, Bd. 99. Springer, Berlin (1959)

[Da] Danicic, I.: An extension of a theorem of Heilbronn. *Mathematika* **5**, 249–256 (1958)

[DP] David, S., Philippon, P.: Minorations des hauteurs normalisées des sou-variétés des tores. *Ann. Sc. Norm. Super. Pisa Cl. Sci IV. Ser.* **28**, 489–543 (1999)

[ES] Evertse, J.-H., Schlickewei, H.P.: A quantitative version of the absolute subspace theorem. *J. Reine Angew. Math.* **548**, 21–127 (2002)

[He] Heilbronn, H.A.: On the distribution of the sequence $n^2\theta(mod 1)$. *Q. J. Math. Oxf. Ser.* **19**, 249–256 (1948)

[Hl] Hlawka, E.: Zur Geometrie der Zahlen. *Math. Z.* **42**, 285–312 (1943)

[Ma] Mahler, K.: Zur Approximation der Exponentialfunktion und des Logarithmus I. *J. Reine Angew. Math.* **166**, 118–136 (1932)

[Pi] Pitman, J.: Cubic inequalities. *J. Lond. Math. Soc.* **43**, 119–126 (1968)

[R1] Roth, K.F.: On irregularities of distribution. *Mathematika* **1**, 73–79 (1954)

[R2] Roth, K.F.: Rational approximations to algebraic numbers. *Mathematika* **2**, 1–20 (1955)

[St] Stepanov, S.A.: An elementary proof of the Hasse–Weil theorem for hyperelliptic curves. *J. Number Theory* **4**, 118–143 (1972)

[Th] Thue, A.: Om en generel i store hele tal uløsbar ligning. *Christiania Vidensk. Selsk. Skr. I Mat. Nat. Kl.* **7** (1908)

[W1] Wirsing, E.: Approximation mit algebraischen Zahlen beschränkten Grades. *J. Reine Angew. Math.* **206**, 67–77 (1961)

[W2] Wirsing, E.: On approximation of algebraic numbers by algebraic numbers of bounded degree. In: *1969 Number Theory Institute – Proceedings of the 1969 Summer Institute on Number Theory – American Mathematical Society.* Proc. Symp. Pure Math., 20, pp. 213–247. American Mathematical Society, Providence, R.I. (1971)

[Zh] Zhang, S.: Positive line bundles on arithmetic surfaces. *Ann. Math.* **136**, 569–587 (1992)

*Diophantine Approximation*
H. P. Schlickewei et al., Editors


# SCHÄFFER'S DETERMINANT ARGUMENT

Roger C. Baker
*Department of Mathematics, Brigham Young University, Provo, Utah 84602, U.S.A.*
baker@math.byu.edu

To Wolfgang M. Schmidt on the occasion of his 70th birthday

## 1 Introduction

Let $\| \ldots \|$ denote distance from the nearest integer. Various versions of the following problem in simultaneous Diophantine approximation have been studied since 1957, beginning with Danicic [5]. Given an integer $h \geq 2$. we seek a number $\theta$ having the following property, for every $\epsilon > 0$ and every pair $\boldsymbol{\alpha} = (\alpha_1, \ldots, \alpha_h)$, $\boldsymbol{\beta} = (\beta_1, \ldots, \beta_h)$ in $\mathbb{R}^h$:

*For $N > C(h, \epsilon)$, there is an integer $n$, $1 \leq n \leq N$, satisfying*

$$\|n^2\alpha_j + n\beta_j\| < N^{-\theta+\epsilon} \quad (j = 1, \ldots, h).$$

It is convenient to say that $\theta$ is *admissible for h quadratic polynomials* if $\theta$ possesses the above property. The best known result for general $h$ is that

$$\frac{1}{h^2+h} \text{ is admissible for } h \text{ quadratic polynomials.} \tag{1.1}$$

Most of the ideas leading to (1.1) occur in the lectures of W. M. Schmidt [7]. In particular [7] contains the corresponding result for the special case $\boldsymbol{\beta} = \mathbf{0}$. The finishing touches for (1.1) are in [1,2]; see also [3]. One should note the correction in [4], which applies equally to Theorem 5.1 of [3]. This theorem is used in proving (1.1) in [3], and again in the present paper.

Schäffer [6] was able to improve (1.1) in the case $h = 2$, showing that $2/11$ is admissible for a pair of quadratic polynomials. The key to his improvement is Lemma 4 of [6], which we need not restate here since it is essentially subsumed under Theorems 2 and 3 below. Schäffer's lemma is an ingenious refinement of the "determinant argument" of Schmidt. This is Lemma 18A of [7], abstracted as Lemma 7.6 in [3] and repeated below as Lemma 4.

Theorems 2 and 3 will be applied to give the following modest improvement of (1.1).

---

*Keywords.* Fractional part, quadratic polynomial, exponential sum, lattice, successive minima.

*2000 Mathematics subject classification.* 11J54.

**Theorem 1.** *Let $h \geq 3$. The number $(h^2 + h - 1/2)^{-1}$ is admissible for $h$ quadratic polynomials.*

We now give a version of Schäffer's lemma for $\mathbb{R}^h$. We write $\boldsymbol{ab}$ for inner product in $\mathbb{R}^h$, and $|\boldsymbol{a}| = (\boldsymbol{aa})^{1/2}$. The constants $C(h, \epsilon)$, $C(h)$ need not be the same at each occurrence. The cardinality of a finite set $\mathcal{E}$ is denoted by $|\mathcal{E}|$.

**Theorem 2.** *Let $h \geq 2$, $\epsilon > 0$, $M > C(h, \epsilon)$, $A \geq 1$, $U \geq 1$, $UA \leq M$ and $0 < V < 1$, with*

$$M^{h-1+\epsilon} AV < 1. \tag{1.2}$$

*Let $\boldsymbol{e} \in \mathbb{R}^h$. Let $\mathcal{A}$ be a subset of $\mathbb{Z}^h$, with*

$$|\mathcal{A}| > M^{2\epsilon} \max(1, (M^h V)^{h/(h+1)}).$$

*Suppose that, for $\boldsymbol{p}$ in $\mathcal{A}$, we have*

$$|\boldsymbol{p}| \leq A, \tag{1.3}$$

*and there are coprime integers $\ell(\boldsymbol{p})$, $w(\boldsymbol{p})$,*

$$0 < \ell(\boldsymbol{p}) \leq U, \tag{1.4}$$

*with*

$$|\ell(\boldsymbol{p})\boldsymbol{pe} - w(\boldsymbol{p})| < V. \tag{1.5}$$

*Then there is a subset $\mathcal{C}$ of $\mathcal{A}$ and a natural number $\ell$ such that*

$$|\mathcal{C}| \geq |\mathcal{A}| M^{-\epsilon} \min(1, (M^h V)^{-h/(h+1)})$$

*and $\ell(\boldsymbol{p}) = \ell$ for all $\boldsymbol{p}$ in $\mathcal{C}$.*

In Theorem 3, we assume a somewhat similar situation but we suppose that there is some "known repetition" among the $\ell(\boldsymbol{p})$. We use this to get a "lot of repetition". The linear span of a set $S$ in $\mathbb{R}^h$ is denoted by Span $S$.

**Theorem 3.** *Let $h \geq 2$, $\epsilon > 0$, $M > C(h, \epsilon)$, $A \geq 1$, $U \geq 1$, $UA \leq M$, $0 < V < 1$ and let $\boldsymbol{e} \in \mathbb{R}^h$. Let $\mathcal{A}$ be a subset of $\mathbb{Z}^h$, $W =$ Span $\mathcal{A}$, dim $W = m$. Suppose that, for each $\boldsymbol{p}$ in $\mathcal{A}$,*

$$A/2 < |\boldsymbol{p}| \leq A, \tag{1.6}$$

*and there exist coprime integers $\ell(\boldsymbol{p})$, $w(\boldsymbol{p})$ satisfying*

$$U/2 < \ell(\boldsymbol{p}) \leq U, \tag{1.7}$$

$$|\ell(\boldsymbol{p})\boldsymbol{pe} - w(\boldsymbol{p})| < V. \tag{1.8}$$

*Suppose that for some integer $n$, $2 \leq n \leq m$ with*

$$C(h) U^{1+m-n} A^m V < 1 \tag{1.9}$$

*for a suitable positive $C(h)$, there are linearly independent $\boldsymbol{p}_1, \ldots, \boldsymbol{p}_n$ in $\mathcal{A}$ with $\ell(\boldsymbol{p}_1) = \ell(\boldsymbol{p}_2) = \cdots = \ell(\boldsymbol{p}_n)$. Then there is a subset $\mathcal{C}$ of $\mathcal{A}$ and a natural number $\ell'$ such that*

$$|\mathcal{C}| > |\mathcal{A}| M^{-\epsilon}$$

*and $\ell(\boldsymbol{p}) = \ell'$ for all $\boldsymbol{p}$ in $\mathcal{C}$.*

Lemma 4 of [6] is essentially equivalent to the cases $h = 2$ of Theorems 3 and 4, taken together.

## 2 Proofs of Theorems 2 and 3

As in [3], the *determinant* of $t$ vectors $\boldsymbol{a}_1, \ldots, \boldsymbol{a}_t$ in $\mathbb{R}^h$, where $1 \le t \le h$, is the $t$-dimensional volume of the parallellepiped

$$\left\{ \sum_{i=1}^{t} y_i \boldsymbol{a}_i : 0 \le y_1, \ldots, y_t \le 1 \right\}$$

and is denoted by $\det(\boldsymbol{a}_1, \ldots, \boldsymbol{a}_t)$. Note that

$$\det(\boldsymbol{a}_1, \ldots, \boldsymbol{a}_t)^2 = \det\{\boldsymbol{a}_i \boldsymbol{a}_j : 1 \le i, j \le t\}$$

is an integer whenever $\boldsymbol{a}_1, \ldots, \boldsymbol{a}_t$ are in $\mathbb{Z}^h$; compare [8, equation (2.1), p. 4]. If $\boldsymbol{a}_1, \ldots, \boldsymbol{a}_t$ are linearly independent, and

$$\Lambda = \left\{ \sum_{i=1}^{t} n_i \boldsymbol{a}_i : n_1, \ldots, n_t \in \mathbb{Z} \right\}$$

is the $t$-dimensional lattice generated by $\boldsymbol{a}_1, \ldots, \boldsymbol{a}_t$, then the *determinant* of $\Lambda$ is defined to be $d(\Lambda) = \det(\boldsymbol{a}_1, \ldots, \boldsymbol{a}_t)$.

The unit ball in $\mathbb{R}^h$ is denoted by $K_0$.

We begin with a few observations from linear algebra.

**Lemma 1.** *Let $\boldsymbol{v}_0, \boldsymbol{v}_1, \ldots, \boldsymbol{v}_h$ be in $\mathbb{R}^h$, $\boldsymbol{v}_0 \ne \boldsymbol{0}$. Then*

$$\det(\boldsymbol{v}_1, \ldots, \boldsymbol{v}_h) \le h \max\left(\frac{|\boldsymbol{v}_1|}{|\boldsymbol{v}_0|}, \ldots, \frac{|\boldsymbol{v}_h|}{|\boldsymbol{v}_0|}\right) \max_i \det(\boldsymbol{v}_0, \boldsymbol{v}_1, \ldots, \boldsymbol{v}_{i-1}, \boldsymbol{v}_{i+1}, \ldots, \boldsymbol{v}_h).$$

*Proof.* Evidently we may suppose that $|\boldsymbol{v}_0| = 1$ and, after applying a linear isometry to $\mathbb{R}^h$, that $\boldsymbol{v}_0 = (1, 0, \ldots, 0)$. Let $\boldsymbol{v}_i = (v_{i1}, \ldots, v_{ih})$, and let $M_i$ be the cofactor of $v_{i1}$ in the matrix $A = [v_{ij} : 1 \le i, j \le h]$. Then

$$\det(\boldsymbol{v}_1, \ldots, \boldsymbol{v}_h) \le \sum_{i=1}^{h} |v_{i1} M_i| \le h \max_i |\boldsymbol{v}_i| \max_i |M_i|. \tag{2.1}$$

Now consider the matrix $A_i$ obtained by replacing row $i$ of $A$ by $\boldsymbol{v}_0$. We have

$$\det(\boldsymbol{v}_0, \boldsymbol{v}_1, \ldots, \boldsymbol{v}_{i-1}, \boldsymbol{v}_{i+1}, \ldots, \boldsymbol{v}_h) = |\det A_i| = |M_i|. \tag{2.2}$$

The lemma follows from (2.1), (2.2). □

**Lemma 2.** *Let $\boldsymbol{x}_1, \ldots, \boldsymbol{x}_h$ be linearly independent in $\mathbb{R}^h$. The distance between parallel hyperplanes*

$$\boldsymbol{c} + a_i \boldsymbol{x}_h + \text{Span}\{\boldsymbol{x}_1, \ldots, \boldsymbol{x}_{h-1}\} \quad (i = 1, 2)$$

*is*

$$|a_1 - a_2| \frac{\det(\boldsymbol{x}_1, \ldots, \boldsymbol{x}_{h-1}, \boldsymbol{x}_h)}{\det(\boldsymbol{x}_1, \ldots, \boldsymbol{x}_{h-1})}.$$

*Proof.* It suffices to show that the distance $d$ from $\boldsymbol{x}_h$ to $\text{Span}\{\boldsymbol{x}_1, \ldots, \boldsymbol{x}_{h-1}\}$ is

$$\frac{\det(\boldsymbol{x}_1, \ldots, \boldsymbol{x}_h)}{\det(\boldsymbol{x}_1, \ldots, \boldsymbol{x}_{h-1})}.$$

We use the Gram–Schmidt process to replace $\boldsymbol{x}_1, \ldots, \boldsymbol{x}_h$ by an orthogonal set

$$\boldsymbol{v}_1 = \boldsymbol{x}_1,\ \boldsymbol{v}_2 = \boldsymbol{x}_2 - \frac{\boldsymbol{x}_2 \cdot \boldsymbol{v}_1}{\boldsymbol{v}_1 \cdot \boldsymbol{v}_1}\,\boldsymbol{v}_1$$

and so on. Note that

$$\det(\boldsymbol{x}_1, \ldots, \boldsymbol{x}_i) = \det(\boldsymbol{v}_1, \ldots, \boldsymbol{v}_i) = |\boldsymbol{v}_1| \ldots |\boldsymbol{v}_i| \quad (i = 1, \ldots, h).$$

Hence

$$d = |\boldsymbol{v}_h| = \frac{\det(\boldsymbol{v}_1, \ldots, \boldsymbol{v}_h)}{\det(\boldsymbol{v}_1, \ldots, \boldsymbol{v}_{h-1})} = \frac{\det(\boldsymbol{x}_1, \ldots, \boldsymbol{x}_h)}{\det(\boldsymbol{x}_1, \ldots, \boldsymbol{x}_{h-1})}.$$

□

**Lemma 3.** *Let $\boldsymbol{x}_1, \ldots, \boldsymbol{x}_h$ be linearly independent points of a hyperplane with equation $\boldsymbol{a}\boldsymbol{x} = b$ in $\mathbb{R}^h$. Then for any $\boldsymbol{x}$ in the hyperplane, we have*

$$\boldsymbol{x} = t_1 \boldsymbol{x}_1 + \cdots + t_h \boldsymbol{x}_h \tag{2.3}$$

*with $t_1 + \cdots + t_h = 1$.*

*Proof.* We may define $t_1, \ldots, t_h$ uniquely via (2.3). On each side of (2.3), take the inner product with $\boldsymbol{a}$:

$$\begin{aligned} b = \boldsymbol{a}\boldsymbol{x} &= t_1 \boldsymbol{a}\boldsymbol{x}_1 + \cdots + t_h \boldsymbol{a}\boldsymbol{x}_h \\ &= (t_1 + \cdots + t_h) b. \end{aligned}$$

Since $b \neq 0$ by independence of $\boldsymbol{x}_1, \ldots, \boldsymbol{x}_h$, the lemma follows. □

Throughout the remainder of the paper, constants implied by $\ll$ depend at most on $h$ and $\epsilon$. We suppose (as we may) that $\epsilon$ is sufficiently small. A quantity of the form $C(h)\epsilon^2$ is denoted by $\delta$.

**Lemma 4.** *Let $N > C(h, \epsilon)$. Let $\Lambda$ be an $h$-dimensional lattice in $\mathbb{R}^h$ with $d(\Lambda) = D \le N^2$ and $\Lambda \cap K_0 = \{\boldsymbol{0}\}$. Let $\Pi$ be the dual lattice of $\Lambda$. Let $\mathcal{A}$ be a subset of $\Pi$ with $|\boldsymbol{p}| \le N$ for all $\boldsymbol{p}$ in $\mathcal{A}$. Suppose that* Span $\mathcal{A}$ *has dimension $t$, and that any $t$ vectors in $\mathcal{A}$ have determinant $\le Z$. Let $\boldsymbol{e} \in \mathbb{R}^h$. Let $U$, $V$ be positive numbers, $U \le N$, such that for any $\boldsymbol{p}$ in $\mathcal{A}$ there are coprime integers $\ell(\boldsymbol{p})$, $w(\boldsymbol{p})$ satisfying*

$$1 \le \ell(\boldsymbol{p}) \le U,\ |\ell(\boldsymbol{p})\boldsymbol{e}\boldsymbol{p} - w(\boldsymbol{p})| < V.$$

*Suppose further that*

$$Z U^t V D N^{\epsilon} \le 1.$$

*Then there is an integer $\ell$ and a subset $\mathcal{C}$ of $\mathcal{A}$ with $|\mathcal{C}| \ge |\mathcal{A}| N^{-\delta}$, $\ell(\boldsymbol{p}) = \ell$ for all $\boldsymbol{p}$ in $\mathcal{C}$.*

*Proof.* As noted above, this is Lemma 7.6 of [3]. □

*Proof of Theorem 2.* There are two cases to consider.

*Case 1.* There is a subset $\mathcal{E}$ of $\mathcal{A}$ with

$$|\mathcal{E}| > |\mathcal{A}| M^{-\epsilon/2} \min(1, (M^h V)^{-h/(h+1)}),$$

such that Span $\mathcal{E}$ has dimension $t \le h - 1$.

We apply Lemma 4 with $\Lambda = \Pi = \mathbb{Z}^h$, $D = 1$, $N = M$ and $\mathcal{E}$, $\epsilon^2$ in place of $\mathcal{A}$, $\epsilon$. Clearly we may take $Z = A^t$. Now

$$\begin{aligned} ZU^t V D N^{\epsilon^2} &\ll A^t U^t V M^{\epsilon^2} \\ &\ll M^{h-1+\epsilon^2} V \ll M^{-\epsilon^2}. \end{aligned}$$

Hence there is a subset $\mathcal{C}$ of $\mathcal{E}$ and a natural number $\ell$ such that

$$|\mathcal{C}| \geq |\mathcal{E}| M^{-\delta} > |\mathcal{A}| M^{-\epsilon} \min(1, (M^h V)^{-h/(h+1)})$$

and $\ell(\boldsymbol{p}) = \ell$ for all $\boldsymbol{p}$ in $\mathcal{C}$.

*Case* 2. Case 1 does not hold.

It is convenient to write

$$f(\boldsymbol{x}) = \boldsymbol{x}\boldsymbol{e},\, R = (M/V)^{1/(h+1)}.$$

By Dirichlet's theorem, there is a point $\boldsymbol{p}_0$ in $\mathbb{Z}^h$ and an integer $w_0$ such that

$$0 < |\boldsymbol{p}_0| \leq R,\ |f(\boldsymbol{p}_0) - w_0| \ll R^{-h}. \tag{2.4}$$

We choose $\boldsymbol{p}_1, \ldots, \boldsymbol{p}_{h-1}$ in $\mathcal{A}$ to maximize

$$C = \det(\boldsymbol{p}_0, \ell(\boldsymbol{p}_1)\boldsymbol{p}_1, \ldots, \ell(\boldsymbol{p}_{h-1})\boldsymbol{p}_{h-1}).$$

Since we are in case 2, we have Span $\mathcal{A} = \mathbb{R}^h$ and $C > 0$. Let us write

$$\ell_j = \ell(\boldsymbol{p}_j),\, w_j = w(\boldsymbol{p}_j) \quad (j = 1, \ldots, h).$$

We note that

$$\det(\boldsymbol{p}_0, \ell_1\boldsymbol{p}_1, \ldots, \ell_{j-1}\boldsymbol{p}_{j-1}, \ell(\boldsymbol{p})\boldsymbol{p}, \ell_{j+1}\boldsymbol{p}_{j+1}, \ldots, \ell_{h-1}\boldsymbol{p}_{h-1}) \leq C \tag{2.5}$$

for all $\boldsymbol{p}$ in $\mathcal{A}$, by choice of $\boldsymbol{p}_1, \ldots, \boldsymbol{p}_{h-1}$, while

$$\det(\ell(\boldsymbol{p})\boldsymbol{p}, \ell_1\boldsymbol{p}_1, \ldots, \ell_{h-1}\boldsymbol{p}_{h-1}) \ll \frac{M}{|\boldsymbol{p}_0|} C \tag{2.6}$$

by Lemma 1, (2.5), (1.3) and (1.4).

It follows from (2.5), (2.6) and Cramer's rule that if we write $\ell(\boldsymbol{p})\boldsymbol{p}$ in the form

$$\ell(\boldsymbol{p})\boldsymbol{p} = y_0\boldsymbol{p}_0 + y_1\boldsymbol{p}_1 + \cdots + y_{h-1}\boldsymbol{p}_{h-1}, \tag{2.7}$$

then

$$|y_i| \leq 1 \quad (i = 1, \ldots, h-1),\ |y_0| \ll \frac{M}{|\boldsymbol{p}_0|}. \tag{2.8}$$

Let $E(\boldsymbol{x})$ be the linear function on $\mathbb{R}^h$ for which

$$E(\boldsymbol{p}_0) = w_0,\ E(\ell_j\boldsymbol{p}_j) = w_j \quad (j = 1, \ldots, h-1).$$

Then $E$ takes the form

$$E(\boldsymbol{x}) = \frac{1}{C}\boldsymbol{B}\boldsymbol{x}$$

with $\boldsymbol{B} = (B_1, \ldots, B_h) \in \mathbb{Z}^h$. Let us write gcd $(B_1, \ldots, B_h, C) = D$ for the greatest common divisor of $B_1, \ldots, B_h$ and $C$.

Now consider the linear function $F = f - E$. We have

$$F(\boldsymbol{p}_0) = f(\boldsymbol{p}_0) - w_0 \ll R^{-h},$$
$$F(\ell_j \boldsymbol{p}_j) = f(\ell_j \boldsymbol{p}_j) - w_j \ll V \quad (j = 1, \ldots, h-1),$$

from (2.4), (1.5). Taking into account (2.7), (2.8),

$$\begin{aligned} F(\ell(\boldsymbol{p})\boldsymbol{p}) &\ll \frac{M}{|\boldsymbol{p}_0|} R^{-h} + V \\ &\ll |\boldsymbol{p}_0|^{-1}(MR^{-h} + RV) \\ &\ll |\boldsymbol{p}_0|^{-1} M^{1/(h+1)} V^{h/(h+1)} \end{aligned} \tag{2.9}$$

for all $\boldsymbol{p}$ in $\mathcal{A}$. It is convenient to define

$$H = |\boldsymbol{p}_0|^{-1} M^{1/(h+1)} V^{h/(h+1)};$$

the above calculation gives $V \ll H$.

We can now give a bound for the integer

$$k(\boldsymbol{p}) = CD^{-1}(E(\ell(\boldsymbol{p})\boldsymbol{p}) - w(\boldsymbol{p})). \tag{2.10}$$

We have

$$\begin{aligned} k(\boldsymbol{p}) &= CD^{-1}(-F(\ell(\boldsymbol{p})\boldsymbol{p}) + f(\ell(\boldsymbol{p})\boldsymbol{p}) - w(\boldsymbol{p})) \\ &\ll CD^{-1}(H + V) \ll CD^{-1}H \end{aligned}$$

from (2.9), (1.5).

Next we distinguish two subcases of case 2.

*Case 2a.* $CD^{-1}H < M^{-\epsilon^2}$. In this case, $k(\boldsymbol{p}) = 0$ for all $\boldsymbol{p}$ in $\mathcal{A}$. The points

$$(\ell(\boldsymbol{p})\boldsymbol{p}, w(\boldsymbol{p}))$$

lie in an $h$-dimensional hyperplane in $\mathbb{R}^{h+1}$. If we fix any $h$ linearly independent points $\boldsymbol{p}_1', \ldots, \boldsymbol{p}_h'$ of $\mathcal{A}$, then for any $\boldsymbol{p}$ in $\mathcal{A}$,

$$\det \begin{bmatrix} \ell(\boldsymbol{p})\boldsymbol{p} & w(\boldsymbol{p}) \\ \ell(\boldsymbol{p}_1')\boldsymbol{p}_1' & w(\boldsymbol{p}_1') \\ \vdots & \\ \ell(\boldsymbol{p}_h')\boldsymbol{p}_h' & w(\boldsymbol{p}_h') \end{bmatrix} = 0.$$

Expanding by the first row,

$$0 = \ell(\boldsymbol{p})G \pm w(\boldsymbol{p}) \det(\ell(\boldsymbol{p}_1')\boldsymbol{p}_1', \ldots, \ell(\boldsymbol{p}_h')\boldsymbol{p}_h')$$

for some integer $G$, so that $\ell(\boldsymbol{p})$ is a divisor of

$$L = \det(\ell(\boldsymbol{p}_1')\boldsymbol{p}_1', \ldots, \ell(\boldsymbol{p}_h')\boldsymbol{p}_h').$$

Since $L \le M^h$, $L$ has at most $M^\epsilon$ divisors, and there is a divisor $\ell$ of $L$ such that $\ell(\boldsymbol{p}) = \ell$ for $\boldsymbol{p}$ in a subset $\mathcal{F}$ of $\mathcal{A}$ with $|\mathcal{F}| > |\mathcal{A}|M^{-\epsilon}$.

*Case 2b.* $CD^{-1}H \ge M^{-\epsilon^2}$. In this case, there are

$$\ll CD^{-1}H + 1 \ll CD^{-1}HM^{\epsilon^2}$$

possible values of $k(\boldsymbol{p})$. There is an integer $k$ and a subset $\mathcal{A}_1$ of $\mathcal{A}$ with

$$|\mathcal{A}_1| \gg |\mathcal{A}|C^{-1}DH^{-1}M^{-\epsilon^2}, \tag{2.11}$$
$$k(\boldsymbol{p}) = k \text{ for all } \boldsymbol{p} \text{ in } \mathcal{A}_1. \tag{2.12}$$

In particular, the subset $S$ of $\mathbb{Z}^h$ consisting of solutions of

$$D^{-1}\boldsymbol{B}\boldsymbol{x} \equiv k \pmod{CD^{-1}}$$

contains $\{\ell(\boldsymbol{p})\boldsymbol{p} : \boldsymbol{p} \in \mathcal{A}_1\}$. Now $S$ is a translate $\Lambda_0 + \boldsymbol{R}$ of the sublattice $\Lambda_0$ of $\mathbb{Z}^h$ consisting of solutions of

$$D^{-1}\boldsymbol{B}\boldsymbol{x} \equiv 0 \pmod{CD^{-1}}.$$

It is easy to see that $\det \Lambda_0 = CD^{-1}$.

The lattice $\Lambda_1$ generated by $\boldsymbol{p}_0, \ell_1\boldsymbol{p}, \ldots, \ell_{h-1}\boldsymbol{p}_{h-1}$ is contained in $\Lambda_0$, since

$$D^{-1}\boldsymbol{B}\boldsymbol{p}_0 = CD^{-1}w_0, D^{-1}B\ell_j\boldsymbol{p}_j = CD^{-1}w_j \quad (j = 1, \ldots, h-1).$$

The index of $\Lambda_1$ in $\Lambda_0$ is

$$\frac{\det \Lambda_1}{\det \Lambda_0} = \frac{C}{CD^{-1}} = D.$$

Hence we can write $\Lambda_0$ as a union of $D$ translates of $\Lambda_1$. We conclude that there is a $\boldsymbol{Q}$ in $\mathbb{Z}^h$ and a subset $\mathcal{A}_2$ of $\mathcal{A}_1$ such that

$$\ell(\boldsymbol{p})\boldsymbol{p} \in \boldsymbol{Q} + \Lambda_1 \quad (\boldsymbol{p} \in \mathcal{A}_2), \tag{2.13}$$
$$|\mathcal{A}_2| \geq |\mathcal{A}_1|D^{-1} \gg |\mathcal{A}|C^{-1}H^{-1}M^{-\epsilon^2}, \tag{2.14}$$

from (2.11).

We now seek a hyperplane that contains many of the points $\ell(\boldsymbol{p})\boldsymbol{p}$ with $\boldsymbol{p}$ in $\mathcal{A}_2$. For $n \in \mathbb{Z}$, let

$$L_n = \boldsymbol{Q} + n\ell_{h-1}\boldsymbol{p}_{h-1} + \text{Span}\,\{\boldsymbol{p}_0, \ell_1\boldsymbol{p}_1, \ldots, \ell_{h-2}\boldsymbol{p}_{h-2}\}.$$

If $n_0$ and $n_1$ are the smallest and largest integers for which $L_n$ meets the ball $MK_0$, then

$$n_0 - n_1 \ll \frac{M \det(\boldsymbol{p}_0, \ell_1\boldsymbol{p}_1, \ldots, \ell_{h-2}\boldsymbol{p}_{h-2})}{C}$$

(by Lemma 2)

$$\ll \frac{M^{h-1}|\boldsymbol{p}_0|}{C}.$$

Since $C \leq |\boldsymbol{p}_0|M^{h-1}$, it follows that there is an $n$ for which

$$|\{\boldsymbol{p} \in \mathcal{A}_2 : \ell(\boldsymbol{p})\boldsymbol{p} \in L_n\}| \gg |\mathcal{A}_2|CM^{-h+1}|\boldsymbol{p}_0|^{-1}$$
$$\gg |\mathcal{A}|H^{-1}M^{-h+1-\delta}|\boldsymbol{p}_0|^{-1}$$

(from (2.14))

$$\gg |\mathcal{A}|V^{-h/(h+1)}M^{-h^2/(h+1)-\delta}$$

from the definition of $H$. Let

$$\mathcal{A}_3 = \{\boldsymbol{p} \in \mathcal{A}_2 : \ell(\boldsymbol{p})\boldsymbol{p} \in L_n\}.$$

Since we are in case 2, Span $\mathcal{A}_3$ is $\mathbb{R}^h$. We select linearly independent points $\boldsymbol{p}'_1, \ldots, \boldsymbol{p}'_h$ in $\mathcal{A}_3$.

Recalling Lemma 3, for any $\boldsymbol{p}$ in $\mathcal{A}_3$, there are real $t_1, \ldots, t_h$ with

$$t_1 + \cdots + t_h = 1, \tag{2.15}$$

$$\ell(\boldsymbol{p})\boldsymbol{p} = t_1\ell(\boldsymbol{p}'_1)\boldsymbol{p}'_1 + \cdots + t_h\ell(\boldsymbol{p}'_h)\boldsymbol{p}'_h. \tag{2.16}$$

Now

$$\det\begin{bmatrix} \ell(\boldsymbol{p})\boldsymbol{p} & w(\boldsymbol{p}) \\ \ell(\boldsymbol{p}'_1)\boldsymbol{p}'_1 & w(\boldsymbol{p}'_1) \\ \vdots & \\ \ell(\boldsymbol{p}'_h)\boldsymbol{p}'_h & w(\boldsymbol{p}'_h) \end{bmatrix} = \det\begin{bmatrix} \ell(\boldsymbol{p})\boldsymbol{p} & -kC^{-1}D \\ \ell(\boldsymbol{p}'_1)\boldsymbol{p}'_1 & -kC^{-1}D \\ \vdots & \\ \ell(\boldsymbol{p}'_h)\boldsymbol{p}'_h & -kC^{-1}D \end{bmatrix}$$

(subtract $B_j/C$ times column $j$ from column $h+1$ for $j = 1, \ldots, h$ and use (2.10), (2.12))

$$= \det\begin{bmatrix} \mathbf{0} & 0 \\ \ell(\boldsymbol{p}'_1)\boldsymbol{p}'_1 & -kC^{-1}D \\ \vdots & \\ \ell(\boldsymbol{p}'_h)\boldsymbol{p}'_h & -kC^{-1}D \end{bmatrix} = 0.$$

For the penultimate step, we subtract $t_1$ times row $2, \ldots, t_h$ times row $h+1$ from row 1 and use (2.15), (2.16).

We can now argue as in case 2a to show that there is a subset $\mathcal{C}$ of $\mathcal{A}_3$, on which $\ell(\boldsymbol{p})$ is constant, say, $\ell(\boldsymbol{p}) = \ell'$, satisfying

$$|\mathcal{C}| \geq |\mathcal{A}_3|M^{-\delta} \geq |\mathcal{A}|M^{-\epsilon}(M^hV)^{-h/(h+1)}.$$

Thus a subset $\mathcal{C}$ of $\mathcal{A}$ with the required properties exists in all cases. □

*Proof of Theorem 3.* Let $\Gamma$ denote the $m$-dimensional lattice $\mathbb{Z}^h \cap W$; let $\boldsymbol{x}_1, \ldots, \boldsymbol{x}_m$ be a basis of $\Gamma$. Let $\boldsymbol{p}_{n+1}, \ldots, \boldsymbol{p}_m$ be chosen in $\mathcal{A}$ so that $\boldsymbol{p}_1, \ldots, \boldsymbol{p}_m$ is a basis of $W$. Let us write $\ell(\boldsymbol{p}_j) = \ell_j$, $w(\boldsymbol{p}_j) = w_j$ $(j = 1, \ldots, m)$. Thus

$$\ell_j = \ell_1 \qquad (j = 2, \ldots, n). \tag{2.17}$$

We now write

$$\boldsymbol{p}_j = p_{j1}\boldsymbol{x}_1 + \cdots + p_{jm}\boldsymbol{x}_m \quad (j = 1, \ldots, m),$$

so that the $p_{jk}$ are integers. Let $P$ be the matrix $[\ell_j p_{jk}]_{1 \leq j,k \leq m}$. Then

$$\det(\ell_1\boldsymbol{p}_1, \ldots, \ell_m\boldsymbol{p}_m) = |\det P| \det(\boldsymbol{x}_1, \ldots, \boldsymbol{x}_m). \tag{2.18}$$

We now imitate the construction in the previous proof. Define the linear function $E_1$ on $W$ by the conditions

$$E_1(\ell_j\boldsymbol{p}_j) = w_j \qquad (j = 1, \ldots, m).$$

Let $A_j = E_1(\boldsymbol{x}_j)$, so that

$$E_1(\alpha_1 \boldsymbol{x}_1 + \cdots + \alpha_m \boldsymbol{x}_m) = A_1\alpha_1 + \cdots + A_m\alpha_m.$$

Since

$$E_1(\ell_j p_{j1} \boldsymbol{x}_1 + \cdots + \ell_j p_{jm} \boldsymbol{x}_m) = E_1(\ell_j \boldsymbol{p}_j) = w_j,$$

we have

$$A_1 \ell_j p_{j1} + \cdots + A_m \ell_j p_{jm} = w_j \quad (j = 1, \ldots, m).$$

If we solve for $A_i$ by Cramer's rule, we obtain

$$|A_i| = \frac{\det P_i}{\det P}, \tag{2.19}$$

where $P_i$ is obtained from $P$ by replacing column $i$ by a column with entries $w_1, \ldots, w_m$. Clearly we may cancel $\ell_1^{n-1}$ from numerator and denominator on the right side of (2.19). This gives

$$A_i = \frac{B_i}{\ell_1^{-n+1} \det P} \qquad (B_i \in \mathbb{Z}),$$

so that

$$\ell_1^{-n+1} (\det P) E_1(\boldsymbol{p}) \in \mathbb{Z} \quad (\boldsymbol{p} \in \mathcal{A}). \tag{2.20}$$

We observe that

$$\begin{aligned} |\det P| &= \frac{\det(\ell_1 \boldsymbol{p}_1, \ldots, \ell_m \boldsymbol{p}_m)}{\det(\boldsymbol{x}_1, \ldots, \boldsymbol{x}_m)} \\ &\ll \det(\ell_1 \boldsymbol{p}_1, \ldots, \ell_m \boldsymbol{p}_m) \end{aligned} \tag{2.21}$$

from (2.18).

Now let $F_1 = f - E_1$. If we write $\ell(\boldsymbol{p})\boldsymbol{p}$ in the form

$$\ell(\boldsymbol{p})\boldsymbol{p} = \alpha_1 \ell_1 \boldsymbol{p}_1 + \cdots + \alpha_m \ell_m \boldsymbol{p}_m,$$

then

$$\begin{aligned} |\alpha_i| &= \frac{\det(\ell_1 \boldsymbol{p}_1, \ldots, \ell_{i-1} \boldsymbol{p}_{i-1}, \ell(\boldsymbol{p})\boldsymbol{p}, \ell_{i+1} \boldsymbol{p}_{i+1}, \ldots, \ell_m \boldsymbol{p}_m)}{\det(\ell_1 \boldsymbol{p}_1, \ldots, \ell_m \boldsymbol{p}_m)} \\ &\ll \frac{A^m}{\det(\boldsymbol{p}_1, \ldots, \boldsymbol{p}_m)} \end{aligned}$$

by (1.6), (1.7). Hence

$$\begin{aligned} F_1(\ell(\boldsymbol{p})\boldsymbol{p}) &\ll \frac{A^m}{\det(\boldsymbol{p}_1, \ldots, \boldsymbol{p}_m)} \max_i |F_1(\ell_i \boldsymbol{p}_i)| \\ &\ll \frac{A^m}{\det(\boldsymbol{p}_1, \ldots, \boldsymbol{p}_m)} V \end{aligned} \tag{2.22}$$

for all $\boldsymbol{p}$ in $\mathcal{A}$, by (1.8) and the definition of $F_1$.

Given $\boldsymbol{p}$ in $\mathcal{A}$, we now estimate the integer

$$k(\boldsymbol{p}) = \ell_1^{-n+1} \det P(E_1(\ell(\boldsymbol{p})\boldsymbol{p}) - w(\boldsymbol{p})).$$

We have

$$|k(\boldsymbol{p})| \le \ell_1^{-n+1} |\det P| (|F_1(\ell(\boldsymbol{p})\boldsymbol{p})| + |f(\ell(\boldsymbol{p})\boldsymbol{p}) - w(\boldsymbol{p})|)$$

$$\ll \ell_1^{-n+1} |\det P| \frac{A^m}{\det(\boldsymbol{p}_1, \ldots, \boldsymbol{p}_m)} V$$

(by (2.22), (1.8))

$$\ll \ell_1^{-n+1} \ell_1 \ldots \ell_m A^m V$$

(by (2.21))

$$\ll U^{m-n+1} A^m V$$

by (1.7). Taking (1.9) into account, with $C(h)$ suitably chosen, we have $|k(\boldsymbol{p})| < 1$, and indeed $k(\boldsymbol{p}) = 0$. We may now complete the proof by the argument in case 2a of the preceding proof. The points $(\ell(\boldsymbol{p})\boldsymbol{p}, w(\boldsymbol{p}))$ lie in an $m$-dimensional subspace of $\mathbb{R}^h$. In the role of the determinant in case 2a, we use

$$\det \begin{bmatrix} \ell(\boldsymbol{p})\boldsymbol{p} & w(\boldsymbol{p}) \\ \ell(\boldsymbol{p}_1)\boldsymbol{p}_1 & w(\boldsymbol{p}_1) \\ & \vdots \\ \ell(\boldsymbol{p}_m)\boldsymbol{p}_m & w(\boldsymbol{p}_m) \end{bmatrix}.$$

□

## 3 A lemma with four alternatives

In the present section we prove a lemma with four alternatives as a stage in the proof of Theorem 1. I have arranged the proof in this way for comparison with the "three-alternatives lemma" (Lemma 17B of [7]). The corresponding result in [3] (formulated a little differently) is Lemma 7.7.

**Lemma 5.** *Let $h \ge 3$, $\epsilon > 0$. Let $N \ge C(h, \epsilon)$. Let $\Delta$ satisfy*

$$1 \le \Delta^{h+1-(1/2h)+\epsilon} \le N. \tag{3.1}$$

*Let $\Lambda = \Delta^{1/h}\mathbb{Z}^h$, $\Pi = \Delta^{-1/h}\mathbb{Z}^h$, and let $\boldsymbol{a}_1, \boldsymbol{a}_2 \in \mathbb{R}^h$. Then either*

(i) *for every $t$, the set $K_0 + \Lambda + t$ contains a point $n^2\boldsymbol{a}_2 + n\boldsymbol{a}_1$ with $1 \le n \le N$; or*

(ii) *there is a primitive point $\boldsymbol{p}$ in $\Pi$ and a natural number $q$ with*

$$|\boldsymbol{p}| < N^\delta, q < N^\delta |\boldsymbol{p}|^{-2}, \|q\boldsymbol{a}_i\boldsymbol{p}\| < N^{\delta-i}|\boldsymbol{p}|^{-1} \quad (i = 1, 2); \tag{3.2}$$

*or*

(iii) *there is a pair of linearly independent points $\boldsymbol{p}_1, \boldsymbol{p}_2$ of $\Pi$, a natural number $q$, and there are numbers $a, B, 0 < a < N^\delta, 1 < B < N$, such that*

$$|\boldsymbol{p}_1|\,|\boldsymbol{p}_2| \ll a^2 N^{\delta-1} B, \tag{3.3}$$

$$q \ll a^{-2}B^{-2}N^{2+\delta}, \tag{3.4}$$

$$|\boldsymbol{p}_j|\,\|q\boldsymbol{p}_k\boldsymbol{a}_i\| \ll a^{-1}B^{-1}N^{1-i+\delta} \quad (i = 1, 2;\ (j,k) = (1,2), (2,1)); \tag{3.5}$$

*or*

(*iv*) *there are three linearly independent points* $\boldsymbol{p}_1, \boldsymbol{p}_2, \boldsymbol{p}_3$ *in* $\Pi$ *with* $|\boldsymbol{p}_j| < N^\delta$ $(j = 1, 2, 3)$ *and a natural number* $q$ *with*

$$q < N^\delta \Delta^2, \quad \|q\boldsymbol{p}_j\boldsymbol{a}_i\| < N^{\delta - i}\Delta^2 \quad (i = 1, 2; \quad j = 1, 2, 3). \tag{3.6}$$

For the proof of Lemma 5, we require the following variant of Lemma 5 of [6].

**Lemma 6.** *Let* $W$ *be a subspace of* $\mathbb{R}^h$, $\dim W = 2$, *such that* $\Gamma = W \cap \mathbb{Z}^h$ *is a two-dimensional lattice. Let* $\mathcal{A}$ *be a set of primitive points* $\boldsymbol{p}$ *of* $\Gamma$, $|\mathcal{A}| \geq 8$. *Suppose that*

$$A/2 < |\boldsymbol{p}| \leq A \qquad (\boldsymbol{p} \in \mathcal{A}) \tag{3.7}$$

*and* $\boldsymbol{e}_1, \boldsymbol{e}_2$ *in* $\mathbb{R}^h$ *and* $V_1, V_2$ *are such that*

$$9A^2V_j < 1 \qquad (j = 1, 2), \tag{3.8}$$

$$|\boldsymbol{p}\boldsymbol{e}_j - v_j(\boldsymbol{p})| < V_j \qquad (j = 1, 2, \ \boldsymbol{p} \in \mathcal{A}), \tag{3.9}$$

*where* $v_j(\boldsymbol{p}) \in \mathbb{Z}$. *Then there are linearly independent points* $\boldsymbol{p}_1, \boldsymbol{p}_2$ *of* $\Gamma$ *for which*

$$|\boldsymbol{p}_1||\boldsymbol{p}_2| \ll A^2|\mathcal{A}|^{-1}, \tag{3.10}$$

$$\max(|\boldsymbol{p}_1| \, \|\boldsymbol{p}_2\boldsymbol{e}_j\|, |\boldsymbol{p}_2| \, \|\boldsymbol{p}_1\boldsymbol{e}_j\|) \ll V_j A|\mathcal{A}|^{-1} \quad (j = 1, 2). \tag{3.11}$$

*Proof.* Let $\boldsymbol{w}_1, \boldsymbol{w}_2$ be an orthonormal basis of $W$. We write each $\boldsymbol{p}$ in $\mathcal{A}$ as

$$\boldsymbol{p} = (r\cos\alpha)\boldsymbol{w}_1 + (r\sin\alpha)\boldsymbol{w}_2, \ r = r(\boldsymbol{p}) > 0, \ \alpha = \alpha(\boldsymbol{p}) \in [0, 2\pi).$$

Now for some $k$, $0 \leq k \leq 3$, there is a subset $\mathcal{A}'$ of $\mathcal{A}$ having

$$|\mathcal{A}'| \geq |\mathcal{A}|/4,$$
$$\alpha(\boldsymbol{p}) \in [k\pi/2, (k+1)\pi/2] \qquad (\boldsymbol{p} \in \mathcal{A}').$$

Let $\boldsymbol{q}_1, \boldsymbol{q}_2, \boldsymbol{r}_1, \boldsymbol{r}_2$ be chosen in $\mathcal{A}'$ so that $\alpha(\boldsymbol{q}_1)$ is least, $\alpha(\boldsymbol{q}_2)$ is greatest, and $\alpha(\boldsymbol{r}_2) - \alpha(\boldsymbol{r}_1)$ is positive and as small as possible. Clearly the $\alpha(\boldsymbol{p})$ $(\boldsymbol{p} \in \mathcal{A}')$ are distinct, and

$$\begin{aligned} 0 < \det(\boldsymbol{r}_1, \boldsymbol{r}_2) &\ll A^2(\alpha(\boldsymbol{r}_2) - \alpha(\boldsymbol{r}_1)) \\ &\ll |\mathcal{A}|^{-1}A^2(\alpha(\boldsymbol{q}_2) - \alpha(\boldsymbol{q}_1)) \\ &\ll |\mathcal{A}|^{-1}\det(\boldsymbol{q}_1, \boldsymbol{q}_2). \end{aligned} \tag{3.12}$$

Let $C$ be the index in $\Gamma$ of the lattice $\Gamma_0$ generated by $\boldsymbol{q}_1, \boldsymbol{q}_2$. Then

$$C\Gamma \subset \Gamma_0. \tag{3.13}$$

We introduce the linear functions $E_j : W \to \mathbb{R}$ defined by

$$E_j(\boldsymbol{q}_1) = v_j(\boldsymbol{q}_1), \quad E_j(\boldsymbol{q}_2) = v_j(\boldsymbol{q}_2)$$

for $j = 1, 2$. We observe that

$$CE_j(\boldsymbol{x}) \in \mathbb{Z} \qquad (\boldsymbol{x} \in \Gamma) \tag{3.14}$$

from (3.13).

Let $f_j(\boldsymbol{x}) = \boldsymbol{x}\boldsymbol{e}_j$ and $F_j = f_j - E_j$. Then

$$|F_j(\boldsymbol{q}_i)| < V_j \qquad (1 \leq i, j \leq 2)$$

from (3.9). Moreover, given $\boldsymbol{p} \in \mathcal{A}'$, $\boldsymbol{p} = x_1\boldsymbol{q}_1 + x_2\boldsymbol{q}_2$, we have

$$|x_1| = \frac{\det(\boldsymbol{p}, \boldsymbol{q}_2)}{\det(\boldsymbol{q}_1, \boldsymbol{q}_2)} \le 4, \quad |x_2| = \frac{\det(\boldsymbol{p}, \boldsymbol{q}_1)}{\det(\boldsymbol{q}_1, \boldsymbol{q}_2)} \le 4$$

by (3.7) and the choice of $\boldsymbol{q}_1, \boldsymbol{q}_2$. Hence

$$|F_j(\boldsymbol{p})| < 8V_j. \tag{3.15}$$

The integer $k_j(\boldsymbol{p}) = C(E_j(\boldsymbol{p}) - v_j(\boldsymbol{p}))$ satisfies

$$|k_j(\boldsymbol{p})| \le C(|f_j(\boldsymbol{p}) - v_j(\boldsymbol{p})| + |F_j(\boldsymbol{p})|) < 9CV_j \qquad (\boldsymbol{p} \in \mathcal{A}')$$

from (3.9), (3.15).

Taking $\boldsymbol{s}_1, \boldsymbol{s}_2$ to be a basis of $\Gamma$, we see that

$$C = \frac{\det(\boldsymbol{q}_1, \boldsymbol{q}_2)}{\det(\boldsymbol{s}_1, \boldsymbol{s}_2)} \le \det(\boldsymbol{q}_1, \boldsymbol{q}_2) \le A^2,$$

and in view of (3.8), $|k_j(\boldsymbol{p})| < 9A^2V_j < 1$. Hence $k_j(\boldsymbol{p}) = 0$. In particular,

$$E_j(\boldsymbol{p}) \in \mathbb{Z} \qquad (j = 1, 2) \tag{3.16}$$

for all $\boldsymbol{p}$ in $\mathcal{A}'$.

The set $\Gamma_1$ of $\boldsymbol{p}$ in $\Gamma$ satisfying (3.16) is clearly a two-dimensional lattice, and indeed $\det \Gamma_1 \le \det(\boldsymbol{r}_1, \boldsymbol{r}_2)$.

By Minkowski's theorem, there are linearly independent points $\boldsymbol{p}_1, \boldsymbol{p}_2$ in $\Gamma_1$ with

$$\begin{aligned} |\boldsymbol{p}_1|\,|\boldsymbol{p}_2| &\ll \det \Gamma_1 \le \det(\boldsymbol{r}_1, \boldsymbol{r}_2) \\ &\ll |\mathcal{A}|^{-1} \det(\boldsymbol{q}_1, \boldsymbol{q}_2) \ll |\mathcal{A}|^{-1} A^2, \end{aligned} \tag{3.17}$$

on taking into account (3.12), (3.7).

Now let $u_{j,i} = E_j(\boldsymbol{p}_i)$. Then $u_{j,i}$ is an integer, and

$$\begin{aligned} |\boldsymbol{p}_1|\,|\boldsymbol{p}_2\boldsymbol{e}_j - u_{j,2}| &= |\boldsymbol{p}_1|\,|F_j(\boldsymbol{p}_2)| \\ &\le |\boldsymbol{p}_1| \left( \frac{\det(\boldsymbol{p}_2, \boldsymbol{q}_2)}{\det(\boldsymbol{q}_1, \boldsymbol{q}_2)} |F_j(\boldsymbol{q}_1)| + \frac{\det(\boldsymbol{p}_2, \boldsymbol{q}_1)}{\det(\boldsymbol{q}_1, \boldsymbol{q}_2)} |F_j(\boldsymbol{q}_2)| \right) \end{aligned}$$

(by the argument leading to (3.15))

$$\begin{aligned} &\le \frac{|\boldsymbol{p}_1|\,|\boldsymbol{p}_2|(|\boldsymbol{q}_2| + |\boldsymbol{q}_1|)V_j}{\det(\boldsymbol{q}_1, \boldsymbol{q}_2)} \\ &\ll |\mathcal{A}|^{-1} A V_j \end{aligned}$$

in view of (3.17). The same bound holds with $\boldsymbol{p}_1, \boldsymbol{p}_2$ interchanged. This completes the proof of Lemma 6. □

*Proof of Lemma 5.* Suppose that alternative (i) does not hold. By a slight variant of the proof of [3, Lemma 7.5], there are numbers $a$ and $B$ such that

$$\Delta^{-1} \ll a \ll N^{\delta}, \tag{3.18}$$
$$B \gg N^{1-\delta} \Delta^{-1} a^{-1}, \tag{3.19}$$

and there is a set $\mathcal{B}$ of primitive points of $\Pi$ with

$$a < |\boldsymbol{p}| \le 2a \qquad (\boldsymbol{p} \in \mathcal{B}), \tag{3.20}$$

$$|\mathcal{B}| \gg NB^{-1}(\log N)^{-2}. \tag{3.21}$$

Further, for each $\boldsymbol{p}$ in $\mathcal{B}$ there are integers $q = q(\boldsymbol{p})$, $v_1 = v_1(\boldsymbol{p})$, $v_2 = v_2(\boldsymbol{p})$ satisfying

$$1 \le q < a^{-2}B^{-2}N^{2+\delta}, \tag{3.22}$$

$$(q, v_1, v_2) = 1,\ (q, v_2) < N^{\delta}a^{-1}, \tag{3.23}$$

$$|q\boldsymbol{a}_i\boldsymbol{p} - v_i| < a^{-1}B^{-2}N^{2-i+\delta} \quad (i = 1, 2). \tag{3.24}$$

Let us write $s = s(\boldsymbol{p}) = (q, v_2)$, $r = r(\boldsymbol{p}) = qs^{-1}$, $v = v(\boldsymbol{p}) = v_2s^{-1}$. Then we note that

$$r \ge 1, s \ge 1, rs < a^{-2}B^{-2}N^{2+\delta}, \tag{3.25}$$

$$s|r\boldsymbol{a}_2\boldsymbol{p} - v| < a^{-1}B^{-2}N^{\delta},\ (r, v) = 1, \tag{3.26}$$

$$|sr\boldsymbol{a}_1\boldsymbol{p} - v_1| < a^{-1}B^{-2}N^{1+\delta},\ (s, v_1) = 1, \tag{3.27}$$

$$s < N^{\delta}a^{-1}. \tag{3.28}$$

There are now two cases to consider. Suppose first that

$$B \ge N^{1-3\epsilon^2}. \tag{3.29}$$

Take any $\boldsymbol{p} \in \mathcal{B}$. Then alternative (ii) holds with this choice of $\boldsymbol{p}$ and $q = q(\boldsymbol{p})$. For

$$q < a^{-2}B^{-2}N^{2+\delta} < a^{-2}N^{\delta} < |\boldsymbol{p}|^{-2}N^{\delta}$$

by (3.22), (3.29), (3.20), while

$$\begin{aligned}\|q\boldsymbol{a}_i\boldsymbol{p}\| &< a^{-1}B^{-2}N^{2-i+\delta} < a^{-1}N^{-i+\delta}\\ &< |\boldsymbol{p}|^{-1}N^{-i+\delta} \quad (i = 1, 2)\end{aligned}$$

by (3.24), (3.29), (3.20).

Now suppose that (3.29) is false. Clearly (3.21) yields a subset $\mathcal{B}'$ of $\mathcal{B}$ with

$$|\mathcal{B}'| \ge |\mathcal{B}|\,N^{-\epsilon^2} \ge N^{2\epsilon^2},\ \ U/2 < r(\boldsymbol{p}) \le U < a^{-2}B^{-2}N^{2+\delta} \quad (\boldsymbol{p} \in \mathcal{B}').$$

We apply Theorem 2 with $\epsilon^2$ in place of $\epsilon$,

$$\mathcal{A} = \Delta^{1/h}\mathcal{B}',\ \boldsymbol{e} = \boldsymbol{a}_2\Delta^{-1/h},\ \ell(\boldsymbol{p}) = r(\boldsymbol{p}),\ w(\boldsymbol{p}) = v(\boldsymbol{p}).$$

Thus we may take

$$\begin{aligned}&A = 2\Delta^{1/h}a, \quad U < a^{-2}B^{-2}N^{2+\delta},\\ &V = a^{-1}B^{-2}N^{\delta}, M = UA,\end{aligned}$$

in view of (3.20), (3.25), (3.26). We must verify (1.2), (1.3). We have

$$\begin{aligned}M^{h-1+\epsilon^2}AV &\ll U^{h-1}A^hVN^{\delta}\\ &\ll a^{-h+1}B^{-2h}N^{2h-2+\delta}\Delta\\ &\ll \Delta^{2h+1}N^{-2+\delta} \ll N^{-\delta}\end{aligned} \tag{3.30}$$

from (3.19), (3.18), (3.1). Moreover,

$$\begin{aligned}|\mathcal{A}|M^{-2\epsilon^2}(M^hV)^{-h/(h+1)} &\gg N^{1-\delta}B^{-1}(a^{-h-1}B^{-2h-2}N^{2h}\Delta)^{-h/(h+1)}\\ &\gg N^{1-2h^2/(h+1)-\delta}B^{2h-1}a^h\Delta^{-h/(h+1)}\\ &\gg N^{2h-2h^2/(h+1)-\delta}\Delta^{-2h+1-h/(h+1)} \gg M^\delta\end{aligned}$$

from (3.21), (3.19), (3.18), (3.1). This establishes that (1.2), (1.3) hold. Thus there is a subset $\mathcal{A}_1$ of $\mathcal{A}$ with

$$|\mathcal{A}_1| \gg M^{2\epsilon^2}$$

and $r(\boldsymbol{p}) = r$ for all $\boldsymbol{p}$ in $\mathcal{A}_1$.

We now use Theorem 3 to find a subset $\mathcal{A}_2$ of $\mathcal{A}$ with

$$|\mathcal{A}_2| \gg |\mathcal{A}|M^{-\delta} \gg N^{1-\delta}B^{-1}$$

and $r(\boldsymbol{p}) = r$ for all $\boldsymbol{p}$ in $\mathcal{A}_2$. We take $\mathcal{A}$, $\boldsymbol{e}$, $\ell(\boldsymbol{p})$, $w(\boldsymbol{p})$, $A$, $U$, $V$ and $M$ as above. We have $2 \le m \le h$. Since $\mathcal{A}_1$ consists of primitive points, we can certainly take $n \ge 2$. It follows that

$$U^{1+m-n}A^mV \ll M^{h-1}AV \ll N^{-\delta}.$$

Having "fixed $r$" on the set $\mathcal{B}_1 = \Delta^{-1/h}\mathcal{A}_2$ in (3.25)–(3.28), we now "fix $s$". In view of (3.20), (3.27), (3.28) we may apply Lemma 4 with $\mathcal{B}_1$ in place of $\mathcal{A}$, $\boldsymbol{e} = r\boldsymbol{a}_1$, $\ell(\boldsymbol{p}) = s(\boldsymbol{p})$, $w(\boldsymbol{p}) = v_1(\boldsymbol{p})$, and with

$$Z = (2a)^t, \quad U = N^\delta a^{-1}, \quad V = a^{-1}B^{-2}N^{1+\delta},$$

where $t$ is the dimension of Span $\mathcal{B}_1$. Now

$$\begin{aligned}ZU^tV\Delta N^\delta &\ll (2a)^t(N^\delta a^{-1})^t a^{-1}B^{-2}N^{1+\delta}\Delta\\ &\ll \Delta^3N^{-1+\delta} \ll N^{-\delta}\end{aligned}$$

from (3.19), (3.18), (3.1). Thus there is a subset $\mathcal{B}_2$ of $\mathcal{B}_1$ with

$$|\mathcal{B}_2| \gg |\mathcal{B}_1|N^{-\delta} \gg N^{1-\delta}B^{-1}, \tag{3.31}$$

with $s(\boldsymbol{p})$, and indeed $q(\boldsymbol{p})$, constant throughout $\mathcal{B}_2$:

$$q(\boldsymbol{p}) = q.$$

If $\mathcal{B}_2$ contains three linearly independent points, it is clear that alternative (iv) of Lemma 5 holds. It remains to consider the case where $W$ = Span $\mathcal{B}_2$ has dimension 2. In that case, we apply Lemma 6 with $\epsilon^2$ in place of $\epsilon$, $\Delta^{1/h}\mathcal{B}_2$ in place of $\mathcal{A}$, taking $\boldsymbol{e}_j = \Delta^{-1/h}q\boldsymbol{a}_j$ $(j = 1, 2)$, so that (3.7)–(3.9) hold with

$$A = 2\Delta^{1/h}a, \quad V_j = a^{-2}B^{-2}N^{2-j+\delta}.$$

The condition (3.8) is satisfied, since

$$\Delta^{2/h}a^2V_j \ll \Delta^{2/h+2}N^{-1+\delta} \ll N^{-\delta} \quad (j = 1, 2)$$

from (3.19), (3.18), (3.1). Let $\boldsymbol{p}_1'$, $\boldsymbol{p}_2'$ be the independent points of $W \cap \mathbb{Z}^h$ provided by Lemma 6, and $\boldsymbol{p}_i = \Delta^{-1/h}\boldsymbol{p}_i'$. Then (3.3), (3.4), (3.5) follow from (3.10), (3.31), (3.22), (3.11). Thus alternative (iii) holds, and the proof of Lemma 5 is complete. □

## 4 Proof of Theorem 1

**Lemma 7.** *Let $h \geq 1$, $\epsilon > 0$, $N > C(h,\epsilon)$. Let $\Lambda$ be an $h$-dimensional lattice in $\mathbb{R}^h$ with*

$$K_0 \cap \Lambda = \{\mathbf{0}\},$$
$$d(\Lambda)^{h+1+\epsilon} \leq N. \tag{4.1}$$

*For any $\boldsymbol{a}_1, \boldsymbol{a}_2$ in $\mathbb{R}^h$, there is a natural number $n \leq N$ such that*

$$n^2 \boldsymbol{a}_2 + n\boldsymbol{a}_1 \in K_0 + \Lambda.$$

*Proof.* This is Theorem 7.2 of [3]. It contains the admissibility of $1/(h^2+h)$ as a special case, as we see on taking $\Lambda = N^{1/(h^2+h)-\epsilon}\mathbb{Z}^h$. (The methods of the present paper do not seem to be strong enough to sharpen Lemma 7 for a general lattice.) □

The following lemma is a refinement of [3, Lemma 7.9]. We give the proof in detail for the convenience of readers. The orthogonal complement of a subspace $T$ in $\mathbb{R}^h$ is denoted by $T^\perp$.

**Lemma 8.** *Let $\Lambda$ be an $h$-dimensional lattice in $\mathbb{R}^h$ with polar lattice $\Pi$. Let $\Pi'$ be a $t$-dimensional lattice contained in $\Pi$, let $T = \operatorname{Span}\Pi'$, and let $\boldsymbol{p}_1, \ldots, \boldsymbol{p}_t$ be a linearly independent set in $\Pi'$. Then there is a natural number c,*

$$c \ll \det(\boldsymbol{p}_1, \ldots, \boldsymbol{p}_t)/d(\Pi'), \tag{4.2}$$

*having the following property. Given $\boldsymbol{a}$ in $\mathbb{R}^h$, $c\boldsymbol{a}$ may be written in the form*

$$c\boldsymbol{a} = \boldsymbol{\ell} + \boldsymbol{s} + \boldsymbol{b}, \tag{4.3}$$

*where $\boldsymbol{\ell} \in \Lambda$, $\boldsymbol{s} \in T^\perp$ and*

$$|\boldsymbol{b}| \ll d(\Pi')^{-1} \max_{1 \leq i \leq t} |\boldsymbol{p}_1| \ldots |\boldsymbol{p}_{i-1}|\, \|\boldsymbol{p}_i \boldsymbol{a}\|\, |\boldsymbol{p}_{i+1}| \ldots |\boldsymbol{p}_t|. \tag{4.4}$$

*Proof.* Let $\lambda_1, \ldots, \lambda_t$ be the successive minima of $\Pi'$ with respect to $K_0$ and let $\boldsymbol{q}_1, \ldots, \boldsymbol{q}_t$ be linearly independent points of $\Pi'$ with $|\boldsymbol{q}_j| = \lambda_j$. By Minkowski's theorem,

$$1 \leq v := \frac{\det(\boldsymbol{q}_1, \ldots, \boldsymbol{q}_t)}{d(\Pi')} \leq \frac{|\boldsymbol{q}_1| \ldots |\boldsymbol{q}_t|}{d(\Pi')} \ll 1. \tag{4.5}$$

Arguing as in the proof of Lemma 7.8 of [3], we find points $\boldsymbol{\ell}_1, \ldots, \boldsymbol{\ell}_t$ of $v^{-1}\Lambda$ such that

$$\boldsymbol{\ell}_i \boldsymbol{q}_j = \begin{cases} 1 & \text{if } i = j \\ 0 & \text{if } i \neq j. \end{cases} \tag{4.6}$$

Let $\boldsymbol{w}_1, \ldots, \boldsymbol{w}_t$ be an orthonormal basis of $T$, and write

$$\boldsymbol{p}_j = p_{j1}\boldsymbol{w}_1 + \cdots + p_{jt}\boldsymbol{w}_t, \quad \boldsymbol{q}_j = q_{j1}\boldsymbol{w}_1 + \cdots + q_{jt}\boldsymbol{w}_t.$$

There are integers $c_{ij}$ such that

$$v\boldsymbol{p}_j = c_{j1}\boldsymbol{q}_1 + \cdots + c_{jt}\boldsymbol{q}_t \qquad (j = 1, \ldots, t).$$

Write $C = [c_{ij}]$, $c = |\det C|$, and let $C_{ij}$ be the cofactor of $c_{ij}$ in $C$. Obviously

$$v^t \det(\boldsymbol{p}_1, \ldots, \boldsymbol{p}_t) = c \det(\boldsymbol{q}_1, \ldots, \boldsymbol{q}_t).$$

Taking (4.5) into account, we obtain (4.2).

We now fix $j$ and solve the $t$ equations

$$c_{j1}q_{1i} + \cdots + c_{jt}q_{ti} = vp_{ji} \qquad (i = 1, \ldots, t)$$

for $c_{js}$ by Cramer's rule. This yields

$$\begin{aligned} c_{js} &\ll \frac{|\boldsymbol{q}_1| \ldots |\boldsymbol{q}_{s-1}|\,|\boldsymbol{p}_j|\,|\boldsymbol{q}_{s+1}| \ldots |\boldsymbol{q}_t|}{\det(\boldsymbol{q}_1, \ldots, \boldsymbol{q}_t)} \\ &\ll |\boldsymbol{p}_j|/|\boldsymbol{q}_s| \end{aligned}$$

by (4.5). It follows that for $1 \le i, r \le t$,

$$\begin{aligned} C_{ir} &\ll \frac{|\boldsymbol{p}_1| \ldots |\boldsymbol{p}_{i-1}|\,|\boldsymbol{p}_{i+1}| \ldots |\boldsymbol{p}_t|}{|\boldsymbol{q}_1| \ldots |\boldsymbol{q}_{r-1}|\,|\boldsymbol{q}_{r+1}| \ldots |\boldsymbol{q}_t|} \\ &\ll \frac{|\boldsymbol{p}_1| \ldots |\boldsymbol{p}_{i-1}|\,|\boldsymbol{p}_{i+1}| \ldots |\boldsymbol{p}_t|\,|\boldsymbol{q}_r|}{d(\Pi')}. \end{aligned} \tag{4.7}$$

We are now ready to deduce the representation (4.3), (4.4). We have

$$v\boldsymbol{p}_j\boldsymbol{a} = vx_j + P_j,$$

where $x_j \in \mathbb{Z}$ and $P_j \ll \|\boldsymbol{p}_j\boldsymbol{a}\|$. That is,

$$c_{j1}\boldsymbol{q}_1\boldsymbol{a} + \cdots + c_{jt}\boldsymbol{q}_t\boldsymbol{a} = vx_j + P_j \quad (j = 1, \ldots, t).$$

For a fixed $i$, we multiply the $j$-th equation by $C_{ji}$ and add to get

$$c\boldsymbol{q}_i\boldsymbol{a} = vy_i + V_i,$$

where $y_i \in \mathbb{Z}$ and

$$\begin{aligned} V_i &\ll \max_j |C_{ji}P_j| \\ &\ll \frac{|\boldsymbol{q}_i|}{d(\Pi')} \max_j |\boldsymbol{p}_1| \ldots |\boldsymbol{p}_{j-1}|\, \|\boldsymbol{p}_j\boldsymbol{a}\|\, |\boldsymbol{p}_{j+1}| \ldots |\boldsymbol{p}_t| \end{aligned} \tag{4.8}$$

in view of (4.7).

Define $\boldsymbol{\ell} = v(y_1\boldsymbol{\ell}_1 + \cdots + y_t\boldsymbol{\ell}_t)$; then $\boldsymbol{\ell} \in \Lambda$ and

$$\boldsymbol{q}_i(c\boldsymbol{a} - \boldsymbol{\ell}) = vy_i + V_i - vy_i = V_i \quad (i = 1, \ldots, t).$$

We now decompose $c\boldsymbol{a} - \boldsymbol{\ell}$ into

$$c\boldsymbol{a} - \boldsymbol{\ell} = \boldsymbol{b} + \boldsymbol{s} \qquad (\boldsymbol{b} \in T, \boldsymbol{s} \in T^{\perp})$$

and give a bound for $|\boldsymbol{b}|$. We have

$$\boldsymbol{q}_i\boldsymbol{b} = \boldsymbol{q}_i(\boldsymbol{b} + \boldsymbol{s}) = V_i \quad (i = 1, \ldots, t)$$

because $\boldsymbol{q}_i \in T$. Writing $\boldsymbol{b} = b_1\boldsymbol{w}_1 + \cdots + b_t\boldsymbol{w}_t$, we have the equations

$$q_{i1}b_1 + \cdots + q_{it}b_t = V_i \quad (i = 1, \ldots, t)$$

for $b_1, \ldots, b_t$. Solving by Cramer's rule,

$$\det(\boldsymbol{q}_1, \ldots, \boldsymbol{q}_t)b_j = \pm(Q_{1j}V_1 + \cdots + Q_{tj}V_t), \tag{4.9}$$

where $Q_{ij}$ is the cofactor of $q_{ij}$ in $[q_{rs}]$. Now

$$|Q_{ij}| \ll \prod_{\ell \neq i} |\boldsymbol{q}_\ell|. \tag{4.10}$$

We obtain

$$|b_j| \ll d(\Pi')^{-1} \sum_{i=1}^{t} \left( \prod_{\ell \neq i} |\boldsymbol{q}_\ell| \right) |\boldsymbol{q}_i| d(\Pi')^{-1} \max_k |\boldsymbol{p}_i| \ldots |\boldsymbol{p}_{k-1}| \, \|\boldsymbol{p}_k \boldsymbol{a}\| \, |\boldsymbol{p}_{k+1}| \ldots |\boldsymbol{p}_t|$$

on combining (4.8)–(4.10) and recalling (4.5). Now the lemma follows on a further application of (4.5). □

*Proof of Theorem 1.* Let $\epsilon > 0$, $h \geq 3$, $N > C(h, \epsilon)$. Take $\boldsymbol{\alpha} = (\alpha_1, \ldots, \alpha_h)$, $\boldsymbol{\beta} = (\beta_1, \ldots, \beta_h) \in \mathbb{R}^h$. Suppose that there is no natural number $n \leq N$ such that

$$\|\alpha_i n^2 + \beta_i n\| < N^{\epsilon - \varphi} \quad (i = 1, \ldots, h), \tag{4.11}$$

where $\varphi^{-1} = h^2 + h - 1/2$. Write $\boldsymbol{a}_2 = N^{\varphi - \epsilon}\boldsymbol{\alpha}$, $\boldsymbol{a}_1 = N^{\varphi - \epsilon}\boldsymbol{\beta}$, $\Lambda = N^{\varphi - \epsilon}\mathbb{Z}^h$. Then there is no natural number $n \leq N$ such that $n^2\boldsymbol{a}_2 + n\boldsymbol{a}_1 \in K_0 + \Lambda$. Moreover, $\Lambda$ satisfies the hypotheses of Lemma 5 with $\Delta = N^{h(\varphi - \epsilon)}$. Hence one of the cases (ii), (iii) or (iv) must hold. We apply Lemma 8, taking $\Pi'$ to be the lattice generated by $\boldsymbol{p}$ in case (ii); by $\boldsymbol{p}_1$, $\boldsymbol{p}_2$ in case (iii); and by $\boldsymbol{p}_1$, $\boldsymbol{p}_2$, $\boldsymbol{p}_3$ in case (iv). Let $\Lambda' = \Lambda \cap T^\perp$. In each case, we have the inequality $d(\Lambda') \ll d(\Pi')\Delta$ whenever $\dim T < h$ [3, Lemma 7.8]. Our choices of $\boldsymbol{a}$ are $\boldsymbol{a}_i = q^i \boldsymbol{a}_i$ for $i = 1, 2$. We obtain the representation

$$cq^i \boldsymbol{a}_i = \boldsymbol{\ell}_i + \boldsymbol{s}_i + \boldsymbol{b}_i \quad (i = 1, 2),$$

where $\boldsymbol{\ell}_i \in \Lambda$, $\boldsymbol{s}_i \in T^\perp$ and

$$c \ll 1, \quad |\boldsymbol{b}_i| \ll |\boldsymbol{p}|^{-1} \|\boldsymbol{p} q^i \boldsymbol{a}_i\| \tag{4.12}$$

in case (ii),

$$c \ll |\boldsymbol{p}_1| \, |\boldsymbol{p}_2| / d(\Pi'), \; |\boldsymbol{b}_i| \ll d(\Pi')^{-1} |\boldsymbol{p}_1| \, \|\boldsymbol{p}_2 q^i \boldsymbol{a}\| \tag{4.13}$$

in case (iii),

$$c \ll |\boldsymbol{p}_1| \, |\boldsymbol{p}_2| \, |\boldsymbol{p}_3| / d(\Pi'), \; |\boldsymbol{b}_i| \ll d(\Pi')^{-1} |\boldsymbol{p}_1| \, |\boldsymbol{p}_2| \, \|\boldsymbol{p}_3 q^i \boldsymbol{a}_i\| \tag{4.14}$$

in case (iv). (We permit renumbering of the $\boldsymbol{p}_i$ in cases (iii), (iv).)

We now apply Lemma 8 in the space $T^\perp$, whose dimension we denote by $t$. We replace $\epsilon$ by $\epsilon^2$, $\Lambda$ by $2\Lambda'$, $\boldsymbol{a}_i$ by $2c^{i-1}\boldsymbol{s}_i$ and $N$ by $d(2\Lambda')^{t+1}N^\delta$. Thus if $t > 0$ there is a natural number $x$,

$$x \leq d(2\Lambda')^{t+1} N^\delta \ll d(\Pi')^{t+1} \Delta^{t+1} N^\delta, \tag{4.15}$$

such that $2x^2 c\boldsymbol{s}_2 + 2x\boldsymbol{s}_1 \in 2\Lambda' + K_0$. This implies

$$x^2 c\boldsymbol{s}_2 + x\boldsymbol{s}_1 \in \Lambda + \frac{1}{2} K_0. \tag{4.16}$$

If $t = 0$, we take $x = 1$. Of course (4.16) holds, since $\boldsymbol{s}_1 = \boldsymbol{s}_2 = \boldsymbol{0}$.

Now let $n = xcq$. We shall show that

$$n \ll N^{1-\delta}, \tag{4.17}$$

$$x^i c^{i-1} |\boldsymbol{b}_i| \ll N^{-\delta} \quad (i = 1, 2). \tag{4.18}$$

Suppose for a moment that (4.17), (4.18) hold. We see that the natural number $n \le N$ satisfies

$$\begin{aligned} n^2 \boldsymbol{a}_2 + n\boldsymbol{a}_1 &= x^2 c(\boldsymbol{\ell}_2 + \boldsymbol{s}_2 + \boldsymbol{b}_2) + x(\boldsymbol{\ell}_1 + \boldsymbol{s}_1 + \boldsymbol{b}_1) \\ &= (x^2 c\boldsymbol{s}_2 + x\boldsymbol{s}_1) + (x^2 c\boldsymbol{b}_2 + x\boldsymbol{b}_1) + \boldsymbol{\ell}, \end{aligned}$$

where $\boldsymbol{\ell} \in \Lambda$. Taking (4.16)–(4.18) into account, $n^2\boldsymbol{a}_2 + n\boldsymbol{a}_1 \in \Lambda + K_0$. This contradicts our hypothesis. Hence there must be a solution of (4.11) after all, and the proof is complete.

It remains to prove (4.17), (4.18). Consider case (ii) first. Here $t = h - 1$,

$$\begin{aligned} n = xcq &\ll d(\Pi')^h \Delta^h q N^\delta \\ &\ll |\boldsymbol{p}|^h \Delta^h |\boldsymbol{p}|^{-2} N^\delta \ll \Delta^h N^\delta \ll N^{1-\delta} \end{aligned}$$

from (4.15), (4.12), (3.2), (3.1). Further

$$\begin{aligned} x^i c^{i-1} |\boldsymbol{b}_i| &\ll d(\Pi')^{hi} \Delta^{hi} |\boldsymbol{p}|^{-1} q^{i-1} \|q\boldsymbol{p}\boldsymbol{a}_i\| \\ &\ll |\boldsymbol{p}|^{hi-1} \Delta^{hi} |\boldsymbol{p}|^{-2i+1} N^{\delta-i} \\ &\ll (\Delta^h N^{-1+\delta})^i \ll N^{-\delta}, \end{aligned}$$

again from (4.15), (4.12), (3.2), (3.1).

Now consider case (iii). Here $t = h - 2$,

$$\begin{aligned} n = xcq &\ll d(\Pi')^{h-1} \Delta^{h-1} |\boldsymbol{p}_1| \, |\boldsymbol{p}_2| d(\Pi')^{-1} a^{-2} B^{-2} N^{2+\delta} \\ &\ll (|\boldsymbol{p}_1| \, |\boldsymbol{p}_2|)^{h-1} \Delta^{h-1} a^{-2} B^{-2} N^{2+\delta} \end{aligned}$$

from (4.15), (4.13), (3.4), and since

$$d(\Pi') \le |\boldsymbol{p}_1| \, |\boldsymbol{p}_2|. \tag{4.19}$$

Recalling (3.3),

$$\begin{aligned} n &\ll (a^2 N^{-1} B)^{h-1} \Delta^{h-1} a^{-2} B^{-2} N^{2+\delta} \\ &\ll a^{2h-4} B^{h-3} N^{-h+3+\delta} \Delta^{h-1} \\ &\ll \Delta^{h-1} N^\delta \ll N^{1-\delta} \end{aligned}$$

since $a < N^\delta$, $B < N$. Similarly,

$$x^i c^{i-1} |\boldsymbol{b}_i| \ll d(\Pi')^{(h-1)i} \Delta^{(h-1)i} N^\delta (|\boldsymbol{p}_1| \, |\boldsymbol{p}_2|)^{i-1} d(\Pi')^{-i} |\boldsymbol{p}_1| q^{i-1} \|q\boldsymbol{p}_2\boldsymbol{a}_i\|$$

(from (4.15), (4.13))

$$\ll (|\boldsymbol{p}_1| \, |\boldsymbol{p}_2|)^{(h-1)i-1} \Delta^{(h-1)i} (a^{-2} B^{-2} N^2)^{i-1} B^{-1} N^{1-i+\delta}$$

(from (4.19), (3.4), (3.5))

$$\begin{aligned} &\ll \Delta^{(h-1)i} (a^2 N^{-1} B)^{(h-1)i-1} (a^{-2} B^{-2} N^2)^{i-1} a^{-1} B^{-1} N^{1-i+\delta} \\ &\ll (\Delta^{h-1} a^{2h-5} B^{h-3} N^{-h+2+\delta})^i \ll (\Delta^{h-1} N^{-1+\delta})^i \\ &\ll N^{-\delta} \end{aligned}$$

from (3.3), (3.1).

Finally, consider case (iv). Here $t = h - 3$. Suppose first that $t > 0$. Then

$$n = xcq \ll d(\Pi')^{h-2}\Delta^{h-2}|\boldsymbol{p}_1|\,|\boldsymbol{p}_2|\,|\boldsymbol{p}_3| d(\Pi')^{-1} N^{\delta}\Delta^2 \\ \ll \Delta^h N^{\delta} \ll N^{1-\delta}$$

from (4.15), (4.14), (3.6) and the bounds

$$d(\Pi') \leq |\boldsymbol{p}_1|\,|\boldsymbol{p}_2|\,|\boldsymbol{p}_3| < N^{\delta}. \tag{4.20}$$

Similarly,

$$x^i c^{i-1}|\boldsymbol{b}_i| \ll d(\Pi')^{(h-2)i}\Delta^{(h-2)i}N^{\delta}(|\boldsymbol{p}_1|\,|\boldsymbol{p}_2|\,|\boldsymbol{p}_3|)^{i-1} d(\Pi')^{-i}|\boldsymbol{p}_1|\,|\boldsymbol{p}_2| q^{i-1}\|q\boldsymbol{p}_3\boldsymbol{a}_i\|$$

(from (4.15), (4.14))

$$\ll \Delta^{(h-2)i+2(i-1)+2}N^{\delta-i}$$

(from (4.20), (3.6))

$$\ll (\Delta^h N^{-1+\delta})^i \ll N^{-\delta}.$$

We argue a little differently in case (iv) if $h = 3$, $t = 0$. We have $\Pi' = \Pi$,

$$n = cq \ll |\boldsymbol{p}_1|\,|\boldsymbol{p}_2|\,|\boldsymbol{p}_3| d(\Pi')^{-1}\Delta^2 N^{\delta} \\ \ll \Delta^3 N^{\delta} \ll N^{1-\delta}$$

from (4.14), (3.6), (4.20). Similarly,

$$c^{i-1}|\boldsymbol{b}_i| \ll (|\boldsymbol{p}_1|\,|\boldsymbol{p}_2|\,|\boldsymbol{p}_3|)^{i-1} d(\Pi')^{-i}|\boldsymbol{p}_1|\,|\boldsymbol{p}_2| q^{i-1}\|q\boldsymbol{p}_3\boldsymbol{a}_i\| \\ \ll \Delta^{i+2(i-1)+2}N^{-i+\delta} \ll N^{-\delta}$$

from (4.14), (3.6), (4.20). We have now obtained (4.17), (4.18) in all cases, and the proof of Theorem 1 is complete. □

## References


1. Baker, R.C.: Fractional parts of several polynomials II. *Mathematika* **25**, 76–93 (1978)
2. Baker, R.C.: Fractional parts of several polynomials III. *Q. J. Math. Oxf. II. Ser.* **31**, 19–36 (1980)
3. Baker, R.C.: *Diophantine Inequalities*. Oxford University Press, Oxford (1986)
4. Baker, R.C.: Correction to 'Weyl sums and Diophantine approximation'. *J. Lond. Math. Soc. II. Ser.* **46**, 202–204 (1992)
5. Danicic, I.: Contributions to number theory, Ph.D. thesis, University of London, London, United Kingdom (1957)
6. Schäffer, S.: Fractional parts of pairs of quadratic polynomials. *J. Lond. Math. Soc. II. Ser.* **51**, 429–441 (1995)
7. Schmidt, W.M.: *Small Fractional Parts of Polynomials*. Regional Conference Series in Mathematics, 32. American Mathematical Society, Providence, R.I. (1977)
8. Schmidt, W.M.: *Diophantine Approximation and Diophantine Equations*. Lect. Notes Math., vol. 1467. Springer, Heidelberg (1991)

*Diophantine Approximation*
H. P. Schlickewei et al., Editors


# ARITHMETIC PROGRESSIONS AND TIC-TAC-TOE GAMES

József Beck
*Department of Mathematics, Rutgers University, Busch Campus, Hill Center, New Brunswick, New Jersey 08903, U.S.A.*
jbeck@math.rutgers.edu

In honor of Wolfgang Schmidt's 70th birthday

## 1 Van der Waerden's theorem

This paper is partly an overview of the subject (see Sections 1–4), in fact, as far as I know, the first attempt to do that, and partly an ordinary research paper containing proofs for new results (Sections 5–8). I use many different sources; to make the reader's life easier, I decided to keep the paper (more-or-less) self-contained – this explains the considerable length.

*Every "irregular" structure, if it is large enough, contains a "regular" substructure of some given size* – this is the motto of Ramsey theory, a main chapter of modern Combinatorics. For example, a well-known puzzle states that in a party of 6 people there is always a group of three who either all know each other or are all strangers to each other. A more difficult version is the following: in a party of 18 people there is always a group of four who either all know each other or are all strangers to each other. In technical terms, given any 2-coloring of the $\binom{18}{2} = 153$ edges of a complete graph on 18 points, there is always a set of 4 points such that the $\binom{4}{2} = 6$ edges joining them all have the same color. (Note that both values of 6 and 18 are best possible.) These puzzles illustrate a very important general theorem discovered by Ramsey in 1930. Ramsey theory was named after Ramsey, but the most influential result of Ramsey theory is not Ramsey's theorem: the most influential result is (arguably) a theorem of van der Waerden concerning *arithmetic progressions*, which, in fact, was proved in 1927, three years before Ramsey's work. (It is a sad fact that Ramsey died at a very young age shortly after his combinatorial paper was published; van der Waerden, on the other hand, moved to Algebra, and never returned to Combinatorics again.)

Van der Waerden's theorem [20], which solved a long-standing conjecture of Schur, goes as follows (in this paper the old theorems are *lettered*, and the new theorems are *numbered*).

*Keywords.* Arithmetic progression, probabilistic method, derandomization, strategy.

*2000 Mathematics subject classification.* 91A24.

**Theorem A** (van der Waerden 1927)**.** *For all positive integers n and k, there exists an integer W such that, if the set of integers* $\{1, 2, \ldots, W\}$ *is k-colored, then there exists a monochromatic n-term arithmetic progression.*

Let $W(n, k)$ be the least such integer; we call it the van der Waerden threshold.

Note that van der Waerden's theorem was originally classified as a result in number theory – see e.g. the wonderful book of Khintchine titled "Three pearls of Number Theory" – and it was only the last few decades when van der Waerden's theorem, with its several generalizations (like Hales–Jewett theorem, Szemerédi's theorem, etc.), became a cornerstone of Combinatorics.

The original van der Waerden proof was based on the idea of studying *iterated* arithmetic progressions, that is, progressions of progressions of progressions of ... progressions of arithmetic progressions. This is nothing else than the *combinatorial structure of the family of n-in-a-line's in the d-dimensional* $n \times n \times \cdots \times n = n^d$ *hypercube*. This observation is precisely formulated in a theorem of Hales and Jewett (see Section 2). To get an idea what is going on here, we include an *outline* of the original proof of van der Waerden.

**The original "double induction" proof of van der Waerden.** First we study the simplest nontrivial case $W(3, 2)$: we show that given any 2-coloring (say, red and blue) of the integers $\{1, 2, \ldots, 325\}$ there is a monochromatic 3-term arithmetic progression. (Of course, 325 is an "accidental" number; the exact value of the threshold is actually known: it is the much smaller value of $W(3, 2) = 9$.) The proof is explained by the following picture:

$$\begin{array}{ccccc} a...a...b & \leftarrow d \rightarrow & a...a...b & \leftarrow d \rightarrow & ?...?...? \\ \mathbf{a}...a...b & \leftarrow d \rightarrow & a...\mathbf{a}...b & \leftarrow d \rightarrow & ?...?...\mathbf{a} \\ a...a...\mathbf{b} & \leftarrow d \rightarrow & a...a...\mathbf{b} & \leftarrow d \rightarrow & ?...?...\mathbf{b} \end{array}$$

What this picture means is the following. It is easy to see that any block of five consecutive integers contains a 3-term arithmetic progression of the color code *a...a...a* or *a...a...b*. The first case is a monochromatic 3-term arithmetic progression (A.P.), and we are done. The second case is a 3-term A.P. where the first two terms have the same color and the third term has the other color. The pigeonhole principle implies that the *same ab*-triplet shows up twice:

$$a...a...b \qquad a...a...b$$

Indeed, divide the $\{1, \ldots, 325\}$ interval into 65 blocks of length 5. Since each block has 5 numbers, and we have 2 colors, there are $2^5 = 32$ ways to 2-color a 5-block. By the pigeonhole principle, among the first 33 blocks there are two which are colored in exactly the same way. Assume that the distance between these two identically colored 5-blocks is $d$, and consider the third 5-block such that the blocks form a 3-term A.P.:

$$a...a...b \quad \leftarrow d \rightarrow \quad a...a...b \quad \leftarrow d \rightarrow \quad ?...?...?$$

There are two possibilities. If the last ? has color $a$, then **a** forms a monochromatic 3-term A.P.:

$$\mathbf{a}...a...b \quad \leftarrow d \rightarrow \quad a...\mathbf{a}...b \quad \leftarrow d \rightarrow \quad ?...?...\mathbf{a}$$

If the last ? has color $b$, then **b** forms a monochromatic 3-term A.P.:

$$a...a...\mathbf{b} \quad \leftarrow d \rightarrow \quad a...a...\mathbf{b} \quad \leftarrow d \rightarrow \quad ?...?...\mathbf{b}$$

This completes the proof of the case of 2 colors. Note that $325 = 5(2 \cdot 2^5 + 1)$. If one breaks up the line

$$a...a...b \quad \leftarrow d \rightarrow \quad a...a...b \quad \leftarrow d \rightarrow \quad ?...?...?$$

as follows

$$\begin{array}{ccc} & & ?...?...? \\ & a...a...b & \\ a...a...b & & \end{array}$$

then the argument above basically means "to play Tic-Tac-Toe with *seven* winning triplets instead of the usual eight":

$$\begin{array}{ccc} (1,3) & (2,3) & (3,3) \\ (1,2) & (2,2) & (3,2) \\ (1,1) & (2,1) & (3,1) \end{array}$$

where the diagonal $\{(1,3), (2,2), (3,1)\}$ does not show up in the argument.

Next consider the case of 3 colors, and again we want a monochromatic 3-term A.P. Repeating the previous argument, we obtain the $ab$-configuration as before:

$$a...a...b \quad a...a...b \quad ?...?...?$$

This time we are not done yet, since the last ? can have the third color:

$$a...a...b \quad a...a...b \quad ?...?...c$$

However, the pigeonhole principle implies that the same $abc$-block shows up twice:

$$a...a...b \;\; a...a...b \;\; ?...?...c \qquad a...a...b \;\; a...a...b \;\; ?...?...c$$

Consider the third block such that the 3 blocks form a 3-term A.P.:

$$a..a..b \;\; a..a..b \;\; ?..?..c \qquad a..a..b \;\; a..a..b \;\; ?..?..c \qquad ?..?..? \;\; ?..?..? \;\; ?..?..?$$

This time there are 3 possibilities. If the last ? has color $a$, then **a** forms a monochromatic 3-term A.P.:

$$\mathbf{a}..a..b \;\; a..a..b \;\; ?..?..c \qquad a..a..b \;\; a..\mathbf{a}..b \;\; ?..?..? \qquad ?..?..? \;\; ?..?..? \;\; ?..?..\mathbf{a}$$

If the last ? has color $b$, then **b** forms a monochromatic 3-term A.P.:

$$a..a..\mathbf{b} \;\; a..a..b \;\; ?..?..c \qquad a..a..b \;\; a..a..\mathbf{b} \;\; ?..?..c \qquad ?..?..? \;\; ?..?..? \;\; ?..?..\mathbf{b}$$

And finally, if the last ? has color $c$, then **c** forms a monochromatic 3-term A.P.:

$$a..a..b \;\; a..a..b \;\; ?..?..\mathbf{c} \qquad a..a..b \;\; a..a..b \;\; ?..?..\mathbf{c} \qquad ?..?..? \;\; ?..?..? \;\; ?..?..\mathbf{c}$$

This is how we can force a monochromatic 3-term A.P. if there are 3 colors. The argument gives the upper bound

$$W(3,3) \leq 7(2 \cdot 3^7 + 1)(2 \cdot 3^{7(2 \cdot 3^7 + 1)} + 1) < 3^{20000}.$$

The case of 3 colors basically means "to play Tic-Tac-Toe on a 3-dimensional $3 \times 3 \times 3 = 3^3$ board" (instead of the usual $3 \times 3 = 3^2$ board).

If there are 4 colors, then we get the same *abc*-configuration as before, but the last ? can have the 4th color:

*a..a..b a..a..b ?..?..c a..a..b a..a..b ?..?..c ?..?..? ?..?..? ?..?..d*

Again by the pigeonhole principle, the following configuration will definitely show up:

*a.a.b a.a.b ?.?.c a.a.b a.a.b ?.?.c ?.?.? ?.?.? ?.?.d*
*a.a.b a.a.b ?.?.c a.a.b a.a.b ?.?.c ?.?.? ?.?.? ?.?.d*
*?.?.? ?.?.? ?.?.? ?.?.? ?.?.? ?.?.? ?.?.? ?.?.? ?.?.?*

What it means is that there are two identical *abcd*-blocks, separated from each other, and ?...? stands for the third block such that the 3 blocks form a 3-term A.P.

There are 4 possibilities. If the last ? has color *a*, then **a** forms a monochromatic 3-term A.P., if the last ? has color *b*, then **b** forms a monochromatic 3-term A.P., if the last ? has color *c*, then **c** forms a monochromatic 3-term A.P., and finally, if the last ? has color *d*, then **d** forms a monochromatic 3-term A.P. (we replace the last ? by •):

**a**.*a*.**b** *a.a.b* ?.?.**c** *a.a.b a.a.b ?.?.c ?.?.? ?.?.?* ?.?.**d**
*a.a.b a.a.b ?.?.c a.a.b a*.**a**.**b** ?.?.**c** *?.?.? ?.?.?* ?.?.**d**
*?.?.? ?.?.? ?.?.? ?.?.? ?.?.? ?.?.? ?.?.? ?.?.?* ?.?.•

This is how we can force a monochromatic 3-term A.P. if there are 4 colors. The case of 4 colors basically means "to play Tic-Tac-Toe on a 4-dimensional $3 \times 3 \times 3 \times 3 = 3^4$ board".

Repeating this argument we get a finite bound for an arbitrary number of colors: $W(3,k) < \infty$. *But how to get a monochromatic 4-term A.P.?* Consider the simplest case of two colors. We recall the (very clumsy) upper bound $W(3,2) \leq 325$. It follows that 2-coloring any block of 500 consecutive integers, there is always a configuration *a...a...a...a* or *a...a...a...b*. In the first case we are done. In the second case we can force the existence of a 3-term A.P. of identical *ab*-blocks:

*a...a...a...b a...a...a...b a...a...a...b*

Indeed, any 500-block has $2^{500}$ possible 2-colorings, so if we take $W(3, 2^{500})$ consecutive 500-blocks, then we get a 3-term A.P. of identical 500-blocks. Consider the 4th block such that the 4 blocks form a 4-term A.P.:

*a...a...a...b a...a...a...b a...a...a...b ?...?...?...?*

Now there are two cases. If the last ? has color *a*, then **a** forms a monochromatic 4-term A.P.:

**a**...*a...a...b a*...**a**...*a...b a...a*...**a**...*b* ?...?...?...**a**

If the last ? has color *b*, then **b** forms a monochromatic 4-term A.P.:

*a...a...a*...**b** *a...a...a*...**b** *a...a...a*...**b** ?...?...?...**b**

This is how we can force a monochromatic 4-term A.P. if there are two colors. The case of 2 colors basically means "to play Tic-Tac-Toe on a 2-dimensional $4 \times 4 = 4^2$ board".

Finally consider the case of 3 colors, and we want a monochromatic 4-term A.P. Repeating the previous argument, every block of $W(3, 3^{3^{20000}})$ consecutive integers contains an $ab$-configuration as before:

$$a...a...a...b \quad a...a...a...b \quad a...a...a...b \quad ?...?...?...?$$

This time we are not done yet, since the last ? can have the third color:

$$a...a...a...b \quad a...a...a...b \quad a...a...a...b \quad ?...?...?...c$$

This $abc$-block has

$$3^{W\left(3,3^{3^{20000}}\right)}$$

3-colorings, so among

$$W\left(3, 3^{W\left(3,3^{3^{20000}}\right)}\right)$$

consecutive blocks of length $W(3, 3^{3^{20000}})$ there are three identically colored blocks which form a 3-term A.P.:

$$\begin{array}{l} a..a..a..b \;\; a..a..a..b \;\; a..a..a..b \;\; ?..?..?..c \\ \quad a..a..a..b \;\; a..a..a..b \;\; a..a..a..b \;\; ?..?..?..c \\ \qquad a..a..a..b \;\; a..a..a..b \;\; a..a..a..b \;\; ?..?..?..c \end{array}$$

Consider the fourth block such that the 4 blocks form a 4-term A.P.:

$$\begin{array}{l} a..a..a..b \;\; a..a..a..b \;\; a..a..a..b \;\; ?..?..?..c \\ \quad a..a..a..b \;\; a..a..a..b \;\; a..a..a..b \;\; ?..?..?..c \\ \qquad a..a..a..b \;\; a..a..a..b \;\; a..a..a..b \;\; ?..?..?..c \\ \qquad\quad ?..?..?..? \;\; ?..?..?..? \;\; ?..?..?..? \;\; ?..?..?..? \end{array}$$

There are 3 possibilities. If the last ? has color $a$, then **a** forms a monochromatic 4-term A.P., if the last ? has color $b$, then **b** forms a monochromatic 4-term A.P., and if the last ? has color $c$, then **c** forms a monochromatic 4-term A.P. (we replace the last ? by $\bullet$):

$$\begin{array}{l} \mathbf{a}..a..a..\mathbf{b} \;\; a..a..a..b \;\; a..a..a..b \;\; ?..?..?..\mathbf{c} \\ \quad a..a..a..b \;\; a..\mathbf{a}..a..\mathbf{b} \;\; a..a..a..b \;\; ?..?..?..\mathbf{c} \\ \qquad a..a..a..b \;\; a..a..a..b \;\; a..a..\mathbf{a}..\mathbf{b} \;\; ?..?..?..\mathbf{c} \\ \qquad\quad ?..?..?..? \;\; ?..?..?..? \;\; ?..?..?..? \;\; ?..?..?..\bullet \end{array}$$

This is how we can force a monochromatic 4-term A.P. if there are 3 colors. The case of 3 colors basically means "to play Tic-Tac-Toe on a 3-dimensional $4 \times 4 \times 4 = 4^3$ board".

Studying these special cases it is easy to see how the *double induction* proof of van der Waerden's theorem goes in the general case. This completes our outline of van der Waerden's proof.

Van der Waerden's proof was adapted by Hales and Jewett (see [16]) to study monochromatic $n$-in-a-line's in an arbitrary $k$-coloring of the $d$-dimensional $n \times n \times \cdots \times n = n^d$ hypercube. The Hales–Jewett theorem, which will be formulated in Section 2, has a wonderful application to the multidimensional generalization of the well-known Tic-Tac-Toe game. I explain this connection in reverse order: first I talk about "Tic-Tac-Toe games" in general, and point out the connection with the Hales–Jewett theorem later.

## 2 Hypercube Tic-Tac-Toe and positional games

Tic-Tac-Toe and its variants are board games which are won (or *lost* in the "Reverse" version) by the first player to complete some kind of winning pattern. The main example of this class is of course Tic-Tac-Toe (arguably the most well-known board game in the world).

**Tic-Tac-Toe** (or Noughts-and-Crosses in the UK). The game board is a big square which is partitioned into $3 \times 3 = 9$ congruent small squares. Whoever moves first puts a cross in one of the nine small squares. The opponent puts a nought into any other small square, and then they alternate cross and nought in the remaining empty squares until one player wins by getting three of his own squares in a line. If neither player gets three squares in a line the play is a draw. There are eight winning triplets: three horizontal, three vertical and two diagonal lines.

Note that Tic-Tac-Toe is a "3-in-a-row" game on a $3 \times 3$ board. The 2-dimensional generalization is the "$n$-in-a-row" game on an $n \times n$ board; we simply call it the $n^2$ game. The multidimensional generalization is $n^d$ Tic-Tac-Toe, or simply the $n^d$ **TTT game.** Two players alternately put their marks in the cells of a $d$-dimensional (hyper)cube of size $n \times \cdots \times n = n^d$. The winner is the player who occupies a full-length-line *first*, i.e., who has $n$ of his marks in a line *first*. If neither player gets $n$-in-a-line, the play is a draw.

More precisely, the board $V$ of the $n^d$ TTT game (or simply "$n^d$ game") is the set of integral $d$-tuples

$$V = \left\{\mathbf{a} = (a_1, a_2, \ldots, a_d) \in \mathbb{Z}^d : 1 \le a_j \le n \text{ for each } 1 \le j \le d\right\}.$$

The winning sets of the $n^d$ game are the $n$-in-a-line sets, that is, the $n$-element sequences

$$\left(\mathbf{a}^{(1)}, \mathbf{a}^{(2)}, \ldots, \mathbf{a}^{(n)}\right)$$

of the board $V$ such that, for each $j$, the sequence $a_j^{(1)}, a_j^{(2)}, \ldots, a_j^{(n)}$ composed of the $j$th coordinates is either $1, 2, 3, \ldots, n$ ("increasing"), or $n, n-1, n-2, \ldots, 1$ ("decreasing"), or a *constant.*

In other words, the winning sets are exactly the $n$-in-a-line's in the $n^d$ hypercube if each elementary "cell" is identified with its own center. Note that in higher dimensions most of the winning lines are some kind of diagonal. The special case $n = 3, d = 2$ gives ordinary Tic-Tac-Toe.

The total number of winning lines in the $n^d$ game is $((n+2)^d - n^d)/2$. Indeed, for each $j$, the sequence $a_j^{(1)}, a_j^{(2)}, \ldots, a_j^{(n)}$ composed of the $j$th coordinates is either *increasing* (i.e., $1, 2, 3, \ldots, n$), or *decreasing* (i.e., $n, n-1, n-2, \ldots, 1$), or a *constant* $c \in \{1, 2, \ldots, n\}$. Since for each coordinate we have $(n+2)$ possibilities, this gives $(n+2)^d$. But we have to subtract $n^d$ since in a line at least one coordinate must change. Finally, we have to divide by 2 because every line has two orientations.

An alternative geometric/intuitive way of getting $((n+2)^d - n^d)/2$ goes as follows. Imagine the board $n^d$ is surrounded by an additional layer of cells, one cell thick. This new object is a cube

$$(n+2) \times (n+2) \times \cdots \times (n+2) = (n+2)^d.$$

It is easy to see that every winning line of the $n^d$-board extends to a uniquely determined *pair* of cells in the new surface layer. So the total number of lines is $((n+2)^d - n^d)/2$.

For later application note that there are at most $(3^d - 1)/2$ $n$-in-a-line's through any point. If $n$ is odd, equality is attained in the center ("center has the maximum degree"); if $n$ is even, the maximum degree of the family of $n$-in-a-line's drops to $2^d - 1$ (which is much smaller than $(3^d - 1)/2$ if $d$ is large).

To prove this elementary fact, consider an arbitrary point $\mathbf{c} = (c_1, c_2, \ldots, c_d)$ of the $n^d$-board and let $L$ be an oriented line containing $\mathbf{c}$. The $j$th coordinates of the points along line $L$ are either

(i) increasing from 1 to $n$, or
(ii) decreasing from $n$ to 1, or
(iii) remain constant $c_j$.

Since every line has two orientations, and at least one coordinate is changing, the maximum degree is $(3^d - 1)/2$, and equality is achieved at the center only. (This suggests that the center is probably the optimum first move for First Player if $n$ is odd.)

If $n$ is even, the maximum degree of the family of winning lines drops to the much smaller value of $2^d - 1$. Indeed, let $\mathbf{c} = (c_1, c_2, \ldots, c_d)$ be an arbitrary point of the $n^d$-board, and consider the family of all lines containing $\mathbf{c}$. Fixing an index-set $I \subset \{1, 2, \ldots, d\}$, where $I$ is a proper subset, there is at most one line $L$ in this family for which the $j$th coordinates of the points along line $L$ remain constant for every $j \in I$, and either increase from 1 to $n$ or decrease from $n$ to 1 for $j \notin I$. So the maximum degree is at most

$$\sum_{i=0}^{d-1} \binom{d}{i} = 2^d - 1.$$

In this case equality occurs for many points of the board: equality occurs if and only if there is a common $c \in \{1, 2, \ldots, n\}$ such that every coordinate $c_j$ is either $c$ or $(n+1-c)$ $(j = 1, 2, \ldots, d)$.

The $n^2$ games are rather dull (with the possible exception of ordinary $3^2$ Tic-Tac-Toe itself): the $2^2$ game is a trivial First Player win, and the $n^2$ game with $n \geq 3$ is a draw. In fact, the longer the winning size $n$, the easier to force a draw – we are going to prove this for $n \geq 4$, see Section 6 – explaining why $3^2$ is the most interesting $n^2$ game.

Far less is known about the $n^d$ game with $d \geq 3$; for example, "Is it true that $5^3$ is a draw game?", or "Is it true that $5^4$ is a First Player win?".

An extremely far-reaching generalization of $n^d$ Tic-Tac-Toe is the class of positional games.

**Positional Games.** Let $(V, \mathcal{F})$ be an arbitrary finite *hypergraph*. What it means is that $V$ is an arbitrary finite set, called the **board** of the game, and $\mathcal{F}$ is an arbitrary family of subsets of $V$, called the family of **winning sets**. The two players, First Player and Second Player, alternately occupy previously unoccupied elements ("points") of board $V$. That player wins who occupies all the elements of some winning set $A \in \mathcal{F}$ *first*; otherwise the play ends in a draw.

Sometimes we just give the family $\mathcal{F}$ of winning sets. Then the board $V$ is the union $\bigcup_{A\in\mathcal{F}} A$ of all winning sets.

We recall the well-known Zermelo's theorem [21].

**Theorem B** (Zermelo 1912). *Every finite perfect-information 2-player game is* **determined**, *which means that either*

*(a) First Player has a winning strategy, or*
*(b) Second Player has a winning strategy, or*
*(c) both of them have a drawing strategy.*

*Remarks.* Alternatives (a), (b), (c) are what we call the three **outcomes of a game**. Of course every *play* has three possible outcomes: either First Player wins, or Second Player wins, or the play ends in a draw, but the outcome of a particular play has nothing to do with the *outcome of the game*. For example, Second Player can easily lose in ordinary Tic-Tac-Toe (e.g., if First Player opens in the center, and Second Player replies on the side), even if Tic-Tac-Toe is a *draw game* (i.e., the game itself belongs to class (c)).

The *existence* **proof** of Zermelo's theorem is a simple application of De Morgan's law. Indeed, there are three alternatives only: either

(a) First Player = **I** has a winning strategy: $\exists x_1 \forall y_1 \exists x_2 \forall y_2 \cdots$ such that **I** wins; or
(b) Second Player = **II** has a winning strategy: $\forall x_1 \exists y_1 \forall x_2 \exists y_2 \cdots$ such that **II** wins;
(c′) or the negation of (a)∨(b):

$$\neg\left((\exists x_1 \forall y_1 \exists x_2 \forall y_2 \cdots \ \mathbf{I}\text{ wins}) \vee (\forall x_1 \exists y_1 \forall x_2 \exists y_2 \cdots \ \mathbf{II}\text{ wins})\right),$$

which, by De Morgan's law, is equivalent to

$$(\forall x_1 \exists y_1 \forall x_2 \exists y_2 \cdots \ \mathbf{I}\text{ loses or draw}) \wedge (\exists x_1 \forall y_1 \exists x_2 \forall y_2 \cdots \ \mathbf{II}\text{ loses or draw}).$$

So the third alternative is that both players have a drawing strategy, which is exactly case (c). This completes the existence proof of Theorem B. □

Every game can be visualized as a "tree of all possible plays" (called game-tree). The exhaustive search of the game-tree gives a *constructive* proof of Zermelo's theorem. It provides an *explicit* winning (or drawing) strategy. The bad news is that exhaustive search is usually impractical. From a complexity point of view a better way to do the same thing is to work with the (usually smaller) position-graph; then "exhaustive search" is called "backward labeling". Unfortunately even "backward labeling" is impractical: it takes "exponential time".

**Strategy Stealing Argument.** It is well known that whoever plays first in a positional game can force at least a draw. In other words, for positional games case (b) cannot occur. Heuristically this is "obvious" because positional games are *symmetric* (i.e., both players want the same thing: to occupy a whole winning set first), and First Player has the *advantage* of the first move which "breaks" the symmetry. I return to this "heuristic proof" later.

**Theorem C** (Strategy Stealing). *Let $(V, \mathcal{F})$ be an arbitrary finite hypergraph. Then playing the positional game on $(V, \mathcal{F})$, First Player can force at least a draw, i.e., a draw or possibly a win.*

*Remark.* Theorem C seems to be *folklore*. "Strategy stealing" was definitely used by J. Nash in the late 1940s (Nash proved that in the well-known game of Hex the first player has a winning strategy), but the first publication of Theorem C (with Theorem E below) is probably in Hales and Jewett [16].

*Proof.* Assume that Second Player (**II**) has a winning strategy *STR*, and one wants to obtain a contradiction. The idea is to see what happens if First Player (**I**) steals and uses *STR*. A winning strategy for a player is a list of instructions telling the player that if the opponent does this, then he does that, so if the player follows the instructions, he will always win. Now **I** can use **II**'s winning strategy *STR* to win as follows. **I** takes an arbitrary first move, and *pretends* to be the second player (he ignores his first move). After **II**'s each move, **I**, as a fake second player, reads the instruction in *STR* to take action. If **I** is told to take a move that is still available, he takes it. If this move was taken by him before as his ignored "arbitrary" first move, then he takes another "arbitrary move". The crucial point here is that an extra move, namely the last "arbitrary move", only *benefits* **I** in a positional game. □

This was a nonconstructive proof: I did *not* construct the claimed drawing strategy. In a nonconstructive proof it is particularly important to be very precise, so the reader is justly wondering: "Where is the precise definition of the concept of *strategy*?". Well, I have to admit my approach was informal: I felt the concept of *strategy* was so natural/intuitive that I could take it for granted, something like "common sense" (I hope the reader agrees on this with me).

Unfortunately "strategy stealing" doesn't supply an *explicit* strategy; this is a major "weakness" of the argument.

**Open Problem 1** *Can the reader find an* **explicit** *First-Player's drawing strategy for the $5^3$ game, and in general, for every single $n^d$ game?*

This seems to be a hopeless problem. I don't know anything beyond "exhaustive search".

**Warning.** I recall the "heuristic proof" of Theorem C: in a positional game the two players have exactly the same goal ("to occupy a whole winning set first"), but First Player has the "first-move-advantage", which breaks the symmetry in favor of him, and guarantees at least a First-Player's drawing strategy.

Unfortunately this is a *faulty argument*. Indeed, if one repeats it for the *Reverse* Positional Game, where that player *loses* who occupies a whole winning set first, then it leads to a *false* conclusion, namely, to the conclusion that "Second Player cannot lose". Indeed, one can argue that, in a Reverse Positional Game, the two players have exactly the same goal, but First Player has the "first-move-disadvantage", which breaks the symmetry in favor of the opponent, implying that Second Player cannot lose. But the conclusion is *false*: it is *not* true that Second Player cannot lose a Reverse

Positional Game. There are infinitely many Reverse Positional Games in which the First Player has a winning strategy, e.g., the Reverse $3^3$-game.

**Theorem D** (Reverse $n^d$ game)**.** *Consider the* **Reverse** $n^d$ *game: the only difference in the rule is that the player who occupies a whole n-in-a-line first is the loser. If n is odd (i.e., the geometric center of the board is a "cell") First Player has an* **explicit** *drawing strategy. If n is even (i.e., the geometric center of the board is not a "cell") Second Player has an* **explicit** *drawing strategy.*

The shockingly simple **proof** is due to Golomb and Hales [13]. If $n$ is *odd*, then First-Player's opening move is the center $C$, and whenever Second Player claims a point (i.e., a cell) $P$, then First Player chooses the *reflection* $P'$ of $P$ with respect to center $C$ (i.e., $P$, $C$, $P'$ are on the same line and the $PC$-distance equals the $CP'$-distance). Assume First Player colors his points red, and second player colors his points blue. We show that First Player cannot lose. Indeed, assume that First Player loses, and $L$ is the *first* $n$-in-a-line owned by a player during the course of a play (i.e., $L$ is a red line). Observe that $L$ cannot contain center $C$. Indeed, every completed $n$-in-a-line containing the center has precisely $(n+1)/2$ red and $(n-1)/2$ blue points. If $L$ doesn't contain the center, then its reflection $L'$ is a complete blue line, and since $L'$ was completed *before* $L$, we get a contradiction.

On the other hand, if $n$ is *even*, then Second Player can use the "reflection strategy", i.e., choosing $P'$ if First-Player's last move is $P$, and achieves at least a draw. □

Note that the $3^3$ board does *not* have a drawing end-position ("easy case-study"), so First-Player's explicit drawing strategy in the Reverse $3^3$ game is automatically "upgraded" to a winning strategy.

**Ramsey Theory: when draw is impossible.** When can First Player win in a positional game? A partial answer (sufficient condition) is the following: First Player has a *winning strategy* in a positional game when *draw is impossible*.

Of course this condition is not necessary. For example, the full-length branches of a binary tree with $n$ levels form an $n$-uniform family of $2^{n-1}$ winning sets such that First Player has an easy win (the players take the vertices of the tree). But this positional game has plenty of drawing end-positions (e.g., all vertices of degree one are red, and the rest of the vertices are all blue).

A drawing end-position in a positional game $(V, \mathcal{F})$ gives a *halving* 2-coloring of the board $V$ such that no winning set $A \in \mathcal{F}$ is monochromatic; I call it a *proper halving 2-coloring* of hypergraph $(V, \mathcal{F})$. (Of course *halving* 2-coloring means to have $\lceil |V|/2 \rceil$ of one color and $\lfloor |V|/2 \rfloor$ of the other color.)

A slightly more general concept is when we allow *arbitrary* 2-colorings, not just halving 2-colorings.

The *chromatic number* $\chi(\mathcal{F})$ of hypergraph $\mathcal{F}$ is the least integer $r \geq 2$ such that the elements of the board $V$ can be colored with $r$ colors yielding no monochromatic $A \in \mathcal{F}$. If the chromatic number of $\mathcal{F}$ is at least three, then draw is impossible. In this case First-Player's (at least) drawing strategy in Theorem C is automatically upgraded to a winning strategy. *Ramsey Theory* is exactly the theory of hypergraphs with chromatic number at least three (see [15]).

**Theorem E** (Win by Ramsey Theory)**.** *Suppose that the board V is finite, and the family $\mathcal{F}$ of winning sets has the property that there is no proper halving 2-coloring;*

*this happens, for example, if $\mathcal{F}$ has chromatic number at least three. Then First Player has a winning strategy in the positional game on $(V, \mathcal{F})$.*

Theorem E doesn't say a word about *how* to win. Unfortunately nobody knows "how to win" (try "exhaustive search" like "backward labeling")! Theorem E describes a subclass of positional games with the remarkable property that one can easily determine the winner without being able to say how one wins.

Now we are ready to formulate the Hales–Jewett theorem which was briefly mentioned at the end of Section 1. The Hales–Jewett theorem – the combinatorial content of van der Waerden's theorem on arithmetic progressions – states that there is a (least) *finite* threshold number $HJ(n)$ such that the family of all "$n$-in-a-lines" in the $n^d$ hypercube (i.e., the family of winning sets in the $n^d$ game) has chromatic number $\geq 3$ if $d \geq HJ(n)$. We call $HJ(n)$ the Hales–Jewett threshold. By the Ramsey Criterion (Theorem E), First Player has a winning strategy in the $n^d$ game if $d \geq HJ(n)$.

**Open Problem 2** *Find an* **explicit** *First-Player's winning strategy in the $n^d$ game when $d \geq HJ(n)$.*

Unfortunately this problem seems to be hopelessly difficult.

Notice that, in view of Theorem E the assumption $d \geq HJ(n)$ for win in the $n^d$ game is a little bit too strong: it suffices to assume that $n^d$ does not have a proper halving 2-coloring. This motivates the following definition: let $HJ_{1/2}(n)$ denote the least integer $d$ such that in each *halving* 2-coloring of $n^d$ there is a monochromatic $n$-in-a-line (i.e., *geometric* line). We call $HJ_{1/2}(n)$ the *halving* version of the Hales–Jewett number. By definition

$$HJ_{1/2}(n) \leq HJ(n).$$

Is there an $n^d$ game for which *strict inequality* holds? Well, I don't know the answer to this question, but I *do* know an "almost $n^d$ game" for which strict inequality holds: it is the "$3^3 \setminus \{center\}$ game" in which the center is removed, and also the 13 3-in-a-line's going through the center are removed. The "$3^3 \setminus \{center\}$ game" – a truncated version of the $3^3$ game – has $3^3 - 1 = 26$ points and $(5^3 - 3^3)/2 - 13 = 49 - 13 = 36$ winning triplets. The "$3^3 \setminus \{center\}$ game" has chromatic number two, but every proper 2-coloring has the type (12, 14) meaning that one color class has 12 points and the other one has 14 points; proper *halving* 2-coloring, therefore, does not exist. (By the way, this implies, in view of Theorem E, that the "$3^3 \setminus \{center\}$ game" is a First Player win.)

The "$3^3 \setminus \{center\}$ game" was a kind of "natural game" example. In the family of *all* hypergraphs it is easy to find examples distinguishing *proper 2-coloring* from *proper halving 2-coloring* in a much more dramatic fashion. We don't even need hypergraphs, it suffices to consider graphs: consider the complete bipartite graph $K_{a,b}$ (i.e., let $A$ and $B$ be disjoint sets where $A$ is $a$-element and $B$ is $b$-element, and take the $ab$ point-pairs such that one point is from $A$ and the other one is from $B$). This graph has chromatic number two, and the *only* proper 2-coloring of the $(a+b)$-element point-set is the $(A, B)$-coloring. If $a = 1$ and $b$ is "large", then the proper 2-coloring is very far from a *halving* 2-coloring.

Let's return to the Hales–Jewett theorem: the Hales–Jewett theorem is about monochromatic $n$-in-a-line's of the $n^d$ hypercube. What the proof actually gives is more: it guarantees the existence of a monochromatic *combinatorial line*. To explain what it means, let $[n] = \{1, 2, \ldots, n\}$. An *x-string* is a finite word $a_1 a_2 a_3 \cdots a_d$ of the symbols $a_i \in [n] \cup \{x\}$ where at least one symbol $a_i$ is $x$. An $x$-string is denoted by $\mathbf{w}(x)$. For every integer $i \in [n]$ and $x$-string $\mathbf{w}(x)$, let $\mathbf{w}(x; i)$ denote the string obtained from $\mathbf{w}(x)$ by replacing each $x$ by $i$. A *combinatorial line* is a set of $n$ strings $\{\mathbf{w}(x; i) : i \in [n]\}$ where $\mathbf{w}(x)$ is an $x$-string.

Every combinatorial line is a geometric line, i.e., $n$-in-a-line, but the converse is not true. Before showing a counterexample, note that a *geometric line* can be described as an *$xx'$-string* $a_1 a_2 a_3 \cdots a_d$ of the symbols $a_i \in [n] \cup \{x\} \cup \{x'\}$ where at least one symbol $a_i$ is $x$ or $x'$. An $xx'$-string is denoted by $\mathbf{w}(xx')$. For every integer $i \in [n]$ and $xx'$-string $\mathbf{w}(xx')$, let $\mathbf{w}(xx'; i)$ denote the string obtained from $\mathbf{w}(xx')$ by replacing each $x$ by $i$ and each $x'$ by $(n+1-i)$. A *directed geometric line* is a *sequence* $\mathbf{w}(xx'; 1)$, $\mathbf{w}(xx'; 2)$, $\mathbf{w}(xx'; 3), \ldots, \mathbf{w}(xx'; n)$ of $n$ strings, where $\mathbf{w}(xx')$ is an $xx'$-string. Note that every geometric line has two orientations.

As we said before, it is *not* true that every geometric line is a combinatorial line. What is more, it is clear from the definition that there are substantially more geometric lines than combinatorial lines: in the $n^d$ game there are $((n+2)^d - n^d)/2$ geometric lines and $(n+1)^d - n^d$ combinatorial lines. Note that the maximum degree of the family of combinatorial lines is $2^d - 1$, and the maximum is attained in the points of the "main diagonal" $(j, j, \ldots, j)$ where $j$ runs from 1 to $n$.

For example, in ordinary Tic-Tac-Toe:

| | | |
|---|---|---|
| (1, 3) | (2, 3) | (3, 3) |
| (1, 2) | (2, 2) | (3, 2) |
| (1, 1) | (2, 1) | (3, 1) |

the "main diagonal" $\{(1,1), (2,2), (3,3)\}$ is a combinatorial line defined by the $x$-string $xx$, $\{(1,1), (2,1), (3,1)\}$ is another combinatorial line defined by the $x$-string $x1$, but the "other diagonal"

$$\{(1,3), (2,2), (3,1)\}$$

is a geometric line defined by the $xx'$-string $xx'$. The "other diagonal" is the only geometric line of the $3^2$ game which is *not* a combinatorial line.

The Hales–Jewett threshold $HJ(n,k)$ is the smallest integer $d$ such that in each $k$-coloring of $[n]^d = n^d$ there is a monochromatic *geometric* line. The modified Hales–Jewett threshold $HJ^c(n,k)$ is the smallest integer $d$ such that in each $k$-coloring of $[n]^d = n^d$ there is a monochromatic *combinatorial* line ("c" stands for "combinatorial"). Trivially

$$HJ(n,k) \leq HJ^c(n,k).$$

In the case of "two colors" ($k = 2$) we write: $HJ(n) = HJ(n,2)$ and $HJ^c(n) = HJ^c(n,2)$; trivially $HJ(n) \leq HJ^c(n)$.

Similarly to $HJ(n)$ we can define the *halving* version of $HJ^c(n)$ as follows: $HJ^c_{1/2}(n)$ is the smallest integer $d$ such that in each *halving* 2-coloring of $[n]^d = n^d$ there is a monochromatic *combinatorial* line. By definition $HJ^c_{1/2}(n) \leq HJ^c(n)$.

In 1961 Hales and Jewett observed that van der Waerden's proof of Theorem A can be adapted to the $n^d$ board, and this way they proved that $HJ^c(n,k) < \infty$ for all positive integers $n$ and $k$ (which, of course, implies that $HJ(n,k) < \infty$ for all positive integers $n$ and $k$).

How large is $HJ(n) = HJ(n,2)$? Unfortunately our present knowledge on the Hales–Jewett threshold number $HJ(n)$ is rather disappointing. The best known upper bound on $HJ(n)$ was proved by Shelah [12] in 1988. It is a primitive recursive function (the *supertower* function), which is much-much better than the original van der Waerden–Hales–Jewett bound. The original "double-induction" argument gave the notorious Ackermann function. Unfortunately Shelah's bound is still enormous for "layman combinatorics".

For a *precise* discussion we have to introduce the so-called Grzegorczyk hierarchy of primitive recursive functions. In fact, we define the *representative* function for each class. (For a more detailed treatment of primitive recursive functions we refer the reader to any monography of Mathematical Logic.)

Let $g_1(n) = 2n$, and for $i > 1$, let $g_i(n) = g_{i-1}\Big(g_{i-1}\big(\dots g_{i-1}(1)\dots\big)\Big)$, where $g_{i-1}$ is taken $n$ times. An equivalent definition is $g_i(n+1) = g_{i-1}\big(g_i(n)\big)$. For example, $g_2(n) = 2^n$ is the exponential function,

$$g_3(n) = 2^{2^{2^{\cdot^{\cdot^{\cdot^{2}}}}}}$$

is the "tower function" of height $n$. The next function $g_4(n+1) = g_3\big(g_4(n)\big)$ is what we call the "Shelah's supertower function" because this is exactly what shows up in Shelah's proof. Note that $g_k(x)$ is the *representative* function of the $(k+1)$st Grzegorczyk class.

Note that the original van der Waerden–Hales–Jewett proof proceeded by a *double induction* on $n$ ("length") and $k$ ("number of colors"), and yielded an extremely large upper bound for $HJ^c(n,k)$. Actually the original argument gave the same upper bound $U(n,k)$ for both $HJ^c(n,k)$ and $W(n,k)$ ("van der Waerden threshold"). We define $U(n,k)$ as follows. For $n = 3$: $U(3,2) = 1000$ and for $k \geq 2$, $U(3,k+1) = (k+1)^{U(3,k)}$. For $n = 4$: $U(4,2) = U\left(3, 2^{U(3,2)}\right)$ and for $k \geq 2$,

$$U(4,k+1) = U\left(3, (k+1)^{U(4,k)}\right).$$

In general, for $n \geq 4$, let

$$U(n,2) = U\left(n-1, 2^{U(n-1,2)}\right)$$

and for $k \geq 2$,

$$U(n,k+1) = U\left(n-1, (k+1)^{U(n,k)}\right).$$

It is easy to see that for every $n \geq 3$ and $k \geq 2$, $U(n,k) > g_n(k)$. It easily follows that the function $U(x,2)$ (i.e., the case of two colors) eventually *majorizes* $g_n(x)$ for every $n$ (we recall that $g_n(x)$ is the representative function of the $(n+1)$th Grzegorczyk class). It follows that $U(x,2)$ is *not* primitive recursive. In fact, $U(x,2)$ behaves like the well-known Ackermann function $A(x) = g_x(x)$, the classical example of a recursive but not primitive recursive function. In plain language, the original Ackermann function upper bound was ENOOOOORMOUSLY LARGE BEYOND IMAGINATION!!!

In 1988 Shelah proved the following much-much better upper bound.

**Theorem F** (Shelah's primitive recursive upper bound for the Hales–Jewett threshold). *For every $n \geq 1$ and $k \geq 1$,*

$$HJ^c(n,k) \leq \frac{1}{(n+1)k} g_4(n+k+2).$$

*That is, given any k-coloring of the hypercube $[n]^d = n^d$ where the dimension $d \geq \frac{1}{(n+1)k} g_4(n+k+2)$, there is always a monochromatic* **combinatorial** *line.*

Consider $HJ(n) = HJ(n,2)$ and $HJ^c(n) = HJ^c(n,2)$: an easy case-study shows that $HJ(3) = HJ^c(3) = 3$, but the numerical value of $HJ(4)$ remains a complete mystery. We know that it is $\geq 5$ (see [13]), and also that it is finite, but no one can prove a "reasonable" upper bound like $HJ(4) \leq 1000$ or even a much weaker bound like $HJ(4) \leq 10^{1000}$. Shelah's proof gives the upper bound

$$HJ(4) \leq HJ^c(4) \leq g_3(24) = 2^{2^{2^{\cdot^{\cdot^{2}}}}}$$

where the "height" of the tower is 24. This upper bound is *absurdly large*; it is rather disappointing that Ramsey theory is unable to provide a "reasonable" upper bound even for the first "nontrivial" value $HJ(4)$ of the Hales–Jewett function $HJ(n)$.

It is a wide open problem whether $HJ(n)$ is less than the "plain" tower function $g_3(n)$.

It seems to be highly unlikely that the game-theoretic "phase transition" between win and draw for the $n^d$ game is anywhere close to the Hales–Jewett number $HJ(n)$, but no method is known for handling this problem. To make it quantitative, we introduce the **Win Number** for the $n^d$ game. Let $\mathbf{w}(n-\text{line})$ denote the least threshold such that for every $d \geq \mathbf{w}(n-\text{line})$ the $n^d$ game is a First-Player's win ("**w**" stands for "win"). Theorem E yields the inequality $\mathbf{w}(n-\text{line}) \leq HJ(n)$.

**Open Problem 3** *Is it true that* $\mathbf{w}(n-\text{line}) < HJ(n)$ *for all sufficiently large values of n? Is it true that*

$$\frac{\mathbf{w}(n-\text{line})}{HJ(n)} \longrightarrow 0 \text{ as } n \to \infty?$$

A winning set of an $n^d$ Tic-Tac-Toe game is an $n$-term arithmetic progression on a straight line. This motivates the "Arithmetic Progression Game": this is a positional game in which the board is the set of the first $N$ integers $[N] = \{1, 2, 3, \ldots, N\}$, and the winning sets are the $n$-term arithmetic progressions in $[N]$. I call this "Arithmetic Progression Game" the $(N, n)$ **van der Waerden Game**. An obvious motivation for the name is the van der Waerden theorem: for every $n$ there is a (least) threshold $W(n) = W(n, 2)$ such that given any 2-coloring of $[N]$ with $N = W(n)$ there is always a monochromatic $n$-term arithmetic progression. $W(n)$ is called the *van der Waerden Number*. If $N \geq W(n)$ then Theorem E applies, and yields that First Player has a winning strategy in the $(N, n)$ **van der Waerden Game**.

Actually, in view of Theorem E, $W(n)$ can be replaced by its *halving* version $W_{1/2}(n)$. $W_{1/2}(n)$ is defined as the least integer $N$ such that each *halving* 2-coloring of the interval $[N]$ yields a monochromatic $n$-term arithmetic progression. Trivially $W_{1/2}(n) \leq W(n)$; is there an $n$ with strict inequality? Well, I don't know.

What do we know about the van der Waerden number $W(n) = W(n, 2)$? First note that Shelah's theorem above (Theorem F) immediately gives the following primitive recursive upper bound for the van der Waerden threshold.

**Theorem G** (Shelah). *For all positive integers n and k,*

$$W(n, k) \le g_4(n + k + 3).$$

*Proof.* It is easy to see that

$$W(n, k) \le n^{HJ(n,k)}.$$

Indeed, we embed the $d$-dimensional cube $[n]^d$ into the interval $\{0, 1, 2, \ldots, n^d - 1\}$ in the following natural 1-to-1 way: given any string $\mathbf{w} = a_1 a_2 \cdots a_d \in [n]^d$, let

$$f(\mathbf{w}) = (a_1 - 1) + (a_2 - 1)n + (a_3 - 1)n^2 + \cdots + (a_d - 1)n^{d-1}.$$

Observe that $f$ maps any $n$-in-a-line ("geometric line") into an $n$-term arithmetic progression. It follows that

$$W(n, k) \le n^{HJ(n,k)}.$$

Therefore, by Shelah's bound,

$$W(n, k) \le n^{HJ(n,k)} \le n^{g_4(n+k+2)} \le g_4(n + k + 3),$$

and Theorem G follows. □

Theorem G was enormously improved by a recent breakthrough of T. Gowers: he pushed $W(n)$ down *well* below the "plain" tower function $g_3(n)$. In fact, Gowers [14] proved much more: he proved a quantitative *Szemerédi theorem*, i.e., a *density* version of the van der Waerden theorem. To formulate the result I use the arrow-notation $a \uparrow b$ for $a^b$, with the obvious convention that $a \uparrow b \uparrow c$ stands for $a \uparrow (b \uparrow c) = a^{b^c}$. The relevant "two-color" special case of Gowers's theorem goes as follows.

**Theorem H** (Gowers's analytic result). *Let*

$$N \ge 2 \uparrow 2 \uparrow 2 \uparrow 2 \uparrow 2 \uparrow (n + 9) = 2^{2^{2^{2^{2^{n+9}}}}},$$

*and let S be an arbitrary subset of* $\{1, 2, \ldots, N\}$ *of size* $\ge N/2$. *Then S contains an n-term arithmetic progression.*

Of course this implies that $W(n) = W(n, 2) \le 2 \uparrow 2 \uparrow 2 \uparrow 2 \uparrow 2 \uparrow (n + 9)$. This bound is a huge improvement to Shelah's supertower function, but it is still far too large for "layman combinatorics". Note that Gowers's paper is extremely complicated: it is 128 pages long and uses deep analytic techniques; Shelah's proof, on the other hand, is surprisingly short and elementary.

Gowers's density theorem implies that, if

$$N \ge 2 \uparrow 2 \uparrow 2 \uparrow 2 \uparrow 2 \uparrow (n + 9),$$

then First Player has a winning strategy in the $(N, n)$ van der Waerden Game. As usual with the applications of Theorem E we have no idea how First-Player's winning strategy actually looks like.

**Open Problem 4** *Consider the* $(N, n)$ *van der Waerden Game where* $N \geq W(n)$*; for example, let*

$$N \geq 2 \uparrow 2 \uparrow 2 \uparrow 2 \uparrow 2 \uparrow (n+9).$$

*Find an* **explicit** *First-Player's winning strategy.*

It seems to be highly unlikely that the "phase transition" between win and draw for the van der Waerden game is anywhere close to the van der Waerden number $W(n)$. The corresponding *Win Number* is defined as follows. Let $\mathbf{w}(n-\text{term A.P.})$ denote the least threshold such that for every $N \geq \mathbf{w}(n-\text{term A.P.})$ the $(N, n)$ van der Waerden game is a First Player win.

Theorem E implies $\mathbf{w}(n-\text{term A.P.}) \leq W(n)$. It is easily seen

$$\mathbf{w}(3-\text{term A.P.}) = 5 < 9 = W(3).$$

I don't know the exact value of $\mathbf{w}(4-\text{term A.P.})$ but I know that it is less than $W(4) = 35$, and similarly, $\mathbf{w}(5-\text{term A.P.})$ is unknown but it is definitely less than $W(5) = 178$.

**Open Problem 5** *Is it true that* $\mathbf{w}(n-\text{term A.P.}) < W(n)$ *for all sufficiently large values of* $n$*? Is it true that*

$$\frac{\mathbf{w}(n-\text{term A.P.})}{W(n)} \longrightarrow 0 \text{ as } n \to \infty?$$

## 3 Win vs. Weak Win

The primary object of game-playing is of course *winning* ("winning is everything" as they say it here in the USA). Unfortunately no one knows how to force a win in a general positional game. The next best thing is *Weak Win.*

**Weak Win and Strong Draw: the Maker–Breaker version.** As we have seen in Theorem C, Second Player has no chance to win a positional game against a perfect First Player. Then, why doesn't he just concentrate on preventing First Player from building a winning set and simply ignore his own desire for building? We thus can name him *Breaker*, and the other one *Maker*. (We usually assume that Maker is the first player and Breaker is the second player, but in most cases this is almost irrelevant.)

In general, on the same hypergraph $(V, \mathcal{F})$, one can play the "symmetric" positional game, and also the "asymmetric" **Maker–Breaker game.** Here Maker's aim is to claim every element of a winning set $A \in \mathcal{F}$, but not necessarily first; and Breaker's aim is to prevent Maker from doing so (i.e., to put his mark in every winning set; Breaker does not want to occupy a winning set). The winner is the one who achieves his goal; so draw is impossible by definition.

If First Player can force a win in the positional game on $(V, \mathcal{F})$, then of course the *same* play gives him, as Maker, a win in Maker–Breaker version (on the same hypergraph). But the converse is *not* true. It is possible that Maker, as the first player, has a winning strategy in the Maker–Breaker version, while the second player can force a draw in the positional game. This happens, for example, in ordinary Tic-Tac-Toe: the $3^2$ game is a draw, but the Maker–Breaker version is a Maker's win (Maker is the first player). This is why one can call a Maker's winning strategy a **Weak Win Strategy**. Breaker's winning strategy can be called a **Strong Draw Strategy**. We know a lot about the complementary concepts of *Weak Win* and *Strong Draw*, but we know nothing about "ordinary win" except Theorem E (which unfortunately doesn't say a word about *how* to win).

Understanding "ordinary win" seems to be beyond the reach of contemporary mathematics. The next best thing is to understand Weak Win, i.e., to draw the line between Weak Win and Strong Draw.

While playing the positional game on a hypergraph, both players have their own threats, and either of them, fending off the other's, may build his own winning set. Therefore, a play is a delicate balancing between threats and counterthreats and can be of very intricate structure even if the hypergraph itself is simple.

The Maker–Breaker version is usually somewhat simpler. Maker doesn't have to waste valueable moves fending off his opponent's threats. Maker can simply concentrate on his own goal of *building*, and Breaker can concentrate on *blocking* the opponent (unlike in a positional game where either player has to *build and block* at the same time). Doing one job at a time is definitely simpler.

For example, consider the following "plausible conjecture" about "ordinary win".

**Open Problem 6** *Is it true that if the $n^d$ game is a First-Player's win, then the $n^D$ game, where $D > d$, is also a win?*

A good reason why Open Problem 6 is so difficult is that one can construct a finite hypergraph (i.e., a positional game) which is a First-Player's win, but adding an extra winning set turns it into a draw game. A more sophisticated "Extra Set Paradox" is the following. One can construct a finite hypergraph which is a draw, but it has an induced sub-hypergraph (i.e., all winning sets in a subset of the board) which is a First-Player's win. One can even construct such a *uniform* hypergraph (i.e., each winning set has the same size).

We challenge the reader to find such examples (not as easy as it seems!).

Note that the Maker–Breaker version of Open Problem 6 is trivial: Maker uses the Weak Win strategy within a $d$-dimensional subcube of the $n^D$ board.

How about the Maker–Breaker version of Open Problem 2? In contrast to positional games, in Maker–Breaker games there *is* an explicit version of Theorem E: Weak Win is guaranteed by a simple *copycat strategy*.

**Theorem I** (Weak Win by Ramsey Theory). *Let $(V, \mathcal{F})$ be a finite hypergraph of chromatic number $\geq 3$, and let $(V', \mathcal{F}')$ be a point-disjoint copy of $(V, \mathcal{F})$. Assume $Y$ contains $V \cup V'$ and $\mathcal{G}$ contains $\mathcal{F} \cup \mathcal{F}'$. Then Maker has an* **explicit** *Weak Win strategy playing on $(Y, \mathcal{G})$.*

Theorem I seems to be *folklore* among Ramsey theorists. An interesting infinite version is in [2].

*Proof.* Let $f : V \to V'$ be the isomorphism between $(V, \mathcal{F})$ and $(V', \mathcal{F}')$. Pick a player and call him "Maker": I show that Maker can force a Weak Win by using the following copycat pairing strategy. If the opponent's last move was $x \in V$ or $x' \in V'$, then Maker's next move is $f(x) \in V'$ or $f^{-1}(x') \in V$ (unless it was already occupied by Maker before; then Maker's next move is arbitrary). Since the chromatic number of $(V, \mathcal{F})$ is at least three, one of the two players will completely occupy a winning set. If this player is Maker, we are done. If the opponent occupies some $A \in \mathcal{F}$, then Maker occupies $f(A) \in \mathcal{F}'$, and we are done again. □

For example, consider the $n^d$ game where $d \geq HJ(n) + 1$ ($HJ(n)$ is the Hales–Jewett number). The condition guarantees that the board contains two disjoint copies of $n^{HJ(n)}$. The copycat strategy of Theorem I supplies an *explicit* Weak Win strategy for either player. Theorem I solves the Weak Win version of Open Problem 2, except for the "boundary case" $d = HJ(n)$.

This doesn't mean, however, that Maker–Breaker games are "easy". Absolutely not! For example, the well-known and notoriously difficult *Hex* is equivalent to a Maker–Breaker game, but to be a Maker–Breaker game does *not* help to "break" Hex. We recall *Hex*: the board is a rhombus of hexagons of size $n \times n$ (the standard size is $n = 11$). The two players, White ("First Player") and Black ("Second Player") take the two pairs of opposite sides of the board. The players alternately put their pieces on unoccupied hexagons (White has white pieces, and Black has black pieces). White wins if his pieces connect his opposite sides of the board, and Black wins if his pieces connect the other pair. Observe that Hex is *not* a positional game: the winning sets for White and Black are (mostly) *different*. In fact the only common winning sets are the chains connecting diagonally opposite corners.

Let *WeakHex* denote the Maker–Breaker game in which the board is the $n \times n$ Hex board, Maker = White, Breaker = Black, and the winning sets are the connecting chains of White. We claim that Hex and *WeakHex* are equivalent. First note that draw in Hex is impossible. Indeed, in order to prevent the opponent from making a connecting chain, one must build a "river", which itself contains a chain connecting the *other* pair of opposite sides (this fact seems plausible, but the precise proof is not completely trivial). This means that Breaker's goal in *WeakHex* (i.e., "blocking") is equivalent to Black's goal in Hex (i.e., "building first"). Here "equivalent" means that Breaker has a winning strategy in *WeakHex* if and only if Black has a winning strategy in Hex. Since draw is impossible, Hex and *WeakHex* are equivalent.

Even if Hex is not a positional game, the Strategy Stealing argument still applies (why? use reflection symmetry!), and yields that White has a winning strategy. This remarkable theorem was discovered by J. Nash in the late 1940s (probably the first application of Strategy Stealing). The fact that Hex is equivalent to a Maker–Breaker (namely, to *WeakHex*) doesn't make it "easy": we don't have the slightest clue how to find an *explicit* winning strategy. It remains an outstanding open problem (the cases $n \leq 7$ are solved by exhaustive search).

Next consider the $(N, n)$ van der Waerden Game, and assume that

$$N \geq 2 \uparrow 2 \uparrow 2 \uparrow 2 \uparrow 2 \uparrow (n + 9)$$

("Gowers's upper bound" for $W(n)$). If First Player just wants a *Weak Win* (i.e., to occupy an $n$-term arithmetic progression, but not necessarily first), then he does *not* need to follow any particular strategy, simply "showing up" is enough. Indeed, at the end of a play he will certainly occupy half of $[N]$ ($N/2$ integers), and by Gowers's density theorem (quantitative Szemerédi theorem) he *must* have an $n$-term arithmetic progression independently of his actual play.

By contrast, to achieve a Weak Win in the $n^d$ game just "showing up" is *not* enough. First Player must do something special, but this "special" is not too sophisticated: a simple "copycat" Pairing Strategy will do the trick (see Theorem I).

Weak Win motivates the introduction of the *Weak* versions of the *Win Numbers*. Let's begin with the $n^d$ game: $\mathbf{ww}(n\text{--line})$ denotes the least threshold such that for every $d \geq \mathbf{ww}(n\text{--line})$ First Player can force a Weak Win in the $n^d$ game ("**ww**" stands for "weak win").

The study of ordinary Tic-Tac-Toe yields $\mathbf{ww}(3\text{--line}) = 2 < 3 = \mathbf{w}(3\text{--line})$. Patashnik's well-known computer-assisted study of the $4^3$ game yields $\mathbf{ww}(4\text{--line}) = \mathbf{w}(4\text{--line}) = 3$ (see [17]).

**Open Problem 7** *(a) Is it true that* $\mathbf{ww}(n\text{--line}) < \mathbf{w}(n\text{--line})$ *for all sufficiently large values of $n$? Is it true that*

$$\frac{\mathbf{ww}(n\text{--line})}{\mathbf{w}(n\text{--line})} \longrightarrow 0 \text{ as } n \to \infty?$$

*(b) Is it true that*

$$\frac{\mathbf{ww}(n\text{--line})}{HJ(n)} \longrightarrow 0 \text{ as } n \to \infty?$$

Next consider the van der Waerden game. Let $\mathbf{ww}(n\text{--term A.P.})$ denote the least threshold such that for every $N \geq \mathbf{ww}(n\text{--term A.P.})$ First Player can force a Weak Win in the $(N, n)$ van der Waerden game.

It is easily seen

$$\mathbf{ww}(3\text{--term A.P.}) = \mathbf{w}(3\text{--term A.P.}) = 5 < 9 = W(3).$$

**Open Problem 8** *(a) Is it true that* $\mathbf{ww}(n\text{--term A.P.}) < \mathbf{w}(n\text{--term A.P.})$ *for all sufficiently large values of $n$? Is it true that*

$$\frac{\mathbf{ww}(n\text{--term A.P.})}{\mathbf{w}(n\text{--term A.P.})} \longrightarrow 0 \text{ as } n \to \infty?$$

*(b) Is it true that*

$$\frac{\mathbf{ww}(n\text{--term A.P.})}{W(n)} \longrightarrow 0 \text{ as } n \to \infty?$$

I conclude this section by recalling the trivial inequalities

$$\mathbf{ww}(n\text{--line}) \leq \mathbf{w}(n\text{--line}) \leq HJ_{1/2}(n) \leq HJ(n),$$
$$\mathbf{ww}(\text{comb. } n\text{--line}) \leq \mathbf{w}(\text{comb. } n\text{--line}) \leq HJ^c_{1/2}(n) \leq HJ^c(n),$$

and

$$\mathbf{ww}(n\text{--term A.P.}) \leq \mathbf{w}(n\text{--term A.P.}) \leq W_{1/2}(n) \leq W(n).$$

## 4 Old lower bounds

The starting point of this section is an old result of Erdős [9] (see also [10]), which was his first step toward developing the so-called Probabilistic Method in Combinatorics (see [1]).

**Theorem J** (Erdős 1947). *Let $\mathcal{F}$ be an n-uniform hypergraph, and assume that $|\mathcal{F}| < 2^{n-1}$. Then*

*(a) there is a Proper 2-Coloring; and what is somewhat more,*
*(b) there also is a Proper Halving 2-Coloring (i.e., Drawing End-Position).*

Both (a) and (b) can be proved by a simple "counting argument". The **proof** of Theorem J (a) goes as follows. Let $N = |V|$ denote the size of the union set ("board") $V$ of hypergraph $\mathcal{F}$. A simple counting argument shows that under the condition $|\mathcal{F}| < 2^{n-1}$ there exists a Proper 2-Coloring. Indeed, there are $2^N$ 2-colorings of board $V$, and for every single winning set $A \in \mathcal{F}$ there exists $2^{N-n+1}$ "bad" 2-colorings which are monochromatic on $A$. By hypothesis $2^N - |\mathcal{F}|2^{N-n+1} > 0$, which implies that throwing out all "bad" 2-colorings, there must remain at least one Proper 2-Coloring (i.e., no $A \in \mathcal{F}$ is monochromatic).

To prove Theorem J (b) we have to find a Drawing End-Position (i.e., a 2-coloring of the board by "colors" X and O such that the two color classes have the same size, and each winning set contains both marks). For notational simplicity assume that $N$ is even. The idea is exactly the same as that of (a), except that we restrict ourselves to the $\binom{N}{N/2}$ Halving 2-Colorings instead of the $2^N$ (arbitrary) 2-colorings. The analogue of $2^N - |\mathcal{F}|2^{N-n+1} > 0$ is the following requirement: $\binom{N}{N/2} - 2|\mathcal{F}|\binom{N-n}{N/2} > 0$. This holds because

$$\frac{\binom{N-n}{N/2}}{\binom{N}{N/2}} = \frac{N/2}{N}\frac{(N/2)-1}{N-1}\frac{(N/2)-2}{N-2}\cdots\frac{(N/2)-n+1}{N-n+1} \leq 2^{-n},$$

and (b) follows. □

The previous argument can be stated in the following slightly different form: The *average number* ("expected value" or "first moment") of winning sets completely occupied by either player is precisely

$$2|\mathcal{F}|\frac{\binom{N-n}{N/2}}{\binom{N}{N/2}}, \quad \text{which is } \textit{less than one}.$$

Since the *minimum* is less or equal to the *average*, and the *average* is *less than one*, *there must exist* at least one *Drawing* End-Position (i.e., no player owns a whole winning set).

This kind of "counting argument", discovered by Erdős, belongs to the same category as Euclid's proof of the existence of infinitely many primes, or Pythagoras's proof of the irrationality of $\sqrt{2}$: astonishingly simple and fundamentally important at the same time.

Erdős's "counting argument" (Theorem J) was developed later in two very different ways: first by Wolfgang M. Schmidt [18] in 1962, and later by L. Lovász in a joint work with Erdős in 1973 (see [11]).

First an observation: The family of winning sets in the $n^d$ game has the important additional feature that the winning sets are on straight lines, and any two straight lines have at most one point in common. A hypergraph with the intersection property that any two hyperedges have at most one point in common is called **Almost Disjoint**.

Schmidt's work can be summarized in the following theorem: part (a) is a general hypergraph result; part (b) is the special case of "arithmetic progressions".

**Theorem K** (Schmidt 1962).

*(a)* *Let $\mathcal{F}$ be an $n$-uniform Almost Disjoint hypergraph. Assume that the Maximum Degree of $\mathcal{F}$ is less than $2^{n-5\sqrt{n\log n}}$ and the size $|\mathcal{F}|$ of the hypergraph is less than $8^n$, then $\mathcal{F}$ has chromatic number two, that is, the hypergraph has a Proper 2-Coloring.*

*(b)* $W(n) = W(n, 2) \geq 2^{n-5\sqrt{n\log n}}$.

Note that part (b) is not a corollary of part (a); the family of $n$-term arithmetic progressions in an interval $[N] = \{1, 2, \ldots, N\}$ is *not* Almost Disjoint, but it is "close enough" in the sense that the proof of (a) can be *adapted* to prove (b).

Note that Schmidt's idea motivated our "game-theoretic decomposition technique" (see Section 7).

The following result is about arbitrary hypergraphs, not just Almost Disjoint (or "nearly" Almost Disjoint) hypergraphs (see [11]).

**Theorem L** (Erdős–Lovász 2-Coloring Theorem 1973). *If $\mathcal{F}$ is an $n$-uniform hypergraph, and its Maximum Neighborhood Size is at most $2^{n-3}$, then the hypergraph has a Proper 2-Coloring (i.e., the points can be colored by two colors so that no hyperedge $A \in \mathcal{F}$ is monochromatic.) In particular, if the Maximum Degree is at most $2^{n-3}/n$, then the hypergraph has a Proper 2-Coloring.*

*Remark.* The very surprising message of the Erdős–Lovász theorem is that the "global size" of hypergraph $\mathcal{F}$ is irrelevant – it can even be infinite! – all what matters is the "local size".

The proof of Theorem L is strikingly short.

*Proof of Theorem L.* This is again the "counting argument", but used in a more sophisticated way than in the proof of Theorem J.

Let $|\mathcal{F}| = M$, let $\mathcal{F} = \{A_1, A_2, \ldots, A_M\}$, let $V$ denote the board, and let $|V| = N$. Let $\mathcal{C}$ denote the set of all $2^N$ possible 2-colorings of $V$. Let $I \subset [M]$ be an arbitrary index-set where $[M] = \{1, 2, \ldots, M\}$, then $\mathcal{C}(I :$ proper$) \subset \mathcal{C}$ denotes the set of 2-colorings of the board $V$ such that no $A_i$, $i \in I$ becomes monochromatic. For arbitrary $I \subset [M]$ and $j \in [M]$ with $j \notin I$, let $\mathcal{C}(I :$ proper $\wedge\, j :$ mono$)$ denote the set of 2-colorings of the board $V$ such that no $A_i$, $i \in I$ becomes monochromatic but $A_j$ *is* monochromatic.

We actually prove a *stronger* statement; it is often easier to prove a stronger statement by induction. The proof of Theorem L is an excellent example of the principle that "to prove more may be less trouble".

**Statement** *Let $I \subset [M]$ and $j \in [M]$ with $j \notin I$. Then*

$$\frac{|\mathcal{C}(I : \text{proper} \wedge j : \text{mono})|}{|\mathcal{C}(I : \text{proper})|} \leq 2 \cdot 2^{-n+1}.$$

*Remark.* Note that $2^{-n+1}$ is the probability that in a random 2-coloring a given $n$-set becomes monochromatic.

*Proof of the Statement.* We prove this *stronger* Statement by induction on $|I|$. If $|I| = 0$ (i.e., $I$ is the empty set) then the Statement reduces to the following triviality:

$$\frac{|\mathcal{C}(j : \text{mono})|}{|\mathcal{C}|} = 2^{-n+1} < 2 \cdot 2^{-n+1}.$$

Next assume that index-set $I$ is not empty. For notational convenience write $I = \{1, 2, \ldots, i\}$, and among the elements of $I$ let $1, 2, \ldots, d$ ($d \leq i$) denote the neighbors of $A_j$ (i.e., $A_1, \ldots, A_d$ intersect $A_j$, but $A_{d+1}, \ldots, A_i$ do not intersect $A_j$). By hypothesis, $d \leq 2^{n-3}$. Since $A_{d+1}, \ldots, A_i$ are disjoint from $A_j$, we trivially have the equality

$$\frac{|\mathcal{C}(\{d+1, d+2, \ldots, i\} : \text{proper} \wedge j : \text{mono})|}{|\mathcal{C}(\{d+1, d+2, \ldots, i\} : \text{proper})|} = 2^{-n+1}. \tag{4.1}$$

Furthermore,

$$\mathcal{C}(\{1, 2, \ldots, i\} : \text{proper}) =$$
$$\mathcal{C}(\{d+1, d+2, \ldots, i\} : \text{proper}) \setminus \bigcup_{k=1}^{d} \mathcal{C}(\{d+1, d+2, \ldots, i\} : \text{proper} \wedge k : \text{mono}), \tag{4.2}$$

and by the induction hypothesis,

$$\frac{|\mathcal{C}(\{d+1, d+2, \ldots, i\} : \text{proper} \wedge k : \text{mono})|}{|\mathcal{C}(\{d+1, d+2, \ldots, i\} : \text{proper})|} \leq 2 \cdot 2^{-n+1}. \tag{4.3}$$

Since $d \cdot 2 \cdot 2^{-n+1} \leq 2^{n-3} \cdot 2 \cdot 2^{-n+1} = 1/2$, by (4.2) and (4.3) we obtain

$$|\mathcal{C}(\{1, 2, \ldots, i\} : \text{proper})| \geq \frac{1}{2} |\mathcal{C}(\{d+1, d+2, \ldots, i\} : \text{proper})|. \tag{4.4}$$

Now the proof of the Statement is straightforward: by (4.1) and (4.4),

$$2^{-n+1} = \frac{|\mathcal{C}(\{d+1, d+2, \ldots, i\} : \text{proper} \wedge j : \text{mono})|}{|\mathcal{C}(\{d+1, d+2, \ldots, i\} : \text{proper})|} \geq$$
$$\frac{|\mathcal{C}(\{1, 2, \ldots i\} : \text{proper} \wedge j : \text{mono})|}{2|\mathcal{C}(\{1, 2, \ldots i\} : \text{proper})|} = \frac{|\mathcal{C}(I : \text{proper} \wedge j : \text{mono})|}{2|\mathcal{C}(I : \text{proper})|},$$

which is exactly the Statement. □

The deduction of Theorem L from the Statement is obvious: indeed, by an *iterated application* of the Statement

$$\frac{|\mathcal{C}(\{1, 2, \ldots M\} : \text{proper})|}{|\mathcal{C}|} \geq \left(1 - 2^{-n+2}\right)^{|\mathcal{F}|} > 0,$$

which proves the *existence* of a Proper 2-Coloring of hypergraph $\mathcal{F}$. Since $|\mathcal{C}| = 2^N$, the total number of Proper 2-Colorings is at least

$$2^N \cdot \left(1 - 2^{-n+2}\right)^{|\mathcal{F}|} \approx 2^N \cdot e^{-|\mathcal{F}|/2^{n-2}}.$$

This completes the proof of Theorem L. □

The Erdős–Lovász 2-coloring theorem implies the lower bound

$$W(n) = W(n, 2) \geq \frac{2^n}{8n},$$

which is somewhat better than Schmidt's lower bound, but it is still in the same range of $W(n) \geq (2 + o(1))^n$. The proof of the Erdős–Lovász 2-coloring theorem was a pure existence argument; Schmidt's proof, on the other hand, is "more constructive".

The *lower bound problem* for the van der Waerden number, however, must be considered a triumph for the hard-core *constructivists*. We start with an example: The inequality $W(4) \geq 35$ is established by the following explicit 2-coloring of the interval $I = \{0, 1, 2, \ldots, 33\}$: the first color-class consists of 0, 11, and the quadratic nonresidues (mod 11) in $I$

$$0, 2, 6, 7, 8, 10, 11, 13, 17, 18, 19, 21, 24, 28, 29, 30, 32,$$

and of course the other color-class is the complement

$$1, 3, 4, 5, 8, 12, 14, 15, 16, 20, 22, 23, 25, 26, 27, 31, 33;$$

it is easy to see that no class contains a 4-term arithmetic progression. This algebraic construction is due to J. Folkman (the inequality is actually an equality: $W(4) = 35$). A similar but more sophisticated explicit algebraic construction was discovered by Berlekamp in 1968; it gives the lower bound $W(n) > (n-1)2^{n-1}$ if $n-1$ is a prime. Berlekamp's construction, just like Folkman's example above, is a Proper *Halving* 2-Coloring (see [7]). It follows that $W_{1/2}(n) > (n-1)2^{n-1}$ if $n-1$ is a *prime*, and because there is always a prime between $n$ and $n - n^{2/3}$ if $n$ is large enough, Berlekamp's construction implies the lower bound $W_{1/2}(n) \geq (2 + o(1))^n$ for *every* $n$.

Next we switch from the van der Waerden number $W(n)$ to the Hales–Jewett number $HJ(n)$. In 1961 Hales and Jewett [16] proved the linear lower bound $HJ(n) \geq n$ by an explicit construction.

**Theorem M** (Hales–Jewett linear lower bound). *The Hales–Jewett number satisfies the linear lower bound $HJ(n) \geq n$.*

*Proof.* We can assume that $n \geq 5$. Indeed, every "reasonable" play of Tic-Tac-Toe leads to a drawing end-position (which solves case $n = 3$), and even if the $4^3$ game is a First Player win, it nevertheless *does* have a drawing end-position – we challenge the reader to find one! – which settles case $n = 4$.

For $n \geq 5$ we are going to define a Proper 2-Coloring of the $n^{n-1}$-hypergraph by using an elegant explicit algebraic construction. The idea of Hales and Jewett is to add up $(n-1)$ 1-dimensional 2-colorings (i.e., 2-colorings of $[n] = \{1, 2, \ldots, n\}$), where the addition is taken (mod 2).

Let $\mathbf{v}_1, \ldots, \mathbf{v}_i = (v_{i,1}, \ldots, v_{i,n}), \ldots, \mathbf{v}_d$ be $d$ $n$-dimensional 0-1 vectors, i.e., $v_{i,j} \in \{0, 1\}$ for all $1 \leq i \leq d, 1 \leq j \leq n$. Each $\mathbf{v}_i$ can be viewed as a 2-coloring of $[n] = \{1, 2, \ldots, n\}$. Now the vector sequence $\mathbf{v}_1, \ldots, \mathbf{v}_d$ defines a 2-coloring $f : [n]^d \to \{0, 1\}$ of the board of the $n^d$-game as follows: for every $(j_1, \ldots, j_d) \in [n]^d$ let

$$f(j_1, \ldots, j_d) \equiv v_{1,j_1} + v_{2,j_2} + \cdots + v_{d,j_d} \pmod 2. \tag{4.5}$$

Which vector sequence $\mathbf{v}_1, \ldots, \mathbf{v}_d$ defines a *Proper* 2-Coloring of the $n^d$-hypergraph? To answer this question, consider an arbitrary winning line $L$. Line $L$ can be parametrized by an "$x$-vector" as follows. The first coordinate of the "$x$-vector" is either a constant $c_1$, or $x$, or $(n+1-x)$, similarly, the second coordinate of the "$x$-vector" is either a constant $c_2$, or $x$, or $(n+1-x)$, and so on:

$$L: \text{(either constant } c_1 \text{ or } x \text{ or } (n+1-x), \ldots, \\ \text{either constant } c_d \text{ or } x \text{ or } (n+1-x)), \tag{4.6}$$

and the $k$th point $\mathbf{P}_k$ on line $L$ is obtained by the substitution $x = k$ in "$x$-vector" (4.6) ($k = 1, 2, \ldots, d$). What is the $f$-color – see (4.5) – of point $\mathbf{P}_k$? To answer this question, for every $i = 1, 2, \ldots, d$ write

$$\varepsilon_i = \begin{cases} 0, & \text{if the } i\text{th coordinate in (4.6) is a constant } c_i; \\ 1, & \text{otherwise.} \end{cases} \tag{4.7}$$

For an arbitrary $n$-dimensional vector $\mathbf{a} = (a_1, a_2, \ldots, a_n)$ define the "reverse":

$$\mathbf{a}^{(\text{rev})} = (a_n, a_{n-1}, \ldots, a_1). \tag{4.8}$$

It follows from (4.5)–(4.8) that the $f$-color of the $k$th point $\mathbf{P}_k$ on line $L$ is

$$f(\mathbf{P}_k) \equiv \sum_{1 \le i \le d:\ \varepsilon_i = 0} v_{i,c_i} + \left( k\text{th coordinate of } \sum_{i=1}^{d} \varepsilon_i \mathbf{w}_i \right) \pmod 2, \tag{4.9}$$

where

$$\mathbf{w}_i = \begin{cases} \mathbf{v}_i, & \text{if the } i\text{th coordinate in (4.6) is } x; \\ \mathbf{v}_i^{(\text{rev})}, & \text{if the } i\text{th coordinate in (4.6) is } (n+1-x). \end{cases} \tag{4.10}$$

It follows from (4.9)–(4.10) that line $L$ is monochromatic if and only if

$$\sum_{i=1}^{d} \varepsilon_i \mathbf{w}_i \equiv \text{ either } \mathbf{0} = (0, \ldots, 0) \text{ or } \mathbf{1} = (1, \ldots, 1) \pmod 2.$$

It suffices, therefore to find $(n-1)$ $n$-dimensional 0-1 vectors $\mathbf{v}_1, \ldots, \mathbf{v}_{n-1}$ such that for each choice of $\mathbf{w}_i \in \{\mathbf{v}_i, \mathbf{v}_i^{(\text{rev})}\}$, $\varepsilon_i \in \{0, 1\}$, $1 \le i \le n-1$, where $(\varepsilon_1, \ldots, \varepsilon_{n-1}) \ne \mathbf{0}$, the vector

$$\varepsilon_1 \mathbf{w}_1 + \varepsilon_2 \mathbf{w}_2 + \cdots + \varepsilon_{n-1} \mathbf{w}_{n-1} \pmod 2 \tag{4.11}$$

is neither $\mathbf{0}$ nor $\mathbf{1}$.

We give the following explicit construction: For $n \ge 5$ let

(1,0,0,0,0,0,.........,0,........,0,0,0,0,0,0)
(0,1,0,0,0,0,.........,0,........,0,0,0,0,0,0)
(0,0,1,0,0,0,.........,0,........,0,0,0,0,0,0)
(0,0,0,1,0,0,.........,0,........,0,0,0,0,0,0)
(0,0,0,0,1,0,.........,0,........,0,0,0,0,0,0)
.......................................................
.......................................................
.......................................................
.......................................................

..................................................

(0,0,0,0,1,0,.........,0,........,0,1,0,0,0,0)

(0,0,0,1,0,0,.........,0,........,0,0,1,0,0,0)

(0,0,1,0,0,0,.........,0,........,0,0,0,1,0,0)

(0,1,0,0,0,0,.........,0,........,0,0,0,0,1,0)

(1,0,0,0,0,0,.........,0,........,0,0,0,0,0,1)

That is, the first $\lfloor n/2 \rfloor$ $n$-dimensional vectors are $\mathbf{e}_i = (0, \ldots, 0, 1, 0, \ldots, 0)$ such that the only nonzero coordinate is 1 at the $i$th place, $1 \le i \le \lfloor n/2 \rfloor$, the rest are symmetric "self-reversed" vectors, and in each vector the $\lceil (n+1)/2 \rceil$th coordinate is zero.

It remains to show that this construction satisfies the requirement (see (4.11)). First we show that the vector

$$\varepsilon_1 \mathbf{w}_1 + \varepsilon_2 \mathbf{w}_2 + \cdots + \varepsilon_{n-1} \mathbf{w}_{n-1} \pmod 2$$

in (4.11) is $\neq \mathbf{1} = (1, \ldots, 1)$. Indeed, the $\lceil (n+1)/2 \rceil$th coordinate of vector (4.11) is always zero.

So assume that vector (4.11) equals $\mathbf{0} = (0, \ldots, 0)$. Then from the first and $n$th coordinates we see that

either $\varepsilon_1 + \varepsilon_{n-1} \equiv \varepsilon_{n-1} \equiv 0 \pmod 2$

or $\varepsilon_1 + \varepsilon_{n-1} \equiv \varepsilon_1 \equiv 0 \pmod 2$.

In both cases we obtain that $\varepsilon_1 = \varepsilon_{n-1} = 0$. Similarly, we have $\varepsilon_2 = \varepsilon_{n-2} = 0$, $\varepsilon_3 = \varepsilon_{n-3} = 0$, and so on. We conclude that all coefficients must be zero: $\varepsilon_1 = \varepsilon_2 = \cdots = \varepsilon_{n-1} = 0$, which is impossible. This completes the proof of Theorem M. □

## 5 New lower bound results

The $n^d$-hypergraph is very far from being Degree-Regular; it is in fact extremely *irregular*. Indeed, the Average Degree of the family of winning sets in $n^d$ is

$$\text{AverageDegree}(n^d) = \frac{n \cdot \text{familysize}}{\text{boardsize}} = \frac{n\left((n+2)^d - n^d\right)/2}{n^d} \approx \frac{n}{2}\left(e^{2d/n} - 1\right).$$

This is *much* smaller than the Maximum Degree $(3^d - 1)/2$ ($n$ odd) and $2^d - 1$ ($n$ even), namely, about (roughly speaking) the $n$th root of the Maximum Degree. It is natural, therefore, to ask the following

*Question A:* Can one reduce the Maximum Degree of an arbitrary $n$-uniform hypergraph close to the order of the Average Degree?

The answer is an easy *yes* if one is allowed to throw out *whole* winning sets. But throwing out a whole winning set means that Breaker loses control over that set, and Maker might completely occupy it. So we cannot throw out whole sets, but we can throw out a few points from each winning set. In other words, we can *partially truncate* the winning sets, but we cannot throw them out entirely. So the right question is

*Question B:* Can one reduce the Maximum Degree of an arbitrary $n$-uniform hypergraph, by *partially* truncating the winning sets, close to the order of the Average Degree?

Well, the answer to Question B is *no* for general $n$-uniform hypergraphs (we leave it to the reader to construct an example), but it is *yes* for the special case of the $n^d$-hypergraphs.

**Theorem 1** (Degree Reduction by Partial Truncation).

*(a)* *Let $\mathcal{F}_{n,d}$ denote the family of n-in-a-line's (i.e., geometric lines) in the $n^d$ board; $\mathcal{F}_{n,d}$ is an n-uniform Almost Disjoint hypergraph. Let $0 < \alpha < 1/2$ be an arbitrary real number. Then for each geometric line $L \in \mathcal{F}_{n,d}$ there is a $2\lfloor(\frac{1}{2}-\alpha)n\rfloor$-element subset $\widetilde{L} \subset L$ such that the truncated family $\widetilde{\mathcal{F}_{n,d}} = \{\widetilde{L} : L \in \mathcal{F}_{n,d}\}$ has Maximum Degree*

$$\text{MaxDegree}\left(\widetilde{\mathcal{F}_{n,d}}\right) < d + d^{\lceil d/\alpha n\rceil - 1}.$$

*(b)* *Let $\mathcal{F}^c_{n,d}$ denote the family of combinatorial lines in the $n^d$ board; $\mathcal{F}^c_{n,d}$ is an n-uniform Almost Disjoint hypergraph. Let $0 < \beta < 1$ be an arbitrary real number. Then for each combinatorial line $L \in \mathcal{F}^c_{n,d}$ there is a $\lfloor(1-\beta)n\rfloor$-element subset $\widetilde{L} \subset L$ such that the truncated family $\widetilde{\mathcal{F}^c_{n,d}} = \{\widetilde{L} : L \in \mathcal{F}^c_{n,d}\}$ has Maximum Degree*

$$\text{MaxDegree}\left(\widetilde{\mathcal{F}^c_{n,d}}\right) < d + d^{\lceil d/\beta n\rceil - 1}.$$

*Remarks.* (1) Let $\alpha > c_0 > 0$, that is, let $\alpha$ be "separated from zero". Then the upper bound $d^{O(d/n)}$ of the Maximum Degree of $\widetilde{\mathcal{F}_{n,d}}$ is not that far from the order of magnitude of the Average Degree $\frac{n}{2}\left(e^{2d/n}-1\right)$ of $\mathcal{F}_{n,d}$. Indeed, what really matters is the exponent, and the two exponents $const \cdot d/n$ and $2d/n$ are the same apart from a constant factor.

(2) In the applications of (a) we always need that $(\frac{1}{2}-\alpha)n \geq 1$, or equivalently $\alpha \leq \frac{1}{2} - \frac{1}{n}$, since otherwise the "pseudo-line" $\widetilde{L}$ becomes empty.

*Proof of Theorem 1.* Case (a). We recall that the maximum degree of the family of winning lines is $(3^d-1)/2$ if $n$ is *odd*, and $2^d-1$ if $n$ is *even*. The maximum is achieved for the center ($n$ is *odd*), and for the points $(c_1, c_2, \ldots, c_d)$ such that there is a $c \in \{1, \ldots, n\}$ with $c_j \in \{c, n+1-c\}$ for every $j = 1, 2, \ldots, d$ ($n$ is *even*). This motivates our basic idea: we define $\widetilde{L} \subset L$ by throwing out the points with "large coordinate-repetition".

Let $\mathbf{P} = (a_1, a_2, a_3, \ldots, a_d)$, $a_i \in \{1, 2, \ldots, n\}$, $1 \leq i \leq d$ be an arbitrary point of the board of the $n^d$ game. We study the *coordinate-repetitions* of $\mathbf{P}$. Let $\ell = \lfloor(n+1)/2\rfloor$, and write $[\ell] = \{1, 2, \ldots, \ell\}$. Let $b \in [\ell]$ be arbitrary. Consider the *multiplicity* of $b$ *and* $(n+1-b)$ in $\mathbf{P}$: let

$$\begin{aligned} m(\mathbf{P}, b) &= |M(\mathbf{P}, b)| \text{ where } M(\mathbf{P}, b) = \{1 \leq i \leq d : a_i = b \text{ or } (n+1-b)\} \\ &= M(\mathbf{P}, n+1-b). \end{aligned}$$

Observe that in the definition of multiplicity we identify $b$ and $(n+1-b)$.

For example, let

$$\mathbf{P} = (3, 7, 3, 5, 1, 3, 5, 4, 3, 1, 5, 3, 3, 5, 5, 1, 4, 2, 6, 5, 2, 7) \in [7]^{22},$$

then $m(\mathbf{P}, 1) = 5$, $m(\mathbf{P}, 2) = 3$, $m(\mathbf{P}, 3) = 12$, $m(\mathbf{P}, 4) = 2$.

For every *n-line* $L$ of the $n^d$ game we choose one of the two orientations. An orientation can be described by an $x$-vector $\mathbf{v} = \mathbf{v}(L) = (v_1, v_2, v_3, \ldots, v_d)$ where the $i$th coordinate $v_i$ is either a constant $c_i$, or variable $x$, or variable $(n+1-x)$, $1 \leq i \leq d$, and for at least one index $i$, $v_i$ is $x$ or $(n+1-x)$.

For example, in the ordinary $3^2$ Tic-Tac-Toe:

$$\begin{array}{ccc} (1,3) & (2,3) & (3,3) \\ (1,2) & (2,2) & (3,2) \\ (1,1) & (2,1) & (3,1) \end{array}$$

$\{(1,1), (2,2), (3,3)\}$ is a winning line defined by the $x$-vector $xx$, $\{(1,2), (2,2), (3,2)\}$ is another winning line defined by the $x$-vector $x2$, and finally $\{(1,3), (2,2), (3,1)\}$ is a winning line defined by the $x$-vector $xx'$, where $x' = (n+1-x)$.

The $k$th point $\mathbf{P}_k$ ($1 \le k \le n$) of line $L$ is obtained by putting $x = k$ in the $x$-vector $\mathbf{v} = \mathbf{v}(L)$ of the line. The sequence $(\mathbf{P}_1, \mathbf{P}_2, \dots, \mathbf{P}_n)$ gives an *orientation* of line $L$. The second (i.e., reversed) orientation comes from $x$-vector $\mathbf{v}^*$ which is obtained from $\mathbf{v} = \mathbf{v}(L)$ by switching coordinates $x$ and $(n+1-x)$ that are *variables*.

Let $b \in [\ell]$ ($\ell = \lfloor (n+1)/2 \rfloor$) be arbitrary, and consider the multiplicity of $b$ and $(n+1-b)$ in $x$-vector $\mathbf{v} = \mathbf{v}(L)$:

$$m(L,b) = |M(\mathbf{v}(L), b)| \text{ where } M(\mathbf{v}(L), b) = \{1 \le i \le d : \ v_i = b \text{ or } (n+1-b)\}.$$

Similarly, consider the multiplicity of $x$ and $(n+1-x)$ in $x$-vector $\mathbf{v} = \mathbf{v}(L)$:

$$m(L,x) = |M(\mathbf{v}(L), x)| \text{ where } M(\mathbf{v}(L), x) = \{1 \le i \le d : \ v_i = x \text{ or } (n+1-x)\}.$$

It follows that $m(\mathbf{P}_k, k) = m(L,k) + m(L,x)$, and $m(\mathbf{P}_k, b) = m(L,b)$ if $k \notin \{b, n+1-b\}$, where $\mathbf{P}_k$ is the $k$th point of line $L$ in the orientation $\mathbf{v} = \mathbf{v}(L)$, and $b \in [\ell]$.

Let $m_0 = \lceil d/\alpha n \rceil$. For every line $L$ define the "index-set"

$$B_L = \{k \in [\ell] : \ m(L,k) < m_0\}.$$

Then

$$d > \sum_{b \in [\ell]} m(L,b) \ge \sum_{b \in [\ell] \setminus B_L} m(L,b) \ge \sum_{b \in [\ell] \setminus B_L} m_0 = m_0(\ell - |B_L|),$$

and so

$$|B_L| > \ell - \frac{d}{m_0} \ge \lfloor (n+1)/2 \rfloor - \alpha n,$$

which implies that

$$|B_L| \ge \left\lfloor \left(\frac{1}{2} - \alpha\right) n \right\rfloor + \{1 \text{ or } 0\}$$

depending on the parity of $n$ (1 if $n$ is *odd*, and 0 if $n$ is *even*). Let

$$B_L^* = \{k : \ k \in B_L \text{ or } (n+1-k) \in B_L\},$$

then clearly

$$|B_L^*| \ge 2 \left\lfloor \left(\frac{1}{2} - \alpha\right) n \right\rfloor + \{1 \text{ or } 0\}$$

depending on the parity of $n$. For every line $L$, the "index-set" $B_L^*$ defines a subset $\widetilde{L} \subset L$ (we call $\widetilde{L}$ a *pseudo-line*) as follows: let $\widetilde{L} = \{\mathbf{P}_k : \ k \in B_L^*\}$ if $n$ is *even*, and

$\widetilde{L} = \{\mathbf{P}_k : k \in B_L^* \setminus \{\ell\}\}$ if $n$ is *odd* (i.e., we throw out the "mid-point" when there is one). Here $\mathbf{P}_k$ is the $k$th point of line $L$ in the chosen orientation.

The above-mentioned definition of the pseudo-line has one trivial formal problem: $\widetilde{L}$ may have too many points, and this indeed happens for lines like the "main diagonal", then it is sufficient to throw out arbitrary points to get to the desired size $2\left\lfloor (\frac{1}{2} - \alpha)n \right\rfloor$.

Now fix an arbitrary point $\mathbf{P} = (c_1, c_2, \ldots, c_d)$ of the $n^d$-board. We have to estimate the number of pseudo-lines through $\mathbf{P}$. To find a line $L$ such that $\mathbf{P} = \mathbf{P}_k$ for some $k \in B_L^*$ (i.e., $\mathbf{P}$ is the $k$th point of line $L$) we must choose a subset $Y$ of $M(\mathbf{P}, k)$ of size $y < m_0$ and for $i \in M(\mathbf{P}, k) \setminus Y$ change $c_i = k$ to $x$ and $c_i = n + 1 - k$ to $n + 1 - x$ (here we use that $k \neq (n + 1 - k)$; indeed, this follows from $k \neq \ell = \lfloor (n + 1)/2 \rfloor$ when $n$ is odd).

Let $K = \{k \in [\ell] : m(\mathbf{P}, k) \neq 0\}$. Then clearly

$$|K| \leq \sum_{k \in K} m(\mathbf{P}, k) \leq d.$$

Thus the number of pseudo-lines through $\mathbf{P}$ is at most

$$\sum_{y=0}^{m_0-1} \sum_{k \in K} \binom{m(\mathbf{P}, k)}{y} \leq |K| + \sum_{y=1}^{m_0-1} \binom{\sum_{k \in K} m(\mathbf{P}, k)}{y} < d + d^{m_0-1},$$

which completes the proof of Theorem 1 (a).

Case (b): *combinatorial* lines. This case is even simpler than that of case (a); we leave the details to the reader. □

The first application of Theorem 1 is to improve on the Hales–Jewett linear lower bound $HJ(n) \geq n$ of the Hales–Jewett number. The improvement comes from combining Theorem 1 (a) with the Erdős–Lovász 2-Coloring Theorem.

Applying the Erdős–Lovász 2-Coloring Theorem to the truncated hypergraph $\widetilde{\mathcal{F}_{n,d}}$, we get a Proper 2-Coloring of the $n^d$-hypergraph if $0 < c_0 < \alpha < c_1 < 1/2$ and

$$d^{d/\alpha n} \leq 2^{(1-2\alpha)n + O(\log n)}.$$

Taking logarithms we obtain the requirement $d \log_2 d/\alpha n \leq (1 - 2\alpha)n + O(\log n)$, which is equivalent to $d \log_2 d \leq \alpha(1 - 2\alpha)n^2(1 + o(1))$. Since $\alpha(1 - 2\alpha)$ attains its maximum at $\alpha = 1/4$, we conclude that $d \log_2 d \leq n^2(1 + o(1))/8$, which is equivalent to $d \leq (\frac{\log 2}{16} + o(1))n^2/\log n$.

How about *combinatorial* lines? Repeating the same calculations for $\widetilde{\mathcal{F}^c_{n,d}}$, we obtain the similar inequality $d \log_2 d \leq \beta(1-\beta)n^2(1+o(1))$. Since $\beta(1-\beta)$ attains its maximum at $\beta = 1/2$, we conclude that $d \log_2 d \leq n^2(1 + o(1))/4$, which is equivalent to $d \leq (\frac{\log 2}{8} + o(1))n^2/\log n$.

This gives the following new lower bound for the two Hales–Jewett numbers.

**Theorem 2.** *We have the nearly quadratic lower bound*

$$HJ(n) \geq \left(\frac{\log 2}{16} + o(1)\right) \frac{n^2}{\log n}$$

*and*

$$HJ^c(n) \geq \left(\frac{\log 2}{8} + o(1)\right) \frac{n^2}{\log n}$$

*where the* $o(1)$ *in either case is in fact* $O(\log\log n/\log n)$. □

Unlike the proof of the linear lower bound $HJ(n) \geq n$, which was an *explicit* algebraic construction, here we cannot provide an explicit Proper 2-Coloring. Indeed, the proof of the Erdős–Lovász 2-Coloring Theorem was an existence argument: it didn't say a word how to find the existing Proper 2-Coloring (try out all possible $2^N$ 2-colorings of the board, where $N$ is the board-size!). A much more efficient "algorithmization" of the Erdős–Lovász 2-Coloring Theorem is developed in Beck [5].

Note that Berlekamp's explicit algebraic construction – $W(n) > (n-1)2^{n-1}$ if $n$ is a prime – was a Proper *Halving* 2-Coloring. The Erdős–Lovász 2-Coloring Theorem, however, does *not* (and probably *cannot*, see Open Problem 9 (f)) provide a Proper *Halving* 2-Coloring, which raises the following natural question. Is it true that $HJ_{1/2}(n) \geq \left(\frac{\log 2}{16} + o(1)\right) \frac{n^2}{\log n}$? And similarly: Is it true that $HJ^c_{1/2}(n) \geq \left(\frac{\log 2}{8} + o(1)\right) \frac{n^2}{\log n}$? Well, the answer is "yes", but I don't know an easy proof. What I can do is a "game-theoretic approach": in view of the trivial inequality

$$\mathbf{ww}(n\text{–line}) \leq \mathbf{w}(n\text{–line}) \leq HJ_{1/2}(n) \leq HJ(n),$$

it is enough to prove that

$$\mathbf{ww}(n\text{–line}) \geq \left(\frac{\log 2}{16} + o(1)\right) \frac{n^2}{\log n}.$$

Similarly, in view of

$$\mathbf{ww}(\text{comb. } n\text{–line}) \leq \mathbf{w}(\text{comb. } n\text{–line}) \leq HJ^c_{1/2}(n) \leq HJ(n),$$

it is enough to prove that

$$\mathbf{ww}(\text{comb. } n\text{–line}) \geq \left(\frac{\log 2}{8} + o(1)\right) \frac{n^2}{\log n}.$$

That is, a Strong Draw strategy – in fact, any drawing strategy! – yields the existence of a drawing end-position, i.e., a proper *halving* 2-coloring. In the following sections we work out the details of this "game-theoretic approach".

## 6 More new lower bounds via games

What the inequality

$$\mathbf{ww}(n\text{–line}) \geq \left(\frac{\log 2}{16} + o(1)\right) \frac{n^2}{\log n}$$

actually means is that if the dimension $d \leq \left(\frac{\log 2}{16} + o(1)\right) \frac{n^2}{\log n}$, then either player can force a Strong Draw in the $n^d$ game. That is, either player can put his own mark in every $n$-in-a-line.

What do we know about Strong Draw? What kind of general criterions are available? Let's start with the simplest case: the simplest possible way to force a strong draw is to apply a *pairing strategy*. Pairing strategy is probably the most common technique in Game Theory; we already used pairing strategy twice: the reflection strategy of Theorem D and the copycat strategy of Theorem I both were *pairing strategy*.

Pairing strategy means a decomposition of the board (or some part of the board) into disjoint pairs, and when your opponent takes one member from a pair, you take the other one. It is applied when the opponent cannot achieve the objective (win or draw) without choosing both points of at least one pair.

I recall that a hypergraph is called **almost disjoint** if any two hyperedges have at most one point in common.

If the family of winning sets is *almost disjoint,* then the question "can pairing strategy work here" is a standard *perfect matching* problem. Indeed, in an almost disjoint family two distinct winning sets cannot share the same pair of points. So pairing strategy works if and only if one can find a family of disjoint 2-element representatives of the hypergraph of winning sets. But to find a family of disjoint 2-element representatives of a given hypergraph is a well-characterized, completely solved problem in Matching Theory ("Bigamy version of Hall's Marriage Theorem").

In general, if the hypergraph of winning sets is *not* necessarily almost disjoint, then the existence of a family of disjoint 2-element representatives is a *sufficient but not necessary* condition for the existence of a pairing strategy. The following two criterions were (probably) first published in Hales and Jewett [16] (and rediscovered independently in several other papers).

**Theorem N** (Pairing Strategy Draw)**.** *Consider the positional game on $(V, \mathcal{F})$, and assume that for every subfamily $\mathcal{G} \subseteq \mathcal{F}$,*

$$\left| \bigcup_{A \in \mathcal{G}} A \right| \geq 2|\mathcal{G}|.$$

*Then either player can force a pairing strategy draw.*

**Theorem O** (Degree Criterion for Pairing Strategy Draw)**.** *Let $\mathcal{F}$ be an $n$-uniform hypergraph, i.e., $|A| = n$ for every $A \in \mathcal{F}$. Further assume that the Maximum Degree is at most $n/2$, that is, every $x \in V$ is contained in at most $n/2$ elements of $\mathcal{F}$. Then playing on $\mathcal{F}$ either player can force a pairing strategy draw.*

Note that the two Pairing Strategy Criterions (Theorems N–O) are very general. They are *local* conditions in the sense that they don't give any restriction on the global size of hypergraph $\mathcal{F}$. Both hold for infinite boards as well.

Hales and Jewett [16] proved, by a pioneering application of Theorem O, that if

$$n \geq 3^d - 1 \ (n \text{ odd}) \quad \text{or} \quad n \geq 2^{d+1} - 2 \ (n \text{ even}), \tag{6.1}$$

then the $n^d$ game is a pairing strategy draw. Indeed, it immediately follows from the elementary fact that, if $n$ is odd, there are at most $(3^d - 1)/2$ winning lines through any point; if $n$ is even, the maximum degree of the family of geometric lines drops to $2^d - 1$ (which is much smaller than $(3^d - 1)/2$ if $d$ is large). For the proof of this elementary fact see the beginning of Section 2.

Consider the special case $d = 2$ in (6.1). Then the bounds

$$n \geq 3^d - 1 \ (n \text{ odd}) \quad \text{and} \quad n \geq 2^{d+1} - 2 \ (n \text{ even})$$

immediately solve the $n^2$ game for all $n \geq 9$, $n$ odd, and $n \geq 6$, $n$ even: the game is a pairing strategy draw. To settle the case $n = 7$ we apply a simple "truncation trick". Note that the center is the only cell with 4 winning lines passing through it. Throwing out the center from these 4 lines and an arbitrary point from each one of the rest of the lines, the size of the winning sets decreases to 6. Since the new maximum degree is 3, and $6 = 2 \cdot 3$, Theorem O applies to the "truncated family", and proves that the $7^2$ game is also a pairing strategy draw.

If $n = 5$, then again Second Player can force a draw by a pairing strategy: whenever First Player occupies a numbered cell, Second Player takes the other cell of the same number:

$$n = 5 \qquad \begin{bmatrix} 11 & 7 & 1 & 1 & 12 \\ 6 & 2 & 8 & 9 & 2 \\ 6 & 3 & * & 3 & 10 \\ 4 & 7 & 8 & 4 & 10 \\ 12 & 5 & 5 & 9 & 11 \end{bmatrix}$$

If First Player takes the *-marked center, Second Player may take any cell, and if the cell he is required to take by the pairing strategy is occupied, he may play anywhere.

If a pairing strategy forces a draw in an almost disjoint hypergraph, then there are at least twice as many points as winning sets. Therefore, if the Point/Line ratio in an $n^d$ game is less than 2, then it is *not* a pairing strategy draw. For example, in the $4^2$ game, pairing strategy *cannot* exist because the number of points ("cells") is less than twice the number of winning lines: $16 = 4^2 < 2(4 + 4 + 2) = 20$.

Note that $4^2$ is nevertheless a draw game; we prove it later in this section.

In 1963 Hales and Jewett [16] made the following elegant conjecture.

**Hales–Jewett Conjecture**

(a) *If there are at least twice as many points ("cells") as winning lines, then the $n^d$ game is always a draw.*

(b) *What is more, if there are at least twice as many points as winning lines, then Theorem N always applies, and the draw is actually a Pairing Strategy Draw.*

Since the total number of $n$-in-a-line's in the $n^d$ board is $((n+2)^d - n^d)/2$, and the number of points is trivially $n^d$, the condition "there are at least twice as many points as winning lines" means

$$n^d \geq (n+2)^d - n^d, \tag{6.2}$$

that is, $2 \geq ((n+2)/n)^d$. Since $((n+2)/n)^d = \left(1 + \frac{2}{n}\right)^d \approx e^{2d/n}$, (6.2) is asymptotically equivalent to

$$n \geq \frac{2d}{\log 2} + O(1) = 2.885d + O(1). \tag{6.3}$$

Observe that the Hales–Jewett conjecture – see (6.3) – is "exponentially better" than (6.1) (which was a straightforward consequence of Theorem O).

In view of the Bigamy Corollary of the well-known Marriage Theorem ("Hall's theorem") a *necessary and sufficient* condition for the existence of a draw-forcing pairing is that for every subfamily of winning lines the union set has at least twice as many points as the number of lines in the subfamily (the phrase "Bigamy" refers to the fact that "every man needs 2 wives"). Consequently, what part (b) of the Hales–Jewett Conjecture really says is that *the Point/Line ratio attains its minimum for the family of all lines, and for any proper subfamily the ratio is greater or equal.* This **Ratio Conjecture** is very compelling, not only when a Pairing Strategy exists but in general for *arbitrary* $n^d$-game. The **Ratio Conjecture**, as a generalization of Hales–Jewett Conjecture (b), was formulated in Patashnik [12]. Unfortunately we know very little about the *Ratio Conjecture* in general.

Now we leave pairing strategy; in the rest of the paper we prove and apply completely different tools, namely *Potential Criterions*, which give much better results than what pairing strategy does.

The first *Potential Criterion* was found by Erdős and Selfridge [12] in 1973: it is a sufficient condition for a Strong Draw. Quite contrary to the pairing strategy and other local approaches, the Erdős–Selfridge theorem is a *global* criterion.

**Theorem P** (Erdős–Selfridge Theorem). *Let $\mathcal{F}$ be an n-uniform hypergraph, and assume that* $|\mathcal{F}| + \text{MaxDeg}(\mathcal{F}) < 2^n$, *where* $\text{MaxDeg}(\mathcal{F})$ *denotes the maximum degree of hypergraph $\mathcal{F}$. Then playing the positional game on $\mathcal{F}$ Second Player can force a Strong Draw.*

*Remark.* The following simple general observation will be used repeatedly in the rest of the paper: If Second Player can force a Strong Draw in a positional game, then First Player can also force a Strong Draw. Indeed, use "strategy stealing".

*Proof.* Let $\mathcal{F} = \{A_1, A_2, \ldots, A_M\}$, where $M < 2^{n-1}$. Assume we are at the stage of the play where First Player already occupied $x_1, x_2, \ldots, x_i$, and Second Player occupied $y_1, y_2, \ldots, y_{i-1}$. The question is how to choose Second-Player's next point $y_i$. Those winning sets which contain at least one $y_j$ $(j \le i-1)$ are "harmless" – we call them "dead sets". The winning sets which are not "dead" are called "survivors". The "survivors" have a chance to be completely occupied by First Player at the end of the play, so they each represent some "danger". What is the total "danger" of the *whole* position? We evaluate the given position by the following expression, called "danger-function": $D_i = \sum_{s \in S} 2^{-u_s}$, where $u_s$ is the number of unoccupied elements of the "survivor" $A_s$ ($s \in S_i$ = "index-set of the survivors") and index $i$ indicates that we are at the stage of choosing the $i$th point $y_i$ of Second Player. A natural choice for $y_i$ is to minimize the "danger" $D_{i+1}$ at the next stage. How to do that? The simple linear structure of the danger-function $D_i$ gives an easy answer to this question. Let $y_i$ and $x_{i+1}$ denote the next two moves. What is the effect of these two points on $D_i$? How do we get $D_{i+1}$ from $D_i$? Well, $y_i$ "kills" all the "survivors" $A_s \ni y_i$, which means we have to subtract the sum

$$\sum_{s \in S_i:\ y_i \in A_s} 2^{-u_s}$$

from $D_i$. On the other hand, $x_{i+1}$ doubles the "danger" of each "survivor" $A_s \ni x_{i+1}$, that is, we have to add the sum $\sum_{s \in S_i:\ x_{i+1} \in A_s} 2^{-u_s}$ back to $D_i$. Warning: If some "survivor" $A_s$ contains both $y_i$ and $x_{i+1}$, then we do not have to give the corresponding term $2^{-u_s}$ back because that $A_s$ was previously "killed" by $y_i$.

The natural choice for $y_i$ is the unoccupied $z$ for which $\sum_{s \in S_i:\, z \in A_s} 2^{-u_s}$ attains its maximum. Then what we subtract is at least as large as what we add back:

$$D_{i+1} \le D_i - \sum_{s \in S_i:\, y_i \in A_s} 2^{-u_s} + \sum_{s \in S_i:\, x_{i+1} \in A_s} 2^{-u_s}$$

$$\le D_i - \sum_{s \in S_i:\, y_i \in A_s} 2^{-u_s} + \sum_{s \in S_i:\, y_i \in A_s} 2^{-u_s} = D_i.$$

In other words, Second Player can force the *decreasing property* $D_1 \ge D_2 \ge \cdots \ge D_{\text{end}}$ of the danger-function.

Second-Player's ultimate goal is to prevent First Player from completely occupying some $A_j \in \mathcal{F}$, that is, to avoid $u_j = 0$. If $u_j = 0$ for some $j$, then $D_{\text{end}} \ge 2^{-u_j} = 1$. By hypothesis

$$D_{\text{start}} = D_1 = \sum_{A:\, x_1 \in A \in \mathcal{F}} 2^{-n+1} + \sum_{A:\, x_1 \notin A \in \mathcal{F}} 2^{-n} \le |\mathcal{F}| 2^{-n+1} < 1,$$

so by the *decreasing property* of the danger-function, $D_{\text{end}} < 1$. This completes the proof of the Erdős–Selfridge theorem. □

*Remarks.* (1) If $\mathcal{F}$ is $n$-uniform, then multiplying the *danger* $2^{-u_s}$ of a *survivor* by $2^n$, the renormalized danger becomes $2^{n-u_s}$. The exponent, $n - u_s$, is the number of First-Player's marks in a *survivor* (i.e., Second-Player-free) set ($u_s$ denotes the number of unoccupied points). This means the following *Power-of-Two Scoring System*. A winning set containing an O (Second-Player's mark) scores zero, a blank winning set scores 1, a set with a single X (First-Player's mark) and no O scores 2, a set with two X's and no O scores 4, a set with three X's and no O scores 8, and so on (i.e., the "values" are integers rather than small fractions). Occupying a whole $n$-element winning set scores $2^n$, i.e., due to the renormalization, the "target value" becomes $2^n$ (instead of 1).

It is just a matter of taste which scoring system one prefers: the first one, where the "scores" were negative powers of 2 and the "target value" was 1, or the second one, where the "scores" were positive powers of 2 and the "target value" was $2^n$.

(2) The most frequently applied special case of the Erdős–Selfridge theorem is the following: *If $\mathcal{F}$ is $n$-uniform and $|\mathcal{F}| < 2^n$ or $< 2^{n-1}$, then playing on $(V, \mathcal{F})$ First or Second Player can force a Strong Draw.*

To have a better understanding of what is going on here, it is worth to study the *randomized game* where both players are "random generators". The calculation somewhat simplifies if we study the *random 2-coloring* instead: the points of the board are colored (say) red and blue independently of each other with probability $p = 1/2$. (This model is a little bit different from considering the halving 2-colorings only: the case which corresponds to the randomized game.) Indeed, in the random 2-coloring model the *expected number* of monochromatic winning sets is clearly $2^{-n+1}|\mathcal{F}|$, which is less than 1 (by the hypothesis of the Erdős–Selfridge theorem; the case of the second player). So there must *exist* an end-position with no monochromatic winning set: a drawing end-position. Now the real meaning of the Erdős–Selfridge theorem becomes clear: it "upgrades" the existing drawing end-position – see Theorem J – to a Drawing Strategy. The proof is an "algorithmization" of the "counting argument".

(3) Theorem P is sharp: the full-length branches of a binary tree with $n$ levels form an $n$-uniform family of $2^{n-1}$ winning sets such that the first player can occupy a full branch in $n$ moves (the players take vertices of the tree).

By using Theorem P it is easy to solve the $4^2$ game (which cannot have a pairing strategy draw), and also the $8^3$ game, without any case-study!

**Corollary.** *In both the* $4^2$ *and* $8^3$ *games Second Player can force a Strong Draw.*

*Proof.* In the $4^2$ game there are 10 winning lines and the maximum degree is 3. Since $3 + 10 < 2^4 = 16$, Theorem P applies, and we are done.

In the $8^3$ game there are $(10^3 - 8^3)/2 = 244$ winning lines and the maximum degree is $2^3 - 1 = 7$. Since $244 + 7 < 2^8 = 256$, Theorem P applies, and we are done again. □

In the $4^2$ game the Point/Line ratio is less than 2, implying that the $4^2$ game is a draw but *not* a pairing strategy draw. In the $8^3$ game, however, there are more than twice as many points ("cells") as winning lines (indeed, $8^3 = 512 > 2 \cdot (10^3 - 8^3)/2 = 488$), so there is a chance to find a draw-forcing pairing strategy. And indeed there is one: a beautiful symmetric pairing (strategy draw), due to S. Golomb, is described in pp. 677–678 of [8] (volume two) or in [13].

Unfortunately I don't know any similar elegant short proof for ordinary $3^2$ Tic-Tac-Toe (of course every child "knows" that it is a draw, but the only known *precise* proof is an unpleasant exhaustive search).

Is Theorem P powerful enough to settle part (a) of the Hales–Jewett conjecture? Unfortunately, the answer is "no"; indeed, the Erdős–Selfridge criterion applies to the $n^d$ game if

$$\frac{(n+2)^d - n^d}{2} < 2^{n-1},$$

which implies that either player can force a Strong Draw if $n > const \cdot d \cdot \log d$. Unfortunately this "superlinear" bound falls short of proving the linear bound $n \geq 2d/\log 2 = 2.885d$ (which is asymptotically equivalent to part (a) of the Hales–Jewett conjecture).

If Theorem P is not powerful enough to "beat" pairing strategy in the $n^d$ game, then why did we include it in this paper? Well, the answer is that Theorem P is the "first step in the right direction": we develop a "variant" of Theorem P, see Theorem 3 below, which will immediately prove part (a) of the Hales–Jewett conjecture (at least in large dimensions), and combining Theorem 3 with Theorem 1 will lead to further improvements. We have to warn the reader that Theorem 3 is somewhat "ugly": it has a "free parameter" $k$ which will be optimized in the applications.

**Theorem 3.** *Let* $\mathcal{F}$ *be an* $m$*-uniform almost disjoint hypergraph. Assume that the Maximum Degree of* $\mathcal{F}$ *is at most* $D$*, that is, every point of the board is contained in at most* $D$ *hyperedges of* $\mathcal{F}$ *("local size"). Moreover assume that the total number of* *winning sets is $|\mathcal{F}| = M$ ("global size"). If there is an integer $k$ with $2 \le k \le m/2$ such that*

$$M\binom{m(D-1)}{k} < 2^{km-k(k+1)-\binom{k}{2}-1}, \tag{6.4}$$

*then Second Player can force a Strong Draw in the positional game on $\mathcal{F}$.*

*Remarks.* Theorem 3 is rather difficult to understand at first sight. One difficulty is the role of parameter $k$; what is the optimal choice for $k$? To answer this question, take $k$th roots of both sides of (6.4): we see that (6.4) holds if

$$M^{1/k} \cdot (D-1) \cdot \frac{4m}{k} < 2^{m-3k/2-1}. \tag{6.5}$$

(6.5) is equivalent to

$$D-1 < k \cdot M^{-1/k} \cdot 2^{-3k/2} \cdot \frac{2^{m-3}}{m}. \tag{6.6}$$

The product $k \cdot M^{-1/k} \cdot 2^{-3k/2}$ in (6.6) attains its maximum by choosing our integral parameter $k$ around $(\frac{2}{3}\log_2 M)^{1/2}$, where $\log_2$ is the base two logarithm ("binary logarithm") (which assumes, in view of the requirement $k \le m/2$, that $M < 2^{3m^2/8}$), and obtain the following result.

**Corollary 1.** *If $\mathcal{F}$ is an $m$-uniform almost disjoint hypergraph with global size $|\mathcal{F}| < 2^{3m^2/8}$, and the maximum degree $D$ of $\mathcal{F}$ is less than*

$$\frac{2^{m-\sqrt{6\log_2|\mathcal{F}|}-3}}{m}, \tag{6.7}$$

*then Second Player can force a Strong Draw in the positional game played on $\mathcal{F}$.*

I call Theorem 3 a "neighborhood principle" type result. The name "neighborhood principle" emphasizes the fact that what really matters here is an exponential upper bound on the "neighborhood size", and the global size $|\mathcal{F}|$ is *almost* irrelevant in the sense that it can be super-exponentially large (like $2^{c\cdot m^2}$, see (6.7)).

Corollary 1 is a justification of the "neighborhood principle" for *almost disjoint* hypergraphs. It raises some very natural questions like:

(i) What happens if $|\mathcal{F}| > 2^{3m^2/8}$?
(ii) What happens if the hypergraph is *not* almost disjoint?

We are going to return to question (i) in Section 8 (question (ii) will be discussed in another paper).

If $|\mathcal{F}| \le m^m$, then Corollary 1 yields

**Corollary 2.** *Let $\mathcal{F}$ be an $m$-uniform family of almost disjoint sets. Assume that $|\mathcal{F}| \le m^m$ and the Maximum Degree of $\mathcal{F}$ is at most $2^{m-3\sqrt{m\log m}}$. If $m > c_0$, i.e., if $m$ is sufficiently large, then Second Player can force a Strong Draw in the positional game on $\mathcal{F}$.*

Note that Corollary 2 follows fairly easily from a 20-year-old result of mine (see Beck [3]). Corollary 2 applies to the $n^d$ game if

$$\frac{(n+2)^d - n^d}{2} \le n^n \text{ and} \tag{6.8}$$

$$\frac{3^d - 1}{2} \le 2^{n-3\sqrt{n\log n}} \ (n \text{ odd}), \quad 2^d - 1 \le 2^{n-3\sqrt{n\log n}} \ (n \text{ even}). \tag{6.9}$$

Inequalities (6.8)–(6.9) trivially hold if

$$n \ge \left(\frac{\log 3}{\log 2} + \varepsilon\right) d \text{ with } d > c_0(\varepsilon) \ (n \text{ odd}), \tag{6.10}$$

$$n \ge (1+\varepsilon)\, d \text{ with } d > c_0(\varepsilon) \ (n \text{ even}). \tag{6.11}$$

Observe that both (6.10) and (6.11) are asymptotically better than (6.3) (note that $\log 3/\log 2 = 1.585$). This means that there are *infinitely many* $n^d$ games in which pairing strategy draw *cannot* exist, but the game is nevetheless a draw (in fact a Strong Draw).

According to my calculations, Theorem 3 proves part (a) of the Hales–Jewett conjecture for all dimensions $d \ge 32$. For example, if $d = 32$, the Hales–Jewett conjecture applies for $n$ at least $2/(2^{1/32} - 1)$, which is about 91.3 so for $n \ge 92$. Theorem 3 settles both “border line” cases “$d = 32$ and $n = 92$” and “$d = 32$ and $n = 93$” with $k = 12$.

Of course, the case $n$ *odd* is always harder, since the maximum degree is much larger. A low-dimensional example where Theorem 3 “beats” Pairing Strategy is the $44^{16}$ game: the Point/Line ratio is less than 2 (so Pairing Strategy cannot force a draw); on the other hand, Theorem 3 applies with $k = 8$, and guarantees a Strong Draw.

Let me recall what we just proved about hypercube Tic-Tac-Toe:

If $n$ is *odd* and $n > (\log 3/\log 2 + \varepsilon)\, d$, or if $n$ is *even* and $n > (1+\varepsilon)\, d$, then the $n^d$ game is a Strong Draw.

This is a “linear” bound; on the other hand, Theorem 2 gave a much stronger (nearly) “quadratic” bound for the *existence* of a Proper 2-Coloring of the $n^d$-hypergraph. Can this Proper 2-Coloring of the $n^d$-hypergraph be “upgraded” to a Drawing Strategy in the $n^d$ game?

**Question 1.** Is it true that, if $d \le \left(\frac{\log 2}{16} + o(1)\right) \frac{n^2}{\log n}$, then the $n^d$ game is a Draw?

Let’s see how close we can get to Question 1 by combining Theorem 1 with *Theorem 3* (instead of the Erdős–Lovász 2-Coloring Theorem). We apply Theorem 3 to the truncated hypergraph $\widetilde{\mathcal{F}_{n,d}}$ (see Theorem 1), which means the requirement

$$d^{d/\alpha n + O(1)} \cdot 2^{3k/2} \cdot n^{d/k} \le 2^{(1-2\alpha)n + O(1)}, \tag{6.12}$$

assuming $k = const \cdot n$. Note in advance that $d = O(n^2/\log n)$, so taking logarithms in (6.12) we obtain

$$\frac{d\log d}{\alpha n \log 2} + \frac{3k}{2} + \frac{d}{k} \cdot \frac{\log d}{2\log 2} + O(\log n) \le (1-2\alpha)n.$$

This inequality is satisfied if $\alpha = 2/13$, $k = \sqrt{d\log d/3\log 2} + O(1)$, $n > 5.5\sqrt{d\log d}$, and $n > c_0$. Note that $n > 5.5\sqrt{d\log d}$ is asymptotically equivalent to $d < \frac{n^2}{60.5\log n}$.

The analogue of (6.12) for *combinatorial lines* goes as follows:

$$d^{d/\beta n+O(1)} \cdot 2^{3k/2} \cdot n^{d/k} \leq 2^{(1-\beta)n+O(1)}. \qquad (6.12')$$

This inequality is satisfied if $\beta = 3/10, k = \sqrt{d \log d / 3 \log 2} + O(1), n > 4.5\sqrt{d \log d}$, and $n > c_0$. Note that $n > 4.5\sqrt{d \log d}$ is asymptotically equivalent to $d < \frac{n^2}{40.5 \log n}$.

Let us summarize what we have learned so far about the $n^d$ game.

**Theorem 4.** *In the $n^d$ game,*

(i) *if $d \geq$ "Shelah's supertower function of n", then First Player can force an (ordinary) Win (but we don't know how);*

(ii) *if $d < \frac{n^2}{60.5 \log n}$ and n is sufficiently large, then Second Player can force a Strong Draw;*

(iii) *if $d \leq n/4$, then the game is a Pairing Strategy Draw.*

(iv) *if $d < \frac{n^2}{40.5 \log n}$ and n is sufficiently large, then Second Player can force a Strong Draw in the "combinatorial lines only" version of the $n^d$ game.*

Of course (i) is not a new result; we just included it for the sake of completeness.

Theorem 4 (iii) easily follows from Theorem 1 and Theorem O. Indeed, if $n \geq 4d$, then applying Theorem 1 (a) with $\alpha = 1/4$ we have

$$\text{MaxDegree}\left(\widetilde{\mathcal{F}_{n,d}}\right) \leq d \leq \lfloor n/4 \rfloor,$$

so Theorem O applies to the $2\lfloor n/4 \rfloor$-uniform $\widetilde{\mathcal{F}_{n,d}}$, and yields a *Pairing Strategy Draw*. Note that the bound $n \geq 4d$ is "exponentially better" than (6.1), but falls short of (6.3). This is how close I can get to part (b) of the Hales–Jewett conjecture.

Theorem 4 (ii) falls short of answering Question 1 above: the constant factor is $1/60.5 = 0.01653$ which is substantially less than $(\log 2)/16 = 0.04332$, but the nearly quadratic order of magnitude is the same.

Question 1 was about a *particular* application of the Erdős–Lovász 2-Coloring Theorem (namely, the application to hypercube Tic-Tac-Toe); the natural *generalization* of Question 1 is the following.

**Question 2.** Can the Erdős–Lovász Proper 2-Coloring in general be *upgraded* to a Drawing Strategy?

To give a precise meaning to the vague Question 2 I need to introduce the concept of the *Maximum Neighborhood Size* of a hypergraph. If $\mathcal{F}$ is a hypergraph and $A \in \mathcal{F}$ is a hyperedge, then the $\mathcal{F}$-neighborhood of $A$ is $\mathcal{F}_A = \{B \in \mathcal{F} : B \cap A \neq \emptyset\}$, that is, the set of elements of $\mathcal{F}$ which intersect $A$, including $A$ itself. Now the *Maximum Neighborhood Size* of $\mathcal{F}$ is the maximum of $|\mathcal{F}_A| = |\{B \in \mathcal{F} : B \cap A \neq \emptyset\}|$, where $A$ runs over all elements of $\mathcal{F}$.

The *Maximum Neighborhood Size* is very closely related to the Maximum Degree. Indeed, if $\mathcal{F}$ is $m$-uniform, its Maximum Degree is $D$, and its Maximum Neighborhood Size is $S$, then $D + 1 \leq S \leq m(D-1) + 1$.

The best possible conjecture would be the following, containing the Erdős–Selfridge theorem as a special case.

**Open Problem 9** *(a) Assume that $\mathcal{F}$ is an n-uniform hypergraph, and its Maximum Neighborhood Size is less than $2^{n-1}$. Is it true that playing on $\mathcal{F}$ Second Player has a Strong Draw?*

Maybe the sharp upper bound $< 2^{n-1}$ is not quite right, and an "accidental" counterexample disproves it. The weaker version (b) below would be equally interesting.

**Open Problem 9** *(b) If (a) is too difficult (or false), then how about if the upper bound on the Maximum Neighborhood Size is replaced by an upper bound $2^{n-c}/n$ on the Maximum Degree, where c is a sufficiently large positive constant?*

*(c) If (b) is still too difficult, then how about a polynomially weaker version where the upper bound on the Maximum Degree is replaced by $n^{-c} \cdot 2^n$, where $c > 1$ is a positive absolute constant?*

*(d) If (c) is still too difficult, then how about an exponentially weaker version where the upper bound on the Maximum Degree is replaced by $c^n$, where $2 > c > 1$ is an absolute constant?*

*(e) How about if we make the extra assumption that the hypergraph is Almost Disjoint (which holds for the $n^d$ game anyway)?*

*(f) How about if we just want a Draw End-Position, i.e., a Proper Halving 2-Coloring?*

In the next section we prove Theorem 3.

## 7 Big Game–Small Game decomposition

*Proof of Theorem 3.* First Player is called "Maker" and Second Player is called "Breaker". The basic idea of the proof is a *decomposition* of the game into *two noninteracting games*; I was motivated by Schmidt's proof of Theorem K (see [18]). The *two noninteracting games* have disjoint boards: we call them the *big game* and the *small game*. This is in Breaker's mind only; Maker does not know anything about the decomposition whatsoever.

*Noninteracting games* means that playing the *big game* Breaker has no knowledge of the happenings in the *small game*, and similarly, playing the *small game* Breaker has no knowledge of the happenings in the *big game*. In other words, we assume that Breaker is "schizophrenic": he has two personalities, one for the *big game* and one for the *small game*, and the two personalities know nothing about each other. I call it the "Iron Curtain Principle". This, at first sight very weird, assumption is crucial in the proof!

Whenever Maker picks a point from the *big board* ("board of the big game"), Breaker responds in the big board, and similarly, whenever Maker picks a point from the *small board* ("board of the small game"), Breaker responds in the small board. In other words, Breaker follows the "Same Board Rule".

The *small game* contains the winning sets that are "dangerous", where Maker is close to winning (i.e., the *small board* is a kind of "Emergency Room"). In the *small game*, Breaker's goal is to block the most dangerous winning sets, and he uses a straightforward *Pairing Strategy*.

Breaker's goal in the *big game* is to prevent too complex winning-set-configurations from graduating into the *small game*. This is how the *big game* ensures that

Breaker's Pairing Strategy in the *small game* will actually *work*. In the *big game* Breaker uses the sophisticated Erdős–Selfridge Power-of-Two Scoring System (i.e., *not* the Erdős–Selfridge theorem itself, but rather its proof-technique – see Lemma 5 below). The key fact is that the number of *big sets* depends primarily on the Maximum Degree $D$ (rather than on the much larger "global size" $M$). The (relatively) small Maximum Degree keeps the family of *big sets* under control, and this is how Breaker can ensure, by using the Power-of-Two Scoring System, that the *small game* remains relatively simple, and a pairing strategy can indeed block every "dangerous" winning set.

The board of the big game ("big board") is going to shrink during a play. Consequently, the board of the small game ("small board"), which is the complement of the big board, keeps growing during a play. At the beginning of the play the big board is equal to the whole board $V$ (i.e., the small game is "not born yet"). Let $V_{\text{BIG}}(i)$ and $V_{\text{small}}(i) = V \setminus V_{\text{BIG}}(i)$ denote the big board and the small board after Maker's $i$th move and before Breaker's $i$th move. Then we have

$$V = V_{\text{BIG}}(0) \supseteq V_{\text{BIG}}(1) \supseteq V_{\text{BIG}}(2) \supseteq V_{\text{BIG}}(3) \supseteq \cdots ,$$
$$\emptyset = V_{\text{small}}(0) \subseteq V_{\text{small}}(1) \subseteq V_{\text{small}}(2) \subseteq V_{\text{small}}(3) \subseteq \cdots .$$

The big game is played on the family of *big sets*, and the small game is played on the family of *small sets*. What are the *big sets* and the *small sets*? Well, it is much simpler (and natural) to define the *small sets* first, and to define the *big sets* later.

Let $x_1, x_2, \ldots, x_i, \ldots$ and $y_1, y_2, \ldots, y_i, \ldots$ denote, respectively, the points of Maker and Breaker. At the beginning of the play, when the board of the small game is empty (i.e., the small game is "not born yet"), Breaker chooses his points $y_1, y_2, y_3, \ldots$ according to the Erdős–Selfridge Power-of-Two Scoring System (see Lemma 5 below) applied to the family $\mathcal{B}$ of big sets (family $\mathcal{B}$ will be defined later). In the course of a play in the big game an $n$-element winning set $A \in \mathcal{F}$ is *dead* when it contains some point of Breaker (for the first time). Note that *dead* elements of $\mathcal{F}$ don't represent danger any more (they are marked by Breaker, so Maker cannot completely occupy them). At a given stage of a play in the big game those elements of $\mathcal{F}$ which are not dead yet are called *survivors*. A *survivor* $A \in \mathcal{F}$ becomes *dangerous* when Maker occupies its $(m - k - 1)$th point; then the unoccupied $(k + 1)$-element part of this dangerous $A \in \mathcal{F}$ becomes an *emergency set*. This is why the *big game* is shrinking: whenever an *emergency set* arises, its points are *removed* from the big board, and at the same time they are added to the small board. In other words, the small board is precisely the union of *all emergency sets* (and consequently, the big board is precisely the complement of the union of all emergency sets). This means that the *small board* is a kind of "Emergency Room" where Breaker takes care of the *emergency sets* (Breaker's goal is to put his mark in every *emergency set*).

What are the *small sets*, that is, the winning sets in the small game? Well, a set $S$ is a *small set* if it satisfies the following two requirements:

(a) $S$ is the intersection of a *survivor* $A \in \mathcal{F}$ and the board of the small game,
(b) $|S| \geq k + 1$.

The *emergency sets* are obviously all *small* sets. Those small sets which are *not* emergency sets are called *secondary sets*. What is the goal of the small game? Well, Breaker wins the small game if he can prevent Maker from completely occupying a small set;

otherwise Maker wins. (In other words, Breaker's goal is to block every small set – he will in fact employ a simple pairing strategy.)

Now we are ready to define the *big sets*. The *big sets* play an auxiliary role in the proof. We use them to guarantee that, when Breaker follows a winning strategy in the big game, the small game is very simple in the following sense:

(i) a secondary set doesn't exist – see Lemma 1; and
(ii) Breaker can block every emergency set by a pairing strategy employing the "private points" – see Lemma 2 below.

To ensure these two requirements we introduce a key definition: a $k$-element subfamily $\mathcal{G} = \{A_1, A_2, \ldots, A_k\} \subset \mathcal{F}$ is called *$\mathcal{F}$-linked* if there is a set $A \in \mathcal{F}$ with $A \notin \mathcal{G}$ such that $A$ intersects each element of family $\mathcal{G}$, i.e., $A \cap A_i \neq \emptyset$, $1 \leq i \leq k$. (Note that parameter $k$ is an integer between 2 and $m/2$.)

The big game is played on the family $\mathcal{B}$ of *big sets*. What are the *big sets*? Well, they are the *union sets* $\bigcup_{i=1}^{k} A_i$ of all possible *$\mathcal{F}$-linked* $k$-element subfamilies $\mathcal{G} = \{A_1, A_2, \ldots, A_k\}$ of $\mathcal{F}$:

$$\mathcal{B} = \left\{ B = \bigcup_{A \in \mathcal{G}} A : \mathcal{G} \subset \mathcal{F},\ |\mathcal{G}| = k,\ \mathcal{G} \text{ is } \mathcal{F}\text{−linked} \right\}.$$

The total number of *big sets* is clearly not more than

$$M\binom{m(D-1)}{k}. \tag{7.1}$$

Indeed, there are $|\mathcal{F}| = M$ ways to choose "linkage" $A$, there are at most $n(D-1)$ other sets intersecting $A$, and we have to choose $k$ sets $A_1, A_2, \ldots, A_k$ among them.

Each big set $B = \bigcup_{i=1}^{k} A_i$ has cardinality $\geq kn - \binom{k}{2}$. Indeed, since $\mathcal{F}$ is almost disjoint,

$$|B| = \left|\bigcup_{i=1}^{k} A_i\right| \geq \sum_{i=1}^{k} |A_i| - \sum_{1 \leq i < j \leq k} |A_i \cap A_j| \geq km - \binom{k}{2}. \tag{7.2}$$

What is *Maker's goal* in the big game? Of course Maker doesn't know about the "big game" (or the "small game"); this whole "decomposition" is in Breaker's mind only, so it is up to Breaker to define "Maker's goal" in the big game. The definition goes as follows: "Maker wins the big game" if he can occupy all but $k(k+1)$ points of some big set $B \in \mathcal{B}$ *in the big board* before Breaker could put his first mark in this $B$ *in the big board*; otherwise Breaker wins the big game. The reason why I had to write "*in the big board*" twice in the previous sentence is that the big board is shrinking, and so the big board does not necessarily contain all big sets. The intersection of a big set with the small board (i.e., the part outside of the big board) is "invisible" in the big game: whatever happens in the small game has no effect in the big game. (For example, if Breaker can block a big set in the small board, that doesn't count as a "blocking in the big game"; this is the curse of the "Iron Curtain Principle".)

We are going to show that Breaker can win the big game by using the Erdős–Selfridge power-of-two scoring system if the total number of big sets is not too large, namely, if

$$|\mathcal{B}| < 2^{km-\binom{k}{2}-k(k+1)-1}. \tag{7.3}$$

Note that the board of the big game is *shrinking* during a play, but it is *not* going to cause any extra difficulty in the argument (see Lemma 5).

**Lemma 1.** *Assume that Breaker has a winning strategy in the big game, and he follows it in a play. Then there is no secondary set in the small game.*

*Proof.* Let $S^* = \{z_1, z_2, z_3, \ldots\}$ be an arbitrary secondary set. Since the small board is the union of the emergency sets, by property (a) above every point $z_i \in S^*$ is contained in some emergency set (say) $S_i$, $1 \leq i \leq |S^*|$. Almost disjointness implies that different points $z_i \in S^*$ are contained in different emergency sets $S_i$. Since $S_i$ is an emergency set, there is a winning set $A_i \in \mathcal{F}$ such that $S_i \subset A_i$ and $A_i \backslash S_i$ was completely occupied by Maker during the play in the big game. Note that Breaker didn't block $A_i$ in the big game, since $S_i$ was removed from the big board (and added to the small board). Again almost disjointness implies that the sets $A_i$, $i = 1, 2, 3, \ldots$ are all different. The union set $\bigcup_{i=1}^{k} A_i$ is a *big set* since $\{A_1, A_2, \ldots, A_k\}$ is $\mathcal{F}$*-linked* by $A^*$ where $A^* \in \mathcal{F}$ is defined by $S^* \subset A^*$ ($A^*$ is the *ancestor* of $S^*$). Since for every $i$, $A_i \backslash S_i$ was completely occupied by Maker, Maker was able to occupy all but $k(k+1)$ points of the particular big set $B = \bigcup_{i=1}^{k} A_i$ during a play in the big game, and Breaker didn't block $B$ in the big game, i.e., Maker wins the big game. This contradicts the assumption that Breaker has a winning strategy in the big game and follows it in the play. This contradiction shows that a secondary set cannot exist. □

A similar argument proves

**Lemma 2.** *If Breaker wins the big game, then every emergency set has at least two "private" points, that is, two points which are never going to be contained in any other emergency set during the whole course of the small game. (In other words, a point of an emergency set is called "private" if it has degree one in the family of all emergency sets.)*

Let $S_1, S_2, S_3, \ldots$ be the complete list of *emergency sets* arising in this order during the course of a play (when a bunch of two or more emergency sets arise at the same time, then the ordering within the bunch is arbitrary). Let $\widetilde{S_1} = S_1$, $\widetilde{S_2} = S_2 \backslash S_1$, $\widetilde{S_3} = S_3 \backslash (S_1 \cup S_2)$, and in general,

$$\widetilde{S_j} = S_j \backslash \left( \bigcup_{i=1}^{j-1} S_i \right).$$

We call $\widetilde{S_j}$ the "disjoint part" of $S_j$. Of course, the "disjoint part" of $S_j$ contains its "private points", so by Lemma 2 every "disjoint part" $\widetilde{S_j}$ has at least 2 elements (exactly what we need for a pairing strategy – see below).

When the first dangerous set $A \in \mathcal{F}$ arises, say, at the $i$th move of Maker, then the whole board $V$ splits into two nonempty parts for the *first time*. The two parts are the big board $V_{BIG}(i)$ and the small board $V_{small}(i)$, where $V = V_{BIG}(i) \cup V_{small}(i)$. Whenever Maker picks a point from the big board, Breaker responds in the big board; whenever Maker picks a point from the small board, Breaker responds in the small board ("Same Board Rule"). This is how the game falls apart into two noninteracting, disjoint games: the shrinking *big game* and the growing *small game*.

During the course of a play in the *small game* Breaker uses the following trivial pairing strategy: when Maker occupies a point of the small board which is contained in the "disjoint part" $\widetilde{S}$ of an emergency set $S$, then Breaker picks another point of the same $\widetilde{S}$ (if he *finds* one; if he *doesn't*, then he makes an arbitrary move). In view of the remark after Lemma 2, Breaker can block every emergency set in the small game *under the condition* that he can win the big game. Since a secondary set cannot exist (see Lemma 1), we obtain

**Lemma 3.** *If Breaker can win the big game, then he can win the small game, i.e., he can block every small set.*

Next we prove

**Lemma 4.** *If Breaker can win the big game, then he can block every winning set $A \in \mathcal{F}$ either in the big game or in the small game.*

*Proof.* Indeed, assume that at the end of a play some $A_0 \in \mathcal{F}$ is completely occupied by Maker. We derive a contradiction as follows. We distinguish two cases.

Case 1. During the course of the big game Maker occupies $(m-k-1)$ points of $A_0$, that is, $|A_0 \cap V_{BIG}(j)| = m-k-1$ for some $j$.

Let $j$ be the first index such that $|A_0 \cap V_{BIG}(j)| = m-k-1$, i.e., in the big game Maker occupied the $(m-k-1)$th point of $A_0$ at his $j$th move. Then $A_0$ becomes a dangerous set, and $S_0 = A_0 \setminus V_{BIG}(j)$ goes to the small game as an emergency set. Since Breaker can block every emergency set in the small game by a "pairing strategy" (see Lemma 3), we have a contradiction.

Case 2. At the end of the big game Maker has less than $(m-k-1)$ points of $A_0$. Then for some $j$, $A_0 \cap V_{small}(j)$ must become a *secondary set*, which contradicts Lemma 1. □

Therefore, the *last step* of the proof of Theorem 3 is to show that Breaker has a winning strategy in the big game. We recall that Maker wins the big game if he can occupy all but $k(k+1)$ points of some big set $B \in \mathcal{B}$ before Breaker could put his first mark in this $B$. In view of (7.3) this means (at least) $km - \binom{k}{2} - k(k+1)$ points of Maker in $B$ (before Breaker could put his first mark in this $B$). What we need here is not the Erdős–Selfridge theorem itself, but the following slightly modified version (we will apply it to the big game with $b = km - \binom{k}{2} - k(k+1)$, where $m$ is from Lemma 5 below).

**Lemma 5** (Modified Erdős–Selfridge)**.** *Let $\mathcal{B}$ be a hypergraph such that every winning set $B \in \mathcal{B}$ has at least $b$ points. There are two players, Maker and Breaker, who alternately occupy previously unoccupied points of the board (Maker starts). Assume that after Maker's each move the unoccupied part of the board may shrink, but the board doesn't change after Breaker's moves. Maker wins the game if he can occupy $b$ points of some winning set $B \in \mathcal{B}$ before Breaker could put his first mark in this $B$; otherwise Breaker wins. Now if $|\mathcal{B}| < 2^{b-1}$, then Breaker has a winning strategy.*

*Proof.* This is basically the Erdős–Selfridge proof. Breaker employs the following Power-of-Two Scoring System: a winning set blocked by Breaker *scores* zero ("dead set"), a blank winning set *scores* 1, a set with a single mark of Maker and no mark of Breaker *scores* 2, a set with two marks of Maker and no mark of Breaker *scores* $2^2 = 4$, and so on (these are the "survivors").

Suppose that we are in the middle of a play, and it is Breaker's turn to choose his $i$th point $y_i$. What is the "danger" of this particular position? We evaluate the position by the total sum, over all winning sets, of the *scores*; we denote it by $D_i$, and call it the "danger-function" (index $i$ indicates that we are at the stage of choosing the $i$th point of Breaker). A natural choice for $y_i$ is to minimize the "danger" $D_{i+1}$ at the next stage of the play. Let $y_i$ and $x_{i+1}$ denote the next moves of Breaker and Maker (in this order). Note that the board doesn't change after choosing $y_i$. What are the effects of these two moves on $D_i$? Well, $y_i$ "kills" all "survivors" $B \ni y_i$ ("survivor" means that the set was not blocked by Breaker before), which means we have to *subtract* the sum of the *scores* of all "survivors" $B \in \mathcal{B}$ with $y_i \in B$ from $D_i$. On the other hand, $x_{i+1}$ doubles the "danger" of each "survivor" $B \ni x_{i+1}$, which means we have to *add* the sum of the *scores* of all "survivors" $B \in \mathcal{B}$ with $x_{i+1} \in B$ back to $D_i$ one more time. If some "survivor" $B$ contains *both* $y_i$ and $x_{i+1}$, then we don't have to give the score back with $x_{i+1}$ because that $B$ was previously "killed" by $y_i$.

So the natural choice for $y_i$ is that unoccupied point $z$ which makes the "biggest damage": for which the sum of the *scores* of all "survivors" $B \in \mathcal{B}$ with $z \in B$ attains the *maximum.* Loosely speaking: $y_i$ is the "biggest damage point", so $x_{i+1}$ is at most the "second biggest damage point". Then what we subtract from $D_i$ is greater or equal to what we add back to it. In other words, Breaker can force the decreasing property $D_{\text{start}} = D_1 \geq D_2 \geq \cdots \geq D_i \geq D_{i+1} \geq \cdots \geq D_{\text{end}}$ of the "danger-function" (the "shrinking" of the unoccupied part of the board doesn't change this key property).

Maker wins the game if he can occupy $b$ points of some set $B \in \mathcal{B}$ before Breaker could put his first mark in this $B$. Assume this happens right before the $j$th move of Breaker. Then this $B$ scores $2^b$, implying $D_j \geq 2^b$; we call $2^b$ the "target value".

On the other hand, the "initial value" ($x_1$ is the first point of Maker)

$$D_{\text{start}} = D_1 = \sum_{B:\, x_1 \in B \in \mathcal{B}} 2^1 + \sum_{B:\, x_1 \notin B \in \mathcal{B}} 2^0 \leq \sum_{B \in \mathcal{B}} 2.$$

By the *decreasing property* of the "danger-function", if Maker wins at his $j$th move, then

$$2^b \leq D_j \leq D_{\text{start}} \leq \sum_{B \in \mathcal{B}} 2. \tag{7.4}$$

Since *no play is possible from an initial position if the target position has a higher danger value than the initial position,* Maker cannot win the play if (7.4) is *violated.* In other words, inequality $|\mathcal{B}| < 2^{b-1}$ (a violation of (7.4)) guarantees that Breaker wins. Breaker's winning strategy is to keep choosing the "biggest damage point". This completes the proof of Lemma 5. □

By (7.2) each big set $B \in \mathcal{B}$ has cardinality $\geq km - \binom{k}{2}$. Therefore, we apply Lemma 5 to the big game with $b = km - \binom{k}{2} - k(k+1)$. We recall the upper bound on the total number of big sets (see (7.1)):

$$|\mathcal{B}| \leq M\binom{m(D-1)}{k}.$$

By Lemma 5 Breaker wins the big game if

$$|\mathcal{B}| \leq M\binom{m(D-1)}{k} < 2^{km-\binom{k}{2}-k(k+1)-1},$$

exactly as we predicted in (7.3). This completes the proof of Theorem 3. □

## 8 How good are the new lower bounds? Strong Draw and Weak Win

Before talking about "how good the new lower bounds are" (the main subject of this section), I recall Question 1 from Section 5:

**Question 1.** Is it true that, if $d \leq \left(\frac{\log 2}{16} + o(1)\right)\frac{n^2}{\log n}$, then the $n^d$ game is a Draw?

By combining Theorem 1 with Theorem 3 we could prove the somewhat weaker result that, if $d \leq \frac{n^2}{60.5 \log n}$, then the $n^d$ game is a Draw (in fact a Strong Draw). This falls short of Question 1 by a constant factor.

In the proof of Theorem 3 we applied the technique of "Big Game–Small Game Decomposition". This technique breaks down in the range $|\mathcal{F}| > 2^{3m^2/8}$, where $\mathcal{F}$ is an $m$-uniform almost disjoint hypergraph.

By developing a more sophisticated decomposition technique we could largely extend the range far beyond $|\mathcal{F}| > 2^{const \cdot m^2}$, and prove the following ("very ugly") theorem.

**Theorem 5.** *If $\mathcal{F}$ is an $m$-uniform Almost Disjoint hypergraph such that*

$$|\mathcal{F}|^{m^2 \cdot 2^{-k/4}} \cdot \max\left\{2^{7k/2} \cdot D, m \cdot 2^{5k/2} \cdot D^{1+1/k-2}\right\} \leq 2^m$$

*holds for some integer $k \geq 8\log_2 m$ ("binary logarithm"), where $D = \text{MaxDeg}(\mathcal{F})$, then either player can force a Strong Draw in the positional game on $\mathcal{F}$.*

By choosing $k$ around $\sqrt{m}$ in Theorem 5 ($m$ is sufficiently large), we obtain the following special case.

**Corollary 1.** *If $\mathcal{F}$ is an $m$-uniform Almost Disjoint hypergraph,*

$$\text{MaxDeg}(\mathcal{F}) < 2^{m-4\sqrt{m}} \text{ and } |\mathcal{F}| < 2^{2^{\sqrt{m}/5}},$$

*then, for $m > c_0$, either player can force a Strong Draw in the positional game on $\mathcal{F}$.*

Combining Theorem 1 with Corollary 1 (instead of Theorem 3) it is easy to obtain an affirmative answer to Question 1. In other words, we can "upgrade" the existing Proper 2-Coloring of the $n^d$-hypergraph to a Drawing Strategy (in fact Strong Draw Strategy), which of course immediately yields the existence of a Proper *Halving* 2-Coloring.

**Corollary 2.** *(a)*

$$\mathbf{ww}(n\text{--line}) \geq \left(\frac{\log 2}{16} + o(1)\right) \frac{n^2}{\log n},$$

*that is, if*

$$d \leq \left(\frac{\log 2}{16} + o(1)\right) \frac{n^2}{\log n},$$

*then either player can force a Strong Draw in the $n^d$ game.*

*(b) As a byproduct, under the condition of (a), we obtain a Proper Halving 2-Coloring of the $n^d$-hypergraph, implying the lower bound*

$$HJ_{1/2}(n) \geq \mathbf{ww}(n\text{--line}) \geq \left(\frac{\log 2}{16} + o(1)\right) \frac{n^2}{\log n}.$$

*(c)*

$$\mathbf{ww}(\text{comb. } n\text{--line}) \geq \left(\frac{\log 2}{8} + o(1)\right) \frac{n^2}{\log n},$$

*that is, if*

$$d \leq \left(\frac{\log 2}{8} + o(1)\right) \frac{n^2}{\log n},$$

*then either player can force a Strong Draw in the "combinatorial lines only" version of the $n^d$ game.*

*(d) As a byproduct, under the condition of (c), we obtain a Proper Halving 2-Coloring of the "combinatorial lines only" hypergraph on the $n^d$ board, implying the lower bound*

$$HJ^c_{1/2}(n) \geq \mathbf{ww}(\text{comb. } n\text{--line}) \geq \left(\frac{\log 2}{8} + o(1)\right) \frac{n^2}{\log n}.$$

The proof of Theorem 5 is rather long (20 pages), and will be discussed in another paper – this paper is already far too long! – titled "Tic-Tac-Toe games" (to be submitted for publication soon).

Now we are ready to discuss the topic described in the title of this section. How good is Corollary 2 (a) of Theorem 5? Is there any room left for further improvement? To see the limitations, we have to study the complement of Strong Draw: *Weak Win* in the $n^d$ game. We apply a general "Potential Weak Win Criterion" that was proved more than 20 years ago (see Theorem 1 in Beck [4]).

**Theorem Q** (Potential Weak Win Criterion). *Let $\mathcal{F}$ be an n-uniform hypergraph, and let $V$ denote the union set (board). Assume that, fixing any two distinct points of $V$, there are no more than $\Delta_2 = \Delta_2(\mathcal{F})$ winning sets $A \in \mathcal{F}$ containing both points. If equality occurs for some point pair, then I call $\Delta_2(\mathcal{F})$ the Pair-Degree of the hypergraph.*

*If*

$$\frac{|\mathcal{F}|}{|V|} > 2^{n-3} \cdot \Delta_2,$$

*then First Player has a Weak Win in $(V, \mathcal{F})$.*

*In particular, for Almost Disjoint hypergraph the condition simplifies to* $|\mathcal{F}| > 2^{n-3}|V|$.

*Remark.* If $\mathcal{F}$ is $n$-uniform, then $\frac{|\mathcal{F}|}{|V|}$ is $\frac{1}{n}$ times the *Average Degree*. Indeed, it follows from the equality $n|\mathcal{F}| = \text{AverDeg}(\mathcal{F})|V|$.

In other words, the hypothesis of Theorem Q is a simple Density Condition: To guarantee Weak Win all what one needs is a sufficiently "dense" hypergraph.

Before proving Theorem Q, let's first apply it to the $n^d$ game. First Player has a Weak Win if

$$\frac{(n+2)^d - n^d}{2} > 2^{n-3} n^d,$$

which is equivalent to

$$\left(1 + \frac{2}{n}\right)^d > 2^{n-2} + 1. \tag{8.1}$$

Inequality (8.1) holds if $d > \frac{1}{2}(\log 2) \cdot n^2$. It follows that Corollary 2 (a) of Theorem 5 is best possible apart from a factor of $\log n$.

I have a suspicion that the order $d \le const \cdot n^2/\log n$ for Strong Draw can be improved to $d \le const \cdot n^2$, that is, $n^2$ is the right order of magnitude for the "phase transition" between Weak Win and Strong Draw in the $n^d$ game, but I don't have the confidence to formulate it as a conjecture.

Considering small values, inequality (8.1) holds for the weak $3^3$, $4^4$, $5^7$, $6^{10}$, $7^{14}$, $8^{19}$, $9^{25}$, $10^{31}$, ... games, so in these games First Player can force a Weak Win. Note that $3^3$ and $4^4$ on the list can be replaced by $3^2$ and $4^3$. Indeed, ordinary Tic-Tac-Toe has an easy Weak Win, and in view of Patashnik's computer-assisted work, the $4^3$ game has an ordinary Win (see [17]). The list of small Weak Win $n^d$ games: $3^2$, $4^3$, $5^7$, $6^{10}$, $7^{14}$, $8^{19}$, ... is complemented by the following list of known small Strong Draw games: $4^2$, $8^3$, $14^4$, $20^5$, $24^6$, $26^7$, .... There is a big gap between the two lists, proving that our knowledge of the small $n^d$ games is very unsatisfactory.

I leave it to the reader to study the case of *combinatorial lines*.

*Proof of Theorem Q.* I basically repeat the proof of the Erdős–Selfridge theorem. I call First Player "Maker" and Second Player "Breaker". Assume we are at the stage of the play where Maker ("First Player") already occupied $x_1, x_2, \cdots, x_i$, and Breaker ("Second Player") occupied $y_1, y_2, \cdots, y_i$. The question is how to choose Maker's next point $x_{i+1}$. Those winning sets which contain at least one $y_j$ $(j \le i)$ are "useless" for Maker. We call them "dead sets". The winning sets which are not "dead" (yet) are called "survivors". The "survivors" have a chance to be completely occupied by Maker ("Weak Win"). What is the total "chance" of the position? We evaluate the given position by the following "chance function": $C_i = \sum_{s \in S_i} 2^{n-u_s}$, where $u_s$ is the number of unoccupied points of the "survivor" $A_s$ ($s \in S_i$ = "index-set of the survivors", and index $i$ indicates that we are at the stage of choosing the $(i+1)$st point $x_{i+1}$ of Maker. Note that the "chance function" can be much greater than 1 (i.e., it is not a *probability*), but it is always nonnegative.

A natural choice for $x_{i+1}$ is to maximize the "chance" $C_{i+1}$ at the next stage. Let $x_{i+1}$ and $y_{i+1}$ denote the next moves of the two players. What is their effect on $C_{i+1}$?

Well, first $x_{i+1}$ doubles the "chances" for each "survivor" $A_s \ni x_{i+1}$, that is, we have to add the sum $\sum_{s\in S_i:\ x_{i+1}\in A_s} 2^{n-u_s}$ to $C_i$.

On the other hand, $y_{i+1}$ "kills" all the "survivors" $A_s \ni y_{i+1}$, which means we have to subtract the sum $\sum_{s\in S_i:\ y_{i+1}\in A_s} 2^{n-u_s}$ from $C_i$.

**Warning:** We have to make a correction to those "survivors" $A_s$ which contain both $x_{i+1}$ and $y_{i+1}$. These 'survivors" $A_s$ were "doubled" first and "killed" second. So what we have subtract from $C_i$ is not

$$\sum_{s\in S_i:\ \{x_{i+1},y_{i+1}\}\subset A_s} 2^{n-u_s}$$

but the twice as large

$$\sum_{s\in S_i:\ \{x_{i+1},y_{i+1}\}\subset A_s} 2^{n-u_s+1}.$$

It follows that

$$C_{i+1} = C_i + \sum_{s\in S_i:\ x_{i+1}\in A_s} 2^{n-u_s} - \sum_{s\in S_i:\ y_{i+1}\in A_s} 2^{n-u_s} - \sum_{s\in S_i:\ \{x_{i+1},y_{i+1}\}\subset A_s} 2^{n-u_s}.$$

Now the natural choice for $x_{i+1}$ is the unoccupied $z$ for which $\sum_{s\in S_i:\ z\in A_s} 2^{n-u_s}$ attains its maximum. Then clearly

$$C_{i+1} \ge C_i - \sum_{s\in S_i:\ \{x_{i+1},y_{i+1}\}\subset A_s} 2^{n-u_s}.$$

We trivally have

$$\sum_{s\in S_i:\ \{x_{i+1},y_{i+1}\}\subset A_s} 2^{n-u_s} \le \frac{\Delta_2}{4}.$$

Indeed, there are at most $\Delta_2$ winning sets $A_s$ containing the given two points $\{x_{i+1}, y_{i+1}\}$, and $2^{n-u_s} \le 2^{n-2}$, since $x_{i+1}$ and $y_{i+1}$ were definitely unoccupied points at the previous stage.

Therefore,

$$C_{i+1} \ge C_i - \Delta_2 2^{n-2}.$$

What happens at the end? Let $\ell$ denote the number of stages, i.e., the $\ell$th stage is the last one. Clearly $\ell = |V|/2$. Inequality $C_\ell = C_{\text{last}} > 0$ means that Breaker could not "kill" (block) all the winning sets. Indeed, at the last stage all points are occupied, so $C_\ell = C_{\text{last}} > 0$ means that Maker was able to completely occupy a winning set, meaning a Weak Win.

So all what we have to check is that $C_\ell = C_{\text{last}} > 0$. But this is trivial. Indeed, $C_{\text{start}} = C_0 = |\mathcal{F}|$, so we have,

$$C_{\text{last}} \ge |\mathcal{F}| - \frac{|V|}{2}\Delta_2 2^{n-2}.$$

It follows that $C_{\text{last}} > 0$ if $|\mathcal{F}| > 2^{n-3}|V|\Delta_2$, which implies that at the end of the play Maker was able to completely occupy a winning set. □

A major difficulty of studying the $n^d$ *hypercube* Tic-Tac-Toe comes from the highly irregular nature of the $n^d$-hypergraph. Indeed, the Average Degree of the $n^d$-hypergraph is about (very roughly) the $n$th root of the Maximum Degree. This is why we desperately needed Theorem 1 (Degree Reduction by Partial Truncation), and that led to an "error factor" of $\log n$ that we couldn't get rid of.

If we switch to the $n^d$ *Torus* and consider the family of all "Torus-Lines" instead of all "geometric lines" in the $n^d$ Hypercube, then we obtain a *Degree-Regular* hypergraph, and there is no need for "wasteful" degree reduction. The Degree-Regular family of "Torus-Lines" forms the family of winning sets of the $n^d$ **Torus Tic-Tac-Toe**. The $8^2$ Torus Tic-Tac-Toe is particularly interesting because it comes up in a natural way in the proof of the amusing fact that the *Unrestricted 9-in-a-row* is a draw game.

**Unrestricted $n$-in-a-row game**. *Unrestricted* means that the game is played on an *infinite* chessboard (infinite in every direction). In the *Unrestricted 5-in-a-row game* the players alternately occupy little squares of an infinite chessboard, the first player marks his squares by X, and the second player marks his squares by O. That player wins who first gets 5 consecutive marks of his own in a row horizontally, or vertically, or diagonally (of slope 1 or $-1$). *Unrestricted n-in-a-row* differs in only one aspect: the winning size is $n$ instead of 5. The Unrestricted 4-in-a-row game is an easy first-player's win. The Unrestricted 5-in-a-row game is conjectured to be a first-player's win, too, but there is no proof. For a "reasonable board size" like $20 \times 20$ there is a first-player's winning strategy (computer-assisted proof; huge case study), but no one knows how to extend it to the whole plane ("curse of the Extra Set Paradox"). In the other direction, Unrestricted 8-in-a-row is known to be a draw game (this is the best what we know). Here I give a simple Pairing Strategy proof of the slightly weaker result that Unrestricted 9-in-a-row is a draw game. Indeed, cover the infinite chessboard with disjoint copies of the following 8 by 8 matrix

$$n = 8 : \quad \begin{bmatrix} \backslash & \backslash & - & - & | & | & / & / \\ - & - & \backslash & \backslash & / & / & | & | \\ - & - & / & / & \backslash & \backslash & | & | \\ / & / & | & | & - & - & \backslash & \backslash \\ \backslash & \backslash & | & | & - & - & / & / \\ | & | & \backslash & \backslash & / & / & - & - \\ | & | & / & / & \backslash & \backslash & - & - \\ / & / & - & - & | & | & \backslash & \backslash \end{bmatrix}$$

(see [8]). What this 8 by 8 matrix represents is a direction-marking of the $4 \cdot 8 = 32$ "Torus-Lines" of the $8 \times 8$ *Torus*. The direction-marks $-$, $|$, $\backslash$, and $/$ mean (respectively) "horizontal", "vertical", "diagonal of slope $-1$", and "diagonal of slope 1". Each one of the 32 "Torus-Lines" contains 2 marks of its own direction. The periodic extension of the 8 by 8 matrix over the whole plane gives a Drawing Strategy for the Unrestricted 9-in-a-row game. Either player responds to the opponent's last move by taking the *nearest* similarly marked square in the direction indicated by the mark in the opponent's last move square.

Let's return to the general $n^d$ *Torus* game. The $n^d$ *Torus* is an abelian group, implying that the $n^d$ torus-hypergraph is translation-invariant (any two points "look the same"). The $n^d$ torus-hypergraph is degree-regular: every point has degree $(3^d - 1)/2$

(which is, by the way, the same as the degree of the center in the $n^d$ hypergraph when $n$ is odd). So the total number of winning sets ("Torus-Lines") is $(3^d - 1)n^{d-1}/2$. Of course, the board-size remains $n^d$.

We owe the reader a formal definition of the concept of "Torus-Line". A *Torus-Line* $L$ is formally defined by a point $P \in L$ and a vector $\mathbf{v} = (a_1, \ldots, a_d)$, where each coordinate $a_i$ is either 0 or +1 or −1 $(1 \le i \le d)$. The $n$ points of line $L$ are $P + k\mathbf{v}$ (mod $n$), where $k = 0, 1, \ldots, n-1$.

The *combinatorial line* version goes as follows: A *Comb-Torus-Line* $L$ is formally defined by a point $P \in L$ and a vector $\mathbf{v} = (a_1, \ldots, a_d)$, where each coordinate $a_i$ is either 0 or +1 $(1 \le i \le d)$. The $n$ points of line $L$ are $P + k\mathbf{v}$ (mod $n$), where $k = 0, 1, \ldots, n-1$.

The $n^d$ comb-torus-hypergraph is degree-regular: every point has degree $2^d - 1$ (which is, by the way, the same as the maximum degree of the family of all combinatorial lines in the $n^d$ hypercube). So the total number of winning sets ("Comb-Torus-Lines") is $(2^d - 1)n^{d-1}$.

A peculiarity of this new "line-concept" is that two different Torus-Lines may have more than one point in common! We recommend the reader to study the $4^2$ torus game, and to find two different Torus-Lines with *two* common points. We show that this "pair-intersection" cannot happen when $n$ is *odd*, and if $n$ is *even*, then it can happen only under very special circumstances.

For Comb-Torus-Lines, however, there is no surprise.

**Statement 1.** (a) Any two different Torus-Lines have at most *one* common point if $n$ is *odd*, and at most *two* common points if $n$ is *even*. In the second case the distance between the two common points along either Torus-Line containing both is always $n/2$.

(b) Any two different Comb-Torus-Lines have at most *one* common point.

The proof of (a) goes as follows. Let $L_1$ and $L_2$ be two different Torus-Lines with (at least) two common points $\mathbf{P}$ and $\mathbf{Q}$. Then there exist $k, l, \mathbf{v}, \mathbf{w}$ with $1 \le k, l \le n-1$, $\mathbf{v} = (a_1, \ldots, a_d)$, $\mathbf{w} = (b_1, \ldots, b_d)$ (where $a_i$ and $b_i$ are either 0 or +1 or −1 $(1 \le i \le d)$), $\mathbf{v} \not\equiv \pm\mathbf{w}$, such that $\mathbf{Q} \equiv \mathbf{P} + k\mathbf{v} \equiv \mathbf{P} + l\mathbf{w}$ (mod $n$). It follows that $k\mathbf{v} \equiv l\mathbf{w}$ (mod $n$), or equivalently, $ka_i \equiv lb_i$ (mod $n$) for every $i = 1, 2, \ldots, d$. Since $a_i$ and $b_i$ are either 0 or +1 or −1 $(1 \le i \le d)$), the only solution is $k = l = n/2$ (no solution if $n$ is *odd*).

We leave the proof of (b) to the reader. □

What can we say about the two-dimensional $n^2$ torus game? Well, we know a lot; we just demonstrated that the $8^2$ torus game is a Pairing Strategy Draw, and this is a sharp result (since the Point/Line ratio of the $7^2$ torus is $49/28 = 7/4$; i.e. less than 2).

The Erdős–Selfridge theorem applies if $4n + 4 < 2^n$, which gives that the $n^2$ torus game is a Strong Draw for every $n \ge 5$. The $4^2$ torus game is also a draw (mid-size "case-study"), but I don't know any elegant proof. On the other hand, the $3^2$ torus game is an easy First-Player's win.

Next consider the three-dimensional $n^3$ torus game. The Erdős–Selfridge theorem applies if $13n + 13 < 2^n$, which gives that the $n^3$ torus game is a Strong Draw for every $n \ge 11$. I am convinced that the $10^3$ torus game is also a draw, but I don't know how to prove it.

How about the four-dimensional $n^4$ torus game? The Erdős–Selfridge theorem applies if $40n + 40 < 2^n$, which gives that the $n^4$ torus game is a Strong Draw for every $n \geq 18$. This $n = 18$ can be improved to $n = 15$ by using an adaptation of the "Big Game–Small Game Decomposition" technique (see Section 7). The $15^4$ torus game is the smallest four-dimensional example that I know to be a Strong Draw.

**Theorem 6.** *We have*

$$(a)\quad \frac{\log 2}{2} n^2 \geq \mathbf{ww}(n\text{–line}) \geq \left(\frac{\log 2}{16} + o(1)\right) \frac{n^2}{\log n},$$

$$(b)\quad (\log 2) n^2 \geq \mathbf{ww}(\text{comb. } n\text{–line}) \geq \left(\frac{\log 2}{8} + o(1)\right) \frac{n^2}{\log n},$$

$$(c)\quad \mathbf{ww}(n\text{–line in torus}) = \left(\frac{\log 2}{\log 3} + o(1)\right) n,$$

$$(d)\quad \mathbf{ww}(\text{comb. } n\text{–line in torus}) = (1 + o(1))n,$$

$$(e)\quad \mathbf{ww}(n\text{–term A.P.}) = (2 + o(1))^n.$$

As we said before, the lower bounds ("Strong Draw") in (a) and (b) both follow from Theorem 5 combined with Theorem 1 (Theorem 3 combined with Theorem 1 gives a weaker constant factor).

The proofs of the lower bounds in (c) and (d) are much simpler: each one is a straightforward application of Theorem 3 (in fact, its Corollary 2). Well, I "cheated" a little bit: if $n$ is *even*, then the $n^d$-torus-hypergraph is *not* Almost Disjoint, so one cannot apply Theorem 3 directly to the case of (c) with even $n$. Nevertheless, any two Torus-Lines have at most two common points, and the *proof-technique* of Theorem 3 can be easily adapted to yield basically the same result as for Almost Disjoint hypergraphs.

A similar adaptation works for the lower bound in (e), see Beck [4].

The upper bounds ("Weak Win") in (a), (b), (c) with *odd* $n$, (d), and (e) all follow easily from Theorem Q. However, the case of "(c) with *even* $n$" is far less easy! Not only that the $n^d$-torus-hypergraph is *not* Almost Disjoint, but the Pair-Degree of this hypergraph is *exponentially large* (namely, $2^{d-1}$), which makes Theorem Q useless for this case. Indeed, let $\mathbf{P}$ be an arbitrary point of the $n^d$ Torus, and let $\mathbf{Q}$ be another point such that the coordinates of $\mathbf{P}$ are all shifted by $n/2$ (modulo $n$, of course): $\mathbf{Q} = \mathbf{P} + \mathbf{n/2}$ (mod $n$), where $\mathbf{n/2} = (n/2, n/2, \ldots, n/2)$; then there are exactly $2^{d-1}$ Torus-Lines containing both $\mathbf{P}$ and $\mathbf{Q}$. This $2^{d-1}$ is in fact the Pair-Degree. An application of Theorem Q gives that, if $|\mathcal{F}|/|V| = 3^{d-1}/n > 2^{n-3} \cdot 2^{d-1}$, then First Player can force a Weak Win; that is,

$$d \geq \left(\frac{\log 2}{\log 3 - \log 2} + o(1)\right) n$$

suffices for Weak Win. This bound is clearly weaker than what we stated in (c); Theorem Q is not powerful enough. Instead of Theorem Q we have to apply a much more complicated Weak Win criterion: see Theorem R below (Beck [6]).

Let $\mathcal{F}$ be an arbitrary finite hypergraph; for arbitrary integers $p \geq 2$ define the "big hypergraph" $\mathcal{F}_2^p$ as follows:

$$\mathcal{F}_2^p = \left\{ \bigcup_{i=1}^{p} A_i : \{A_1, \ldots, A_p\} \in \binom{\mathcal{F}}{p}, \left| \bigcap_{i=1}^{p} A_i \right| \geq 2 \right\}.$$

In other words, $\mathcal{F}_2^p$ is the family of all union sets $\bigcup_{i=1}^p A_i$ where $\{A_1, \ldots, A_p\}$ runs over all unordered $p$-tuples of distinct elements of $\mathcal{F}$ having at least 2 points in common. Note that even if $\mathcal{F}$ is an ordinary uniform hypergraph, i.e., a set has multiplicity 0 or 1 only and every set has the same size, $\mathcal{F}_2^p$ may become a *nonuniform multi-hypergraph* (i.e., a "big set" may have arbitrary *multiplicity*, not just 0 and 1, and the "big hypergraph" fails to remain uniform). More precisely, if $\{A_1, \ldots, A_p\}$ is an unordered $p$-tuple of distinct elements of $\mathcal{F}$ and $\{A'_1, \ldots, A'_p\}$ is another unordered $p$-tuple of distinct elements of $\mathcal{F}$, $|\bigcap_{i=1}^p A_i| \geq 2$, $|\bigcap_{j=1}^p A'_j| \geq 2$, and $\bigcup_{i=1}^p A_i = \bigcup_{j=1}^p A'_j$, i.e., $\bigcup_{i=1}^p A_i$ and $\bigcup_{j=1}^p A'_j$ are equal as *sets*, then they still represent *distinct* hyperedges of the "big hypergraph" $\mathcal{F}_2^p$.

For an arbitrary hypergraph $(V, \mathcal{H})$ write

$$T(\mathcal{H}) = \sum_{A \in \mathcal{H}} 2^{-|A|}.$$

If a set has multiplicity (say) $M$, then of course it shows up $M$ times in the summation.

**Theorem R** ($p$-power criterion for Weak Win). *If there exists a positive integer $p \geq 2$ such that*

$$\frac{T(\mathcal{F})}{|V|} > p + 4p \left(T(\mathcal{F}_2^p)\right)^{1/p},$$

*then in the Positional Game on hypergraph $(V, \mathcal{F})$ First Player can force a Weak Win.*

At first sight this criterion is completely "out of the blue", without any motivation ("deus ex machina"), hopelessly incomprehensible. The reader is wondering:

(i) Where did this criterion come from?
(ii) What was the motivation to conjecture it in the first place?
(iii) How to prove it?
(iv) Why is this complicated criterion so useful?

Well, one thing is clear, which gives at least a *partial* answer to question (ii): the lower index "2" in $\mathcal{F}_2^p$ is responsible for "controlling the Pair-Degree", and the hypothesis of the criterion means that a kind of "generalized Pair-Degree" is subtantially less than the Average Degree – this explains why Theorem R is a kind of "more sophisticated" version of Theorem Q.

To get a satisfying answer to questions (i)–(iii) above I refer the reader to Beck [6]; here I only answer question (iv) by an application: I apply Theorem R to the $n^d$-torus-hypergraph $\mathcal{F}_{n,d}^t$. Trivially

$$T(\mathcal{F}_{n,d}^t) = \frac{3^d - 1}{2} \cdot n^{d-1} \cdot 2^{-n} \quad \text{and} \quad |V| = n^d.$$

To estimate the more complicated sum $T\left((\mathcal{F}^t_{n,d})^p_2\right)$, where $p \geq 2$, I begin with a simple observation (I recommend the reader to study the proof of Statement 1 again).

*Observation:* Let **P** and **Q** be two arbitrary points of the $n^d$ Torus, and let $k$ denote the number of coordinates of $\mathbf{P} - \mathbf{Q}$ (mod $n$) which are different from zero. If these $k$ coordinates are all equal to $n/2$, then the number of Torus-Lines containing both **P** and **Q** is exactly $2^{k-1}$; otherwise there is at most one Torus-Lines containing both **P** and **Q**.

It follows from this *Observation* that

$$T\left((\mathcal{F}^t_{n,d})^p_2\right) \leq n^d \left(\sum_{k=1}^{d} \binom{d}{k} \binom{2^{k-1}}{p} 2^{-pn+2\binom{p}{2}}\right)$$

$$< \frac{n^d}{2^{pn-p^2}} \left(\sum_{k=0}^{d} \binom{d}{k} (2^p)^k\right) = \frac{n^d}{2^{pn-p^2}} (1+2^p)^d < \frac{3^{pd} n^d}{2^{p(n-p)}}. \tag{8.2}$$

Let $d_0$ be the least integer such that $3^{d_0} \geq n^2 2^{n+1}$; this means $d_0 = (\log 2/\log 3)n + O(\log n)$. By choosing $p = p_0 = (2/\log 3)\log n + O(1)$ and $d = d_0$, in view of (8.2) we obtain that both inequalities below

$$\frac{T(\mathcal{F}^t_{n,d})}{|V|} \geq n \tag{8.3}$$

and

$$\left(T\left((\mathcal{F}^t_{n,d})^p_2\right)\right)^{1/p} \leq \frac{3n^{d/p}}{2^{n-p}} < 1 \tag{8.4}$$

hold at the same time. (8.3) and (8.4) together imply that Theorem R applies, and guarantees a First Player's Weak Win in the $n^{d_0}$ Torus game. Finally note that, if $d > d_0$, then the application of Theorem R is even simpler; I leave the trivial calculations to the reader. This completes the proof of the missing part "(c) with *even n*: upper bound" of Theorem 6.

This concludes our long journey from arithmetic progressions to Tic-Tac-Toe-like games.

## References

1. Alon, N., Spencer, J.: *The Probabilistic Method*. Academic Press, New York (1992)
2. Baumgartner, J., Galvin, F., Laver, R., McKenzie, R.: Game theoretic versions of partition relations. In: Hajnal, A., Rado, R., Sós, V. T. (eds.) *Infinite and Finite Sets*. Colloq. Math. Soc. János Bolyai, vol. 10, pp. 131–135. North-Holland, Amsterdam (1973)
3. Beck, J.: On positional games. *J. Comb. Theory, Ser. A* **30**, 117–133 (1981)
4. Beck, J.: Van der Waerden and Ramsey type games. *Combinatorica* **2**, 103–116 (1981)
5. Beck, J.: An algorithmic approach to the Lovász Local Lemma. I. *Random Struct. Algorithms* **2**, 343–365 (1991)
6. Beck, J.: Positional games and the second moment method. *Combinatorica* **22**, 169–216 (2002)
7. Berlekamp, E.R.: A construction for partitions which avoid long arithmetic progressions. *Can. Math. Bull.* **11**, 409–414 (1968)
8. Berlekamp, E.R., Conway, J.H., Guy, R.K.: *Winning Ways*, vol. I and II. Academic Press, London (1982)
9. Erdős, P.: Some remarks on the theory of graphs. *Bull. Am. Math. Soc.* **53**, 292–294 (1947)
10. Erdős, P.: On a combinatorial problem, I. *Nordisk Mat. Tidskr.* **11**, 5–10 (1963)

11. Erdős, P., Lovász, L.: Problems and results on 3-chromatic hypergraphs and some related questions. In: Hajnal, A., Rado, R., Sós, V. T. (eds.) *Infinite and Finite Sets*. Colloq. Math. Soc. Janos Bolyai, vol. 10, pp. 609–627. North-Holland, Amsterdam (1975)
12. Erdős, P., Selfridge, J.: On a combinatorial game. *J. Comb. Theory Ser. A* **14**, 298–301 (1973)
13. Golomb, S.W., Hales, A.W.: Hypercube tic-tac-toe. In: Nowakowski, R. J. (ed.) *More on Games of No Chance*. Math. Sci. Res. Inst. Publ., vol. 42, pp. 167–182. Cambridge University Press, Cambridge (2002)
14. Gowers, T.: A new proof of Szemerédi's theorem. *Geom. Funct. Anal.*, **11**, 465–588 (2001)
15. Graham, R.L., Rothschild, B.L., Spencer, J.H.: *Ramsey Theory*. Wiley-Interscience Series in Discrete Mathematics. Wiley, New York (1980)
16. Hales, A.W., Jewett, R.I.: On regularity and positional games. *Trans. Am. Math. Soc.* **106**, 222–229 (1963)
17. Patashnik, O.: Qubic: $4 \times 4 \times 4$ tic-tac-toe. *Math. Mag.* **53**, 202–216 (1980)
18. Schmidt, W.M.: Two combinatorial theorems on arithmetic progressions. *Duke Math. J.* **29**, 129–140 (1962)
19. Shelah, S.: Primitive recursive bounds for van der Waerden numbers. *J. Am. Math. Soc.* **1**, 683–697 (1988)
20. van der Waerden, B.L.: Beweis einer Baudetschen Vermutung. *Nieu Arch. Wiskd.* **15**, 212–216 (1927)
21. Zermelo, E.: Über eine Anwendung der Mengenlehre und der Theorie des Schachspiels. In: *Proceedings of the Fifth International Congress of Mathematicians*, vol. 2, pp. 501–504. Cambridge University Press, Cambridge (1913)

*Diophantine Approximation*
H. P. Schlickewei et al., Editors


# METRIC DISCREPANCY RESULTS FOR SEQUENCES $\{n_k x\}$ AND DIOPHANTINE EQUATIONS

István Berkes[1], Walter Philipp[2], and Robert F. Tichy[3]
[1] *Institut für Statistik, Technische Universität Graz, Steyrergasse 17, 8010 Graz, Austria*
berkes@tugraz.at
[2] *Department of Statistics, University of Illinois at Urbana-Champaign, Champaign, Illinois, USA*
[3] *Institut für Mathematik A, Technische Universität Graz, Steyrergasse 30, 8010 Graz, Austria*
tichy@tugraz.at

Dedicated to Professor Wolfgang M. Schmidt on the occasion of his 70th birthday.

## 1 Introduction

Let $(n_k)$ be an increasing sequence of positive integers. For $0 \le x \le 1$, set

$$\eta_k = \eta_k(x) := n_k x \pmod 1 . \tag{1}$$

The discrepancy of the first $N$ elements of the sequence $(\eta_k)$ is defined as

$$D_N = D_N(x) := \sup_{0 \le s \le 1} \left| \frac{1}{N} \operatorname{card}(k \le N : \eta_k(x) \le s) - s \right| . \tag{2}$$

In his fundamental paper on uniform distribution mod 1, H. Weyl [23] proved, among many other things, that $D_N(x) \to 0$ for almost all $x \in (0, 1)$, i.e., that $(n_k x)$ is uniformly distributed mod 1 for all $x \in (0, 1)$ except for a set of Lebesgue measure zero. This was later improved independently by Cassels [5] and by Erdős and Koksma [7] who proved that for almost all $x \in (0, 1)$

$$N D_N(x) = O(N^{1/2} (\log N)^{5/2+\varepsilon}), \qquad \varepsilon > 0.$$

The best result so far has been achieved by R. C. Baker [1], who reduced the exponent $\frac{5}{2}$ of the logarithm to $\frac{3}{2}$. The exact exponent of the logarithm is still an open problem, except for the fact that it cannot be less than $\frac{1}{2}$, as was shown by Berkes and Philipp [4].

*Keywords.* Discrepancy, lacunary sequences, Diophantine equations, law of iterated algorithm.

*2000 Mathematics subject classification.* 11K38, 11D45, 42A55, 60F15.

Determining the exact order of magnitude of $D_N(x)$ for a concrete sequence $(n_k)$ is generally a hard problem and a satisfactory solution exists only in a few special cases. For $n_k = k$, Kesten [15] proved that

$$ND_N(x) \sim \frac{2}{\pi^2} \log N \log\log N$$

in measure. (For the remainder term, see Schoissengeier [20].) Another important case when a sharp bound for the magnitude of $D_N(x)$ is known is the case when $(n_k)$ is a lacunary sequence. Philipp [17,18] proved that if $(n_k)$ satifies the Hadamard gap condition

$$n_{k+1}/n_k \geq 1 + \rho, \quad \rho > 0, \ k = 1, 2, \ldots, \tag{3}$$

then we have for almost all $x$

$$\frac{1}{4} \leq \limsup_{N\to\infty} \frac{ND_N(x)}{\sqrt{N \log\log N}} \leq C(\varrho), \tag{4}$$

where $C(\varrho) \ll \frac{1}{\varrho}$. This result has an obvious probabilistic flavor. If $(x_n)$ is a sequence of independent random variables uniformly distributed (in the probabilistic sense) over $(0, 1)$, then by the classical Chung–Smirnov law of the iterated logarithm for empirical distribution functions (see, e.g., Shorack and Wellner [21, p. 504]), the discrepancy $D_N^*$ of $(x_n, n \leq N)$ satisfies

$$\limsup_{N\to\infty} \frac{ND_N^*}{\sqrt{N \log\log N}} = \frac{1}{\sqrt{2}} \tag{5}$$

with probability one. Thus, roughly speaking, under the Hadamard gap condition (3) the sequence $n_k x$ (mod 1) behaves like a sequence of independent random variables. This heuristics, which plays an important role in harmonic analysis and is the key for understanding a number of interesting phenomena (see, e.g., Kac [13]), should be used, however, with great care. Berkes and Philipp [4] have constructed sequences $(n_k)$ satisfying (3) for which the lower bound $\frac{1}{4}$ in (4) can be improved to $c \log\log \frac{1}{\varrho}$ with an absolute constant $c$. Hence there cannot be an upper bound (4), independent of $\varrho$, that works for all sequences $(n_k)$ satisfying a Hadamard gap condition (3). A deeper analysis of the problem shows that under the Hadamard gap condition (3), the behavior of $D_N$ is determined by a delicate interplay between the speed of growth and the number-theoretic properties of $(n_k)$. If $(n_k)$ grows extremely rapidly, then $\{n_k x\}$ is indeed a nearly i.i.d. (independent, identically distributed) sequence of random variables as one can easily see from the mixing relation

$$\lim_{n\to\infty} |x \in (\alpha, \beta) : \{nx\} \leq t| = (\beta - \alpha)t \qquad (0 \leq \alpha < \beta \leq 1),$$

where $|\cdot|$ stands for the Lebesgue measure. See Philipp [18, Lemma 4.2.1], where a remainder term is also given. Specifically, if

$$\sum_{k=1}^{\infty} n_k/n_{k+1} < \infty,$$

then the limsup in (4) is $1/\sqrt{2}$, in accordance with (5). (This follows easily using the approximation method in Berkes [2].) If, however, $n_{k+1}/n_k$ is bounded, then the arithmetic structure of $(n_k)$ comes into play. The number-theoretic effect becomes particularly clear if in (2) we compute the right hand side for a single $s$ only (without the sup), i.e., if we study the behavior of the sum

$$\sum_{k \le N} f(n_k x), \tag{6}$$

where $f = I_{(0,s)} - s$, extended with period 1. (Here, and in the sequel, $I_A(\cdot)$ denotes the indicator function of the set $A$.) In the case $n_k = 2^k$ the sum in (6) is asymptotically normally distributed, as shown by Kac [12]. The corresponding LIL (law of the iterated logarithm)

$$\limsup_{N\to\infty} \frac{1}{\sqrt{N \log\log N}} \sum_{k \le N} f(n_k x) = \gamma \quad \text{a.e.} \tag{7}$$

is also valid, where $\gamma = \gamma(s)$ is an explicitly computable constant (see, e.g., Berkes and Philipp [3]). Thus in this case the behavior of (6) is the same as that of sums of independent random variables. On the other hand, Erdős and Fortet showed (see [13, p. 646]) that for $n_k = 2^k - 1$ both the central limit theorem and the LIL (7) break down; in fact, the limsup in (7) is not any more a constant almost everywhere. This interesting phenomenon was cleared up by Gaposhkin [11], who showed that the sum in (6) satisfies the central limit theorem for all "nice" functions $f$ if and only if for any fixed nonzero integers $a, b, c$ the number of solutions of the Diophantine equation

$$an_\nu + bn_\mu = c$$

is bounded by a constant $C = C(a, b)$. For concrete sequences $(n_k)$, this criterion is usually not easy to verify, but one can give simple sufficient criteria for its validity. For example, the above Diophantine condition is satisfied if for any rational $r$ and any positive integer sequences $k_m, l_m$ tending to infinity, the limit relation

$$\lim_{m\to\infty} n_{k_m}/n_{l_m} = r$$

can hold only if the fraction on the left side equals $r$ for $m \ge m_0$. In particular, the criterion is satisfied in each of the following cases:

(a) $\lim_{k\to\infty} n_{k+1}/n_k = \infty$
(b) $n_k | n_{k+1}$ for any $k$
(c) $\lim_{k\to\infty} n_{k+1}/n_k = \alpha$, where $\alpha^r$ is irrational for $r = 1, 2, \ldots$

In the case when Gaposhkin's Diophantine condition is satisfied, the corresponding LIL (7) is also valid (see Berkes and Philipp [3]).

The previous results give a fairly complete picture on the discrepancy $D_N(x)$ and the underlying probabilistic structure of $n_k x$ mod 1 in the case when $(n_k)$ satisfies the Hadamard gap condition (3). The purpose of this paper is to study the same problem for sub-Hadamard sequences, i.e., when $n_{k+1}/n_k \to 1$. This problem is considerably harder than the Hadamard case and very few results are known here. Philipp [19] proved, verifying a conjecture of R. C. Baker, that the LIL (4) holds for all Hardy–Littlewood–Pólya sequences $(n_k)$. These are defined as follows. Let $(q_1, q_2, \ldots, q_\tau)$

be a finite set of coprime positive integers and let $(n_k)$ be the multiplicative semigroup generated by $(q_1, q_2, \ldots, q_\tau)$ and arranged in increasing order. Thus

$$(n_k)_{k=1}^{\infty} = (q_1^{\alpha_1} q_2^{\alpha_2} \ldots q_\tau^{\alpha_\tau},\ \alpha_i \geq 0,\ 1 \leq i \leq \tau).$$

Then relation (4) holds with a constant $C(r)$ on the right side depending only on the number $r$ of primes involved in the prime factorization of $q_1, \ldots, q_\tau$. Note that Hardy–Littlewood–Pólya sequences grow fairly rapidly: they are subexponential but satisfy the gap condition

$$n_{k+1} - n_k \geq \frac{n_k}{(\log n_k)^\gamma} \qquad (k = 1, 2, \ldots)$$

for some $\gamma > 0$. (See Tijdeman [22].) The proof of Lemma 3 below will also show (cf. relation (15)) that $n_k \geq \exp(c_1 k^\alpha)$ for some $0 < \alpha < 1$.

In the opposite direction one can show (see Berkes and Philipp [4]) that for any $\varepsilon_k \to 0$, there exists a sequence $(n_k)$ of integers satisfying

$$n_{k+1}/n_k \geq 1 + \varepsilon_k, \quad k = 1, 2, \ldots$$

such that

$$\limsup_{N\to\infty} \frac{N D_N(x)}{\sqrt{N \log\log N}} = \infty$$

for almost every $x$. Hence no subexponential speed of growth can guarantee, by itself, the law of the iterated logarithm (4) for $D_N(x)$ and thus in the sub-Hadamard domain the LIL is an individual affair: its validity depends on the specific properties of the sequence $(n_k)$. While the Hardy–Littlewood–Pólya sequences are the only known examples for LIL behavior (4) in the subexponential domain, in this paper we will see that they represent the rule rather than the exception. Indeed, we will show that the "majority" of sequences $(n_k)$ (in a suitable statistical sense) with a minimal subexponential speed of growth satisfies the LIL, and this is the case for all "sufficiently irregular" $(n_k)$. These results will follow from Theorem 1 below, the main result of our paper, which gives explicit Diophantine conditions on $(n_k)$ guaranteeing the validity of (4). To formulate the theorem, we introduce the following conditions.

We will say that a sequence $(n_k)$ satisfies

*condition A*, if for any fixed nonzero integers $a, b, c$ the number of solutions of the Diophantine equation $an_\nu + bn_\mu = c$ is bounded by a constant $C = C(a, b)$;
*condition A**, if $(n_k)$ satisfies condition A with a constant $C(a, b)$ independent of $a, b$;
*condition B*, if there exist constants $0 < \alpha < \frac{1}{2}$ and $C > 0$ such that for each positive integer $b$ and for each $R \geq 1$ the number of solutions $(h, n_\nu)$ of the Diophantine equation $hn_\nu = b$ with $h \in \mathbb{N}$, $1 \leq h \leq R$ does not exceed $CR^\alpha$;
*condition C*, if there exist constants $0 < \beta < \frac{3}{2}$ and $C > 0$ such that for each $N \geq 1$ and for all fixed integers $h_i$ with $0 < |h_i| \leq N^4$, for $i = 1, 2, 3, 4$, the number of nondegenerate solutions of the Diophantine equation

$$h_1 n_{\nu_1} + h_2 n_{\nu_2} + h_3 n_{\nu_3} + h_4 n_{\nu_4} = 0 \tag{8}$$

subject to

$$1 \leq \nu_i \leq N, \qquad i = 1, 2, 3, 4 \tag{9}$$

does not exceed $CN^\beta$;

*condition G*, if there exists a constant $0 < \eta < 1$ such that for all $k \geq k_0 = k_0(\eta)$ we have

$$n_{k+k^{1-\eta}}/n_k \geq k. \tag{10}$$

Here, and in the sequel, $n_j$ is meant as $n_{[j]}$ if $j$ is not an integer. Condition A is Gaposhkin's necessary and sufficient condition for the nearly independent behavior of $\{n_k x\}$ in the Hadamard case; condition A* is the uniform version of A. Conditions B and C are analogous Diophantine conditions which play a basic role in the subexponential domain. Condition G is the growth condition we will assume throughout this paper; it implies $n_k \geq \exp(k^\eta)$ and thus it restricts our investigations to a zone under the exponential speed. It is easy to see that condition G is implied by the gap condition

$$n_{k+1}/n_k \geq 1 + ck^{-\alpha} \qquad (k = 1, 2, \ldots) \tag{11}$$

for any $0 < \eta < 1 - \alpha$ but, unlike (11), relation (10) does not require that all individual gaps $n_{k+1} - n_k$ are large.

With the above notations, we can formulate now our main result.

**Theorem 1.** *Assume $(n_k)$ is an increasing sequence of positive integers satisfying conditions B, C and G. Then there is a constant D, depending only on the constants $\alpha, \beta, \eta$ and C appearing in these conditions such that for almost all $x \in (0, 1)$*

$$\frac{1}{4} \leq \limsup_{N\to\infty} \frac{ND_N(x)}{\sqrt{N \log\log N}} \leq D.$$

In Section 2 we give some comments on the conditions B, C and G. We will show that they are satisfied by a wide class of sequences including the Hardy–Littlewood–Pólya sequences as well as "almost all" sequences (in some natural sense). This requires application of a recent version of the subspace theorem due to Evertse, Schlickewei and Schmidt [8]. The details of the proof including the whole probabilistic machinery, such as martingale inequalities and chaining arguments will be given in a subsequent paper which will also extend and simplify the methods used in Philipp [19].

The Diophantine condition in Theorem 1 involves equations of the type

$$a_1 n_{\nu_1} + \cdots + a_p n_{\nu_p} = b \tag{12}$$

with arbitrary nonzero coefficients $a_1, \ldots, a_p$. Equations of this type with coefficients $\pm 1$ play an important role in the study of lacunary exponential sums $\sum e^{2\pi i h n_k x}$; the idea goes back to Sidon (see, e.g., Kahane [14], Gaposhkin [10]). Bounds on $\sum e^{2\pi i h n_k x}$ lead, in turn, to bounds on the discrepancy $D_N(x)$, in view of the Erdős–Turán inequality. Hence bounding the number of solutions of the Diophantine equation (12) with coefficients $\pm 1$ will also lead to bounds for the discrepancy $D_N(x)$ of $n_k x$, although these bounds are necessarily cruder than the LIL (4), due to the "usual defect" of the Erdős–Turán inequality. However, for comparison with classical Sidon theory, we formulate here one such result.

**Theorem 2.** *Let $(n_k)$ be an increasing sequence of positive integers and let $p \geq 2$. Assume that there exists a constant $C > 0$, depending on p and the sequence $(n_k)$, such that the number of solutions of the Diophantine equation*

$$\pm n_{\nu_1} \pm \cdots \pm n_{\nu_p} = b$$

*is at most C for any $b \neq 0$. Then*

$$ND_N(x) = O\left(N^{1/2}(\log N)^{1+1/p}\right)$$

*for almost every $x$.*

As a byproduct of the proof of Theorem 1, we can get precise asymptotic results for lacunary sums (6) not only for centered indicator functions $f$ but for any function $f$ of bounded variation (BV). In particular, it is easy to get the central limit theorem and the law of the iterated logarithm (with a precise constant) for such sums. These results can be proved much simpler than Theorem 1 itself. For example, the LIL for the sums (6) expresses the LIL for the discrepancy of $(n_k x)$, where we consider a single interval $(0, s)$, instead of taking the sup for all intervals in (2). This special case can be handled relatively easily; the main difficulty in the proof of Theorem 1 is to get the uniformity over all subintervals of $(0, 1)$. Since sums $\sum f(n_k x)$ received considerable attention in harmonic analysis, it is worth formulating such a corollary of Theorem 1 and comparing it with the earlier theory.

**Theorem 3.** *Let $(n_k)$ be an increasing sequence of positive integers satisfying conditions C and G. Assume that*

$$f(x+1) = f(x), \quad \int_0^1 f(x)\,\mathrm{d}x = 0, \quad f \in BV(0,1)$$

*and*

$$\int_0^1 \left(\sum_{k \le N} f(n_k x)\right)^2 \mathrm{d}x \sim \sigma^2 N$$

*for some $\sigma > 0$. Then*

$$\frac{1}{\sqrt{N}} \sum_{k \le N} f(n_k x) \to \mathcal{N}(0, \sigma^2) \quad \text{in distribution}$$

*and*

$$\limsup_{N \to \infty} \frac{1}{(N \log\log N)^{1/2}} \sum_{k \le N} f(n_k x) = \sigma\sqrt{2} \quad a.e.$$

As we noted before, the asymptotic behavior of $\sum_{k \le N} f(n_k x)$ is completely known in the case when $(n_k)$ satisfies the Hadamard gap condition (3). In particular, a necessary and sufficient condition for the CLT (Central Limit Theorem) is the Diophantine condition A. On the other hand, practically nothing is known in the sub-Hadamard case, with the exception of a recent CLT of Fukuyama and Petit [9] concerning the Hardy–Littlewood–Pólya sequence. Theorem 3 gives a fairly complete description of the asymptotics of $\sum_{k \le N} f(n_k x)$ under the sub-Hadamard growth condition G. Note that instead of the two-term Diophantine condition A we assumed here the four-term condition C, but it is easy to see that Theorem 3 remains valid if condition C is replaced by A*. Hence a condition very close to A is essentially the right Diophantine assumption for the CLT also in the subexponential domain. A more detailed investigation of the situation will be given elsewhere.

In conclusion we make some comments on the permutation invariance of our results. Changing the order of terms of a sequence $(x_n)$ leads generally to a drastic change of its discrepancy (see, e.g., Kuipers and Niederreiter [16]). Thus it is unclear what happens in Theorem 1 if we permute the terms of the sequence $(n_k)$. On the other hand, the usual heuristics behind our theorems says that the sequence $n_k x \mod 1$ behaves like a sequence of independent, identically distributed random variables. As the i.i.d. character of a sequence is permutation-invariant, it is natural to expect that our results remain valid after any permutation of $(n_k)$. In the case when $(n_k)$ satisfies the Hadamard gap condition (3), this is indeed the case: the proof in Philipp [17] uses the multiplicative orthogonality of lacunary trigonometric series and thus it is permutation-invariant. However, the martingale method in the proof of Theorem 1 uses the increasing character of $(n_k)$ in an essential way, and thus the permutation invariance of Theorem 1 remains open. Using different methods, permutation invariance can be proved under a stronger form of the Diophantine condition C; we shall prove this in a subsequent paper.

## 2 Comments on conditions B, C and G

We first show that the Hardy–Littlewood–Pólya sequences satisfy conditions B, C and G.

**Lemma 1.** *Let $(n_k)$ be a Hardy–Littlewood–Pólya sequence. Then there is a constant $C > 0$ such that for each positive integer $b$ and for each $R \geq 1$ the number of solutions $(h, n_\nu)$ of the Diophantine equation*

$$hn_\nu = b \tag{13}$$

*with $h \in \mathbb{N}$, $1 \leq h \leq R$ does not exceed $C(\log R)^r$. Here $r$ is the number of primes involved in the prime factorization of $q_1, \dots, q_\tau$.*

*Proof.* Let $p_1, \dots, p_r$ be the primes appearing in the prime factorization of $q_1, \dots, q_\tau$ and write $b = p_1^{\alpha_1} \dots p_r^{\alpha_r} M$, where $M$ is not divisible by $p_1, \dots, p_r$. Then $n_\nu$ in (13) has the form $n_\nu = p_1^{\beta_1} \dots p_r^{\beta_r}$ with integers $\beta_i \geq 0$ and thus (13) implies that $\beta_i \leq \alpha_i$, $i = 1, \dots, r$ and

$$h = p_1^{\alpha_1 - \beta_1} \dots p_r^{\alpha_r - \beta_r} M = p_1^{\delta_1} \dots p_r^{\delta_r} M$$

with integers $\delta_i \geq 0$. Now $h \leq R$ implies $p_1^{\delta_1} \dots p_r^{\delta_r} \leq R/M \leq R$ and consequently $\delta_i \leq \log R / \log 2$, $i = 1, \dots, r$. Thus the number of $r$-tuples $(\delta_1, \dots, \delta_r)$ and consequently the number of $h$'s that can possibly yield a candidate $h$ for a solution $(h, n_\nu)$ of (13) is at most $(1 + \log R/\log 2)^r$. As $h$ in (13) determines $\nu$ uniquely, Lemma 1 is proved. □

**Lemma 2.** *A Hardy–Littlewood–Pólya sequence satisfies condition C with $\beta = 1$ and*

$$C = \exp(18^9(\tau + 1)),$$

*where $\tau$ denotes the number of generating elements of the sequence.*

*Proof.* The number of choices for $\nu_4$ in (8) is $N$ and thus the lemma follows from Theorem 1.1 of Evertse et al. [8] upon fixing $\nu_4$ and dividing (8) by $h_4 n_{\nu_4}$. □

**Lemma 3.** *A Hardy–Littlewood–Pólya sequence* $(n_k)$ *satisfies condition G.*

*Proof.* Let $q_1, \dots, q_\tau$ be the generating elements of $(n_k)$. Clearly, an element $n_j = q_1^{\delta_1} \dots q_\tau^{\delta_\tau}$ of the sequence $(n_k)$ satisfies $n_j \le R$ iff

$$\delta_1 \log q_1 + \dots + \delta_\tau \log q_\tau \le \log R$$

and thus the number $A(R)$ of elements of $(n_k)$ in $[0, R]$ equals the number of lattice points $(\delta_1, \dots, \delta_\tau)$ in the $\tau$-dimensional "tetrahedron"

$$x_1 \log q_1 + \dots + x_\tau \log q_\tau \le \log R, \qquad x_1 \ge 0, \dots, x_\tau \ge 0.$$

The volume of the tetrahedron is $c_1(\log R)^\tau$, where

$$c_1 = c_1(\tau) = \frac{1}{\tau! \log q_1 \dots \log q_\tau},$$

and thus by a well-known argument in analysis we have, as $R \to \infty$,

$$A(R) = c_1(\log R)^\tau + O((\log R)^{\tau-1}). \tag{14}$$

From (14) and the trivial relation $A(n_k) = k$ we get

$$\log n_k \sim \left(\frac{k}{c_1}\right)^{1/\tau}. \tag{15}$$

Formulas (14) and (15) and $\log kn_k \sim \log n_k$ imply that the number of $n_j$'s in the interval $[n_k, kn_k]$ is

$$c_1[(\log kn_k)^\tau - (\log n_k)^\tau] + O((\log kn_k)^{\tau-1}) \sim c_1\tau(\log k)(\log n_k)^{\tau-1}$$
$$\sim c_2 k^{(\tau-1)/\tau} \log k$$

as $k \to \infty$. Thus for $k \ge k_0$ we have

$$n_{k+2c_2k^{(\tau-1)/\tau} \log k} \ge kn_k$$

and consequently (10) holds with any $\eta < 1/\tau$. □

We now show that, in some sense, almost all sequences $(n_k)$ growing like a polynomial with a fixed large degree will satisfy conditions B and C. We shall construct these sequences by induction. Let $n_1 = 1$ and suppose that $n_1 < n_2 < \dots < n_{k-1}$ have already been constructed and satisfy

$$(j-1)^{50} < n_j \le j^{50} \qquad j = 1, 2, \dots, k-1. \tag{16}$$

Then the cardinality of the set of integers of the form

$$a_1 n_{\mu_1} + a_2 n_{\mu_2} + a_3 n_{\mu_3}$$

with $1 \le \mu_1, \mu_2, \mu_3 \le k-1$, $|a_1|, |a_2|, |a_3| \le k^{11}$ is at most $(2k^{11}+1)^3(k-1)^3 = O(k^{36})$. Hence the cardinality of the set of integers included in the set of rational numbers

$$\frac{1}{a}(a_1 n_{\mu_1} + a_2 n_{\mu_2} + a_3 n_{\mu_3}), \quad a \in \mathbb{Z} - \{0\}, \ |a| \le k^{11} \tag{17}$$

subject to $1 \le \mu_1, \mu_2, \mu_3 \le k-1$, $|a_1|, |a_2|, |a_3| \le k^{11}$ is $O(k^{47})$. Thus, the interval $((k-1)^{50}, k^{50}]$ contains at most that many integers of the form (17). This number is at most $O(1/k^2)$ times the total number of integers in the interval. Calling these numbers "bad", we choose now $n_k$ from the "good" integers (which constitute an overwhelming majority for $k$ large), and note that (16) is satisfied also for $j = k$. This construction yields an infinite increasing sequence $(n_k)$ with the property that for $k \geq k_0$ the Diophantine equation

$$a_1 n_{\mu_1} + a_2 n_{\mu_2} + a_3 n_{\mu_3} + a_4 n_{\mu_4} = 0 \tag{18}$$

with $1 \leq \mu_i \leq k$, $i = 1, 2, 3, 4$ and $\max(|a_i|, i = 1, 2, 3, 4) \leq k^{11}$ has no solution if one of the indices, say, $\mu_4$, equals $k$ and the corresponding factor $a_4 \neq 0$, while the other three indices $\mu_i$ are strictly less than $k$. Call this property NS (for "no solutions").

We now show that the constructed sequence $(n_k)$ satisfies condition C. Let $N \geq N_0$ be given and consider (8) subject to $0 < |h_i| \leq N^4$, $i = 1, 2, 3, 4$, fixed and $1 \leq \nu_1 \leq \nu_2 \leq \nu_3 \leq \nu_4 \leq N$. We can assume without loss of generality that $\nu_4 > 3N^{4/11}$, since otherwise (8) can have only $(3N^{4/11})^4 = O(N^\beta)$ solutions, where $\beta = 16/11 < 3/2$. We now distinguish cases regarding the relative size of the indices $\nu_i$. If $\nu_4 > \nu_3$, we set $k = \nu_4$. Then using property NS it follows that (8) has no solutions subject to $0 < |h_i| \leq N^4$, since then $|h_i| \leq N^4 < k^{11}$ by $k = \nu_4 > 3N^{4/11}$. (Note that the validity of NS has been established only for $k \geq k_0$, but by $k = \nu_4 > 3N^{4/11}$ this is satisfied if $N \geq N_0$. For the finitely many remaining values $1 \leq N < N_0$, condition C is trivially satisfied.) If $\nu_4 = \nu_3 > \nu_2$, then (8) reduces to

$$h_1 n_{\nu_1} + h_2 n_{\nu_2} + h^* n_{\nu_4} = 0 \tag{19}$$

with $h^* = h_3 + h_4 \neq 0$ since otherwise the proper subsum $h_3 n_{\nu_3} + h_4 n_{\nu_4}$ would vanish and thus the solution $(n_{\nu_1}, n_{\nu_2}, n_{\nu_3}, n_{\nu_4})$ of (8) would be degenerate. In (18) we set $a_1 = h_1, a_2 = h_2, a_3 = 0$, $a_4 = h^*$ and $\mu_4 = k$, and conclude that by property NS, (19) has no solutions since $|h^*| \leq 2N^4 < k^{11}$. If $\nu_4 = \nu_3 = \nu_2 > \nu_1$, then (8) reduces to

$$h_1 n_{\nu_1} + h^{**} n_{\nu_4} = 0 \tag{20}$$

with $h^{**} = h_2 + h_3 + h_4 \neq 0$ since otherwise we would have $h_1 = 0$, contrary to the assumption. In (18) we set $a_1 = h_1, a_2 = a_3 = 0, a_4 = h^{**}$ and $\mu_4 = k$, and conclude that (20) cannot have a solution since $|h^{**}| \leq 3N^4 < k^{11}$. Finally, if $\nu_4 = \nu_3 = \nu_2 = \nu_1$, then there are only $N$ possibilities for the 4-tuple $(\nu_1, \nu_2, \nu_3, \nu_4)$ and thus for fixed $h_i$ the number of solutions of (8) is at most $N$, regardless of the restrictions on $h_i$.

To verify that $(n_k)$ also satisfies condition B, let $R \geq 1, b \geq 1$ be given and consider the equation $hn_\nu = b$ with $h \in \mathbb{N}$, $1 \leq h \leq R$. If this equation has a solution $(h, n_\nu)$ at all, let $(l, n_\mu)$ denote its solution with the largest $\mu$. This means we have to study the Diophantine equation

$$hn_\nu - ln_\mu = 0 \tag{21}$$

subject to $\nu \leq \mu$ and $1 \leq h \leq R$. Set $k = \mu$. If $k \leq R^{1/4}$, then $\nu \leq R^{1/4}$, and since $\nu$ uniquely determines $h$ in (21), in this case the number of solutions $(h, n_\nu)$ of (21) does not exceed $R^{1/4}$, regardless of the restriction on $h$. If $k > R^{\frac{1}{4}}$, then by property NS, equation (21) has no solutions other than $h = l, \nu = \mu$, since from $\nu < \mu$, $1 \leq h \leq R$ it follows $1 \leq l \leq h \leq R \leq k^4 \leq k^{11}$. Thus the impossibility of a solution follows by setting in (18) $a_1 = a_2 = 0, a_3 = h$ and $a_4 = -l$.

If instead of (16) we require $n_j \in I_j$, where $I_1, I_2, \ldots$ are disjoint intervals on the positive line, each lying to the right of the preceding one, then the same construction will work as long as the length $|I_j|$ of the interval $I_j$ satisfies $|I_j| \geq j^{49}$. Specifically, if $n_1 < n_2 < \cdots < n_{k-1}$ are given, the number of "bad" choices for $n_k$ in the interval $I_k$ is $O(1/k^2)$ times the total number of integers in the interval, and thus if we choose $n_k$ at random, uniformly among all integers in the interval $I_k$, then the Borel–Cantelli lemma shows that with probability one, all choices for $k \geq k_0$ will be "good". Hence the above construction yields the following

**Corollary.** Let $I_1, I_2, \ldots$ be disjoint intervals on the positive line, each lying to the right of the preceding one, such that $|I_k| \geq k^{49}$, $k = 1, 2, \ldots$ and let $(n_k)$ be a random sequence such that $n_k$ is uniformly distributed over the integers of the interval $I_k$. Then $(n_k)$ satisfies conditions B and C with probability one.

With a proper choice of the intervals $I_k$ we can "regulate" the speed of growth of $(n_k)$; in fact, we can guarantee an arbitrarily prescribed speed of growth provided this exceeds that of $k^\gamma$, $\gamma$ large. Specifically, if $\phi(k)$, $k \geq 0$ is a sequence of integers with $\phi(0) = 0$ and $\phi(k) - \phi(k-1) > 2k^{49}$, $k = 1, 2, \ldots$, then choosing $I_k = [\phi(k) - k^{49}, \phi(k) + k^{49}]$ will imply $n_k \sim \phi(k)$. In particular, we can guarantee the validity of condition G as well.

**Acknowledgments.** Research by I.B. was supported by FWF grant S9603-N13 and OTKA grants T 43037, K 61052, and K 67961. Research by R.F.T. was supported by FWF grant S9603-N13.

**Note.** Walter Philipp passed away on 19 July 2006 at the age of 69 near Graz, Austria.

## References

1. Baker, R.C.: Metric number theory and the large sieve. *J. Lond. Math. Soc. Ser. II* **24**, 34–40 (1981)
2. Berkes, I.: On almost i.i.d. subsequences of the trigonometric system. In: Odell, E.W., Rosenthal, H.P. (eds.) *Functional Analysis: Proceedings of the Seminar at the University of Texas at Austin*. Lect. Notes Math., vol. 1332, pp. 54–63. Springer, Heidelberg (1987)
3. Berkes, I., Philipp, W.: An a.s. invariance principle for lacunary series $f(n_k x)$. *Acta Math. Acad. Sci. Hung.* **34**, 141–155 (1979)
4. Berkes, I., Philipp, W.: The size of trigonometric and Walsh series and uniform distribution mod 1. *J. Lond. Math. Soc. Ser. II.* **50**, 454–464 (1994)
5. Cassels, J.W.S.: Some metrical theorems in Diophantine approximation III. *Proc. Camb. Philos. Soc.* **46**, 219–225 (1950)
6. Drmota, M., Tichy, R.F.: *Sequences, Discrepancies and Applications*. Lect. Notes Math., vol. 1651. Springer, Heidelberg (1997)
7. Erdős, P., Koksma, J.F.: On the uniform distribution modulo 1 of sequences $(f(n, \vartheta))$. *Proc. K. Ned. Akad. Wet.* **52**, 851–854 (1949)
8. Evertse, J.-H., Schlickewei, R.H.-P., Schmidt, W.M.: Linear equations in variables which lie in a multiplicative group. *Ann. Math. (2)* **155**, 807–836 (2002)
9. Fukuyama, K., Petit, B.: Le théorème limite central pour les suites de R. C. Baker. *Ergodic Theory Dyn. Syst.* **21**, 479–492 (2001)
10. Gaposhkin, V.F.: Lacunary series and independent functions (in Russian). *Usp. Mat. Nauk* **21**(6), 3–82 (1966)
11. Gaposhkin, V.F.: On the central limit theorem for some weakly dependent sequences (in Russian). *Teor. Verojatn. Primen.* **15**, 666–684 (1970)
12. Kac, M.: On the distribution of values of sums of type $\sum f(2^k t)$. *Ann. Math.* **47**, 33–49 (1947)
13. Kac, M.: Probability methods in some problems of analysis and number theory. *Bull. Am. Math. Soc.* **55**, 641–665 (1949)
14. Kahane, J.: *Some Random Series of Functions*, 2nd edn. Cambridge University Press, Cambridge (1985)

15. Kesten, H.: The discrepancy of random sequences $\{kx\}$. *Acta Arith.* **10**, 183–213 (1964)
16. Kuipers, L., Niederreiter, H.: *Uniform Distribution of Sequences.* Wiley, New York (1974)
17. Philipp, W.: Limit theorems for lacunary series and uniform distribution mod 1. *Acta Arith.* **26**, 241–251 (1975)
18. Philipp, W.: A functional law of the iterated logarithm for empirical distribution functions of weakly dependent random variables. *Ann. Probab.* **5**, 319–350 (1977)
19. Philipp, W.: Empirical distribution functions and strong approximation theorems for dependent random variables. A problem of Baker in probabilistic number theory. *Trans. Am. Math. Soc.* **345**, 707–727 (1994)
20. Schoissengeier, J.: A metrical result on the discrepancy of $(n\alpha)$. *Glasg. Math. J.* **40**, 393–425 (1998)
21. Shorack, R., Wellner, J.: *Empirical Processes with Applications to Statistics.* Wiley, New York (1986)
22. Tijdeman, R.: On integers with many small prime factors. *Compos. Math.* **26**, 319–330 (1973)
23. Weyl, H.: Über die Gleichverteilung von Zahlen mod. Eins. *Math. Ann.* **77**, 313–352 (1916)

*Diophantine Approximation*
H. P. Schlickewei et al., Editors


# MAHLER'S CLASSIFICATION OF NUMBERS COMPARED WITH KOKSMA'S, II

Yann Bugeaud
*Université Louis Pasteur, U. F. R. de mathématiques, 7, rue René Descartes, 67084 Strasbourg, France*
bugeaud@math.u-strasbg.fr

À Wolfgang Schmidt, pour son soixante-dixième anniversaire

## 1 Introduction

Mahler [8], in 1932, and Koksma [7], in 1939, introduced two related measures of the degree of approximation of a complex transcendental number $\xi$ by algebraic numbers. Following Mahler [8], for any integer $n \geq 1$, we denote by $w_n(\xi)$ the supremum of the exponents $w$ for which

$$0 < |P(\xi)| < \mathrm{H}(P)^{-w}$$

has infinitely many solutions in integer polynomials $P(X)$ of degree at most $n$. Here, $\mathrm{H}(P)$ stands for the naïve height of the polynomial $P(X)$, that is, the maximum of the absolute values of its coefficients. Following Koksma [7], for any integer $n \geq 1$, we denote by $w_n^*(\xi)$ the supremum of the exponents $w$ for which

$$0 < |\xi - \alpha| < \mathrm{H}(\alpha)^{-w-1}$$

has infinitely many solutions in complex algebraic numbers $\alpha$ of degree at most $n$. Here, $\mathrm{H}(\alpha)$ stands for the naïve height of $\alpha$, that is, the naïve height of its minimal defining polynomial over $\mathbf{Z}$. Clearly, the functions $w_1$ and $w_1^*$ coincide.

For any integer $n \geq 2$ and any complex transcendental number $\xi$ we have

$$w_n^*(\xi) \leq w_n(\xi) \leq w_n^*(\xi) + n - 1. \tag{1}$$

The first inequality in (1) is easy (see, e.g., [4, Section 3.4]), and the second one is due to Wirsing [14]. Furthermore, Sprindžuk [13] established that $w_n(\xi) = w_n^*(\xi) = (n-1)/2$ holds for all $n \geq 1$ and almost all $\xi$ (in the sense of the Lebesgue measure on the complex plane). This raises the question whether there exist complex numbers $\xi$ such that $w_n^*(\xi) < w_n(\xi)$ for some integer $n \geq 2$. In 1976, R. C. Baker [1] gave a positive answer to this problem by proving that for any integer $n \geq 2$ the function $w_n - w_n^*$ can take any value in the interval $[0, (n-1)/n]$. This has been subsequently improved upon by Bugeaud [2], who showed that, for any integer

---

*Keywords.* Classification of complex numbers, roots of polynomial.

*2000 Mathematics subject classification.* 11J04.

$n \geq 3$, the set of values taken by the function $w_n - w_n^*$ contains the interval $[0, n/4]$. Like Baker's, his approach originates in two papers by Schmidt [10,11], where the existence of $T$-numbers is established (these are transcendental numbers $\xi$ for which $\limsup_{n\to+\infty} w_n(\xi)/n = +\infty$ and $w_n(\xi)$ is finite for every $n \geq 1$). The main novelty introduced in [2] is the use in the inductive construction of integer polynomials having two zeros very close to each other.

The above quoted results from [1] and [2] have been obtained by constructing suitable *real* numbers $\xi$ for which $w_n(\xi)$ and $w_n^*(\xi)$ differ. In the present paper, we are mainly interested in the approximation of *complex nonreal* transcendental numbers $\xi$. In this case, (1) can be replaced by the sharper inequalities

$$w_n^*(\xi) \leq w_n(\xi) \leq w_n^*(\xi) + \frac{n-1}{2}, \tag{2}$$

see [14] or Section 9.1 of [4]. Furthermore, we have $w_n(\xi) = w_n^*(\xi)$ for $1 \leq n \leq 3$. Now, we may ask whether there are complex nonreal numbers $\xi$ such that $w_n^*(\xi) < w_n(\xi)$ for some integer $n \geq 4$. A positive answer has been given by Baker [1] when $n$ is even, but his method does not seem to extend to the case of odd $n$. In the present paper, we show that the approach followed in [2] can be adapted to prove, for any odd integer $n \geq 5$, the existence of complex nonreal transcendental numbers $\xi$ with $w_n^*(\xi) < w_n(\xi)$. This confirms a guess made at the end of [2]. Our main tool is the construction of families of irreducible integer polynomials of odd degree having two complex nonreal roots very close to each other.

We further show how to use the integer polynomials introduced in [5] to improve upon some of the results established in [2] on the set of values taken by the functions $w_n - w_n^*$ on the set of real numbers.

## 2 Results

We begin by pointing out a direct consequence of our main result, given in Theorem 2 below.

**Theorem 1.** *Let $n \geq 4$ be an integer. Then there exist complex nonreal numbers $\xi$ such that $w_n^*(\xi) < w_n(\xi)$.*

For even values of $n$, Theorem 1 is due to Baker [1]. However, his method does not seem to extend (at least straightforwardly) to the case of odd values of $n$.

Theorem 1 is an immediate consequence of [1] and Theorem 2 below, which asserts the existence of complex nonreal numbers with special properties.

**Theorem 2.** *Let $n \geq 5$ be an odd integer and set $F(n) = (5n^3 + 5n^2 + 5n - 3)/2$. Let $w_n$ and $w_n^*$ be real numbers such that*

$$\begin{aligned} w_n^* + \frac{1}{2} &\leq w_n \leq w_n^* + \frac{n+5}{16}, \quad w_n > F(n), \quad (n = 5, 7), \\ w_n^* + 1 - \frac{9}{2n} &\leq w_n \leq w_n^* + \frac{n+5}{16}, \quad w_n > F(n), \quad (n \geq 9). \end{aligned} \tag{3}$$

*Then there exist complex nonreal numbers $\xi$ such that*

$$w_n^*(\xi) = w_n^* \text{ and } w_n(\xi) = w_n.$$

The main tool for establishing Theorem 2 is the construction of integer polynomials of odd degree having two complex nonreal roots very close to each other, and not close to the real axis (thus, two pairs of complex nonreal roots). Then, denoting by $\gamma_j$ one of these roots, we construct $\xi$ as the limit of a suitable sequence $\xi_j = (c_j + \gamma_j)/g_j$, where the $c_j$'s and the $g_j$'s are positive integers tending to infinity. To ensure that $\xi_j$ is sufficiently far away from the real axis, we have to choose $\gamma_j$ with a strong dependence on $g_j$. The integer polynomials we are using are the polynomials

$$P_{n,a,b}(X) = X\left(X^2 - 2X + b^2 + 1\right)^{(n-1)/2} \\ +2\left(a^2X^2 - 2a(a+1)X + 2a + 1 + a^2 + a^2b^2\right)^2$$

defined in Lemma 1 below, which have two roots very close to $1+ib+a^{-1}$. They have been constructed by suitably modifying the polynomials $X^n - 2(aX-1)^2$ used in [2]. As far as we know, no example of integer polynomials with two complex nonreal roots very close to each other appeared previously in the literature.

A result similar to Theorem 2 can be proved for even integers $n \geq 6$ by using in the inductive construction the polynomials

$$\left(X^2 - 2X + b^2 + 1\right)^{n/2} + 2\left(a^2X^2 - 2a(a+1)X + 2a + 1 + a^2 + a^2b^2\right)^2,$$

instead of the polynomials $P_{n,a,b}(X)$. However, a slightly sharper result follows by combining ideas from [1,2] and results from [5], see Theorem 4 below.

We take the opportunity of the present paper to point out how the results from [2] can be improved by using families of polynomials introduced in [5], where Theorem A below is established.

**Theorem A.** *Let $n \geq 6$ and $a$ be integers with $n$ even. Set*

$$\tilde{P}_{n,a}(X) := (X^{n/2} - aX + 1)^2 - 2X^{n-2}(aX-1)^2.$$

*If $a$ is large enough, then the polynomial $\tilde{P}_{n,a}(X)$ is irreducible and has two real roots in the disc of center $a^{-1} + a^{-1-n/2}$ and of radius $3a^{-n}$.*

Using Theorem A instead of Lemma 3 from [2], it is possible to improve Theorem 1 from [2] as follows.

**Theorem 3.** *Let $n \geq 6$ be an even integer. Let $\Delta$ be in $[1 - 1/n, n/2)$. Set $\mu := (n\Delta - n + 1)/(n - 2\Delta)$ and $G(n) = n(n+1)(n+2\mu) + 3n - 1$. Let $w_n$ be a real number with $w_n > G(n)$. Then there exist real numbers $\xi$ such that*

$$w_n(\xi) = w_n \quad \text{and} \quad w_n(\xi) = w_n^*(\xi) + \Delta.$$

The proof of Theorem 3 follows exactly the same lines as that of Theorem 3 from [2] combined with Section 6 of that paper. However, we work with algebraic numbers $\gamma_j$ having large heights, thus we have to modify the lower bound given in inequality (6) from [2]. This is the reason why $G(n)$ in Theorem 3 above is much larger than $F(n)$ occurring in Theorem 1 from [2].

The following corollary is an immediate consequence of our Theorem 3 and Theorem 1 from [1].

**Corollary 1.** *For any even integer $n \geq 6$, the set of values taken by the function $w_n - w_n^*$ contains the interval* $[0, n/2)$.

Using similar ideas as in [1], it is easy to adapt the proof of [2] to construct complex nonreal numbers $\xi$ for which $w_n(\xi)$ and $w_n^*(\xi)$ differ for some *even* integer $n \geq 6$. Indeed, we construct $\xi$ as a limit of algebraic numbers of degree $n$ of the form $\xi_j = (c_j + id_j + \gamma_j)/g_j$, where $i^2 = -1$, the $c_j$'s, $d_j$'s and $g_j$'s are positive integers and the $\gamma_j$'s are real algebraic numbers of degree $n/2$. As in [2], we can take for the $\gamma_j$'s roots of the polynomials $X^{n/2} - 2(aX-1)^2$ or $X^{n/2} - 2a^{n/2}$, for a suitable positive integer $a$. To improve upon the results obtained in this way when $n$ is divisible by 4 (as Theorem 3 above improves Theorem 1 from [2]), we can take for the $\gamma_j$'s roots of the polynomials $\tilde{P}_{n/2,a}(X)$. Baker [1] established that, for any even integer $n \geq 4$, the set of values taken by the function $w_n - w_n^*$ on complex nonreal numbers includes the interval $[0, (n-2)/(2n)]$. Proceeding as described above, we improve upon Baker's result for $n \geq 6$.

**Theorem 4.** *For any even integer $n \geq 6$, the set of values taken by the function $w_n - w_n^*$ on complex nonreal numbers contains the interval* $[0, n/16]$. *If, moreover, $n$ is divisible by* 4, *then this set contains the interval* $[0, n/8]$.

Theorems 2 to 4 show that inequalities (1) and (2) are close to be best possible. However, we are unable to solve completely the following question.

**Problem 1.** *Let $n \geq 2$ be an integer, and let $\Delta$ be in* $[0, n-1]$. *Does there exist a real number $\xi$ with $w_n(\xi) = w_n^*(\xi) + \Delta$? Let $n \geq 4$ be an integer, and let $\Delta$ be in* $[0, (n-1)/2]$. *Does there exist a complex nonreal number $\xi$ with $w_n(\xi) = w_n^*(\xi) + \Delta$?*

Problem 1 has been (nearly completely) solved for $n = 2$ and $n = 3$. Namely, it is established in [3] that the set of values taken by the function $w_2 - w_2^*$ (resp. $w_3 - w_3^*$) contains the half-open interval $[0, 1)$ (resp. $[0, 2)$). The key tool for the case $n = 3$ is a result of Evertse [6] on integer cubic polynomials having two real roots very close to each other. Evertse's result has been subsequently refined by Schönhage [12], but this improvement does not seem to imply (at least straightforwardly) the existence of real numbers $\xi$ with $w_3(\xi) = w_3^*(\xi) + 2$.

Presumably, it is possible to extend the range of values in (3) and to prove that, when $n \geq 6$ is congruent to 2 modulo 4, the set of values taken by $w_n - w_n^*$ on complex nonreal numbers include the interval $[0, n/8]$. Indeed, by a suitable modification of the polynomials $\tilde{P}_{n,a}(X)$ defined in Theorem A above, we can construct families of integer polynomials of even degree having two complex nonreal roots very close to each other, even closer than in Lemma 1 below. However, we failed to prove that these polynomials are irreducible.

It is apparent from [1,2] and the present work that Schmidt's construction offers much flexibility to confirm the existence of transcendental numbers with special properties of approximation. In this respect, there remain many interesting unanswered problems, including the following one. Recall that, for almost all real numbers $\xi$, the quality of approximation to $\xi$ by algebraic numbers of degree at most $n$ is the same as the quality of approximation to $\xi$ by algebraic *integers* of degree at most $n+1$.

**Problem 2.** *To construct real numbers which are strictly better approximable by algebraic integers of degree $n+1$ than by algebraic numbers of degree at most $n$. To construct real numbers which are strictly better approximable by algebraic numbers of degree at most $n$ than by algebraic integers of degree $n+1$.*

We point out that Roy [9] recently constructed real numbers that are unexpectedly badly approximable by cubic integers, but very well approximable by quadratic numbers.

The sequel of the paper is organized as follows. In Section 3, we establish the key lemma. Afterwards, in Section 4, we formulate Theorem 5, which is the main step in our inductive construction. Its proof is given in the same section. In Section 5 we derive Theorem 2 from Theorem 5. Finally, Section 6 is devoted to the proof of Theorem 3, whereas Theorem 4 is established in Section 7.

## 3 An auxiliary result

The key ingredient for the proof of Theorem 2 is the following lemma which asserts the existence of irreducible, monic integer polynomials having two (pairs of) complex roots very close to each other.

**Lemma 1.** *Let $n \geq 5$ be an odd integer. Let $a$ and $b$ be positive integers with $a \geq b/10$ and $b$ odd. If $a$ is sufficiently large, then the polynomial*

$$P_{n,a,b}(X) := X\left(X^2 - 2X + b^2 + 1\right)^{(n-1)/2} + 2\left(a^2X^2 - 2a(a+1)X + 2a + 1 + a^2 + a^2b^2\right)^2$$

*is irreducible and has two complex nonreal roots very close to each other, namely,*

$$\delta^+(n,a,b) := 1 + ib + a^{-1} + \frac{\sqrt{1+ib}}{2\sqrt{2}}(1+i)^{(n-1)/2} a^{-(n+7)/4} b^{(n-5)/4} + \varepsilon^+(n,a,b)$$

*and*

$$\delta^-(n,a,b) := 1 + ib + a^{-1} - \frac{\sqrt{1+ib}}{2\sqrt{2}}(1+i)^{(n-1)/2} a^{-(n+7)/4} b^{(n-5)/4} + \varepsilon^-(n,a,b),$$

*where $|\varepsilon^+(n,a,b)|, |\varepsilon^-(n,a,b)| \leq c_1(n)\, a^{-(n+9)/4} b^{(n-5)/4}$ for some constant $c_1(n)$, depending only on $n$. Furthermore, we have*

$$\left|P'_{n,a,b}\left(\delta^+(n,a,b)\right)\right| \asymp a^{-(n-9)/4} b^{(n+5)/4}. \tag{4}$$

*Proof.* Since $b$ is odd, the irreducibility of $P_{n,a,b}(X)$ follows from the Eisenstein criterion applied with the prime number 2. Then, we study the function $x \mapsto P_{n,a,b}(1 + ib + a^{-1} + x)$ in a neighbourhood of the origin and we use the inequality $a \geq b/10$ to show that $P_{n,a,b}(X)$ has two roots which can be expressed as stated above. The estimate (4) is a straightforward calculation. □

Actually, Lemma 1 holds under a slightly weaker condition than $a \geq b/10$, namely, it is enough to assume that $a \geq c_2(n) b^{(n-3)/(n+3)}$, for a suitable positive constant $c_2(n)$. Taking this into consideration yields a (very) slight improvement of Theorem 2. However, to avoid additional technical difficulties, we choose to keep the weaker assumption $a \geq b/10$.

We direct the reader to our previous work [2] for the other auxiliary results used in the proof of Theorem 2.

## 4 The inductive construction

Theorem 5 below gives an explicit inductive construction of sequences $(\xi_j)_{j \geq 1}$ of complex nonreal algebraic numbers of odd degree $n$. It will be proved in Section 5 that such sequences converge to complex nonreal numbers having the property stated in Theorem 2. We use in Theorem 5 the same notation as in Lemma 1, namely, we denote by $\delta^+(n, a, b)$ the root of the polynomial $P_{n,a,b}(X)$ defined in this lemma.

For any real numbers $a, b, c, d$ with $a < b$ and $c < d$, the set of complex points

$$\{x + iy : a < x < b, c < y < d\}$$

will often be called the rectangle $(a, b) \times (c, d)$ and will be denoted by $(a, b) \times (c, d)$.

The norm of an algebraic number means the product of its conjugates over $\mathbf{Q}$.

**Theorem 5.** *Let $n \geq 5$ be an odd integer. Let $\mu$, $\nu$ be real numbers with $1 \leq \mu \leq n-1$ and $\nu > 5$. Set $H(n) = (5n^3 + 5n^2 + 3n + 1)/2$ and let $\chi > H(n)$ be a real number. Then, there exist a positive real number $\lambda < 1/3$, prime numbers $g_1 \geq 11, g_2, \ldots$, positive integers $c_1, c_2, \ldots$, and positive even integers $d_1, d_2, \ldots$ such that the following conditions are satisfied. Writing $\gamma_j := \delta^+(n, [g_j^\mu], g_j + d_j)$ for $j = 1, 2, \ldots$, we have*

($I_j$) *$g_j$ does not divide the norm of $c_j + \gamma_j$ for $j \geq 1$.*
($II_1$) *$\xi_1 = (c_1 + \gamma_1)/g_1 \in (1, 2) \times (5, 6)$.*
($II_j$) *$\xi_j = (c_j + \gamma_j)/g_j$ belongs to the rectangle*

$$I_{j-1} = \left(\Re e\, \xi_{j-1} + \frac{1}{2} g_{j-1}^{-\nu}, \Re e\, \xi_{j-1} + \frac{5}{8} g_{j-1}^{-\nu}\right) \times \left(\Im m\, \xi_{j-1} + \frac{1}{2} g_{j-1}^{-\nu}, \Im m\, \xi_{j-1} + \frac{5}{8} g_{j-1}^{-\nu}\right).$$

($III_1$) *$|\xi_1 - \alpha| \geq 2\lambda\, \mathrm{H}(\alpha)^{-\chi}$ for any algebraic number $\alpha \neq \xi_1$ of degree $\leq n$.*
($III_j$) *$|\xi_j - \alpha| \geq \lambda\, \mathrm{H}(\alpha)^{-\chi}$ for any algebraic number $\alpha \notin \{\xi_1, \ldots, \xi_j\}$ of degree $\leq n$ for $j \geq 2$.*

Observe that, under the assumption of Theorem 5, we have $g_j + d_j \leq 10[g_j^\mu]$ and $g_j + d_j$ is odd for any $j \geq 1$. Thus, we can indeed apply the results established in Lemma 1. Furthermore, setting $s = 4n$, we check that there exists a positive constant $c_3(n)$ such that

$$\mathrm{H}(\gamma_j) \leq c_3(n)\, g_j^s, \qquad \text{for any } j \geq 1. \tag{5}$$

Since we will also deal in the present work with other families of algebraic numbers playing the same role as the $\gamma_j$'s, it is convenient to introduce the parameter $s$, see Section 6 and the remark at the end of the present section.

To simplify the notation, in what follows we denote by $\alpha$ a complex algebraic number of degree less than or equal to $n$. By the definitions of $H(n)$ and of $s$, there exists a positive real number $\varepsilon$ such that

$$2\chi > n(n+1)(n+s) + 3n + 1 + 5n^2 s\varepsilon. \tag{6}$$

In order to prove Theorem 5, we add three extra conditions $(IV_j)$, $(V_j)$ and $(VI_j)$, which should be satisfied by the numbers $\xi_j$. We denote by Leb the Lebesgue measure on the complex plane.

Let $J_j$ denote the subset of $I_j$ consisting of the complex numbers $z = x + iy \in I_j$ satisfying

$$\max\{|x - \Re e\, \alpha|, |y - \Im m\, \alpha|\} \geq 2\lambda\, \mathrm{H}(\alpha)^{-\chi}$$

for any algebraic number $\alpha$ of degree $\leq n$, distinct from $\xi_1, \ldots, \xi_j, z$ and of height $\mathrm{H}(\alpha)$ satisfying

$$\mathrm{H}(\alpha) \geq (\lambda g_j^{\nu})^{1/\chi}.$$

The supplementary conditions are the following.

$(IV_j)$ $\xi_j \in J_{j-1}$ $(j \geq 2)$.
$(V_j)$ If $\mathrm{H}(\alpha) \leq g_j^{2/(n+1+\varepsilon)}$, then we have $|\xi_j - \alpha| \geq 1/g_j$ for $j \geq 1$.
$(VI_j)$ The measure of $J_j$ satisfies $\mathrm{Leb}(J_j) \geq \mathrm{Leb}(I_j)/2$ for $j \geq 1$.

We construct the sequences $c_1, c_2, \ldots, d_1, d_2, \ldots, g_1, g_2, \ldots$ by induction. At the $j$-th stage, there are two distinct steps. Step $(A_j)$ consists in building an algebraic number

$$\xi_j = \frac{c_j + \gamma_j}{g_j}$$

of degree $n$ satisfying conditions $(I_j)$ to $(V_j)$. In step $(B_j)$, we show that the number $\xi_j$ constructed in $(A_j)$ satisfies $(VI_j)$ as well, provided that $g_j$ is chosen large enough in terms of

$$n, \mu, \nu, \chi, \varepsilon, \lambda, \xi_1, \ldots, \xi_{j-1}. \tag{7}$$

The symbols $o$, $\gg$ and $\ll$ used throughout steps $(A_j)$ and $(B_j)$ mean that the numerical implicit constants depend (at most) on the quantities (7). Furthermore, the symbol $o$ implies "as $g_j$ tends to infinity".

Step $(A_1)$ is rather easy. Let $g_1 > \max\{11, n\}$ be a prime number. There are $\gg g_1^2$ numbers $\xi_1 = (c_1 + \gamma_1)/g_1$ in the rectangle $(1, 2) \times (5, 6)$, since for any fixed $d_1$ there are $\gg g_1$ suitable choices for $c_1$. Observe that condition $(I_1)$ is satisfied if, and only if, $g_1$ does not divide $P(-c_1)$, where $P(X)$ denotes the minimal defining polynomial of $\gamma_1$. Thus, by Lemma 5 from [2], there are $\gg g_1^2$ numbers $\xi_1 = (c_1 + \gamma_1)/g_1$ in the rectangle $(1, 2) \times (5, 6)$ that satisfy condition $(I_1)$. These $\gg g_1^2$ numbers have mutual distances at least $g_1^{-1}$ and, since there are only $o(g_1^2)$ algebraic numbers $\alpha$ of degree at most $n$ satisfying $\mathrm{H}(\alpha) \leq g_1^{2/(n+1+\varepsilon)}$, one can choose $\xi_1$ such that $(V_1)$ is satisfied. Furthermore, we point out that there are $\gg g_1^2$ suitable choices for the pair $(c_1, d_1)$, where the constant implicit in $\gg$ depends only on $n$. Further, by Lemma 2 from [2], we have $|\xi_1-\alpha|\geq 2\lambda\,\mathrm{H}(\alpha)^{-n}$, with $\lambda=(n+1)^{-2(n+1)}\,\mathrm{H}(\xi_1)^{-n}/2$, for any algebraic number $\alpha\neq\xi_1$ of degree at most $n$. Thus $(I_1)$, $(II_1)$, $(III_1)$ and $(V_1)$ are satisfied.

Let $j\geq 2$ be an integer and assume that $c_1,\ldots,c_{j-1},d_1,\ldots,d_{j-1},g_1,\ldots,g_{j-1}$ have been constructed. Step $(A_j)$ is much harder to verify, since we have no control on the set $J_{j-1}$. Thus, it seems difficult to check that the condition $(IV_j)$ holds. To overcome this problem, we follow Schmidt's argument [11], also used by Baker [1]. We set $\xi_j=(c_j+\gamma_j)/g_j$ for some positive integers $c_j$, $d_j$ (recall that the definition of $\gamma_j$ requires an integer $d_j$) and $g_j>8g_{j-1}$ and we introduce the set $J'_{j-1}$ formed by the complex numbers $z=x+iy\in I_{j-1}$ satisfying

$$\max\{|x-\Re e\,\alpha|,|y-\Im m\,\alpha|\}\geq 2\lambda\,\mathrm{H}(\alpha)^{-\chi}$$

for any algebraic number $\alpha$ of degree $\leq n$, distinct from $\xi_1,\ldots,\xi_j,z$, and whose height $\mathrm{H}(\alpha)$ satisfies the inequalities

$$(\lambda\,g_{j-1}^{\nu})^{1/\chi}\leq \mathrm{H}(\alpha)\leq (c_4(n)\,g_j^{n(n+s)})^{1/(\chi-n)}, \tag{8}$$

for a suitable constant $c_4(n)$ which will be defined just after inequality (13). Since, by (6), we have

$$\chi-n>n(n+1)(n+s)/2, \tag{9}$$

the exponent of $g_j$ in the right member of (8) is strictly less than $2/(n+1)$. Thus, there are $o(g_j^2)$ algebraic numbers $\alpha$'s satisfying (8), and we observe that, unlike for $J_{j-1}$, the complement in $I_{j-1}$ of the set $J'_{j-1}$ is a finite union of very small rectangles, and, more precisely, a union of $o(g_j^2)$ rectangles.

We will prove that for $g_j$ large enough we have $\gg g_j^2$ choices for the pair $(c_j,d_j)$ in order that conditions $(I_j)$ to $(V_j)$ are fulfilled. We stress that if $\xi'_j=(c'_j+\gamma'_j)/g_j$ for some positive integers $c'_j$ and $d'_j$, then we have

$$\xi'_j-\xi_j=\frac{c'_j-c_j}{g_j}+i\,\frac{d'_j-d_j}{g_j}+O\left(g_j^{-7/2}\right), \tag{10}$$

by Lemma 1.

Let $\alpha$ be an algebraic number of degree $\leq n$. Since $g_j+d_j\leq 10g_j$, we infer from (5) and Lemmas 2 and 4 from [2] that there exist positive constants $c_5(n)$ and $c_6(n)$ such that

$$|\xi_j-\alpha|\geq c_5(n)\,\mathrm{H}(\xi_j)^{-n}\,\mathrm{H}(\alpha)^{-n}\geq c_6(n)\,g_j^{-n(n+s)}\,\mathrm{H}(\alpha)^{-n}. \tag{11}$$

In particular, using that $2\sqrt{2}\lambda<1$, we have

$$|\xi_j-\alpha|\geq 2\sqrt{2}\lambda\,\mathrm{H}(\alpha)^{-\chi} \tag{12}$$

as soon as

$$\mathrm{H}(\alpha)^{\chi-n}\geq c_6(n)^{-1}\,g_j^{n(n+s)}. \tag{13}$$

We take $c_4(n)=c_6(n)^{-1}$.

By $(VI_{j-1})$ and $J'_{j-1}\supset J_{j-1}$, we have $\mathrm{Leb}(J'_{j-1})\gg 1$. Since the complement in $I_{j-1}$ of the set $J'_{j-1}$ is the union of $o(g_j^2)$ rectangles, if $g_j$ is a sufficiently large prime number, then, using (10) and Lemma 5 from [2] as in step $(A_1)$, we get that there exist $\gg g_j^2$ numbers $\xi_j = (c_j + \gamma_j)/g_j$ in $J'_{j-1}$ such that $(I_j)$ is satisfied. Such $\xi_j$'s also belong to $J_{j-1}$, since (13) implies (12), and condition $(IV_j)$ is satisfied.

Thus, we are left with $\gg g_j^2$ suitable algebraic numbers $\xi_j$, mutually distant by at least $g_j^{-1}$, as follows from (10). Only $o(g_j^2)$ algebraic numbers $\alpha$ of degree at most $n$ satisfy

$$\mathrm{H}(\alpha) \leq g_j^{2/(n+1+\varepsilon)}, \tag{14}$$

thus one can choose $\xi_j$ in such a way that $|\xi_j - \alpha| \geq 1/g_j$ for the numbers $\alpha$ satisfying (14). Consequently, there are $\gg g_j^2$ algebraic numbers $\xi_j$ satisfying $(I_j)$, $(II_j)$, $(IV_j)$ and $(V_j)$.

To prove that such a $\xi_j$ also satisfies $(III_j)$, we argue exactly as in [2]. We omit the details. Thus, the proof of step $(A_j)$ is completed.

Let $j \geq 1$ be an integer. For the proof of step $(B_j)$, we first establish that if $g_j$ is large enough and if $z$ lies in $I_j$, then we have

$$|z - \alpha| \geq 2\sqrt{2}\lambda \mathrm{H}(\alpha)^{-\chi} \tag{15}$$

for any algebraic number $\alpha \neq \xi_j$ such that

$$(\lambda g_j^{\nu})^{1/\chi} \leq \mathrm{H}(\alpha) \leq g_j^{\nu/(\chi-(n+1)/2-\varepsilon)}. \tag{16}$$

Let then $\alpha \neq \xi_j$ be an algebraic number of degree $\leq n$ satisfying (16) and let $z = x + iy$ be in $I_j$, that is, such that

$$\begin{aligned} 1/2\, g_j^{-\nu} &< x - \Re e\, \xi_j < 5/8\, g_j^{-\nu} \\ 1/2\, g_j^{-\nu} &< y - \Im m\, \xi_j < 5/8\, g_j^{-\nu}. \end{aligned} \tag{17}$$

If $g_j^{\nu/(\chi-(n+1)/2-\varepsilon)} \leq g_j^{2/(n+1+\varepsilon)}$, then $\mathrm{H}(\alpha) \leq g_j^{2/(n+1+\varepsilon)}$ and it follows from $(V_j)$, (16) and (17) that

$$\begin{aligned} |z - \alpha| &\geq |\xi_j - \alpha| - |\xi_j - z| \\ &\geq g_j^{-1} - g_j^{-\nu} \geq 2\sqrt{2} g_j^{-\nu} \geq 2\sqrt{2}\lambda \mathrm{H}(\alpha)^{-\chi}. \end{aligned} \tag{18}$$

Otherwise, we have

$$g_j^{\nu/(\chi-(n+1)/2-\varepsilon)} > g_j^{2/(n+1+\varepsilon)}, \tag{19}$$

and, by (11), we get

$$\begin{aligned} |z - \alpha| &\geq |\xi_j - \alpha| - |\xi_j - z| \\ &\geq c_6(n)\, g_j^{-n(n+s)}\, \mathrm{H}(\alpha)^{-n} - g_j^{-\nu} \\ &\geq c_6(n)\, g_j^{-n(n+s)}\, \mathrm{H}(\alpha)^{-n}/2, \end{aligned} \tag{20}$$

To check the last inequality, we have to verify that

$$2g_j^{-\nu} \leq c_6(n)\, g_j^{-n(n+s)}\, \mathrm{H}(\alpha)^{-n}. \tag{21}$$

In view of (16), inequality (21) is true as soon as

$$2g_j^{n\nu/(\chi-(n+1)/2-\varepsilon)} \leq c_6(n)\, g_j^{\nu}\, g_j^{-n(n+s)},$$

which, by (19), holds for $g_j$ large enough when

$$\frac{n}{\chi-(n+1)/2-\varepsilon}<1-\frac{n(n+1+\varepsilon)(n+s)}{2\chi-n-1-2\varepsilon}, \tag{22}$$

in particular when $\chi$ satisfies (6).

Moreover, we have

$$c_6(n)\, g_j^{-n(n+s)}\, \mathrm{H}(\alpha)^{-n} \geq 4\sqrt{2}\lambda \mathrm{H}(\alpha)^{-\chi}. \tag{23}$$

Indeed, by (16), $\lambda<1$ and (19), we get

$$\begin{aligned}\mathrm{H}(\alpha)^{\chi-n} &\geq (\lambda g_j^{\nu})^{(\chi-n)/\chi}\\ &\geq \lambda\, g_j^{(\chi-n)(2\chi-n-1-2\varepsilon)/(\chi n+\chi+\chi\varepsilon)}\\ &\geq 4\sqrt{2}\lambda\, c_6(n)^{-1}\, g_j^{n(n+s)},\end{aligned}$$

since we infer from (6) that

$$(\chi-n)(2\chi-n-1-2\varepsilon)>\chi\, n\,(n+1+\varepsilon)\,(n+s). \tag{24}$$

Combining (20) and (23), we have checked that we have

$$|z-\alpha| \geq 2\sqrt{2}\lambda \mathrm{H}(\alpha)^{-\chi},$$

when (19) holds; hence, by (18), (15) is true if $\alpha \neq \xi_j$ satisfies (16). Consequently, if $g_j$ is large enough, then the complement $J_j^c$ of $J_j$ in $I_j$ is contained in the union of the rectangles

$$\begin{aligned}&\left(\Re e\, \alpha-2\lambda \mathrm{H}(\alpha)^{-\chi}, \Re e\, \alpha+2\lambda \mathrm{H}(\alpha)^{-\chi}\right)\\ &\quad\times\left(\Im m\, \alpha-2\lambda \mathrm{H}(\alpha)^{-\chi}, \Im m\, \alpha+2\lambda \mathrm{H}(\alpha)^{-\chi}\right),\end{aligned}$$

where $\alpha$ runs over the set of algebraic numbers of degree $\leq n$ and with height greater than $g_j^{\nu/(\chi-(n+1)/2-\varepsilon)}$. The Lebesgue measure of $J_j^c$ is then

$$\ll \sum_{H>g_j^{\nu/(\chi-(n+1)/2-\varepsilon)}} H^{n-2\chi} = o(g_j^{-2\nu}) = o(\mathrm{Leb}(I_j)).$$

Thus, we conclude that we can find $g_j$ large enough such that $\mathrm{Leb}(J_j) \geq \mathrm{Leb}(I_j)/2$. This completes step $(B_j)$ as well as the proof of Theorem 5.

*Remark.* Observe that the size of the function $n \mapsto H(n)$ occurring in the statement of Theorem 5 is implied by the conditions (9), (22), and (24), the most constraining one being (22).

Likewise, we can rework the proof of Theorem 3 from [2] using the upper bound $\mathrm{H}(\gamma_j) \leq c_3(n)\, g_j^s$. Then, modifying accordingly the inequalities displayed on p. 97, l.-5, on p. 99, l.-4, and on p. 100, l.5 from [2], we see that $\chi$ has to satisfy simultaneously

$$\begin{aligned}\chi-n &> n(n+1)(n+s),\\ n &< \chi-n-1-\varepsilon-n(n+s)(n+1+\varepsilon),\\ (\chi-n)(\chi-n-1-\varepsilon) &> \chi n(n+s)(n+1+\varepsilon).\end{aligned}$$

The most constraining inequality is the second one, which gives

$$\chi > n(n+s)(n+1) + 2n + 1 \tag{25}$$

when we omit $\varepsilon$. This observation will be used in Section 6.

## 5 Completion of the proof of Theorem 2

For $n = 5$ or 7, let $\Delta$ be in $[7/16, (n+5)/16]$ and set

$$\mu = \frac{n + 16\Delta - 7}{n - 16\Delta + 7}.$$

Observe that $\mu$ is in $[1, n-1]$. For $n \geq 9$, let $\Delta$ be in $[1 - 9/(2n), (n+5)/16]$. If $\Delta \in [1 - 9/(2n), (n-1)/16]$, then we set

$$\mu = \frac{4n\Delta - 3n + 9}{n - 9},$$

and we observe that $\mu$ is in $[1, (n-4)/4]$. If $\Delta \in [(n-1)/16, (n+5)/16]$, then we set

$$\mu = \frac{n + 16\Delta - 7}{n - 16\Delta + 7},$$

and we observe that $\mu$ is in $[(n-4)/4, n-1]$. Let $w_n > (5n^3 + 5n^2 + 5n - 3)/2$ and set $w_n^* = w_n - \Delta$. Set $\nu = n(w_n^* + 1)$ and $\chi = w_n - n + 2$, in such a way that $\chi > (5n^3 + 5n^2 + 3n + 1)/2$. The sequence $(\xi_j)_{j \geq 1}$ obtained by applying Theorem 5 with these parameters is a Cauchy sequence, thus it converges towards a complex number denoted by $\xi$. Our choice for the $\gamma_j$'s implies that $\xi$ is nonreal.

We write $A \ll B$ if there exists a constant $c(n)$, depending only on $n$, such that $|A| < c(n)B$, and we write $A \asymp B$ if we have both $A \ll B$ and $B \ll A$.

By the definition of $\gamma_j$, the minimal defining polynomial of $\xi_j$ is

$$\begin{aligned} Q_j(X) := {} & (g_j X - c_j)\Big((g_j X - c_j)^2 - 2(g_j X - c_j) + (g_j + d_j)^2 + 1\Big)^{(n-1)/2} + \\ & + 2\Big([g_j^{\mu}]^2 (g_j X - c_j)^2 - 2[g_j^{\mu}]([g_j^{\mu}] + 1)(g_j X - c_j) + \\ & + 2[g_j^{\mu}] + 1 + [g_j^{\mu}]^2 + [g_j^{\mu}]^2 (g_j + d_j)^2\Big)^2. \end{aligned}$$

This polynomial is indeed irreducible and primitive by $(I_j)$ and the first statement of Lemma 1.

Furthermore, for any $j \geq 1$ we have

$$g_j^{-\nu}/2 \leq |\xi - \xi_j| \leq 2 g_j^{-\nu},$$

and we deduce that

$$|\xi - \xi_j| \asymp \mathrm{H}(\xi_j)^{-\nu/n} \asymp \mathrm{H}(\xi_j)^{-w_n^* - 1}. \tag{26}$$

Further, if $\alpha$ is of degree $\leq n$ and is not equal to one of the $\xi_j$'s, then we have

$$|\xi - \alpha| \geq \lambda \mathrm{H}(\alpha)^{-\chi},$$

whence

$$|\xi - \alpha| \geq \mathrm{H}(\alpha)^{-w_n^*-1}, \tag{27}$$

since $\chi \leq w_n^* + 1$, by (1). It follows from (26) and (27) that

$$w_n^*(\xi) = w_n^*.$$

It now remains for us to prove that $w_n(\xi) = w_n$. Denote by $\xi_j = \beta_{j1}, \beta_{j2}, \ldots, \beta_{jn}$ the roots of the polynomial $Q_j(X)$. Observe that, for any $k \geq 1$, we have

$$\beta_{jk} = \frac{c_j + \delta_k}{g_j}$$

for a suitable root $\delta_k$ of the polynomial $P_{n,[g_j^\mu],g_j+d_j}(X)$. By (26) and Lemma 1, we get

$$\begin{aligned}
|Q_j(\xi)| &= g_j^n \cdot |\xi - \xi_j| \cdot \prod_{2\leq k\leq n} |\xi - \beta_{jk}| \\
&\asymp g_j^n \cdot |\xi - \xi_j| \cdot \prod_{2\leq k\leq n} |\xi_j - \beta_{jk}| \\
&\asymp g_j \cdot \mathrm{H}(\xi_j)^{-w_n^*-1} \cdot \left| P'_{n,[g_j^\mu],g_j+d_j}\left(\delta^+(n, [g_j^\mu], g_j + d_j)\right)\right| \\
&\asymp \mathrm{H}(\xi_j)^{-w_n^*-1} \cdot g_j^{(n+9)/4-\mu(n-9)/4}.
\end{aligned}$$

Assume first that $1 \leq \mu \leq (n-4)/4$. Then, we have $\mathrm{H}(\xi_j) = \mathrm{H}(Q_j) \asymp g_j^n$, which yields

$$|Q_j(\xi)| \asymp \mathrm{H}(Q_j)^{-w_n^*-1+(n+9)/(4n)-\mu(n-9)/(4n)},$$

hence

$$w_n(\xi) \geq w_n^* + \frac{3}{4} - \frac{9}{4n} + \frac{\mu}{4} - \frac{9\mu}{4n} = w_n^* + \Delta, \tag{28}$$

by the definition of $\Delta$. Assume now that $(n-4)/4 \leq \mu \leq n-1$. Then, we have $\mathrm{H}(\xi_j) = \mathrm{H}(Q_j) \asymp g_j^{4(\mu+1)}$, which yields

$$w_n(\xi) \geq w_n^* + \frac{7}{16} + \frac{n}{16} \cdot \frac{\mu - 1}{\mu + 1} = w_n^* + \Delta, \tag{29}$$

by the definition of $\Delta$. In order to show that the inequalities in (28) and (29) are indeed equalities, we argue exactly as Baker [1] did (see also [2]). We omit the details. This completes the proof of the theorem.

## 6 Proof of Theorem 3

The proof of Theorem 3 follows step by step that of Theorem 1 from [2]. Instead of working with the integer polynomials given in Lemma 3 from [2], we use Theorem A stated in Section 2. Let $n \geq 6$ be an even integer. Let $\Delta$ be in $[(n-1)/n, n/2)$ and set

$$\mu = \frac{n(\Delta - 1) + 1}{n - 2\Delta}.$$

We observe that $\mu$ is in $[0, +\infty)$. Keeping the notation from [2, Section 6], we take for $\gamma_j$ the root of

$$\tilde{P}_{n,a}(X) := (X^{n/2} - aX + 1)^2 - 2X^{n-2}(aX - 1)^2$$

nearest to $a^{-1} + a^{-1-n/2} + 2\sqrt{2}\,a^{-n}$, where $a = [g_j^{\mu}]$. Then, the minimal defining polynomial of $\xi_j$ is

$$Q_j(X) = \left((g_jX - c_j)^{n/2} - a(g_jX - c_j) + 1\right)^2$$
$$-2(g_jX - c_j)^{n-2}\left(a(g_jX - c_j) - 1\right)^2.$$

Observe further that

$$\mathrm{H}(\xi_j) = \mathrm{H}(Q_j) \asymp g_j^{n+2\mu}. \tag{30}$$

The inequality $\mathrm{H}(\gamma_j) \leq 2g_j^{n-2}$ used in [2, page 98, line 3] does not hold anymore, thus we have to modify accordingly inequality (3) from [2] (i.e., we have to assume that $\chi$ is sufficiently large) in order to be able to argue as in [2]. The only consequence is that the function $n \mapsto F(n)$ defined in Theorem 1 from [2] must be replaced by a larger one. By the remark at the end of Section 4, we have to take $s = n + 2\mu$ in (25), thus, we have to assume $\chi > 2n(n+\mu)(n+1) + 2n + 1$.

We argue as in Section 6 from [2] and as in the above Section 5. Since

$$|\tilde{P}'_{n,[g_j^{\mu}]}(\gamma_j)| \asymp a^{-n+2},$$

we get

$$|Q_j(\xi)| \asymp g_j\,\mathrm{H}(Q_j)^{-w_n^*-1}\,g_j^{-\mu(n-2)}.$$

By (30), this implies the lower estimate

$$w_n(\xi) \geq w_n^* + 1 + \frac{\mu(n-2) - 1}{n + 2\mu}.$$

By definition of $\mu$, we obtain

$$w_n(\xi) \geq w_n^* + \Delta. \tag{31}$$

We prove that there is equality in (31) exactly as in [1] or [2]. □

## 7 Proof of Theorem 4

Let $n \geq 6$ be an even integer. As in Theorem 2 of [1], the number $\xi$ is obtained as the limit of a sequence of algebraic numbers of the form

$$\xi_j = \frac{c_j + id_j + \gamma_j}{g_j},$$

where the $\gamma_j$'s are suitable real algebraic numbers of degree $n/2$ and the $c_j$'s, $d_j$'s and $g_j$'s satisfy $g_j < c_j < 2g_j$ and $5g_j < d_j < 6g_j$. We omit the details of the construction of the $\xi_j$'s, since it is very similar to that in Theorem 5 above. Set $m = n/2$. Let $\mu$, $\mu'$ and $\mu''$ be real numbers in $[0, (m-2)/2]$, in $[0, 1]$ and in $[0, +\infty)$, respectively. Set $a = [g_j^{\mu}]$, $a' = [g_j^{\mu'}]$ and $a'' = [g_j^{\mu''}]$. We choose for $\gamma_j$ roots of the polynomials

$$\hat{P}_{m,a}(X) = X^m - 2(aX-1)^2, \qquad \check{P}_{m,a'}(X) = X^m - 2a'^m$$

or, when $n$ is divisble by 4 (that is, when $m$ is even),

$$\tilde{P}_{m,a''}(X) := (X^{m/2} - a''X + 1)^2 - 2X^{m-2}\,(a''X-1)^2.$$

We observe that the $\xi_j$'s are of degree $n$ and roots of polynomials of the form either $\hat{P}_{m,a}(g_jX - c_j - id_j) \times \hat{P}_{m,a}(g_jX - c_j + id_j)$, or $\check{P}_{m,a'}(g_jX - c_j - id_j) \times \check{P}_{m,a'}(g_jX - c_j + id_j)$, or $\tilde{P}_{m,a''}(g_jX - c_j - id_j) \times \tilde{P}_{m,a''}(g_jX - c_j + id_j)$, whose heights are $\asymp g_j^{2m}$, $\asymp g_j^{2m}$, and $\asymp g_j^{2(m+2\mu)}$, respectively. Furthermore, we have

$$\left| \frac{d}{dx}\left(\hat{P}_{m,a}(X - id_j) \times \hat{P}_{m,a}(X + id_j)\right)(\gamma_j + id_j) \right| \ll g_j^{m-\mu(m-2)/2},$$

$$\left| \frac{d}{dx}\left(\check{P}_{m,a'}(X - id_j) \times \check{P}_{m,a'}(X + id_j)\right)(\gamma_j + id_j) \right| \ll g_j^{m+\mu'(m-1)},$$

and

$$\left| \frac{d}{dx}\left(\tilde{P}_{m,a''}(X - id_j) \times \tilde{P}_{m,a''}(X + id_j)\right)(\gamma_j + id_j) \right| \ll g_j^{m-\mu''(m-4)}.$$

These estimates imply that, firstly, working with polynomials of the form $\hat{P}_{m,a}(g_jX - c_j - id_j) \times \hat{P}_{m,a}(g_jX - c_j + id_j)$, we construct a nonreal complex number $\xi$ such that

$$w_n(\xi) = w_n^*(\xi) + 1 - \frac{m+1}{2m} + \mu \cdot \frac{m-2}{4m}.$$

Secondly, working with polynomials of the form $\check{P}_{m,a'}(g_jX - c_j - id_j) \times \check{P}_{m,a'}(g_jX - c_j + id_j)$, we construct a nonreal complex number $\xi$ such that

$$w_n(\xi) = w_n^*(\xi) + 1 - \frac{m+1}{2m} - \mu' \cdot \frac{m-1}{2m}.$$

Thirdly, working with polynomials of the form $\tilde{P}_{m,a''}(g_jX - c_j - id_j) \times \tilde{P}_{m,a''}(g_jX - c_j + id_j)$, we construct a nonreal complex number $\xi$ such that

$$w_n(\xi) = w_n^*(\xi) + 1 - \frac{m+1}{2(m+2\mu'')} + \frac{\mu''(m-4)}{2(m+2\mu'')}.$$

Recalling that $n = 2m$ and letting $\mu$, $\mu'$ and $\mu''$ vary in their respectives ranges of values, this proves that the set of values taken by the function $w_n - w_n^*$ contains the intervals

$$\left[\frac{1}{2} - \frac{1}{n}, \frac{n}{16}\right] \quad \text{and} \quad \left[0, \frac{1}{2} - \frac{1}{n}\right],$$

and, if $n$ is divisible by 4, the interval

$$\left[\frac{1}{2} - \frac{1}{n}, \frac{n}{8}\right].$$

This completes the proof of Theorem 4. □

## References

1. Baker, R.C.: On approximation with algebraic numbers of bounded degree. *Mathematika* **23**, 18–31 (1976)
2. Bugeaud, Y.: Mahler's classification of numbers compared with Koksma's. *Acta Arith.* **110**, 89–105 (2003)
3. Bugeaud, Y.: Mahler's classification of numbers compared with Koksma's, III. *Publ. Math. (Debrecen)* **65**, 305–316 (2004)
4. Bugeaud, Y.: *Approximation by Algebraic Numbers*. Cambridge Tracts in Mathematics, vol. 160. Cambridge University Press, Cambridge (2004)
5. Bugeaud, Y., Mignotte, M.: On the distance between roots of an integer polynomial. *Proc. Edinb. Math. Soc.* **47**, 553–556 (2004)
6. Evertse, J.-H.: Distances between the conjugates of an algebraic number. *Publ. Math. (Debrecen)* **65**, 323–340 (2004)
7. Koksma, J.F.: Über die Mahlersche Klasseneinteilung der transzendenten Zahlen und die Approximation komplexer Zahlen durch algebraische Zahlen. *Monatsh. Math. Phys.* **48**, 176–189 (1939)
8. Mahler, K.: Zur Approximation der Exponentialfunktionen und des Logarithmus. I, II. *J. Reine Angew. Math.* **166**, 118–150 (1932)
9. Roy, D.: Approximation to real numbers by cubic algebraic numbers, II. *Ann. Math.* **158**, 1081–1087 (2003)
10. Schmidt, W.M.: $T$-numbers do exist. In: *Symposia Mathematica su Teoria dei Numeri, Istituto Nazionale di Alta Mathematica, Rome 1968*. Symp. Math., 4, pp. 3–26. Academic Press, London (1970)
11. Schmidt, W.M.: Mahler's $T$-numbers. In: *1969 Number Theory Institute*. Proc. Symp. Pure Math., vol. 20, pp. 275–286. American Mathematical Society, Providence (1971)
12. Schönhage, A.: Polynomial root separation examples. *J. Symb. Comput.* **41**, 1080–1090 (2006)
13. Sprindžuk, V.G.: *Mahler's Problem in Metric Number Theory*. American Mathematical Society, Providence (1969)
14. Wirsing, E.: Approximation mit algebraischen Zahlen beschränkten Grades. *J. Reine Angew. Math.* **206**, 67–77 (1961)

*Diophantine Approximation*
H. P. Schlickewei et al., Editors


# RATIONAL APPROXIMATIONS TO A $q$-ANALOGUE OF $\pi$ AND SOME OTHER $q$-SERIES

Peter Bundschuh[1] and Wadim Zudilin[2]
[1] *Mathematical Institute, University of Cologne, Weyertal 86–90, 50931 Cologne, Germany*
pb@mi.uni-koeln.de
[2] *Department of Mechanics and Mathematics, Moscow Lomonosov State University, Vorobiovy Gory, GSP-1, 119991 Moscow, Russia*
wadim@mi.ras.ru

To Wolfgang M. Schmidt on the occasion of his 70th birthday

## 1 Introduction

One of the famous mathematical constants is $\pi$, Archimedes' constant. There are several analytic ways to define it, e.g., by the (slowly convergent) series

$$\pi = 4\sum_{\nu=0}^{\infty} \frac{(-1)^\nu}{2\nu+1}, \tag{1}$$

or by the (Gaussian probability density) integral

$$\pi = \left( \int_{-\infty}^{\infty} e^{-x^2}\,\mathrm{d}x \right)^2; \tag{2}$$

for a comprehensive exposition of different representations and bibliography we refer the reader to [Fi, Section 1.4].

It does not seem very surprising that the number $\pi$ has a natural $q$-analogue

$$\pi_q = 1 + 4\sum_{\nu=0}^{\infty} \frac{(-1)^\nu q^{2\nu+1}}{1-q^{2\nu+1}} = 1 + 4\sum_{\mu=1}^{\infty} \frac{q^\mu}{1+q^{2\mu}}, \qquad |q| < 1, \tag{3}$$

since computations on a basis of (1) show that

$$\lim_{\substack{q\to 1\\ |q|<1}} (1-q)\pi_q = \pi$$

*Keywords.* Irrationality, $q$-analogues of mathematical constants, basic hypergeometric series, $q$-binomial theorem.

*2000 Mathematics subject classification.* Primary 11J72; Secondary 11J82, 33D15.

(cf. [Zu1]). In addition, the well-known Jacobi relation of the series in (3) with the modular theta series,

$$\pi_q = \left( \sum_{n=-\infty}^{\infty} q^{n^2} \right)^2 \tag{4}$$

(which may be interpreted as a $q$-analogue of (2)), together with Nesterenko's theorem [Ne] immediately implies the transcendence of $\pi_q$ at algebraic points $q$, $0 < |q| < 1$. Unfortunately, algebraic independence measures from [Ne] do not give any reasonable (i.e., Liouville type) estimates for the irrationality measure of $\pi_q$. Recall that Liouville type estimates are known for its ordinary version $\pi$; the best known one is due to Hata [Ha2].

The present paper is aimed at deducing fairly good irrationality measures for the values $\pi_q$ and the more general series

$$\ell_p(x, z) = x \sum_{\nu=1}^{\infty} \frac{z^\nu}{p^\nu - x} \tag{5}$$

in the case $p = q^{-1} \in \mathbb{Z} \setminus \{0, \pm 1\}$. In particular, we prove the irrationality of $\ell_p(x, z)$ provided some natural restrictions for the real numbers $x$ and $z$ are satisfied.

## 2 Main results and reduction

**Theorem 1.** *Assume that* $p \in \mathbb{Z} \setminus \{0, \pm 1\}$ $\notin$; $x, z \in \mathbb{Q} \setminus \{0\}$, $x \notin \{p, p^2, p^3, \dots\}$, $|z| < |p|$ *and that one of the following two conditions hold*:

(a) $z^k = \pm p^l$, *where the integers* $k, l$ *are coprime, or*
(b) $x^{k_1}/z^{k_2} = p^l$, *where* $k_1, k_2$ *are positive and coprime integers,* $l$ *is integer.*

*Then the value of the series* $\ell_p(x, z)$ *in* (5) *is irrational.*

Our (constructive) proof of the theorem also implies estimates for the irrationality exponents of the values $\ell_p(x, z)$, which depend on the "multiplicative dependence" factor (i.e., on the integers $k, l$ in case (a) or $k_1, k_2, l$ in case (b)). We recall that the *irrationality exponent* of a real irrational number $\gamma$ is defined by the relation

$$\begin{aligned} \mu = \mu(\gamma) = \inf\{c \in \mathbb{R} : &\text{ the inequality } |\gamma - a/b| \leqslant |b|^{-c} \text{ has} \\ &\text{only finitely many solutions in } (a, b) \in \mathbb{Z}^2 \text{ with } b \neq 0\}. \end{aligned}$$

We do not present estimates for $\mu(\ell_p(x, z))$ in general, under the assumptions of Theorem 1, but we are particularly interested in the following cases:

(a) $x = \sqrt{p} \in \mathbb{Z}$, $z = -1$ (the case corresponds to the series in (3)),
(b) $z = -1$, and
(c) $x = z$.

**Theorem 2.** *The irrationality exponent of* $\pi_q$ *satisfies the estimate*

$$\mu(\pi_q) \leqslant 10.31789151\dots,$$

*where* $q = p^{-1}$ *and* $p \in \mathbb{Z} \setminus \{0, \pm 1\}$.

**Theorem 3.** *Let $p \in \mathbb{Z}\setminus\{0, \pm 1\}$ and $x \in \mathbb{Q}\setminus\{0, p, p^2, p^3, \dots\}$. Then the irrationality exponent of $\ell_p(x, -1)$ satisfies the estimate*

$$\mu(\ell_p(x,-1)) = \mu\left(\sum_{\nu=1}^{\infty} \frac{(-1)^{\nu-1}}{p^\nu - x}\right) \leqslant \frac{6\pi^2}{\pi^2 - 6} = 15.30327658\dots.$$

**Theorem 4.** *Let $p \in \mathbb{Z} \setminus \{0, \pm 1\}$ and $x \in \mathbb{Q}$, $0 < |x| < |p|$. Then the irrationality exponent of $\ell_p(x, x)$ satisfies the estimate*

$$\mu(\ell_p(x,x)) = \mu\left(\sum_{\nu=1}^{\infty} \frac{x^\nu}{p^\nu - x}\right) \leqslant 6,$$

*and also*

$$\mu(\ell_p(-1,-1)) = \mu\left(\sum_{\nu=1}^{\infty} \frac{(-1)^{\nu-1}}{p^\nu + 1}\right) \leqslant \frac{3\pi^2}{\pi^2 - 3} = 4.31011910\dots$$

*in the special case $x = -1$.*

If $x = 1$ (or $z = 1$), the series in (5) is a $q$-analogue of the logarithm

$$-\log(1-z) = \sum_{\nu=1}^{\infty} \frac{z^\nu}{\nu}.$$

Therefore, irrationality results in this case (provided $p \in \mathbb{Z} \setminus \{0, \pm 1\}$) are known from [Be], [Bo1], [BV], and the best estimates for the irrationality exponents are given in [Zu2] and [MVZ]. The case $z = -1$ is treated in [Bo2, Theorem 2] but without an explicit indication of an estimate for the irrationality exponent of $\ell_p(x, -1)$; we include a result connecting our construction and those of the note [Bo2], in Section 4. The irrationality of the series $\ell_p(x, x)$ was previously obtained in [Du1], but under some very unnatural conditions on the rational number $x$ and only in a qualitative form.

From (4) and Theorem 2, it follows the estimate

$$\mu\left(\sum_{n=-\infty}^{\infty} q^{n^2}\right) \leqslant 20.63578302\dots$$

for the irrationality exponent of the theta series; this estimate is essentially worse than the known one

$$\mu\left(\sum_{n=-\infty}^{\infty} q^{n^2}\right) \leqslant \frac{3+\sqrt{5}}{2} = 2.61803398\dots$$

obtained in [Bu] by another method.

Theorems 1–4 remain valid in some cases when the parameter $p$ is not necessarily integer, like it is treated in [Du2]. In these cases, however, we should introduce several new technical conditions and make our proofs more complicated. Therefore, we do all our results in "readable" settings.

To simplify our further considerations, we mention that the functional equation

$$\ell_p(x,z) = \frac{xz}{p-x} + xz\sum_{\nu=1}^{\infty} \frac{z^\nu}{p^{\nu+1} - x} = \frac{xz}{p-x} + z\ell_p\left(\frac{x}{p}, z\right) \tag{6}$$

allows us to scale the variable $x$ by integer powers of $p$ and hence to reduce the situation to the case $1 \leqslant |x| < |p|$. On the other hand, the symmetry

$$\ell_p(x,z) = x\sum_{\nu=1}^{\infty}\frac{z^\nu}{p^\nu - x} = z\sum_{\nu=1}^{\infty}\frac{x^\nu}{p^\nu - z} = \ell_p(z,x) \tag{7}$$

gives a way to scale the variable $z$ as well, in order to reduce it to the case $1 \leqslant |z| < |p|$.

## 3 Hypergeometric construction

In order to construct rational approximations to the values of the series (5), we adopt the construction given in [MVZ]. Throughout the paper, we use standard $q$-notations [GR]:

$$(a;q)_0 = 1, \qquad (a;q)_n = \prod_{\nu=1}^{n}(1 - aq^{\nu-1}) \quad \text{for } n = 1, 2, \dots, \infty,$$

$$\begin{bmatrix} n \\ k \end{bmatrix}_q = \frac{(q;q)_n}{(q;q)_k \cdot (q;q)_{n-k}} \quad \text{for } k = 0, 1, \dots, n \quad \text{and} \quad n = 0, 1, 2, \dots.$$

$${}_2\phi_1\left(\begin{matrix} a, \; b \\ c \end{matrix}\,\middle|\, q, y\right) = \sum_{n=0}^{\infty}\frac{(a;q)_n(b;q)_n}{(q;q)_n(c;q)_n}y^n.$$

In particular, the series in (5) may be written as

$$\ell_p(x,z) = \frac{xz}{p-x}\cdot {}_2\phi_1\left(\begin{matrix} p, \; p/x \\ p^2/x \end{matrix}\,\middle|\, p, z\right) = \frac{qxz}{1-qx}\cdot {}_2\phi_1\left(\begin{matrix} q, \; qx \\ q^2x \end{matrix}\,\middle|\, q, qz\right),$$

where $p = 1/q$.

Let $n$ and $m$ be positive integers, $m \leqslant n$. Consider the rational (in terms of $T$) function

$$\widetilde{R}(q,x;T) = \frac{(qT;q)_n \cdot x^n}{(q^{n+1}xT;q)_{n+1}} = \frac{\prod_{k=1}^{n}(1 - q^kT)}{\prod_{k=0}^{n}(1 - q^{n+k+1}xT)}\cdot x^n,$$

which is of order $O(T^{-1})$ as $T \to \infty$, hence may be decomposed in the sum of partial fractions:

$$\widetilde{R}(q,x;T) = \sum_{k=0}^{n}\frac{A_k(q)}{1 - q^{n+k+1}xT}. \tag{8}$$

**Lemma 1.** *For $k = 0, 1, \dots, n$, we have the following explicit formula for the coefficients in the decomposition* (8)*:*

$$A_k(q) = (-1)^n p^{n(n+1)/2}\frac{(qx;q)_n}{(q;q)_n}\cdot\frac{(q^{-n};q)_k(q^{n+1}x;q)_k}{(q;q)_k(qx;q)_k}q^k. \tag{9}$$

*In addition, we have the formula*

$$\begin{aligned} A_k(q)p^{(n+k+1)(m+1)} &= (-1)^k p^{(2n+1)(m+1)-m(m+1)/2} \\ &\quad\times p^{(n-m-k)(n-m-k-1)/2}\frac{\prod_{j=1}^{n}(p^{k+j} - x)}{(p;p)_k(p;p)_{n-k}} \end{aligned} \tag{10}$$

*and the estimate*

$$|A_k(q)p^{(n+k+1)(m+1)}| \leqslant |p|^{n^2+2nm+O(n)} \tag{11}$$

*valid for all* $k = 0, 1, \dots, n$.

*Proof.* The standard procedure of determining coefficients in a partial fraction decomposition gives us

$$\begin{aligned}
A_k(q) &= \left(\widetilde{R}(q,x;T)(1-q^{n+k+1}xT)\right)\big|_{T=q^{-(n+k+1)}x^{-1}} \\
&= \frac{\prod_{j=1}^{n}(1-q^{-(k+j)}x^{-1})}{\prod_{j=1}^{k}(1-q^{-j})\cdot\prod_{j=1}^{n-k}(1-q^{j})}\cdot x^n \\
&= \frac{(-1)^n q^{-n(n+1)/2-nk}(q^{k+1}x;q)_n}{(-1)^k q^{-k(k+1)/2}(q;q)_k(q;q)_{n-k}} \\
&= (-1)^{n+k}p^{n(n+1)/2+nk-k(k+1)/2}\frac{(q^{k+1}x;q)_n}{(q;q)_k(q;q)_{n-k}};
\end{aligned} \tag{12}$$

in particular, we derive the estimate (11) by applying $|(q^A x;q)_B| \leqslant (2x)^B$ for $A, B > 0$. Since

$$\begin{aligned}
\frac{(q^{-n};q)_k}{(q;q)_k} &= (-1)^k q^{-nk+k(k-1)/2}\frac{(q;q)_n}{(q;q)_k(q;q)_{n-k}}, \\
\frac{(q^{n+1}x;q)_k}{(qx;q)_k} &= \frac{(qx;q)_{n+k}/(qx;q)_n}{(qx;q)_k} = \frac{(q^{k+1}x;q)_n}{(qx;q)_n},
\end{aligned}$$

going on in (12) we obtain (9); on the other hand, from (12) we also derive

$$A_k(q)p^{(n+k+1)(m+1)} = (-1)^k p^{(n+k+1)(m+1)+(n-k)(n-k+1)/2}\frac{\prod_{j=1}^{n}(p^{k+j}-x)}{(p;p)_k(p;p)_{n-k}}$$

leading to representation (10). □

Take now $R(q,x;T) = \widetilde{R}(q,x;T)\cdot T^{m+1}$ and consider the quantity

$$I_{n,m}(q,x,z) = I(q,x,z) = z^{n+1}\sum_{t=0}^{\infty} z^t R(q,x;T)\big|_{T=q^t}. \tag{13}$$

**Lemma 2.** *The quantity* (13) *admits the following representation:*

$$I(q,x,z) = A(p)\ell_p(x,z) - A'(p) - A''(p), \tag{14}$$

*where*

$$A(p) = x^{-m-1} \sum_{k=0}^{n} A_k(q) p^{(n+k+1)(m+1)} z^{-k}, \tag{15}$$

$$A'(p) = x^{-m} \sum_{k=0}^{n} A_k(q) p^{(n+k+1)(m+1)} z^{-k} \sum_{l=1}^{k} \frac{z^l}{p^l - x}, \tag{16}$$

$$A''(p) = p^{(n+1)(m+1)+n(n+1)/2} z \sum_{l=0}^{m-1} \frac{x^{-(l+1)}}{p^{m-l} - z} \times \sum_{k=0}^{l} (-1)^k \begin{bmatrix} n \\ k \end{bmatrix}_p \begin{bmatrix} n+l-k \\ n \end{bmatrix}_p p^{(n-k)(n-k+1)/2} x^k. \tag{17}$$

*Proof.* Going on with the right-hand side of (13) we obtain

$$\begin{aligned} I(q,x,z) &= \sum_{t=-n}^{\infty} z^{t+n+1} \sum_{k=0}^{n} \frac{A_k(q) q^{t(m+1)}}{1 - q^{t+n+k+1} x} \\ &= \sum_{k=0}^{n} A_k(q) q^{-(n+k+1)(m+1)} z^{-k} \sum_{t=-n}^{\infty} \frac{q^{(t+n+k+1)(m+1)} z^{t+n+k+1}}{1 - q^{t+n+k+1} x} \\ &= \sum_{k=0}^{n} A_k(q) p^{(n+k+1)(m+1)} z^{-k} \sum_{l=k+1}^{\infty} \frac{q^{l(m+1)} z^l}{1 - q^l x}. \end{aligned}$$

The last inner sum may be computed as follows:

$$\begin{aligned} \sum_{l=k+1}^{\infty} \frac{q^{l(m+1)} z^l}{1 - q^l x} &= x^{-m} \sum_{l=k+1}^{\infty} \frac{q^l z^l}{1 - q^l x} - x^{-m} \sum_{l=k+1}^{\infty} \frac{(q^l - q^{l(m+1)} x^m) z^l}{1 - q^l x} \\ &= x^{-m} \sum_{l=1}^{\infty} \frac{q^l z^l}{1 - q^l x} - x^{-m} \sum_{l=1}^{k} \frac{q^l z^l}{1 - q^l x} - x^{-m} \sum_{l=k+1}^{\infty} q^l z^l \sum_{j=1}^{m} (q^l x)^{j-1} \\ &= x^{-m} \sum_{l=1}^{\infty} \frac{q^l z^l}{1 - q^l x} - x^{-m} \sum_{l=1}^{k} \frac{q^l z^l}{1 - q^l x} - x^{-m} \sum_{j=1}^{m} x^{j-1} \frac{(q^j z)^{k+1}}{1 - q^j z} \\ &= x^{-m} \sum_{l=1}^{\infty} \frac{z^l}{p^l - x} - x^{-m} \sum_{l=1}^{k} \frac{z^l}{p^l - x} - z^{k+1} \sum_{j=1}^{m} \frac{p^{-jk} x^{-(m+1-j)}}{p^j - z}; \end{aligned}$$

therefore, we get representation (14) with $A(p)$, $A'(p)$ given in (15), (16) and

$$\begin{aligned} A''(p) &= z \sum_{k=0}^{n} A_k(q) p^{(n+k+1)(m+1)} \sum_{j=1}^{m} \frac{p^{-jk} x^{-(m+1-j)}}{p^j - z} \\ &= z p^{(n+1)(m+1)} \sum_{j=1}^{m} \frac{x^{-(m+1-j)}}{p^j - z} \sum_{k=0}^{n} A_k(q) p^{k(m+1-j)} \end{aligned}$$

$$= (-1)^n p^{(n+1)(m+1)+n(n+1)/2} z \sum_{l=0}^{m-1} \frac{x^{-(l+1)}}{p^{m-l} - z}$$
$$\times \frac{(qx;q)_n}{(q;q)_n} \sum_{k=0}^{n} \frac{(q^{-n};q)_k (q^{n+1}x;q)_k}{(q;q)_k (qx;q)_k} q^{-lk},$$

where we use formula (9). The inner series in the representation of $A''(p)$ is a basic hypergeometric ${}_2\phi_1$-series. Applying to it Heine's transformation formula [GR, formula (1.4.1)],

$${}_2\phi_1\left(\begin{matrix} a, b \\ c \end{matrix} \,\middle|\, q, y\right) = \frac{(b;q)_\infty (ay;q)_\infty}{(c;q)_\infty (y;q)_\infty} \cdot {}_2\phi_1\left(\begin{matrix} c/b, y \\ ay \end{matrix} \,\middle|\, q, b\right),$$

we obtain

$$A''(p) = (-1)^n p^{(n+1)(m+1)+n(n+1)/2} z \sum_{l=0}^{m-1} \frac{x^{-(l+1)}}{p^{m-l} - z}$$
$$\times \frac{(q^{-(n+l)};q)_n}{(q;q)_n} \sum_{k=0}^{l} \frac{(q^{-n};q)_k (q^{-l};q)_k}{(q;q)_k (q^{-(n+l)};q)_k} q^{(n+1)k} x^k$$
$$= p^{(n+1)(m+1)+n(n+1)/2} z \sum_{l=0}^{m-1} \frac{x^{-(l+1)}}{p^{m-l} - z}$$
$$\times \sum_{k=0}^{l} (-1)^k \begin{bmatrix} n \\ k \end{bmatrix}_q \begin{bmatrix} n+l-k \\ n \end{bmatrix}_q q^{-n(n+1)/2 - n(l-k) + k(k+1)/2} x^k,$$

which, after necessary manipulations, becomes the required formula (17). □

To deduce arithmetic properties from the coefficients (15)–(17) in representation (14), we will write the rational numbers $x$, $z$ in the form $x = x_1/x_2$, $z = z_1/z_2$, where $x_1, x_2, z_1, z_2$ are all integers.

For a given $x = x_1/x_2 \in \mathbb{Q} \setminus \{0\}$, by $D_n(p, x)$ denote the least common multiple of the polynomials $p - x, p^2 - x, \dots, p^n - x$ in the ring $\mathbb{Q}[p]$; throughout the paper, any such common multiple in $\mathbb{Q}[p]$ is normalized by the condition that its leading coefficient is 1. Since the polynomial $D_n(p, x)$ divides the product $\prod_{j=1}^{n}(p^j - x)$, we always have

$$x_2^n D_n(p, x_1/x_2) \in \mathbb{Z}[p], \qquad x_1, x_2 \in \mathbb{Z} \setminus \{0\}.$$

In fact, both the polynomials $D_n(p, x)$ and $\prod_{j=1}^{n}(p^j - x)$ coincide except for cases when $x$ is a root of unity [MVZ]; for instance, if $x = 1$,

$$D_n(p, 1) = \prod_{j=1}^{n} \Phi_j(p),$$

where $\Phi_j(p)$ are cyclotomic polynomials (see [BV, Section 2]).

**Lemma 3.** *Let $x = x_1/x_2$ and $z = z_1/z_2$ be as above, and let $\widehat{D}_{n,m}(p, x, z)$ denote the least common multiple of the polynomials $(p; p)_n D_n(p, x)$ and $D_m(p, z)$, and*

$$M = (2n+1)(m+1) - \frac{m(m+1)}{2}.$$

*Then we have the inclusions*

$$\begin{aligned}
&(x_1x_2z_1z_2)^{n+m} p^{-M} \widehat{D}_{n,m}(p, x, z) \cdot A(p) \in \mathbb{Z}[p],\\
&(x_1x_2z_1z_2)^{n+m} p^{-M} \widehat{D}_{n,m}(p, x, z) \cdot A'(p) \in \mathbb{Z}[p],\\
&(x_1x_2z_1z_2)^{n+m} p^{-M} \widehat{D}_{n,m}(p, x, z) \cdot A''(p) \in \mathbb{Z}[p].
\end{aligned} \tag{18}$$

*Proof.* In accordance with (10), (15), (16) we deduce

$$x_1^n x_2^{m+1} z_1^n p^{-(2n+1)(m+1)+m(m+1)/2}(p; p)_n \cdot A(p) \in \mathbb{Z}[p],$$
$$x_1^n x_2^m z_1^n p^{-(2n+1)(m+1)+m(m+1)/2}(p; p)_n \cdot x_2^n D_n(p, x) \cdot A'(p) \in \mathbb{Z}[p];$$

from (17) we have

$$x_1^m p^{-(n+1)(m+1)-n(n+1)/2} \cdot z_2^m D_m(p, z) \cdot A''(p) \in \mathbb{Z}[p].$$

Since

$$\begin{aligned}
M &= (n+1)(m+1) + \frac{n(n+1)}{2} - \frac{(n-m)(n-m-1)}{2}\\
&\leqslant (n+1)(m+1) + \frac{n(n+1)}{2},
\end{aligned}$$

we arrive at the desired inclusions (18). □

**Lemma 4.** *The following estimate holds:*

$$|A(p)| \leqslant |p|^{n^2+2nm+O(n)}.$$

*Proof.* The required estimate is a consequence of (15) and (11). □

**Lemma 5.** *As $m \to \infty$, we have*

$$I(q, x, z) = x^n z^{n+1} c_0(q)\big(1 + O(|q|^m)\big), \tag{19}$$

*where $c_0(q) = \prod_{k=1}^{\infty}(1 - q^k)$; in particular,*

$$I(q, x, z) = |p|^{O(n)} \qquad as \quad n \to \infty. \tag{20}$$

*Proof.* Since

$$\frac{z^{t+1}R(q, x; q^{t+1})}{z^t R(q, x; q^t)} = zq^{m+1} \cdot \frac{1 - q^{t+n+1}}{1 - q^{t+1}} \cdot \frac{1 - q^{t+n+1}x}{1 - q^{t+2n+2}x} = O(|q|^m)$$

uniformly in $t = 0, 1, 2, \ldots$, from (13) we deduce

$$\begin{aligned}
I(q, x, z) &= z^{n+1} \cdot z^0 R(q, x; q^0)\big(1 + O(|q|^m)\big)\\
&= x^n z^{n+1} \frac{\prod_{k=1}^{n}(1 - q^k)}{\prod_{k=0}^{n}(1 - q^{n+k+1}x)}\big(1 + O(|q|^m)\big),
\end{aligned}$$

which yields the desired asymptotic formula (19) since $n \geqslant m$. □

## 4 Integral construction

Recall that $|p| > 1$ and $|z| < |p|$ (thanks to the reduction procedure already applied in Section 1). For the same positive integers $n \geqslant m$ as in the previous section, we now introduce the complex integral

$$F(p, x, z) = \frac{1}{2\pi i} \oint_{|T|=|p|} \frac{\prod_{j=1}^{n}(x - p^j/T)}{\prod_{j=0}^{n}(1 - p^j T)} \sum_{\nu=1}^{\infty} \frac{z^\nu}{x - p^\nu/T} \cdot \frac{\mathrm{d}T}{T^{m-n+1}} \tag{21}$$

prompted by P. Borwein's construction in [Bo2]. Clearly, the integrand is a meromorphic function of $T$ and we claim

**Lemma 6.** *Let $p, x, z \in \mathbb{C}^*$ satisfy $|p| > 1$ and $|z| < |p|$. Then*

$$\begin{aligned} F(p, x, z) &= \sum_{k=0}^{n}(-1)^k z^{-k} p^{k(k+1)/2+k(m-n)} \frac{\prod_{j=1}^{n}(x - p^{k+j})}{(p; p)_k (p; p)_{n-k}} \sum_{l=k+1}^{\infty} \frac{z^l}{p^l - x} \\ &\quad + \frac{1}{(m-1)!} \frac{\mathrm{d}^{m-1}}{\mathrm{d}T^{m-1}} \left( \frac{\prod_{j=1}^{n}(xT - p^j)}{\prod_{j=0}^{n}(1 - p^j T)} \sum_{\nu=1}^{\infty} \frac{z^\nu}{xT - p^\nu} \right) \Bigg|_{T=0}. \end{aligned} \tag{22}$$

*Proof.* On the contour $|T| = |p|$, the integrand on the right-hand side of (21) is holomorphic and has poles in $|T| < |p|$ at the points $T = p^0, p^{-1}, \dots, p^{-n}$ (all simple) and at $T = 0$ (of exact order $m$). The first poles $T = p^{-k}$ contribute, for $k = 0, 1, \dots, n$,

$$\begin{aligned} &-\frac{\prod_{j=1}^{n}(x - p^{j+k})}{p^k \cdot \prod_{j=0, j\neq k}^{n}(1 - p^{j-k})} \cdot \sum_{\nu=1}^{\infty} \frac{z^\nu}{x - p^{\nu+k}} \cdot p^{k(m-n+1)} \\ &= (-1)^k z^{-k} p^{k(k+1)/2+k(m-n)} \frac{\prod_{j=1}^{n}(x - p^{k+j})}{(p; p)_k (p; p)_{n-k}} \sum_{l=k+1}^{\infty} \frac{z^l}{p^l - x}, \end{aligned}$$

while the contribution of the pole $T = 0$ equals

$$\frac{1}{(m-1)!} \frac{\mathrm{d}^{m-1}}{\mathrm{d}T^{m-1}} \left( \frac{\prod_{j=1}^{n}(xT - p^j)}{\prod_{j=0}^{n}(1 - p^j T)} \sum_{\nu=1}^{\infty} \frac{z^\nu}{xT - p^\nu} \right) \Bigg|_{T=0}.$$

□

Defining

$$B(p) = B(p, x, z) = x^{-1} \sum_{k=0}^{n}(-1)^k z^{-k} p^{k(k+1)/2+k(m-n)} \frac{\prod_{j=1}^{n}(x - p^{k+j})}{(p; p)_k (p; p)_{n-k}} \tag{23}$$

and

$$B'(p) = \sum_{k=0}^{n}(-1)^k z^{-k} p^{k(k+1)/2+k(m-n)} \frac{\prod_{j=1}^{n}(x - p^{k+j})}{(p; p)_k (p; p)_{n-k}} \sum_{l=1}^{k} \frac{z^l}{p^l - x}, \tag{24}$$

$$B''(p) = -\frac{1}{(m-1)!} \frac{\mathrm{d}^{m-1}}{\mathrm{d}T^{m-1}} \left( \frac{\prod_{j=1}^{n}(xT - p^j)}{\prod_{j=0}^{n}(1 - p^j T)} \sum_{\nu=1}^{\infty} \frac{z^\nu}{xT - p^\nu} \right) \Bigg|_{T=0}, \tag{25}$$

and using the fact

$$\sum_{l=k+1}^{\infty}\frac{z^l}{p^l-x}=\sum_{l=1}^{\infty}\frac{z^l}{p^l-x}-\sum_{l=1}^{k}\frac{z^l}{p^l-x}=x^{-1}\ell_p(x,z)-\sum_{l=1}^{k}\frac{z^l}{p^l-x},$$

we can rewrite formula (22) as

$$F(p,x,z)=B(p)\ell_p(x,z)-B'(p)-B''(p). \tag{26}$$

**Lemma 7.** *Let $p, x, z$ be as in Lemma 6, $q=p^{-1}$, and let the quantities $I(q,x,z)$ and $F(p,x,z)$ be defined as in (13) and (21), respectively. Then*

$$I(q,x,z)=(-1)^n x^{-m} p^{(n+1)(m+1)+n(n+1)/2}F(p,x,z). \tag{27}$$

*Proof.* We will use the representations (14) and (26) for the given quantities. Verification of the equalities

$$\begin{aligned}A(p)&=(-1)^n x^{-m} p^{(n+1)(m+1)+n(n+1)/2}B(p),\\ A'(p)&=(-1)^n x^{-m} p^{(n+1)(m+1)+n(n+1)/2}B'(p)\end{aligned} \tag{28}$$

is straightforward due to formulae (10), (15), (16) and (23), (24).

To derive a closed expression for (25), we first mention the formula

$$\begin{aligned}&\frac{1}{j!}\frac{\mathrm{d}^j}{\mathrm{d}T^j}\left(\sum_{\nu=1}^{\infty}\frac{z^\nu}{xT-p^\nu}\right)\bigg|_{T=0}\\ &=(-1)^j x^j\sum_{\nu=1}^{\infty}\frac{z^\nu}{(xT-p^\nu)^{j+1}}\bigg|_{T=0}\\ &=-x^j\sum_{\nu=1}^{\infty}\left(\frac{z}{p^{j+1}}\right)^\nu=-\frac{x^jz}{p^{j+1}-z},\qquad j=0,1,2,\dots.\end{aligned}$$

Therefore, if

$$\frac{\prod_{j=1}^n(xT-p^j)}{\prod_{j=0}^n(1-p^jT)}=\sum_{l=0}^{\infty}a_lT^l, \tag{29}$$

then by (25)

$$B''(p)=z\sum_{l=0}^{m-1}a_l\frac{x^{m-l-1}}{p^{m-l}-z}. \tag{30}$$

It remains to develop the power series in (29). With this aim, we apply the $q$-binomial theorem [GR, Section 1.3]

$$\frac{(ay;q)_\infty}{(y;q)_\infty}=\sum_{\nu=0}^{\infty}\frac{(a;q)_\nu}{(q;q)_\nu}y^\nu$$

in two different ways:

$$\prod_{j=1}^{n}(xT-p^j)=(xT)^n\frac{(p/xT;p)_\infty}{(p^{n+1}/xT;p)_\infty}=(xT)^n\sum_{\nu=0}^{\infty}(-1)^\nu p^{\nu(\nu+1)/2}\begin{bmatrix}n\\ \nu\end{bmatrix}_p(xT)^{-\nu}$$

$$=\sum_{k=0}^{n}(-1)^{n-k}p^{(n-k)(n-k+1)/2}\begin{bmatrix}n\\ k\end{bmatrix}_p(xT)^k,$$

$$\frac{1}{\prod_{j=0}^{n}(1-p^jT)}=\frac{(p^{n+1}T;p)_\infty}{(T;p)_\infty}=\sum_{\nu=0}^{\infty}\begin{bmatrix}n+\nu\\ n\end{bmatrix}_pT^\nu.$$

Then

$$\sum_{l=0}^{\infty}a_lT^l=\sum_{k=0}^{n}(-1)^{n-k}p^{(n-k)(n-k+1)/2}\begin{bmatrix}n\\ k\end{bmatrix}_p(xT)^k\sum_{\nu=0}^{\infty}\begin{bmatrix}n+\nu\\ n\end{bmatrix}_pT^\nu$$

$$=(-1)^n\sum_{l=0}^{n}T^l\sum_{k=0}^{\min\{l,n\}}(-1)^kp^{(n-k)(n-k+1)/2}\begin{bmatrix}n\\ k\end{bmatrix}_p\begin{bmatrix}n+l-k\\ n\end{bmatrix}_px^k,$$

whence

$$a_l=(-1)^n\sum_{k=0}^{l}(-1)^kp^{(n-k)(n-k+1)/2}\begin{bmatrix}n\\ k\end{bmatrix}_p\begin{bmatrix}n+l-k\\ n\end{bmatrix}_px^k \tag{31}$$

for $l\leqslant n$. Substituting the result (31) into (30) and comparing with (17) we conclude that

$$A''(p)=(-1)^nx^{-m}p^{(n+1)(m+1)+n(n+1)/2}B''(p). \tag{32}$$

Finally, equalities (28) and (32) exactly mean that (27) holds. □

*Remark.* Lemma 7 explains that the two analytic constructions of rational approximations to $\ell_p(x,z)$ are the same, even looking quite differently.

## 5 Proofs

*Proof of Theorem 1.* Clearly, the polynomial $\widehat{D}_{n,m}(p,x,z)$ divides the product

$$\prod_{j=1}^{n}(p^j-1)\cdot\prod_{j=1}^{n}(p^j-x)\cdot\prod_{j=1}^{m}(p^j-z), \tag{33}$$

which has the total degree $n(n+1)+m(m+1)/2$ in $p$. Since

$$\begin{aligned}n(n+1)+\frac{m(m+1)}{2}&=(2n+1)(m+1)-\frac{m(m+1)}{2}\\&\quad+(n-m)(n-m-1)-(m+1)\\&\geqslant M-(m+1),\end{aligned}$$

it is impossible to get an irrationality result on $\ell_p(x,z)$ using just the product (33) (cf. (14), (18) and (20)). On the other hand, when $p, x=x_1/x_2$ and $z=z_1/z_2$ are multiplicatively dependent in accordance with cases (a) or (b), we always have an essentially better choice than $(x_1x_2z_1z_2)^{n+m}p^{-M}\widehat{D}_{n,m}(p,x,z)$ in (18) for the denominators of the numbers $A(p)$, $A'(p)$ and $A''(p)$, hence the irrationality of $\ell_p(x,z)$ follows. Till the end of the proof we assume that $m=n$.

(a) The numbers $k,l$ in the relation $z^k=\pm p^l$ are not both even; using, if necessary, scaling $z\mapsto zp$ (which is the consequence of (6) and (7)) we may assume that $l$ is odd and $k,l$ are of the same sign, hence positive. Write $z^k=\pm p^l$ in the form $z=\varepsilon_0 p_0^l$, where $p_0=|p|^{1/k}$ must be integer (since $z$ is rational and $p$ is integer) and $\varepsilon_0\in\{\pm1\}$.

For any $j=0,1,2,\dots$ and $\varepsilon\in\{\pm1\}$, we have

$$(p_0^{kj-1}-\varepsilon)\mid(p_0^{l(kj-1)}-\varepsilon)$$

and

$$(p_0^{kj-1}-\varepsilon)\mid(p_0^{2(kj-1)}-1)\mid(p_0^{2ki}-1)\qquad\text{with}\quad i=kj-1.$$

Therefore, the integers

$$\prod_{i=1}^{n}(p^i-1)=\prod_{\substack{i=1\\ i\text{ odd}}}^{n}(p^i-1)\cdot\prod_{i=1}^{\lfloor n/2\rfloor}(p_0^{2ki}-1)$$

and

$$\prod_{j=1}^{\lfloor n/l\rfloor}z_2(p^{lj}-z)=\pm\prod_{j=1}^{\lfloor n/l\rfloor}z_2(p_0^{lkj}-\varepsilon p_0^l)=\pm\prod_{j=1}^{\lfloor n/l\rfloor}z_2p_0^l(p_0^{l(kj-1)}-\varepsilon)$$

($\lfloor\cdot\rfloor$ denotes the integer part of a real number) have the common divisor

$$E_n=\prod_{j=1}^{N}(p_0^{kj-1}-\varepsilon),\qquad N=\min\left\{\left\lfloor\frac{n}{l}\right\rfloor,\left\lfloor\frac{n}{2k}\right\rfloor\right\},$$

satisfying

$$\lim_{n\to\infty}\frac{\log|E_n|}{n^2\log|p|}=\frac{1}{2\max\{l^2,4k^2\}}>0.$$

(b) Scaling $x\mapsto x/p$ in (6) allows us to assume that $l\neq0$ in $x^{k_1}/z^{k_2}=p^l$. Using the symmetry (7) we may reduce the general case to the following one: $k_1>0$, $k_2>0$ and $l>0$.

Since $k_1$ and $k_2$ are coprime and positive, the linear equation $k_1i-k_2j=1$ possesses a solution $(i_0,j_0)\in\mathbb{Z}^2$ with $i_0>0$ and $j_0>0$, but also the pairs $i=i_0+k_2t$ and $j=j_0+k_1t$ involving positive integers for $t=0,1,2,\dots$, form other solutions of the equation. We would like to find a common divisor of the integers

$$\prod_{\substack{i=1\\ i\equiv i_0\ (\mathrm{mod}\ k_2)}}^{\lfloor n/l\rfloor}x_2(p^{li}-x)\quad\text{and}\quad\prod_{\substack{j=1\\ j\equiv j_0\ (\mathrm{mod}\ k_1)}}^{\lfloor n/l\rfloor}z_2(p^{lj}-z).\tag{34}$$

We have

$$\begin{aligned}x_2(p^{li}-x) &= x_1(x^{k_1 i-1}z^{-k_2 i}-1) = x_1(x^{k_1(i_0+k_2t)-1}z^{-k_2(i_0+k_2t)}-1)\\ &= x_1\big((x^{k_1}/z^{k_2})^{k_2t}\cdot x^{k_2 j_0}/z^{k_2 i_0}-1\big)\\ &= x_1\big((p^{lt}x^{j_0}/z^{i_0})^{k_2}-1\big) \quad \text{for} \quad i=i_0+k_2t, \quad t=0,1,2,\dots,\end{aligned}$$

and

$$\begin{aligned}z_2(p^{lj}-z) &= z_1(x^{k_1 j}z^{-k_2 j-1}-1) = z_1(x^{k_1(j_0+k_1t)}z^{-k_2(j_0+k_1t)-1}-1)\\ &= z_1\big((x^{k_1}/z^{k_2})^{k_1t}\cdot x^{k_1 j_0}/z^{k_1 i_0}-1\big)\\ &= z_1\big((p^{lt}x^{j_0}/z^{i_0})^{k_1}-1\big) \quad \text{for} \quad j=j_0+k_1t, \quad t=0,1,2,\dots;\end{aligned}$$

moreover, the numbers $x_1\big((p^{lt}x^{j_0}/z^{i_0})^{k_2}-1\big)$ and $z_1\big((p^{lt}x^{j_0}/z^{i_0})^{k_1}-1\big)$ have the divisor $y(p^{lt}x^{j_0}/z^{i_0}-1)$, where $y=x_2^{j_0}z_1^{i_0}$ is the denominator of $x^{j_0}/z^{i_0}$, for $t=0,1,2,\dots$. Therefore, the numbers (34) are divisible by

$$E_n=\prod_{t=0}^{N} y(p^{lt}x^{j_0}/z^{i_0}-1), \qquad N=\min\Big\{\Big\lfloor\frac{n-i_0}{k_2}\Big\rfloor, \Big\lfloor\frac{n-j_0}{k_1}\Big\rfloor\Big\},$$

satisfying

$$\lim_{n\to\infty}\frac{\log|E_n|}{n^2\log|p|}=\frac{l}{2\max\{k_1^2,k_2^2\}}>0.$$

In both cases (a) and (b) we proved the existence of the integer $E_n$ such that the least common denominator of the numbers $A(p)$, $A'(p)$ and $A''(p)$ in (14) divides

$$\frac{1}{E_n}\cdot(x_1x_2z_1z_2)^{2n}p^{-(3n+2)(n+1)/2}\prod_{j=1}^{n}(p^j-1)\prod_{j=1}^{n}(p^j-x)\prod_{j=1}^{n}(p^j-z)$$

and

$$\lim_{n\to\infty}\frac{\log|E_n|}{n^2\log|p|}>0.$$

Finally, using Lemmas 2 and 5 we conclude with the irrationality of the quantity $\ell_p(x,z)$. □

Deriving estimates for irrationality exponents requires the following general assertion.

**Proposition 1.** *Let $p,x,z$ be as in Theorem 1, and let a real parameter $\alpha$ lie in the interval $0<\alpha\leqslant 1$; denote*

$$C_0(\alpha)=2\alpha-\frac{1}{2}\alpha^2-C_2(\alpha), \qquad C_1(\alpha)=1+\frac{1}{2}\alpha^2+C_2(\alpha),$$

*where*

$$C_2(\alpha)=\lim_{n\to\infty}\frac{\deg_p \widehat{D}_{n,\lfloor\alpha n\rfloor}(p,x,z)}{n^2}.$$

*Then, if* $C_0(\alpha) > 0$,

$$\mu\big(\ell_p(x,z)\big) \leqslant 1 + \frac{C_1(\alpha)}{C_0(\alpha)}. \tag{35}$$

*Remark.* The additional real parameter $\alpha$ gives a possibility to minimize the value of the right-hand side in (35). The final result is, therefore, as follows:

$$\mu\big(\ell_p(x,z)\big) \leqslant 1 + \min_{\substack{0<\alpha\leqslant 1\\ C_0(\alpha)>0}} \frac{C_1(\alpha)}{C_0(\alpha)}.$$

*Proof.* Take $m = \lfloor \alpha n \rfloor$. If $x = x_1/x_2$ and $z = z_1/z_2$, where $x_1, x_2, z_1, z_2 \in \mathbb{Z} \setminus \{0\}$, then by Lemmas 2 and 3

$$(x_1x_2z_1z_2)^{n+m} p^{-M} \widehat{D}_{n,m}(p,x,z) \cdot I(q,x,z) \in \mathbb{Z}[p]\ell_p(x,z) + \mathbb{Z}[p]. \tag{36}$$

The growth of the coefficients of $\ell_p(x,z)$ in the linear forms (36) is determined by the quantity

$$\limsup_{n\to\infty} \frac{\log|(x_1x_2z_1z_2)^{n+m} p^{-M} \widehat{D}_{n,m}(p,x,z)A(p)|}{n^2\log|p|} \leqslant C_1(\alpha)$$

by Lemma 4, while the linear forms themselves behave as follows:

$$\lim_{n\to\infty} \frac{\log|(x_1x_2z_1z_2)^{n+m} p^{-M} \widehat{D}_{n,m}(p,x,z)I(q,x,z)|}{n^2\log|p|} = -C_0(\alpha)$$

by Lemma 5. Combining this asymptotic information and standard arguments (see, e.g., [Ha1, Lemma 3.1]) we conclude with the required estimate (35). □

*Proof of Theorem 2.* It follows from definition (3) that

$$\pi_q = 1 - 4\ell_{p^2}(p,-1) = 1 - 4\ell_{p^2}(-1,p).$$

The polynomial $D_n(p^2,p)$ is the least common multiple of the polynomials $p^2 - p, p^4 - p, \dots, p^{2n} - p$, hence

$$D_n(p^2,p) = p\prod_{j=1}^{n} \Phi_{2j-1}(p),$$

where $\Phi_l(x)$ are the cyclotomic polynomials. It is proved in [MVZ] that

$$D_m(p^2,-1) = \prod_{j=1}^{m} \Phi_{2j}(p^2);$$

in particular, $D_n(p^2,p)$ and $D_m(p^2,-1)$ are coprime. On the other hand, the product

$$(p^2;p^2)_n = \prod_{l=1}^{n}(1-p^{2l}) = \pm\prod_{l=1}^{n}\prod_{j|l}\Phi_j(p^2)$$

contains all cyclotomic factors $\Phi_{2j}(p^2)$ with $2j \leqslant n$; thus

$$\widehat{D}_{n,m}(p^2,p,-1) = p\cdot(p^2;p^2)_n \cdot \prod_{j=1}^{n}\Phi_{2j-1}(p) \cdot \prod_{j=\lfloor n/2\rfloor+1}^{m} \Phi_{2j}(p^2)$$

with the estimate (when $m \geqslant n/2$)

$$\begin{aligned}\deg_p \widehat{D}_{n,m}(p^2, p, -1) &= 2 \cdot \frac{n^2}{2} + \frac{8}{\pi^2}n^2 + 2 \cdot \frac{4}{\pi^2}\left(m^2 - \left(\frac{n}{2}\right)^2\right) + O(n \log n) \\ &= \left(1 + \frac{6}{\pi^2}\right)n^2 + \frac{8}{\pi^2}m^2 + O(n \log n) \qquad (37)\end{aligned}$$

(see [MVZ, formula (11)]). Take $m = \lfloor \alpha n \rfloor$, where $1/2 \leqslant \alpha \leqslant 1$. From Lemmas 2 and 3 we have

$$p^{m+1}p^{-2M}\widehat{D}_{n,m}(p^2, p, -1) \cdot I(q^2, p, -1) \in \mathbb{Z}[p]\ell_{p^2}(p, -1) + \mathbb{Z}[p],$$

whence

$$\begin{aligned}&-\lim_{n\to\infty} \frac{\log|p^{m+1}p^{-2M}\widehat{D}_{n,m}(p^2, p, -1) \cdot I(q^2, p, -1)|}{n^2 \log|p|} \\ &\quad = 4\alpha - \alpha^2 - \left(1 + \frac{6}{\pi^2} + \frac{8}{\pi^2}\alpha^2\right) = C_0(\alpha)\end{aligned}$$

and

$$\begin{aligned}&\limsup_{n\to\infty} \frac{\log|p^{m+1}p^{-2M}\widehat{D}_{n,m}(p^2, p, -1) \cdot A(p^2)|}{n^2 \log|p|} \\ &\quad \leqslant 2 + \alpha^2 + \left(1 + \frac{6}{\pi^2} + \frac{8}{\pi^2}\alpha^2\right) = C_1(\alpha).\end{aligned}$$

This leads to the estimate

$$\mu(\pi_q) = \mu\big(\ell_{p^2}(p, -1)\big) \leqslant 1 + \frac{C_1(\alpha)}{C_0(\alpha)}$$

for the irrationality exponent. The optimal choice $\alpha = 0.99756567\ldots$ gives the estimate

$$\mu(\pi_q) \leqslant 10.31789151\ldots,$$

while taking simply $\alpha = 1$ we obtain only $\mu(\pi_q) \leqslant 10.31808189\ldots$.

It is worth mentioning that taking $x = -1$, $z = p$ (and not $x = p$, $z = -1$ as before) we get

$$\widehat{D}_{n,m}(p^2, -1, p) = p \cdot (p^2; p^2)_n \cdot \prod_{j=1}^{n} \Phi_{2j}(p^2) \cdot \prod_{j=\lfloor n/2 \rfloor+1}^{m} \Phi_{2j-1}(p)$$

with exactly the same estimate for degree as in (37) (and, hence, with the same estimate for $\mu(\pi_q)$). □

*Proof of Theorem 3.* In [MVZ, Lemma 1], it is proved that $D_m(p, -1) = \prod_{l=1}^{m} \Phi_{2l}(p)$. All cyclotomic polynomials $\Phi_{2l}(p)$ with $2l \leqslant n$ divide $(p; p)_n = \prod_{j=1}^{n}(1 - p^j)$, whence the least common multiple $\widehat{D}_{n,m}(p, x, -1)$ divides the polynomial

$$(p; p)_n \prod_{j=1}^{n}(p^j - x) \cdot \prod_{l=\lfloor n/2 \rfloor+1}^{m} \Phi_{2l}(p)$$

of the total degree in $p$ (when $m \geqslant n/2$)

$$\frac{1}{2}n^2 + \frac{1}{2}n^2 + \sum_{l=\lfloor n/2\rfloor+1}^{m} \varphi(2l) = n^2 + \frac{4}{\pi^2}\left(m^2 - \left(\frac{n}{2}\right)^2\right) + O(n \log n)$$
$$= \left(1 - \frac{1}{\pi^2}\right)n^2 + \frac{4}{\pi^2}m^2 + O(n \log n).$$

Therefore, if $\alpha \geqslant 1/2$,

$$C_2(\alpha) = 1 - \frac{1}{\pi^2} + \frac{4}{\pi^2}\alpha^2,$$

and Proposition 1 applied with the (optimal) choice $\alpha = 1$ yields

$$\mu\big(\ell_p(x, -1)\big) \leqslant \frac{6\pi^2}{\pi^2 - 6} = 15.30327658\ldots.$$

□

*Proof of Theorem 4.* In the case $x = z$ the polynomial $\widehat{D}_{n,m}(p, x, z)$ divides the product

$$\prod_{j=1}^{n}(p^j - 1) \cdot \prod_{j=1}^{n}(p^j - x)$$

and, in fact, coincides with $(p; p)_n D_n(p, x)$. In the notations of Proposition 1, taking $\alpha = 1$ we obtain $C_2(\alpha) = 1$,

$$C_0(\alpha) = \tfrac{3}{2} - C_2(\alpha) = \tfrac{1}{2}, \qquad C_1(\alpha) = \tfrac{3}{2} + C_2(\alpha) = \tfrac{5}{2}.$$

This leads to the estimate

$$\mu\big(\ell_p(x, x)\big) \leqslant 1 + \frac{5/2}{1/2} = 6$$

for the irrationality exponent. In the case $x = -1$ the estimate may be improved by taking the better estimate for $\deg_p \widehat{D}_{n,n}(p, -1, -1)$ (cf. the proof of Theorem 3):

$$\mu\big(\ell_p(-1, -1)\big) = \mu\left(\sum_{\nu=1}^{\infty} \frac{(-1)^{\nu-1}}{p^\nu + 1}\right) \leqslant 1 + \frac{2 + 3/\pi^2}{1 - 3/\pi^2} = 4.31011910\ldots.$$

□

**Acknowledgements.** We thank Yu. Nesterenko kindly for his valuable comments and notification of a serious misprint in an earlier version of the paper. The work was supported by an Alexander von Humboldt research fellowship and partially supported by grant nr. 03-01-00359 of the Russian Foundation for Basic Research.

## References

[Be] Bézivin, J.-P.: Indépendance linéaire des valeurs des solutions transcendantes de certaines équations fonctionnelles. *Manuscr. Math.* **61**, 103–129 (1988)

[Bo1] Borwein, P.: On the irrationality of $\sum(1/(q^n + r))$. *J. Number Theory* **37**, 253–259 (1991)

[Bo2] Borwein, P.B.: On the irrationality of certain series. *Math. Proc. Camb. Philos. Soc.* **112**, 141–146 (1992)

[Bu] Bundschuh, P.: Verschärfung eines arithmetischen Satzes von Tschakaloff. *Port. Math.* **33**, 1–17 (1974)

[BV] Bundschuh, P., Väänänen, K.: Arithmetical investigations of a certain infinite product. *Compos. Math.* **91**, 175–199 (1994)

[Du1] Duverney, D.: Sur l'irrationalité de $\sum_{n=1}^{+\infty} r^n/(q^n - r)$. *C. R. Acad. Sci. Paris Sér. I* **320**, 1–4 (1995)

[Du2] Duverney, D.: A propos de la série $\sum_{n=1}^{+\infty} \frac{x^n}{q^n-1}$. *J. Théor. Nombres Bordx.* **8**, 173–181 (1996)

[Fi] Finch, S.R.: *Mathematical Constants*. Encyclopedia of Mathematics and Its Applications, vol. 94. Cambridge University Press, Cambridge (2003)

[GR] Gasper, G., Rahman, M.: *Basic Hypergeometric Series*. Encyclopedia of Mathematics and Its Applications, vol. 35. Cambridge Cambridge University Press, Cambridge (1990)

[Ha1] Hata, M.: Legendre type polynomials and irrationality measures. *J. Reine Angew. Math.* **407**, 99–125 (1990)

[Ha2] Hata, M.: Rational approximations to $\pi$ and some other numbers. *Acta Arith.* **63**, 335–349 (1993)

[MVZ] Matalo-Aho, T., Väänänen, K., Zudilin, W.: New irrationality measures for $q$-logarithms. *Math. Comput.* **75**, 879–889 (2006)

[Ne] Nesterenko, Yu.V.: Modular functions and transcendence questions. *Mat. Sb.* **187**, 65–96 (1996)

[Zu1] Zudilin, W.: Diophantine problems for $q$-zeta values. *Mat. Zametki* **72**, 936–940 (2002)

[Zu2] Zudilin, W.: Heine's basic transform and a permutation group for $q$-harmonic series. *Acta Arith.* **111**, 153–164 (2004)

*Diophantine Approximation*
H. P. Schlickewei et al., Editors


# ORTHOGONALITY AND DIGIT SHIFTS IN THE CLASSICAL MEAN SQUARES PROBLEM IN IRREGULARITIES OF POINT DISTRIBUTION

William W. L. Chen[1] and Maxim M. Skriganov[2]
[1] *Department of Mathematics, Macquarie University, Sydney, NSW 2109, Australia*
wchen@maths.mq.edu.au
[2] *Steklov Mathematical Institute, Fontanka 27, St. Petersburg 191011, Russia*
skrig@pdmi.ras.ru

To Wolfgang Schmidt on the occasion of his 70th birthday

## 1 Introduction

Suppose that $\mathcal{A}_N$ is a distribution of $N > 1$ points, not necessarily distinct, in the $n$-dimensional unit cube $U^n = [0,1)^n$, where $n \geq 2$. We consider the $L_2$-discrepancy

$$\mathcal{L}_2[\mathcal{A}_N] = \left( \int_{U^n} |\mathcal{L}[\mathcal{A}_N; Y]|^2 \, \mathrm{d}Y \right)^{1/2},$$

where for every $Y = (y_1, \ldots, y_n) \in U^n$, the local discrepancy $\mathcal{L}[\mathcal{A}_N; Y]$ is given by

$$\mathcal{L}[\mathcal{A}_N; Y] = \#(\mathcal{A}_N \cap B_Y) - N \operatorname{vol} B_Y .$$

Here

$$B_Y = [0, y_1) \times \ldots \times [0, y_n) \subseteq U^n$$

is a rectangular box of volume $\operatorname{vol} B_Y = y_1 \ldots y_n$, and $\#(\mathcal{S})$ denotes the number of points of a set $\mathcal{S}$, counted with multiplicity.

Roth [12, 13] established a lower bound for the $L_2$-discrepancy in any given dimension $n \geq 2$, as well as a corresponding upper bound. More precisely, for every distribution $\mathcal{A}_N$ of $N$ points in the unit cube $U^n$, we have

$$\mathcal{L}_2[\mathcal{A}_N] > c_n (\log N)^{(n-1)/2},$$

where the positive constant $c_n$ depends only on the dimension $n$. Furthermore, there exist distributions $\mathcal{B}_N$ of $N$ points in the unit cube $U^n$ such that

*Keywords.* Irregularities of distribution, orthogonality, digit shift, coding theory.

*2000 Mathematics subject classification.* 11K38.

$$\mathcal{L}_2[\mathcal{B}_N] < C_n(\log N)^{(n-1)/2}, \tag{1}$$

where the positive constant $C_n$ again depends only on the dimension $n$.

However, Roth's upper bound technique involves a probabilistic argument, as are the subsequent arguments of Chen [1,2], Frolov [7], Dobrovol'skiĭ [4] and Skriganov [15,16], and no explicit distribution $\mathcal{B}_N$ is given in any of these papers.

The first explicit constructions of such distributions in any dimension $n \geq 2$ can be found in a recent paper of Chen and Skriganov [3], where it is shown that for every integer $N > 1$, a distribution $\mathcal{D}_N$ of $N$ points in the unit cube $U^n$ can be constructed explicitly to satisfy the inequality

$$\mathcal{L}_2[\mathcal{D}_N] < 2^{n+1} p^{2n} \left( \frac{\log N}{\log p} + 2n + 1 \right)^{(n-1)/2}, \tag{2}$$

where $p \geq 2n^2$ is a prime. An important concept of the approach in [3] is the use of suitably generalized Walsh functions which form an orthonormal basis of the space $L_2(U^n)$. One then shows that under suitable conditions, a collection of functions that arise as coefficients of the Fourier–Walsh series of approximations to the characteristic functions of rectangular boxes $B_Y$ is *quasi-orthonormal*.

In this paper, we shall make use of the fact that under suitable conditions, many relevant subcollections of these functions are in fact orthogonal, as has been observed by Skriganov [18] in his recent work on $L_q$-discrepancy. This enables us to substantially simplify a major aspect of the proof in [3]. We establish the following improvement of the inequality (2).

**Theorem 1.** *Let $p \geq 2n^2$ be a prime. Then for every $N > 1$, a distribution $\mathcal{D}_N$ of $N$ points in the unit cube $U^n$ can be constructed explicitly to satisfy the inequality*

$$\mathcal{L}_2[\mathcal{D}_N] < 2^{1-n} p^{2n} \left( \frac{\log N}{\log p} + 2n + 1 \right)^{(n-1)/2}. \tag{3}$$

Note that the constant in the inequality (3) is not best possible, but it represents a savings of a factor $4^n$ from that in the inequality (2) nevertheless. We remark that it is not the main purpose of our work here to improve such constants.

Upper bounds of the form (1) have been given in Roth [13] or in Chen [2] with the constants $C_n$ given implicitly or inductively. In particular, this is achieved in [2] by the use of digit shifts. Here we shall show that digit shifts are in fact intimately related to the orthogonality property mentioned above. We establish the following existence result with explicitly given values for $C_n$, again not sharp.

**Theorem 2.** *Let $p \geq n - 1$ be a prime. Then for every $N > 1$, there exists a distribution $\mathcal{D}_N$ of $N$ points in the unit cube $U^n$ which satisfies the inequality*

$$\mathcal{L}_2[\mathcal{D}_N] < 2^{1-n} p^{n+1/2} \left( \frac{\log N}{\log p} + 2 \right)^{(n-1)/2}. \tag{4}$$

Throughout, the letter $p$ denotes a prime number. We shall be concerned with point sets that possess the structure of vector spaces over the finite field

$$\mathbf{F}_p = \{0, 1, \ldots, p-1\}$$

of residues modulo $p$. We shall discuss these point sets in Section 2, together with an inner product and two special metrics central to our argument. In particular, we shall state a number of results concerning these point sets, which we shall combine in Section 3 with crucial results from our work in [3] to establish Theorem 1. We then deduce Theorem 2 in Section 4, where we need a crucial result of Faure [5].

The remainder of the paper is devoted to the establishment of all the results stated in Section 2, and is organized as follows. In Section 5, we recall necessary facts on generalized Walsh functions. In Section 6, we extend the two special metrics introduced in Section 2 to $n$-tuples of nonnegative integers in order to cement the intimate relationship between our point sets and the generalized Walsh functions. In Section 7, we consider a suitable approximation of the discrepancy function which can be described by a suitably truncated Fourier–Walsh series. By making use of the roles played by orthogonality and digit shifts, we establish an expression for certain mean squares of this approximation in terms of integrals over Fourier–Walsh coefficients. Finally, we deduce Theorem 5 in Section 8 and Theorems 3 and 4 in Section 9.

For convenience, $\mathbf{N}$ denotes the set of all positive integers, $\mathbf{N}_0$ denotes the set of all nonnegative integers, $\mathbf{Z}$ denotes the set of all integers, $\mathbf{Q}$ denotes the set of all rational numbers, and $\mathbf{C}$ denotes the set of all complex numbers. If $z \in \mathbf{C}$, then $\overline{z} \in \mathbf{C}$ denotes its complex conjugate. Finally, if $\mathcal{S}$ is a finite set, then $\#(\mathcal{S})$ denotes the number of elements of $\mathcal{S}$.

## 2 Linear distributions

We shall be concerned with a class of sets $D \subset U^n$ which possess the structure of vector spaces over the finite field $\mathbf{F}_p$. For any $s \in \mathbf{N}_0$, let

$$\mathbf{Q}(p^s) = \{mp^{-s} : m = 0, 1, \ldots, p^s - 1\}.$$

Observe that any $x \in \mathbf{Q}(p^s)$ can be represented uniquely in the form

$$x = \sum_{i=1}^{s} \eta_i(x) p^{-i},$$

where the coefficients $\eta_i(x) \in \mathbf{F}_p$ for every $i = 1, \ldots, s$.

For any two vectors $X = (x_1, \ldots, x_n)$ and $Y = (y_1, \ldots, y_n)$ in $\mathbf{Q}^n(p^s)$ and any two scalars $\alpha, \beta \in \mathbf{F}_p$, we write

$$\alpha X \oplus \beta Y = (\alpha x_1 \oplus \beta y_1, \ldots, \alpha x_n \oplus \beta y_n) \in \mathbf{Q}^n(p^s) \tag{5}$$

by setting

$$\eta_i(\alpha x_j \oplus \beta y_j) = \alpha \eta_i(x_j) + \beta \eta_i(y_j) \pmod{p}$$

for every $i = 1, \ldots, s$ and $j = 1, \ldots, n$. It is easy to see that with respect to the arithmetic operations (5), the set $\mathbf{Q}^n(p^s)$ forms a vector space of dimension $ns$ over the finite field $\mathbf{F}_p$.

We say that a subset $D \subset \mathbf{Q}^n(p^s)$ is a linear distribution (in base $p$) if $D$ is a subspace of the vector space $\mathbf{Q}^n(p^s)$.

Suppose now that $s \in \mathbf{N}_0$ is chosen and fixed. Then any $x \in \mathbf{Q}(p^s)$ can also be represented in the form

$$x = \sum_{i=1}^{s} \xi_i(x) p^{i-s-1}, \tag{6}$$

where $\xi_i(x) = \eta_{s+1-i}(x) \in \mathbf{F}_p$ for every $i = 1, \ldots, s$. Using this representation, we can define an inner product on the space $\mathbf{Q}^n(p^s)$ as follows. For every $x, y \in \mathbf{Q}(p^s)$, we let

$$\langle x, y \rangle = \langle y, x \rangle = \sum_{i=1}^{s} \xi_i(x) \xi_{s+1-i}(y).$$

For vectors $X = (x_1, \ldots, x_n)$ and $Y = (y_1, \ldots, y_n)$ in $\mathbf{Q}^n(p^s)$, we write

$$\langle X, Y \rangle = \langle Y, X \rangle = \sum_{j=1}^{n} \langle x_j, y_j \rangle.$$

For any linear distribution $D \subseteq \mathbf{Q}^n(p^s)$, where $s \in \mathbf{N}_0$, we now define the dual distribution $D^\perp \subseteq \mathbf{Q}^n(p^s)$ by

$$D^\perp = \{X \in \mathbf{Q}^n(p^s) : \langle X, Y \rangle = 0 \text{ for every } Y \in D\}.$$

It is easy to check that $D^\perp$ is a subspace of $\mathbf{Q}^n(p^s)$, and is therefore also a linear distribution. Furthermore, we have $(D^\perp)^\perp = D$, so that $D$ and $D^\perp$ are mutually dual subspaces of $\mathbf{Q}^n(p^s)$.

Following [3] and [17], we next introduce two metrics on the vector space $\mathbf{Q}^n(p^s)$. For any $x \in \mathbf{Q}(p^s)$, the Hamming weight $\kappa(x)$ is the number of nonzero coefficients $\xi_i(x)$ in the representation (6), while the Rosenbloom–Tsfasman weight is defined by

$$\rho(x) = \begin{cases} 0 & \text{if } x = 0, \\ \max\{i = 1, \ldots, s : \xi_i(x) \neq 0\} & \text{if } x \neq 0; \end{cases}$$

see [11]. For $X = (x_1, \ldots, x_n) \in \mathbf{Q}^n(p^s)$, we now let

$$\kappa(X) = \sum_{j=1}^{n} \kappa(x_j) \quad \text{and} \quad \rho(X) = \sum_{j=1}^{n} \rho(x_j).$$

It is easy to check that $\kappa(X) = \rho(X) = 0$ if and only if $X = 0$. One can also easily check the triangle inequalities for both weights. These give rise to metrics (or distances) on the vector space $\mathbf{Q}^n(p^s)$.

If a linear distribution $D \subseteq \mathbf{Q}^n(p^s)$ contains at least two points, then we can consider the Hamming weight

$$\kappa(D) = \min\{\kappa(X) : X \in D \setminus \{0\}\},$$

and the Rosenbloom–Tsfasman weight

$$\rho(D) = \min\{\rho(X) : X \in D \setminus \{0\}\}.$$

We shall establish the following improvements of Lemma 2D of [3].

**Theorem 3.** *Suppose that $p \geq 2n^2$ is a prime, and that $D \subset \mathbf{Q}^n(p^s)$ is a linear distribution of $p^s$ points, with dual linear distribution $D^\perp \subset \mathbf{Q}^n(p^s)$ that satisfies $\kappa(D^\perp) \geq 2n+1$ and $\rho(D^\perp) \geq s+1$. Then*

$$\mathcal{L}_2[D] < 2^{1-n}p^n(s+1)^{(n-1)/2}. \tag{7}$$

**Theorem 4.** *Suppose that $p \geq 2n^2$ is a prime, and that $D \subset \mathbf{Q}^n(p^s)$ is a linear distribution of $p^s$ points, with dual linear distribution $D^\perp \subset \mathbf{Q}^n(p^s)$ that satisfies $\kappa(D^\perp) \geq 2n+1$ and $\rho(D^\perp) \geq s+1$. Then there exists an approximation $\mathcal{M}[D;Y]$ of the discrepancy function $\mathcal{L}[D;Y]$, with error*

$$|\mathcal{L}[D;Y] - \mathcal{M}[D;Y]| \leq n \quad \textit{for every } Y \in U^n, \tag{8}$$

*such that the quantity*

$$\mathcal{M}_2[D] = \left(\int_{U^n} |\mathcal{M}[D;Y]|^2 \,\mathrm{d}Y\right)^{1/2} \tag{9}$$

*can be evaluated precisely. Furthermore, we have*

$$\left|(\mathcal{L}_2[D])^2 - (\mathcal{M}_2[D])^2\right| \leq 2n\mathcal{M}_2[D] + 3n^2. \tag{10}$$

The approximation $\mathcal{M}[D;Y]$ will be given explicitly in Section 7.

For any linear distribution $D \subseteq \mathbf{Q}^n(p^s)$, we can consider cosets of the form

$$D \oplus T = \{X \oplus T : X \in D\}, \tag{11}$$

obtained by applying the same digit shift $T \in \mathbf{Q}^n(p^s)$ to every point of $D$.

**Theorem 5.** *Suppose that $p \geq n-1$ is a prime, and that $D \subset \mathbf{Q}^n(p^s)$ is a linear distribution of $p^s$ points, with dual linear distribution $D^\perp \subset \mathbf{Q}^n(p^s)$ that satisfies $\rho(D^\perp) \geq s+1$. Then there exists a digit shift $T \in \mathbf{Q}^n(p^s)$ such that*

$$\mathcal{L}_2[D \oplus T] < 2^{1-n}p^n(s+1)^{(n-1)/2}. \tag{12}$$

The restrictions imposed on the prime $p$ in terms of the dimension $n$ in all our results here can be relaxed if we work with linear distributions with *deficiencies*. More precisely, for each prime $p$, including $p = 2$, we can explicitly construct linear distributions $D \subset \mathbf{Q}^n(p^s)$ of $p^s$ points with corresponding dual linear distributions $D^\perp \subset \mathbf{Q}^n(p^s)$ satisfying $\rho(D^\perp) \geq s+1-\delta$, where the deficiency $\delta$ is a nonnegative integer which depends only on the dimension $n$ and satisfies the bound $\delta = O(n \log n)$. This approach will lead only to a renormalization of the constants in our bounds (7) and (12), but will not necessarily improve the estimates, and so appears to be not worth persuing. However, the benefit of this approach is that with the choice $p = 2$, it is possible to obtain very precise information on the discrepancy of linear distributions in terms of the distribution of the Rosenbloom–Tsfasman weight within the dual linear distributions; see the paragraph after the proof of Lemma 7.5. This leads in turn to very accurate estimates for the mean squares discrepancy of such linear distributions.

## 3 Deduction of Theorem 1

We shall proceed along the lines of [3], but shall omit some of the details by summarizing lengthy steps as lemmas and giving references where appropriate. Let $g = 2n$, and let $p \geq gn = 2n^2$ be a prime. Given any natural number $N > 1$, we choose $\sigma \in \mathbf{N}$ such that

$$p^{g(\sigma-1)} < N \leq p^{g\sigma}, \tag{13}$$

and consider first of all a linear distribution of $p^{g\sigma}$ points in $U^n$.

The following result is essentially Lemma 2E of [3].

**Lemma 3.1.** *Suppose that $p \geq gn$ is a prime, where $g = 2n$. Then for every $\sigma \in \mathbf{N}$, a linear distribution $D(g,\sigma) \subset \mathbf{Q}^n(p^{g\sigma})$ of $p^{g\sigma}$ points can be constructed explicitly, with dual linear distribution $(D(g,\sigma))^\perp \subset \mathbf{Q}^n(p^{g\sigma})$ satisfying*

$$\kappa((D(g,\sigma))^\perp) \geq g+1 \quad \textit{and} \quad \rho((D(g,\sigma))^\perp) \geq g\sigma + 1.$$

Clearly the hypotheses of Theorem 3 are satisfied with $g = 2n$ and $s = g\sigma$, and so it follows immediately that

$$\mathcal{L}_2[D(g,\sigma)] < 2^{1-n} p^n (g\sigma + 1)^{(n-1)/2}.$$

Our next task is to select a subset of $D(g,\sigma)$ and rescale. Consider the subset

$$\mathcal{D}_N^*(g) = D(g,\sigma) \cap ([0, Np^{-g\sigma}) \times U^{n-1}) \subseteq D(g,\sigma).$$

To guarantee that $\mathcal{D}_N^*(g)$ contains exactly $N$ points, we need the following special case of Lemma 2C of [3] or Theorem 4.2 of [17]. It relates the uniform distribution of the points of a linear distribution $D$ and the spacing of the points of the dual linear distribution $D^\perp$ with respect to the Rosenbloom–Tsfasman metric $\rho$.

**Lemma 3.2.** *Suppose that $D \subset \mathbf{Q}^n(p^s)$ is a linear distribution of $p^s$ points, with dual linear distribution $D^\perp \subset \mathbf{Q}^n(p^s)$. Then the following statements are equivalent:*

(i) *The Rosenbloom–Tsfasman weight $\rho(D^\perp) \geq s + 1$.*
(ii) *Every rectangular box of the type*

$$\prod_{j=1}^{n} [m_j p^{-s_j}, (m_j+1)p^{-s_j}) \subset U^n,$$

*where $m_1, \ldots, m_n, s_1, \ldots, s_n \in \mathbf{N}_0$ satisfy $s_1 + \ldots + s_n = s$, contains precisely one point of the linear distribution $D$.*

To see that $\mathcal{D}_N^*(g)$ contains exactly $N$ points, we observe simply that every rectangular box of the form $[mp^{-g\sigma}, (m+1)p^{-g\sigma}) \times U^{n-1} \subset U^n$, where $m \in \mathbf{N}_0$, contains exactly one point of $D(g,\sigma)$.

We next rescale $\mathcal{D}_N^*(g)$ to obtain

$$\mathcal{D}_N = \mathcal{D}_N(g) = \{(N^{-1}p^{g\sigma}x_1, x_2, \ldots, x_n) : (x_1, x_2, \ldots, x_n) \in \mathcal{D}_N^*(g)\}.$$

Then in view of (13) and noting that $g = 2n$, we have

$$\int_{U^n} |\mathcal{L}[\mathcal{D}_N; Y]|^2 \,\mathrm{d}Y = N^{-1} p^{g\sigma} \int_{[0,Np^{-g\sigma})\times U^{n-1}} |\mathcal{L}[D(g,\sigma); Y]|^2 \,\mathrm{d}Y$$

$$\leq N^{-1} p^{g\sigma} \int_{U^n} |\mathcal{L}[D(g,\sigma); Y]|^2 \,\mathrm{d}Y < 4^{1-n} N^{-1} p^{g\sigma} p^{2n} (g\sigma + 1)^{n-1}$$

$$< 4^{1-n} p^g p^{2n} \left(\frac{\log N}{\log p} + g + 1\right)^{n-1} = 4^{1-n} p^{4n} \left(\frac{\log N}{\log p} + 2n + 1\right)^{n-1}.$$

The inequality (3) now follows on taking square roots.

## 4 Deduction of Theorem 2

Let $p \geq n-1$ be a prime. Given any natural number $N > 1$, we choose $s \in \mathbf{N}$ such that

$$p^{s-1} < N \leq p^s, \tag{14}$$

and consider first of all a linear distribution of $p^s$ points in $U^n$.

The following result is due to Faure [5]. We remark that the condition $p \geq n-1$ cannot be relaxed, as observed by Chen [2].

**Lemma 4.1.** *Suppose that $p \geq n-1$ is a prime. Then for every $s \in \mathbf{N}$, a linear distribution $D \subset \mathbf{Q}^n(p^s)$ of $p^s$ points can be constructed explicitly such that condition (ii) of Lemma 3.2 is satisfied, so that the dual linear distribution $D^\perp \subset \mathbf{Q}^n(p^s)$ has Rosenbloom–Tsfasman weight $\rho(D^\perp) \geq s+1$.*

It follows from Theorem 5 that there exists a digit shift $T \in \mathbf{Q}^n(p^s)$ such that the inequality (12) holds. Next, observe that condition (ii) of Lemma 3.2 remains valid if we replace the linear distribution $D$ by its coset $D \oplus T$, and so the subset

$$\mathcal{D}_N^* = (D \oplus T) \cap ([0, Np^{-s}) \times U^{n-1}) \subseteq D \oplus T$$

contains exactly $N$ points. We now rescale $\mathcal{D}_N^*$ to obtain

$$\mathcal{D}_N = \{(N^{-1} p^s x_1, x_2, \ldots, x_n) : (x_1, x_2, \ldots, x_n) \in \mathcal{D}_N^*\}.$$

Then in view of (14), we have

$$\int_{U^n} |\mathcal{L}[\mathcal{D}_N; Y]|^2 \,\mathrm{d}Y = N^{-1} p^s \int_{[0,Np^{-s})\times U^{n-1}} |\mathcal{L}[D \oplus T; Y]|^2 \,\mathrm{d}Y$$

$$\leq N^{-1} p^s \int_{U^n} |\mathcal{L}[D \oplus T; Y]|^2 \,\mathrm{d}Y < 4^{1-n} N^{-1} p^s p^{2n} (s+1)^{n-1}$$

$$< 4^{1-n} p^{2n+1} \left(\frac{\log N}{\log p} + 2\right)^{n-1}.$$

The inequality (4) now follows on taking square roots.

## 5 Walsh functions

Every $\ell \in \mathbf{N}_0$ can be written uniquely in the form

$$\ell = \sum_{i=1}^{\infty} \lambda_i(\ell) p^{i-1}, \tag{15}$$

where the coefficients $\lambda_i(\ell) \in \mathbf{F}_p$ for every $i \in \mathbf{N}$.

For any two vectors $L = (\ell_1, \ldots, \ell_n)$ and $K = (k_1, \ldots, k_n)$ in $\mathbf{N}_0^n$ and any two scalars $\alpha, \beta \in \mathbf{F}_p$, we write

$$\alpha L \oplus \beta K = (\alpha \ell_1 \oplus \beta k_1, \ldots, \alpha \ell_n \oplus \beta k_n) \in \mathbf{N}_0^n \tag{16}$$

by setting

$$\lambda_i(\alpha \ell_j \oplus \beta k_j) = \alpha \lambda_i(\ell_j) + \beta \lambda_i(k_j) \pmod{p}$$

for every $i \in \mathbf{N}$ and $j = 1, \ldots, n$. It is easy to see that with respect to the arithmetic operations (16), the set $\mathbf{N}_0^n$ forms a vector space over the finite field $\mathbf{F}_p$.

On the other hand, every $x \in U$ can be represented in the form

$$x = \sum_{i=1}^{\infty} \eta_i(x) p^{-i}, \tag{17}$$

where the coefficients $\eta_i(x) \in \mathbf{F}_p$ for every $i \in \mathbf{N}$, and this representation is unique if we agree that the series in (17) is finite if

$$x \in \mathbf{Q}(p^\infty) = \bigcup_{s=0}^{\infty} \mathbf{Q}(p^s).$$

In this case, for any two vectors $X = (x_1, \ldots, x_n)$ and $Y = (y_1, \ldots, y_n)$ in $\mathbf{Q}^n(p^\infty)$ and any two scalars $\alpha, \beta \in \mathbf{F}_p$, we can extend (5) to

$$\alpha X \oplus \beta Y = (\alpha x_1 \oplus \beta y_1, \ldots, \alpha x_n \oplus \beta y_n) \in \mathbf{Q}^n(p^\infty) \tag{18}$$

by setting

$$\eta_i(\alpha x_j \oplus \beta y_j) = \alpha \eta_i(x_j) + \beta \eta_i(y_j) \pmod{p}$$

for every $i \in \mathbf{N}$ and $j = 1, \ldots, n$. It is easy to see that with respect to the arithmetic operations (18), the set $\mathbf{Q}^n(p^\infty)$ forms a vector space over the finite field $\mathbf{F}_p$.

For every $\ell \in \mathbf{N}_0$ and every $x \in U$, we let

$$w_\ell(x) = e_p\left(\sum_{i=1}^{\infty} \lambda_i(\ell) \eta_i(x)\right), \tag{19}$$

where $e_p(z) = \mathrm{e}^{2\pi \mathrm{i} z/p}$ for every real number $z$, and where the coefficients $\lambda_i(\ell)$ and $\eta_i(x)$ are given by (15) and (17) respectively. The functions $w_\ell$ are known as the Walsh functions if $p = 2$ and the Chrestenson or Chrestenson–Levy functions if $p > 2$. For simplicity, we refer to them all as Walsh functions here. A detailed study of such functions can be found in [8] or [14]. Here it suffices to mention that while $w_0(x) = 1$ for every $x \in U$, we have

$$\int_U w_\ell(x)\,\mathrm{d}x = 0 \quad \text{for every } \ell \in \mathbf{N}. \tag{20}$$

For every $L = (\ell_1, \dots, \ell_n) \in \mathbf{N}_0^n$ and every $X = (x_1, \dots, x_n) \in U^n$, we let

$$W_L(X) = \prod_{j=1}^{n} w_{\ell_j}(x_j). \tag{21}$$

It is well known that

$$W_{L \oplus K}(X) = W_L(X) W_K(X) \quad \text{for every } X \in U^n \text{ and } L, K \in \mathbf{N}_0^n, \tag{22}$$

and that

$$W_L(X \oplus Y) = W_L(X) W_L(Y) \quad \text{for every } X, Y \in \mathbf{Q}^n(p^\infty) \text{ and } L \in \mathbf{N}_0^n. \tag{23}$$

Furthermore, for every $K, L \in \mathbf{N}_0^n$, we have

$$\int_{U^n} W_K(X)\overline{W_L(X)}\,\mathrm{d}X = \int_{U^n} W_{K \ominus L}(X)\,\mathrm{d}X = \begin{cases} 1 & \text{if } K = L, \\ 0 & \text{if } K \neq L. \end{cases}$$

Indeed, the Walsh functions form an orthonormal basis of the Hilbert space $L_2(U^n)$ of square-integrable functions on the $n$-dimensional unit cube $U^n$. For each $f \in L_2(U^n)$, we have the Fourier–Walsh expansion

$$f(X) \simeq \sum_{L \in \mathbf{N}_0^n} \widetilde{f}_L \overline{W_L(X)},$$

where the symbol $\simeq$ denotes that the series converges in the $L_2$-norm, and where the Fourier–Walsh coefficients are given by

$$\widetilde{f}_L = \int_{U^n} W_L(X) f(X)\,\mathrm{d}X.$$

Known results on characters of abelian groups (cf. [9]) can often be restated in terms of Walsh functions. Here we need the following result. Let

$$\mathbf{N}_0^n(p^s) = \{L = (\ell_1, \dots, \ell_n) \in \mathbf{N}_0^n : 0 \le \ell_j < p^s \text{ for every } j = 1, \dots, n\}.$$

The mapping

$$\theta : \mathbf{Q}^n(p^s) \to \mathbf{N}_0^n(p^s) : (x_1, \dots, x_n) \mapsto (p^s x_1, \dots, p^s x_n) \tag{24}$$

is clearly an isomorphism of vector spaces.

**Lemma 5.1.** *For every linear distribution $D \subseteq \mathbf{Q}^n(p^s)$ and every $L \in \mathbf{N}_0^n(p^s)$, we have*

$$\sum_{X \in D} W_L(X) = \begin{cases} \#(D) & \text{if } L \in \theta(D^\perp), \\ 0 & \text{if } L \notin \theta(D^\perp), \end{cases}$$

*where $\theta(D^\perp) = \{\theta(Y) : Y \in D^\perp\}$ denotes the image under the mapping* (24) *of the dual linear distribution $D^\perp \subseteq \mathbf{Q}^n(p^s)$.*

A special case of this is the useful orthogonality result below.

**Lemma 5.2.** *For every $L', L'' \in \mathbf{N}_0^n(p^s)$, we have*

$$\sum_{T\in\mathbf{Q}^n(p^s)} \overline{W_{L'}(T)}\, W_{L''}(T) = \begin{cases} p^{ns} & \text{if } L' = L'', \\ 0 & \text{otherwise.} \end{cases} \tag{25}$$

## 6 More weights and metrics

In Section 2, we considered Hamming and Rosenbloom–Tsfasman weights defined on elements in $\mathbf{Q}^n(p^s)$. The purpose of this section is to consider their analogues on $\mathbf{N}_0^n$.

For any $\ell \in \mathbf{N}_0$, the Hamming weight $\kappa(\ell)$ denotes the number of non-zero coefficients $\lambda_i(\ell)$ in the representation (15), while the Rosenbloom–Tsfasman weight is defined by

$$\rho(\ell) = \begin{cases} 0 & \text{if } \ell = 0, \\ \max\{i \in \mathbf{N} : \lambda_i(\ell) \neq 0\} & \text{if } \ell \in \mathbf{N}. \end{cases}$$

Note that for every $\ell \in \mathbf{N}$, we have

$$p^{\rho(\ell)-1} \leq \ell < p^{\rho(\ell)}.$$

For $L = (\ell_1, \ldots, \ell_n) \in \mathbf{N}_0^n$, we now let

$$\kappa(L) = \sum_{j=1}^n \kappa(\ell_j) \quad \text{and} \quad \rho(L) = \sum_{j=1}^n \rho(\ell_j). \tag{26}$$

It is easy to check that $\kappa(L) = \rho(L) = 0$ if and only if $L = 0$. One can also easily check the triangle inequalities for both weights. These give rise to metrics (or distances) on the vector space $\ell \in \mathbf{N}_0^n$.

These metrics are intimately related to those defined in Section 2 on elements in $\mathbf{Q}^n(p^s)$. It is not difficult to see that the mapping (24) is an isomorphism that preserves the metrics $\kappa$ and $\rho$. More precisely, for every $X \in \mathbf{Q}^n(p^s)$, we have

$$\kappa(X) = \kappa(\theta(X)) \quad \text{and} \quad \rho(X) = \rho(\theta(X)). \tag{27}$$

## 7 Approximation of the discrepancy function

For any $Y = (y_1, \ldots, y_n) \in U^n$, we consider the characteristic function $\chi(Y, X)$ of the rectangular box $B_Y = [0, y_1) \times \ldots \times [0, y_n)$, so that

$$\chi(Y, X) = \begin{cases} 1 & \text{if } X \in B_Y, \\ 0 & \text{if } X \notin B_Y. \end{cases}$$

It is clear that if $X = (x_1, \ldots, x_n)$, then

$$\chi(Y, X) = \prod_{j=1}^n \chi(y_j, x_j),$$

where for every $j = 1, \dots, n$,

$$\chi(y_j, x_j) = \begin{cases} 1 & \text{if } x_j \in [0, y_j), \\ 0 & \text{if } x_j \notin [0, y_j). \end{cases}$$

For any linear distribution $D \subset \mathbf{Q}^n(p^s)$ of $p^s$ points, we have

$$\mathcal{L}[D; Y] = \sum_{X \in D} \chi(Y, X) - p^s y_1 \dots y_n.$$

The function $\chi(y, x)$ has a Fourier–Walsh expansion of the form

$$\chi(y, x) \simeq \sum_{\ell=0}^{\infty} \widetilde{\chi}_\ell(y) \overline{w_\ell(x)},$$

where for every $\ell \in \mathbf{N}_0$, the Fourier–Walsh coefficients are defined by

$$\widetilde{\chi}_\ell(y) = \int_0^y w_\ell(x)\, \mathrm{d}x.$$

In particular, we have $\widetilde{\chi}_0(y) = y$.

Following [3], for given $s \in \mathbf{N}_0$, we approximate $\chi(y, x)$ by the truncated series

$$\chi_s(y, x) = \sum_{\ell=0}^{p^s-1} \widetilde{\chi}_\ell(y) \overline{w_\ell(x)}, \tag{28}$$

the characteristic function $\chi(Y, X)$ by the product

$$\chi_s(Y, X) = \prod_{j=1}^{n} \chi_s(y_j, x_j), \tag{29}$$

and the discrepancy function $\mathcal{L}[D; Y]$ by

$$\mathcal{M}[D; Y] = \sum_{X \in D} \chi_s(Y, X) - p^s y_1 \dots y_n. \tag{30}$$

The following estimate, essentially Lemma 6A of [3], gives a bound for the error of this approximation process.

**Lemma 7.1.** *Suppose that $D \subset \mathbf{Q}^n(p^s)$ is a linear distribution of $p^s$ points, with dual linear distribution $D^\perp \subset \mathbf{Q}^n(p^s)$ satisfying $\rho(D^\perp) \geq s + 1$. Then for every $Y \in U^n$, we have $|\mathcal{L}[D; Y] - \mathcal{M}[D; Y]| \leq n$.*

For every $L = (\ell_1, \dots, \ell_n) \in \mathbf{N}_0^n$ and $Y = (y_1, \dots, y_n) \in U^n$, write

$$\widetilde{\chi}_L(Y) = \widetilde{\chi}_{\ell_1}(y_1) \dots \widetilde{\chi}_{\ell_n}(y_n). \tag{31}$$

In view of (28)–(31), (21) and Lemma 5.1, we have

$$\begin{aligned}\mathcal{M}[D;Y] &= \sum_{X\in D}\sum_{L\in\mathbf{N}_0^n(p^s)} \widetilde{\chi}_L(Y)\,\overline{W_L(X)} - p^s\widetilde{\chi}_0(y_1)\ldots\widetilde{\chi}_0(y_n)\\ &= \sum_{L\in\mathbf{N}_0^n(p^s)}\left(\sum_{X\in D}\overline{W_L(X)}\right)\widetilde{\chi}_L(Y) - p^s\widetilde{\chi}_0(Y)\\ &= p^s\sum_{L\in\theta(D^\perp)\setminus\{0\}}\widetilde{\chi}_L(Y). \end{aligned} \tag{32}$$

The result below follows immediately.

**Lemma 7.2.** *Suppose that $D\subset\mathbf{Q}^n(p^s)$ is a linear distribution of $p^s$ points. Suppose further that for any distinct $L', L''\in\theta(D^\perp)\setminus\{0\}$, the functions $\widetilde{\chi}_{L'}$ and $\widetilde{\chi}_{L''}$ are orthogonal to each other. Then*

$$\int_{U^n}|\mathcal{M}[D;Y]|^2\,\mathrm{d}Y = p^{2s}\sum_{L\in\theta(D^\perp)\setminus\{0\}}\int_{U^n}|\widetilde{\chi}_L(Y)|^2\,\mathrm{d}Y. \tag{33}$$

Next, we consider digit shifts.

**Lemma 7.3.** *Suppose that $D\subset\mathbf{Q}^n(p^s)$ is a linear distribution of $p^s$ points. Then*

$$\frac{1}{p^{ns}}\sum_{T\in\mathbf{Q}^n(p^s)}\int_{U^n}|\mathcal{M}[D\oplus T;Y]|^2\,\mathrm{d}Y = p^{2s}\sum_{L\in\theta(D^\perp)\setminus\{0\}}\int_{U^n}|\widetilde{\chi}_L(Y)|^2\,\mathrm{d}Y. \tag{34}$$

*Proof.* For every fixed $T\in\mathbf{Q}^n(p^s)$, the coset $D\oplus T$, obtained by applying the same digit shift to every point of $D$ and given by (11), satisfies

$$\begin{aligned}\mathcal{M}[D\oplus T;Y] &= \sum_{X\in D}\sum_{L\in\mathbf{N}_0^n(p^s)} \widetilde{\chi}_L(Y)\,\overline{W_L(X\oplus T)} - p^s\widetilde{\chi}_0(y_1)\ldots\widetilde{\chi}_0(y_n)\\ &= \sum_{L\in\mathbf{N}_0^n(p^s)}\overline{W_L(T)}\left(\sum_{X\in D}\overline{W_L(X)}\right)\widetilde{\chi}_L(Y) - p^s\widetilde{\chi}_0(Y)\\ &= p^s\sum_{L\in\theta(D^\perp)\setminus\{0\}}\overline{W_L(T)}\,\widetilde{\chi}_L(Y), \end{aligned}$$

in view of (23). It follows that

$$\begin{aligned}
&\sum_{T\in\mathbf{Q}^n(p^s)} |\mathcal{M}[D\oplus T;Y]|^2\\
&= p^{2s}\sum_{T\in\mathbf{Q}^n(p^s)}\left|\sum_{L\in\theta(D^\perp)\setminus\{0\}} \overline{W_L(T)}\,\widetilde{\chi}_L(Y)\right|^2\\
&= p^{2s}\sum_{T\in\mathbf{Q}^n(p^s)}\sum_{L',L''\in\theta(D^\perp)\setminus\{0\}} \overline{W_{L'}(T)}\,W_{L''}(T)\,\widetilde{\chi}_{L'}(Y)\,\overline{\widetilde{\chi}_{L''}(Y)}\\
&= p^{2s}\sum_{L',L''\in\theta(D^\perp)\setminus\{0\}}\left(\sum_{T\in\mathbf{Q}^n(p^s)} \overline{W_{L'}(T)}\,W_{L''}(T)\right)\widetilde{\chi}_{L'}(Y)\,\overline{\widetilde{\chi}_{L''}(Y)}\\
&= p^{ns}p^{2s}\sum_{L\in\theta(D^\perp)\setminus\{0\}} |\widetilde{\chi}_L(Y)|^2,
\end{aligned}$$

in view of Lemma 5.2. The identity (34) then follows immediately on integrating over $Y\in U^n$. □

Note that the right hand sides of (33) and (34) are identical. The former is a consequence of the orthogonality of the Fourier–Walsh coefficients, while the latter is a consequence of the orthogonality condition (25) brought into play by the digit shifts. Noting (31), we see that to progress further, we clearly need to study the integral

$$\int_{U^n} |\widetilde{\chi}_L(Y)|^2\,\mathrm{d}Y = \prod_{j=1}^{n}\int_U |\widetilde{\chi}_{\ell_j}(y_j)|^2\,\mathrm{d}y_j. \tag{35}$$

**Lemma 7.4.** *We have*

$$\int_U |\widetilde{\chi}_0(y)|^2\,\mathrm{d}y = \frac{1}{4}+\frac{1}{4(p^2-1)}\sum_{j=1}^{p-1}\csc^2\frac{\pi j}{p}. \tag{36}$$

*Furthermore, for every $\ell\in\mathbf{N}$, we have*

$$\int_U |\widetilde{\chi}_\ell(y)|^2\,\mathrm{d}y = p^{-2\rho(\ell)}\left(\frac{1}{2}\csc^2\frac{\pi\lambda(\ell)}{p}-\frac{1}{4}+\frac{1}{4(p^2-1)}\sum_{j=1}^{p-1}\csc^2\frac{\pi j}{p}\right), \tag{37}$$

*where $\lambda(\ell)=\lambda_{\rho(\ell)}(\ell)$ denotes the leading coefficient in the $p$-ary expansion* (15) *of $\ell$.*

*Proof.* We have the Fine–Price formula, that for every $\ell\in\mathbf{N}_0$,

$$\widetilde{\chi}_\ell(y) = p^{-\rho(\ell)}u_\ell(y), \tag{38}$$

where

$$u_0(y) = \frac{1}{2}w_0(y)+\sum_{i=1}^{\infty}p^{-i}\sum_{j=1}^{p-1}\zeta^j(1-\zeta^j)^{-1}w_{jp^{i-1}}(y), \tag{39}$$

and where for every $\ell \in \mathbf{N}$,

$$u_\ell(y) = (1 - \zeta^{\lambda(\ell)})^{-1} w_{\tau(\ell)}(y) + \left(\frac{1}{2} - (1 - \zeta^{\lambda(\ell)})^{-1}\right) w_\ell(y)$$
$$+ \sum_{i=1}^{\infty} p^{-i} \sum_{j=1}^{p-1} \zeta^j (1 - \zeta^j)^{-1} w_{\ell + jp^{\rho(\ell)+i-1}}(y). \tag{40}$$

Here $\tau(\ell) = \ell - \lambda(\ell) p^{\rho(\ell)-1}$, and $\zeta = \mathrm{e}^{2\pi \mathrm{i}/p}$ is a primitive $p$-th root of unity. For details, see Fine [6] and Price [10]. The right hand side of (40) is a linear combination of distinct Walsh functions. It follows that for every $\ell \in \mathbf{N}$, we have

$$\int_U |u_\ell(y)|^2 \, \mathrm{d}y = \frac{1}{(1 - \zeta^{\lambda(\ell)})(1 - \zeta^{-\lambda(\ell)})} + \left(\frac{1}{2} - \frac{1}{1 - \zeta^{\lambda(\ell)}}\right)\left(\frac{1}{2} - \frac{1}{1 - \zeta^{-\lambda(\ell)}}\right)$$
$$+ \sum_{i=1}^{\infty} p^{-2i} \sum_{j=1}^{p-1} |1 - \zeta^j|^{-2}$$
$$= 2|1 - \zeta^{\lambda(\ell)}|^{-2} - \frac{1}{4} + \frac{1}{p^2 - 1} \sum_{j=1}^{p-1} |1 - \zeta^j|^{-2}. \tag{41}$$

The identity (37) follows on combining (38) and (41) with the observation

$$|1 - \zeta^j|^2 = \left(1 - \cos\frac{2\pi j}{p}\right)^2 + \sin^2\frac{2\pi j}{p} = 4 \sin^2 \frac{\pi j}{p}. \tag{42}$$

Similarly, we have

$$\int_U |u_0(y)|^2 \, \mathrm{d}y = \frac{1}{4} + \sum_{i=1}^{\infty} p^{-2i} \sum_{j=1}^{p-1} |1 - \zeta^j|^{-2} = \frac{1}{4} + \frac{1}{p^2 - 1} \sum_{j=1}^{p-1} |1 - \zeta^j|^{-2}. \tag{43}$$

The identity (36) follows on combining (38), (42) and (43). □

**Lemma 7.5.** *For every $L \in \mathbf{N}_0^n$, we have*

$$\int_{U^n} |\widetilde{\chi}_L(Y)|^2 \, \mathrm{d}Y \leq \frac{p^{2n - 2\rho(L)}}{4^n}.$$

*Proof.* In view of (35), it suffices to show that for every $\ell \in \mathbf{N}_0$, we have

$$\int_U |\widetilde{\chi}_\ell(y)|^2 \, \mathrm{d}y \leq \frac{p^{2 - 2\rho(\ell)}}{4}.$$

Suppose first of all that $\ell \neq 0$. Then using the inequality that

$$\csc^2 \frac{\pi j}{p} \leq \frac{p^2}{4} \quad \text{for every } j = 1, \ldots, p - 1,$$

we see from (37) that

$$\int_U |\widetilde{\chi}_\ell(y)|^2\,\mathrm{d}y \le p^{-2\rho(\ell)}\left(\frac{p^2}{8}+\frac{1}{4}+\frac{p^2(p-1)}{16(p^2-1)}\right) \le \frac{p^{2-2\rho(\ell)}}{4}.$$

On the other hand, it follows similarly from (36) that

$$\int_U |\widetilde{\chi}_0(y)|^2\,\mathrm{d}y \le \frac{1}{4}+\frac{p^2(p-1)}{16(p^2-1)} \le \frac{p^2}{4} = \frac{p^{2-2\rho(0)}}{4}.$$

□

Here we take a digression and make a brief comment on the case $p=2$, where it is easy to show that

$$\int_{U^n} |\widetilde{\chi}_L(Y)|^2\,\mathrm{d}Y = \frac{4^{-\rho(L)}}{3^n}.$$

Suppose that $D$ is a linear distribution of $2^s$ points. Then it follows immediately from Lemma 7.3 that

$$\frac{1}{2^{ns}}\sum_{T\in\mathbf{Q}^n(2^s)}\int_{U^n} |\mathcal{M}[D\oplus T;Y]|^2\,\mathrm{d}Y = \frac{4^s}{3^n}\sum_{L\in\theta(D^\perp)\setminus\{0\}} 4^{-\rho(L)}.$$

Furthermore, it follows immediately from Lemma 7.2 that

$$\int_{U^n} |\mathcal{M}[D;Y]|^2\,\mathrm{d}Y = \frac{4^s}{3^n}\sum_{L\in\theta(D^\perp)\setminus\{0\}} 4^{-\rho(L)}, \tag{44}$$

provided that the functions $\widetilde{\chi}_{L'}$ and $\widetilde{\chi}_{L''}$, where $L', L''\in\theta(D^\perp)\setminus\{0\}$, are orthogonal to each other. The formula (44) shows that the $L_2$-norm of the approximation $M[D;Y]$ coincides with a Rosenbloom–Tsfasman enumerator for the subspace $D^\perp$.

We now return to our main discussion. The following estimate is given by Lemma 6D of [3].

**Lemma 7.6.** *Suppose that $D\subset\mathbf{Q}^n(p^s)$ is a linear distribution of $p^s$ points, with dual linear distribution $D^\perp\subset\mathbf{Q}^n(p^s)$ satisfying $\rho(D^\perp)\ge s+1$. Then*

$$\sum_{L\in\theta(D^\perp)\setminus\{0\}} p^{2s-2\rho(L)} < (s+1)^{n-1}.$$

Combining Lemmas 7.5 and 7.6, we conclude that

$$p^{2s}\sum_{L\in\theta(D^\perp)\setminus\{0\}}\int_{U^n} |\widetilde{\chi}_L(Y)|^2\,\mathrm{d}Y \le \frac{p^{2n}}{4^n}(s+1)^{n-1}. \tag{45}$$

## 8 Deduction of Theorem 5

Suppose that $p\ge n-1$ is a prime. Suppose further that $D\subset\mathbf{Q}^n(p^s)$ is a linear distribution of $p^s$ points, with dual linear distribution $D^\perp\subset\mathbf{Q}^n(p^s)$ that satisfies

$\rho(D^\perp) \geq s+1$. Combining Lemma 7.3 and the inequality (45), we conclude that

$$\frac{1}{p^{ns}} \sum_{T \in \mathbf{Q}^n(p^s)} \int_{U^n} |\mathcal{M}[D \oplus T; Y]|^2 \,\mathrm{d}Y \leq \frac{p^{2n}}{4^n}(s+1)^{n-1}.$$

It follows that there exists a digit shift $T \in \mathbf{Q}^n(p^s)$ such that

$$\int_{U^n} |\mathcal{M}[D \oplus T; Y]|^2 \,\mathrm{d}Y \leq \frac{p^{2n}}{4^n}(s+1)^{n-1}.$$

It is not too difficult to check that the conclusion of Lemma 7.1 remains valid for this coset $D \oplus T$, so that

$$\begin{aligned}\int_{U^n} |\mathcal{L}[D \oplus T; Y]|^2 \,\mathrm{d}Y &\leq 2 \int_{U^n} |\mathcal{M}[D \oplus T; Y]|^2 \,\mathrm{d}Y + 2n^2 \\ &\leq \frac{2p^{2n}}{4^n}(s+1)^{n-1} + 2n^2 \leq \frac{4p^{2n}}{4^n}(s+1)^{n-1}, \end{aligned} \tag{46}$$

where the last inequality is valid with the possible exception of the cases

$$\begin{cases} s = 0, \\ s = 1,\ p = 2,\ n = 2, \\ s = 1,\ p = 2,\ n = 3, \\ s = 2,\ p = 2,\ n = 2. \end{cases} \tag{47}$$

The inequality (12) follows immediately from the inequality (46) on taking square roots. On the other hand, the inequality (12) holds trivially for each of the first three exceptional cases in (47). For the remaining case, we simply note that the uniformity of the linear distribution $D$ implies that $|\mathcal{L}[D \oplus T; Y]| \leq 3$ for every $Y \in U^2$, so that the inequality (12) follows again. This completes the proof of Theorem 5.

## 9 Deduction of Theorems 3 and 4

The following crucial orthogonality relationship arises from the very recent work of Skriganov [18] on $L_q$-discrepancy.

**Lemma 9.1.** *Suppose that $D \subset \mathbf{Q}^n(p^s)$ is a linear distribution of $p^s$ points, with dual linear distribution $D^\perp \subset \mathbf{Q}^n(p^s)$ satisfying $\kappa(D^\perp) \geq 2n+1$. Then for any distinct $L', L'' \in \theta(D^\perp) \setminus \{0\}$, the functions $\widetilde{\chi}_{L'}$ and $\widetilde{\chi}_{L''}$ are orthogonal to each other.*

Suppose that $p \geq 2n^2$ is a prime. Suppose further that $D \subset \mathbf{Q}^n(p^s)$ is a linear distribution of $p^s$ points, with dual linear distribution $D^\perp \subset \mathbf{Q}^n(p^s)$ that satisfies

$$\kappa(D^\perp) \geq 2n+1 \quad \text{and} \quad \rho(D^\perp) \geq s+1.$$

The assertions (8) and (9) of Theorem 4 follow immediately from Lemmas 7.1, 7.2, 9.1 and 7.4, together with the identity (35). The assertion (10) of Theorem 4 follows as a simple consequence of the inequality (8).

Furthermore, combining Lemmas 9.1 and 7.2 with the inequality (45), we have

$$\int_{U^n} |\mathcal{M}[D; Y]|^2 \,\mathrm{d}Y \leq \frac{p^{2n}}{4^n}(s+1)^{n-1}.$$

It then follows from Lemma 7.1 that

$$\int_{U^n} |\mathcal{L}[D;Y]|^2\,\mathrm{d}Y \leq 2\int_{U^n} |\mathcal{M}[D;Y]|^2\,\mathrm{d}Y + 2n^2$$

$$\leq \frac{2p^{2n}}{4^n}(s+1)^{n-1} + 2n^2 \leq \frac{4p^{2n}}{4^n}(s+1)^{n-1}.$$

The inequality (7) follows immediately on taking square roots. This completes the proof of Theorem 3.

We complete this paper by giving the very short proof of Lemma 9.1. Note first that a consequence of (31) is the identity

$$\int_{U^n} \widetilde{\chi}_{L'}(Y)\,\overline{\widetilde{\chi}_{L''}(Y)}\,\mathrm{d}Y = \prod_{j=1}^{n}\int_{U} \widetilde{\chi}_{\ell'_j}(y_j)\,\overline{\widetilde{\chi}_{\ell''_j}(y_j)}\,\mathrm{d}y_j.$$

On the other hand, the condition $\kappa(D^{\perp}) \geq 2n+1$ and the relationship (27) imply that for any distinct $L', L'' \in \theta(D^{\perp})\setminus\{0\}$, we must have $\kappa(L'\ominus L'') \geq 2n+1$. It follows from (26) and the pigeonhole principle that $\kappa(\ell'_j \ominus \ell''_j) \geq 3$ for some $j = 1,\ldots,k$. Hence Lemma 9.1 is an immediate consequence of the following one-dimensional result.

**Lemma 9.2.** *Suppose that $\ell', \ell'' \in \mathbf{N}_0$ and $\kappa(\ell'\ominus\ell'') \geq 3$. Then $\widetilde{\chi}_{\ell'}$ and $\widetilde{\chi}_{\ell''}$ are orthogonal to each other.*

*Proof.* The Fine–Price formula (38)–(40) can be rewritten in the following form. For every $\ell \in \mathbf{N}_0$, we have

$$\widetilde{\chi}_{\ell}(y) = p^{-\rho(\ell)} w_{\ell}(y) v_{\ell}(y), \tag{48}$$

where

$$v_0(y) = \frac{1}{2}w_0(y) + \sum_{i=1}^{\infty} p^{-i}\sum_{j=1}^{p-1} \zeta^j(1-\zeta^j)^{-1} w_{jp^{i-1}}(y), \tag{49}$$

and where for every $\ell \in \mathbf{N}$,

$$v_{\ell}(y) = (1-\zeta^{\lambda(\ell)})^{-1} w_{\lambda(\ell)p^{\rho(\ell)-1}}(y) + \left(\frac{1}{2} - (1-\zeta^{\lambda(\ell)})^{-1}\right) w_0(y)$$

$$+\sum_{i=1}^{\infty} p^{-i}\sum_{j=1}^{p-1} \zeta^j(1-\zeta^j)^{-1} w_{jp^{\rho(\ell)+i-1}}(y). \tag{50}$$

Note first of all from (49) and (50) that for every $\ell \in \mathbf{N}_0$, there exists a set $\mathcal{K} = \mathcal{K}(\ell)$ of nonnegative integers, depending only on $\ell$, such that

$$v_{\ell}(y) = \sum_{k\in\mathcal{K}} c_k w_k(y),$$

where for every $k \in \mathcal{K}$, the $p$-ary expansion of $k$ has at most one nonzero coefficient. Suppose now that $\ell', \ell'' \in \mathbf{N}_0$ are distinct. Then there exist two sets $\mathcal{K}' = \mathcal{K}'(\ell')$ and $\mathcal{K}'' = \mathcal{K}''(\ell'')$ of nonnegative integers such that

$$v_{\ell'}(y)\,\overline{v_{\ell''}(y)} = \sum_{k'\in\mathcal{K}'}\sum_{k''\in\mathcal{K}''} c_{k'}\overline{c_{k''}} w_{k'\ominus k''}(y),$$

where for every $k' \in \mathcal{K}'$ and $k'' \in \mathcal{K}''$, the $p$-ary expansion of $k' \ominus k''$ has at most two nonzero coefficients. Combining this with (48), we conclude that

$$\begin{aligned}\widetilde{\chi}_{\ell'}(y)\,\overline{\widetilde{\chi}_{\ell''}(y)} &= p^{-\rho(\ell')-\rho(\ell'')} w_{\ell'\ominus\ell''}(y) \sum_{k'\in\mathcal{K}'}\sum_{k''\in\mathcal{K}''} c_{k'}\overline{c_{k''}} w_{k'\ominus k''}(y) \\ &= p^{-\rho(\ell')-\rho(\ell'')} \sum_{k'\in\mathcal{K}'}\sum_{k''\in\mathcal{K}''} c_{k'}\overline{c_{k''}} w_{\ell'\ominus\ell''\oplus k'\ominus k''}(y). \qquad (51)\end{aligned}$$

On the other hand, the condition $\kappa(\ell' \ominus \ell'') \geq 3$ ensures that the $p$-ary expansion of $\ell' \ominus \ell''$ has at least three nonzero coefficients. It follows that for every $k' \in \mathcal{K}'$ and $k'' \in \mathcal{K}''$, the $p$-ary expansion of $\ell' \ominus \ell'' \oplus k' \ominus k''$ has at least one nonzero coefficient, so that $\ell' \ominus \ell'' \oplus k' \ominus k''$ is nonzero. It follows from (51) and (20) that

$$\begin{aligned}&\int_U \widetilde{\chi}_{\ell'}(y)\,\overline{\widetilde{\chi}_{\ell''}(y)}\,\mathrm{d}y \\ &\quad = p^{-\rho(\ell')-\rho(\ell'')} \sum_{k'\in\mathcal{K}'}\sum_{k''\in\mathcal{K}''} c_{k'}\overline{c_{k''}} \int_U w_{\ell'\ominus\ell''\oplus k'\ominus k''}(y)\,\mathrm{d}y = 0.\end{aligned}$$

□

**Acknowledgments.** The research of the second author has been supported by RFFI Project No. 02-01-00086 and INTAS Grant No. 00-429.

## References

1. Chen, W.W.L.: On irregularities of distribution. *Mathematika* **27**, 153–170 (1980)
2. Chen, W.W.L.: On irregularities of distribution II. *Q. J. Math. Oxf.* **34**, 257–279 (1983)
3. Chen, W.W.L., Skriganov, M.M.: Explicit constructions in the classical mean squares problem in irregularities of point distribution. *J. Reine Angew. Math.* **545**, 67–95 (2002)
4. Dobrovol'skiĭ, N.M.: An effective proof of Roth's theorem on quadratic dispersion. *Usp. Mat. Nauk* **39**, 155–156 (1984); *Russ. Math. Surv.* **39**, 117–118 (1984)
5. Faure, H.: Discrépance de suites associées à un système de numération (en dimension $s$). *Acta Arith.* **41**, 337–351 (1982)
6. Fine, N.J.: On the Walsh functions. *Trans. Am. Math. Soc.* **65**, 373–414 (1949)
7. Frolov, K.K.: An upper bound for the discrepancy in the $L_p$-metric. *Dokl. Akad. Nauk SSSR* **252**, 805–807 (1980)
8. Golubov, B.I., Efimov, A.V., Skvorčov, V.A.: *The Walsh Series and Transformations: Theory and Applications*. Kluwer, Dordrecht (1991)
9. Hewitt, E., Ross, K.A.: *Abstract Harmonic Analysis*, vol. 1. Springer, Heidelberg (1963)
10. Price, J.J.: Certain groups of orthonormal step functions. *Can. J. Math.* **9**, 413–425 (1957)
11. Rosenbloom, M.Yu., Tsfasman, M.A.: Codes in the $m$-metric. *Probl. Peredachi Inf.* **33**, 55–63 (1997); *Probl. Inf. Transm.* **33**, 45–52 (1997)
12. Roth, K.F.: On irregularities of distribution. *Mathematika* **11**, 73–79 (1954)
13. Roth, K.F.: On irregularities of distribution IV. *Acta Arith.* **37**, 67–75 (1980)

14. Schipp, F., Wade, W.R., Simon, P.: *Walsh Series: An Introduction to Dyadic Harmonic Analysis*. Hilger, Bristol (1990)
15. Skriganov, M.M.: Lattices in algebraic number fields and uniform distribution modulo 1. *Algebra Anal.* **1**, 207–228 (1989); *Leningr. Math. J.* **1**, 535–558 (1990)
16. Skriganov, M.M.: Constructions of uniform distributions in terms of geometry of numbers. *Algebra Anal.* **6**, 200–230 (1994); *St. Petersbg. Math. J.* **6**, 635–664 (1995)
17. Skriganov, M.M.: Coding theory and uniform distributions. *Algebra Anal.* **13**, 191–239 (2001); *St. Petersbg. Math. J.* **13**, 301–337 (2002)
18. Skriganov, M.M.: Harmonic analysis on totally disconnected groups and irregularities of point distributions. *J. Reine Angew. Math.* **600**, 25–49 (2006)

*Diophantine Approximation*
H. P. Schlickewei et al., Editors


# APPLICATIONS OF THE SUBSPACE THEOREM TO CERTAIN DIOPHANTINE PROBLEMS

## A survey of some recent results

Pietro Corvaja[1] and Umberto Zannier[2]
[1] *Dipartimento di Matematica e Informatica, Università di Udine, Via delle Scienze 206, 33100 Udine, Italy*
corvaja@dimi.uniud.it
[2] *Scuola Normale Superiore, Piazza dei Cavalieri 7, 56100 Pisa, Italy*
u.zannier@sns.it

To Wolfgang M. Schmidt on the occasion of his 70th birthday

## Introduction

One of the cornerstones of modern Diophantine Approximation is the Schmidt Subspace Theorem. Its original form was obtained by Wolfgang Schmidt around 1970, as an evolution of slightly special cases related to an analogue of Roth's Theorem for simultaneous rational approximations to several algebraic numbers. While Roth's Theorem considers rational approximations to a given algebraic point on the line, the Subspace Theorem deals with approximations to given hyperplanes in higher dimensional space, defined over the field of algebraic numbers, by means of rational points in that space.

Schmidt's proofs introduced several substantial ideas with respect to the pattern used since Thue, until Roth. Also, Schmidt soon obtained spectacular applications, for instance to the problem of simultaneous approximations (which was one of the original motivations), of the approximations to an algebraic number by numbers of bounded degree and also concerning the so-called norm-form equations; in all of these topics Schmidt obtained in a sense final answers via his Subspace Theorem.

An important evolution of the Subspace Theorem occurred with the versions by H.P. Schlickewei, where the approximations were measured simultaneously with respect to all the absolute values in a prescribed finite set $S$ of places of a given number field. (A slightly special case was obtained at about the same time by E. Dubois and G. Rhin.) These versions were soon applied (by J.-H. Evertse, A. van der Poorten and

---

*Keywords.* Diophantine approximation, subspace theorem, linear recurrence sequences, irregular points on varieties.

*2000 Mathematics subject classification.* 11J25, 11G35, 11D57.

Schlickewei, M. Laurent and others) to diophantine equations of new type, like the so-called $S$-unit equation $X_1 + \cdots + X_n = 1$, to be solved in a finitely generated subgroup of the multiplicative group of a number field (Siegel had considered only the case $n = 2$); in turn, this was applied to linear recurrences. Also, the Subspace Theorem motivated rather broad conjectures (e.g., by P. Vojta) concerning the distribution of integral points on algebraic varieties. We do not further pause on these developments, nor on the quantitative point of view, and refer instead, e.g., to [S], [V] and [Z] for several references on the whole subject.

More recently the present authors have given new applications to well-known diophantine problems with linear recurrences and about integral points on algebraic varieties. The object of the present paper is to describe in a single survey this more recent work.

## The quotient problem

This concerns linear recurrences over $\mathbf{C}$, expressed as is well known by formulas of the type $f(n) = \sum_{i=1}^{r} P_i(n)a_i^n$, for nonzero polynomials $P_i \in \mathbf{C}[X]$ and distinct complex numbers $a_i$, called the *roots*. The recurrence is called *simple* or *power-sum* if the $P_i$ are constants and *nondegenerate* if no ratio of distinct roots is a root of unity.

Throughout we shall deal with algebraic data; the general case may be often reduced to this by specialization (see [R]).

Given recurrences $f, g$, the so-called *Hadamard quotient* $f(n)/g(n)$ is not a recurrence in general. A necessary condition for being a recurrence is of course that *all* the values $f(n)/g(n)$, $n \in \mathbf{N}$, lie in a finitely generated ring (we agree that for $g(n) = 0$ this means that $f(n) = 0$). It was Pisot who conjectured the converse implication, while it was van der Poorten [vdP2] who obtained a general proof (see also [R]), after an incomplete argument by Pourchet [Po].

The general case of this theorem is rather delicate, and the ingenious proof by Pourchet and van der Poorten relies on an intricate auxiliary construction and on certain $p$-adic estimates. However, even such a method leaves open the natural question of the *infinitude* of the set of $n \in \mathbf{N}$ such that $f(n)/g(n)$ lies in $\mathbf{Z}$, or more generally in a prescribed finitely generated ring $\mathcal{R}$: to assume that *all* the values (not merely an infinity of them) lie in $\mathbf{Z}$ (or $\mathcal{R}$) is crucial for those proofs.

For nondegenerate power-sums $f, g$ defined over $\mathbf{Q}$, the problem is solved in [CZ1]. In this case, with the aid of the Subspace Theorem it is established that (see Thm. 1 there):

**Theorem A.** *If $f, g$ are power-sums with positive rational roots and if the ratio $f(n)/g(n)$ is in $\mathbf{Z}$ for infinitely many $n \in \mathbf{N}$, then $f/g$ is a recurrence.*

We stress that the conclusion is easy to test in practice, and can in fact be reduced to verifying the divisibility between certain polynomials (see for this [CZ1] or [CZ4, Lemma 2.1] or [vdP]). Also, the restriction to "positive" roots is immaterial (just a normalization) and the method of [CZ1] often works even over $\overline{\mathbf{Q}}$ and for nonsimple recurrences, as noted in that paper.

In general however, it is crucial for that method that $g$ admits a *dominant root* (which is automatic for positive roots in $\mathbf{R}$). By this we mean that *there exists an absolute value $v$ of $\overline{\mathbf{Q}}$ such that $g$ has a unique root which is maximal for $v$*. It may well happen that such dominant-root assumption, often crucial in the whole theory, is not satisfied even in the nondegenerate case. (Consider, e.g., the recurrence $g(n+3)+g(n+2)+g(n+1)=g(n)$.)

Coming back to Theorem A, we shall illustrate its proof just with the special case when $f(n)=a^n-1$, $g(n)=b^n-1$, for integers $a,b>1$. Namely, we shall give a proof of the claim:

*Claim: If* $(a^n-1)/(b^n-1)\in\mathbf{Z}$ *for infinitely many* $n\in\mathbf{N}$, *then* $a$ *is a power of* $b$.

We shall deduce this from a lemma, illustrative of the general strategy, which shall be useful also later.

We take this opportunity to state a version of Schmidt's Subspace Theorem due to H.P. Schlickewei; we have borrowed it from [S, Thm. 1E, p.178].

**Subspace Theorem.** *Let* $S$ *be a finite set of absolute values of* $\mathbf{Q}$, *including the infinite one and normalized in the usual way* (i.e., $|p|_v=p^{-1}$ *if* $v|p$). *Extend each* $v\in S$ *to* $\overline{\mathbf{Q}}$ *in some way. For* $v\in S$ *let* $L_{1,v},\dots,L_{N,v}$ *be* $N$ *linearly independent linear forms in* $N$ *variables with algebraic coefficients and let* $\epsilon>0$. *Then the solutions* $\mathbf{x}:=(x_1,\dots,x_N)\in\mathbf{Z}^N$ *to the inequality*

$$\prod_{v\in S}\prod_{i=1}^{n}|L_{i,v}(\mathbf{x})|_v<||\mathbf{x}||^{-\epsilon}$$

*where* $||\mathbf{x}||:=\max\{|x_i|\}$, *are contained in finitely many proper subspaces of* $\mathbf{Q}^N$.

Actually, the statement in [S] is more precise and quantifies the number of relevant subspaces (and a finite number of remaining exceptions) in terms of the linear forms. Substantial work has been recently done from the quantitative point of view, mainly by Evertse, Schlickewei, Schmidt; however, we shall not be concerned here with this.

We also mention a new, more general, geometric version of the Subspace Theorem, obtained by Faltings and Wüstholz [FW], in which the approximating points are restricted to lie on an algebraic subvariety.

**Lemma 1.** *Let* $\xi$ *be a power-sum with positive roots in* $\mathbf{Q}$ *and coefficients in a number field* $K$ *(embedded in* $\mathbf{C}$*). For* $n$ *in an infinite set* $\Sigma$ *of natural numbers, let* $z(n)$ *be integers such that* $|z(n)-\xi(n)|\ll l^n$, *where* $0<l<1$. *Then there exists a power-sum* $\zeta$ *with integer roots and rational coefficients such that* $z(n)=\zeta(n)$ *for all but finitely many* $n\in\Sigma$. *Moreover, any root of* $\zeta$ *is a root of* $\xi$.

*Proof.* Write $\xi(n)=\sum_{i=1}^{h}c_i(a_i/b)^n$ where $c_i\in K^*$ and the $a_i,b$ are positive integers, such that the $a_i/b$ are distinct. Let $S$ be the set of absolute values of $\mathbf{Q}$ consisting of $\infty$ and all primes dividing some of the $a_i$ or $b$. Extend each value in $S$ to $K$ in some way, the infinite value being extended so to coincide with the complex absolute value in the given embedding of $K$ in $\mathbf{C}$. Define the linear forms $L_{i,v}$ for $v\in S$ and $i=0,1,\dots,h$ as follows: $L_{0,\infty}=L:=X_0-\sum_{i=1}^{h}c_iX_i$, $L_{i,\infty}=X_i$ for $i=1,\dots,h$, while for $v\in S$, $v\neq\infty$, put $L_{i,v}=X_i$ for $i=0,1,\dots,h$. Consider, for $n\in\Sigma$, the vector $\mathbf{x}_n=(b^nz(n),a_1^n,\dots,a_h^n)\in\mathbf{Z}^{h+1}$. We have

$$\prod_{v\in S}\prod_{i=0}^{h}|L_{i,v}(\mathbf{x}_n)|_v=\prod_{i=1}^{h}\left(\prod_{v\in S}|a_i^n|_v\right)\cdot|L(\mathbf{x}_n)|\prod_{v\in S\setminus\{\infty\}}|b^nz(n)|_v.$$

By the product formula and the choice of $S$ we have $\prod_{v\in S}|a_i^n|_v = 1$ for $i = 1,\dots,h$. By our assumptions we have $|L(\mathbf{x}_n)| \ll (bl)^n$. Finally, $S$ includes all finite absolute values nontrivial on $b$ and the $z(n)$ are integers, so, by the product formula,

$$\prod_{v\in S\setminus\{\infty\}} |b^n z(n)|_v \le \prod_{v\in S\setminus\{\infty\}} |b^n|_v = |b|^{-n}.$$

On the other hand, $||\mathbf{x}_n|| \ll A^n$ for some $A > 0$ independent of $n$. Combining such estimates we get

$$\prod_{v\in S}\prod_{i=0}^{h} |L_{i,v}(\mathbf{x}_n)|_v < ||\mathbf{x}_n||^{-\epsilon}$$

for large $n \in \Sigma$, provided $\epsilon < \log(1/l)/\log A$. By the Subspace Theorem there exist finitely many nonzero rational linear forms $\Lambda_j(X_0,\dots,X_h)$ such that each vector $\mathbf{x}_n$ as above is a zero of some $\Lambda_j$. Suppose first $\Lambda_j$ does not depend on $X_0$. Then, if $\Lambda_j(\mathbf{x}_n) = 0$ we have a nontrivial relation $\sum_{i=1}^h u_i(a_i/b)^n = 0$, for constant rational $u_i$; however, this can hold for a finite number of $n$ at most, since the $a_i/b$ are distinct and positive. Hence there is some rational linear form $\Lambda$, depending on $X_0$, such that $\Lambda(\mathbf{x}_n) = 0$ holds for infinitely many $n \in \Sigma$. We may write

$$\Lambda(X_0,\dots,X_h) = X_0 - \sum_{i=1}^{h} v_i X_i, \qquad v_i \in \mathbf{Q}$$

whence $z(n) = \zeta(n) := \sum_{i=1}^h v_i(a_i/b)^n \in \mathcal{E}_{\mathbf{Q}}$ for $n$ lying in an infinite subset $\Sigma_1$ of $\Sigma$. Since $|z(n) - \xi(n)| \ll l^n$ for $n \in \Sigma$, where $l < 1$, we see that $v_i = c_i$ for $i$ such that $|a_i/b| \ge 1$. By definition any root of $\zeta$ is then a root of $\xi$. Now, $\zeta$ takes integer values on $\Sigma_1$; an application of the Subspace Theorem similar to and simpler than the above one then proves that the roots of $\zeta$ must be integral. (If not, let $p$ be a prime appearing in some denominator; the integrality of the values implies that a suitable linear form in the $a_i^n$ is $p$-adically small and the above pattern applies; see [CZ1].) In particular $v_i = 0$ if $|a_i/b| < 1$. We conclude that the linear form $\Lambda$ is determined in terms of $\xi$ only and the lemma follows. □

To deduce the Claim, fix an integer $r$ such that $b^r > a$, put $z(n) = (a^n-1)/(b^n-1)$ and observe the identity $(b^{rn}-1)z(n) = (a^n-1)(1+b^n+\dots+b^{(r-1)n})$ which may be rewritten as

$$z(n) + \sum_{(i,j)\in A} (-1)^i a^{in} b^{(j-r)n} = \frac{z(n)}{b^{rn}}, \qquad A = \{0,1\}\times\{0,1,\dots,r-1\}.$$

Since the right side tends to zero exponentially, we may apply Lemma 1 obtaining some nontrivial equation

$$\gamma_0 b^{rn}(a^n-1) + (b^n-1)\left(\sum_{(i,j)\in A} \gamma_{ij} a^{in} b^{jn}\right) = 0 \tag{1}$$

holding for all $n$ in a suitable infinite set $\Sigma'$, where the coefficients $\gamma_0, \gamma_{ij}$ are rationals not all zero. Now, this implies that $a, b$ are multiplicatively dependent, for otherwise the functions $n \mapsto a^n, n \mapsto b^n, n \in \Sigma'$ would be algebraically independent (because $\Sigma'$ is infinite), whence (1) would lead to the identity $\gamma_0 V^r(U-1) + (V-1)(\sum_{(i,j)\in A} \gamma_{ij} U^i V^j) = 0$ in the variables $U, V$. However it is immediate to check that this yields the vanishing of all the coefficients $\gamma_0, \gamma_{ij}$, a contradiction. Hence we may write $a = c^p$, $b = c^q$ for some integer $c > 1$ and some positive integers $p, q$ and we then have just to show that $q$ divides $p$. Write $p = mq + s$ for integers $m, s$ with $0 \le s < q$. Then $a^n = (c^{qn})^m \cdot c^{sn} \equiv c^{sn} \pmod{c^{qn} - 1}$. Therefore, since $c^{qn} - 1 = b^n - 1$ divides $a^n - 1$ for all $n$ in a certain infinite set, we find that $c^{qn} - 1$ divides in fact $c^{sn} - 1$ for all $n$ in the set. This is however plainly impossible if $0 < s < q$, whence $s = 0$ as desired.

We pause to point out another application of the above lemma to rational approximations of special type to algebraic numbers. We recall here Ridout's extension of Roth's Theorem, proving for instance that rational approximations whose denominators are of the form, say, $2^n$, satisfy a lower bound $2^{-n(1+\epsilon)}$, which supersedes Roth's bound $2^{-n(2+\epsilon)}$. For the sake of example, we note here that Ridout's bound holds for denominators, say, $2^n + 1$. In fact, if $z(n)/(2^n + 1)$, $n \in \Sigma$, are rational approximations to the algebraic number $\alpha$, then the power-sum $\xi(n) = \alpha 2^n + \alpha$ is "close" to the integer $z(n)$, and Lemma 1 easily leads to the sought conclusion. Naturally, this gives a sufficient condition for transcendency (see, e.g., [TrZ] for an explicit application). Transcendency applications of similar ideas appear also in the paper [CZ3] (see Thm. 3 and corollaries). For instance, Corollary 3 therein immediately yields:

**Theorem B.** *Let $\{m_i\}$ be an increasing sequence of integers satisfying*

$$\sup_N \limsup_n (m_{n+N}/m_n) = \infty.$$

*Let also $a_i$ be positive real algebraic numbers taken from a finite set. Then the function defined in* (0, 1) *by the series $\sum_{i=1}^{\infty} a_i x^{m_i}$ takes transcendental values at all algebraic points in* (0, 1).

This applies in particular to the Fredholm series $\sum_{i=1}^{\infty} x^{2^i}$, and we get back results obtained by K. Mahler by means of his method relying on certain functional equations. Theorem B equally applies to lacunary series $\sum_{i=1}^{\infty} x^{m_i}$, where $m_{i+1}/m_i > l$, for a fixed $l > 1$, even in the absence of suitable functional equations.

Going back to Theorem A, we again remark that the given arguments work in far greater generality; however, we need the mentioned "dominant root" assumption in order to apply Lemma 1; this amounts to construct a nonobvious linear form which is small at the relevant vectors. (The "obvious" forms are just the variables $X_i$, whose values at the $\mathbf{x}_n$ are small at suitable $p$-adic places.) In the general case, a small linear form may be constructed out of several dominant roots (instead of a single dominant root), but the inequality so obtained turns out to be too weak for an application of the Subspace Theorem. Nevertheless, a somewhat surprising device allows to produce many small linear forms out of a single one (one multiplies the initial linear form by many monomials in the dominant roots) and this procedure finally enables to eliminate the annoying assumption about the dominant root. The arguments appear in [CZ4] in full detail; since the construction is a bit technical we do not reproduce it here, but rather just state the main theorem of [CZ4]:

**Theorem C.** *Let $f, g$ be recurrences defined over a number field $k$. Suppose that $S$ is a finite set of places of $k$ such that $f(n)/g(n) \in \mathcal{O}_{k,S}$ for infinitely many $n \in \mathbf{N}$. Then there exist a nonzero polynomial $P(n)$ and positive integers $q, r$ such that both $P(n)f(qn+r)/g(qn+r)$ and $g(qn+r)/P(n)$ are recurrences.*

In practice the conclusion says that, over a suitable arithmetic progression, the recurrence $g$ divides $f$ up to a polynomial factor $P(n)$. Often one can take $P = 1$: this occurs, e.g., when $g$ is a power-sum, and then we get a substantial extension of Theorem A. However it is not generally the case that $P$ can be taken constant, as shown by examples like $(2^n - 2)/n$; now the quotient is integral whenever $n$ is a prime, hence for a fairly dense set in $\mathbf{N}$. In an appendix to [CZ4] we show that *if the polynomial denominator cannot be eliminated, then the set of $n$ such that $f(n)/g(n)$ is integer has zero density*; this in particular immediately yields a sharpening of van der Poorten's Theorem. Still in other words, these results say that a divisibility relation between infinitely many pairs of values $f(n), g(n)$ may in a sense be explained by algebraic identities. This is rather easy to prove for polynomials $f, g$; we may view the above statements as analogues for polynomial-exponential functions.

Actually, the method of [CZ1] (or [CZ4]) yields, more precisely, a nontrivial bound for the cancellation in the quotient $f(n)/g(n)$, i.e., for the g.c.d.$(f(n), g(n))$. In some cases, like $(a^n - 1)/(b^n - 1)$ of the Claim, it is possible with more effort to get a nearly best-possible conclusion in this direction; in [BCZ] the following is proved:

**Theorem D.** *If $a, b \in \mathbf{Z}$ are multiplicatively independent, then for all $\epsilon > 0$ we have the estimate*

$$\gcd(a^n - 1, b^n - 1) \ll_\epsilon \exp(\epsilon n).$$

Note that the relevant g.c.d. may often be quite large; in fact, the lower bound $(a^n - 1, b^n - 1) > \exp(\exp(c \log n / \log\log n))$ (some $c > 0$) is valid for infinitely many integers $n$ (see Prop. 10 in Adelman, Pomerance, Rumely, *Ann. Math.* **117**, 173–206, 1983).

The methods of [BCZ] combined with the Lang-Liardet results about points on the intersection of curves in $\mathbf{G}_m^2$ with finitely generated groups actually lead to a similar, more general, conclusion, when $a^n, b^n$ are replaced by arbitrary $S$-units $u, v \in \mathbf{Z}$ (see [CZ5, Remark 1] and [Z, Thm. IV.3]); namely, for a fixed finite set $S$ of places of $\mathbf{Q}$ one can prove that, *for any positive $\epsilon$, $\gcd(u-1, v-1) \ll \max(|u|, |v|)^\epsilon$ for multiplicatively independent $S$-units $u, v \in \mathbf{Z}$.* In [CZ6] we extend this to algebraic numbers and pairs of functions of $u, v$ more general than $u - 1, v - 1$. Also, we formulate the result in terms of Weil heights, e.g., as in the following statement:

**Theorem E.** *Let $S$ be a finite set of places in a number field $k$. Then, for multiplicatively independent $S$-units $u, v \in k$ we have the asymptotic $h(u-1 : v-1) \sim h(1 : u : v)$ (for $h(u) + h(v) \to \infty$).*

We owe to J. Silverman the observation that this is equivalent to a special case of the so-called Vojta's conjecture (Conj. 3.4.3 in [V]), related to integral points off divisors on algebraic varieties; namely the case when the variety is the blow up of $\mathbf{G}_m^2$ at $(1, 1)$ and the relevant $D$ is the sum of the exceptional divisor with the divisor at infinity of $\mathbf{G}_m^2$ in its embedding in $\mathbf{P}_1^2$ (see also [Si]).

This kind of results also admit applications to lower bounds for the order of an integral matrix modulo an integer $N$ tending to infinity, as in the paper [CRZ]; as remarked by Z. Rudnick, this is related with certain dynamical systems.

Also, the results yield in particular a proof of a (sharp form of a) conjecture by Győry, Sarkozy and Stewart; in [CZ5] we prove that:

**Theorem F.** *For positive integers $a \geq b > c > 0$, the greatest prime factor of $(ab+1)(ac+1)$ tends to infinity as $a$ tends to infinity.*

The original conjecture predicted the same conclusion for $(ab+1)(ac+1)(bc+1)$. The result may be seen as a uniform version of the well-known theorem (due to Pólya) that the greatest prime factor of the values at integers of a quadratic polynomial with distinct rational roots tends to infinity. The link with the previous context is provided by the observation that, if $u := ab+1$ and $v := ac+1$ have all their prime factors in a prescribed finite set $S$, then $u, v$ are $S$-units such that $\gcd(u-1, v-1) \geq a$ is "large". Then Theorem F follows from Theorem E.

Further applications of the methods, e.g., to study the length of the continued fraction for quotients $f(n)/g(n)$ of power-sums over $\mathbf{Q}$, have been given in [CZ7]. For instance we have:

**Theorem G.** *If $a, b$ are multiplicatively independent positive integers, then the length of the euclidean algorithm for $(a^n - 1) : (b^n - 1)$ tends to infinity as $n$ tends to infinity.*

Note that Theorems D and G express in different terms the complexity of the relevant rational fractions. Theorem G appears as Corollary 3 in [CZ7], obtained therein as an application of a general statement for arbitrary pairs of power sums with rational coefficients and roots. (See also [Z, I,IV]; the case $a^n : b^n$ was a result by Pourchet, after a question by Mendès-France; see [CZ7].)

## The $d$-th root problem

In addition to the mentioned quotient conjecture solved by van der Poorten, Pisot formulated a "$d$-th-root conjecture": *If all the values $f(n)$ of a recurrence are $d$-th powers in a given number field $k$, then $f$ is identically a $d$-th power of a recurrence.* After some partial results by several authors, a complete proof was given in [Z2] by means of congruence considerations; one applies the Lang-Weil bound for points on varieties over finite fields, but first one has to reduce exponential congruences to polynomial ones.

In analogy with the case of the Pisot Quotient-conjecture, the arguments in [Z2] do not help in establishing the more fundamental question of the *finiteness* of the solutions of $f(n) = y^d$, $n \in \mathbf{Z}$, $y \in k$, for a given integer $d \geq 2$ and a general recurrence $f$ satisfying appropriate necessary assumptions. In special cases (e.g., for binary recurrences or when $f$ has a dominant root and $d$ is large enough with respect to $f$) this has been worked out by several authors, like Pethö, Schinzel, Shorey, Stewart, Tijdeman (see, e.g., [ShSt] and [ShT]); they used Baker's method, obtaining, whenever the arguments applied, effective conclusions.[1] However, when $f$ has three or more

1. On the contrary the present arguments do not lead to effective conclusions; it is however rather easy to estimate the number of solutions, using suitable quantitative versions of the Subspace Theorem.

roots, such considerations seem not to extend to the case of general $d$, e.g., to the case $d = 2$.

For unrestricted (but fixed) $d$, the first finiteness results valid for any number of roots have been obtained in [CZ1], in the general case of power-sums defined over $\mathbf{Q}$. That paper actually considers arbitrary algebraic equations $F(y, f(n)) = 0$, where $F$ is a polynomial and where $f$ is a power-sum over $\mathbf{Q}$. It is also observed that the same arguments often apply to simple recurrences over $\overline{\mathbf{Q}}$ with the sole assumption of a dominant root. A result in this direction appears as Theorem 2 in [CZ3]. Here we shall just sketch a brief deduction from Lemma 1 of the following result of [CZ1]:

**Theorem H.** *Let $f$ be a power-sum with positive rational roots. If $f(n)$ is a square in $\mathbf{Q}$ for infinitely many integers $n$, then we have an identity $f(n) = a^{n+e}g(n)^2$ for suitable $e \in \{0, 1\}$, $a \in \mathbf{N}$ and power sum $g$ with rational coefficients and roots.*

The principle will be to approximate a square root of $f(n)$ by means of a power sum. In this step the dominant root plays a crucial role. In fact, let us write $f(n) = \sum_{i=1}^{r} c_i a_i^n$, where for our purposes the $c_i$, $a_i$ may be assumed to be nonzero integers, with $0 < a_r < \cdots < a_1$; then, on writing

$$\sqrt{f(n)} = \sqrt{c_1}(\sqrt{a_1})^n\sqrt{1 + \rho(n)},$$

where $\rho(n)$ is a power sum with positive roots $< 1$, we may expand with the binomial theorem in a series

$$\sqrt{f(n)} = \sqrt{c_1}(\sqrt{a_1})^n\sigma(n),$$

where $\sigma(n) = \sum_{i=1}^{\infty} d_i b_i^n$ is an "infinite power sum" with rational coefficients and roots, which converges absolutely for large enough $n$. Truncating this series it is easy to see that there exists a power sum $\sigma^*(n)$ such that

$$\sqrt{f(n)} = \sqrt{c_1}(\sqrt{a_1})^n\sigma^*(n) + O(l^n)$$

for some $l < 1$. Now, writing $n = 2m + e$, $e \in 0, 1$, we obtain

$$\sqrt{f(n)} = \sqrt{c_1 a_1^e}a_1^m\sigma^*(2m + e) + O(l^{2m}).$$

Naturally, these formulas hold for appropriate choices of the signs. Applying Lemma 1 separately for $e = 0, 1$, with $m$ in place of $n$, $z(m) = \sqrt{f(2m + e)}$, $\xi(m) = \sqrt{c_1 a_1^e}a_1^m\sigma^*(2m + e)$, we get, for the relevant integers $m$, an expression of $z(m)$ as a power sum; it is then easy to see that this leads to an identity holding for all integers $m$, yielding the sought conclusion.

Even the special case when, say, $f(n) = 3^n + 2^n + 1$ was not known; note that now the conclusion implies the finiteness of the perfect squares in the sequence $3^n + 2^n + 1$; in fact, an identity $3^n + 2^n + 1 = a^{n+e}g(n)^2$ would easily imply (see [vdP]) that the polynomial $X^2 + Y^2 + 1$ is a square in $\mathbf{C}[X, Y]$, which is not the case.

It is tempting to modify these questions by adding a further variable, namely to consider the perfect squares of the form $3^m + 2^n + 1$, say. To our knowledge, to prove their finiteness is an open problem. (Sometimes congruence arguments may be helpful, but they cannot answer the question if "perfect square" is interpreted in an arbitrary number field.)

We now illustrate how such problem happens to be less artificial than it perhaps appears. Actually, it represents a typical instance of the problem of the $S$-integral points on a variety $\mathbf{P}_2 \setminus D$, where $D$ is a divisor which is the sum of two lines and a conic. To see the link, say that the lines and conic are given by $X_0 = 0$, $X_1 = 0$ and $X_2^2 = X_0^2 + X_0X_1$, and that $S = \{\infty, 2, 3\}$. Then the $S$-integral points $(x_0 : x_1 : x_2)$ are those such that both $u := x_1/x_0$ and $v := (x_2/x_0)^2 - 1 - (x_1/x_0)$ are $S$-units. Namely, they correspond to $S$-units $u, v$ such that $1+u+v$ is a perfect square. On the other hand, $u, v$ have the shape $\pm 2^a 3^b$, and the connection with the above becomes clear.

Now, a broad conjecture by Lang and Vojta (see [HS, p. 486]) implies that in these cases the set of $S$-integral points is not Zariski dense. This is known when $D$ is the sum of four lines in general position; the next simplest cases, presently unknown, occur just when $D$ is as above the sum of two lines and a conic. On the other hand, known methods prove that the solutions $(y, 2^n, 3^m)$ to $y^2 = 1 + u + v$ either are Zariski dense or are finite in number; in conclusion, the finiteness in question follows from the Lang–Vojta conjecture and the above problem well illustrates one of the simplest cases of it.

Our methods cannot answer such questions. However they lead to results on the distribution of solutions; for instance, *for every infinite sequence of integer solutions* $(y, m, n) \in \mathbf{Z}^3$ *for the equation* $y^2 = 3^m + 2^n + 1$, *the ratio* $m \log 3/n \log 2$ *converges to* 1. Unconditional finiteness results can be obtained for the similar equation $y^2 = 6^m + 2^n + 1$ or more generally for $f(a^m, y) = b^n$ where $a, b$ are non-coprime integers and $f$ is a polynomial verifying suitable assumptions (see [CZ2]). The principle of the proof is to use the mentioned distributional constraint and to combine it with an analogue constraint relative to a $p$-adic absolute value in place of the usual one. For the method to work, a relevant prime $p$ must divide both $a$ and $b$.

## Integral points on certain affine varieties

These methods have proved useful also to study the integral points on affine varieties $X \subset \mathbf{A}^n$ whose divisor $D$ at infinity has suitable properties, like being highly reducible. Here $D$ is defined as the sum of the components of the divisor $\tilde{X} \setminus X$, where $\tilde{X}$ is the projective completion of $X$ in $\mathbf{P}_n$. We at once mention that to our knowledge the conditions on $D$ which we shall meet may well depend on the embedding of $X$ in some affine space.

A simple instance of the methods occurs with a proof of Siegel's Theorem on integral points which avoids any recourse to the arithmetic of Jacobians, which heavily appeared in Siegel's arguments. We recall that this result asserts the finiteness of integral points (over a number field $k$) on affine curves $X$ which either have positive genus or have at least three points at infinity; in the above language this last condition amounts to $D$ having at least three components. We remark at once that we may work under this assumption; in fact, when $X$ has positive genus we may first go to an unramified cover $X'$ of $X$, of degree $\geq 3$, and then apply the result to $X'$; note that by the Chevalley–Weil Theorem, $X'$ will continue to have infinitely many integral points if $X$ does. (Here it may be necessary to increase $k$ and also to allow more denominators; see [CZ8] for more details on this.)

To explain the principle of this proof, we first recall that special cases of Siegel's Theorem were known, and proved without recourse to the Jacobian, only for curve of particular shape; this was the case in Thue's original papers, where he dealt with equations $f(x, y) = c$ for a homogeneous polynomial $f$. The new method is to change the embedding of the curve, to get an advantageous induced metric; in practice, this amounts to the existence of (many) linear spaces with high order contact with the curve at a point at infinity. To achieve this purpose, it proves necessary to increase freely the dimension of the ambient space; it is here that Roth's Theorem no longer suffices, being necessary a multidimensional extension of it, represented precisely by the Subspace Theorem. We illustrate this method by sketching the proof of the following

**Theorem.** *Let $\tilde{C}$ be a nonsingular projective curve over a number field $k$; let $D = Q_1 + \ldots + Q_r$ for distinct points $Q_i$, where $r \geq 3$. Then $C = \tilde{C} \setminus D$ has only finitely many $S$-integral points.*

In this argument we suppose that $k$ is a sufficiently large number field. For a large integer $N$ let us consider the vector space $V = V_N$ of the rational functions $\varphi \in k(\tilde{C})$ with $\operatorname{div}(\varphi) \geq -ND$. Denote by $d = d_N$ its dimension and let $\varphi_1, \ldots, \varphi_d$ be a basis (over $k$). By Riemann-Roch we have $d \geq Nr + O(1)$. Let $\{P_n\}$ be an infinite sequence of $S$-integer points of $C$. By going to an infinite subsequence we may assume that, for each $v \in S$, $P_n$ converges to a point $P^v \in \tilde{C}(k_v)$. Denote by $S'$ the subset of $S$ consisting of the places $v$ such that $P^v$ is one of the points $Q_i$ at infinity and put $S'' = S \setminus S'$. For every $v \in S'$ there exists a basis $\{L_{1v}, \ldots, L_{dv}\}$ of $V$ such that for each $j = 1, \ldots, d$ the function $L_{jv}$ vanishes at $P^v$ with a multiplicity $\geq j - 1 - N$. For every $v \in S''$ let $\{L_{1v}, \ldots, L_{dv}\}$ be any basis of $V$. Let $t_v \in k(C)$ be a local parameter at $P^v$, for $v \in S'$. We have, for $v \in S'$,

$$|L_{jv}(P_n)|_v \ll |t_v(P_n)|_v^{j-1-N}, \qquad j = 1, \ldots, d.$$

For $v \in S''$ we have $|L_{jv}(P_N)|_v \ll 1$. Then

$$\begin{aligned}\prod_{v\in S}\prod_{j=1}^{d} |L_{jv}(P_n)|_v &\ll \Bigg(\prod_{v\in S'} |t_v(P_n)|_v\Bigg)^{(d/2)(d-2N-1)} \\ &\ll \Bigg(\prod_{v\in S'} |t_v(P_n)|_v\Bigg)^{d((r-2)N+O(1))/2}.\end{aligned}$$

Thanks to the integrality assumption on the $P_n$, the projective height of the point $(\varphi_1(P_n) : \ldots : \varphi_d(P_n))$ is bounded by $\prod_{v\in S'} |t_v(P_n)|_v^{-N}$. The above inequality gives the estimation

$$\prod_{v\in S}\prod_{j=1}^{d} |L_{jv}(P_n)|_v \ll H(\varphi_1(P_n) : \ldots : \varphi_d(P_n))^{-d((r-2)N+O(1))/2N}.$$

Since the $L_{iv}$ are linear forms in $\varphi_1, \ldots, \varphi_d$, we may apply the Subspace Theorem for large enough $N$, which concludes the proof.

As remarked above, the case of positive genus and fewer than three points at infinity reduces to this statement, after taking a suitable cover and applying the Chevalley–Weil Theorem.

The advantage of this method is not only of methodological nature, since it provides for instance sharper quantitative versions of Siegel's Theorem. In [CZ9] we have applied a uniform quantitative version of the Subspace Theorem, due to Evertse [E], to the estimation of the number of integral points on algebraic curves, especially in the case of three or more points at infinity. For instance it follows that, *for a fixed irreducible affine curve C with three points at infinity, defined over a number field* $k_0$, *the number of its integral points over a variable extension field* $k/k_0$ *is bounded only in terms of the degree* $[k : k_0]$.

Other applications of these principles concern varieties of higher dimension. In the paper [CZ10] we study the case of an affine surface $X \subset \mathbf{A}^n$ such that its closure $\tilde{X} \subset \mathbf{P}_n$ is nonsingular. We denote by $D_1, \dots, D_r$ the irreducible components of $\tilde{X} \setminus X$. The following result appears as Theorem 1 there, where we have denoted by $(A.B)$ the intersection product of the divisors $A$, $B$ in $\tilde{X}$.

**Theorem I.** [CZ10]. *Let* $X$, $\tilde{X}$ *be as above and assume that no three of the* $D_i$ *share a common point. Assume also that there exist positive integers* $p_1, \dots, p_r, c$, *with either*

*(a)* $r \geq 4$ *and* $p_i p_j (D_i.D_j) = c$ *for all* $i, j$, *or*

*(b)* $r \geq 5$ *and* $D_i^2 = 0$, $p_i p_j (D_i.D_j) = c$ *for* $i \neq j$.

*Then the S-integral points are not Zariski dense in X.*

We briefly sketch two applications of this result. First, consider a *double Pell's equation* like $y^2 = 2x^2 + 1$, $z^2 = 3x^2 + 1$. This represents an affine curve $C$ of genus 1 with four points at infinity. By Siegel's Theorem it has only finitely many $S$-integral points. However it has three infinite families of *quadratic-integral points*, namely integral points defined over a *variable* number field, of degree $\leq 2$ over the ground field. To obtain the families, for instance solve in $\mathbf{Z}^2$ the first Pell's equation and then define $z := \sqrt{3x^2 + 1}$ and proceed similarly to obtain two further families. Now, Theorem I implies that only finitely many quadratic-integral points escape from this description (see [CZ10], also for a more general analysis). In fact, consider the *symmetric square* $X$ of $C$, i.e., the surface $C \times C / \sim$ where the equivalence $\sim$ identifies $(P, Q)$ and $(Q, P)$. It may be checked that it has four divisors at infinity to which Theorem I(a) applies. Now, a quadratic integral point $P$ on $C$ gives rise to the point $Q := (P, P')/\sim$ on $X$, where $P'$ is the nontrivial conjugate of $P$. Then $Q$ is an ordinary integral point on $X$, hence lies on a certain fixed curve on $X$, by the conclusion of Theorem I; this is the crucial deduction and a little more work gives the stated claim.

Secondly, we note that part (b) applies in particular to certain "generic" surfaces in affine 5-space: we start with a surface $X \subset \mathbf{A}^5$ defined by three equations $f_i(x_1, \dots, x_5) = 0$, $i = 1, 2, 3$, where $f_i$ are polynomials of degree $d$ in each variable. By embedding $\mathbf{A}^5$ in the compactification $\mathbf{P}_1^5$ one obtains a complete surface $\tilde{X}$, which we suppose to be smooth, with five divisors at infinity $D_1, \dots, D_5$, namely the inverse images of the points at infinity of $\mathbf{P}_1$ under the five natural projections.

These divisors in general satisfy assumption (b); the self-intersections vanish because they are fibers of morphisms to $\mathbf{P}_1$ and, for $i \neq j$, the product $(D_i.D_j)$ will be (for general choice of the $f_i$) equal to $(3d)^3$. Again it follows from Theorem I that such surfaces $X$ have a set of integral points which is not Zariski-dense. (See [CZ10] and [Z3] for other applications.)

Theorem I seems not to apply to the natural case of hypersurfaces in $\mathbf{A}^3$. However the basic principles of the method may be adapted also in these cases, actually in any number of variables. In [CZ11] we prove for instance the following generalization of Thue's theorem for the hypersurface $X$ defined by an equation $f_1 \cdots f_r = g$; here the $f_i, g$ are polynomials in $n$ variables for which we denote with a bar their homogeneizations:

**Theorem J.** *Suppose that the set of common zeros (in $\mathbf{P}_n$) of $X_0\overline{g}$ and any $n-1$ of the forms $\overline{f}_i$ is finite and that no $n$ of the $\overline{f}_i$ have a common zero at infinity. Assume also that*

$$\sum_{i=1}^{r} \deg f_i > n \max(\deg f_i) + \deg g.$$

*Then $X(\mathcal{O}_S^n)$ is not Zariski-dense in $X$.*

Note that Thue's equation is the extremely special case when the $f_i$ are linear in two variables, $r \geq 3$ and $g$ is constant.

We should remark that this idea of choosing suitable embeddings for the relevant varieties occurs also in joint work by Evertse and Ferretti; for instance in the paper [EF] they use embedding techniques to obtain a quantitative version of the above quoted theorem of Faltings and Wüstholz, and actually in this way they surprisingly recover that result from the original one by Schmidt and Schlickewei.

The same methods of proof of Theorem J yield also a version of the Subspace Theorem for arbitrary polynomials in place of linear forms. For instance in [CZ11] we sketch a proof of the following result:

**Theorem K.** *For $v \in S$, let $f_{iv}$, $i = 1, \ldots, n-1$, be polynomials in $k[X_1, \ldots, X_n]$ of degrees $\delta_{iv} > 0$. Put $\delta_v = \max_i \delta_{iv}$ and $\mu := \min_{v \in S} \sum_{i=1}^{n-1} \delta_{iv}/\delta_v$. Fix $\epsilon > 0$ and consider the Zariski closure $\mathcal{H}$ in $\mathbf{P}_n$ of the set of solutions $\mathbf{x} \in \mathcal{O}_S^n$ of*

$$\prod_{v \in S} \prod_{i=1}^{n-1} |f_{iv}(\mathbf{x})|_v^{1/\delta_v} \leq H(\mathbf{x})^{\mu-n-\epsilon}.$$

*Suppose that, for $v \in S$, $X_0$ and the $\overline{f}_{iv}$, $i = 1, \ldots, n-1$, define a variety of dimension 0. Then $\dim \mathcal{H} \leq n-1$. Moreover, if $\mathcal{H}'$ is a component of $\mathcal{H}$ of dimension $n-1$, there exists $v \in S$ such that the $\overline{f}_{iv}$ determine in $\mathcal{H}'$ a variety of dimension 1.*

A special case of this, involving a single polynomial, is the following result, appearing as Corollary 2 in [CZ11]:

**Theorem L.** *Let $f \in \overline{\mathbf{Q}}[X_1, \ldots, X_n]$ have degree $\delta > 0$ and let $\epsilon > 0$. The vectors $\mathbf{x} \in \mathbf{Z}^n$ such that*

$$0 < |f(\mathbf{x})| < H(\mathbf{x})^{-\delta(n-1)-\epsilon}$$

*are all contained in a single variety of dimension $\leq n-2$.*

Here the absolute value is understood in the classical sense, viewing $\overline{\mathbf{Q}}$ as embedded in $\mathbf{C}$. A similar result holds for any absolute value (with the exponent $-\delta n - \epsilon$ if the value is finite).

This corollary goes in the direction of questions raised by Schmidt in [S3, §11.3]. Note that the exponent on the right is best possible, as may be shown by taking $f$ to be the power of a suitable linear form with algebraic coefficients (the solutions of the corresponding inequality may be Zariski-dense in $\mathbf{A}^n$). The conclusion about the dimension is however not generally best-possible: in the very recent preprint [CZ12] we show that $n-2$ may be replaced by 0.

These results are in the same spirit of those in the papers of Faltings and Wüstholz [FW] and Evertse and Ferretti [EF]; therein however the results involve certain geometric quantities which seem not straighforward to compute, even in concrete cases. It is only very recently that Evertse and Ferretti [EF2] have applied their techniques to recover and also sharpen Theorem K.

## References


[BCZ] Bugeaud, Y., Corvaja, P., Zannier, U.: An upper bound for the G.C.D. of $a^n - 1$ and $b^n - 1$. *Math. Z.* **243**, 79–84 (2003)

[CRZ] Corvaja, P., Rudnick, Z., Zannier, U.: A lower bound for periods of matrices. *Commun. Math. Phys.* **252**, 535–541 (2004)

[CZ1] Corvaja, P., Zannier, U.: Diophantine equations with power sums and universal Hilbert sets. *Indag. Math., N.S.* **9**, 317–332 (1998)

[CZ2] Corvaja, P., Zannier, U.: On the diophantine equation $f(a^m, y) = b^n$. *Acta Arith.* **94**, 25–40 (2000)

[CZ3] Corvaja, P., Zannier, U.: Some new applications of the subspace theorem. *Compos. Math.* **131**, 319–340 (2002)

[CZ4] Corvaja, P., Zannier, U.: Finiteness of integral values for the ratio of two linear recurrences. *Invent. Math.* **149**, 431–451 (2002)

[CZ5] Corvaja, P., Zannier, U.: On the greatest prime factor of $(ab+1)(ac+1)$. *Proc. Am. Math. Soc.* **131**, 1705–1709 (2003)

[CZ6] Corvaja, P., Zannier, U.: A lower bound for the height of a rational function at $S$-unit points. *Monatsh. Math.* **144**, 203–224 (2005)

[CZ7] Corvaja, P., Zannier, U.: On the length of the continued fraction for values of quotients of power sums. *J. Théor. Nombres Bordx.* **17**, 737–748 (2005)

[CZ8] Corvaja, P., Zannier, U.: A subspace theorem approach to integral points on curves. *C. R. Acad. Sci. Paris, Ser. I* **334**, 267–271 (2002)

[CZ9] Corvaja, P., Zannier, U.: On the number of integral points on algebraic curves. *J. Reine Angew. Math.* **565**, 27–42 (2003)

[CZ10] Corvaja, P., Zannier, U.: On integral points on surfaces. *Ann. Math. (2)* **160**, 705–726 (2004)

[CZ11] Corvaja, P., Zannier, U.: On a general Thue's equation. *Am. J. Math.* **126**, 1033–105 (2004)

[CZ12] Corvaja, P., Zannier, U.: Addendum to the paper "On a general Thue's equation". *Am. J. Math.* **128**, 1057–1066 (2006)

[DZ] Dèbes, P., Zannier, U.: Universal Hilbert subsets. *Math. Proc. Camb. Philos. Soc.* **124**, 127–134 (1998)

[E] Evertse, J.-H.: An improvement of the quantitative subspace theorem. *Compos. Math.* **101**, 225–311 (1996)

[EF] Evertse, J.-H., Ferretti, R.: Diophantine inequalities on projective varieties. *Int. Math. Res. Not.* **25**, 1295–1330 (2002)

[EF2] Evertse, J.-H., Ferretti, R.: A generalization of the subspace theorem with polynomials of higher degree. (this volume)

[FW] Faltings, G., Wüstholz, G.: Diophantine approximations on projective varieties. *Invent. Math.* **116**, 109–138 (1994)

[HS] Hindry, M., Silverman, J.H.: *Diophantine Geometry*. Springer, Heidelberg (2000)

[PeZ] Perelli, A., Zannier, U.: Arithmetic properties of certain recurrent sequences. *J. Aust. Math. Soc. Ser. A* **37**, 4–16 (1984)

[vdP] van der Poorten, A.J.: Some facts that should be better known, especially about rational functions. In: Mollin, R.A. (ed.) *Number Theory and Applications*, NATO ASI Series C Mathematical and Physical Sciences, vol. 265, pp. 497–528. Kluwer Academic, Dordrecht (1989)

[vdP2] van der Poorten, A.J.: Solution de la conjecture de Pisot sur le quotient de Hadamard de deux fractions rationnelles. *C. R. Acad. Sci. Paris Sér. I* **306**, 97–102 (1988)

[Po] Pourchet, Y.: Solution du problème arithmétique du quotient de Hadamard de deux fractions rationnelles. *C. R. Acad. Sci. Paris Sér. A* **288**, 1055–1057 (1979)

[R] Rumely, R.: Note on van der Poorten's proof of the Hadamard quotient theorem I, II. In: *Séminaire de Théorie des Nombres de Paris 1986–87*. Prog. Math., vol. 75, pp. 349–409. Birkhäuser, Boston (1988)

[S] Schmidt, W.M.: *Diophantine Approximations and Diophantine Equations*. Lect. Notes Math., vol. 1467. Springer, Heidelberg (1991)

[S2] Schmidt, W.M.: Linear recurrence sequences and polynomial-exponential equations. In: Amoroso, F., Zannier, U. (eds.) *Diophantine Approximation, Lectures Given at the C.I.M.E. Summer School Held in Cetraro, Italy, June 28 – July 6, 2000*. Lect. Notes Math., vol. 1819, pp. 171–247. Springer, Heidelberg (2003)

[S3] Schmidt, W.M.: Approximations to algebraic numbers. *Enseign. Math. II. Sér.* **17**, 187–253 (1971)

[ShSt] Shorey, T.N., Stewart, C.L.: Pure powers in recurrence sequences and some related diophantine equations. *J. Number Theory* **27**, 324–352 (1987)

[ShT] Shorey, T.N., Tijdeman, R.: *Exponential Diophantine Equations*. Cambridge University Press, Cambridge (1986)

[Si] Silverman, J.: Generalized greatest common divisors, divisibility sequences and Vojta's conjecture for blow-ups, arXiv:math.NT/0407415 v3, (2004)

[TrZ] Troi, G., Zannier, U.: Note on the density constant in the distribution of self-numbers II. *Boll. Unione Mat. Ital. Sez. B Artic. Ric. Mat (8)* **2**, 397–399 (1999)

[V] Vojta, P.: *Diophantine Approximations and Value Distribution Theory*. Lect. Notes Math., vol. 1239. Springer, Heidelberg (1987)

[Z] Zannier, U.: *Some applications of diophantine approximation to diophantine equations*. Editrice Forum, Udine (2003)

[Z2] Zannier, U.: A proof of Pisot $d$th root conjecture. *Ann. Math. (2)* **151**, 375–383 (2000)

[Z3] Zannier, U.: On the integral points on the complement of ramification divisors. *J. Inst. Math. Jussieu* **4**, 317–330 (2005)

*Diophantine Approximation*
H. P. Schlickewei et al., Editors


# A GENERALIZATION OF THE SUBSPACE THEOREM WITH POLYNOMIALS OF HIGHER DEGREE

Jan-Hendrik Evertse[1] and Roberto G. Ferretti[2]
[1] *Mathematisch Instituut, Universiteit Leiden, Postbus 9512, 2300 RA Leiden, The Netherlands*
evertse@math.leidenuniv.nl
[2] *Università della Svizzera Italiana, Via Buffi 23, 6900 Lugano, Switzerland*
roberto.ferretti@lu.unisi.ch

To Professor Wolfgang Schmidt on his 70th birthday

## 1 Introduction

1.1 The Subspace Theorem can be stated as follows. Let $K$ be a number field (assumed to be contained in some given algebraic closure $\overline{\mathbb{Q}}$ of $\mathbb{Q}$), $n$ a positive integer, $0 < \delta \leqslant 1$ and $S$ a finite set of places of $K$. For $v \in S$, let $L_0^{(v)}, \dots, L_n^{(v)}$ be linearly independent linear forms in $\overline{\mathbb{Q}}[x_0, \dots, x_n]$. Then the set of solutions $\mathbf{x} \in \mathbb{P}^n(K)$ of

$$\log\left(\prod_{v\in S}\prod_{i=0}^{n}\frac{|L_i^{(v)}(\mathbf{x})|_v}{\|\mathbf{x}\|_v}\right) \leqslant -(n+1+\delta)h(\mathbf{x}) \tag{1.1}$$

is contained in the union of finitely many proper linear subspaces of $\mathbb{P}^n$.

Here, $h(\cdot)$ denotes the absolute logarithmic height on $\mathbb{P}^n(\overline{\mathbb{Q}})$, $|\cdot|_v$, $\|\cdot\|_v$ ($v \in S$) denote normalized absolute values on $K$ and normalized norms on $K^{n+1}$, and each $|\cdot|_v$ has been extended to $\overline{\mathbb{Q}}$ (see §1.4). The Subspace Theorem was first proved by Schmidt [14, 15] for the case that $S$ consists of the archimedean places of $K$, and then later extended by Schlickewei [13] to the general case.

1.2 We state a generalization of the Subspace Theorem in which the linear forms $L_i^{(v)}$ are replaced by homogeneous polynomials of arbitrary degree, and in which the solutions are taken from an $n$-dimensional projective subvariety of $\mathbb{P}^N$, where $N \geqslant n \geqslant 1$.

By a projective subvariety of $\mathbb{P}^N$ we mean a geometrically irreducible Zariski-closed subset of $\mathbb{P}^N$. For a Zariski-closed subset $X$ of $\mathbb{P}^N$ and for a field $\Omega$, we denote by $X(\Omega)$ the set of $\Omega$-rational points of $X$. For homogeneous polynomials $f_1, \dots, f_r$ in the variables $x_0, \dots, x_N$ we denote by $\{f_1 = 0, \dots, f_r = 0\}$ the Zariski-closed subset of $\mathbb{P}^N$ given by $f_1 = 0, \dots, f_r = 0$.

*Keywords.* Diophantine approximation, subspace theorem.

*2000 Mathematics subject classification.* 11J68, 11J25.

Then our result reads as follows:

**Theorem 1.1.** *Let $K$ be a number field, $S$ a finite set of places of $K$ and $X$ a projective subvariety of $\mathbb{P}^N$ defined over $K$ of dimension $n \geqslant 1$ and degree $d$. Let $0 < \delta \leqslant 1$. Further, for $v \in S$ let $f_0^{(v)}, \dots, f_n^{(v)}$ be a system of homogeneous polynomials in $\overline{\mathbb{Q}}[x_0, \dots, x_N]$ such that*

$$X(\overline{\mathbb{Q}}) \cap \left\{ f_0^{(v)} = 0, \dots, f_n^{(v)} = 0 \right\} = \emptyset \quad \text{for } v \in S. \tag{1.2}$$

*Then the set of solutions $\mathbf{x} \in X(K)$ of the inequality*

$$\log \left( \prod_{v \in S} \prod_{i=0}^{n} \frac{|f_i^{(v)}(\mathbf{x})|_v^{1/\deg f_i^{(v)}}}{\|\mathbf{x}\|_v} \right) \leqslant -(n+1+\delta)h(\mathbf{x}) \tag{1.3}$$

*is contained in a finite union $\bigcup_{i=1}^{u} (X \cap \{G_i = 0\})$, where $G_1, \dots, G_u$ are homogeneous polynomials in $K[x_0, \dots, x_N]$ not vanishing identically on $X$ of degree at most*

$$(8n+6)(n+2)^2 d\Delta^{n+1}\delta^{-1} \quad \text{with } \Delta := \operatorname{lcm}\left( \deg f_i^{(v)} : v \in S,\ 0 \leqslant i \leqslant n \right).$$

It should be noted that if $N = n$, $X = \mathbb{P}^n$ and $f_0^{(v)}, \dots, f_n^{(v)}$ are linear forms, then condition (1.2) means precisely that $f_0^{(v)}, \dots, f_n^{(v)}$ are linearly independent.

We give an immediate consequence:

**Corollary 1.2.** *Let $f_0, \dots, f_n$ be homogeneous polynomials in $\overline{\mathbb{Q}}[x_0, \dots, x_n]$ such that*

$$\left\{ \mathbf{x} \in \overline{\mathbb{Q}}^{n+1} : f_0(\mathbf{x}) = \cdots = f_n(\mathbf{x}) = 0 \right\} = \{\mathbf{0}\}.$$

*Let $0 < \delta \leqslant 1$. Then the set of solutions $\mathbf{x} = (x_0, \dots, x_n) \in \mathbb{Z}^{n+1}$ of*

$$\prod_{i=0}^{n} |f_i(\mathbf{x})|^{1/deg f_i} \leqslant \left( \max_{0 \leqslant i \leqslant n} |x_i| \right)^{-\delta}$$

*is contained in some finite union of hypersurfaces $\{G_1 = 0\} \cup \cdots \cup \{G_u = 0\}$, where each $G_i$ is a homogeneous polynomial in $\mathbb{Q}[x_0, \dots, x_n]$ of degree at most $(8n+6)(n+2)^2\Delta^{n+1}\delta^{-1}$ with $\Delta := \operatorname{lcm}(\deg f_i : 0 \leqslant i \leqslant n)$.*

1.3 In their paper [6], Faltings and Wüstholz introduced a new method to prove the Subspace Theorem and gave some examples showing that their method enables to prove extensions of the Subspace Theorem with higher degree polynomials instead of linear forms, and with solutions from an arbitrary projective variety. Ferretti [7,8] observed the role of Mumford's degree of contact [10] (or the Chow weight, see §2.3) in the work of Faltings and Wüstholz and worked out several other cases. Evertse and Ferretti [4] showed that the extensions of the Subspace Theorem as proposed by Faltings and Wüstholz in [6] can be deduced directly from the Subspace Theorem itself.

Recently, Corvaja and Zannier [2, Theorem 3] obtained a result similar to our Theorem 1.1 with $X = \mathbb{P}^n$. (More precisely, Corvaja and Zannier gave an essentially equivalent affine formulation, in which the polynomials $f_i^{(v)}$ need not be homogeneous and in which the solutions $\mathbf{x}$ have $S$-integer coordinates.) In fact, Corvaja and Zannier showed that the set of solutions of (1.3) is contained in a finite union of hypersurfaces in $\mathbb{P}^n$ and gave some further information about the structure of these hypersurfaces, on the other hand they did not provide an explicit bound for their degrees. Corvaja and Zannier stated their result only for the case $X = \mathbb{P}^n$, but with their methods this may be extended to the case that $X$ is a complete intersection. In contrast, our result is valid for arbitrary projective subvarieties $X$ of $\mathbb{P}^N$.

In their paper [2], Corvaja and Zannier proved also finiteness results for several classes of Diophantine equations. It is likely, that similar results can be deduced by means of our approach, but we have not gone into this.

1.4 Below we state a quantitative version of Theorem 1.1. We first introduce the necessary notation. All number fields considered in this paper are contained in a given algebraic closure $\overline{\mathbb{Q}}$ of $\mathbb{Q}$. Let $K$ be a number field and denote by $G_K$ the Galois group of $\overline{\mathbb{Q}}$ over $K$. For $\mathbf{x} = (x_0, \dots, x_N) \in \overline{\mathbb{Q}}^{N+1}$, $\sigma \in G_K$ we write $\sigma(\mathbf{x}) = (\sigma(x_0), \dots, \sigma(x_N))$. Denote by $M_K$ the set of places of $K$. For $v \in M_K$, choose an absolute value $|.|_v$ normalized such that the restriction of $|.|_v$ to $\mathbb{Q}$ is $|.|^{[K_v:\mathbb{R}]/[K:\mathbb{Q}]}$ if $v$ is archimedean and $|.|_p^{[K_v:\mathbb{Q}_p]/[K:\mathbb{Q}]}$ if $v$ lies above the prime number $p$. Here $|.|$ is the ordinary absolute value, and $|.|_p$ is the $p$-adic absolute value with $|p|_p = p^{-1}$. These absolute values satisfy the product formula $\prod_{v \in M_K} |x|_v = 1$ for $x \in K^*$.

Given $\mathbf{x} = (x_0, \dots, x_N) \in K^{N+1}$ we put $\|\mathbf{x}\|_v := \max(|x_0|_v, \dots, |x_N|_v)$ for $v \in M_K$. Then the absolute logarithmic height of $\mathbf{x}$ is defined by $h(\mathbf{x}) = \log(\prod_{v \in M_K} \|\mathbf{x}\|_v)$. By the product formula, $h(\lambda\mathbf{x}) = h(\mathbf{x})$ for $\lambda \in K^*$. Moreover, $h(\mathbf{x})$ depends only on $\mathbf{x}$ and not on the choice of the particular number field $K$ containing $x_0, \dots, x_N$. Thus, this function $h$ gives rise to a height on $\mathbb{P}^N(\overline{\mathbb{Q}})$.

Given a system $f_0, \dots, f_m$ of polynomials with coefficients in $\overline{\mathbb{Q}}$ we define $h(f_0, \dots, f_m) := h(\mathbf{a})$, where $\mathbf{a}$ is a vector consisting of the nonzero coefficients of $f_0, \dots, f_m$. Further by $K(f_0, \dots, f_m)$ we denote the extension of $K$ generated by the coefficients of $f_0, \dots, f_m$. The height of a projective subvariety $X$ of $\mathbb{P}^N$ defined over $\overline{\mathbb{Q}}$ is defined by $h(X) := h(F_X)$, where $F_X$ is the Chow form of $X$ (see §2.3 below).

For every $v \in M_K$ we choose an extension of $|\cdot|_v$ to $\overline{\mathbb{Q}}$ (this amounts to extending $|\cdot|_v$ to the algebraic closure $\overline{K}_v$ of $K_v$ and choosing an embedding of $\overline{\mathbb{Q}}$ into $\overline{K}_v$). Further for $v \in M_K$, $\mathbf{x} = (x_0, \dots, x_N) \in \overline{\mathbb{Q}}^{N+1}$ we put $\|\mathbf{x}\|_v := \max(|x_0|_v, \dots, |x_N|_v)$.

1.5 Schmidt [16] was the first to obtain a quantitative version of the Subspace Theorem, giving an explicit upper bound for the number of subspaces containing all solutions with "large" height. Since then his basic result has been improved and generalized in various directions. Evertse and Schlickewei [5, Theorem 3.1] deduced a quantitative version of the Absolute Subspace Theorem, dealing with solutions in $\mathbb{P}^n(\overline{\mathbb{Q}})$ of some absolute extension of (1.1). Their result can be stated as follows.

Let again $K$ be a number field, and $S$ a finite set of places of $K$ of cardinality $s$. Let $n \geqslant 1$, $0 < \delta \leqslant 1$. For $v \in S$, let $L_0^{(v)}, \dots, L_n^{(v)}$ be linearly independent linear forms in $\overline{\mathbb{Q}}[x_0, \dots, x_n]$. Put $\mathcal{D} := \prod_{v \in S} |\det(L_0^{(v)}, \dots, L_n^{(v)})|_v$ and assume that $[K(L_i^{(v)}) : K] \leqslant C$ for $v \in S$, $i = 0, \dots, n$. Then the set of $\mathbf{x} \in \mathbb{P}^n(\overline{\mathbb{Q}})$ with

$$\log\left(\mathcal{D}^{-1}\prod_{v\in S}\prod_{i=0}^{n}\max_{\sigma\in G_K}\frac{|L_i^{(v)}(\sigma(\mathbf{x}))|_v}{\|\sigma(\mathbf{x})\|_v}\right)\leqslant -(n+1+\delta)h(\mathbf{x}),$$

$$h(\mathbf{x})\geqslant 9(n+1)\delta^{-1}\log(n+1)+\max\left(h(L_i^{(v)}):\ v\in S,\ 0\leqslant i\leqslant n\right)$$

is contained in the union of not more than

$$(3n+3)^{(2n+2)s}8^{(n+10)^2}\delta^{-(n+1)s-n-5}\log(4C)\log\log(4C)$$

proper linear subspaces of $\mathbb{P}^n(\overline{\mathbb{Q}})$ which are all defined over $K$.

Typically, the lower bound for $h(\mathbf{x})$ depends on the linear forms $L_i^{(v)}$, while the upper bound for the number of subspaces does not depend on the $L_i^{(v)}$.

1.6 We now state an analogue for inequalities with higher degree polynomials instead of linear forms. We first list some notation: $\delta$ is a real with $0<\delta\leqslant 1$, $K$ is a number field, $S$ is a finite set of places of $K$ of cardinality $s$, $X$ is a projective subvariety of $\mathbb{P}^N$ defined over $K$ of dimension $n\geqslant 1$ and degree $d$, $f_0^{(v)},\dots,f_n^{(v)}$ $(v\in S)$ are systems of homogeneous polynomials in $\overline{\mathbb{Q}}[x_0,\dots,x_N]$,

$$\begin{cases} C:=\max\left([K(f_i^{(v)}):K]:\ v\in S,\ i=0,\dots,n\right),\\ \Delta:=\operatorname{lcm}\left(\deg f_i^{(v)}:\ v\in S,\ i=0,\dots,n\right),\end{cases}\tag{1.4}$$

$$\begin{cases} A_1:=(20n\delta^{-1})^{(n+1)s}\cdot\exp\left(2^{12n+16}n^{4n}\delta^{-2n}d^{2n+2}\Delta^{n(2n+2)}\right)\\ \qquad\qquad\cdot\log(4C)\log\log(4C),\\ A_2:=(8n+6)(n+2)^2d\Delta^{n+1}\delta^{-1},\\ A_3:=\exp\left(2^{6n+20}n^{2n+3}\delta^{-n-1}d^{n+2}\Delta^{n(n+2)}\log(2Cs)\right),\\ H:=\log(2N)+h(X)+\max\left(h(1,f_i^{(v)}):\ v\in S,\ 0\leqslant i\leqslant n\right).\end{cases}\tag{1.5}$$

**Theorem 1.3.** *Assume that*

$$X(\overline{\mathbb{Q}})\cap\left\{f_0^{(v)}=0,\dots,f_n^{(v)}=0\right\}=\emptyset\quad \textit{for } v\in S.\tag{1.2}$$

*Then there are homogeneous polynomials $G_1,\dots,G_u\in K[x_0,\dots,x_N]$ with*

$$u\leqslant A_1,\quad \deg G_i\leqslant A_2\ \textit{for } i=1,\dots,u$$

*which do not vanish identically on $X$, such that the set of $\mathbf{x}\in X(\overline{\mathbb{Q}})$ with*

$$\log\left(\prod_{v\in S}\prod_{i=0}^{n}\max_{\sigma\in G_K}\frac{|f_i^{(v)}(\sigma(\mathbf{x}))|_v^{1/\deg f_i^{(v)}}}{\|\sigma(\mathbf{x})\|_v}\right)\leqslant -(n+1+\delta)h(\mathbf{x}),\tag{1.6}$$

$$h(\mathbf{x})\geqslant A_3\cdot H\tag{1.7}$$

*is contained in $\bigcup_{i=1}^{u}(X\cap\{G_i=0\})$.*

Clearly, the bounds in Theorem 1.3 are much worse than those in the result of Evertse and Schlickewei. It would be very interesting if one could replace $A_1$, $A_3$ by quantities which are at most exponential in (some power of) $n$ and which are polynomial in $\delta^{-1}$, $d$, $\Delta$. Further, we do not know whether the dependence of $A_2$ on $\delta$ is needed.

1.7 Our starting point is a result for twisted heights on $\mathbb{P}^n$ (a quantitative version of the Absolute Parametric Subspace Theorem), due to Evertse and Schlickewei [5, Theorem 2.1] (see also Proposition 3.1 in Section 3 below). From this, we deduce an analogous result for twisted heights on arbitrary projective varieties; the statement of this result is in Section 2 (Theorem 2.1) and its proof in Section 3. The proof involves some arguments from Evertse and Ferretti [4], in particular an explicit lower bound of the normalized Chow weight of a projective variety in terms of the $m$-th normalized Hilbert weight of that variety. In Section 4 we give some height estimates; here we use heavily Rémond's exposé [12]. Then in Section 5 we deduce Theorem 1.3. Using that $\mathbb{P}^N(K)$ has only finitely many points with height below any given bound, Theorem 1.1 follows at once from Theorem 1.3.

## 2 Twisted heights

2.1 The quantitative version of the Absolute Parametric Subspace Theorem of Evertse and Schlickewei mentioned in the previous section deals with a class of twisted heights defined on $\mathbb{P}^n(\overline{\mathbb{Q}})$ parametrized by a real $Q \geqslant 1$. Roughly speaking, this result states that there are a finite number of proper linear subspaces of $\mathbb{P}^n$ such that for every sufficiently large $Q$, the set of points in $\mathbb{P}^n(\overline{\mathbb{Q}})$ with small $Q$-height is contained in one of these subspaces. Theorem 2.1 stated below is an analogue in which the points are taken from an arbitrary projective variety instead of $\mathbb{P}^n$. Loosely speaking, Theorem 1.3 stated in the previous section is proved by defining a suitable finite morphism $\varphi$ from $X$ to a projective variety $Y \subset \mathbb{P}^R$ and a finite number of classes of twisted heights on $Y$ as above, and applying Theorem 2.1 to each of these classes.

2.2 Let $K$ be a number field. For finite extensions of $K$ we define normalized absolute values similarly as for $K$. Thus, if $L$ is a finite extension of $K$, $w$ is a place of $L$, and $v$ is the place of $K$ lying below $w$, then

$$|x|_w = |x|_v^{d(w|v)} \quad \text{for } x \in K, \text{ with } d(w|v) := \frac{[L_w : K_v]}{[L : K]}, \tag{2.1}$$

where $K_v$, $L_w$ denote the completions at $v$, $w$, respectively.

We denote points on $\mathbb{P}^R$ by $\mathbf{y} = (y_0, \dots, y_R)$. For $v \in M_K$, let $\mathbf{c}_v = (c_{0v}, \dots, c_{Rv})$ be a tuple of reals such that $c_{0v} = \cdots = c_{Rv} = 0$ for all but finitely many places $v \in M_K$ and put $\mathbf{c} = (\mathbf{c}_v : v \in M_K)$. Further, let $Q$ be a real $\geqslant 1$. We define a twisted height on $\mathbb{P}^R(\overline{\mathbb{Q}})$ as follows. First put

$$H_{Q,\mathbf{c}}(\mathbf{y}) := \prod_{v \in M_K} \max_{0 \leqslant i \leqslant R} \left(|y_i|_v Q^{c_{iv}}\right) \quad \text{for } \mathbf{y} = (y_0, \dots, y_R) \in \mathbb{P}^R(K);$$

by the product formula, this is well-defined on $\mathbb{P}^R(K)$. For any finite extension $L$ of $K$ we put

$$c_{iw} := c_{iv} \cdot d(w|v) \quad \text{for } w \in M_L, \tag{2.2}$$

where $M_L$ is the set of places of $L$ and $v$ the place of $K$ lying below $w$. Then for $\mathbf{y} \in \mathbb{P}^R(\overline{\mathbb{Q}})$, we define

$$H_{Q,\mathbf{c}}(\mathbf{y}) := \prod_{w \in M_L} \max_{0 \leqslant i \leqslant R} \left( |y_i|_w Q^{c_{iw}} \right) \tag{2.3}$$

where $L$ is any finite extension of $K$ such that $\mathbf{y} \in \mathbb{P}^R(L)$. In view of (2.1) this definition does not depend on $L$.

2.3 Let $Y$ be a (by definition irreducible) projective subvariety of $\mathbb{P}^R$ of dimension $n$ and degree $D$, defined over $K$. We recall that up to a constant factor there is a unique polynomial $F_Y(\mathbf{u}^{(0)}, \dots, \mathbf{u}^{(n)})$ with coefficients in $K$ in blocks of variables $\mathbf{u}^{(0)} = (u_0^{(0)}, \dots, u_R^{(0)}), \dots, \mathbf{u}^{(n)} = (u_0^{(n)}, \dots, u_R^{(n)})$, called the *Chow form* of $Y$, with the following properties: $F_Y$ is irreducible over $\overline{\mathbb{Q}}$; $F_Y$ is homogeneous in each block $\mathbf{u}^{(h)}$ $(h = 0, \dots, n)$; and $F_Y(\mathbf{u}^{(0)}, \dots, \mathbf{u}^{(n)}) = 0$ if and only if $Y$ and the hyperplanes $\sum_{i=0}^R u_i^{(h)} y_i = 0 (h = 0, \dots, n)$ have a $\overline{\mathbb{Q}}$-rational point in common. It is well known that the degree of $F_Y$ in each block $\mathbf{u}^{(h)}$ is $D$.

Let $\mathbf{c} = (c_0, \dots, c_R)$ be a tuple of reals. Introduce an auxiliary variable $t$ and substitute $t^{c_i} u_i^{(h)}$ for $u_i^{(h)}$ in $F_Y$ for $h = 0, \dots, n$, $i = 0, \dots, R$. Thus we obtain an expression

$$\begin{aligned} &F_Y(t^{c_0} u_0^{(0)}, \dots, t^{c_R} u_R^{(0)}; \dots; t^{c_0} u_0^{(n)}, \dots, t^{c_R} u_R^{(n)}) \\ &\quad = t^{e_0} G_0(\mathbf{u}^{(0)}, \dots, \mathbf{u}^{(n)}) + \dots + t^{e_r} G_r(\mathbf{u}^{(0)}, \dots, \mathbf{u}^{(n)}), \end{aligned} \tag{2.4}$$

with $G_0, \dots, G_r \in K[\mathbf{u}^{(0)}, \dots, \mathbf{u}^{(n)}]$ and $e_0 > e_1 > \dots > e_r$. Now we define the *Chow weight* of $Y$ with respect to $\mathbf{c}$ by[1]

$$e_Y(\mathbf{c}) := e_0. \tag{2.5}$$

2.4 We formulate our main result for twisted heights. Below, $Y$ is a projective subvariety of $\mathbb{P}^R$ of dimension $n \geqslant 1$ and degree $D$, defined over $K$, and $\mathbf{c}_v = (c_{0v}, \dots, c_{Rv})$ $(v \in M_K)$ are tuples of reals such that

$$c_{iv} \geqslant 0 \text{ for } v \in M_K, i = 0, \dots, R; \tag{2.6}$$

$$c_{0v} = \dots = c_{Rv} = 0 \text{ for all but finitely many } v \in M_K; \tag{2.7}$$

$$\sum_{v \in M_K} \max(c_{0v}, \dots, c_{Rv}) \leqslant 1. \tag{2.8}$$

Put

$$E_Y(\mathbf{c}) := \frac{1}{(n+1)D} \left( \sum_{v \in M_K} e_Y(\mathbf{c}_v) \right). \tag{2.9}$$

1. The Chow weight was introduced in [4], and named such because of its relation to the Chow form. It is an adaptation of the *degree of contact* earlier introduced by Mumford [10], so perhaps the naming "Mumford weight" would have been a happier choice. Roughly speaking, the degree of contact of $Y$ with respect to $\mathbf{c}$ is defined for integer tuples $\mathbf{c}$ and it is equal to $e_r$ instead of $e_0$.

Further, let $0 < \delta \leqslant 1$, and put

$$\begin{cases} B_1 := \exp\left(2^{10n+4}\delta^{-2n}D^{2n+2}\right)\cdot\log(4R)\log\log(4R), \\ B_2 := (4n+3)D\delta^{-1}, \\ B_3 := \exp\left(2^{5n+4}\delta^{-n-1}D^{n+2}\log(4R)\right). \end{cases} \tag{2.10}$$

**Theorem 2.1.** *There are homogeneous polynomials $F_1, \dots, F_t \in K[y_0, \dots, y_R]$ with*

$$t \leqslant B_1, \quad \deg F_i \leqslant B_2 \quad \textit{for } i = 1, \dots, t,$$

*which do not vanish identically on $Y$, such that for every real number $Q$ with*

$$\log Q \geqslant B_3 \cdot (h(Y) + 1)$$

*there is $F_i \in \{F_1, \dots, F_t\}$ with*

$$\left\{\mathbf{y} \in Y(\overline{\mathbb{Q}}) : H_{Q,\mathbf{c}}(\mathbf{y}) \leqslant Q^{E_Y(\mathbf{c})-\delta}\right\} \subset Y \cap \{F_i = 0\}. \tag{2.11}$$

## 3 Proof of Theorem 2.1

3.1 We first recall the quantitative version of the Absolute Parametric Subspace Theorem of Evertse and Schlickewei. As before, $K$ is an algebraic number field and $R, n$ are integers with $R \geqslant n \geqslant 1$. We denote the coordinates on $\mathbb{P}^n$ by $(x_0, \dots, x_n)$. Given an index set $I = \{i_0, \dots, i_n\}$ with $i_0 < \dots < i_n$ and linear forms $L_j = \sum_{i=0}^n a_{ij}x_i$ ($j \in I$) we write $\det(L_j : j \in I) := \det(a_{i,i_j})_{i,j=0,\dots,n}$.

Let $L_0, \dots, L_R$ be linear forms in $K[x_0, \dots, x_n]$ with $\operatorname{rank}\{L_0, \dots, L_R\} = n+1$. Further, let $I_v$ ($v \in M_K$) be subsets of $\{0, \dots, R\}$ of cardinality $n+1$ such that

$$\operatorname{rank}\{L_i : i \in I_v\} = n+1 \quad \text{for } v \in M_K. \tag{3.1}$$

Define

$$\mathcal{H} := \prod_{v \in M_K} \max_I |\det(L_i : i \in I)|_v, \quad \mathcal{D} := \prod_{v \in M_K} |\det(L_i : i \in I_v)|_v; \tag{3.2}$$

here the maximum is taken over all subsets $I$ of $\{0, \dots, R\}$ of cardinality $n+1$. According to [4, Lemma 7.2] we have

$$\mathcal{D} \geqslant \mathcal{H}^{1-\binom{R+1}{n+1}}. \tag{3.3}$$

Let $\mathbf{d}_v = (d_{iv} : i \in I_v)$ ($v \in M_K$) be tuples of reals such that

$$d_{iv} = 0 \ \text{ for } i \in I_v \text{ and for all but finitely many } v \in M_K, \tag{3.4}$$

$$\sum_{v \in M_K}\sum_{i \in I_v} d_{iv} = 0, \tag{3.5}$$

$$\sum_{v \in M_K} \max(d_{iv} : i \in I_v) \leqslant 1 \tag{3.6}$$

and write $\mathbf{d} = (\mathbf{d}_v : v \in M_K)$.

We define a twisted height on $\mathbb{P}^n(\overline{\mathbb{Q}})$ as follows. For any real number $Q \geqslant 1$ we first put

$$H^*_{Q,\mathbf{d}}(\mathbf{x}) = \prod_{v\in M_K} \left(\max_{i\in I_v} |L_i(\mathbf{x})|_v Q^{-d_{iv}}\right) \quad \text{for } \mathbf{x} \in \mathbb{P}^n(K).$$

More generally, if $L$ is any finite extension of $K$, put

$$d_{iw} := d(w|v)d_{iv}, \quad I_w := I_v, \tag{3.7}$$

where $v$ is the place of $K$ lying below $w$. Then for $\mathbf{x} \in \mathbb{P}^n(\overline{\mathbb{Q}})$ we define

$$H^*_{Q,\mathbf{d}}(\mathbf{x}) = \prod_{w\in M_L} \left(\max_{i\in I_w} |L_i(\mathbf{x})|_v Q^{-d_{iw}}\right), \tag{3.8}$$

where $L$ is any finite extension of $K$ such that $\mathbf{x} \in \mathbb{P}^n(L)$. This is independent of the choice of $L$.

Now the result of Evertse and Schlickewei [5, Theorem 2.1] is as follows:

**Proposition 3.1.** *Let* $I_v$ $(v \in M_K)$, $\mathbf{d} = (\mathbf{d}_v : v \in M_K)$, *satisfy* (3.1), (3.4), *respectively, and let* $0 < \varepsilon \leqslant 1$.

*There are proper linear subspaces* $T_1, \dots, T_t$ *of* $\mathbb{P}^n$, *defined over* $K$, *with*

$$t \leqslant 4^{(n+9)^2} \varepsilon^{-n-5} \log(3R) \log\log(3R), \tag{3.9}$$

*such that for every real number* $Q$ *with*

$$Q \geqslant \max\left(\mathcal{H}^{1/\binom{R+1}{n+1}}, (n+1)^{2/\varepsilon}\right) \tag{3.10}$$

*there is* $T_i \in \{T_1, \dots, T_t\}$ *with*

$$\{\mathbf{x} \in \mathbb{P}^n(\overline{\mathbb{Q}}) : H^*_{Q,\mathbf{d}}(\mathbf{x}) \leqslant \mathcal{D}^{1/(n+1)} Q^{-\varepsilon}\} \subset T_i\,. \tag{3.11}$$

3.2 We recall some results from [4]. As in Section 2, we denote the coordinates on $\mathbb{P}^R$ by $(y_0, \dots, y_R)$. Let $Y$ be a projective variety of $\mathbb{P}^R$ defined over $K$ of dimension $n$ and degree $D$. Let $I_Y$ be the prime ideal of $Y$, i.e., the ideal of polynomials from $\overline{\mathbb{Q}}[y_0, \dots, y_R]$ vanishing identically on $Y$. For $m \in \mathbb{N}$, denote by $\overline{\mathbb{Q}}[y_0, \dots, y_R]_m$ the vector space of homogeneous polynomials in $\overline{\mathbb{Q}}[y_0, \dots, y_R]$ of degree $m$, and put $(I_Y)_m := \overline{\mathbb{Q}}[y_0, \dots, y_R]_m \cap I_Y$. Then the Hilbert function of $Y$ is defined by

$$H_Y(m) := \dim_{\overline{\mathbb{Q}}} \left(\overline{\mathbb{Q}}[y_0, \dots, y_R]_m / (I_Y)_m\right).$$

The scalar product of $\mathbf{a} = (a_0, \dots, a_R)$, $\mathbf{b} = (b_0, \dots, b_R) \in \mathbb{R}^{R+1}$ is given by $\mathbf{a} \cdot \mathbf{b} := a_0 b_0 + \cdots + a_R b_R$. For $\mathbf{a} = (a_0, \dots, a_R) \in (\mathbb{Z}_{\geqslant 0})^{R+1}$, denote by $\mathbf{y}^{\mathbf{a}}$ the monomial $y_0^{a_0} \cdots y_R^{a_R}$. Then the $m$-th Hilbert weight of $Y$ with respect to a tuple $\mathbf{c} = (c_0, \dots, c_R) \in \mathbb{R}^{R+1}$ is defined by

$$s_Y(m, \mathbf{c}) := \max \left(\sum_{i=1}^{H_Y(m)} \mathbf{a}_i \cdot \mathbf{c}\right), \tag{3.12}$$

where the maximum is taken over all sets of monomials $\{\mathbf{y}^{\mathbf{a}_1}, \dots, \mathbf{y}^{\mathbf{a}_{H_Y(m)}}\}$ whose residue classes modulo $(I_Y)_m$ form a basis of $\overline{\mathbb{Q}}[y_0, \dots, y_R]_m / (I_Y)_m$.

We recall Evertse and Ferretti [4, Theorem 4.1]:

**Proposition 3.2.** *Let* $\mathbf{c} = (c_0, \dots, c_R)$ *be a tuple of nonnegative reals. Let* $m > D$ *be an integer. Then*

$$\tfrac{1}{mH_Y(m)} \cdot s_Y(m, \mathbf{c}) \geqslant \tfrac{1}{(n+1)D} \cdot e_Y(\mathbf{c}) - \tfrac{(2n+1)D}{m} \cdot \max(c_0, \dots, c_R). \tag{3.13}$$

Let $m$ be a positive integer. Put

$$n_m := H_Y(m) - 1, \quad R_m := \tbinom{R+m}{m} - 1,$$

and let $\mathbf{y}^{\mathbf{a}_0}, \dots, \mathbf{y}^{\mathbf{a}_{R_m}}$ be the monomials of degree $m$ in $y_0, \dots, y_R$, in some order. Denote by $\varphi_m$ the Veronese map of degree $m$, $\mathbf{y} \mapsto (\mathbf{y}^{\mathbf{a}_0}, \dots, \mathbf{y}^{\mathbf{a}_{R_m}})$. Lastly, denote by $Y_m$ the smallest linear subspace of $\mathbb{P}^{R_m}$ containing $\varphi_m(Y)$.

**Lemma 3.3.** (i) $Y_m$ *is defined over* $K$*;*

(ii) $\dim Y_m = n_m \leqslant D\binom{m+n}{n}$*;*

(iii) $h(Y_m) \leqslant Dm\binom{m+n}{n}\left(D^{-1}h(Y) + (3n+4)\log(R+1)\right)$.

*Proof.* (i), (iii) [4, Lemma 8.3]; (ii) Chardin [1, Théorème 1]. □

3.3 Let $\mathbf{c}_v \in \mathbb{R}^R$ $(v \in M_K)$ be tuples with (2.6) and (2.8). For a suitable value of $m$, we link the twisted height $H_{Q,\mathbf{c}}$ from Theorem 2.1 to a twisted height on $\mathbb{P}^{n_m}$ to which Proposition 3.1 is applicable. Put

$$m := [(4n+3)D\delta^{-1}]. \tag{3.14}$$

Then by Proposition 3.2 and (2.6) we have

$$\frac{1}{mH_Y(m)} \cdot \left(\sum_{v \in M_K} s_Y(m, \mathbf{c}_v)\right) \geqslant \frac{1}{(n+1)D} \cdot \left(\sum_{v \in M_K} e_Y(\mathbf{c}_v)\right) - \frac{\delta}{2}. \tag{3.15}$$

Denote as before the coordinates on $\mathbb{P}^R$ by $\mathbf{y} = (y_0, \dots, y_R)$, those on $\mathbb{P}^{n_m} = \mathbb{P}^{H_Y(m)-1}$ by $\mathbf{x} = (x_0, \dots, x_{n_m})$, and those on $\mathbb{P}^{R_m} = \mathbb{P}^{\binom{R+m}{m}-1}$ by $\mathbf{z} = (z_0, \dots, z_{R_m})$. Since $Y_m$ is an $n_m$-dimensional linear subspace of $\mathbb{P}^{R_m}$ defined over $K$, there are linear forms $L_0, \dots, L_{R_m} \in K[x_0, \dots, x_{n_m}]$ such that the map

$$\psi_m : \mathbf{x} \mapsto (L_0(\mathbf{x}), \dots, L_{R_m}(\mathbf{x}))$$

is a linear isomorphism from $\mathbb{P}^{n_m}$ to $Y_m$. Thus, $\psi_m^{-1}\varphi_m$ is an injective map from $Y$ into $\mathbb{P}^{n_m}$.

For $v \in M_K$ there is a subset $I_v$ of $\{0, \dots, R_m\}$ of cardinality $n_m + 1 = H_Y(m)$ such that $\{\mathbf{y}^{\mathbf{a}_i} : i \in I_v\}$ is a basis of $\overline{\mathbb{Q}}[y_0, \dots, y_R]_m/(I_Y)_m$ and

$$s_Y(m, \mathbf{c}_v) = \sum_{i \in I_v} \mathbf{a}_i \cdot \mathbf{c}_v. \tag{3.16}$$

Now define the tuples $\mathbf{d}_v = (d_{iv}, i \in I_v)$ $(v \in M_K)$ by

$$\begin{aligned} d_{iv} &= -\frac{1}{m} \cdot \mathbf{a}_i \cdot \mathbf{c}_v + \frac{1}{m(n_m+1)}\left(\sum_{j \in I_v} \mathbf{a}_j \cdot \mathbf{c}_v\right) \\ &= -\frac{1}{m} \cdot \mathbf{a}_i \cdot \mathbf{c}_v + \frac{1}{mH_Y(m)} \cdot s_Y(m, \mathbf{c}_v), \end{aligned} \tag{3.17}$$

and put $\mathbf{d} = (\mathbf{d}_v : v \in M_K)$. Similarly to (3.2) we define

$$\mathcal{H} := \prod_{v \in M_K} \max_I |\det(L_i : i \in I)|_v, \quad \mathcal{D} := \prod_{v \in M_K} |\det(L_i : i \in I_v)|_v,$$

where the maximum is taken over all subsets $I$ of $\{0, \ldots, R_m\}$ of cardinality $n_m + 1$. Then by, e.g., [4, page 1300] we have

$$\log \mathcal{H} = h(Y_m). \tag{3.18}$$

We define in a usual manner a twisted height on $\mathbb{P}^{n_m}(\overline{\mathbb{Q}})$ by putting

$$H^*_{Q,\mathbf{d}}(\mathbf{x}) = \prod_{w \in M_L} \max_{i \in I_w} \left( |L_i(\mathbf{x})|_w Q^{-d_{iw}} \right)$$

for $\mathbf{x} \in \mathbb{P}^{n_m}(\overline{\mathbb{Q}})$, where $L$ is any finite extension of $K$ such that $\mathbf{x} \in \mathbb{P}^{n_m}(L)$, $Q \geqslant 1$ is a real number, and $d_{iw} = d(w|v)d_{iv}$, $I_w = I_v$ with $v$ the place of $K$ below $w$. It follows at once from (2.7) that $d_{iv} = 0$ for all but finitely many $v$ and for $i \in I_v$. Therefore this height is well-defined.

**Lemma 3.4.** *Assume that*

$$Q \geqslant \mathcal{D}^{6/\delta m(n_m+1)}. \tag{3.19}$$

*Let* $\mathbf{y} \in Y(\overline{\mathbb{Q}})$ *be such that*

$$H_{Q,\mathbf{c}}(\mathbf{y}) \leqslant Q^{E_Y(\mathbf{c})-\delta}, \tag{3.20}$$

*where* $E_Y(\mathbf{c}) = \frac{1}{(n+1)D}\left(\sum_{v \in M_K} e_Y(\mathbf{c}_v)\right)$. *Let* $\mathbf{x} = \psi_m^{-1}\varphi_m(\mathbf{y})$. *Then*

$$H^*_{Q^m,\mathbf{d}}(\mathbf{x}) \leqslant \mathcal{D}^{1/(n_m+1)}(Q^m)^{-\delta/3}. \tag{3.21}$$

*Proof.* Put $s_v := \frac{1}{mH_Y(m)} s_Y(m, \mathbf{c}_v)$, $s := \sum_{v \in M_K} s_v$. We first show that

$$H^*_{Q^m,\mathbf{d}}(\mathbf{x}) \leqslant Q^{-ms} \left(H_{Q,\mathbf{c}}(\mathbf{y})\right)^m. \tag{3.22}$$

Take a finite extension $L$ of $K$ such that $\mathbf{y} \in Y(L)$. We have $\mathbf{x} \in \mathbb{P}^{n_m}(L)$ and $L_i(\mathbf{x}) = \mathbf{y}^{\mathbf{a}_i}$ for $i = 0, \ldots, R_m$. So for $w \in M_L$ we have (putting $s_w := d(w|v)s_v$, with $v$ the place of $K$ below $w$),

$$\max_{i \in I_w} \left( |L_i(\mathbf{x})|_w (Q^m)^{-d_{iw}} \right) = \max_{i \in I_w} \left( |\mathbf{y}^{\mathbf{a}_i}|_w Q^{\mathbf{a}_i \cdot \mathbf{c}_w - ms_w} \right)$$

$$\leqslant \max_{i=0,\ldots,R_m} \left( |\mathbf{y}^{\mathbf{a}_i}|_w Q^{\mathbf{a}_i \cdot \mathbf{c}_w - ms_w} \right) \leqslant \left( Q^{-s_w} \max_{i=0,\ldots,R} \left( |y_i|_w Q^{c_{iw}} \right) \right)^m.$$

By taking the product over all $w \in M_L$, (3.22) follows.

Now a successive application of (3.19), (3.22), (3.20), (3.15) gives

$$H^*_{Q^m,\mathbf{d}}(\mathbf{x}) \leqslant \mathcal{D}^{1/(n_m+1)} Q^{m\delta/6} \cdot Q^{-ms} Q^{mE_Y(\mathbf{c})-m\delta} \leqslant \mathcal{D}^{1/(n_m+1)}(Q^m)^{-\delta/3}.$$

□

3.4 To complete the proof of Theorem 2.1 we apply Proposition 3.1 to (3.21); that is, we apply Proposition 3.1 with $n = n_m$, $R = R_m$, $\varepsilon = \delta/3$, and with $Q^m$ in place of $Q$. For the moment we assume

$$\log Q \geqslant \frac{6}{(n_m+1)m\delta}(R_m+1)^{n_m+1}(h(Y_m)+1). \tag{3.23}$$

In view of (3.18), this is precisely (3.10) with $R = R_m, n = n_m, \varepsilon = \delta/3$ and with $Q^m$ in place of $Q$.

We have to verify that (3.1), (3.4), (3.5), (3.6) are satisfied with $n_m, R_m$ in place of $n, R$. First, (3.1) follows at once from the definition of $I_v$ and the fact that $\psi_m$ is a linear isomorphism. Secondly, (3.4) follows from (2.7) and (3.17). Thirdly, (3.5) follows from (3.17), (3.16). Finally, (3.6) is a consequence of (2.6), (2.8) and the fact that $\frac{1}{mH_Y(m)} \cdot s_Y(m, \mathbf{c}_v)$ can be expressed as a maximum of linear forms in $c_{0v}, \ldots, c_{Rv}$ whose coefficients are nonnegative and have sum equal to 1.

Thus, there are proper linear subspaces $T_1, \ldots, T_t$ of $\mathbb{P}^{n_m}$, defined over $K$, with

$$t \leqslant 4^{(n_m+9)^2}(3/\delta)^{n_m+5}\log(3R_m)\log\log(3R_m) \tag{3.24}$$

such that for every $Q$ with (3.23) there is $T_i \in \{T_1, \ldots, T_t\}$ with

$$\{\mathbf{x} \in \mathbb{P}^{n_m}(\overline{\mathbb{Q}}) : H^*_{Q^m,\mathbf{d}}(\mathbf{x}) \leqslant \mathcal{D}^{1/(n_m+1)}(Q^m)^{-\delta/3}\} \subset T_i\,.$$

For each space $T_i$ there is a linear form $L_i \in K[z_0, \ldots, z_{R_m}]$ vanishing identically on $\psi_m(T_i)$ but not on $Y_m$. Since by definition, $Y_m$ is the smallest linear subvariety of $\mathbb{P}^{R_m}$ containing $\varphi_m(Y)$, the linear form $L_i$ does not vanish identically on $\varphi_m(Y)$. Replacing in $L_i$ the coordinate $z_j$ by $\mathbf{y}^{\mathbf{a}_j}$ for $j = 0, \ldots, R_m$, we obtain a homogeneous polynomial $F_i \in K[y_0, \ldots, y_R]$ of degree $m$ not vanishing identically on $Y$ such that if $\mathbf{x} = \psi_m^{-1}\varphi_m(\mathbf{y}) \in T_i$, then $F_i(\mathbf{y}) = 0$.

It is easily seen that assumption (3.23), together with (3.18) and (3.3), implies (3.19); hence Lemma 3.4 is applicable. Thus, we infer that there are homogeneous polynomials $F_1, \ldots, F_t \in K[y_0, \ldots, y_R]$ of degree $m$, with $t$ satisfying (3.24), such that for every $Q$ with (3.23) there is $F_i \in \{F_1, \ldots, F_t\}$ with

$$\{\mathbf{y} \in Y(\overline{\mathbb{Q}}) : H_{Q,\mathbf{c}}(\mathbf{y}) \leqslant Q^{E_Y(\mathbf{c})-\delta}\} \subset Y \cap \{F_i = 0\}\,.$$

By (3.14) we have $m \leqslant (4n+3)D\delta^{-1}$, which is the quantity $B_2$ from (2.10). So to complete the proof of Theorem 2.1, it suffices to show that the right-hand side of (3.24) is at most $B_1$ and that the right-hand side of (3.23) is at most $B_3 \cdot (h(Y)+1)$, where $B_1, B_3$ are given by (2.10).

Using $m \geqslant 7$ and the inequality

$$\binom{x+y}{y} \leqslant \frac{(x+y)^{x+y}}{x^x y^y} = \left(1+\frac{y}{x}\right)^x \cdot \left(1+\frac{x}{y}\right)^y \leqslant \left(e\left(1+\frac{x}{y}\right)\right)^y \tag{3.25}$$

for positive integers $x, y$, we infer

$$R_m = \binom{R+m}{m} - 1 \leqslant \left(e\left(1+\frac{R}{m}\right)\right)^m \leqslant (4R)^m\,. \tag{3.26}$$

So by (3.14),

$$\begin{aligned} \log(3R_m)\log\log(3R_m) &\leqslant 2m^2\log(4R)\log\log(4R) \\ &\leqslant 2(8n+6)^2D^2\delta^{-2}\log(4R)\log\log(4R)\,. \end{aligned}$$

Further, by Lemma 3.3, (ii),

$$n_m \leqslant D\binom{m+n}{n} \leqslant D\left(e\left(1+\frac{m}{n}\right)\right)^n$$
$$\leqslant D\left(e(1+7D\delta^{-1})\right)^n \leqslant 2^{5n}\delta^{-n}D^{n+1}. \tag{3.27}$$

Hence the right-hand side of (3.24) is at most

$$4^{(2^{5n}\delta^{-n}D^{n+1}+9)2}(3\delta^{-1})^{2^{5n}\delta^{-n}D^{n+1}+5}$$
$$\times 2(8n+6)^2D^2\delta^{-2}\log(4R)\log\log(4R)$$
$$\leqslant \exp\left(2^{10n+4}\delta^{-2n}D^{2n+2}\right)\cdot\log(4R)\log\log(4R) = B_1\,,$$

while by Lemma 3.3, (3.14), (3.26), (3.27), the right-hand side of (3.23) is at most

$$\frac{6}{(n_m+1)m\delta}\left((4R)^m+1\right)^{n_m+1}$$
$$\times\left(1+Dm\binom{m+n}{n}\right)\left(D^{-1}h(Y)+(3n+4)\log(R+1)\right)$$
$$\leqslant \delta^{-1}\left((4R)^{(4n+3)D\delta^{-1}}+1\right)^{2^{5n}\delta^{-n}D^{n+1}+1}$$
$$\times 2^{5n}\delta^{-n}D^{n+1}(3n+1)\log(R+1)\cdot(h(Y)+1)$$
$$< \exp\left(2^{5n+4}\delta^{-n-1}D^{n+2}\log(4R)\right)\cdot(h(Y)+1) = B_3\cdot(h(Y)+1)\,.$$

This completes the proof of Theorem 2.1. □

## 4 Height estimates

4.1 In this section we deduce some height estimates, using results from Rémond's paper [12].

Let $K$ be a number field. Denote as before the set of places of $K$ by $M_K$, and denote the sets of archimedean and non-archimedean places of $K$ by $M_K^\infty$ and $M_K^0$, respectively. We use the normalized absolute values $|\cdot|_v$ introduced in §1.4. Recall that for each of these absolute values we have chosen an extension to $\overline{\mathbb{Q}}$. In particular, for each $v \in M_K^\infty$ there is an isomorphic embedding $\sigma_v : \overline{\mathbb{Q}} \hookrightarrow \mathbb{C}$ such that $|x|_v = |\sigma_v(x)|^{[K_v:\mathbb{R}]/[K:\mathbb{Q}]}$ for $x \in \overline{\mathbb{Q}}$.

We represent polynomials as $f = \sum_{\mathbf{m}\in M_f} c_f(\mathbf{m})\mathbf{m}$, where the symbol $\mathbf{m}$ denotes a monomial, $M_f$ is a finite set of monomials, and $c_f(\mathbf{m})$ ($\mathbf{m} \in M_f$) are the coefficients. For any map $\sigma$ on the field of definition of $f$ we put $\sigma(f) := \sum_{\mathbf{m}\in M_f} \sigma(c_f(\mathbf{m}))\mathbf{m}$.

We define norms for polynomials $f_i = \sum_{\mathbf{m}\in M_{f_i}} c_{f_i}(\mathbf{m})\mathbf{m}$ ($i = 1,\dots,r$) with complex coefficients:

$$\|f_1,\dots,f_r\| := \max\left(|c_{f_i}(\mathbf{m})| : 1 \leqslant i \leqslant r,\ \mathbf{m}\in M_{f_i}\right)\,,$$
$$\|f_1,\dots,f_r\|_1 := \sum_{i=1}^{r}\sum_{\mathbf{m}\in M_{f_i}} |c_{f_i}(\mathbf{m})|$$

and for polynomials $f_1, \ldots, f_r$ with coefficients in $\overline{\mathbb{Q}}$:

$$\begin{aligned}
&\|f_1, \ldots, f_r\|_v := \max\left(|c_{f_i}(\mathbf{m})|_v \,:\, 1 \leqslant i \leqslant r,\ \mathbf{m} \in M_{f_i}\right) \quad (v \in M_K),\\
&\|f_1, \ldots, f_r\|_{v,1} := \|\sigma_v(f_1), \ldots, \sigma_v(f_r)\|_1^{[K_v:\mathbb{R}]/[K:\mathbb{Q}]} \quad (v \in M_K^\infty),\\
&\|f_1, \ldots, f_r\|_{v,1} := \|f_1, \ldots, f_r\|_v \quad (v \in M_K^0).
\end{aligned} \tag{4.1}$$

Lastly, for polynomials $f_1, \ldots, f_r$ with coefficients in $K$ we define heights

$$h(f_1, \ldots, f_r) := \log\left(\prod_{v \in M_K} \|f_1, \ldots, f_r\|_v\right),$$

$$h_1(f_1, \ldots, f_r) := \log\left(\prod_{v \in M_K} \|f_1, \ldots, f_r\|_{v,1}\right).$$

More generally, for polynomials $f_1, \ldots, f_r$ with coefficients in $\overline{\mathbb{Q}}$ we define $h(f_1, \ldots, f_r)$, $h_1(f_1, \ldots, f_r)$ by choosing a number field $K$ containing the coefficients of $f_1, \ldots, f_r$ and using the above definitions; this is independent of the choice of $K$.

We state without proof some easy inequalities. First, for $\mathbf{x} \in \overline{\mathbb{Q}}^{n+1}$ and $f \in \overline{\mathbb{Q}}[x_0, \ldots, x_n]$ homogeneous of degree $D$ we have

$$\|f(\mathbf{x})\|_v \leqslant \|f\|_{v,1}\|\mathbf{x}\|_v^D \quad \text{for } v \in M_K. \tag{4.2}$$

Secondly, for $\mathbf{x} \in \mathbb{P}^N(\overline{\mathbb{Q}})$ and $f_0, \ldots, f_r \in \overline{\mathbb{Q}}[x_0, \ldots, x_N]$ homogeneous of degree $D$ we have

$$h(\mathbf{y}) \leqslant Dh(\mathbf{x}) + h_1(f_0, \ldots, f_r), \tag{4.3}$$

where $\mathbf{y} = (f_0(\mathbf{x}), \ldots, f_r(\mathbf{x}))$.

Thirdly, if $f \in \overline{\mathbb{Q}}[x_0, \ldots, x_n]$ is homogeneous of degree $D$, and if $g_0, \ldots, g_n \in \overline{\mathbb{Q}}[x_0, \ldots, x_m]$ are homogeneous of equal degree, then for the polynomial $f(g_0, \ldots, g_n)$, obtained by substituting the polynomial $g_i(x_0, \ldots, x_m)$ for $x_i$ in $f$ for $i = 0, \ldots, n$, we have

$$h_1(f(g_0, \ldots, g_n)) \leqslant h_1(f) + Dh_1(g_0, \ldots, g_n). \tag{4.4}$$

Finally, for $f_1, \ldots, f_r \in \overline{\mathbb{Q}}[x_1, \ldots, x_n]$ we have

$$h(f_1, \ldots, f_r) \leqslant h_1(f_1, \ldots, f_r) \leqslant h(f_1, \ldots, f_r) + \log M, \tag{4.5}$$

where $M$ is the number of nonzero coefficients in $f_1, \ldots, f_r$.

4.2 We define another height for multihomogeneous polynomials. Given a field $\Omega$ and tuples of nonnegative integers $\mathbf{l} = (l_0, \ldots, l_m)$, we write $\Omega[\mathbf{l}]$ for the set of polynomials with coefficients in $\Omega$ in blocks of variables $\mathbf{z}^{(0)} = (z_0^{(0)}, \ldots, z_{l_0}^{(0)}), \ldots, \mathbf{z}^{(m)} = (z_0^{(m)}, \ldots, z_{l_m}^{(m)})$ which are homogeneous in block $\mathbf{z}^{(h)}$ for $h = 0, \ldots, m$. For $f \in \Omega[\mathbf{l}]$ we denote by $\deg_h f$ the degree of $f$ in block $\mathbf{z}^{(h)}$.

Let

$$\begin{aligned}
&S(l+1) := \{(z_0, \ldots, z_l) \in \mathbb{C}^{l+1} : |z_0|^2 + \cdots + |z_l|^2 = 1\},\\
&S(\mathbf{l}) := S(l_0+1) \times \cdots \times S(l_m+1).
\end{aligned}$$

Denote by $\mu_{l+1}$ the unique $U(l+1,\mathbb{C})$-invariant measure on $S(l+1)$ normalized such that $\mu_{l+1}(S(l+1))=1$, and let $\mu_{\mathbf{l}}=\mu_{l_0+1}\times\cdots\times\mu_{l_m+1}$ be the product measure on $S(\mathbf{l})$. Then for $f\in\mathbb{C}[\mathbf{l}]$ we set

$$m(f):=\int_{S(\mathbf{l})}\log|f(\mathbf{z}^{(0)},\dots,\mathbf{z}^{(m)})|\cdot\mu_{\mathbf{l}}+\frac{1}{2}\sum_{h=0}^{m}\deg_h f\left(\sum_{j=1}^{l_h}\frac{1}{2j}\right). \tag{4.6}$$

Given a number field $K$, we define for $f\in K[\mathbf{l}]$,

$$h^*(f):=\sum_{v\in M_K^\infty}\frac{[K_v:\mathbb{R}]}{[K:\mathbb{Q}]}m(\sigma_v(f))+\sum_{v\in M_K0}\log\|f\|_v\,. \tag{4.7}$$

Again, this does not depend on the choice of the number field $K$ containing the coefficients of $f$, so it defines a height on $\overline{\mathbb{Q}}[\mathbf{l}]$. It is not difficult to verify that

$$h^*(f_1\cdots f_r)=\sum_{i=1}^{r}h^*(f_i)\quad\text{for } f_1,\dots,f_r\in\overline{\mathbb{Q}}[\mathbf{l}]. \tag{4.8}$$

**Lemma 4.1.** *Let $\mathbf{l}=(l_0,\dots,l_m)$ be a tuple of nonnegative integers, and $f\in\overline{\mathbb{Q}}[\mathbf{l}]$, $f\neq 0$. Then*

$$|h^*(f)-h_1(f)|\leqslant\sum_{h=0}^{m}(\deg_h f)\log(l_h+1)\,.$$

*Proof.* Put $A:=\prod_{h=0}^{m}(l_h+1)^{\deg_h f}$. According to the definitions of $h^*$ and $h_1$, it suffices to prove that for $f\in\mathbb{C}[\mathbf{l}]$,

$$|m(f)-\log\|f\|_1|\leqslant\log A. \tag{4.9}$$

Using $|f(\mathbf{z}^{(0)},\dots,\mathbf{z}^{(m)})|\leqslant\|f\|_1$ for $(\mathbf{z}^{(0)},\dots,\mathbf{z}^{(m)})\in S(\mathbf{l})$ we obtain at once

$$m(f)\leqslant\log\|f\|_1+\frac{1}{2}\sum_{h=0}^{m}\deg_h f\left(\sum_{j=1}^{l_h}\frac{1}{2j}\right)\leqslant\log\|f\|_1+\log A\,.$$

To prove the inequality in the other direction, write $f=\sum_{\mathbf{m}\in M_f}c(\mathbf{m})\mathbf{m}$, where the sum is over a finite number of monomials $\mathbf{m}=\prod_{h=0}^{m}\prod_{j=0}^{l_h}(z_j^{(h)})^{a_{hj}}$ with $\sum_{j=0}^{l_h}a_{hj}=\deg_h f$ for $h=0,\dots,m$. For each such monomial we put

$$\alpha(\mathbf{m}):=\prod_{h=0}^{m}\frac{(\deg_h f)!}{a_{h0}!\cdots a_{h,l_h}!}\,.$$

Then by an argument on [12, pp. 111, 112],

$$\left(\sum_{\mathbf{m}\in M_f}\alpha(\mathbf{m})^{-1}|c(\mathbf{m})|^2\right)^{1/2}\leqslant A^{1/2}\exp(m(f))\,.$$

On combining this with the Cauchy–Schwarz inequality and $\sum_{\mathbf{m}} \alpha(\mathbf{m}) \leqslant A$, we obtain

$$\begin{aligned}\|f\|_1 = \sum_{\mathbf{m}\in M_f} |c(\mathbf{m})| &\leqslant \Big(\sum_{\mathbf{m}\in M_f} \alpha(\mathbf{m})\Big)^{1/2} \cdot \Big(\sum_{\mathbf{m}\in M_f} \alpha(\mathbf{m})^{-1}|c(\mathbf{m})|^2\Big)^{1/2} \\ &\leqslant A\exp(m(f)) .\end{aligned}$$

This proves $\log\|f\|_1 \leqslant m(f) + \log A$, hence (4.9). □

**Lemma 4.2.** *Let $f_1, \dots, f_r \in \overline{\mathbb{Q}}[\mathbf{l}]$ and $f = \prod_{i=1}^r f_i$. Then*

$$h_1(f) \leqslant \sum_{i=1}^r h_1(f_i) \leqslant h_1(f) + 2\sum_{h=0}^m (\deg_h f)\log(l_h + 1) .$$

*Proof.* The first inequality is straightforward, while the second follows from Lemma 4.1 and (4.8). □

4.3 In this subsection, $X$ is a projective subvariety of $\mathbb{P}^N$ of dimension $n \geqslant 1$ and degree $d$ defined over $\overline{\mathbb{Q}}$.

Let $\Delta$ be a positive integer. Denote by $M_\Delta$ the collection of all monomials of degree $\Delta$ in the variables $x_0, \dots, x_N$. Let $\mathbf{u}^{(h)} = (u_{\mathbf{m}}^{(h)} : \mathbf{m} \in M_\Delta)$ $(h = 0, \dots, n)$ be blocks of variables. Up to a constant factor there is a unique, irreducible polynomial $F_{X,\Delta} \in \overline{\mathbb{Q}}[\mathbf{u}^{(0)}, \dots, \mathbf{u}^{(n)}]$, called the $\Delta$-Chow form of $X$, having the following property (see [11]):

$F_{X,\Delta}(\mathbf{u}^{(0)}, \dots, \mathbf{u}^{(n)}) = 0$ if and only if there is a $\overline{\mathbb{Q}}$-rational point in the intersection of $X$ and the hypersurfaces $\sum_{\mathbf{m}\in M_\Delta} u_{\mathbf{m}}^{(h)}\mathbf{m} = 0$ $(h = 0, \dots, n)$.

Notice that $F_{X,1}$ is none other than the Chow form $F_X$ of $X$. The form $F_{X,\Delta}$ corresponds to the Chow form $F_{\varphi_\Delta(X)}$ of the image of $X$ under the Veronese embedding $\varphi_\Delta$ of degree $\Delta$. It is known that $F_{X,\Delta}$ is homogeneous of degree $\Delta^n d$ in $\mathbf{u}^{(h)}$ for $h = 0, \dots, n$.

For a monomial $\mathbf{m} = x_0^{a_0} \cdots x_N^{a_N}$ of degree $\Delta$, put $\beta(\mathbf{m}) = \Delta!/a_0! \cdots a_N!$. Then the modified Chow form $G_{X,\Delta}(\mathbf{u}^{(0)}, \dots, \mathbf{u}^{(n)})$ is obtained by substituting $\beta(\mathbf{m})^{1/2}u_{\mathbf{m}}^{(h)}$ for the variable $u_{\mathbf{m}}^{(h)}$ in the polynomial $F_{X,\Delta}(\mathbf{u}^{(0)}, \dots, \mathbf{u}^{(n)})$. Notice that $G_{X,1} = F_{X,1} = F_X$. Further, using the estimates $|\beta(\mathbf{m})| \leqslant \Delta!$, $|\beta(\mathbf{m})|_p \geqslant |\Delta!|_p$ for each prime number $p$, one easily obtains

$$\begin{aligned}|h_1(F_{X,\Delta}) - h_1(G_{X,\Delta})| &\leqslant \tfrac{1}{2}(n+1)d\Delta^n \log(\Delta!) \\ &\leqslant \tfrac{1}{2}(n+1)d\Delta^{n+1}\log\Delta . \qquad (4.10)\end{aligned}$$

The following is a special case of a fundamental result of Rémond [12, Thm. 2, pp. 99, 100]:

**Lemma 4.3.** $h^*(G_{X,\Delta}) = \Delta^{n+1}h^*(G_{X,1}) = \Delta^{n+1}h^*(F_X)$.

From this we deduce:

**Lemma 4.4.** $h_1(F_{X,\Delta}) \leqslant \Delta^{n+1}h(F_X) + 5(n+1)d\Delta^{n+1}\log(N + \Delta)$.

*Proof.* Recall that $F_{X,\Delta}$ and $G_{X,\Delta}$ are homogeneous of degree $\Delta^n d$ in each block of variables $\mathbf{u}^{(h)}(h = 0, \dots, n)$ and that each of these blocks has $\binom{N+\Delta}{\Delta} \leqslant (N+\Delta)^\Delta$ variables (that is, the number of coefficients of a homogeneous polynomial of degree $\Delta$ in $N+1$ variables). So by (4.10) and Lemma 4.1,

$$\begin{aligned} h_1(F_{X,\Delta}) &\leqslant h_1(G_{X,\Delta}) + \tfrac{1}{2}(n+1)d\Delta^{n+1}\log\Delta \\ &\leqslant h^*(G_{X,\Delta}) + \tfrac{1}{2}(n+1)d\Delta^{n+1}\log\Delta + (n+1)d\Delta^n \log\tbinom{N+\Delta}{\Delta} \\ &\leqslant h^*(G_{X,\Delta}) + \tfrac{3}{2}(n+1)d\Delta^{n+1}\log(N+\Delta)\,. \end{aligned}$$

Then using Lemma 4.3, again Lemma 4.1 and inequality (4.5) we obtain

$$\begin{aligned} h_1(F_{X,\Delta}) &\leqslant \Delta^{n+1}h^*(F_X) + \tfrac{3}{2}(n+1)d\Delta^{n+1}\log(N+\Delta) \\ &\leqslant \Delta^{n+1}h_1(F_X) + \tfrac{5}{2}(n+1)d\Delta^{n+1}\log(N+\Delta) \\ &\leqslant \Delta^{n+1}h(F_X) + \tfrac{5}{2}(n+1)d\Delta^{n+1}\log(N+\Delta) + \Delta^{n+1}\log M\,, \end{aligned}$$

where $M$ is the number of nonzero coefficients of $F_X$. Since $F_X$ is a polynomial in $n+1$ blocks of $N+1$ variables, and homogeneous of degree $d$ in each block, we have, using (3.25)

$$\begin{aligned} M &\leqslant \binom{N+d}{d}^{n+1} \leqslant (e(N+1))^{(n+1)d} \\ &\leqslant \exp\left(\tfrac{5}{2}(n+1)d\log(N+\Delta)\right)\,. \end{aligned}$$

By inserting this into the last inequality, our lemma follows. □

We arrive at the following:

**Proposition 4.5.** *Let $g_0, \dots, g_R$ be homogeneous polynomials of degree $\Delta$ in $\overline{\mathbb{Q}}[x_0, \dots, x_N]$ such that*

$$X(\overline{\mathbb{Q}}) \cap \{g_0 = 0, \dots, g_R = 0\} = \emptyset\,.$$

*Let $Y = \varphi(X)$, where $\varphi$ is the morphism on $X$ given by $\mathbf{x} \mapsto (g_0(\mathbf{x}), \dots, g_R(\mathbf{x}))$. Then*

$$\begin{aligned} h(Y) \leqslant{}& \Delta^{n+1}h(X) + (n+1)d\Delta^n h_1(g_0, \dots, g_R) \\ &+ 5(n+1)d\Delta^{n+1}\log(N+\Delta) + 3(n+1)d\Delta^n\log(R+1)\,. \end{aligned}$$

*Proof.* For $j = 0, \dots, R$ write $y_j$ for $g_j(\mathbf{x})$ and denote by $\mathbf{g}_j$ the vector of coefficients of $g_j$, i.e., $g_j = \sum_{\mathbf{m}\in M_\Delta} c_{g_j}(\mathbf{m})\mathbf{m}$ and $\mathbf{g}_j = (c_{g_j}(\mathbf{m}) : \mathbf{m} \in M_\Delta)$. Introduce blocks of variables $\mathbf{v}^{(h)} = (v_0^{(h)}, \dots, v_R^{(h)})$ $(h = 0, \dots, n)$ and define the polynomial

$$G(\mathbf{v}^{(0)}, \dots, \mathbf{v}^{(n)}) := F_{X,\Delta}\left(\sum_{j=0}^{R} v_j^{(0)}\mathbf{g}_j, \dots, \sum_{j=0}^{R} v_j^{(n)}\mathbf{g}_j\right)\,.$$

Then $G(\mathbf{v}^{(0)}, \dots, \mathbf{v}^{(n)}) = 0$ if and only if $X$ and the hypersurfaces $\sum_{j=0}^R v_j^{(h)} g_j = 0$ $(h = 0, \dots, n)$ have a $\overline{\mathbb{Q}}$-rational point in common, if and only if $Y$ and the hyperplanes $\sum_{j=0}^R v_j^{(h)} y_j = 0$ $(h = 0, \dots, n)$ have a $\overline{\mathbb{Q}}$-rational point in common, if and only if $F_Y(\mathbf{v}^{(0)},\dots,\mathbf{v}^{(n)})=0$, where $F_Y$ is the Chow form of $Y$. Therefore, $G$ is up to a constant factor equal to a power of $F_Y$.

Put $A:=(n+1)d\Delta^{n+1}\log(N+\Delta)$, $B:=(n+1)d\Delta^n\log(R+1)$. Notice that $G$ has degree $d\Delta^n$ in each block $\mathbf{v}^{(h)}$. Further, by (4.4) we have $h_1(G)\leqslant h_1(F_{X,\Delta})+(n+1)d\Delta^n h_1(g_0,\dots,g_R)+B$. Together with Lemma 4.2, Lemma 4.1, this implies

$$\begin{aligned}h(Y)=h(F_Y)&\leqslant h_1(F_Y)\leqslant h_1(G)+2B\\&\leqslant h_1(F_{X,\Delta})+(n+1)d\Delta^n h_1(g_0,\dots,g_R)+3B\\&\leqslant \Delta^{n+1}h(X)+(n+1)d\Delta^n h_1(g_0,\dots,g_R)+5A+3B\,,\end{aligned}$$

proving our proposition. □

## 5 Proof of Theorem 1.3

5.1 We start with some auxiliary results. We denote the coordinates of $\mathbb{P}^R$ by $\mathbf{y}=(y_0,\dots,y_R)$.

**Lemma 5.1.** *Let $Y$ be a projective subvariety of $\mathbb{P}^R$ of dimension $n\geqslant 1$ and degree $D$, defined over $\overline{\mathbb{Q}}$. Let $\mathbf{c}=(c_0,\dots,c_R)$ be a tuple of reals. Let $\{i_0,\dots,i_n\}$ be a subset of $\{0,\dots,R\}$ such that*

$$Y(\overline{\mathbb{Q}})\cap\{y_{i_0}=0,\dots,y_{i_n}=0\}=\emptyset\,. \tag{5.1}$$

*Then*

$$e_Y(\mathbf{c})\geqslant D(c_{i_0}+\cdots+c_{i_n}). \tag{5.2}$$

*Proof.* For a subset $I=\{k_0,\dots,k_n\}$ of $\{0,\dots,R\}$ with $k_0<k_1<\cdots<k_n$, define the *bracket*

$$[I]=[I](\mathbf{u}^{(0)},\dots,\mathbf{u}^{(n)}):=\det\left(u^{(i)}_{k_j}\right)_{i,j=0,\dots,n},$$

where again $\mathbf{u}^{(h)}$ denotes the block of variables $(u^{(h)}_0,\dots,u^{(h)}_R)$. Let $I_1,\dots,I_S$ with $S=\binom{R+1}{n+1}$ be all subsets of $\{0,\dots,R\}$ of cardinality $n+1$. Then the Chow form $F_Y$ of $Y$ can be written as a homogeneous polynomial of degree $D$ in $[I_1],\dots,[I_S]$ :

$$F_Y=\sum_{\mathbf{a}\in A}C(\mathbf{a})[I_1]^{a_1}\cdots[I_S]^{a_S}, \tag{5.3}$$

where $A$ is the set of tuples of nonnegative integers $\mathbf{a}=(a_1,\dots,a_S)$ with $a_1+\cdots+a_S=D$ and where $C(\mathbf{a})\in\overline{\mathbb{Q}}$ for $\mathbf{a}\in A$ [9, p. 41, Theorem IV]. For each bracket $[I]$ we have

$$[I](t^{c_0}u^{(0)}_0,\dots,t^{c_R}u^{(0)}_R;\dots;t^{c_0}u^{(n)}_0,\dots,t^{c_R}u^{(n)}_R)=t^{\sum_{i\in I}c_i}[I],$$

therefore,

$$\begin{aligned}&F_Y(t^{c_0}u^{(0)}_0,\dots,t^{c_R}u^{(0)}_R;\dots;t^{c_0}u^{(n)}_0,\dots,t^{c_R}u^{(n)}_R)\\&\quad=\sum_{\mathbf{a}\in A}C(\mathbf{a})t^{\sum_{j=1}^S a_j(\sum_{i\in I_j}c_i)}[I_1]^{a_1}\cdots[I_S]^{a_S}.\end{aligned} \tag{5.4}$$

Put $\mathbf{e}_0 := (1, 0, \dots, 0)$, $\mathbf{e}_1 := (0, 1, \dots, 0)$, …, $\mathbf{e}_R := (0, 0, \dots, 1)$. Write $\{i_0, \dots, i_n\} =: I_1$. By (5.1) we have $F_Y(\mathbf{e}_{i_0}, \dots, \mathbf{e}_{i_n}) \neq 0$. Further,

$$[I_1](\mathbf{e}_{i_0}, \dots, \mathbf{e}_{i_n}) = 1, \ \ [I](\mathbf{e}_{i_0}, \dots, \mathbf{e}_{i_n}) = 0 \ \text{ for } I \neq I_1.$$

Hence in expression (5.3) there is a term $C \cdot [I_1]^D$ with $C \in \overline{\mathbb{Q}}^*$, and if we substitute $\mathbf{u}^{(j)} = \mathbf{e}_{i_j}$ $(j = 0, \dots, n)$ in (5.4) we obtain $C \cdot t^{D(c_{i_0} + \cdots + c_{i_n})}$. That is, one of the numbers $e_i$ in (2.4) is equal to $D\left(c_{i_0} + \cdots + c_{i_n}\right)$. This implies (5.2) at once. □

In addition, we need the following combinatorial lemma which is a consequence of [3, Lemma 4].

**Lemma 5.2.** *Let $\theta$ be a real with $0 < \theta \leqslant \frac{1}{2}$ and let $q$ be a positive integer. Then there exists a set $\mathcal{W}$ of cardinality at most $(e/\theta)^{q-1}$, consisting of tuples $(c_1, \dots, c_q)$ of nonnegative reals with $c_1 + \cdots + c_q = 1$, with the following property:*

*for every set of reals $A_1, \dots, A_q$ and $\Lambda$ with $A_j \leqslant 0$ for $j = 1, \dots, q$ and $\sum_{j=1}^{q} A_j \leqslant -\Lambda$, there exists a tuple $(c_1, \dots, c_q) \in \mathcal{W}$ such that*

$$A_j \leqslant -c_j(1-\theta)\Lambda \ \text{ for } j = 1, \dots, q.$$

5.2 In what follows, $K$ is a number field, $S$ a finite set of places of $K$, and $X$, $N$, $n$, $d$, $s$, $C$, $f_i^{(v)}$ $(v \in S, i = 0, \dots, n)$, $C$, $\Delta$, $A_1$, $A_2$, $A_3$, $H$ are as in Theorem 1.3. We denote the coordinates on $\mathbb{P}^N$ by $\mathbf{x} = (x_0, \dots, x_N)$.

Let $f_0, \dots, f_R$ be the distinct polynomials among $\sigma(f_j^{(v)})$ $(v \in S, j = 0, \dots, n, \sigma \in G_K)$. Then by (1.4),

$$R \leqslant C(n+1)s - 1\,. \tag{5.5}$$

Let $K'$ be the extension of $K$ generated by the coefficients of $f_0, \dots, f_R$. Put $g_i := f_i^{\Delta/\deg f_i}$ for $i = 0, \dots, R$. Thus, $g_0, \dots, g_R$ are homogenenous polynomials in $K'[x_0, \dots, x_N]$ of degree $\Delta$. Define

$$\varphi : \mathbf{x} \mapsto (g_0(\mathbf{x}), \dots, g_R(\mathbf{x})), \quad Y := \varphi(X)\,.$$

By assumption (1.2), $\varphi$ is a finite morphism on $X$, and $Y$ is a projective subvariety of $\mathbb{P}^R$ defined over $K'$. We have

$$\dim Y = n, \quad \deg Y =: D \leqslant d\Delta^n\,. \tag{5.6}$$

We denote places on $K'$ by $v'$ and define normalized absolute values $|\cdot|_{v'}$ on $K'$ similarly to §1.4. Further, for every $v' \in M_{K'}$ we choose an extension of $|\cdot|_{v'}$ to $\overline{\mathbb{Q}}$. Since $K'/K$ is a normal extension, for every $v' \in M_{K'}$ there is $\tau_{v'} \in G_K$ such that

$$|x|_{v'} = |\tau_{v'}(x)|_v^{1/g(v)} \quad \text{for } x \in \overline{\mathbb{Q}} \tag{5.7}$$

where $v \in M_K$ is the place below $v'$ and $g(v)$ is the number of places of $K'$ lying above $v$. For each $v' \in M_{K'}^{\infty}$ there is an isomorphic embedding $\sigma_{v'} : K' \hookrightarrow \mathbb{C}$ such that $|x|_{v'} = |\sigma_{v'}(x)|^{[K_{v'}:\mathbb{R}]/[K':\mathbb{Q}]}$ for $x \in \overline{\mathbb{Q}}$. We define norms $\|\cdot\|_{v'}$, $\|\cdot\|_{v',1}$ for polynomials similarly as in (4.1), with $K'$, $v'$, $\sigma_{v'}$ in place of $K$, $v$, $\sigma_v$.

5.3 For later purposes we estimate from above $h_1(1, g_0, \ldots, g_R)$ and $h(Y)$. By a straightforward computation we have for $v' \in M_{K'}^{\infty}$,

$$
\begin{aligned}
&\|1, \sigma_{v'}(g_0), \ldots, \sigma_{v'}(g_R)\|_1 \\
&= 1 + \sum_{i=0}^{R} \|\sigma_{v'}(g_i)\|_1 \leqslant 1 + \sum_{i=0}^{R} \|\sigma_{v'}(f_i)\|_1^{\Delta/\deg f_i} \\
&\leqslant 1 + \sum_{i=0}^{R} \left( \binom{\deg f_i + N}{\deg f_i} \|\sigma_{v'}(f_i)\| \right)^{\Delta/\deg f_i} \\
&\leqslant (R+2)(N+\Delta)^{\Delta} \|1, \sigma_{v'}(f_0), \ldots, \sigma_{v'}(f_R)\|^{\Delta} \\
&\leqslant (R+2)(N+\Delta)^{\Delta} \prod_{i=0}^{R} \|1, \sigma_{v'}(f_i)\|^{\Delta} .
\end{aligned}
$$

So for $v' \in M_{K'}^{\infty}$ we have

$$
\|1, g_0, \ldots, g_R\|_{v',1} \leqslant \left((R+2)(N+\Delta)^{\Delta}\right)^{[K'_{v'}:\mathbb{R}]/[K':\mathbb{Q}]} \cdot \prod_{i=0}^{R} \|1, f_i\|_{v'}^{\Delta}.
$$

In an easier manner one obtains for $v' \in M_{K'}^{0}$,

$$
\|1, g_0, \ldots, g_R\|_{v',1} \leqslant \prod_{i=0}^{R} \|1, f_i\|_{v'}^{\Delta}.
$$

So by taking the product over $v' \in M_{K'}$, substituting (5.5), and using that polynomials with conjugate sets of coefficients have the same height,

$$
\begin{aligned}
h_1(1, g_0, \ldots, g_R) &\leqslant \Delta \left( \sum_{i=0}^{R} h(1, f_i) \right) + \Delta \log\left((R+2)(N+\Delta)^{\Delta}\right) \\
&\leqslant \Delta C \left( \sum_{v \in S} \sum_{j=0}^{n} h(1, f_j^{(v)}) \right) + \Delta \log(N+\Delta) + \log(3Cns) ,
\end{aligned}
$$

and by inserting this estimate into Proposition 4.5 we infer

$$
\begin{aligned}
h(Y) \leqslant{}& \Delta^{n+1} h(X) + (n+1) d \Delta^{n+1} C \sum_{v \in S} \sum_{j=0}^{n} h(1, f_j^{(v)}) \\
&+ 6(n+1) d \Delta^{n+1} \log(N+\Delta) + 4(n+1) d \Delta^{n} \log(3Cns) .
\end{aligned}
$$

A straightforward computation gives the more tractable estimates

$$
h_1(g_0, \ldots, g_R) \leqslant 6\Delta^2 Cns \cdot H , \tag{5.8}
$$
$$
h(Y) \leqslant 25 n^2 d \Delta^{n+2} Cs \cdot H , \tag{5.9}
$$

where $H$ is defined by (1.5).

5.4 We reduce (1.6) to a finite number of systems of inequalities and then show that each such system leads to an inequality involving a twisted height.

Let $\mathbf{x} \in X(\overline{\mathbb{Q}})$ be a solution of (1.6). For $v \in S$, let $I_v$ be the subset of $\{0, \dots, R\}$ such that $\{f_j^{(v)} : j=0, \dots, n\} = \{f_i : i \in I_v\}$. Put $G_v := \|1, g_0, \dots, g_R\|_{v,1}$ for $v \in S$. Then

$$\sum_{v\in S}\sum_{i\in I_v} \log\left(\max_{\sigma\in G_K} \frac{|g_i(\mathbf{x})|_v}{G_v\|\sigma(\mathbf{x})\|_v^{\Delta}}\right) \leqslant -(n+1+\delta)\Delta h(\mathbf{x})\,.$$

By (4.2), the terms in the sum are $\leqslant 0$. We apply Lemma 5.2 with $q = (n+1)s$ and $\theta = \frac{\delta}{(2n+2+2\delta)} = 1 - \frac{n+1+\delta/2}{n+1+\delta}$. We infer that there is a set $\mathcal{W}$ with

$$\#\mathcal{W} \leqslant \left(\frac{e(2n+2+2\delta)}{\delta}\right)^{(n+1)s-1} \leqslant (17n\delta^{-1})^{(n+1)s-1} \tag{5.10}$$

consisting of tuples of nonnegative reals $(c_{iv} : v \in S,\ i \in I_v)$ with

$$\sum_{v\in S}\sum_{i\in I_v} c_{iv} = 1\,, \tag{5.11}$$

such that for every solution $\mathbf{x} \in X(\overline{\mathbb{Q}})$ of (1.6) there is a tuple $(c_{iv} : v \in S,\ i \in I_v) \in \mathcal{W}$ with

$$\log\left(\max_{\sigma\in G_K} \frac{|g_i(\sigma(\mathbf{x}))|_v}{G_v\cdot\|\sigma(\mathbf{x})\|_v^{\Delta}}\right) \leqslant -c_{iv}\left(n+1+\frac{\delta}{2}\right)\Delta h(\mathbf{x}) \quad (v \in S, i \in I_v). \tag{5.12}$$

Denote by $S'$ the set of places of $K'$ lying above the places in $S$. Notice that each element of $G_K$ acts as a permutation on $g_0, \dots, g_R$. Let $v' \in S'$. Write $v$ for the place of $K$ lying below $v'$ and let $\tau_{v'} \in G_K$ be given by (5.7). Then we define $I_{v'} \subset \{0, \dots, R\}$, $c_{i,v'}$ $(i \in I_{v'})$ by

$$\begin{aligned} &\{g_i : i \in I_{v'}\} = \{\tau_{v'}^{-1}(g_j) : j \in I_v\} \quad \text{for } v' \in S', \\ &c_{i,v'} := c_{jv}/g(v) \quad \text{for } v' \in S', i \in I_{v'}, \end{aligned}$$

where $j \in I_v$ is the index such that $g_i = \tau_{v'}^{-1}(g_j)$. Further, we put

$$G_{v'} := \|1, g_0, \dots, g_R\|_{v',1} \quad \text{for } v' \in M_{K'}.$$

Then in view of (5.7), we can rewrite system (5.12) as

$$\log\left(\max_{\sigma\in G_K} \frac{|g_i(\sigma(\mathbf{x}))|_{v'}}{G_{v'}\cdot\|\sigma(\mathbf{x})\|_{v'}^{\Delta}}\right) \leqslant -c_{i,v'}\left(n+1+\frac{\delta}{2}\right)\Delta h(\mathbf{x}) \quad (v' \in S', i \in I_{v'}). \tag{5.13}$$

Invoking (5.10), (5.11) we obtain the following:

**Lemma 5.3.** *There is a set $\mathcal{W}'$ of cardinality at most $(17n\delta^{-1})^{(n+1)s-1}$, consisting of tuples of nonnegative reals $(c_{i,v'} : v' \in S',\ i \in I_{v'})$ with*

$$\sum_{v\in S'}\sum_{i\in I_{v'}} c_{i,v'} = 1\,, \tag{5.14}$$

*with the property that for every* $\mathbf{x} \in X(\overline{\mathbb{Q}})$ *with* (1.6) *there is a tuple in* $\mathcal{W}'$ *such that* $\mathbf{x}$ *satisfies* (5.13).

We consider the solutions of a fixed system (5.13). Put

$$c_{i,v'} = 0 \quad \text{for } v' \in S', i \in \{0, \ldots, R\} \backslash I_{v'} \\ \text{and } v' \in M_{K'} \backslash S', i = 0, \ldots, R \tag{5.15}$$

and put $\mathbf{c}_{v'} := (c_{0,v'}, \ldots, c_{R,v'})$ for $v' \in M_{K'}$, $\mathbf{c} := (\mathbf{c}_{v'} : v' \in M_{K'})$. Denote by $\mathbf{y} = (y_0, \ldots, y_R)$ the coordinates of $\mathbb{P}^R$. We define $H_{Q,\mathbf{c}}(\mathbf{y})$, $E_Y(\mathbf{c})$ similarly as (2.3), (2.9), respectively, but with $K'$ in place of $K$.

**Lemma 5.4.** *Let* $\mathbf{x} \in X(\overline{\mathbb{Q}})$ *be a solution of* (5.13) *satisfying* (1.7) *and let* $\sigma \in G_K$. *Put*

$$\mathbf{y} := \varphi(\sigma(\mathbf{x})), \quad Q := \exp\big((n + 1 + \delta/2)\Delta h(\mathbf{x})\big).$$

*Then*

$$H_{Q,\mathbf{c}}(\mathbf{y}) \leqslant Q^{E_Y(\mathbf{c}) - \delta/2(n+2)^2}. \tag{5.16}$$

*Proof.* We first estimate from below $E_Y(\mathbf{c})$. Let $v' \in S'$ and write $I_{v'} = \{i_0, \ldots, i_n\}$. From assumption (1.2), and from the fact that $X$ is defined over $K$ and that $g_{i_0}, \ldots, g_{i_n}$ are conjugate over $K$ to powers of $f_0^{(v)}, \ldots, f_n^{(v)}$, where $v \in S$ is the place below $v'$, it follows that $X(\overline{\mathbb{Q}}) \cap \{g_{i_0} = 0, \ldots, g_{i_n} = 0\} = \emptyset$. Since $Y = \varphi(X)$, for $\mathbf{y} \in Y(\overline{\mathbb{Q}})$ there is $\mathbf{x} \in X(\overline{\mathbb{Q}})$ with $y_i = g_i(\mathbf{x})$ for $i = 0, \ldots, R$. Hence

$$Y(\overline{\mathbb{Q}}) \cap \{y_{i_0} = 0, \ldots, y_{i_n} = 0\} = \emptyset.$$

Now Lemma 5.1 implies

$$\frac{1}{(n+1)D} \cdot e_Y(\mathbf{c}_{v'}) \geqslant \frac{1}{n+1}(c_{i_0,v'} + \cdots + c_{i_n,v'}) = \frac{1}{n+1} \cdot \sum_{i \in I_{v'}} c_{i,v'}.$$

This holds for $v' \in S'$. For $v' \notin S'$ we have $e_Y(\mathbf{c}_{v'}) = 0$ by (5.15). By summing over $v' \in S'$ and using (5.14), we arrive at

$$E_Y(\mathbf{c}) \geqslant \frac{1}{n+1}. \tag{5.17}$$

Now let $\mathbf{x} \in X(\overline{\mathbb{Q}})$ be a solution of (5.13) with (1.7) and let $\sigma \in G_K$. Then $\sigma(\mathbf{x})$ is also a solution of (5.13). In fact, by (5.15), $\sigma(\mathbf{x})$ satisfies (5.13) for $v \in M_K$, $i = 0, \ldots, R$. Write $\mathbf{y} = \varphi(\sigma(\mathbf{x}))$ so that $y_i = g_i(\sigma(\mathbf{x}))$ for $i = 0, \ldots, R$. Let $L$ be a finite normal extension of $K'$ such that $\sigma(\mathbf{x}) \in X(L)$. Pick $w \in M_L$ and let $v'$ be the place of $K'$ below $w$. Then there is $\tau_w \in \mathrm{Gal}(\overline{\mathbb{Q}}/K')$ such that $|x|_w = |\tau_w(x)|_{v'}^{d(w|v')}$ for $x \in L$, where $d(w|v') = [L_w : K'_{v'}]/[L : K']$. Hence for $i = 0, \ldots, R$, with the usual notation $c_{iw} = d(w|v')c_{i,v'}$,

$$\begin{aligned} |y_i|_w Q^{c_{iw}} &= |g_i(\sigma(\mathbf{x}))|_w Q^{c_{iw}} = \big(|g_i(\tau_w\sigma(\mathbf{x}))|_{v'} Q^{c_{i,v'}}\big)^{d(w|v')} \\ &\leqslant \big(G_{v'} \|\tau_w\sigma(\mathbf{x})\|_{v'}^{\Delta}\big)^{d(w|v')} = G_{v'}^{d(w|v)} \|\sigma(\mathbf{x})\|_w^{\Delta}. \end{aligned}$$

By taking the product over $w \in M_L$ and using $h(\sigma(\mathbf{x})) = h(\mathbf{x})$ we obtain

$$H_{Q,\mathbf{c}}(\mathbf{y}) \leqslant \exp\big(h_1(1, g_0, \ldots, g_R)\big) \cdot Q^{1/(n+1+\delta/2)}.$$

Now (5.16) follows by observing that by (5.17), assumption (1.7), and (5.8),

$$\begin{aligned}
&\left(E_Y(\mathbf{c}) - \frac{\delta}{2(n+2)^2} - \frac{1}{n+1+\delta/2}\right)\log Q \\
&\geqslant \left(\frac{1}{n+1} - \frac{\delta}{2(n+2)^2} - \frac{1}{n+1+\delta/2}\right)\log Q \\
&= \frac{\delta(4n+6-\delta(n+1))}{4(n+1)(n+2)^2}\cdot \Delta h(\mathbf{x}) \geqslant \frac{\delta\Delta}{2(n+2)^2}A_3 H \\
&\geqslant 6\Delta^2 CnsH \geqslant h_1(1, g_0, \dots, g_R)\,.
\end{aligned}$$

□

5.5 We finish the proof of Theorem 1.3. We apply Theorem 2.1 with $K'$, $\delta/(2(n+2))^2$ in place of $K$, $\delta$ and, in view of (5.5) and (5.6), with $D \leqslant d\Delta^n$ and $R = C(n+1)s - 1$. Notice that by (5.14),(5.15), the conditions (2.6), (2.7), (2.8) (with $K'$ in place of $K$) are satisfied. Denote by $B_1'$, $B_2'$, $B_3'$ the quantities obtained by substituting $\delta/(2(n+2))^2$ for $\delta$, $C(n+1)s - 1$ for $R$, and $d\Delta^n$ for $D$ in the quantities $B_1$, $B_2$, $B_3$, respectively, defined by (2.10). Recall that if $\mathbf{x}$ satisfies (1.7), then Lemma 5.4 is applicable. Moreover,

$$\begin{aligned}
\log Q &= \left(n+1+\frac{\delta}{2}\right)\Delta h(\mathbf{x}) \geqslant A_3 H \\
&= \exp\left(2^{6n+20}n^{2n+3}\delta^{-n-1}d^{n+2}\Delta^{n(n+2)}\log(2Cs)\right)\cdot H \\
&\geqslant \exp\left(2^{5n+4}(2(n+2)^2\delta^{-1})^{n+1}(d\Delta^n)^{n+2}\log(4C(n+1)s)\right) \\
&\quad\cdot\left(26n^2 d\Delta^{n+2}Cs\right)\cdot H \\
&= B_3'\cdot\left(26n^2 d\Delta^{n+2}Cs\right)\cdot H \geqslant B_3'(h(Y)+1)\,,
\end{aligned}$$

where the last inequality follows from (5.9). Hence Theorem 2.1 is applicable.

Now Theorem 2.1 and Lemma 5.4 imply that there are homogeneous polynomials $F_1, \dots, F_t \in K'[y_0, \dots, y_R]$ not vanishing identically on $Y$, with $t \leqslant B_1'$ and $\deg F_i \leqslant B_2'$ for $i = 1, \dots, t$, with the property that for every solution $\mathbf{x} \in X(\overline{\mathbb{Q}})$ of (5.13) with (1.7), there is $F_i \in \{F_1, \dots, F_t\}$ such that $F_i(\varphi(\sigma(\mathbf{x}))) = 0$ for every $\sigma \in G_K$. (In fact, taking $Q = \exp((n+1+\delta/2)\Delta h(\mathbf{x}))$ it follows from Theorem 2.1 that there is $F_i$ with $F_i(\mathbf{y}) = 0$ for every $\mathbf{y} \in Y(\overline{\mathbb{Q}})$ with $H_{Q,\mathbf{c}}(\mathbf{y}) \leqslant Q^{E_Y(\mathbf{c})-\delta/2(n+2)^2}$, and then by Lemma 5.4 this holds in particular for all points $\mathbf{y} = \varphi(\sigma(\mathbf{x}))$, $\sigma \in G_K$.)

This means that $\tilde{F}_i(\sigma(\mathbf{x})) = 0$ for $\sigma \in G_K$, where $\tilde{F}_i$ is the polynomial obtained by substituting $g_j$ for $y_j$ in $F_i$ for $j = 0, \dots, R$. Notice that $\tilde{F}_i \in K'[x_0, \dots, x_N]$, $\deg \tilde{F}_i \leqslant B_2'\Delta$, and that $\tilde{F}_i$ does not vanish identically on $X$. Write $\tilde{F}_i = \sum_{k=1}^{M}\omega_k\tilde{F}_{ik}$, where $\omega_1, \dots, \omega_M$ is a $K$-basis of $K'$, and the $\tilde{F}_{ik}$ are polynomials with coefficients in $K$. We can choose $G_i \in \{\tilde{F}_{ik} : k = 1, \dots, M\}$ which does not vanish identically on $X$. Now $\sigma(\tilde{F}_i)(\mathbf{x}) = 0$ for $\sigma \in G_K$. Since the polynomials $\tilde{F}_{ik}$ are linear combinations of the polynomials $\sigma(\tilde{F}_i)$ ($\sigma \in G_K$), it follows that $\tilde{F}_{ik}(\mathbf{x}) = 0$ for $k = 1, \dots, M$, so in particular $G_i(\mathbf{x}) = 0$.

It follows that there are homogeneous polynomials $G_1, \dots, G_t \in K[x_0, \dots, x_N]$ with $t \leqslant B_1'$ and $\deg G_i \leqslant B_2'\Delta$ for $i = 1, \dots, t$, not vanishing identically on $X$, such that the set of $\mathbf{x} \in X(\overline{\mathbb{Q}})$ with (5.13) and with (1.7) is contained in $\bigcup_{i=1}^{t} (X \cap \{G_i = 0\})$.

According to Lemma 5.3, there are at most $T := \left(17n\delta^{-1}\right)^{(n+1)s-1}$ different systems (5.13), such that every solution $\mathbf{x} \in X(\overline{\mathbb{Q}})$ of (1.6) satisfies one of these systems. Consequently, there are homogeneous polynomials $G_1, \dots, G_u \in K[x_0, \dots, x_N]$ not vanishing identically on $X$, with $u \leqslant B_1'T$ and with $\deg G_i \leqslant B_2'\Delta$ for $i = 1, \dots, u$, such that the set of $\mathbf{x} \in X(\overline{\mathbb{Q}})$ with (1.6), (1.7) is contained in $\bigcup_{i=1}^{u} (X \cap \{G_i = 0\})$.

Now the proof of Theorem 1.3 is completed by observing that in view of (2.10),

$$B_2'\Delta = (4n+3)(d\Delta^n)(2(n+2)^2\delta^{-1})\Delta = (8n+6)(n+2)^2 d\Delta^{n+1}\delta^{-1} = A_2$$

and

$$\begin{aligned}
B_1'T &\leqslant \exp\left(2^{10n+4}(2(n+2)^2)^{2n}\delta^{-2n}(d\Delta^n)^{2n+2}\right) \\
&\quad \cdot \log(4(n+1)Cs)\log\log(4(n+1)Cs) \cdot \left(17n\delta^{-1}\right)^{(n+1)s-1} \\
&\leqslant \exp\left(2^{12n+16}n^{4n}\delta^{-2n}d^{2n+2}\Delta^{n(2n+2)}\right) \\
&\quad \cdot (20n\delta^{-1})^{(n+1)s} \cdot \log(4C)\log\log(4C) \\
&= A_1 .
\end{aligned}$$

□

**Note added in proof.** After the submission of our manuscript and independently of our work, P.-C. Hu and C.-C. Yang (Isr. J. Math. 157: 47–61, 2007, Theorem 1.1) proved a result similar to our Theorem 1.1 and similar to the result of Corvaja and Zannier [2]. But Hu and Yang restrict themselves to the case $X = \mathbb{P}^N$ and they do not give an upper bound for the degrees of the polynomials $G_i$.

## References


1. Chardin, M.: Une majoration de la fonction de Hilbert et ses conséquences pour l'interpolation algébrique. *Bull. Soc. Math. F.* **117**, 305–318 (1989)
2. Corvaja, P., Zannier, U.: On a general Thue's equation. *Am. J. Math.* **126**, 1033–1055 (2004)
3. Evertse, J.-H.: On equations in $S$-units and the Thue-Mahler equation. *Invent. Math.* **75**, 561–584 (1984)
4. Evertse, J.-H., Ferretti, R.G.: Diophantine inequalities on projective varieties. *Int. Math. Res. Not.* **2002**, 1295–1330 (2002)
5. Evertse, J.-H., Schlickewei, H.P.: A quantitative version of the absolute subspace theorem. *J. Reine Angew. Math.* **548**, 21–127 (2002)
6. Faltings, G., Wüstholz, G.: Diophantine approximations on projective spaces. *Invent. Math.* **116**, 109–138 (1994)
7. Ferretti, R.G.: Mumford's degree of contact and Diophantine approximations. *Compos. Math.* **121**, 247–262 (2000)
8. Ferretti, R.G.: Diophantine approximations and toric deformations. *Duke Math. J.* **118**, 493–522 (2003)
9. Hodge, W.V.D., Pedoe, D.: *Methods of Algebraic Geometry,* vol. II, Cambridge University Press, Cambridge (1952)
10. Mumford, D.: Stability of projective varieties. *Enseign. Math. II. Sér.* **23**, 39–110 (1977)

11. Rémond, G.: Élimination multihomogène. In: Nesterenko, Yu.V., Philippon, P. (eds.) *Introduction to Algebraic Independence Theory*. Lect. Notes Math., vol. 1752, pp. 53–81. Springer, Heidelberg (2001)
12. Rémond, G.: Géometrie diophantienne multiprojective. In: Nesterenko, Yu.V., Philippon, P. (eds.) *Introduction to Algebraic Independence Theory*. Lect. Notes Math., vol. 1752, pp. 95–131. Springer, Heidelberg (2001)
13. Schlickewei, H.P.: The $\wp$-adic Thue-Siegel-Roth-Schmidt theorem. *Arch. Math.* **29**, 267–270 (1977)
14. Schmidt, W.M.: Norm form equations. *Ann. Math.* **96**, 526–551 (1972)
15. Schmidt, W.M.: Simultaneous approximation to algebraic numbers by elements of a number field. *Monatsh. Math.* **79**, 55–66 (1975)
16. Schmidt, W.M.: The subspace theorem in diophantine approximation. *Compos. Math.* **96**, 121–173 (1989)

*Diophantine Approximation*
H. P. Schlickewei et al., Editors


# ON THE DIOPHANTINE EQUATION $G_n(x) = G_m(y)$ WITH $Q(x, y) = 0$

Clemens Fuchs[1], Attila Pethő[2,3], and Robert F. Tichy[4]
[1] *Departement Mathematik, ETH Zürich, Rämistrasse 101, 8092 Zürich, Switzerland*
clemens.fuchs@math.ethz.ch
[2] *Faculty of Informatics, University of Debrecen, PO Box 12, 4010 Debrecen, Hungary*
pethoe@inf.unideb.hu
[3] *Number Theory Research Group, Hungarian Academy of Sciences, PO Box 12, 4010 Debrecen, Hungary*
[4] *Institut für Analysis und Computational Number Theory, TU Graz, Steyrergasse 30/II, 8010 Graz, Austria*
tichy@tugraz.at

Dedicated to Wolfgang Schmidt on his 70th birthday

## 1 Introduction

Let $\mathbf{K}$ denote an algebraically closed field of characteristic 0, and let $A_0, \ldots, A_{d-1}, G_0, \ldots, G_{d-1} \in \mathbf{K}[X]$ and $(G_n(X))_{n=0}^{\infty}$ be a sequence of polynomials defined by the $d$-th order linear recurring relation

$$G_{n+d}(X) = A_{d-1}(X)G_{n+d-1}(X) + \cdots + A_0(X)G_n(X), \quad \text{for } n \geq 0. \tag{1}$$

Furthermore, let $P(X) \in \mathbf{K}[X]$, $\deg P \geq 1$. Recently, we investigated the question, what can be said about the number of solutions of the Diophantine equation

$$G_n(X) = G_m(P(X)). \tag{2}$$

The problem was motivated by properties of families of orthogonal polynomials. For example, the Chebyshev polynomials of the first kind, which are defined by

$$T_n(X) = \cos(n \arccos X),$$

have the well known property that $T_{2n}(X) = T_n(2X^2 - 1)$ for all integers $n$. Let us mention that all orthogonal polynomials satisfy a second order linear recurring sequence, e.g., for the Chebyshev polynomials we have $T_0(X) = 1$, $T_1(X) = X$ and $T_{n+2}(X) = 2XT_{n+1}(X) - T_n(X), n = 0, 1, 2, \ldots$.

Recently, we [8] were able to formulate conditions for sequences of polynomials satisfying a second order linear recurrence under which we could conclude that (2) has

*Keywords.* Diophantine equations, linear recurring sequences, $S$-unit equations.

*2000 Mathematics subject classification.* Primary 11D45; Secondary 11D04, 11D61, 11B37.

only finitely many solutions $m, n \in \mathbb{Z}, m, n \geq 0, m \neq n$. For the proof we used the Main Theorem on $S$-unit equations over finitely generated fields of characteristic zero [3,5]. Furthermore, we were able to quantify our results by transforming our problem in the function field generated by the characteristic root of the recurrence over the rational function field $\mathbf{K}(X)$.

The first author gave suitable extensions of the above results for third order linear recurring sequences (cf. [6]). Later on, we generalized our results to linear recurring sequences $G_n(X)$ of arbitrary order [9]. The conditions are somewhat complicated to be stated, essentially, they ensure that there exist valuations in the underlying function field which have special properties. Let

$$\mathcal{G}(X,T) = T^d - A_{d-1}(X)T^{d-1} - \cdots - A_0(X) \in \mathbf{K}[X][T]$$

denote the characteristic polynomial of the sequence $(G_n(X))_{n=0}^{\infty}$ and $D(X)$ be the discriminant of $\mathcal{G}(X,T)$. We write $\alpha_1, \ldots, \alpha_r$ for the distinct roots of the characteristic polynomial $\mathcal{G}(X,T)$ in the splitting field $L$ of $\mathcal{G}(X,T)$. It is well known that $(G_n(X))_{n=0}^{\infty}$ has a nice "analytic" representation. More precisely, there exist polynomials $C_1(T), \ldots, C_r(T) \in L[T]$ such that

$$G_n(X) = C_1(n)\alpha_1^n + \cdots + C_r(n)\alpha_r^n, \tag{3}$$

holds for all $n \geq 0$. Assuming that $\mathcal{G}(X,T)$ has no multiple roots, i.e., $D(X) \neq 0$, we obtain $C_i(T) = c_i$ for all $i = 1, \ldots, r$, i.e., all the $C_i$ are constant and hence $r = d$. In this case we will call the recurrence $(G_n(X))_{n=0}^{\infty}$ simple (i.e., the characteristic polynomial has only simple roots). Assume now that the $d$-th order ($d \geq 2$) linear recurring sequence $(G_n(X))_{n=0}^{\infty}$ and the polynomial $P \in \mathbf{K}[X]$ satisfy the following conditions:

(i) None of the roots and the quotients of distinct roots of the characteristic polynomial of $(G_n(X))_{n=0}^{\infty}$ is an element of $\mathbf{K}^*$,
(ii) $\deg P \geq 1$, and $\deg D \geq 1$,
(iii) $\deg A_0 \geq 1$, $R(A_0, \ldots, A_d, G_0, \ldots, G_d) \neq 0$ (for details on the polynomial $R$ we refer to [9]), and
(iv) the set of zeros of $A_0$ is not equal to that of $A_0(P)$.

Then equation

$$G_n(X) = c\, G_m(P(X)), \tag{4}$$

where $c \in \mathbf{K}^* = \mathbf{K}\backslash\{0\}$ is variable, has at most

$$C(d, A_0, D, P) := \\ = e^{(6d)^{4d}} \left(\log\left(d^{2d^2} \deg D(\deg P + 1)\right)\right)^{2d^2} (2ed)^{30d^3 d!^2 \deg A_0 \deg P}$$

solutions $(n,m) \in \mathbb{Z}^2$ with $n, m \geq 0, n \neq m$. We also obtained the result under different conditions (see [9, Theorem 2.3]). For the special case of equation (2) we could even show more: assuming the conditions from above with

(i′) None of the roots and the quotients of distinct roots of the characteristic polynomial of $(G_n(X))_{n=0}^{\infty}$ is a root of unity, instead of (i), we proved that equation (2) has at most

$$e^{(12d)^{6d}}$$

solutions $(n, m) \in \mathbb{Z}^2$ with $n, m \geq 0, n \neq m$.

Very recently, U. Zannier used elementary method from the theory of function fields to improve on these results. In fact, he was able to completely describe the matter: suppose that $\deg P \geq 2$ and that the recurring sequence $G_n(X)$ is simple with characteristic roots $\alpha_1, \ldots, \alpha_d$ satisfying that no ratio $\alpha_i/\alpha_j$, $i \neq j$, lies in $\mathbf{K}$. Then, if there are only finitely many solutions $m, n$ of

$$G_n(X) = c\, G_m(P(X)), \quad m, n \in \mathbb{N},$$

where $c = c(m, n) \in \mathbf{K}^*$ may depend on $m, n$, their number is at most $8d^6$. If there are infinitely many solutions, then for suitable $r, s \in \mathbb{N}$ we have an identity

$$G_{sn+v_0}(P(X)) = \eta \xi^n G_{rn+u_0}(X), \quad n \in \mathbb{N}, \quad |r| = |s| \deg P > 0,$$

for suitable $\xi, \eta \in \mathbf{K}^*$, and two cases may occur, which he calls the "cyclic" (which denotes essentially the case $G_n(X) = X^n$, $P(X) = X^p$) and the "Chebyshev" case (which is essentially the example from the motivation, i.e., $G_n(X) = T_n(X)$, $P(X) = T_p(X)$). More precisely we have:

*Cyclic case*: $P$ is of the form $\lambda' \circ X^p \circ \lambda$ for suitable $\lambda, \lambda' \in \mathrm{PGL}_2(\mathbf{K})$. Also, the $\alpha_i$ are in $\mathbf{K}(X)$, of the form $c_i X^{\delta_i} \circ \lambda$, for integers $\delta_i$ and $c_i \in \mathbf{K}$,
*Chebyshev case*: $P(X) = \lambda' \circ T_p \circ \lambda$, $\lambda, \lambda'$ as above. The $\alpha_i$ are quadratic over $\mathbf{K}(X)$ and of the form $c_i(X \pm \sqrt{X^2 - 1})^{\delta_i} \circ \lambda$.

Our aim is to generalize this result to the equation

$$G_n(x) = c\, G_m(y), \tag{5}$$

where $c = c(m, n) \in \mathbf{K}^*$ may vary with $m, n$ and where $x, y$ are algebraically dependent, i.e., a relation $Q(x, y) = 0$ holds for some polynomial $Q(X, Y) \in \mathbf{K}[X, Y]$. Moreover, we want to consider arbitrary linear recurring sequences $(G_n(X))_{n=0}^{\infty}$ (and not only simple linear recurrences as before). This equation can be understood as an identity in $\mathbf{K}[X, Y]/(Q(X, Y))$, which denotes the residue class ring of the curve $Q(x, y) = 0$. Using the ideas introduced in [15], we want to give necessary conditions under which the more general problem has at most finitely many solutions.

Observe that we may assume without loss of generality that $Q(X, Y)$ is absolutely irreducible and we will assume this for the rest of the paper.

## 2 Results

Before we state our results, let us start with a small discussion about the polynomial $Q(X, Y)$ which is assumed to be absolutely irreducible. We can also assume that the leading coefficient of $Y$ in $Q(X, Y)$ belongs to $\mathbf{K}^*$, or equivalently that $y$ is integral over $\mathbf{K}(x)$. Otherwise, there exists a valuation $\nu$ in the function field $\mathbf{K}(x, y)$ which corresponds to a pole of $y$. But this implies by our equation $G_n(x) = c\, G_m(y)$ that

$$0 \leq \nu(G_n(x)) = \nu(G_m(y)) \leq \nu(y) < 0,$$

which is a contradiction. This argument is only true if $G_n(x) \notin \mathbf{K}$, which we may assume if $m, n$ are large enough (cf. [7, Corollary 3]). Observe that clearly the $G_n(x)$

are integral (they are polynomials). By symmetry, we can also assume that $x$ is integral over $\mathbf{K}(y)$ and therefore that the leading coefficient of $X$ in $Q(X,Y)$ does not depend on $Y$. Therefore, we have

$$Q(X,Y) = Y^{\deg Q_Y} + q_1(X)Y^{\deg Q_Y - 1} + \cdots + q_r' X^{\deg Q_X} + q_r(X),$$

with $q_r' \in \mathbf{K}^*, q_i(X) \in \mathbf{K}[X], i = 1, \ldots, r$ and $\deg q_r(X) < \deg Q_X =: r$.

We do not immediately start with our special case: first we study the general situation of intersections of two linear recurrences defined over a function field. The following theorem is a generalization of [15, Corollary 2] to the case of arbitrary (also non-simple) linear recurring sequences $G_n$ and $H_n$ given by

$$G_n = C_1(n)\alpha_1^n + C_2(n)\alpha_2^n + \cdots + C_p(n)\alpha_p^n, \tag{6}$$
$$H_n = D_1(n)\beta_1^n + D_2(n)\beta_2^n + \cdots + D_q(n)\beta_q^n, \tag{7}$$

where $\alpha_i, \beta_j \in L^*$ and $0 \neq C_i, D_i \in L[X]$ and $L$ is a function field in one variable over $\mathbf{K}$, and which is therefore of interest on its own.

**Theorem 1.** *Assume that none of the $\alpha_i$, none of the $\beta_j$ and none of the ratios $\alpha_i/\alpha_j$ and $\beta_i/\beta_j$, $i \neq j$ $(i \neq j)$ lies in $\mathbf{K}^*$. Then one of the following two alternatives (i) or (ii) is true:*

(i) *The equation $G_n = c\,H_m$, $c = c(n,m) \in \mathbf{K}^*$ has at most*

$$C(\operatorname{ord} G_n, \operatorname{ord} H_n) := 9d^4(3d^2 + e^{e^{e^{20d}}} + rd^2)$$

*solutions $(m,n) \in \mathbb{Z}^2$. Here $d = \max\{\operatorname{ord} G_n, \operatorname{ord} H_n\}$, whereas $r$ is the rank of the multiplicative group generated by the $\alpha_i$ and $\beta_j$.*

(ii) *There are integers $n_0, m_0, r, s$, with $rs \neq 0$, there are elements $\xi, \eta \in \mathbf{K}^*$, and there are polynomials $0 \neq P, Q \in \mathbf{K}[X]$ such that we have*

$$G_{n_0+rm} = \frac{P(m)}{Q(m)}\eta\xi^m H_{m_0+sm}$$

*holds for $m \in \mathbb{Z}$.*

*Moreover, when alternative (ii) is true, then in (6), (7) we have $p = q$ and there exist a permutation $\pi$ of $\{1, \ldots, p\}$ and polynomials $S_1, \ldots, S_p \in L[X]$ such that*

$$C_i(n_0 + rX) = \eta\alpha_i^{-n_0} P(X)S_i(X) \text{ and } D_{\pi(i)}(m_0 + sX) = \beta_{\pi(i)}^{-m_0} Q(X)S_i(X),$$

*and $\alpha_i^r/\beta_{\pi(i)}^s \in \mathbf{K}$ for $i = 1, \ldots, p = q$.*

The proof of this result follows the line of proof from [15, Corollary 2] and uses a result due to Shorey and Tijdeman (see [13, pp. 84–85] and [4, Lemma 3]). Some more remarks are in order.

*Remark 1.* First of all, it is quite clear that there can exist infinitely many solutions and that the statement about the polynomials $P, Q$ is necessary. Because, if we assume that

$$G_n = P(n)S_n, \quad H_n = Q(n)S_n,$$

where $P, Q \in \mathbf{K}[X]$ and $(S_n)_{n=0}^{\infty}$ is a linear recurring sequence defined over $L$, then we have $G_n = c\,H_n$ with $c = \frac{P(n)}{Q(n)}$ for all $n \in \mathbb{Z}$.

*Remark 2.* We want to mention that such a conclusion also appears in a similar context about arbitrary (nonsimple) linear recurring sequences. Namely in [2], Corvaja and Zannier proved that if $G_n/H_n$ is an integer for infinitely many $n$, then there exists a polynomial $P$ such that $P(n)G_n/H_n$ is a linear recurring sequence for all $n$ in an arithmetic progression.

*Remark 3.* If we are interested in solutions of the equation $G_n = H_m$, then infinitely many solutions can come only from an identity of the form $G_{n_0+rm} = H_{m_0+sm}$ for all $m \in \mathbb{Z}$, which means that

$$C_i(n_0 + rX)\alpha_i^{n_0} = \eta D_{\pi(i)}(m_0 + sX)\beta_{\pi(i)}^{m_0}, \ \eta \in \mathbf{K} \quad \text{and} \quad \alpha_i^r/\beta_{\pi(i)}^s \in \mathbf{K}$$

for $i = 1, \ldots, p = q$ and where $\pi$ is permutation of $\{1, \ldots, p\}$.

*Remark 4.* We mention that the largest part of the upper bound $C$ (the last two summands in the brackets) comes from the fact that the problem reduces to estimate the number of zeros of a linear recurring sequence of the form $P(n) = \alpha^n Q(n)$, where $\alpha \in \mathbf{K}$ and $P(X), Q(X) \in \mathbf{K}[X]$. Of course this bound can be considerably improved if $\mathbf{K} = \mathbb{R}$ or if $\mathbf{K}$ is an algebraic number field (in this case the upper bound will also depend on the degree of the number field). In the general case, however, we know no better upper bound than the general one (cf. [11,12]).

Now, we are ready to come to our special case, where $G_n = G_n(x)$ and $H_n = G_n(y)$. By Theorem 1 it follows at once that either equation (5) has at most $C(\operatorname{ord} G_n, \operatorname{ord} G_n)$ many solutions or an identity of the type dealt with in alternative (ii) must hold. We investigate the latter case in this more special situation and we prove the following theorem. As usual $\operatorname{Res}_Y(f, g)$ denotes the resultant of the two polynomials $f, g$ with respect to $Y$.

**Theorem 2.** *Assume that the $d$-th order ($d \geq 1$) linear recurring sequence $(G_n(X))_{n=0}^{\infty}$ and the irreducible polynomial $Q(X, Y) \in \mathbf{K}[X, Y]$ satisfy the following conditions:*

*(i)* *None of the $\alpha_i$ and the ratios $\alpha_i/\alpha_j$, $i \neq j$ is an element of $\mathbf{K}^*$,*
*(ii)* $\deg C_i + 1$ *is equal to the multiplicity of $\alpha_i$ for all $i = 1, \ldots, r$, and*
*(iii)* *the set of zeros of the polynomial $A_0(X)$ is not equal to that of* $\operatorname{Res}_Y(A_0(Y), Q(X, Y))$.

*Then there are at most $\tilde{C}(\operatorname{ord} G_n)$ pairs $(m, n) \in \mathbb{Z}^2$ for which equation (5) holds, where*

$$\tilde{C}(d) := 9d^4(3d^2 + e^{e^{e^{20d}}} + rd^2)$$

*and where $r$ is the rank of the multiplicative group generated by the $\alpha_i$.*

The question now is the following: do there occur infinite families of solutions other than those in the cyclic and Chebyschev case from above, when we consider curves $Q(x, y) = 0$, which are not of the form $y = P(x)$?

*Remark 5.* First of all, it is clear that additional infinite families of solutions may appear. For example, we have for

$$Q(x, y) = acx^m - ay^m - b(1 - c) = 0$$

with $a, b, c \in \mathbf{K}, m \geq 3$ and $P(X) = aX^m + b$ that $P(y) = c\,P(x)$. Therefore, we get for $G_n(x) = P(x)^n, n \in \mathbb{Z}$ that

$$G_n(y) = P(y)^n = (cP(x))^n = c^n P(x)^n = c^n G_n(x)$$

for all $n \in \mathbb{Z}$. By [14, VI.3.3. Example, page 197] the genus of $Q(x, y) = 0$ is $g = \frac{(m-1)(m-2)}{2} > 0$. This example shows that at least in the case of positive genus also other infinite families may occur.

*Remark 6.* Observe that condition (ii) is not too restricitive. It just means that the recurrence uses its "full" power and can be assured by assuming that $d$ is the minimal length of a recurrence which is satisfied by $(G_n(X))_{n=0}^{\infty}$.

*Remark 7.* We may mention that condition (iii) also naturally appears in the context of the conditions given in our previous papers (see [8,6,9]). Namely, it is easy to see that we have

$$\mathrm{Res}_Y(A_0(Y), Q(X, Y)) = (\mathrm{lc}A_0)^{\deg_Y Q} X^{\deg_X Q + \deg A_0} + \cdots,$$

where $\mathrm{lc}A_0$ denotes the leading coefficient of $A_0$. If we additionally assume that $\deg_X Q \geq 2$, we therefore have a valuation $\nu$ with $\nu(D(y)) > \nu(D(x))$, which was the main point in our previous considerations.

We mention that from the proof we see that we must exclude that $A_0^r(y) = cA_0(x)^s$ for some $r, s \in \mathbb{N}, c \in \mathbf{K}^*$. Whenever, we can find

$$Q(X, Y) \mid A_0(Y)^r - cA_0(X)^s,$$

we have other infinite families as described above (observe that the example in Remark 4 was constructed with the trivial case $Q(X, Y) = A_0(Y) - cA_0(X)^s$). It follows from Schinzel [10, page 58] that if $A_0(X)$ is indecomposable over $\mathbf{K}$, which means that if $A_0(X) = F_1(F_2(X)), F_1, F_2 \in \mathbf{K}[X]$, then $\deg F_1 = 1$ or $\deg F_2 = 1$, and $\deg A_0 > 31$, then $A_0(Y) - cA_0(X)^s$ is irreducible over $\mathbf{K}$. We conjecture that $F(X, Y) = A_0(Y) - cA_0(X)^s$ with $A_0$ indecomposable (and it is clear that this is needed) and $A_0(y) \neq B(y)^t$ or $-4B(y)^4$ is always irreducible. Schinzel mentioned to us that this conjecture – if true – lies deeper than Capelli's theorem (e.g., see [10, Theorem 19, page 92]), since it depends on the characteristic of $\mathbf{K}$, while Capelli's theorem does not.

*Remark 8.* The motivation to look at this generalisation is the following: if it would be possible to prove that $G_n(x) = c\,G_m(y)$ with $Q(x, y) = 0$ has no solution unless we have a trivial infinite family, then it would be possible to handle the Diophantine equation $G_n(X) = G_m(Y)$ in integers $X, Y$ by the method of Bilu and Tichy [1].

## 3 Proof of Theorem 1

We start by rewriting our equation $G_n = c\,H_m$ as

$$G_n - c\,H_m = \sum_{i=1}^{p} C_i(n)\alpha_i^n 1^m - \sum_{i=1}^{q} cD_i(n)1^n\beta_i^m = 0.$$

We define vectors $A_i = (\alpha_i, 1) \in (L^*)^2$ for $i = 1, \dots, p$, $A_{p+i} = (1, \beta_i) \in (L^*)^2$ for $i = 1, \dots, q$ and polynomials $P_i = C_i, i = 1, \dots, p$, $P_{p+i} = D_i, i = 1, \dots, q$, respectively.

Now we apply the following lemma due to Zannier (see [15, Theorem 1] and also [15, Definition 1]):

**Lemma 3.** *Let $A_1, \dots, A_h \in (L^*)^r$ and let $P_1, \dots, P_h \in L[X_1, \dots, X_r] = L[\mathbf{X}]$ satisfy $\deg P_i \le d_i$. Then the set*

$$S = \{\mathbf{m} \in \mathbb{Z}^r \ : \ P_i(\mathbf{m})A_i^{\mathbf{m}}, i = 1, \dots, h \text{ are linearly independent over } \mathbf{K}\},$$

*(here for $A = (\alpha_1, \dots, \alpha_r)$ we define $A^{\mathbf{m}} = \alpha_1^{m_1} \cdots \alpha_r^{m_r}$) may be expressed as a union of no more than*

$$\left(d_1 + \cdots + d_h + \binom{h}{2}\right)^r$$

*classes, where we say that $S' \subset \mathbb{Z}^r$ is a class relative to a nonempty subset $B$ of $\{1, \dots, h\}$, if*

(i) *for every $\mathbf{m} \in S'$ the elements $P_i(\mathbf{m})A_i^{\mathbf{m}}, i \in B$ are linearly independent over $\mathbf{K}$ and*
(ii) *for some $\mathbf{m}_0 \in S'$ the set $S'$ is made up by all $\mathbf{m}$ satisfying* (i) *and such that for $i, j \in B$ we have $(A_i A_j^{-1})^{\mathbf{m}-\mathbf{m}_0} \in \mathbf{K}^*$.*

Applying Lemma 3 we see that the set of solutions $(m, n) \in \mathbb{Z}^2$ of our equation is contained in the union of at most

$$\left(\operatorname{ord} G_n + \operatorname{ord} H_n + \binom{p+q}{2}\right)^2 \le \left(3 \max\{\operatorname{ord} G_n, \operatorname{ord} H_n\}^2\right)^2$$

classes (for a definition of classes see Lemma 3 or [15, Definition 2]).

We are going to estimate the number of solutions in each class $\Omega$, corresponding to the subset $B = B_\Omega \subset \{1, \dots, p+q\}$. As in the proof of [15, Corollary 2] it is easy to see by [15, Corollary 1(b)] that there are at most $\max\{\operatorname{ord} G_n, \operatorname{ord} H_n\} + \binom{p+q}{2}$ solutions in every class containing distinct integers $i, j$ in $[1, p]$ or $[p+1, p+q]$, respectively, having $C_1(n) \cdots C_p(n) \ne 0$ or $D_1(n) \cdots D_q(n) \ne 0$, respectively. Since these cases appear for at most $\max\{\operatorname{ord} G_n, \operatorname{ord} H_n\}$ many $n$, we get that the number of solutions is bounded by

$$2 \max\{\operatorname{ord} G_n, \operatorname{ord} H_n\} + \binom{p+q}{2} \le 3 \max\{\operatorname{ord} G_n, \operatorname{ord} H_n\}^2.$$

In the case that $B$ contains integers $i_0, j_0 + p$ with $1 \le i_0 \le p, 1, \le j_0 \le q$ it is also plain (by the proof of [15, Corollary 2]) that the solutions in the class $\Omega$ correspond to integers $m$ such that

$$G_{n_0+rm} = c\, H_{m_0+sm} \tag{8}$$

with integers $n_0, m_0, r, s, rs \ne 0$.

In this case we first group together in a single $\gamma^m$ two exponentials $\alpha_i^{rm}$ and $\beta_j^{sm}$, which are linearly dependent over $\mathbf{K}$. Namely, if $\alpha_i^r = \gamma_{(i,j)}, \beta_j^s = \delta_{(i,j)}\gamma_{(i,j)}$ with $\delta_{(i,j)} \in \mathbf{K}^*$ we have

$$C_i(n_0 + rm)\alpha_i^{n_0}\gamma_{(i,j)}^m - c D_j(m_0 + sm)\beta_j^{m_0}\delta_{(i,j)}^m\gamma_{(i,j)}^m. \tag{9}$$

Now, we write

$$C_i(n_0 + rX)\alpha_i^{n_0} = \sum_{l=1}^{u} \rho_l Q_{il}(X),$$

$$D_j(m_0 + sX)\beta_j^{m_0} = \sum_{l=1}^{u} \rho_l \tilde{Q}_{jl}(X),$$

where the $\rho_l \in L^*, l = 1, \ldots, u$ are linearly independent over $\mathbf{K}$ and the $Q_{il}$, $\tilde{Q}_{jl}$ lie in $\mathbf{K}[X]$ for each $i, j, l$. Clearly this is possible for some $u$ with

$$u \leq (p+q)\max\{\deg C_1, \ldots, \deg C_p, \deg D_1, \ldots, \deg D_q\}.$$

Thus, (9) becomes

$$\sum_{l=1}^{u} \left( Q_{il}(m) - c\delta_{(i,j)}^m \tilde{Q}_{jl}(m) \right) \rho_l \gamma_{(i,j)}^m.$$

Up to now we have rewritten (8) as a $\mathbf{K}$-linear combination of expressions of the from $\rho_l \gamma_i^m$, where all these expressions are linearly independent and where $\gamma_i = \gamma_{(i,j)}$ or $\alpha_i, \beta_j$, respectively, depending on whether they could be paired with some other term in (8) or not. We have two possible cases: those $m$ for which all coefficients vanish and those for which not all coefficients vanish. In the latter case the elements $\rho_l \gamma_i^m$ are linearly dependent over $\mathbf{K}$, which can happen (by [15, Lemma 2]) for at most $\binom{2(\operatorname{ord} G_n + \operatorname{ord} H_n) - 1}{2}$ many $m$.

In the first case all terms in $G_{n_0+rm}$ must be paired with the terms in $H_{m_0+sm}$, so we have $p = q$ and $u \leq 2\operatorname{ord} G_n$. Moreover, there exists a permutation $\pi$ of the set $\{1, \ldots, p\}$ which pairs each $\alpha_i$ with some $\beta_j = \beta_{\pi(i)}$ such that (8) can be rewritten as

$$\sum_{i=1}^{p} \sum_{l=1}^{u} \left( Q_{il}(m) - c\delta_i^m \tilde{Q}_{\pi(i)l}(m) \right) \rho_l \gamma_i^m = 0.$$

For simplicity we have written here $\gamma_i, \delta_i$ instead of $\gamma_{(i,\pi(i))}, \delta_{(i,\pi(i))}$, respectively. Observe that there are at most $\max\{\operatorname{ord} G_n, \operatorname{ord} H_n\}$ many $m$ for which $C_i(n_0 + rm)$ or $D_j(m_0 + sm) = 0$. For all other $m$ we have

$$\frac{Q_{il}(m)}{\tilde{Q}_{\pi(i)l}(m)} \delta_i^{-m} = c$$

(recall that $c$ here may depend on $m$) for all $i, l$ or

$$\frac{Q_{il}(m)}{\tilde{Q}_{\pi(i)l}(m)} \frac{\tilde{Q}_{\pi(j)r}(m)}{Q_{jr}(m)} \left( \frac{\delta_j}{\delta_i} \right)^m = 1 \tag{10}$$

for all $i, l, j, r$.

Now, we pause for a moment to cite the following result from [4, page 148].

**Lemma 4.** *Let $P \in \overline{\mathbb{Q}}(X)$ be a rational function with no poles outside the disc $\{z \in \mathbb{C} : |z| \leq A\}$ and let $\alpha \in \overline{\mathbb{Q}}$. If there are infinitely many pairs of integers $m, n$ with*

$$m > n \geq A, \quad P(m)\alpha^m = P(n)\alpha^n,$$

*then $P$ is constant and $\alpha$ is a root of unity.*

We use a specialization argument to reduce our case to Lemma 4. For this let $U \subset \mathbf{K}$ be a finite set consisting of all transcendental elements from the $Q_{il}$, $\tilde{Q}_{\pi(i)l}$ and $\delta_i$ for all $i, l$, together with all possible differences and all multiplicative inverses of these elements. Then by [5, Lemma 3.1] there exists a ring homomorphism $\varphi : \overline{\mathbb{Q}}[U] \longrightarrow \overline{\mathbb{Q}}$ whose restriction to $\overline{\mathbb{Q}}$ is the identity. Applying this map to (10) leads to

$$\frac{\varphi(Q_{ij}(m))}{\varphi(\tilde{Q}_{\pi(i)j}(m))} \frac{\varphi(\tilde{Q}_{\pi(j)r}(m))}{\varphi(Q_{jr}(m))} \left(\frac{\varphi(\delta_j)}{\varphi(\delta_i)}\right)^m = 1. \tag{11}$$

Now, if there are infinitely many such $m$, then there are infinitely many $m, n$ such that

$$\frac{\varphi(Q_{ij}(m))}{\varphi(\tilde{Q}_{\pi(i)j}(m))} \frac{\varphi(\tilde{Q}_{\pi(j)r}(m))}{\varphi(Q_{jr}(m))} \left(\frac{\varphi(\delta_j)}{\varphi(\delta_i)}\right)^m = \frac{\varphi(Q_{ij}(n))}{\varphi(\tilde{Q}_{\pi(i)j}(n))} \frac{\varphi(\tilde{Q}_{\pi(j)r}(n))}{\varphi(Q_{jr}(n))} \left(\frac{\varphi(\delta_j)}{\varphi(\delta_i)}\right)^n.$$

Therefore, $\max\{m, n\} \longrightarrow \infty$ and the above lemma implies that $\varphi(\delta_j/\delta_i)$ and therefore also $\delta_j/\delta_i$ is a root of unity. Moreover,

$$\frac{Q_{il}(X)}{\tilde{Q}_{\pi(i)l}(X)} = \frac{Q_{ir}(X)}{\tilde{Q}_{\pi(i)r}(X)} \quad \text{and} \quad \frac{Q_{jl}(X)}{\tilde{Q}_{\pi(j)l}(X)} = \frac{Q_{jr}(X)}{\tilde{Q}_{\pi(j)r}(X)}$$

differ just by a constant (in fact again a root of unity) for all $i \neq j, l \neq r$ (observe that the equalities follow from (11) at once). It follows that there exist polynomials $P, Q \in \mathbf{K}[X]$ such that

$$P(X)S'_{il}(X) = \eta_i Q_{il}(X), \quad Q(X)S'_{il}(X) = \tilde{\eta}_{\pi(i)} \tilde{Q}_{\pi(i)l}(X)$$

for all $i, l$ with $\eta_i, \tilde{\eta}_{\pi(i)} \in \mathbf{K}$ (independent of $l$) and for some polynomials $S'_{il}(X)$. From this discussion it follows that this case can only hold for all $m$ in the intersection of certain arithmetic progressions, which is either empty or again an arithmetic progression. Moreover, we see that in this case we have $\tilde{\eta}_{\pi(i)}/\eta_i = \eta$ with $\eta$ a suitable root of unity. Therefore, also the second part of the conclusion of Theorem 1 follows from this.

Further, equation (11) can have finitely many solutions in the following two cases: either $\varphi(\delta_j)/\varphi(\delta_i)$ is a root of unity or not. In the second case the number of $m$ satisfying (11) can be bounded by the zero multiplicity of the underlying linear recurring sequence, hence by

$$\exp(\exp(\exp(20(\operatorname{ord} G_n + \operatorname{ord} H_n)))$$

by [11,12], since the degrees of the polynomials are bounded by the order of the recurrences. On the other hand, if $\varphi(\delta_j)/\varphi(\delta_i)$ is a root of unity of order $\ell$ say, then in the $\ell$ arithmetic progressions $m = k\ell + r, 0 \leq r \leq \ell - 1$, we can bound the number of $m$'s by the degrees of $Q_{il}(X)\tilde{Q}_{jr}(X)$ and $\tilde{Q}_{il}(X)Q_{jr}(X)$, respectively. Therefore, we can bound the number of solutions coming from this case by the rank of the multiplicative group generated by $\alpha_1, \ldots, \alpha_p, \beta_1, \ldots, \beta_p$, which is an upper bound for $\ell$, times $\operatorname{ord} G_n + \operatorname{ord} H_n$.

Altogether, we see that there are at most $C(\operatorname{ord} G_n, \operatorname{ord} H_n)$ solutions, which do not come from a trivial relation, which finishes the proof. □

## 4 Proof of Theorem 2

We already know that we have to study the equation

$$G_{n_0+mr}(x) = \frac{P(m)}{Q(m)} \eta \xi^m G_{m_0+ms}(y) \tag{12}$$

with $\xi, \eta \in \mathbf{K}^*$, $P(X), Q(X) \in \mathbf{K}[X]$ and with integers $n_0, m_0, r, s, rs \neq 0$, which is an identity in the function field $\mathbf{K}(x, y)$. Moreover, we have

$$G_n(x) = C_1(n)\alpha_1^n + \cdots + C_p(n)\alpha_p^n,$$
$$G_n(y) = D_1(n)\beta_1^n + \cdots + D_p(n)\beta_p^n,$$

where $\alpha_i$ are the zeros of $\mathcal{G}(x, T)$ and $\beta_j$ are the zeros of $\mathcal{G}(y, T)$, respectively. Obviously, the field $L_0 := \mathbf{K}(x, \alpha_1, \ldots, \alpha_p)$ and $L_1 := \mathbf{K}(y, \beta_1, \ldots, \beta_p)$ are isomorphic over $\mathbf{K}$ and we denote the isomorphism (which sends $x \mapsto y$ and $\alpha_i \mapsto \beta_i$) by $\psi : L_0 \to L_1$.

From Theorem 1 we know that there are polynomials $S_1(X), \ldots, S_p(X)$ and a permutation $\pi$ of $\{1, \ldots, p\}$ such that $C_i(n_0+rX) = \eta\alpha_i^{-n_0}P(X)S_i(X)$, $D_{\pi(i)}(m_0+sX) = \beta_{\pi(i)}^{-m_0}Q(X)S_i(X)$ and $\alpha_i^r/\beta_{\pi(i)}^s \in \mathbf{K}$ for $i = 1, \ldots, p$. Since by the above isomorphism $\psi$, we have that $\alpha_i$ and $\beta_i$ have the same multiplicity and therefore we have $\deg C_i = \deg D_i$ for all $i = 1, \ldots, p$, it follows that

$$\deg S_{\pi(i)} = \deg S_i + \deg Q - \deg P$$

for all $i = 1, \ldots, p$. This implies

$$\deg S_{\pi^k(i)} = \deg S_i + k(\deg Q - \deg P)$$

for every $k \in \mathbb{N}$, where $\pi^k$ denotes as usual the $k$-th iterate of the map $\pi$. Let $\ell$ be the order of $\pi$. Then, we get $\deg S_i = \deg S_i + \ell(\deg Q - \deg P)$ and therefore $\deg Q = \deg P$. This means that $\deg C_i = \deg D_{\pi(i)}$ and by condition (ii) that $\alpha_i$ and $\beta_{\pi(i)}$ have the same multiplicity as roots of the characteristic polynomial $\mathcal{G}(x, T)$ and $\mathcal{G}(y, T)$, respectively.

The proof of Theorem 2 now follows easily from what we have proved up to now. Namely, by assuming that our equation has infinitely many solutions we have that the characteristic roots of $G_n(x)$ and $G_n(y)$ satisfy $\alpha_i^r = c\beta_{\pi(i)}^s$, where $\pi$ is a permutation of the set $\{1, \ldots, p\}$ and $c \in \mathbf{K}^*$ (here $c$ may depend on i). Moreover, the multiplicities of $\alpha_i$ and $\beta_{\pi(i)}$ are the same. By multiplying all these relations according the multiplicities, we therefore get

$$\begin{aligned} A_0(x)^r &= \prod_{i=1}^{p} \prod_{j=1}^{\deg C_i+1} \alpha_i^r = \prod_{i=1}^{p} \prod_{j=1}^{\deg C_i+1} c\beta_{\pi(i)}^s \\ &= \tilde{c}\left( \prod_{i=1}^{p} \prod_{j=1}^{\deg D_{\pi(i)}} \beta_{\pi(i)} \right)^s = \tilde{c}A_0(y)^s, \end{aligned}$$

where $A_0$ is the constant polynomial in the linear recurring equation. But now, condition (iii) of our assumptions excludes that this equation can hold. Therefore, we obtain a contradiction, which shows the finiteness of the number of solutions in this case. □

**Acknowledgements.** The first and the third author were supported by the Austrian Science Foundation FWF, grants S8307-MAT, J2407-N12 and NFN S9611. The second author was supported by the Hungarian National Foundation for Scientific Research Grant nr. 38225 and 42985.

# References

1. Bilu, Yu., Tichy, R.F.: The Diophantine equation $f(x) = g(y)$. *Acta Arith.* **95**, 261–288 (2000)
2. Corvaja, P., Zannier, U.: Finiteness of integral values for the ratio of two linear recurrences. *Invent. Math.* **149**, 431–451 (2002)
3. Evertse, J.-H., Győry, K..: On the number of solutions of weighted unit equations. *Compos. Math.* **66**, 329–354 (1988)
4. Evertse, J.-H., Győry, K., Stewart, C.L., Tijdeman, R.: $S$-Unit equations and their applications. In: Baker, A. (ed.) *New Advances in Transcendence Theory*, pp. 110–174. Cambridge University Press, Cambridge (1988)
5. Evertse, J.-H., Schlickewei, H.-P., Schmidt, W.M.: Linear equations in variables which lie in a multiplicative group. *Ann. Math.* **155**, 1–30 (2002)
6. Fuchs, C.: On the equation $G_n(x) = G_m(P(x))$ for third order linear recurring sequences. *Port. Math., N.S.* **61**, 1–24 (2004)
7. Fuchs, C., Pethő, A.: Effective bounds for the zeros of linear recurrences in function fields. *J. Théor. Nombres Bordx.* **17**, 749–766 (2005)
8. Fuchs, C., Pethő, A., Tichy, R.F.: On the Diophantine equation $G_n(x) = G_m(P(x))$. *Monatsh. Math.* **137**, 173–196 (2002)
9. Fuchs, C., Pethő, A., Tichy, R.F.: On the Diophantine equation $G_n(x) = G_m(P(x))$: Higher-order recurrences. *Trans. Am. Math. Soc.* **355**, 4657–4681 (2003)
10. Schinzel, A.: *Polynomials with Special Regard to Reducibility.* Cambridge University Press, Cambridge (2000)
11. Schmidt, W.M.: The zero multiplicity of linear recurrence sequences. *Acta Math.* **182**, 243–282 (1999)
12. Schmidt, W.M.: Zeros of linear recurrence sequences. *Publ. Math. (Debrecen)* **56**, 609–630 (2000)
13. Shorey, T.N., Tijdeman, R.: *Exponential Diophantine Equations.* Cambridge University Press, Cambridge (1986)
14. Stichtenoth, H.: *Algebraic Function Fields and Codes.* Springer, Heidelberg (1993)
15. Zannier, U.: On the integer solutions of exponential equations in function fields. *Ann. Inst. Fourier* **54**, 849–874 (2004)

*Diophantine Approximation*
H. P. Schlickewei et al., Editors


# A CRITERION FOR POLYNOMIALS TO DIVIDE INFINITELY MANY $k$-NOMIALS

Lajos Hajdu[1,2] and Robert Tijdeman[3]
[1] *Number Theory Research Group, Hungarian Academy of Sciences, P. O. Box 12, 4010 Debrecen, Hungary*
[2] *Institute of Mathematics, University of Debrecen, P.O. Box 12, 4010 Debrecen, Hungary*
hajdul@math.klte.hu
[3] *Mathematical Institute, Leiden University, P.O. Box 9512, 2300 RA Leiden, The Netherlands*
tijdeman@math.leidenuniv.nl

To Wolfgang M. Schmidt on the occasion of his 70th birthday

## 1 Introduction

A polynomial $Q \in \mathbb{Q}[x]$ of the form

$$Q(x) = \sum_{i=1}^{k} a_i x^{m_i} \text{ with } m_1 > \ldots > m_{k-1} > m_k = 0 \text{ and } a_1 = 1$$

is called a standard $k$-nomial. It is worth to mention that the restriction to monic $k$-nomials is only for convenience. We may replace every standard $k$-nomial by any of its constant multiples, and the theorems would still be valid. We call $(m_1, \ldots, m_k)$ the exponent $k$-tuple of $Q$. Note that if $Q$ is a standard $k$-nomial, but not a standard $(k-1)$-nomial, then its exponent $k$-tuple is uniquely determined. Let

$$\mathrm{PR}_k = \{P \in \mathbb{Q}[x] \;:\; \exists\, Q \in \mathbb{Q}[x] \text{ and } r \in \mathbb{Z} \text{ with } \deg(Q) < k \text{ and } r \geq 1 \text{ such that } P(x) \mid Q(x^r) \text{ over } \mathbb{Q}\}.$$

In 1965 Posner and Rumsey observed (see [5], pp. 339 and 348) that $P \in \mathrm{PR}_k$ implies that $P$ divides infinitely many standard $k$-nomials over $\mathbb{Q}$. They conjectured that the converse is also true, that is, if a polynomial $P \in \mathbb{Q}[x]$ divides infinitely many standard $k$-nomials over $\mathbb{Q}$, then $P \in \mathrm{PR}_k$. For $k = 2$ the conjecture obviously holds.

In [2] Győry and Schinzel verified the conjecture in a quantitative form for $k = 3$. They proved that if $P$ divides more than $C_1$ standard trinomials over $\mathbb{Q}$, then $P \in \mathrm{PR}_3$. Here $C_1$ is a number depending on the degree of $P$ and some other parameters, and it is explicitly given in [2]. Later, Schlickewei and Viola [6] provided a value for $C_1$ which depends only on the degree of $P$.

However, the authors of [2] disproved the conjecture for every $k \geq 4$. For every $k \geq 2$ they gave a polynomial $P \in \mathbb{Q}[x]$ that divides infinitely many standard quadri-

*Keywords.* Polynomials, $k$-nomials, quintinomials, transitivity, subspace theorem.

*2000 Mathematics subject classification.* 11C08 (11D57).

nomials over $\mathbb{Q}$ with $P \notin \mathrm{PR}_k$. In fact the quadrinomials have a zero constant term and have therefore only three nonzero terms. In case of polynomials with nonzero constant terms, the problem is more difficult. For every $k \geq 2$ Győry and Schinzel [2] provided a $P \notin \mathrm{PR}_k$ which divides infinitely many standard quintinomials over $\mathbb{Q}$ with nonzero constant terms. They proposed the following problem instead of the disproved conjecture of Posner and Rumsey.

Let $k$ be an integer with $k \geq 4$. Is it true that a polynomial $P \in \mathbb{Q}[x]$ with $P(0) \neq 0$ divides infinitely many standard $k$-nomials with nonzero constant terms if and only if either $P \in \mathrm{PR}_k$, or $P$ divides a standard $\left[\frac{k+1}{2}\right]$-nomial?

For $k \geq 6$ Hajdu [3] gave a negative answer to this question by providing other kinds of counterexamples. He proposed to modify the problem of Győry and Schinzel as follows.

Let $k$ be an integer with $k \geq 4$. Is it true that a polynomial $P \in \mathbb{Q}[x]$ with $P(0) \neq 0$ divides infinitely many standard $k$-nomials with nonzero constant terms if and only if either $P \in \mathrm{PR}_k$ or $P$ divides a standard $(k-2)$-nomial which divides infinitely many standard $k$-nomials over $\mathbb{Q}$?

Schlickewei and Viola [7] described a so-called proper family $\mathcal{F}_k$ of standard $k$-nomials such that if a polynomial $P$ having only simple zeros divides more than $C_2(k)$ elements of $\mathcal{F}_k$, then $P \in \mathrm{PR}_k$.

In [4] Hajdu and Tijdeman gave necessary and sufficient conditions for a polynomial $P \in \mathbb{Q}[x]$ having only simple zeros to divide infinitely many standard quadrinomials or standard quintinomials over $\mathbb{Q}$. Moreover, for $k = 5$ they presented a polynomial which yields negative answers to the problems stated by Győry and Schinzel, and by Hajdu.

The aim of this paper is to extend the results of [4] to polynomials dividing standard $k$-nomials for arbitrary $k \geq 4$. For this purpose, we impose a new type of assumption. More precisely, we assume that the polynomial $P$ dividing infinitely many $k$-nomials is irreducible over $\mathbb{Q}$, and also that its Galois group is sufficiently large. We note that as "almost all" polynomials in $\mathbb{Q}[x]$ are irreducible and have the whole symmetric group as Galois group, we exclude only a minor part of polynomials from our investigations. The new results indicate that the conditions (i) and (ii) of Theorem 1 of [4] (which are the same as in Theorem 1 below) are the "right ones" to characterize polynomials dividing infinitely many standard $k$-nomials over $\mathbb{Q}$. The proofs rely on the Subspace Theorem based on Schmidt's fundamental work.

## 2 The main results

Let $P \in \mathbb{Q}[x]$ be an irreducible polynomial of degree $n$, with Galois group $\mathcal{G}$ and with splitting field $\mathbb{K}$ over $\mathbb{Q}$. We keep this notation for the whole paper. For any $t \in \{1, \dots, n\}$ we say that the Galois group of $P$ is $t$-times transitive, if for all ordered $t$-tuples $(\alpha_{i_1}, \dots, \alpha_{i_t})$ and $(\alpha_{j_1}, \dots, \alpha_{j_t})$ consisting of zeros of $P$ there exists an automorphism $\sigma$ of $\mathbb{K}$ such that $\sigma(\alpha_{i_l}) = \alpha_{j_l}$ for $l = 1, \dots, t$. It is well known that the Galois group of any irreducible polynomial is transitive (with $t = 1$). Note that if $P$ is $t$-times transitive, then it is $s$-times transitive for any integer $s$ with $1 \leq s \leq t$.

**Theorem 1.** *Let $k$ be an integer with $k \geq 4$. An irreducible polynomial $P \in \mathbb{Q}[x]$ with $[2k/3]$-times transitive Galois group $\mathcal{G}$ divides infinitely many standard $k$-nomials with* *nonzero constant terms over $\mathbb{Q}$ if and only if one of the following conditions holds:*

*(i)* $P \in \mathrm{PR}_k$,

*(ii)* *P divides over $\mathbb{Q}$ two different standard k-nomials with the same exponent k-tuple.*

We note that for $k = 4$ the statement follows from Theorem 1 of [4]. We shall derive the following simple corollaries.

**Corollary 1.** *Let P be as in Theorem 1. Then P divides infinitely many standard k-nomials over $\mathbb{Q}$ if and only if either $P \in \mathrm{PR}_k$ or P divides a standard $(k-1)$-nomial over $\mathbb{Q}$.*

**Corollary 2.** *Let P be as in Theorem 1, with the further assumption that* $\deg(P) \geq k$. *Then condition (i) can be replaced by*

*(i′)* *P divides a standard binomial over $\mathbb{Q}$.*

The following statement shows that the conditions (i) and (ii) in Theorem 1 are independent.

**Proposition.** *For every $k \geq 5$ there exist polynomials $P_1, P_2 \in \mathbb{Q}[x]$ such that both divide infinitely many standard k-nomials over $\mathbb{Q}$, (i) holds for $P_1$ but not for $P_2$, and conversely, (ii) holds for $P_2$ but not for $P_1$.*

*Remark 1.* The Proposition, together with Theorem 1, strongly suggests that the conditions (i) and (ii) are necessary and sufficient to characterize polynomials dividing infinitely many standard $k$-nomials over $\mathbb{Q}$.

*Remark 2.* Following the proof of Theorem 1, one can easily see that there is an effectively computable constant $C_3(k)$ depending only on $k$, such that if $P$ divides more than $C_3(k)$ standard $k$-nomials over $\mathbb{Q}$, then the conclusion of the theorem is still valid.

In case of $k = 5$ we need only double transitivity to have the same conclusion as in Theorem 1.

**Theorem 2.** *An irreducible polynomial $P \in \mathbb{Q}[x]$ with doubly transitive Galois group $\mathcal{G}$ divides infinitely many standard quintinomials with nonzero constant terms over $\mathbb{Q}$ if and only if condition (i) or (ii) in Theorem 1 with $k = 5$ holds.*

*Remark 3.* In Theorem 2 of [4] we proved that a polynomial $P \in \mathbb{Q}[x]$ with only simple zeros and with $P(0) \neq 0$ divides infinitely many standard quintinomials with nonzero constant terms over $\mathbb{Q}$ if and only if (i), (ii) or the next condition holds:

(iii) there exist integers $M_1, M_2, M_3, M_4$ such that $P$ divides over $\mathbb{Q}$ infinitely many standard quintinomials $Q_m$ of the form

$$Q_m(x) = x^{M_1+2m} + a_m x^{M_2+m} + b_m x^{M_3+m} + c_m x^{M_4+m} + d_m$$

with $m \in \mathbb{N}$ and $a_m, b_m, c_m, d_m \in \mathbb{Q}$.

The proof of Theorem 2 shows that if $k = 5$ and $P$ has a doubly transitive Galois group, then condition (iii) implies (i) or (ii).

## 3 Basic lemmas

Two algebraic numbers $\beta_1$ and $\beta_2$ are called equivalent, if for some root of unity $\varepsilon$ we have $\beta_1\varepsilon = \beta_2$. Hence we have a partition of the algebraic numbers into equivalence classes.

**Lemma 1.** *Let $k \in \mathbb{Z}$ with $k \geq 2$, and let $P \in \mathbb{Q}[x]$ be a polynomial having only simple zeros. Then $P \in \mathrm{PR}_k$ if and only if the zeros of $P$ belong to the union of at most $k-1$ equivalence-classes defined above.*

*Proof.* The statement is a reformulation of Proposition 2.1 of [7]. □

**Lemma 2.** *Let $\alpha_1, \dots, \alpha_k$ be nonzero elements of a field of characteristic zero, such that $\alpha_i/\alpha_j$ is not a root of unity $(1 \leq i < j \leq k)$. Then the equation*

$$\begin{vmatrix} \alpha_1^{X_1} & \dots & \alpha_k^{X_1} \\ \vdots & \vdots & \vdots \\ \alpha_1^{X_k} & \dots & \alpha_k^{X_k} \end{vmatrix} = 0$$

*has at most $\exp((6k!)^{3k!})$ solutions in $(X_1, \dots, X_k) \in \mathbb{Z}^k$ with $X_k = 0$ for which the above determinant has no vanishing subdeterminant.*

*Proof.* This is a reformulation of Theorem 1.1 in [8]. □

Let $\mathbb{L}$ be an algebraic number field and $\alpha_{ij} \in \mathbb{L}^*$ for $1 \leq i \leq m$, $1 \leq j \leq n$, where $m, n$ are positive integers. Moreover, let $a_i \in \mathbb{L}$ $(1 \leq i \leq m)$. For $i = 1, \dots, m$ and $\underline{x} \in \mathbb{Z}^n$ with $\underline{x} = (x_1, \dots, x_n)$ write $\underline{\alpha_i}^{\underline{x}} = \alpha_{i1}^{x_1} \dots \alpha_{in}^{x_n}$ for brevity. Consider the equation

$$\sum_{i=1}^{m} a_i \underline{\alpha_i}^{\underline{x}} = 0 \quad \text{in} \quad \underline{x} \in \mathbb{Z}^n. \tag{1}$$

Let $\mathcal{P}$ be a partition of the set $\Lambda = \{1, \dots, m\}$, and consider the system of equations

$$(1.\mathcal{P}) \qquad \sum_{i\in\lambda} a_i \underline{\alpha_i}^{\underline{x}} = 0 \quad (\lambda \in \mathcal{P}) \qquad \text{in} \quad \underline{x} \in \mathbb{Z}^n,$$

which is a refinement of (1). Let $\mathcal{S}(\mathcal{P})$ denote the set of those solutions of $(1.\mathcal{P})$ which are not solutions of any $(1.\mathcal{Q})$ where $\mathcal{Q}$ is a proper refinement of $\mathcal{P}$. Set $i_1 \overset{\mathcal{P}}{\sim} i_2$, if $i_1$ and $i_2$ are in the same class of $\mathcal{P}$, and put

$$G(\mathcal{P}) = \{\underline{z} \in \mathbb{Z}^n \;:\; \underline{\alpha_{i_1}}^{\underline{z}} = \underline{\alpha_{i_2}}^{\underline{z}} \text{ for any } i_1, i_2 \text{ with } i_1 \overset{\mathcal{P}}{\sim} i_2\}.$$

Denote the cardinality of the set $A$ by $|A|$.

**Lemma 3.** *Using the above notation, there exists an explicitly computable constant $C(m,n)$ depending only on $m$ and $n$ such that if $\mathcal{P}$ is any partition of $\Lambda$ with*

$$|\mathcal{S}(\mathcal{P})| \geq C(m,n)$$

*then there are different solutions $\underline{z}'$ and $\underline{z}''$ of $(1.\mathcal{P})$ such that $\underline{z}' - \underline{z}'' \in G(\mathcal{P})$.*

*Proof.* The statement follows from Theorem 1.1 of [1] by a simple induction argument. □

**Lemma 4.** *Let $P \in \mathbb{Q}[x]$ be an irreducible polynomial with doubly transitive Galois group $\mathcal{G}$. Then either all the zeros of $P$ are equivalent or no pair of zeros of $P$ is equivalent.*

*Proof.* Suppose $\alpha_1, \alpha_2, \alpha_i, \alpha_j$ are zeros of $P$ such that $\alpha_1 \neq \alpha_2$, $\alpha_i \neq \alpha_j$ and $\alpha_1/\alpha_2$ is a root of unity. Choose a $\sigma \in \mathcal{G}$ such that $\sigma(\alpha_1) = \alpha_i$ and $\sigma(\alpha_2) = \alpha_j$. Then we obtain that $\alpha_i/\alpha_j$ is also a root of unity. □

## 4 Proofs

As the proof of Theorem 2 is more concrete, we give it first. Thereafter we present the proofs of Theorem 1 and Corollaries 1 and 2. The verification of the Proposition is the final item of the section.

*Proof of Theorem 2.* As we mentioned in the Introduction, (i) is sufficient by a result of Posner and Rumsey (see [5], pp. 339 and 348). The sufficiency of (ii) follows by considering suitable linear combinations of the two polynomials. To prove necessity, in view of Remark 3, we may assume that there exist integers $M_1, M_2, M_3, M_4$ such that $P$ divides over $\mathbb{Q}$ infinitely many standard quintinomials $Q_m$ of the form

$$Q_m(x) = x^{M_1+2m} + a_m x^{M_2+m} + b_m x^{M_3+m} + c_m x^{M_4+m} + d_m$$

with $m \in \mathbb{N}$ and $a_m, b_m, c_m, d_m \in \mathbb{Q}$.

Let $A$ be an infinite set of such quintinomials. We may suppose that $n = \deg(P) \geq 5$, otherwise (i) holds by Lemma 1. Let $\alpha_1, \ldots, \alpha_n$ be the zeros of $P$. If any two of these zeros are equivalent, then by Lemmas 4 and 1 we are done. So we may assume that $\alpha_i/\alpha_j$ is not a root of unity whenever $i \neq j$. Observe that the equation

$$\begin{vmatrix} \alpha_{i_1}^{M_1+2m} & \alpha_{i_2}^{M_1+2m} & \alpha_{i_3}^{M_1+2m} & \alpha_{i_4}^{M_1+2m} & \alpha_{i_5}^{M_1+2m} \\ \alpha_{i_1}^{M_2+m} & \alpha_{i_2}^{M_2+m} & \alpha_{i_3}^{M_2+m} & \alpha_{i_4}^{M_2+m} & \alpha_{i_5}^{M_2+m} \\ \alpha_{i_1}^{M_3+m} & \alpha_{i_2}^{M_3+m} & \alpha_{i_3}^{M_3+m} & \alpha_{i_4}^{M_3+m} & \alpha_{i_5}^{M_3+m} \\ \alpha_{i_1}^{M_4+m} & \alpha_{i_2}^{M_4+m} & \alpha_{i_3}^{M_4+m} & \alpha_{i_4}^{M_4+m} & \alpha_{i_5}^{M_4+m} \\ 1 & 1 & 1 & 1 & 1 \end{vmatrix} = 0 \tag{2}$$

has infinitely many solutions in $m$ for any $i_1, \ldots, i_5$ with $1 \leq i_1 < \ldots < i_5 \leq n$. Thus by Lemma 2 the determinant in (2) must have a vanishing subdeterminant for infinitely many $m$. If there is a vanishing subdeterminant of type $2 \times 2$, then the corresponding zeros are equivalent, which is a contradiction. Thus we may assume that

$$D_{u_1u_2u_3} := \begin{vmatrix} \alpha_{u_1}^{M_2} & \alpha_{u_2}^{M_2} & \alpha_{u_3}^{M_2} \\ \alpha_{u_1}^{M_3} & \alpha_{u_2}^{M_3} & \alpha_{u_3}^{M_3} \\ \alpha_{u_1}^{M_4} & \alpha_{u_2}^{M_4} & \alpha_{u_3}^{M_4} \end{vmatrix} = 0$$

for some $u_1, u_2, u_3$ with $1 \le u_1 < u_2 < u_3 \le n$, otherwise by Lemma 2 we get a contradiction. Note that $D_{u_1u_2u_3}$ does not have a $2 \times 2$ vanishing subdeterminant, otherwise we obtain two equivalent zeros, which is a contradiction again.

Suppose first that $D_{u_1u_2u_3} = 0$ for each choice of $u_1, u_2, u_3$. Then there are $r_3, r_4 \in \mathbb{K}$ such that $P$ divides $x^{M_2} + r_3x^{M_3} + r_4x^{M_4}$ over $\mathbb{K}$. Therefore, for an appropriate choice of $m$, $P$ divides both $Q_m$ and the polynomial

$$x^{M_1+2m} + (a_m+1)x^{M_2+m} + (b_m+s_3)x^{M_3+m} + (c_m+s_4)x^{M_4+m} + d_m$$

over $\mathbb{Q}$, where $s_i = \text{trace}(r_i)$ $(i = 3, 4)$. Thus we have (ii), and the theorem follows in this case.

So we may assume that $D_{123} = 0$ and $D_{124} \neq 0$. Then, by the double transitivity of $\mathcal{G}$, there is an automorphism $\sigma$ of $\mathbb{K}$ such that $\sigma(\alpha_1) = \alpha_1$ and $\sigma(\alpha_2) = \alpha_4$. Observe that by $D_{124} \neq 0$, $\sigma(\alpha_3) \neq \alpha_2$. Moreover, $\sigma(\alpha_3) = \alpha_3$ is impossible, since $D_{123} = 0$ and $D_{134} = 0$ yield $D_{124} = 0$. Hence without loss of generality we may assume that $\sigma(\alpha_3) = \alpha_5$, whence $D_{123} = D_{145} = 0$ and $D_{124} \neq 0$. It is easy to check that $D_{j_1j_2j_3} = 0$ with $1 \le j_1 < j_2 < j_3 \le 5$ only if $(j_1, j_2, j_3) = (1, 2, 3)$ or $(1, 4, 5)$.

Consider now (2), with $(i_1, i_2, i_3, i_4, i_5) = (1, 2, 3, 4, 5)$. Expanding the determinant by its middle three rows, after dividing by $(\alpha_1\alpha_2\alpha_3\alpha_4\alpha_5)^m$, we obtain

$$\sum_{\substack{\{i_1,i_2,i_3,i_4,i_5\}=\{1,2,3,4,5\} \\ i_3<i_4<i_5}} (-1)^{i_1+i_2+1} \cdot \text{sgn}(i_2 - i_1) \cdot D_{i_3i_4i_5} \cdot \alpha_{i_1}^{M_1}(\alpha_{i_1}/\alpha_{i_2})^m = 0. \qquad (3)$$

Observe that, by $D_{123} = 0$ and $D_{145} = 0$, (3) is an exponential equation in $\mathbb{K}$ with exactly 16 nonzero terms. Choose a system $\mathcal{P}$ of subsums of the left hand side of (3) such that each subsum in $\mathcal{P}$ vanishes simultaneously for the exponent quintuples corresponding to polynomials in an infinite subset $A_1$ of $A$, but all the proper subsums of each of these subsums do not vanish. Without loss of generality we may assume that $A = A_1$. Applying Lemma 3 to the partition $\mathcal{P}$, we obtain $G(\mathcal{P}) \neq \{0\}$. Note that each class of $\mathcal{P}$ contains at least two elements. There exists a $z \in \mathbb{Z}$ with $z \neq 0$ such that for all $(i_1, i_2), (j_1, j_2)$ we have that if $(\alpha_{i_1}/\alpha_{i_2})^m$ and $(\alpha_{j_1}/\alpha_{j_2})^m$ occur in the same class of $\mathcal{P}$, then

$$(\alpha_{i_1}/\alpha_{i_2})^z = (\alpha_{j_1}/\alpha_{j_2})^z$$

holds. Thus we obtain many multiplicative relations among the $\alpha_i$'s. If $(\alpha_1/\alpha_2)^z = (\alpha_2/\alpha_1)^z$, $(\alpha_1/\alpha_i)^z$ or $(\alpha_i/\alpha_2)^z$ for some $i$ with $3 \le i \le 5$, then we obtain that two zeros of $P$ are equivalent, which is a contradiction. If $(\alpha_1/\alpha_2)^z$ equals $(\alpha_i/\alpha_1)^z$ or $(\alpha_2/\alpha_i)^z$ for some $i$ with $3 \le i \le 5$, then we get

$$\alpha_{j_1}^z \alpha_{j_2}^z \alpha_{j_3}^{-2z} = 1 \qquad (4)$$

for some distinct $j_1, j_2, j_3$ with $1 \le j_1, j_2, j_3 \le 5$. Suppose that $(\alpha_1/\alpha_2)^z = (\alpha_3/\alpha_4)^z$. Checking the possible elements of the class of $(\alpha_1/\alpha_5)^z$, we find that in each case two zeros of $P$ are equivalent or some relation (4) holds. Thus, since the former case is excluded, it remains to prove that (4) is impossible. Assume that (4) holds for some distinct $j_1, j_2, j_3$ with $1 \le j_1, j_2, j_3 \le 5$. Write $\beta_i = \alpha_{j_i}$ for $i = 1, 2, 3$. By the double transitivity of $\mathcal{G}$, there exists an automorphism $\sigma_1$ of $\mathbb{K}$, such that $\sigma_1(\beta_1) = \beta_2$, $\sigma_1(\beta_2) = \beta_3$. Write $\beta_4 = \sigma_1(\beta_3)$. We observe that if $\beta_1 = \beta_4$, then from (4) we get $\beta_1^{3z} = \beta_3^{3z}$, which is a contradiction. So assume that $\beta_1 \ne \beta_4$, and choose inductively automorphisms $\sigma_i$ of $\mathbb{K}$ such that $\sigma_i(\beta_i) = \beta_{i+1}, \sigma_i(\beta_{i+1}) = \beta_{i+2}$, and write $\beta_{i+3} = \sigma_i(\beta_{i+2})$. As $P$ has $n$ zeros, after $j$ steps with $j \le n-3$, we get that $\beta_{j+3} = \beta_l$ with some $l \le j$. Without loss of generality we may assume that $j$ is minimal with this property and that $l = 1$. Define the numbers $\lambda_i$ for $i = 1, \dots, j+1$ in the following way. Put $\lambda_1 = 1, \lambda_2 = -1$, and let $\lambda_{i+2} = 2\lambda_i - \lambda_{i+1}$ $(i = 1, \dots, j-1)$. A simple calculation yields $\lambda_i = (1-(-2)^i)/3$ $(i = 1, \dots, j+1)$. Observe that by (4) and the definition of the $\beta_i$ and $\lambda_i$ we have

$$(\beta_{j+1}^z \beta_{j+2}^z \beta_1^{-2z})^{\lambda_{j+1}} \prod_{i=1}^{j} (\beta_i^z \beta_{i+1}^z \beta_{i+2}^{-2z})^{\lambda_i} = \beta_1^{z(\lambda_1 - 2\lambda_{j+1})} \beta_{j+2}^{z(-2\lambda_j + \lambda_{j+1})} = 1.$$

By induction it is easy to see that $-2\lambda_j + \lambda_{j+1} = -\lambda_1 + 2\lambda_{j+1}$. As clearly $\lambda_1 \ne 2\lambda_{j+1}$, we find that $\beta_1$ and $\beta_{j+2}$ are equivalent. However, by the minimality of $j$ we have $\beta_1 \ne \beta_{j+2}$. This is a contradiction, and the theorem follows.

*Proof of Theorem 1.* The sufficiency of (i) and (ii) just follows as in the proof of Theorem 2. To prove necessity, suppose that $P \in \mathbb{Q}[x]$ of degree $n$ divides infinitely many standard $k$-nomials, and that $P$ is irreducible with $[2k/3]$-times transitive Galois group $\mathcal{G}$. If $n < k$, then (i) holds by Lemma 1 and we are done. Moreover, if two zeros of $P$ are equivalent, then the theorem follows from Lemmas 4 and 1. Thus without loss of generality we may assume that $n \ge k$ and that the zeros of $P$ are pairwise nonequivalent. Let $A$ be an infinite set of $k$-nomials divisible by $P$. Observe that $P \in \mathrm{PR}_{k-1}$ implies that $P \in \mathrm{PR}_k$. Moreover, if $P$ divides two standard $(k-1)$-nomials with the same exponent $(k-1)$-tuple, then either these polynomials are also standard $k$-nomials or $P$ divides a polynomial of degree less than $k$. Hence, as the statement is true for $k = 4$ (cf. Theorem 1 of [4]), by induction we may assume that $A$ does not contain any $(k-1)$-nomial. Let $\alpha_1, \dots, \alpha_n$ be the zeros of $P$. If $P$ divides a standard $k$-nomial $x^{m_1} + a_2 x^{m_2} + \cdots + a_k$ over $\mathbb{Q}$, then for any $i_1, \dots, i_k$ with $1 \le i_1 < \dots < i_k \le n$ we have

$$\begin{vmatrix} \alpha_{i_1}^{m_1} & \dots & \alpha_{i_k}^{m_1} \\ \vdots & \vdots & \vdots \\ \alpha_{i_1}^{m_k} & \dots & \alpha_{i_k}^{m_k} \end{vmatrix} = 0 \tag{5}$$

with $m_k = 0$. We may assume that the set of such $k$-tuples $(m_1, \dots, m_k)$ is infinite, otherwise (ii) holds. Thus, by Lemma 2 we get that for any $i_1, \dots, i_k$ the determinant in (5) must have a proper subdeterminant which vanishes for infinitely many $k$-tuples

$(m_1, \ldots, m_k)$. Choose such a subdeterminant of size $t \times t$ with some $1 \leq u_1 < \ldots < u_t \leq n$ and $0 \leq m_{j_t} < \ldots < m_{j_1} \leq m_1$ such that

$$\begin{vmatrix} \alpha_{u_1}^{m_{j_1}} & \ldots & \alpha_{u_t}^{m_{j_1}} \\ \vdots & \vdots & \vdots \\ \alpha_{u_1}^{m_{j_t}} & \ldots & \alpha_{u_t}^{m_{j_t}} \end{vmatrix} = 0 \tag{6}$$

for infinitely many $(m_1, \ldots, m_k)$, and $t$ is minimal with this property. Observe that $3 \leq t \leq k-1$, since in case of $t = 2$, $P$ has two equivalent zeros, which is a contradiction.

Suppose first that $t \leq 2k/3$. Take any standard $k$-nomial $Q_1$ from $A$ with exponent $k$-tuple $(m_1, \ldots, m_k)$ for which (6) is valid. Observe that as $\mathcal{G}$ is $[2k/3]$-times transitive (6) holds for any system of $t$ zeros of $P$. Hence there are numbers $r_{j_1}, \ldots, r_{j_t}$ from $\mathbb{K}$, one of them being 1, such that $P$ divides $r_{j_1}x^{m_{j_1}} + \cdots + r_{j_t}x^{m_{j_t}}$ over $\mathbb{K}$. Therefore, $P$ divides the nonzero polynomial $Q_2(x) = s_{j_1}x^{m_{j_1}} + \cdots + s_{j_t}x^{m_{j_t}}$ over $\mathbb{Q}$, where $s_{j_l} = \text{trace}(r_{j_l})$ $(l = 1, \ldots, t)$. Then $P$ divides the standard $k$-nomial $Q_1 + Q_2$ (or rather $(1/2)Q_1 + (1/2s_{j_1})Q_2$ if $\deg(Q_1) = m_{j_1}$ and $s_{j_1} \neq 0$) over $\mathbb{Q}$. This implies (ii), and the theorem follows in this case.

Assume now that $t > 2k/3$. Without loss of generality we may assume that there is no $k$-nomial in $A$ for which there is a vanishing subdeterminant in (5) with some $i_1, \ldots, i_k$ of size smaller than $t \times t$, and by the minimality of $t$ that the set $A$ is infinite. As in (6) there are no vanishing subdeterminants, we obtain from Lemma 2 that there exist integers $M_{j_1} > \ldots > M_{j_t} \geq 0$ such that for infinitely many $k$-nomials from $A$ we have $m_{j_1} - m_{j_l} = M_{j_1} - M_{j_l}$ $(l = 2, \ldots, t)$. Again, we may assume that all the $k$-nomials in $A$ have this property.

Now by a simple process we are going to separate the exponents (more precisely, the indices of the exponents) of the polynomials in $A$ into certain sets. Let $I_1 \supset \{j_1, \ldots, j_t\}$ be a maximal subset of $\{1, \ldots, k\}$ such that there exists an infinite subset $A_1$ of $A$ with the following property: for each $i \in I_1$ there exists an integer $C_i$ such that for each polynomial $Q \in A_1$ the exponent tuple satisfies

$$m_{j_1} - m_i = C_i \quad (i \in I_1).$$

Suppose that $I_\gamma$ and $A_\gamma$ with some integer $\gamma \geq 1$ have already been defined. If $\{1, \ldots, k\} \backslash (I_1 \cup \ldots \cup I_\gamma)$ is nonempty, let $I_{\gamma+1}$ be a maximal subset of $\{1, \ldots, k\} \backslash (I_1 \cup \ldots \cup I_\gamma)$ such that there exists an infinite subset $A_{\gamma+1}$ of $A_\gamma$ with the following property: for each pair $(j, k)$ with $j, k \in A_{\gamma+1}$ there is an integer $C_{jk}$ such that each polynomial $Q \in A_{\gamma+1}$ has an exponent tuple satisfying

$$m_j - m_k = C_{jk}$$

(where $C_{jk}$ is independent of $Q$). We continue this process as far as we can. By this method in finitely many, say, $\Gamma$, steps we get an infinite set $A_\Gamma$ and a partition of $\{1, \ldots, k\}$ into disjoint subsets $I_\gamma$ $(\gamma = 1, \ldots, \Gamma)$. Note that the sets $I_\gamma$ $(\gamma = 1, \ldots, \Gamma)$ are connected in the sense that if $s_1, s_2 \in I_\gamma$ and $s$ is an integer with $s_1 < s < s_2$, then $s$ also belongs to $I_\gamma$. Moreover, without loss of generality we may assume that $A = A_\Gamma$. Then for any $Q, Q' \in A$ with exponent $k$-tuples $(m_1, \ldots, m_k)$ and $(m'_1, \ldots, m'_k)$, respectively, we have $m_{s_1} - m_{s_2} = m'_{s_1} - m'_{s_2}$ if and only if $s_1$ and $s_2$ belong to the same $I_\gamma$ for some $\gamma \in \{1, \ldots, \Gamma\}$. Hence there exist integers

$M_i$ $(i = 1, \dots, k)$ such that if $(m_1, \dots, m_k)$ is the exponent $k$-tuple of a standard $k$-nomial from $A$, then $i \in I_\gamma$ implies $m_i = M_i + m^{(\gamma)}$ with some positive integers $m^{(\gamma)}$ $(\gamma = 1, \dots, \Gamma)$ where $m^{(\gamma)}$ depends only on $\gamma$ and not further on $i$. For each $\gamma \in \{1, \dots, \Gamma\}$ put $l_\gamma = |I_\gamma|$, and for any $u_1, \dots, u_{l_\gamma}$ with $1 \le u_1 < \dots < u_{l_\gamma} \le n$ write

$$D^{(\gamma)}_{u_1 \dots u_{l_\gamma}} = |\alpha_{u_r}^{M_i}|_{\substack{i \in I_\gamma \\ r = 1, \dots, l_\gamma}} .$$

Note that $l_1 \ge t > 2k/3$, and consequently $l_\gamma < k/3$ for each $\gamma \in \{2, \dots, \Gamma\}$. Thus by the minimality of $t$ and our assumptions on $A$, we obtain that $D^{(\gamma)}_{u_1 \dots u_{l_\gamma}} \neq 0$ for all $\gamma \ge 2$ and $u_1, \dots, u_{l_\gamma}$ with $1 \le u_1 < \dots < u_{l_\gamma} \le n$. Further, if $D^{(1)}_{u_1 \dots u_{l_1}} = 0$ for all $u_1, \dots, u_{l_1}$ with $1 \le u_1 < \dots < u_{l_1} \le n$, then by a similar argument as in case of $t \le 2k/3$, we obtain (ii), and we are done. So, without loss of generality we may assume that $i_1 = 1, \dots, i_k = k$ in (5), and that $D^{(1)}_{q_1 \dots q_{l_1}} \neq 0$ for some $q_1, \dots, q_{l_1}$ with $1 \le q_1 < \dots < q_{l_1} \le k$. Expanding the determinant in equation (5) by the rows corresponding to the elements of $I_1$, and then dividing by $(\alpha_1 \dots \alpha_k)^{m^{(1)}}$, we obtain an exponential equation in $\mathbb{K}$ of the form

$$\sum (-1)^\varepsilon \left( \prod_{\gamma=1}^{\Gamma} D^{(\gamma)}_{v_{\gamma 1} \dots v_{\gamma l_\gamma}} \right) \prod_{\gamma=2}^{\Gamma} \left( \alpha_{v_{\gamma 1}} \cdots \alpha_{v_{\gamma l_\gamma}} \right)^{m^{(\gamma)} - m^{(1)}} = 0. \tag{7}$$

Here the summation is taken over all partitions $H_\gamma = \{v_{\gamma 1}, \dots, v_{\gamma l_\gamma}\}$ of $\{1, \dots, k\}$ such that $\bigcup_{\gamma=1}^{\Gamma} H_\gamma = \{1, \dots, k\}$, and $v_{\gamma 1} < \dots < v_{\gamma l_\gamma}$ for each $\gamma$. The exponent $\varepsilon$ of $(-1)$ depends only on the choice of the partition $H_\gamma$ $(\gamma = 1, \dots, \Gamma)$. Further, in view of the previous considerations, the coefficients $\prod_{\gamma=1}^{\Gamma} D^{(\gamma)}_{v_{\gamma 1} \dots v_{\gamma l_\gamma}}$ are not all zero. Recall that if $m'^{(\gamma)}$ and $m''^{(\gamma)}$ $(\gamma = 1, \dots, \Gamma)$ correspond to the exponent $k$-tuples of the standard $k$-nomials $Q'$ and $Q''$ in $A$, respectively, then by the definition of $I_\gamma$ we have

$$m'^{(\gamma)} - m'^{(1)} \neq m''^{(\gamma)} - m''^{(1)} \quad (\gamma = 2, \dots, \Gamma).$$

Hence equation (7) is satisfied by infinitely many distinct exponent tuples $(m^{(2)} - m^{(1)}, \dots, m^{(\Gamma)} - m^{(1)})$. Thus, by Lemma 3 there exist integers $z_2, \dots, z_\Gamma$ such that

$$\prod_{\gamma=2}^{\Gamma} \left( \alpha_{v'_{\gamma 1}} \cdots \alpha_{v'_{\gamma l_\gamma}} \right)^{z_\gamma} = \prod_{\gamma=2}^{\Gamma} \left( \alpha_{v''_{\gamma 1}} \cdots \alpha_{v''_{\gamma l_\gamma}} \right)^{z_\gamma} \tag{8}$$

for some different partitions $\{H'_\gamma\}_{\gamma=1}^{\Gamma}$ and $\{H''_\gamma\}_{\gamma=1}^{\Gamma}$ of the set $\{1, \dots, k\}$ with $H'_\gamma = \{v'_{\gamma 1}, \dots, v'_{\gamma l_\gamma}\}$ and $H''_\gamma = \{v''_{\gamma 1}, \dots, v''_{\gamma l_\gamma}\}$ $(\gamma = 1, \dots, \Gamma)$, where

$$z_\gamma = (m'^{(\gamma)} - m'^{(1)}) - (m''^{(\gamma)} - m''^{(1)})$$

for certain $m'^{(\gamma)}, m'^{(1)}, m''^{(\gamma)}, m''^{(1)}$ corresponding to two distinct $k$-nomials in $A$. In particular, by the definition of $I_\gamma$ we have $z_{\gamma_1} \neq z_{\gamma_2}$ whenever $\gamma_1 \neq \gamma_2$ $(\gamma_1, \gamma_2 \in \{2, \dots, \Gamma\})$. Equation (8) leads to an equation of the form

$$\alpha_{w_1}^{\lambda_1} \dots \alpha_{w_h}^{\lambda_h} = 1 \tag{9}$$

with $2 \le h \le 2(k - |I_1|)$, $1 \le w_1 < \ldots < w_h \le k$ and nonzero integers $\lambda_1, \ldots, \lambda_h$. As $|I_1| > 2k/3$, we have $2 \le h \le 2k/3$. Since $\mathcal{G}$ is $[2k/3]$-times transitive, there exists an automorphism $\sigma$ of $\mathbb{K}$ such that $\sigma(\alpha_{w_1}) = \alpha_{w_2}, \sigma(\alpha_{w_2}) = \alpha_{w_1}$, and $\sigma(\alpha_{w_p}) = \alpha_{w_p}$ for $p = 3, \ldots, h$. Together with (9) this yields that $\alpha_{w_1}$ and $\alpha_{w_2}$ are equivalent. It contradicts an earlier assumption. □

*Proof of Corollary 1.* Suppose that (ii) holds, and $P$ divides the standard $k$-nomials

$$Q(x) = \sum_{i=1}^{k} a_i x^{m_i} \quad \text{and} \quad Q'(x) = \sum_{i=1}^{k} b_i x^{m_i}$$

where $m_1 > \ldots > m_{k-1} > m_k = 0$, $a_1 = b_1 = 1$ and $a_i \ne b_i$ for some $i$ with $2 \le i \le k-1$. Then $P$ divides the standard $(k-1)$-nomial $(b_i Q - a_i Q')/(b_i - a_i)$.

On the other hand, if $P$ divides a standard $(k-1)$-nomial $Q$, then $P$ divides the standard $k$-nomials $x^l Q$ for any nonnegative integer $l$. Hence the statement follows. □

*Proof of Corollary 2.* As a binomial can be considered as a linear polynomial in some $x^r$, (i′) implies $P \in \mathrm{PR}_2$, whence (i) follows. On the other hand, if (i) holds, then as $\deg(P) \ge k$, by Lemmas 1 and 4 we get that any two zeros of $P$ are equivalent, which yields (i′). □

*Proof of the proposition.* Fix any $k$ with $k \ge 5$. Then by Lemma 3 of [3] there exists a polynomial $P_1 \in \mathbb{Q}[x]$ of degree $k-1$ such that $P_1$ does not divide any standard $(k-1)$-nomial over $\mathbb{Q}$. Then by definition, (i) is valid for $P_1$, and $P_1$ divides infinitely many standard $k$-nomials. Moreover, (ii) cannot hold for $P_1$, as in that case $P_1$ would divide a standard $(k-1)$-nomial over $\mathbb{Q}$.

On the other hand, the proposition in [4] in case of $k = 5$ and the Theorem together with Lemma 1 and its proof in [3] when $k \ge 6$ guarantees the existence of a polynomial $P_2 \in \mathbb{Q}[x]$ such that (ii) is valid for $P_2$ but (i) is not. □

**Acknowledgements.** We are grateful to J.-H. Evertse and K. Győry for their useful remarks and to the referee for his valuable and helpful suggestions. The research was supported in part by the Netherlands Organization for Scientific Research (NWO). The first author was further supported by the János Bolyai Research Fellowship of the Hungarian Academy of Sciences, by the grants F034981, T042985 and T048791 of the Hungarian National Foundation for Scientific Research and by the FKFP grant 3272-13/066/2001.

## References

1. Evertse, J.-H., Schlickewei, H.P., Schmidt, W.M.: Linear equations in variables which lie in a multiplicative group. *Ann. Math.* **155**, 1–30 (2002)
2. Győry, K., Schinzel, A.: On a conjecture of Posner and Rumsey. *J. Number Theory* **47**, 63–78 (1994)
3. Hajdu, L.: On a problem of Győry and Schinzel concerning polynomials. *Acta Arith.* **78**, 287–295 (1997)
4. Hajdu, L., Tijdeman, R.: Polynomials dividing infinitely many quadrinomials or quintinomials. *Acta Arith.* **107**, 381–404 (2003)
5. Posner, E.C., Rumsey, H. Jr.: Polynomials that divide infinitely many trinomials. *Mich. Math. J.* **12**, 339–348 (1965)
6. Schlickewei, H.P., Viola, C.: Polynomials that divide many trinomials. *Acta Arith.* **78**, 267–273 (1997)
7. Schlickewei, H.P., Viola, C.: Polynomials that divide many $k$-nomials. In: Győry, K., Iwaniec, H., Urbanowicz, J. (eds.) *Number Theory in Progress*, vol. 1, pp. 445–450. de Gruyter, Berlin (1999)
8. Schlickewei, H.P., Viola, C.: Generalized Vandermonde determinants. *Acta Arith.* **95**, 123–137 (2000)

*Diophantine Approximation*
H. P. Schlickewei et al., Editors


# APPROXIMANTS DE PADÉ DES $q$-POLYLOGARITHMES

Christian Krattenthaler[1,*] et Tanguy Rivoal[2,**]
[1] *Institut Girard Desargues, Université Claude Bernard Lyon-I, 21 Avenue Claude Bernard, 69622 Villeurbanne Cedex, France*
[2] *Laboratoire de Mathématiques Nicolas Oresme, CNRS UMR 6139, Université de Caen, BP 5186, 14032 Caen cedex, France*

Dédié à Wolfgang Schmidt, pour son soixante-dixième anniversaire

## 1 Introduction

Considérons la série

$$\zeta_q(s) = \sum_{k=1}^{\infty} k^{s-1} \frac{q^k}{1-q^k},$$

qui converge pour tout complexe $|q| < 1$ et tout entier $s \geq 1$. La notation $\zeta_q$ est justifiée par le fait que cette fonction est un $q$-analogue de la fonction zêta de Riemann $\zeta(s)$ au sens suivant (voir [5, paragraphe 4.1], [3, Theorem 2] ou [8]),

$$\lim_{q\to 1} (1-q)^s \zeta_q(s) = (s-1)! \sum_{k=1}^{\infty} \frac{1}{k^s} = (s-1)!\, \zeta(s).$$

Dans [5], les deux auteurs et W. Zudilin ont montré que la dimension de l'espace vectoriel engendré sur $\mathbb{Q}$ par 1, $\zeta_q(3)$, $\zeta_q(5)$,..., $\zeta_q(A)$ ($A \geq 3$ impair) est minorée par $\frac{\pi+o(1)}{2\sqrt{\pi^2+12}}\sqrt{A}$ lorsque $1/q \in \mathbb{Z} \setminus \{\pm 1\}$. La démonstration utilise les fonctions $q$-polylogarithmes, définies pour tout entier $s \geq 1$ par

$$\mathrm{Li}_s(z; q) = \sum_{k=1}^{\infty} \frac{q^k}{(1-q^k)^s} z^k, \tag{1.1}$$

*Mots clefs.* Approximants de Padé, $q$-analogue du logarithme, $q$-analogues des polylogarithmes, confluence.

*2000 Classification mathématiques par sujets.* Primaire 41A21; Secondaire 33D15.

* Adresse actuelle : Fakultät für Mathematik, Universität Wien, Nordbergstrasse 15, 1090 Vienna, Austria. christian.krattenthaler@univie.ac.at

** Adresse actuelle : Institut Fourier, CNRS UMR 5582, Université Grenoble 1, 100 rue des Maths, BP 74, 38402 Saint-Martin d'Hères cedex, France. tanguy.rivoal@ujf-grenoble.fr

où $z$ et $q$ désignent des nombres complexes tels que $|q| < 1$ et $|zq| < 1$. Ces fonctions constituent des $q$-analogues des polylogarithmes usuels $\mathrm{Li}_j(z)$ au sens suivant :

$$\lim_{q\to 1}(1-q)^s\,\mathrm{Li}_s(z;q)=\sum_{k=1}^{\infty}\frac{z^k}{k^s}=\mathrm{Li}_s(z).$$

Notons que les polylogarithmes sont aussi utilisés au cours de la démonstration du théorème suivant (dont celui rappelé ci-dessus est un $q$-analogue) : la dimension de l'espace vectoriel engendré sur $\mathbb{Q}$ par $1$, $\zeta(3)$, $\zeta(5)$,..., $\zeta(A)$ ($A \geq 3$ impair) est minorée par $\frac{1+o(1)}{1+\log(2)}\log(A)$ (voir [2,6]). Dans les deux cas, la démonstration est en fait basée sur une étude très fine d'une série ($q$-)hypergéométrique bien choisie que l'on commence par exprimer comme une combinaison linéaire polynomiale en les ($q$-)polylogarithmes. Plus précisément, soient $A, n, r$ des entiers positifs tels que $0 \leq r \leq A/2$. Définissons les factorielles décalées (ou symboles de Pochhammer) $(\alpha)_m = \alpha(\alpha+1)\cdots(\alpha+m-1)$ et les factorielles $q$-décalées $(\alpha;q)_m = (1-\alpha)(1-\alpha q)\cdots(1-\alpha q^{m-1})$, avec la convention usuelle que les produits vides pour $m = 0$ valent 1. On pose alors

$$S_n(z;q)=(q;q)_n^{A-2r}\sum_{k=1}^{\infty}q^k\frac{(q^{k-rn};q)_{rn}\,(q^{k+n+1};q)_{rn}}{(q^k;q)_{n+1}^A}\,q^{(k-1/2)(A-2r)n/2}z^{-k},$$

avec $|q| < 1 \leq |z|$ et $A$ pair, ainsi que

$$S_n(z)=n!^{A-2r}\sum_{k=1}^{\infty}\frac{(k-rn)_{rn}\,(k+n+1)_{rn}}{(k)_{n+1}^A}\,z^{-k},$$

avec $|z| \geq 1$. Il est alors facile de montrer l'existence de deux familles de polynômes $P_{j,n}(z;q) \in \mathbb{C}(q)[z]$ et $P_{j,n}(z) \in \mathbb{C}[z]$, de degré au plus $n$, tels que

$$S_n(z;q)=P_{0,n}(z;q)+\sum_{j=1}^{A}P_{j,n}(z;q)\,\mathrm{Li}_j(1/z;q) \tag{1.2}$$

et

$$S_n(z)=P_{0,n}(z)+\sum_{j=1}^{A}P_{j,n}(z)\,\mathrm{Li}_j(1/z). \tag{1.3}$$

Par ailleurs, les numérateurs des sommandes dans les définitions de $S_n(z;q)$ et $S_n(z)$ s'annulent pour les indices $k \in \{1,\ldots,rn\}$, ce qui assure que l'ordre en $z = 0$ des deux séries est exactement $rn+1$ : les équations (1.2) et (1.3) peuvent donc être vues comme des problèmes d'approximations de type Padé pour les séries entières 1 et $\mathrm{Li}_j(z;q)$, respectivement 1 et $\mathrm{Li}_j(z)$. L'information n'est cependant pas suffisante pour affirmer qu'il n'existe, à constante multiplicative près, qu'une seule fonction $S_n(z;q)$, resp. $S_n(z)$, vérifiant (1.2), resp. (1.3), et qui s'annule à l'ordre $rn+1$.

Dans [4], sont énoncées des conditions supplémentaires, de type Padé, portant sur des objets liés à la série $S_n(z)$ et qui suffisent à assurer que $S_n(z)$ est bien la seule solution de (1.3) pour des polynômes $P_{j,n}(z)$ de degré au plus $n$. Voici l'énoncé précis.

*Étant donnés des entiers $A \geq 1$, $n \geq 0$, $\rho \geq 0$, $\sigma \geq 0$ tels que $\rho+\sigma+2 \leq A(n+1)$, on cherche à résoudre le problème d'approximations simultanées de Padé suivant :*

*déterminer des polynômes (dépendants de A, n, ρ, σ)* $P_0(z)$, $\overline{P}_0(z)$ *et* $P_j(z)$ *(pour* $j = 1, \ldots, A$*), de degré au plus n et à coefficients dans* $\mathbb{Q}$*, tels que*

$$\begin{cases} S(z) = P_0(z) + \sum_{j=1}^{A} P_j(z)\,\mathrm{Li}_j(1/z) = \mathcal{O}(z^{-\rho-1}) \quad \text{quand} \quad z \to \infty\,; \\ \overline{S}(z) = \overline{P}_0(z;q) + \sum_{j=1}^{A} P_j(z)\,\mathrm{Li}_j(z) = \mathcal{O}(z^{\sigma+n+1}) \quad \text{quand} \quad z \to 0\,; \\ I(z) = \sum_{j=1}^{A} P_j(z)\dfrac{\log^{j-1}(1/z)}{(j-1)!} = \mathcal{O}((z-1)^{A(n+1)-\rho-\sigma-2}) \quad \text{quand} \quad z \to 1. \end{cases} \tag{1.4}$$

(Ici et dans toute la suite, la fonction logarithme est définie avec sa branche principale : $\log(z) = \log|z| + i\arg(z)$, avec $-\pi < \arg(z) \leq \pi$. On notera $\mathbb{R}_-$ l'ensemble des réels négatifs.) On a alors le résultat suivant, qui résout conplètement ce problème.

**Théorème 1.** *Dans les conditions ci-dessus, le problème* (1.4) *a une solution unique, à une constante multiplicative près. En choisissant cette constante égale à* 1*, on a*

$$S(z) = \sum_{k=1}^{\infty} \frac{(k-\rho)_\rho\,(k+n+1)_\sigma}{(k)_{n+1}^A}\, z^{-k} \quad \text{et}$$

$$I(z) = \frac{1}{2i\pi} \int_C \frac{(s-\rho)_\rho\,(s+n+1)_\sigma}{(s)_{n+1}^A}\, z^{-s}\,\mathrm{d}s,$$

*où* $C$ *est n'importe quelle courbe fermée orientée dans le sens direct qui entoure les pôles de l'intégrande, i.e.,* $0, -1, \ldots, -n$.

Le but de cette note est de prouver un $q$-analogue du théorème précédent. On suppose dorénavant que $q \notin \mathbb{R}_-$ ce qui permet de définir $\log_q(z) = \log(z)/\log(q)$.

*Étant donnés des entiers* $A \geq 1, n \geq 0, \rho \geq 0, \sigma \geq 0$ *et* $\nu \geq 0$ *tels que* $\rho+\sigma+\nu+2 \leq A(n+1)$*, on cherche à résoudre le problème d'approximations simultanées de Padé suivant : déterminer des polynômes (dépendants de A, n, ρ, σ et ν)* $P_0(z;q)$, $\overline{P}_0(z;q)$ *et* $P_j(z;q)$ *(pour* $j = 1, \ldots, A$*) en la variable z, de degré au plus n et à coefficients dans* $\mathbb{Q}(q)$*, tels que*

$$\begin{cases} S(z;q) = P_0(z;q) + \sum_{j=1}^{A} P_j(z;q)\,\mathrm{Li}_j(1/z;q) = \mathcal{O}(z^{-\rho-1}) \quad \text{quand} \quad z \to \infty\,; \\ \overline{S}(z;q) = \overline{P}_0(z;q) + \sum_{j=1}^{A} P_j(z;q)\,\mathrm{Li}_j(z;1/q) = \mathcal{O}(z^{\sigma+n+1}) \quad \text{quand} \quad z \to 0\,; \\ I(z;q) = -\sum_{j=1}^{A} P_j(zq^{1-j};q)\dfrac{\left(-\log_q(1/z)\right)_{j-1}}{(j-1)!} = \mathcal{O}(z-q^{-\ell}) \quad \text{quand} \quad z \to q^{-\ell} \\ \quad \text{pour tout } \ell \in \{-\nu, -\nu+1, \ldots, A(n+1)-\rho-\sigma-\nu-2\}. \end{cases} \tag{1.5}$$

**Théorème 2.** *Dans les conditions ci-dessus, le problème* (1.5) *a une solution unique, à une constante multiplicative près. En choisissant cette constante égale à* 1, *on a*

$$S(z;q) = \sum_{k=1}^{\infty} q^k \frac{(q^{k-\rho};q)_\rho \, (q^{k+n+1};q)_\sigma}{(q^k;q)_{n+1}^A} q^{\nu k} z^{-k}$$

*et*

$$I(z;q) = \frac{1}{2i\pi} \int_C \frac{(sq^{-\rho};q)_\rho \, (sq^{n+1};q)_\sigma}{(s;q)_{n+1}^A} s^{\nu - \log_q(z)} \, \mathrm{d}s,$$

*où $C$ est n'importe quelle courbe fermée orientée dans le sens direct qui entoure les pôles de l'intégrande,* i.e., $1, q^{-1}, \ldots, q^{-n}$, *sans traverser la coupure $\mathbb{R}_-$.*

Avant de passer à la démonstration du Théorème 2, faisons quelques remarques : Les Théorèmes 1 et 2 sont formellement très similaires, à ceci près que le paramètre « $q$-analogique » $\nu$ n'a pas d'équivalent dans le cas classique. La différence majeure se situe dans l'énoncé des conditions d'annulation des fonctions $I(z)$ et $I(z;q)$ : cela n'a cependant rien surprenant, puisqu'il est fréquent dans ce genre de situation que des singularités en certaines puissances de $q$ confluent vers une unique singularité en 1 (avec une certaine multiplicité) lorsque $q \to 1$. Nous explicitons plus en détail la « convergence » du Théorème 2 vers le Théorème 1 au paragraphe 3.

Il est à noter l'utilisation, naturelle dans notre contexte, de $\log(z)/\log(q)$ comme $q$-analogue de la fonction logarithme. Cet analogue, qui possède donc une monodromie non-triviale en 0, est un choix historiquement classique : voir [1]. Certaines théories géométriques récentes (étudiant l'analogie entre équations aux $q$-différences et équations différentielles) ont mis en avant un $q$-analogue différent du logarithme : J. Sauloy [7] utilise comme $q$-logarithme la fonction $\ell_q(z) = z\theta_q'(z)/\theta_q(z)$, avec $\theta_q(z) = \sum_{n\in\mathbb{Z}}(-1)^n q^{-n(n-1)/2} z^n$, qui est méromorphe sur $\mathbb{C}$ et dont les pôles confluent le long d'une spirale lorsque $q \to 1$, cette spirale agissant alors comme une coupure du plan pour le logarithme usuel.

## 2 Démonstration du Théorème 2

Remarquons tout d'abord que les polynômes $P_0(z;q)$ et $\overline{P}_0(z;q)$ sont déterminés de façon unique, une fois connus les autres polynômes du problème (1.5). De plus, celui-ci se traduit par un système linéaire dont les inconnues sont les $A(n+1)$ coefficients des polynômes $P_j(z;q)$ $(j \geq 1)$ et dont le nombre d'équations est $A(n+1)-1$ : il y a donc au moins une solution non-triviale (i.e., non identiquement nulle).

On utilise temporairement la notation $P_j(z;q)$ pour désigner des polynômes génériques de degré au plus $n$ en $z$ et à coefficients dans $\mathbb{Q}(q)$, sans présager qu'il s'agisse des solutions du problème (1.5). On pose

$$P_j(z;q) = \sum_{t=0}^{n} p_{j,t}(q) z^t, \tag{2.1}$$

où $p_{j,t}(q) \in \mathbb{Q}(q)$, de telle sorte que

$$\sum_{j=1}^{A} P_j(z;q)\,\mathrm{Li}_j(1/z;q) = \sum_{k=1-n}^{\infty} q^k z^{-k} \sum_{j=1}^{A} \sum_{t=\max(0,1-k)}^{n} \frac{q^t p_{j,t}(q)}{(1-q^{k+t})^j}.$$

Il est utile à ce point d'introduire la fraction rationnelle (qui dépend aussi de A) :

$$R(s;q) = \sum_{j=1}^{A}\sum_{t=0}^{n} \frac{q^t p_{j,t}(q)}{(1-sq^t)^j} \tag{2.2}$$

$$= \frac{\Pi(s;q)}{(s;q)_{n+1}^{A}}, \tag{2.3}$$

où $\Pi(s;q)$ est un polynôme en $s$, à coefficients dans $\mathbb{Q}(q)$, et de degré $< A(n+1)$ puisque la décomposition en éléments simples (2.2) de $R(s;q)$ est sans partie principale. Il est clair que la connaissance de $\Pi(s;q)$ détermine *de facto* les polynômes $P_j(z;q)$ $(j \geq 1)$. On en déduit que si l'on connaît $\Pi(s;q)$, alors on a

$$S(z;q) = \sum_{k=1}^{\infty} q^k R(q^k;q) z^{-k}$$

et aussi, par un simple calcul de résidus utilisant l'expression (2.2),

$$I(z;q) = \frac{1}{2i\pi}\int_{\mathcal{C}} R(s;q) s^{-\log_q(z)}\,\mathrm{d}s,$$

où $\mathcal{C}$ est n'importe quelle courbe fermée entourant dans le sens direct les pôles de $R(z;q)$ et qui ne traverse pas la coupure $\mathbb{R}_-$. Nous allons maintenant montrer que la solution du problème de Padé (1.5) est unique, à une constante multiplicative près, et la déterminer en explicitant le polynôme « codant » $\Pi(s;q)$. Pour cela, nous interprétons chacune des conditions de (1.5) par quatre lemmes : il en découlera alors que

$$\Pi(s;q) = (sq^{-\rho};q)_\rho\,(sq^{n+1};q)_\sigma\, s^\nu,$$

à une constante multiplicative près.

**Lemme 1.** *Les polynômes $P_1, \ldots, P_A$ vérifient la première condition de* (1.5) *si, et seulement si,*

$$\prod_{i=1}^{\rho}(1-sq^{-i}) = (sq^{-\rho};q)_\rho \quad \text{divise} \quad \Pi(s;q).$$

*Démonstration.* La première condition de (1.5) se traduit par l'annulation des coefficients de Taylor de $S(z;q)$ d'indices $1, 2, \ldots, \rho$, ce qui équivaut à l'annulation de la fonction $k \mapsto \Pi(q^k;q)$ en $k = 1, 2, \ldots, \rho$. Cela équivaut en fait à l'annulation du polynôme $\Pi(s;q)$ en $s = q, q^2, \ldots, q^\rho$, d'où l'assertion. □

**Lemme 2.** *Les polynômes $P_1, \ldots, P_A$ vérifient la deuxième condition de* (1.5) *si, et seulement si,*

$$\prod_{i=n+1}^{n+\sigma}(1-sq^{i}) = (sq^{n+1};q)_\sigma \quad \text{divise} \quad \Pi(s;q).$$

*Démonstration.* Dans la deuxième condition de (1.5), on change $z$ en $1/z$, puis on multiplie par $z^n$, de telle sorte que $z^n\overline{S}(1/z;q) = \mathcal{O}(z^{-\sigma-1})$. On a

$$z^n \sum_{j=1}^{A} P_j(1/z;q)\,\mathrm{Li}_j(1/z;1/q) = \sum_{k=1-n}^{\infty} q^{-k}z^{-k} \sum_{j=1}^{A} \sum_{t=\max(0,1-k)}^{n} \frac{q^{-t}p_{j,n-t}(q)}{(1-q^{-k-t})^j}$$
$$= q^{-n} \sum_{k=1-n}^{\infty} q^{-k}z^{-k} \sum_{j=1}^{A} \sum_{t=0}^{\min(n,n+k-1)} \frac{q^{t}p_{j,t}(q)}{(1-q^{t-k-n})^j}.$$

Pour $k \geq 1$, le coefficient de $z^{-k}$ dans la série $z^n\overline{S}(1/z;q)$ est donc donné par $q^{-n-k}R(q^{-k-n};q)$. L'annulation des coefficients de Taylor de $z^n\overline{S}(1/z;q)$ d'indices $1, 2, \dots, \sigma$ équivaut donc à l'annulation du polynôme $\Pi(s;q)$ en $s = q^{-n-1}, \dots, q^{-n-\sigma}$, d'où l'assertion. □

**Lemme 3.** *La condition $I(q^{-j};q) = 0$ pour $j \in \{-\nu, \dots, -1\}$ équivaut à*

$$s^\nu \text{ divise } \Pi(s;q).$$

*Démonstration.* Rappelons que

$$I(z;q) = \frac{1}{2i\pi} \int_C \frac{\Pi(s;q)}{(s;q)_{n+1}^A} s^{-\log_q(z)}\,\mathrm{d}s.$$

où $\mathcal{C}$ est n'importe quelle courbe fermée orientée dans le sens direct et entourant les pôles de l'intégrande, i.e., $1, q^{-1}, \dots, q^{-n}$, sans traverser la coupure $\mathbb{R}_-$.

Soit $-n \leq j \leq -1$, et supposons que $I(q^{-j}, q) = 0$. Alors, pour tout contour $\mathcal{C}$ entourant les points $1, q^{-1}, \dots, q^{-n}$ mais ne traversant pas la coupure, donc n'entourant pas 0, on a

$$0 = I(q^{-j};q) = \frac{1}{2i\pi} \int_C \frac{\Pi(s;q)}{s^{-j}(s;q)_{n+1}^A}\,\mathrm{d}s,$$

où l'intégrande est une fraction rationnelle dont 0 est peut-être un pôle : nous allons montrer que ce n'est (éventuellement) le cas que si $j < -\nu$, ce qui prouvera que $s^\nu$ divise $\Pi(s;q)$.

Pour cela, notons que, par le théorème des résidus et parce que le résidu à l'infini de l'intégrande $F(s;q)$ de $I(q^{-j};q)$ est nul (car le degré du numérateur de $F(s;q)$ est $\leq A(n+1)-1$, et celui de son dénominateur $\geq A(n+1)+1$), on a

$$0 = -\mathrm{Res}_\infty(F) = \frac{1}{2i\pi} \int_{C'} F(s;q)\,\mathrm{d}s$$
$$= \frac{1}{2i\pi} \int_C F(s;q)\,\mathrm{d}s + \frac{1}{2i\pi} \int_{C_0} F(s;q)\,\mathrm{d}s = \frac{1}{2i\pi} \int_{C_0} F(s;q)\,\mathrm{d}s,$$

où $\mathcal{C}'$ est un cercle entourant tous les pôles de $F$ ainsi que 0, et $\mathcal{C}_0$ un cercle entourant 0 et aucun autre pôle de $F$, les deux orientés dans le sens direct. Donc

$$0 = \frac{1}{2i\pi} \int_{C_0} F(s;q)\,\mathrm{d}s = \frac{1}{(-j-1)!} \Big( (s;q)_{n+1}^{-A} \Pi(s;q) \Big)^{(-j-1)} \Big|_{s=0}. \tag{2.4}$$

Puisque la fonction $s \mapsto (s;q)_{n+1}^{-A}$ ne s'annule pas en $s = 0$, on déduit de (2.4), par récurrence sur $j \in \{-1, -2, \dots, -\nu\}$, que $0 = \Pi(0;q) = \Pi^{(1)}(0;q) = \cdots = \Pi^{(\nu-1)}(0;q)$, ce qui prouve que $s^\nu$ divise $\Pi(s;q)$. La réciproque se montre facilement en renversant cet argument. □

**Lemme 4.** *La condition* $I(q^{-j};q) = 0$ *pour* $j \in \{0, \dots, A(n+1) - \rho - \sigma - \nu - 2\}$ *équivaut à*

$$\deg(\Pi) \le \rho + \sigma + \nu.$$

*Démonstration.* Développons $\Pi(s;q)/(s;q)_{n+1}^A$ en série entière en $s = \infty$ :

$$\frac{\Pi(s;q)}{(s;q)_{n+1}^A} = \sum_{k=\omega}^{\infty} \frac{c_k}{s^k}$$

où $\omega = A(n+1) - \deg(\Pi)$ est l'ordre de cette fraction rationnelle à l'infini, et les $c_k$ sont des nombres complexes. Notons que cette série converge au moins pour $|s|$ assez grand puisque $\omega \ge 1$, disons pour $|s| \ge S$. On choisit alors un cercle suffisamment grand pour que l'on puisse intégrer cette série terme à terme, disons le cercle $\overline{\mathcal{C}} = \{z : |z| = S+1\}$. On a alors

$$I(q^{-j};q) = \frac{1}{2i\pi} \int_{\overline{\mathcal{C}}} \frac{\Pi(s;q)}{(s;q)_{n+1}^A} s^j \, \mathrm{d}s = \sum_{k=\omega}^{\infty} c_k \int_{\overline{\mathcal{C}}} s^{j-k} \, \mathrm{d}s = c_{j+1}.$$

L'annulation de $I(q^{-j};q)$ pour $j \in \{0, \dots, A(n+1) - \rho - \sigma - \nu - 2\}$ équivaut donc à

$$\omega \ge A(n+1) - \rho - \sigma - \nu,$$

ce qui équivaut plus simplement à $\deg(\Pi) \le \rho + \sigma + \nu$. □

*Démonstration du Théorème 2.* Puisque les polynômes $s^\nu$, $(sq^{-\rho};q)_\rho$ et $(sq^{n+1};q)_\sigma$ n'ont pas de racines communes, les trois premiers lemmes montrent que $s^\nu(sq^{-\rho};q)_\rho (sq^{n+1};q)_\sigma$ divise $\Pi(s;q)$. Or le dernier lemme montre que $\deg(\Pi) \le \rho + \sigma + \nu$, ce qui achève la démonstration. □

## 3 Confluence du Théorème 2 vers le Théorème 1

Dans ce paragraphe, nous explicitons le sens précis en lequel le Théorème 2 « tend » vers le Théorème 1.

Pour commencer, on remarque que, évidemment, on a

$$\lim_{q\to 1} (1-q)^{A(n+1)-\sigma-\rho} S(z;q) = S(z) \quad \text{et} \quad \lim_{q\to 1} (1-q)^{A(n+1)-\sigma-\rho} \overline{S}(z;q) = \overline{S}(z).$$

De plus, en faisant la substitution $s \to q^t$ dans l'intégrale définissant $I(z;q)$, on voit que

$$\lim_{q\to 1} (1-q)^{A(n+1)-\sigma-\rho-1} I(z;q) = -I(q).$$

D'autre part, la définition (2.1) des polynômes $P_j(z;q)$ est donnée par leurs coefficients $p_{j,t}(q)$ qui figurent dans (2.2), avec $\Pi(s;q)=(sq^{-\rho};q)_\rho\,(sq^{n+1};q)_\sigma\,s^\nu$. Explicitement, les $p_{j,t}(q)$ sont donnés par

$$p_{j,t}(q)=\frac{(-1)^{A-j}q^{t(A-j-1)}}{(A-j)!}\frac{\partial^{A-j}}{\partial s^{A-j}}\left(\frac{(1-sq^t)^A\Pi(s;q)}{(s;q)_{n+1}^A}\right)\bigg|_{s=q^{-t}}.$$

Par conséquent, $p_{j,t}(q)$ est une fraction rationnelle en $q$ ; si cette fraction rationnelle est écrite sous forme réduite, la plus grande puissance de $1-q$ qui divise le dénominateur est $(1-q)^{A(n+1)-\sigma-\rho-j}$. En particulier, la limite $\lim_{q\to 1}(1-q)^{A(n+1)-\sigma-\rho-j}P_j(z;q)$ existe : c'est un polynôme en $z$, que l'on notera $Q_j(z)$. De façon similaire, la limite $\lim_{q\to 1}(1-q)^{A(n+1)-\sigma-\rho}\overline{P}_0(z;q)$ existe et sera notée $\overline{Q}_0(z)$.

Si l'on combine ces remarques avec (1.1) et le fait que

$$\lim_{q\to 1}(1-q)\log_q(z)=-\log(z),$$

on en déduit que, en multipliant les deux premières conditions dans (1.5) par $(1-q)^{A(n+1)-\sigma-\rho}$, en multipliant la troisième par $(1-q)^{A(n+1)-\sigma-\rho-1}$, puis en faisant finalement tendre $q$ vers 1, on obtient

$$\begin{cases} S(z)=Q_0(z)+\displaystyle\sum_{j=1}^{A}Q_j(z)\,\mathrm{Li}_j(1/z)=\mathcal{O}(z^{-\rho-1}) \quad \text{quand} \quad z\to\infty\,; \\ \overline{S}(z)=\overline{Q}_0(z)+\displaystyle\sum_{j=1}^{A}Q_j(z)\,\mathrm{Li}_j(z)=\mathcal{O}(z^{\sigma+n+1}) \quad \text{quand} \quad z\to 0\,; \\ I(z)=\displaystyle\sum_{j=1}^{A}Q_j(z)\frac{\log^{j-1}(1/z)}{(j-1)!}=\mathcal{O}(?) \quad \text{quand} \quad z\to 1. \end{cases} \tag{3.1}$$

Il nous reste à montrer que l'on peut mettre $\mathcal{O}((z-1)^{A(n+1)-\rho-\sigma-2})$ à la place de $\mathcal{O}(?)$. Pour le faire, supposons donnée une fonction $f(z;q)$ analytique en $z$ dans un voisinage ouvert suffisamment grand de 1, et telle que

$$f(z;q)=\mathcal{O}(z-q^{-\ell}) \quad \text{quand} \quad z\to q^{-\ell} \tag{3.2}$$

pour tout $\ell\in\{-m,-m+1,\ldots,p\}$. Pour simplifier, on considère le cas où $p=0$, car l'argument qui va suivre se généralise sans difficulté au cas où $p$ est quelconque. On développe tout d'abord $f(z;q)$ en série de Taylor autour de $z=1$ :

$$f(z;q)=\sum_{k=1}^{\infty}f_k(q)(z-1)^k, \tag{3.3}$$

ce développement ne contenant pas de terme constant à cause de la condition (3.2) pour $\ell=0$. Nous supposons alors une condition supplémentaire : la limite $\lim_{q\to 1}f_k(q)$ existe pour tout $k$ et

$$\lim_{q\to 1}\sum_{k=1}^{\infty} f_k(q)(z-1)^k = \sum_{k=1}^{\infty}\lim_{q\to 1} f_k(q)(z-1)^k.$$

Cette condition est satisfaite dans notre cas, c'est-à-dire pour

$$f(z;q) = -(1-q)^{A(n+1)-\sigma-\rho-1}\sum_{j=1}^{A} P_j(zq^{1-j};q)\frac{\left(-\log_q(1/z)\right)_{j-1}}{(j-1)!}. \tag{3.4}$$

La condition (3.2) pour les valeurs non-nulles de $\ell$ implique le système d'équations

$$0 = f(q^{-\ell};q) = \sum_{k=1}^{\infty} f_k(q)(q^{-\ell}-1)^k, \quad \ell \in \{-m, -m+1, \dots, -1\}. \tag{3.5}$$

On multiplie la $\ell$-ième équation par

$$c_\ell = \frac{1}{(q^{-\ell}-1)\prod_{h=1,h\neq -\ell}^{m}(q^{-\ell}-q^h)}$$

et en faisant la somme de ces équations multipliées par le facteur $c_\ell$ correspondant sur $\ell \in \{-m, -m+1, \dots, -1\}$, on obtient

$$\begin{aligned} 0 &= \sum_{\ell=-m}^{-1} c_\ell f(q^{-\ell};q) \\ &= \sum_{\ell=-m}^{-1} c_\ell \sum_{k=1}^{\infty} f_k(q)(q^{-\ell}-1)^k \\ &= \sum_{k=1}^{\infty} f_k(q) \sum_{\ell=-m}^{-1} c_\ell (q^{-\ell}-1)^k. \end{aligned} \tag{3.6}$$

À ce point, on note que le choix des coefficients $c_\ell$ implique que les coefficients de $f_k(q)$ dans la somme (3.6) sont nuls pour $k = 1, 2, \dots, m-1$, et que le coefficient de $f_m(q)$ est exactement 1. Dans le résultat

$$0 = f_m(q) + \sum_{k\geq m+1}^{\infty} f_k(q) \sum_{\ell=-m}^{-1} \frac{(q^{-\ell}-1)^k}{(q^{-\ell}-1)\prod_{h=1,h\neq -\ell}^{m}(q^{-\ell}-q^h)},$$

on fait maintenant tendre $q$ vers 1 : comme la somme extérieure porte sur les $k > m$, la limite du sommande de la somme intérieure est toujours zéro. Par conséquent, on obtient bien que $f_m(1) = 0$. De plus, si l'on applique le même argument pour $\ell \in \{-\bar{m}, -\bar{m}+1, \dots, -1\}$ avec $\bar{m} = m-1, m-2, \dots, 1$, alors on obtient que $f_{\bar{m}}(1) = 0$ pour tout $\bar{m} \in \{1, 2, \dots, m\}$, ce qui prouve que

$$f(z;1) = \mathcal{O}\big((z-1)^m\big).$$

Cet argument, appliqué à (3.4), montre que l'on peut bien remplacer $\mathcal{O}(?)$ dans (3.1) par $\mathcal{O}((z-1)^{A(n+1)-\rho-\sigma-2})$, comme annoncé. Le problème d'approximation (3.1) est donc exactement le problème (1.4). Comme il est démontré dans [4] que ce problème a une solution unique, on a forcément l'égalité $Q_j(z) = P_j(z)$ pour tout $j$.

**Remerciements.** Recherche partiellement supportée par le Programme « Accroître le potentiel humain de recherche » de la Commission Européenne, contrat HPRN-CT-2001-00272, « Algebraic Combinatorics in Europe », et par la Fondation Autrichienne de Recherche FWF, contrat S9607-N13, dans le cadre du Réseau National de Recherche « Analytic Combinatorics and Probabilistic Number Theory ».

**Abstract Pade approximants of $\alpha$-polylogarithms.** We solve a Padé-type problem of approximating three specific functions simultaneously by $q$-analogues of polylogarithms, respectively by powers of the logarithm. This problem is intimately related to recent results of the authors and Wadim Zudilin (J. Inst. Math. Jussieu 5, 53–79, 2006) on the dimension of the vector space generated by $q$-analogues of values of the Riemann zeta function at integers. We also show that our result can be considered as a $q$-analogue of a result of Stéphane Fischler and the second author (J. Math. Pures Appl. 82, 1369–1394, 2003).

## Références

1. Adams, C.R.: On the linear ordinary $q$-difference equations. *Ann. Math.* **30**, 195–205 (1929)
2. Ball, K., Rivoal, T.: Irrationalité d'une infinité de valeurs de la fonction zêta aux entiers impairs. *Invent. Math.* **146**, 193–207 (2001)
3. Kaneko, M., Kurokawa, N., Wakayama, M.: A variation of Euler's approach to values of the Riemann zeta function. *Kyushu J. Math.* **57**, 175–192 (2003)
4. Fischler, S., Rivoal, T.: Approximants de Padé et séries hypergéométriques équilibrées. *J. Math. Pures Appl.* **82**, 1369–1394 (2003)
5. Krattenthaler, C., Rivoal, T., Zudilin, W.: Séries hypergéométriques basiques, fonction $q$-zêta et séries d'Eisenstein. *J. Inst. Math. Jussieu* **5**, 53–79 (2006)
6. Rivoal, T.: La fonction Zêta de Riemann prend une infinité de valeurs irrationnelles aux entiers impairs. *C. R. Acad. Sci. Paris Sér. I Math.* **331**, 267–270 (2000)
7. Sauloy, J.: Systèmes aux $q$-différences singuliers réguliers: classification, matrice de connexion et monodromie. *Ann. Inst. Fourier* **50**, 1021–1071 (2000)
8. Zudilin, W.: Diophantine problems for $q$-zeta values. *Mat. Zametki* **72**, 936–940 (2002); *Math. Notes* **72**, 858–862 (2002)

*Diophantine Approximation*
H. P. Schlickewei et al., Editors


# THE SET OF SOLUTIONS OF SOME EQUATION FOR LINEAR RECURRENCE SEQUENCES

Viktor Losert
*Institut für Mathematik, Universität Wien, Nordbergstrasse 15, 1090 Wien, Austria*
viktor.losert@univie.ac.at

To Wolfgang M. Schmidt on the occasion of his 70th birthday

In [SS1] Schlickewei and Schmidt studied the solutions of various linear equations involving members of recurrence sequences. Most of them are of the form

$$F_1(x_1) + \cdots + F_n(x_n) = 0 \tag{A}$$

with $x_i \in \mathbb{Z}$, where $F_j(x) = \sum_{i=0}^{r_j} f_{ji}(x)\,\alpha_{ji}^x$ $(j = 1, \ldots, n)$, $r_j > 0$, with given polynomials $f_{ji}$ and nonzero numbers $\alpha_{ji}$ (thus for each $j$, $(F_j(x))_{x\in\mathbb{Z}}$ is a linear recurrence sequence, see also [ST, Sec. C]). The general assumption of [SS1, p. 220] is that $\alpha_{j0}$ is a root of unity and that $f_{ji} \neq 0$ for $i > 0$ ($f_{j0}$ may be zero), $j = 1, \ldots, n$. Furthermore, they restrict to nondegenerate sequences, i.e., $\alpha_{ji}/\alpha_{jh}$ is not a root of unity for $h \neq i$.

After studying several equations of the type above, the following conjecture was made ([SS1, p. 226]):

The set of solutions $\mathbf{x} = (x_1, \ldots, x_n) \in \mathbb{Z}^n$ of an equation (A) consists of the union of a finite set of points together with a finite set of families $\mathcal{G}_1, \ldots, \mathcal{G}_l$, such that the elements of $\mathcal{G}_j$ can be parametrized as follows:

$$\mathbf{x}(\mathbf{s}) = \mathbf{p}_j(\mathbf{s}) = (p_{j1}(\mathbf{s}), \ldots, p_{jn}(\mathbf{s})) \qquad (\mathbf{s} \in \mathbb{N}^{m(j)})$$

where $m(j) \geq 1$ and each $p_{jk}$ is of polynomial-exponential type, i.e., it is of the form $\sum_{i=1}^{w} u_i(\mathbf{s})\boldsymbol{\beta}_i^{\mathbf{s}}$, where the $u_i$ are polynomials, $\boldsymbol{\beta}_i^{\mathbf{s}} = \beta_{i1}^{s_1} \cdots \beta_{im}^{s_m}$ (for $\mathbf{s} = (s_1, \ldots, s_m)$) with $\beta_{ih} \neq 0$ $(u_i, \boldsymbol{\beta}_i$ possibly depending on $j, k)$.

Our aim is to give a counterexample to this conjecture (see also the Remarks at the end).

*Example.* Consider the following equation

$$2^h + n\,2^n = q\,2^q \tag{B}$$

The set of integer solutions $(h, n, q)$ of (B) looks as follows:

*Keywords.* Linear recurrence sequences, polynomial-exponential type.

*2000 Mathematics subject classification.* 11D61, 11B37.

1. four "isolated" points
   (1.1) $(-2,-1,-4)$, (1.2) $(-3,-1,-3)$, $(-2,-2,-4)$, (1.3) $(-3,-2,-3)$.

Then five infinite families 2–6

| | | | |
|---|---|---|---|
| 2. $q = 0$, | $n = -2^{\mu}$, | $h = \mu - 2^{\mu}$ | $(\mu \geq 0)$. |
| 3. $q = -2\,(2^j+1)$, | $n = q+1$, | $h = n + j$ | $(j \geq 0)$. |
| 4. $q = -8\,\dfrac{2^{2j+1}+1}{3}$, | $n = q+2$, | $h = n + 2j + 2$ | $(j \geq 0)$. |
| 5. $q = -8\,\dfrac{2^{3j+2}+3}{7}$, | $n = q+3$, | $h = n + 3j + 2$ | $(j \geq 0)$. |
| 6. $q = 2^{\mu}\,\dfrac{2^{(j+1)\,2^{\mu}}-1}{2^{2^{\mu}}-1}$, | $n = q - 2^{\mu}$, | $h = n + \mu + (j+1)\,2^{\mu}$ | $(j \geq 0,\ \mu \geq 0)$. |

*Proof.* We put $h = h_1\,2^{\kappa}$, $n = n_1\,2^{\lambda}$, $q = q_1\,2^{\mu}$ with $\kappa, \lambda, \mu \geq 0$ and $h_1, n_1, q_1$ odd integers or zero. First we consider solutions of (B) where one of the three terms of the sum vanishes. There are two possibilities.

a. If $n = 0$, then $q_1 = 1$, hence $q = 2^{\mu}$ and $h = \mu + 2^{\mu}$ with $\mu \geq 0$, giving 6 for $j = 0$.

b. If $q = 0$, then $n_1 = -1$, hence $n = -2^{\lambda}$ and $h = \lambda - 2^{\lambda}$ with $\lambda \geq 0$, giving 2.
In the remaining cases, we assume that the three parts of the sum are nonzero. Then, comparing the exponents of 2 in these parts (they are $h,\ \lambda + n,\ \mu + q$), it follows that two of them have to be equal and the third one bigger than the other two. This gives three cases.

c. $h = \mu + q < \lambda + n$. Then $2^h = (1/q_1)\,q\,2^q$ and Eq. (B) is equivalent to $(1 - 1/q_1)\,q\,2^q = n\,2^n$. This gives $q_1 \neq 1$ and the monotonicity properties of $|n|\,2^n$ imply that $|n| < |q|$.

If $q > 0$, then $n > 0$, hence $n < q$ and Eq. (B) implies $(1 - 1/q_1)\,q/n = 2^{n-q} < 1$. On the other hand, since $q_1 \geq 3$, we have $\frac{1}{2} < \frac{2}{3} \leq 1 - 1/q_1 < (1 - 1/q_1)\,q/n$. This contradicts the condition that $(1 - 1/q_1)q/n$ should be an integer power of 2.

Thus $q < 0$, hence $n < 0$ and $-n < -q$. This implies $(1 - 1/q_1)q/n = 2^{n-q} > 1$. To get an estimate from above, note that $1 - 1/q_1 \leq 2$, consequently $2^n \leq |n|\,2^n \leq 2\,|q|\,2^q$. Taking logarithms, it follows that $n \leq q + 1 + \ln|q|/\ln 2$. Since $2\ln|q| \leq |q|\ln 2$, we arrive at $n \leq q/2 + 1$ and this implies $q/n \leq 2\,(1 - 1/n)$. Thus we get $2^{n-q} = (1 - 1/q_1)\,q/n \leq (1 - 1/q_1)\,(1 - 1/n)\,2 \leq 2^3$ and therefore $n - q \in \{1, 2, 3\}$.

For $n = 3 + q$, we must have equality in the last estimate, hence $n = -1,\ q = -4$, giving (1.1).

For $n = 2 + q$, Eq. (B) becomes $1 - 1/q_1 = 4\,(1 + 2/q)$. Then $q \in \{-3, -4\}$, since otherwise either $1 + 2/q < 0$ or $1 + 2/q > 1/2$. This gives (1.2).

For $n = 1 + q$, Eq. (B) becomes $1 - 1/q_1 = 2\,(1 + 1/q)$. Then $q_1 = -1$ is impossible and $q \leq q_1 \leq -5$ would give $1 + 1/q \geq \frac{4}{5}$ and $1 - 1/q_1 \leq \frac{6}{5}$ which is also impossible. Hence only $q_1 = -3$ remains, then $q = -3$, giving (1.3).

d. $h = \lambda + n < \mu + q$. Then $2^h = (1/n_1)\,n\,2^n$ and Eq. (B) is equivalent to $(1 + 1/n_1)\,n\,2^n = q\,2^q$. As before, this implies $|n| < |q|$.

If $n > 0$, then $n < q$ and Eq. (B) gives $(1 + 1/n_1)\,n/q = 2^{q-n} \geq 2$. On the other hand, $1 + 1/n_1 \leq 2$, consequently $(1 + 1/n_1)\,n/q < 2$, providing a contradiction.

If $n < 0$, then (using Eq. (B)) $n_1 < -1$, hence ($n_1$ being odd) $n_1 \le -3$, and this implies $1 + 1/n_1 \ge \frac{2}{3}$. Then Eq. (B) gives $|q/n| \ge \frac{2}{3} 2^{n-q}$. Since $|q/n| = (|n| + n - q)/|n| \le 1 + n - q$, it follows that $n - q \in \{1, 2\}$.

For $n = 1 + q$, Eq. (B) becomes $(1 + 1/n_1)2 = 1 - 1/n$. Then (see above) $n = -3$ $(= n_1)$, giving 3 for $j = 0$.

For $n = 2 + q$, Eq. (B) becomes $(1 + 1/n_1)4 = 1 - 2/n$, which contradicts $1 + 1/n_1 \ge \frac{2}{3}$.

e. $\lambda + n = \mu + q < h$. Put $c = h - (\lambda + n)$, then $c > 0$ and Eq. (B) is equivalent to $2^c + n_1 = q_1$. In addition, we have $\lambda + n_1 2^\lambda = \mu + q_1 2^\mu$, in particular $\mu \ne \lambda$.

e1. $\mu > \lambda$. Write $n_1 = q_1 2^{\mu-\lambda} + (\mu - \lambda)/2^\lambda$ and put $\mu_0 = (\mu - \lambda)/2^\lambda$. Then $\mu_0 \in \mathbb{N}$ and $\mu > \lambda$, $n_1$ odd implies that $\mu_0$ has to be odd. Put $d = \mu_0 2^\lambda$. Then $\mu = \lambda + d$, $n_1 = q_1 2^d + \mu_0$ and Eq. (B) becomes $-q_1 = (2^c + \mu_0)/(2^d - 1)$. Conversely, any integers $c, d, \mu_0 > 0$ with $\mu_0$ odd and such that $(2^c + \mu_0)/(2^d - 1)$ is an integer, define a solution $(h, n, q)$.

Put $j = [c/d]$. Then $0 \le c - jd < d$ and since for $j > 0$, one knows $(2^c - 2^{c-jd})/(2^d - 1) = 2^{c-d} + \cdots + 2^{c-jd} \in \mathbb{N}$, we must have $(2^{c-jd} + \mu_0)/(2^d - 1) \in \mathbb{N}$ (for $j = 0$, this is trivial).

For $d = 1$, we get $\mu_0 = 1$, $\lambda = 0$, $j = c$, hence $\mu = 1$, $q_1 = -2^j - 1$, $q = 2q_1$ leading to 3 (with $j > 0$).

If $d \ge 2$, then $2^d - 1 \ge 2^{d-1} \ge d \ge \mu_0$, implying $(2^{c-jd} + \mu_0)/(2^d - 1) < 2$. It follows that $2^{c-jd} + \mu_0 = 2^d - 1$, hence $\mu_0 + 1 \ge 2^{d-1}$, giving $d \le 3$.

For $d = 2$, we get $\mu_0 = 1$, $\lambda = 1$, $\mu = 3$, $2^d - 1 - \mu_0 = 2$ which implies $c - jd = 1$. Thus $c = 2j + 1$ with $j \ge 0$, leading to 4.

For $d = 3$, we get $\mu_0 = 3$, $\lambda = 0$, $\mu = 3$, $2^d - 1 - \mu_0 = 4$ which implies $c - jd = 2$. Thus $c = 3j + 2$ with $j \ge 0$, leading to 5.

e2. $\mu < \lambda$. Write $q_1 = n_1 2^{\lambda-\mu} + (\lambda - \mu)/2^\mu$ and put $\lambda_0 = (\lambda - \mu)/2^\mu$. Then (using $\lambda > \mu$ and $q_1$ odd) $\lambda_0$ has to be an odd natural number. Put $d = \lambda_0 2^\mu$. Then $\lambda = \mu + d$, $q_1 = n_1 2^d + \lambda_0$ and Eq. (B) becomes $n_1 = (2^c - \lambda_0)/(2^d - 1)$. Put $j = [c/d]$. Then, as in e1, we must have $(2^{c-jd} - \lambda_0)/(2^d - 1) \in \mathbb{Z}$.

For $d = 1$, we get $\lambda_0 = 1$, $\mu = 0$, $j = c$, hence $\lambda = 1$, $q_1 = 2n_1 + 1$ $(= q)$ leading to 6 for $\mu = 0$, $j > 0$.

If $d \ge 2$, then $\lambda_0 \le d < 2^d - 1$ implies $2^{c-jd} - \lambda_0 = 0$, hence ($\lambda_0$ is odd!) $c = jd$, $\lambda_0 = 1$, giving $j > 0$, $d = 2^\mu$. Then $d \ge 2$ implies $\mu \ge 1$ and we get $\lambda = \mu + 2^\mu$, $n_1 = (2^{j2^\mu} - 1)/(2^{2^\mu} - 1)$, $q_1 = n_1 2^\mu + 1 = (2^{(j+1)2^\mu} - 1)/(2^{2^\mu} - 1)$.

These are the remaining cases of 6 ($\mu > 0$ and $j > 0$). □

*Remarks.*

1. The families 2–5 of the solutions of (B) are given by functions of polynomial-exponential type. But 6 contains double exponential terms, giving a different shape. For each fixed $\mu$, one has a (one parameter) polynomial-exponential family. But the double exponential term causes that these (infinitely many) parts move to infinity more rapidly.

Explicitly, some differences in the asymptotic behaviour can be described as follows. For a discrete subset $\mathcal{D}$ of real numbers let $d_n$ be the cardinality of the set $\{x \in \mathcal{D} : |x| \le 2^n\}$ (this means that $d_n$ does not count multiple representations). Then the following holds:

(a) If $\mathcal{D} = \{p(s) : s \in \mathbb{N}\}$ where $p$ is of polynomial-exponential type (see the introduction), $\mathcal{D}$ is unbounded and $|\beta_i| \leq 1$ holds for all of the roots, then $\liminf_{n\to\infty} (\ln d_n/n) > 0$.
(b) If $\mathcal{D} = \{p(s) : s \in \mathbb{N}\} \subseteq \mathbb{Q}$ where $p$ is of polynomial-exponential type and $|\beta_i| > 1$ holds for at least one of the roots, then there exists $c > 0$ such that $d_n \sim c\,n$ for $n \to \infty$. Moreover, if there is a dominating root, one gets $\limsup_{n\to\infty} (d_{n+t} - d_n) - c\,t \leq 1$ when $t$ is large enough.
(c) If $\mathcal{D} = \{p(\mathbf{s}) : \mathbf{s} \in \mathbb{N}^m\} \subseteq \mathbb{Q}$ where $p$ is of polynomial-exponential type, $m \geq 2$ and $\mathcal{D}$ is unbounded but has no parametrization as in (b), then $\liminf_{n\to\infty}(d_n/n^2) > 0$.
(d) If $\mathcal{D}$ is the set of the $q$-components of the solutions of (B), then $d_n \sim c_1 n$ for $n \to \infty$ (with $c_1 = \frac{23}{6}$), but $\limsup_{n\to\infty} (d_{n+t} - d_n) - c_1 t = \infty$ for all $t > 0$ (similarly, when considering just the family 6; here $c_1 = 2$).

For a function $p$ of polynomial-exponential type let $d'_n$ be the cardinality of the set $\{\mathbf{s} \in \mathbb{N}^m : |p(\mathbf{s})| \leq 2^n\}$ (i.e., the counterpart of $d_n$, taking care of multiplicities). Then in (a) one can show easily that $d'_n$ grows at least exponentially. The assumption $\mathcal{D} \subseteq \mathbb{Q}$ implies that in (b) and (c) we have algebraic recurrence sequences (in the sense of [ST] p. 35). Then in (b) (assuming first that the sequence is nondegenerate, by splitting it into finitely many parts), the corresponding relation for $d'_n$ follows from the result of van der Poorten and Schlickewei [ST, p. 96]. For passage to $d_n$, recall the results on the intersection of linear recurrence sequences (e.g., [La, Th. 2], [SS1, Prop. 1, 2, 3]). They imply that for a nondegenerate sequence the multiplicities are 1, apart of finitely many exceptions (observe that we restrict $s$ to natural numbers). Thus the asymptotic behaviour of $d_n$ and $d'_n$ is the same for nondegenerate sequences. Intersections of nondegenerate sequences (assuming that both have a root which is not a root of unity) can occur only at arithmetic progressions in $\mathbb{N}$ or at exponentially growing subsets. In this way taking finite unions gives (b) (of course, the constant $c$ will be influenced if two sequences coincide at some arithmetic progression). For comparison, in the case of a pair of conjugate dominant roots, the lim sup is still finite, but the bound will be different in general.

For (c), lower estimates of $d'_n$ are again elementary, but here multiple representations are less easy to handle (e.g., infinite multiplicities occur in $p(s_1, s_2) = 2^{s_1} - s_2$). We indicate the argument for $m = 2$. According to the general results of [La], [SS2], the set of solutions of an equation $p(\mathbf{s}) = x_0$ (a constant) decomposes into finitely many parts $\mathcal{S}(\mathcal{P})$ associated to partitions $\mathcal{P}$ of the index set (of the sum representing $p$). For those partitions $\mathcal{P}$ where $\mathcal{G}(\mathcal{P})$ is trivial (notation of [SS2]), $\mathcal{S}(\mathcal{P})$ is finite and by Th. 1 of [SS2], there is a bound for $|\mathcal{S}(\mathcal{P})|$ independent of $x_0$ (it depends just on the degree of the number field and the degrees of the polynomials in $p$). If the partition $\mathcal{P}$ has at least two sets, one of them will not contain the constant term, hence the solutions in $\mathcal{S}(\mathcal{P})$ satisfy some polynomial-exponential equation that is independent of $x_0$. The case where $s_1$ or $s_2$ are not present in the exponentials is settled at the end, otherwise (again assuming nondegeneracy), we get for fixed $s_2$ (with finitely many exceptions) a nonzero recurrence sequence in $s_1$ and as above, a uniform bound for the number of its zeros. Thus $\mathcal{S}(\mathcal{P})$ causes a reduction of the estimate for $d_n$ of order at most $n$. Finally, there is the case where $\mathcal{G}(\mathcal{P})$ is nontrivial and $\mathcal{P}$ is the trivial partition into one set. Then we can relate $p$ (by a transformation of variables) to a polynomial-exponential function with $\beta_{i2} = 1$ for all $i$ (i.e., $s_2$ does no longer appear in the exponentials). If $s_2$ disappears completely, we have case (a) or (b). Otherwise, we get exponential growth of $d_n$ as in (a).

For (d) the estimates are elementary. The families coming from 3–6 have only finitely many terms in common. Denoting e.g.,the values appearing in 6 by $q(j, \mu)$, we have $q(j, \mu) \sim 2^{\mu + j\, 2^{\mu}}$.

It follows that 6 and the whole set of solutions of (B) cannot be parametrized as prescribed by the conjecture: (c) excludes functions with two or more parameters, (a) one-parameter functions whose roots are bounded by one. Hence, apart of finitely many points, the $q$-components of the solutions would be covered by nondegenerate sequences as in (b). The results on intersection of sequences mentioned above imply that the roots have to be nonnegative powers of 2, giving dominant roots. The growth of the numbers $d_n$ arising from 6 is like that for functions in (b), but the distribution on the annuli $\{x : 2^n < |x| \leq 2^{n+t}\}$ is less uniform. As an alternative argument observe that an integer recurrence sequence whose roots are powers of 2 has at most one accumulation point in the space of 2-adic integers. But in 6 we have countably many accumulation points: 0 and $2^{\mu}(1 - 2^{2^{\mu}})^{-1}$ for $\mu = 0, 1, \ldots$ (more generally, one can show that given a nondegenerate integer recurrence sequence, the set of accumulation points in the space of $p$-adic integers is either finite or an uncountable perfect subset).

In [SS1, p. 227], other examples have been given of equations of polynomial-exponential type, whose solutions are parametrized by double exponential (or multi-exponential) terms. But these were not of the form Eq. (A).

2. Solutions as in 6 appear also when the base 2 in the example is replaced by some other rational integer $a \neq 0, \pm 1$. But the situation changes when considering numbers $a$ that are not roots of integers. In particular, our example does not apply to the second part of the conjecture of [SS1].

Similarly, when considering examples of equations (A) containing nonlinear polynomials. For example, for the special case in [Lo] (1.1), where $n = 3$ and the $F_j$ are multiples of different powers of a fixed sequence with $r = 1$, $f_{j0} = 0$, one obtains (by a similar analysis and parametrization as in the proof above) additional algebraic equations for the parameters which possibly exclude the appearance of families as in 6.

**Acknowledgements.** I thank the referee for pointing out an irritating misprint in the first version of this paper.

## References

[La] Laurent, M.: Équations exponentielles-polynômes et suites récurrentes linéaires. II. *J. Number Theory* **31**, 24–53 (1989)

[Lo] Losert, V.: Two equations for linear recurrence sequences. *Acta Arith.* **119**, 109–147 (2005)

[SS1] Schlickewei, H.P., Schmidt, W. M.: Linear equations in members of recurrence sequences. *Ann. Sc. Norm. Super. Pisa Cl. Sci. IV. Ser.* **20**, 219–246 (1993)

[SS2] Schlickewei, H.P., Schmidt, W.M.: The number of solutions of polynomial-exponential equations. *Compos. Math.* **120**, 193–225 (2000)

[ST] Shorey, T.N., Tijdeman, R.: *Exponential Diophantine Equations*. Cambridge University Press, Cambridge (1986)

*Diophantine Approximation*
H. P. Schlickewei et al., Editors


# COUNTING ALGEBRAIC NUMBERS WITH LARGE HEIGHT I

David Masser[1] and Jeffrey D. Vaaler[2]

[1] *Mathematisches Institut, Universität Basel, Rheinsprung 21, 4051 Basel, Switzerland*
David.Masser@unibas.ch

[2] *Department of Mathematics, University of Texas at Austin, 1 University Station C1200, Austin, TX 78712, USA*
vaaler@math.utexas.edu

Dedicated to Professor Wolfgang Schmidt

Let $\mathbb{Q}$ denote the field of rational numbers, $\overline{\mathbb{Q}}$ an algebraic closure of $\mathbb{Q}$, and $H : \overline{\mathbb{Q}} \to [1, \infty)$ the absolute, multiplicative, Weil height. For each positive integer $d$ and real number $\mathcal{H} \geq 1$, it is well known that the number $\overline{\mathbb{Q}}_d(\mathcal{H})$ of points $\alpha$ in $\overline{\mathbb{Q}}$ having degree $d$ over $\mathbb{Q}$ and satisfying $H(\alpha) \leq \mathcal{H}$ is finite. This is the one-dimensional case of Northcott's Theorem [8] (see also [5, page 59]). The systematic study of the counting function $\overline{\mathbb{Q}}_d(\mathcal{H})$, and that of related functions in higher dimensions, was begun by Schmidt [10]. It is relatively easy to prove the existence of a positive constant $C = C(d)$ such that

$$\overline{\mathbb{Q}}_d(\mathcal{H}) \leq C\mathcal{H}^{d(d+1)}, \tag{1}$$

and also the existence of positive constants $c = c(d)$ and $\mathcal{H}_0 = \mathcal{H}_0(d)$ such that

$$c\mathcal{H}^{d(d+1)} \leq \overline{\mathbb{Q}}_d(\mathcal{H}), \quad \text{whenever} \quad \mathcal{H}_0 \leq \mathcal{H}. \tag{2}$$

Care must be taken with the lower bound (2); for Kronecker's Theorem [3] (see also [5, page 54]) implies that if $d > 1$, then $\overline{\mathbb{Q}}_d(1)$ counts the set of all roots of unity of degree $d$. Thus $\overline{\mathbb{Q}}_d(1) = 0$ if $d$ is not a value of Euler's totient function, and in particular $\overline{\mathbb{Q}}_d(1) = 0$ if $d > 1$ is odd. On the other hand $\overline{\mathbb{Q}}_d(2^{1/d})$ must count $2^{1/d}$ and its conjugates for all $d$.

Regarding (1) Schmidt [10, equation (1.4)] proved that

$$\overline{\mathbb{Q}}_d(\mathcal{H}) \leq 2^{2d^2+14d+11}\mathcal{H}^{d(d+1)}, \tag{3}$$

and regarding (2) he proved [10, equation (1.8)] that

$$6^{-d(d+1)}\mathcal{H}^{d(d+1)} \leq \overline{\mathbb{Q}}_d(\mathcal{H}), \quad \text{when} \quad 2 \leq \mathcal{H}^d. \tag{4}$$

---

*Keywords.* Mahler measure, height.

*2000 Mathematics subject classification.* 11R04.

The latter inequality was slightly improved by Loher [6, page 33] to

$$2^{-d-2}(d+1)^{-(d+1)/2}\mathcal{H}^{d(d+1)} \leq \overline{\mathbb{Q}}_d(\mathcal{H}), \quad \text{when} \quad 2 \leq \mathcal{H}^d. \tag{5}$$

But as far as we know there is no similar improvement on the upper bound (3), and indeed no upper bound at all of the form $C(d)\mathcal{H}^{d(d+1)}$ with

$$\lim_{d\to\infty} d^{-2}\log C(d) = 0.$$

See, however, the paper of Dubickas and Konyagin [2] on polynomials.

The asymptotic behaviour of $\overline{\mathbb{Q}}_d(\mathcal{H})$ as $\mathcal{H} \to \infty$ is known up to now only in the cases $d = 1$ and $d = 2$. In the first case it is a classical result (see Schanuel [9] or Lang [5, page 71], in one dimension over $\mathbb{Q}$) that

$$\overline{\mathbb{Q}}_1(\mathcal{H}) = \frac{12\mathcal{H}^2}{\pi^2} + O(\mathcal{H}\log\mathcal{H}).$$

And in the latter case, Schmidt [11, pages 362–364] proved that

$$\overline{\mathbb{Q}}_2(\mathcal{H}) = \frac{8\mathcal{H}^6}{\zeta(3)} + O(\mathcal{H}^4\log\mathcal{H}),$$

where $\zeta$ is the Riemann zeta-function. The paper [11] also contains asymptotic results in higher dimensions.

The object of the present note is to point out that recent work of Chern and the second author [1] quickly implies an asymptotic estimate for $\overline{\mathbb{Q}}_d(\mathcal{H})$ and all $d \geq 1$. Let

$$\gamma(d) = 2^{d+1}(d+1)^e \prod_{j=1}^{e} \frac{(2j)^{d-2j}}{(2j+1)^{d-2j+1}}, \tag{6}$$

where $e = [(d-1)/2]$ and an empty product is 1. Also, write $\theta(1) = \theta(2) = 1$ and $\theta(d) = 0$ for $d \geq 3$.

**Theorem.** *As $\mathcal{H} \to \infty$ we have*

$$\overline{\mathbb{Q}}_d(\mathcal{H}) = \frac{d\gamma(d)\mathcal{H}^{d(d+1)}}{2\zeta(d+1)} + O\left(\mathcal{H}^{d^2}(\log\mathcal{H})^{\theta(d)}\right), \tag{7}$$

*where the constant implicit in the O-notation depends only on d.*

For example, as $\mathcal{H} \to \infty$ we find that

$$\overline{\mathbb{Q}}_3(\mathcal{H}) = \frac{1920\mathcal{H}^{12}}{\pi^4} + O(\mathcal{H}^9),$$

$$\overline{\mathbb{Q}}_4(\mathcal{H}) = \frac{1280\mathcal{H}^{20}}{27\zeta(5)} + O(\mathcal{H}^{16}),$$

and

$$\overline{\mathbb{Q}}_5(\mathcal{H}) = \frac{86016\mathcal{H}^{30}}{\pi^6} + O(\mathcal{H}^{25}).$$

The constant $\gamma(d)$ already appears in [1], where it is shown that

$$\lim_{d\to\infty} \frac{\log\gamma(d)}{d\log d} = -\frac{1}{2}.$$

This implies that Loher's inequality (5) is close to best possible in the sense that any lower bound of the form $d^{-\eta d}\mathcal{H}^{d(d+1)}$ for $0 < \eta < \frac{1}{2}$, even under restrictions $\mathcal{H} \geq \mathcal{H}_0(d)$ and $d \geq d_0$, is false. For this reason we provide a sketch, with Loher's permission, of his proof of (5), at least up to absolute constants. It is also interesting that (5) holds for $\mathcal{H}^d \geq 2$; of course this is useful only if $\mathcal{H}^d$ is slightly bigger than $2\sqrt{d+1}$. In [1, Theorem 5], there are also some explicit lower and upper bounds and furthermore these bounds are asymptotically correct, but there $\mathcal{H}^d$ must be larger than $8^d d^{2d}$.

In preparation for the proof of our theorem we describe the relevant results of [1]. Let $F(x)$ be a polynomial of degree $L$ in $\mathbb{Z}[x]$ and assume that $F$ factors in $\mathbb{C}[x]$ as

$$F(x) = a_0 \prod_{l=1}^{L} (x - \alpha_l).$$

Then the Mahler measure of $F$ is the nonnegative real number

$$M(F) = \exp\left\{ \int_0^1 \log|F(e^{2\pi i t})|\, \mathrm{d}t \right\} = |a_0| \prod_{l=1}^{L} \max\{1, |\alpha_l|\}, \tag{8}$$

where the second equality in (8) follows from Jensen's formula. For $d$ a positive integer and $T$ a positive real number, let $\mathcal{M}(d, T)$ denote the number of polynomials $F(x)$ in $\mathbb{Z}[x]$ having degree at most $d$ and satisfying $M(F) \leq T$. It is easy to see that $\mathcal{M}(d, T)$ is finite for all positive real $T$. In [1, Theorem 3], the authors proved that

$$\mathcal{M}(d, T) = \gamma(d)T^{d+1} + O(T^d), \tag{9}$$

as $T \to \infty$, where the constant implicit in the $O$-notation depends only on $d$. Now let $\mathcal{N}(d, T)$ denote the number of polynomials $F(x)$ in $\mathbb{Z}[x]$ having degree equal to $d$, satisfying $M(F) \leq T$, and having relatively prime coefficients.

**Lemma 1.** *As $T \to \infty$, we have*

$$\mathcal{N}(d, T) = \frac{\gamma(d)}{\zeta(d+1)} T^{d+1} + O\left(T^d (\log T)^{\theta_1(d)}\right), \tag{10}$$

*where $\theta_1(1) = 1$ and $\theta_1(d) = 0$ if $d \geq 2$.*

*Proof.* Let $\mathcal{L}(d, T)$ denote the number of polynomials $F(x)$ in $\mathbb{Z}[x]$ having degree equal to $d$ and satisfying $M(F) \leq T$. Then we have

$$\mathcal{M}(d, T) = \mathcal{L}(d, T) + \mathcal{M}(d-1, T), \tag{11}$$

where $\mathcal{M}(0, T) = 2[T] + 1$ is the number of integers $a_0$ satisfying $|a_0| \leq T$. As

$$\mathcal{M}(d-1, T) = O(T^d),$$

we get the estimate

$$\mathcal{L}(d, T) = \gamma(d)T^{d+1} + O(T^d) \tag{12}$$

from (9) and (11). For each positive integer $n$ let $\mathcal{L}_n(d,T)$ denote the number of polynomials $F(x)$ in $\mathbb{Z}[x]$ having degree equal to $d$, satisfying $M(F)\le T$, and such that the greatest common divisor of the coefficients of $F$ is $n$. Obviously we have

$$\mathcal{L}(d,T)=\sum_{1\le n\le T}\mathcal{L}_n(d,T). \tag{13}$$

If the greatest common divisor of the coefficients of $F$ is $n$, we can write $F(x)=nG(x)$ where $G(x)$ in $\mathbb{Z}[x]$ has relatively prime coefficients. Then $M(F)\le T$ is equivalent to $M(G)\le T/n$. It follows that

$$\mathcal{L}_n(d,T)=\mathcal{L}_1(d,T/n)=\mathcal{N}(d,T/n),$$

and therefore (13) can be written as

$$\mathcal{L}(d,T)=\sum_{1\le n\le T}\mathcal{N}(d,T/n).$$

Then by Möbius inversion we get

$$\mathcal{N}(d,T)=\sum_{1\le n\le T}\mu(n)\mathcal{L}(d,T/n). \tag{14}$$

To complete the proof we use the estimate (12) in the right-hand side of (14). In this way we find that

$$\begin{aligned}\mathcal{N}(d,T)&=\gamma(d)T^{d+1}\sum_{1\le n\le T}\mu(n)n^{-d-1}+O\left(T^d\sum_{1\le n\le T}n^{-d}\right)\\&=\frac{\gamma(d)}{\zeta(d+1)}T^{d+1}+O\left(T^{d+1}\sum_{T<n}n^{-d-1}\right)+O\left(T^d(\log T)^{\theta_1(d)}\right)\\&=\frac{\gamma(d)}{\zeta(d+1)}T^{d+1}+O\left(T^d(\log T)^{\theta_1(d)}\right),\end{aligned}$$

as claimed. □

Next we write

$$\mathcal{N}(d,T)=\mathcal{N}_I(d,T)+\mathcal{N}_R(d,T), \tag{15}$$

where $\mathcal{N}_R(d,T)$ denotes the number of reducible polynomials $F(x)$ in $\mathbb{Z}[x]$ having degree equal to $d$, satisfying $M(F)\le T$, and having relatively prime coefficients, and $\mathcal{N}_I(d,T)$ counts the corresponding number of irreducible polynomials in $\mathbb{Z}[x]$.

**Lemma 2.** *As $T\to\infty$, we have*

$$\mathcal{N}_I(d,T)=\frac{\gamma(d)}{\zeta(d+1)}T^{d+1}+O\left(T^d(\log T)^{\theta(d)}\right). \tag{16}$$

*Proof.* If $d=1$, then $\mathcal{N}_R(1,T)=0$ and the result follows from (10) and (15). Therefore we assume throughout the remainder of the proof that $d\ge 2$.

Let $F(x)$ be a reducible polynomial in $\mathbb{Z}[x]$ having degree equal to $d$, satisfying $M(F)\le T$, and having relatively prime coefficients. Then there exists a pair of positive integers $(d_1, d_2)$ such that $1 \le d_1 \le d_2 \le d-1$ and $d_1 + d_2 = d$, and a corresponding pair of polynomials $F_1(x)$ and $F_2(x)$ in $\mathbb{Z}[x]$ such that

$$F(x) = F_1(x)F_2(x), \quad d_1 = \deg F_1, \quad \text{and} \quad d_2 = \deg F_2. \tag{17}$$

The polynomial $F_1(x)$ determines a unique positive integer $k$ such that

$$2^{k-1} \le M(F_1) < 2^k.$$

As $1 \le M(F) = M(F_1)M(F_2) \le T$, we find that

$$1 \le k \le K, \quad \text{where} \quad K = \left[\frac{\log T}{\log 2}\right] + 1,$$

and

$$M(F_2) \le 2^{1-k}T.$$

From (9) the number of possible polynomials $F_1$ in such a factorization is at most

$$\ll 2^{k(d_1+1)}.$$

Here the constant implicit in $\ll$ depends only on $d$. Similarly, the number of possible polynomials $F_2$ in such a factorization is at most

$$\ll \left(2^{1-k}T\right)^{d_2+1}.$$

It follows that the number of reducible polynomials in $\mathbb{Z}[x]$ having $M(F) \le T$ and a factorization of the type (17) is at most

$$\ll \sum_{k=1}^{K} 2^{k(d_1+1)} \left(2^{1-k}T\right)^{d_2+1} = (2T)^{d_2+1} \sum_{k=1}^{K} 2^{k(d_1-d_2)}. \tag{18}$$

If $d = 2$, then $d_1 = d_2 = 1$ and (18) is $\ll T^2 \log T$. If $3 \le d$, then (18) is easily seen to be $\ll T^d$.

The number of pairs $(d_1, d_2)$ is obviously $\ll d$. Thus we have shown that

$$\mathcal{N}_R(d, T) = O\left(T^d (\log T)^{\theta(d)}\right).$$

When this estimate is combined with (10) and (15), the estimate in Lemma 2 follows. (Note that Schmidt [11, page 363] gives a similar irreducibility argument and Specht [12] obtains related asymptotic results.) □

*Proof of the theorem.* If $\alpha$ is an algebraic number of degree $d$ over $\mathbb{Q}$, if $f_\alpha(x)$ is an irreducible polynomial in $\mathbb{Z}[x]$ having $\alpha$ as a root, then we have

$$H(\alpha)^d = M(f_\alpha). \tag{19}$$

Of course, $\alpha$ and its conjugates each determine exactly two such irreducible polynomials of degree $d$ in $\mathbb{Z}[x]$. And each pair $\pm F(x)$ of irreducible polynomials in $\mathbb{Z}[x]$ of degree $d$ determines a unique collection of $d$ distinct conjugate algebraic numbers in $\overline{\mathbb{Q}}$ each having degree $d$ over $\mathbb{Q}$. From (19) and these remarks we conclude that

$$2\overline{\mathbb{Q}}_d(\mathcal{H}) = d\mathcal{N}_I(d, \mathcal{H}^d). \tag{20}$$

The theorem follows from the estimate (16) and (20).

We now sketch the argument that leads to the lower bound (5), assuming for simplicity that $d \geq 2$. For each $\boldsymbol{b}$ in $\mathbb{Z}^{d+1}$ let $G_{\boldsymbol{b}}(x)$ be the polynomial

$$G_{\boldsymbol{b}}(x) = (2b_0 + 1)x^d + 2b_1 x^{d-1} + 2b_2 x^{d-2} + \cdots + 2b_{d-1}x + 4b_d + 2. \quad (21)$$

Clearly $G_{\boldsymbol{b}}$ is a polynomial of degree $d$ in $\mathbb{Z}$. Applying Eisenstein's criterion as in Schmidt [10, page 173], with the prime 2, we find that each polynomial $G_{\boldsymbol{b}}$ is irreducible in $\mathbb{Q}[x]$. For $X \geq 2$, let $\mathcal{P}(d, X)$ denote the number of polynomials $G_{\boldsymbol{b}}(x)$ of the form (21) such that

$$\|G_{\boldsymbol{b}}\|_\infty = \max\{|2b_0+1|, |2b_1|, |2b_2|, \ldots, |2b_{d-1}|, |4b_d+2|\} \leq X \quad (22)$$

as well as

$$\gcd\{2b_0 + 1, 2b_1\} = 1. \quad (23)$$

Then

$$\mathcal{P}(d, X) = 2f(X)\left(2\left[\frac{1}{2}X\right]+1\right)^{d-2}\left[\frac{1}{4}X + \frac{1}{2}\right] \geq 2^{-d} f(X) X^{d-1},$$

where $f(X)$ is the number of pairs $(b_0, b_1)$ with (23) and

$$\max\{|2b_0+1|, |2b_1|\} \leq X.$$

Now it is easy to show that $f(X) > 0$ is asymptotic to $8\pi^{-2}X^2$, and so there is an absolute constant $c > 0$ such that $f(x) \geq cX^2$ for all $X \geq 2$ (in fact $c = .46$ is permissible). We deduce that

$$\mathcal{P}(d, X) \geq 2^{-d} c X^{d+1}. \quad (24)$$

Using Jensen's inequality and Parseval's identity (see also [4] and [5, page 60]), we get the well-known bound

$$M(G_{\boldsymbol{b}}) \leq (d+1)^{1/2} \max\{|2b_0+1|, |2b_1|, |2b_2|, \ldots, |2b_{d-1}|, |4b_d+2|\}. \quad (25)$$

It follows by (25) that

$$\mathcal{P}(d, X) \leq N_I(d, (d+1)^{1/2}X),$$

and then from (20) we get

$$d\mathcal{P}(d, (d+1)^{-1/2}\mathcal{H}^d) \leq 2\overline{\mathbb{Q}}_d(\mathcal{H}). \quad (26)$$

Now if $\mathcal{H}^d \geq 2\sqrt{d+1}$, then (24) and (26) can be combined to give a lower bound like (5) that actually improves on (5) if $d$ is large enough (thanks to the factor of $d$). Of course, (5) is trivial if $\mathcal{H}^d < 2\sqrt{d+1}$ in view of $\alpha = 2^{1/d}$.

In a sequel to this work [7] we extend our theorem to obtain an asymptotic estimate for the number of algebraic numbers $\alpha$ having degree $d$ over a fixed number field $K$ and $H(\alpha) \leq \mathcal{H}$. In fact we formulate everything in terms of certain generalized heights on $K^{d+1}$, so that we can recover the $(d+1)$-dimensional version of Schanuel's Theorem. For $K = \mathbb{Q}$ an amusing consequence in the style of Lemma 1 is the following: if $r = 1$ or $r = d$, then as $T \to \infty$ there are asymptotically

$$\frac{\delta_r(d) T^{d+1}}{2\zeta(d+1)}$$

connected $r$-dimensional algebraic subgroups of the multiplicative group $\mathbb{G}_m^{d+1}$ whose degree in the projective completion $\mathbb{P}^{d+1}(\mathbb{C})$ does not exceed $T$. Here we find that

$$\delta_1(d) = d + 1, \quad \text{and} \quad \delta_d(d) = \frac{(2d+1)!}{((d+1)!)^3}.$$

**Acknowledgment.** Research of J.D.V. is supported in part by the National Science Foundation (DMS-00-88915).

## References

1. Chern, S.-J., Vaaler, J.D.: The distribution of values of Mahler's measure. *J. Reine Angew. Math.* **540**, 1–47 (2001)
2. Dubickas, A., Konyagin, S.V.: On the number of polynomials of bounded measure. *Acta Arith.* **86**, 325–342 (1998)
3. Kronecker, L.: Zwei Sätze über Gleichungen mit ganzzahligen Coefficienten. *J. Reine Angew. Math.* **53**, 173–175 (1857)
4. Landau, E.: Sur quelques théorémes de M. Petrovic relatifs aux zéros des fonctions analytiques. *Bull. Soc. Math. Fr.* **33**, 251–261 (1905)
5. Lang, S.: *Fundamentals of Diophantine Geometry*. Springer, Heidelberg (1983)
6. Loher, T.: Counting points of bounded height. Ph.D. thesis, University of Basel, Basel, Switzerland (2001)
7. Masser, D., Vaaler, J.D.: Counting algebraic numbers with large height II. *Trans. Am. Math. Soc.* **359**, 427–445 (2007)
8. Northcott, D.G.: An inequality in the theory of arithmetic on algebraic varieties. *Proc. Camb. Philos. Soc.* **45**, 502–509, 510–518 (1949)
9. Schanuel, S.H.: Heights in number fields. *Bull. Soc. Math. Fr.* **107**, 433–449 (1979)
10. Schmidt, W.M.: Northcott's theorem on heights I. A general estimate. *Monatsh. Math.* **115**, 169–181 (1993)
11. Schmidt, W.M.: Northcott's theorem on heights II. The quadratic case. *Acta Arith.* **74**, 343–375 (1995)
12. Specht, W.: Zur Zahlentheorie der Polynome IV. *Math. Z.* **57**, 291–335 (1953)

*Diophantine Approximation*
H. P. Schlickewei et al., Editors


# CLASS NUMBER CONDITIONS FOR THE DIAGONAL CASE OF THE EQUATION OF NAGELL AND LJUNGGREN

Preda Mihăilescu
*Mathematisches Institut, Universität Göttingen, Bunsenstrasse 3–5, 37073 Göttingen, Germany*
preda@uni-math.gwdg.de

L'ancien va droit comme un mur,
Il parle peu, il se souvient.
Les gens l'ont toujours connu
On le salue,
Forcément, c'est l'ancien!*

To W. Schmidt on the occasion of his 70th birthday

## 1 Introduction

The diagonal case of the Nagell–Ljunggren equation is

$$\frac{x^p - 1}{x - 1} = p^e \cdot y^p \quad \text{with } x, y \in \mathbb{Z} \quad e \in \{0, 1\}, \tag{1}$$

and $p$ an odd prime. The only known nontrivial solution is

$$\frac{18^3 - 1}{18 - 1} = 7^3, \tag{2}$$

and it is conjectured to be also the only such solution. However, it is not even proved that (1) has only finitely many solution.

Currently the most effective methods used for the analysis of (1) are tools of Diophantine approximation and linear forms in logarithms and the *Diagonal Case* under investigation is considered to be the hardest case of the Ljunggren–Nagell equation. In a recent paper which treats this equation in general [BHM], Bugeaud, Hanrot and Mignotte prove with these methods that (1) has no nontrivial solutions for $p < 17$, except for (2). We also refer to this paper and Ribenboim's book [Ri] for detailed references on the history of research on the equation under consideration.

One easily verifies that if $x, y$ is solution for a given prime $p$, then $e = 1$ when $x \equiv 1 \mod p$, and $e = 0$ otherwise. For reasons which will be explained below, we

*Keywords.* Exponential diophantine equation, diagonal case of Nagell and Ljunggren.

*2000 Mathematics subject classification.* 11D61, 11D45.

* "L'ancien" by Hugues Aufray. The rimes may not all fit the jubilee as a glove. But they describe a nice way of coming of age in a community, and this *does* fit him.

denote by the Second Case of (1) the set of those (eventual) solutions for which $x \equiv s \mod p$ with $s \in S = \{-1, 0, 1\}$. All remaining solutions belong to the First Case. This case may be treated with methods reminiscent of the same case of Fermat's Last Theorem.

Concretely, we prove in this paper the following theorems with classical flavor and methods.

**Theorem 1.** *Suppose that $x, y, z$ are coprime integers and $p$ a prime such that*

$$\frac{x^p + y^p}{x + y} = z^p, \quad xy(x^2 - y^2) \not\equiv 0 \mod p. \tag{3}$$

*Then the $p$-rank of the relative class group of the $p$-th cyclotomic extension is $r > \sqrt{p} - 1$.*

We shall see that if the above equation has a solution, then all prime factors of $z$ split completely in the $p$-th cyclotomic extension and are thus $\ell \equiv 1 \mod p$. In particular, $z \equiv 1 \mod p$ and a fortiori $z \not\equiv 0 \mod p$.

**Theorem 2.** *If* (3) *has a solution, then*

$$2^{p-1} \equiv 1 \mod p^2 \text{ and } 3^{p-1} \equiv 1 \mod p^2. \tag{4}$$

The equation (3) generalizes both the First Case of Fermat's Equation and of the Nagell–Ljunggren one. The proof of the above theorems follows closely the counterparts for the FLT case, although the present proof of (4) uses an approach which is more direct than that described in [Ri1].

Restricting generality to (1), we then give some explicit upper bounds for solutions of this equations, as follows:

**Theorem 3.** *Suppose that $x, y$ are integers verifying* (1) *and $p \geq 17$. Then there is a $B \in \mathbb{R}_+$ such that $|x| < B$. The values of $B$ in the various cases of the equation are the following:*

$$B = \begin{cases} 4 \cdot (p - 3/2)^{(p+2)/2} & \text{in the First Case} \\ (4p)^{(p-1)/2} & \text{in the Second Case, if } s = 0, \\ 4 \cdot (p-2)^p & \text{otherwise.} \end{cases} \tag{5}$$

It follows in particular that (1) has at most finitely many solutions for fixed $p$. Although the bounds in (5) are not too large, we could not prove so far sufficient lower bounds in general.

For the Second Case, where the classical methods inherited from the study of Fermat's Last Theorem cannot be adapted, we can however prove the condition $p \mid h_p^+$, which is more than was known for FLT:

**Theorem 4.** *If $p \geq 17$ and* (1) *has a solution in the Second Case, then*

$$p \mid h_p^+, \tag{6}$$

*the class number of the maximal real extension of the $p$-th cyclotomic field.*

The plan of this paper is as follows. In the second section we introduce notation and the general notions of cyclotomy which we use subsequently. In the third section we derive the classical results on (3) and prove Theorems 1 and 2. In section four we generalize some methods used in the proof of Catalan's conjecture and thus derive the upper bounds in Theorem 3. Finally, in section five we assume $p \nmid h_p^+$ in the Second Case and derive lower bounds which contradict (5), thus proving Theorem 4.

## 2 Cyclotomic fields

Let $p$ be an odd prime. The study of equations (1) and (3) is intimately related to properties of the $p$-th cyclotomic field. We develop in this section the notation used subsequently and present some classical results on the Stickelberger module and annihilation of $p$-parts of the class group of this field. Furthermore, we deduce some classical results which explicitly relate the equations under investigation to the properties of the cyclotomic field.

### 2.1 Prerequisites and notations

Let $\mathbb{K} = \mathbb{Q}(\zeta) = \mathbb{Q}[X]/(\Phi_p(X))$ be the $p$-th cyclotomic extension, with $\Phi_p(X) = (X^p - 1)/(X - 1)$. We let $P = \{1, 2, \ldots, p-1\}$, $\overline{P} = P \cup \{0\}$ and $\sigma_c$ be the automorphism of $\mathbb{Q}(\zeta)$ with $\zeta \longrightarrow \zeta^c$, for $c \in P$. The Galois group of $\mathbb{K}$ is $G = \text{Gal}(\mathbb{Q}[\zeta]/\mathbb{Q}) = \{\sigma_c : c \in P\}$. Complex conjugation is denoted by $\jmath \in G$, so $\overline{\alpha} = \sigma_{p-1}(\alpha) = \jmath\alpha$. We write $\lambda = (1 - \zeta)$ for an algebraic integer generating the unique ramified prime ideal above $p$ in $\mathbb{K}$. If $\Theta = \sum_{c \in P} n_c \sigma_c \in \mathbb{Z}[G]$, we denote its *weight* by $w(\Theta) = \sum_{c \in P} n_c$. Note the following useful expansions:

$$\zeta^a = (1-\lambda)^a = 1 - a\lambda + \binom{a}{2}\lambda^2 + O(\lambda^3), \tag{7}$$

$$\sigma_a(\lambda) = 1 - \zeta^a = a\lambda \cdot \left(1 - \frac{a-1}{2}\lambda + \frac{(a-1)(a-2)}{6}\lambda^2\right) + O\left(\lambda^4\right). \tag{8}$$

Let $\mathbb{Q}_p$ be the $p$-adic field and $\mathbb{K}_p = \mathbb{Q}_p[X]/(\Phi_p(X))$ the $p$-th cyclotomic extension of $\mathbb{Q}_p$. Then $\mathbb{K}_p/\mathbb{Q}_p$ is a totally ramified extension of degree $p - 1$ and $\text{Gal}(\mathbb{K}_p/\mathbb{Q}_p) \cong G$; we shall denote the $p$-th root of unity in $\mathbb{K}_p$ – i.e., the image of the independent variable $X$ in $\mathbb{K}_p$ – by $\zeta_p$, thus suggesting an embedding $\iota : \mathbb{K} \to \mathbb{K}_p, \zeta \in \mathbb{K} \mapsto \zeta_p \in \mathbb{K}_p$.

Note that $\alpha \in \mathbb{K}$ is a $p$-adic $p$-th power, in the sense that there exists a $\beta \in \mathbb{K}_p$ such that $\iota(\alpha) = \beta^p$, iff

$$\exists \gamma \in \mathbb{K} \quad \text{such that} \quad \alpha \equiv \gamma^p \mod p\lambda^2. \tag{9}$$

We say that $\alpha$ is ***p*-primary**, if (9) is verified. By class field theory (e.g., [Wa, Chapter 9]), if $\alpha$ is a $p$-primary number which is either a unit or such that there is an ideal $\mathfrak{B}$ of order $p$ in $\mathbb{K}$ with $(\alpha) = \mathfrak{B}^p$ (in such case, $\alpha$ is sometimes called $p$-primary singular), then $\mathbb{K}[\alpha^{1/p}]$ is an unramified Abelian extension of degree $p$ (class field). In particular the existence of such a $p$-primary singular number implies $p|h_p$, the class number of $\mathbb{K}$. We shall expand this observation in the section on group rings.

The following lemma provides some means for identifying $p$-adic $p^k$-th powers:

**Lemma 1.** *Let $\alpha \in \mathbb{Z}[\zeta]$, $(\alpha, p) = 1$ and $0 < k \le p$. Then $\iota(\alpha)$ is a $p^k$-th power iff $\alpha \equiv a_0^p \mod p^k\lambda^2$, with $a_0 \in \mathbb{Z}$.*

*Proof.* We start by developing $\lambda^{p-1}$ under use of (7) and Wilson's Theorem:

$$\lambda^{p-1}/p = \prod_{c=1}^{p-1} \frac{1-\zeta}{1-\zeta^c} \equiv 1/(p-1)! \equiv -1 \mod \lambda.$$

Thus $\lambda^{p-1} = -p + O(p\lambda)$. Next note that for any $\beta \in \mathbb{Z}[\zeta]$ we have $\beta = a_0 + \lambda \cdot \gamma$, with $\gamma \in \mathbb{Z}[\zeta]$ and $a_0 \in \mathbb{Z}$, $a_0 \equiv \beta \mod \lambda$.

Let now $k = 1$, $\iota(\alpha) = \beta^p$ and $\beta_N \in \mathbb{Z}[\zeta]$ be such that $\beta \equiv \beta_N \mod p^N\mathbb{Z}[\zeta]$ for some sufficiently large $N > 1$, so $\iota(\alpha) \equiv \beta_N^p \mod p^N$. If $\beta_N = a_0 + \lambda\gamma$ and $\gamma \equiv c \mod \lambda$, then

$$\begin{aligned}\alpha &\equiv (a_0 + \lambda\gamma)^p \equiv a_0^p + a_0^{p-1} p\lambda\gamma + \lambda^p\gamma^p \\ &\equiv a_0^p + pc\lambda + c\lambda^{p-1}\lambda \equiv a_0^p + pc\lambda - pc\lambda \equiv a_0^p \mod p\lambda^2.\end{aligned}$$

For $k > 1$, using the same notation as in the case $k = 1$, we have $\alpha = (a_0 + \lambda\gamma)^{p^k}$. The valuation of the binomial coefficients is $v_p\left(\binom{p^k}{n}\right) = k - v_p(n)$, for $n > 0$. The only terms of the binomial expansion of $(a_0 + \lambda\gamma)^{p^k}$ which do not vanish modulo $p^k\lambda^2$ are the terms zero, one and $p$ and it follows that

$$\alpha \equiv a_0^{p^k} + a_0^{p^k-1}(c\lambda) + \binom{p^k}{p} a_0^{p^k-p}(c\lambda)^p \equiv a_0^{p^k} + c\lambda - c\lambda \equiv a_0^{p^k} \mod p^k\lambda^2.$$

This completes the proof. □

## 2.2 Jacobi sums and the Stickelberger module

Let[1] $\ell$ be a prime such that $p|(\ell - 1)$, $\rho \in \mathbb{C}$ a primitive $\ell$-th root of unity and $\chi : \mathbb{Z}/(\ell \cdot \mathbb{Z}) \to <\zeta>$ a character of conductor $\ell$ and order $p$ with $\chi(0) = 0$. Let $a, b \in P$ with $a + b \not\equiv 0 \mod p$. The Gauss sum $\tau(\chi)$ and Jacobi sum $j(\chi^a, \chi^b)$ are defined as sums of characters by

$$\begin{aligned}\tau(\chi) &= \sum_{x \in \mathbb{Z}/(\ell\cdot\mathbb{Z})} \chi(x) \cdot \rho^x \\ j(\chi^a, \chi^b) &= - \sum_{x \in \mathbb{Z}/(\ell\cdot\mathbb{Z})} \chi^a(x) \cdot \chi^b(1-x)\end{aligned} \tag{10}$$

The Jacobi sums are defined according to Lang [La], thus with a sign change with respect to their classical definition. They are related to Gauss sums by

$$j(\chi^a, \chi^b) = -\frac{\tau(\chi^a) \cdot \tau(\chi^b)}{\tau(\chi^{a+b})}.$$

It is well known that the Gauss sum verifies $\tau(\chi) \cdot \overline{\tau(\chi)} = \ell$. If $\mathfrak{L} \supset (\ell)$ is a prime above $\ell$ in $\mathbb{Z}[\zeta]$, then the ideal $(j(\chi^a, \chi^b))$ will be expressed as the action of a group

1. Part of this section was used in another paper of the author.

ring element $\psi(a, b) \in \mathbb{Z}[G]$ upon $\mathfrak{L}$. The theorem of Stickelberger ([IR], [Wa]) computes this element: for a certain $\mathbb{Z}[\zeta_\ell] \supset \mathfrak{L} \supset (\ell)$,

$$\left(j(\chi^a, \chi^b)\right) = \mathfrak{L}^{\psi(a,b)} \text{ with} \qquad (11)$$

$$\psi(a, b) = \sum_{c \in P} \left( \left[ \frac{c(a+b)}{p} \right] - \left[ \frac{ca}{p} \right] - \left[ \frac{cb}{p} \right] \right) \cdot \sigma_c^{-1}.$$

Here the brackets denote, as usual, the *integer part*: $[x] = \max\{a \in \mathbb{Z} : a \leq x\}$. The ideal $I \subset \mathbb{Z}[G]$ which is generated by such elements is called the *Stickelberger* ideal and will be presented subsequently in some detail.

We define by multiplicativity the set of *Jacobi integers* to be the subset $\mathbf{J} \subset \mathbb{Z}[\zeta]$ of products of Jacobi sums according to the above definition, and $\mathfrak{J} = \{(\mathbf{j}) : \mathbf{j} \in \mathbf{J}\}$ the subset of principal ideals generated by Jacobi integers. Let $\mathfrak{A} \subset \mathbb{Z}[\zeta]$ be an ideal with $\mathbf{N}(\mathfrak{A}) = t$, such that $t$ factors into powers of primes $\ell \equiv 1 \mod p$. Then Stickelberger's theorem implies in particular that $\mathfrak{A}^\Theta \in \mathfrak{J}, \ \forall \Theta \in I$.

The following lemma is a special-case adaptation of proposition 1.2 of [Jh], which relates $\mathfrak{J}$ to $\mathbf{J}$.

**Lemma 2.** *Let $\iota$ be the natural map $\iota : \mathbf{J} \to \mathfrak{J}$ given by $\mathbf{j} \mapsto (\mathbf{j})$. Then $\iota$ is injective. In particular, a principal ideal can be generated by at most one Jacobi integer and if for some $\alpha \in \mathbb{Z}[\zeta]$ with $\alpha \cdot \overline{\alpha} \in \mathbb{Z}$ the equality $(\alpha) = \mathfrak{j} \in \mathfrak{J}$ holds, then there is a unique Jacobi integer* $\mathbf{a}$ *with*

$$\alpha = \pm \zeta^n \cdot \mathbf{a}, \quad n \in \mathbb{Z}. \qquad (12)$$

*Proof.* Let $\alpha$ generate the principal ideal $\mathfrak{j} \in \mathfrak{J}$ and let $\mathbf{a} \in \mathbf{J}$ with $(\alpha) = (\mathbf{a})$: such a Jacobi integer exists by definition of $\mathfrak{J}$. The principal ideals being equal, there is a unit $\varepsilon \in \mathbb{Z}[\zeta]$ such that $\alpha = \varepsilon \cdot \mathbf{a}$. Furthermore, $\mathbf{a} \cdot \overline{\mathbf{a}} \in \mathbb{N}$ follows by multiplicativity from the property of Gauss sums and since $\alpha \cdot \overline{\alpha} \in \mathbb{Z}$, it follows that $\varepsilon \cdot \overline{\varepsilon} = 1$. By Kronecker's unit theorem, $\varepsilon$ is a root of unity. This proves the identity (12).

We still have to prove that the Jacobi integer $\mathbf{a}$ is unique. Iwasawa shows in [Iw] (see also [IR, p. 226, Ex. 13]) that Jacobi integers $\mathbf{j}$ verify in general

$$\mathbf{j} \equiv 1 \mod (1 - \zeta)^2 \mathbb{Z}[\zeta]. \qquad (13)$$

This property is useful for establishing the power $n$ in (12). In particular, assuming there is a second Jacobi integer $\mathbf{a}'$, with $(\alpha) = (\mathbf{a}')$, we would have by (12) that $\alpha = \pm \zeta^{n'} \mathbf{a}'$ for some $n' \in \overline{P}$. By Iwasawa's relation, $\alpha \equiv s\zeta^n \equiv s'\zeta^{n'} \mod \lambda^2$, where $s, s'$ are the implicit signs for $\mathbf{a}, \mathbf{a}'$. In particular $s\zeta^n \equiv s \equiv s'\zeta^{n'} \equiv s' \mod \lambda$ and thus $s = s'$. Also, $1 - \zeta^n \equiv n\lambda \equiv 1 - \zeta^{n'} \equiv n'\lambda \mod \lambda^2$, so $n \equiv n' \mod p$. Consequently, $\mathbf{a} = \mathbf{a}'$, which completes the proof. □

The notions of Gauss and Jacobi sums are easily generalized to composite orders and to characters of prime power conductor [La]. The Jacobi integers are generalized accordingly and their natural extension to *Jacobi fractions* is reflecting the extension of integral to fractional ideals [Jh]. However, we shall not need such generalizations below.

### 2.3 Stickelberger ideal

We introduce in this section some general and computable results about the Stickelberger ideal. We recommend, besides the classics by Washington [Wa] and Lang [La], the very rich and appealing treatment of the subject by Jha [Jh]. We shall write $d = (p-1)/2$ for convenience.

Let $\vartheta = \sum_{c=1}^{p-1} \{c/p\} \cdot \sigma_c^{-1} \in \mathbb{Q}[G]$ be the Stickelberger element. Stickelberger's theorem states precisely that $(\tau(\chi)) = (\mathfrak{L}^{\vartheta})$ as ideals, where $\mathfrak{L}$ lies above the conductor $\tau(\chi) \cdot \overline{\tau}(\chi)$ in $\mathbb{Q}(\zeta)$, while $\mathfrak{L}^{\vartheta}$ is an ideal of an extension of degree $p$ of $\mathbb{Q}(\zeta_p)$. The interpretation of the action of the fractional $\vartheta$ as an element of the group ring of an extension is known and we refer to [La], [Wa] for details.

The Stickelberger ideal is defined by $I = \mathbb{Z}[G] \cap \vartheta\mathbb{Z}[G] \subset \mathbb{Z}[G]$. Let $I' \subset \mathbb{Z}[G]$ be the ideal generated by elements of the form $n - \sigma_n$, $(n, p) = 1$. Then $I = \vartheta I'$ [Wa, §6.2]. The elements $\Theta_n = (n - \sigma_n)\vartheta \in I$ are often called the *Fuchsian* elements. We also write $\Theta_p = p \cdot \vartheta = \sum_{c=1}^{p-1} c \cdot \sigma_c^{-1}$. The differences $\psi_n = \Theta_{n+1} - \Theta_n$, $n \geq 2$ and $\psi_1 = \Theta_2$ are the *Fueter* elements. For the group ring elements defined in the previous section one finds: $\psi(a, b) = \sigma_a \psi_{b/a}$, where the fractional indexes are taken modulo $p$. Both $\Theta_n$, $\psi_n$ form for $n = 1, 2, \ldots, d$ a $\mathbb{Z}$-base of the Stickelberger ideal [Jh]. We have for all $n$ with $(n, p) = 1$:

$$\Theta_n = (n - \sigma_n) \cdot \vartheta = \sum_{c=1}^{p-1} \left( \left[\frac{nc}{p}\right] - n\left[\frac{c}{p}\right] \right) \cdot \sigma_c^{-1} = \sum_{c=1}^{p-1} \left[\frac{nc}{p}\right] \cdot \sigma_c^{-1};$$

$$\psi_n = \sum_c \nu_c[n]\sigma_c^{-1} = \sum_c \left( \left[\frac{(n+1)c}{p}\right] - \left[\frac{nc}{p}\right] \right) \cdot \sigma_c^{-1}, \tag{14}$$

where $\nu_c[n] \geq 0$ and $\nu_c[n] + \nu_{p-c}[n] = 1$.

Note that the last relation implies $0 \leq \nu[n]_c \leq 1$ and

$$(1 + \jmath) \cdot \psi_n = \mathbf{N}_{\mathbb{Q}}^{\mathbb{Q}(\zeta)}, \tag{15}$$

as elements of the group ring. Since the Fueter elements generate $I$ as a $\mathbb{Z}$-module, the identity implies herewith that the *weight*

$$w(\Theta) = w\left( \sum_c n_c \sigma_c^{-1} \right) = \sum_c n_c \equiv 0 \mod (p-1)/2, \quad \forall \Theta \in I.$$

The Stickelberger ideal acts also on roots of unity and there is an additive map $\varphi : I \rightarrow \overline{P}$ such that

$$\zeta_p^{\varphi(\Theta)} = \Theta(\zeta_p), \quad \forall\, \Theta \in I. \tag{16}$$

In fact the map $\varphi$ is determined by the following relation, which motivates the name *Fermat quotient map*:

$$\varphi(\Theta_n) = \frac{n^p - n}{p} \in \overline{P}; \tag{17}$$

this follows from the Voronoi identities [IR]. Note that $\varphi I :\longrightarrow \mathbb{Z}/(p \cdot \mathbb{Z})$ is a linear functional; let $I_f \subset I$ be its kernel. We denote $I_f = \{\theta \in I : \zeta^{\theta} = 1\}$ by *Fermat module* and let

$$I_0 = \left\{ \theta = \sum_{c \in P} n_c \sigma_c^{-1} \in I_f : n_c \geq 0 \right\} \tag{18}$$

be the set of *positive* elements of the Fermat module. Note that if $\theta_p \in I_f/(p \cdot I_f)$, there always is a lift of $\theta_p$ to some positive $\theta \in I_0$, thus $\theta \equiv \theta_p \mod pI_f$. The following lemma ascertains that there always are elements of small weight in the Fermat module.

**Lemma 3.** *Let $p > 7$ be a prime and $I_0$ be the set of positive elements from the associated Fermat module. Then there are $\theta_1, \theta_2 \in I_0$ with $w(\theta_i) = p - 1$, $i = 1, 2$ and such that for all $\sigma \in G$, $\theta_2 \neq \sigma\theta_1$. If $p = 7$, then $\psi_2 \in I_0$ is an element of weight $(p - 1)/2$. For $p = 11$ one has $\theta_1 = \Theta_3 \in I_0$ and $\theta_2 = \Theta_{1,4} = \psi_1 + \sigma_2\psi_4 \in I_0$.*

*Proof.* Let $\varphi_i = \varphi(\psi_i)$, $i = 1, 2, \ldots, (p - 1)/2$ be the Fermat quotients associated to the Fueter base of $I$. Then there are for each pair $(i, j)$ two integers $a_i, b_j \in P$ such that

$$a_i\varphi_i + b_j\varphi_j \equiv 0 \mod p \quad \text{and thus} \quad \varphi(\Theta_{i,j}) = \varphi\left(\sigma_{a_i}\psi_i + \sigma_{b_j}\psi_j\right) = 0.$$

Thus $\Theta_{i,j} \in I_0$ and have weight $(p - 1)$. We still have to show that among these there are two elements which are Galois independent. A simple counting argument shows that there are $\binom{(p-1)/2}{2}$ elements $\Theta_{i,j}$ and each of them has exactly $(p - 1)$ conjugates. Since for $p > 11$ the inequality $\binom{(p-1)/2}{2} > p - 1$ holds, we still have to check the cases $p = 7$ and $p = 11$ (note that the $\Theta_{i,j}$ are not necessarily *all* $\Theta \in I_0$ with weight $p - 1$!). For $p = 7$ we find that $\varphi(\Theta_3) = \varphi(\Theta_2) = 4$ and thus $\varphi(\Theta_3 - \Theta_2) = \varphi(\psi_2) = 0$ and $\psi_2 \in I_0$ is an element of weight $(p - 1)/2$. The Galois independence is verified by writing the two elements as multiples of $\vartheta$; the question then reduces to one of independence in the group ring. Since $w(\theta_i) = p - 1$ for $i = 1, 2$, this completes the proof of the lemma. □

## 2.4 Gauss sums

Let $p$ be an odd prime like above and $\zeta$ a primitive $p$-th root of unity. We now give a definition of Gauss integers by analogy to the one of Jacobi integers above. Let thus $\ell \equiv 1 \mod p$ be a prime, $\zeta_\ell$ a primitive $\ell$-th root of unity and $\mathfrak{L} \subset \mathbb{Z}[\zeta]$ be a prime ideal above $\ell$. Then for any Gauss sum of conductor $\ell$ and order $p$ one has $(\tau(\chi)) = \mathfrak{L}^\vartheta$ as principal ideals. As mentioned above $\tau(\chi) \in \mathbb{L}$, where $\mathbb{L}/\mathbb{Q}(\zeta)$ is a cyclic extension of degree $p$, abelian over $\mathbb{Q}$. It is in fact, a subfield of the cyclotomic extension $\mathbb{Q}(\zeta, \zeta_\ell)$. The Gauss integers are defined by

**Lemma 4.** *Let $y \in \mathbb{Z}$ split in primes $\ell \equiv 1 \mod p$ and $\mathfrak{Y} \subset \mathbb{Z}[\zeta]$ an ideal with norm $\mathbf{N}(\mathfrak{Y}) = y$. Then there is a Gauss sum $G = \tau(\chi)$ of conductor $y' \mid y$ and a Jacobi integer $J$ such that*

$$\mathfrak{L}^\vartheta = (\tau(\chi) \cdot J). \tag{19}$$

*Proof.* Let $y = \prod_{i=1}^k \ell_i^{e_i}$ with $\ell_i \equiv 1 \mod p$ and $e_i > 0$. If $\mathfrak{L}_i \subset \mathbb{Z}[\zeta]$ is a prime which lies above $\ell_i$, then $\mathfrak{L}_i^{p \cdot \vartheta} = (J_i)$ is generated by some Jacobi integer $J_i$, as a consequence of the fact that $p\vartheta = \Theta_p \in I$. We may thus, after eventual division by a Jacobi integer $J_0$, assume that $0 < e_i < p$ for $1 \leq i \leq k$.

In this case, let us first assume that for each $\ell_i$ there is exactly one prime $\mathfrak{L}_i | (\ell_i, \mathfrak{Y})$. Let $\theta_i = (e_i - \sigma_{e_i})\vartheta \in I$; then

$$\mathfrak{L}_i^{e_i \vartheta} = \mathfrak{L}_i^{\theta_i} \cdot \sigma_{e_i} (\mathfrak{L}_i)^{\vartheta} = (J_i \cdot \tau_i(\chi))$$

is a principal ideal generated by a product of a Gauss sum by a Jacobi integer. It is a common fact that a Gauss sum of composite conductor $y'$ reduces to Gauss sums of prime conductors $\ell' \mid y'$ modulo Jacobi sums. Consequently, the claim of the lemma is in this case proved.

Let us now assume that there are two primes $\mathfrak{L}_i, \mathfrak{L}_i'$ which both divide $(\ell_i, \mathfrak{Y})$, with respective exponents, say, $a$ and $b$; thus $\left(\mathfrak{L}_i, \mathfrak{Y}/\mathfrak{L}_i^a\right) = 1$ and $\left(\mathfrak{L}_i', \mathfrak{Y}/\mathfrak{L}_i'^{b}\right) = 1$. The above argument shows then that there will be two Gauss sums $\tau(\chi), \tau(\chi')$ of conductor $\ell_i$ and such that $\left(\mathfrak{L}_i^a \cdot \mathfrak{L}_i'^{b}\right)^{\vartheta} = \left(\tau(\chi) \cdot \tau(\chi') \cdot J\right)$. Either $\frac{\tau(\chi) \cdot \tau(\chi')}{\tau(\chi \cdot \chi')}$ is a Jacobi integer, in the case $\chi \cdot \chi' \neq 1$, or $\tau(\chi) \cdot \tau(\chi') = \ell_i$, if $\chi' = \overline{\chi}$. We can thus reduce these cases to one single generating Gauss sum. The argument can be repeated inductively with respect to the number of distinct primes above $\ell_i$ and dividing $\mathfrak{Y}$, thus completing the proof of the lemma. □

Note that the conductor $y' \mid y$ in the lemma can very well be a proper divisor of $y$. This is, for instance, most simply, the fact when some of the exponents $e_i = p$.

## 2.5 The group ring $\mathbb{F}_p[G]$

If $p$ is an odd prime and $\mathbb{K} = \mathbb{Q}(\zeta)$, the $p$-th cyclotomic extension with Galois group $G$, then its group ring is $\mathbb{Z}[G]$ and it acts naturally on Abelian groups such as the class group or groups of units. Let $\mathbb{F}_p[G] = \mathbb{Z}[G]/(p \cdot \mathbb{Z}[G])$ be the group ring modulo $p$, which acts on the subgroups of elements of order $p$ in the Abelian groups above. The Dirichlet characters of $\mathbb{K}$ with image in $\mathbb{F}_p$ are generated by the Teichmüller (or cyclotomic) character $\omega : G \rightarrow \mathbb{F}_p$ given by $\omega(\sigma_c) = c$.

With this, one defines the following set of orthogonal idempotents on $\mathbb{F}_p[G]$ [Wa, Chapter 6]

$$\varepsilon_i = \frac{1}{p-1} \sum_{\sigma \in G} \omega^i(\sigma) \cdot \sigma^{-1} \in \mathbb{F}_p[G], \quad i = 1, \ldots, p-1.$$

Note that $\varepsilon_{p-1} = \mathbf{N}$ and $\varepsilon_i \cdot \varepsilon_j = \delta_{i,j}\varepsilon_i$, the last relation explaining the name of orthogonal idempotents. Furthermore one has the useful isomorphism of algebras:

$$\mathbb{F}_p[G] \cong \mathbb{F}_p[X]/\left(X^{p-1} - 1\right) : \quad \sigma \mapsto X \mod (X^{p-1} - 1),$$

where we denote as usual a generator of the cyclic group $G$ by $\sigma$. We let $\mathbf{N} = \mathbf{N}_{\mathbb{Q}(\zeta)/\mathbb{Q}}$ and consider it as an element of the group ring $\mathbb{F}_p[G]$; let $\mathbf{R} = \mathbb{F}_p[G]/(\mathbf{N})$ which we identify with $\sum_{i=1}^{p-2} \varepsilon_i \cdot \mathbb{F}_p[G]$. If $A$ is an Abelian group (written multiplicatively) on which $\mathbf{R}$ or $\mathbb{F}_p[G]$ acts, then the idempotents induce a splitting of $A$ in components $A_i = \varepsilon_i \cdot A$, such that $A = \sum_i A_i$ and $A_i \cap A_j = \{1\}$. If $\jmath \in G$ denotes the complex conjugation, one writes

$$\mathbb{F}_p[G]^- = (1 - \jmath)\mathbb{F}_p[G], \quad \mathbb{F}_p[G]^+ = (1 + \jmath)\mathbb{F}_p[G]$$

for the *minus and plus* parts or the group ring (or imaginary and real parts); when working with $\mathbb{Z}[G]$, the factor $1 \pm \jmath$ requires a denominator 2, which is irrelevant in odd characteristic. Idempotents with odd index correspond to the minus parts and idempotents with even index to the plus part. We define the *support and annihilator* of $A$ by

$$\text{supp}\,(A) = \sum_{i \in P:\ \varepsilon_i A \neq \{1\}} \varepsilon_i \mathbf{R}, \quad \text{ann}(A) = \sum_{i \in P:\ \varepsilon_i A = \{1\}} \varepsilon_i \mathbf{R}.$$

Let $B_n$ be the Bernoulli numbers and $\mathcal{A} = \{x \in \mathcal{C} : x^p = 1\}$ the subgroup of classes of order dividing $p$ in the class group $\mathcal{C}$ of $\mathbb{K}$; furthermore $\mathcal{A}^- = \mathbb{F}_p[G]^- \cdot \mathcal{A}$ and $\mathcal{A}_i = \varepsilon_i \mathcal{A}$. It is a fundamental fact of cyclotomy [Wa, Theorem 6.2] that $\mathcal{A}_i^- \neq \{1\}$ iff $B_{p-i} \equiv 0 \mod p$ (note that $i$ must be odd and thus $p - i$ even; Bernoulli numbers of odd index vanish trivially, except for $B_1$). Let $\mathcal{B} = \{i \in P : B_{p-i} \equiv 0 \mod p, \text{ and } i \equiv 1 \mod 2\}$ be the set of *irregular indices* and $\mathcal{B}' = \{j : j \in \mathcal{B} \text{ or } p - j \in \mathcal{B}\}$. Classical bounds on the number $h_p^-$ [Wa] imply that $\sharp\mathcal{B} \leq (p+1)/4 < (p-1)/2$. We define

$$\mathbf{D}_0 = \sum_{i \in \mathcal{B}} \varepsilon_i \mathbb{F}_p[G] \subset \mathbf{R}, \quad \mathbf{A}_0 = \sum_{j \notin \mathcal{B}} \varepsilon_j \mathbb{F}_p[G] \subset \mathbf{R}$$

$$\mathbf{D}_1 = \sum_{i \in \mathcal{B}'} \varepsilon_i \mathbb{F}_p[G] \subset \mathbf{R}, \quad \mathbf{A}_1 = \sum_{j \notin \mathcal{B}'} \varepsilon_j \mathbb{F}_p[G]. \tag{20}$$

By properties of the idempotents, $\mathbf{A}_j \cdot \mathbf{D}_j = 0$ for $j = 0, 1$, while $\mathbf{A}_j \oplus \mathbf{D}_j = \mathbf{R}$. Note that the previous estimate for $\sharp\mathcal{B}$ implies in particular that $\mathbf{A}_j \neq \{1\}$. Furthermore, the modules $\mathbf{D}_j$ are connected to the class group and unit group by:

**Lemma 5.** *Let* $\mathbf{D}_0, \mathbf{D}_1$ *be defined by* (20). *Then*

$$\mathbf{D}_0 = \text{supp}(\mathcal{A}^-),$$

*and if* $I_p$ *denotes the image of the Stickelberger ideal in* $\mathbb{F}_p[G]$, *we have*

$$(1 - \jmath)I_p^- = \mathbf{A}_0 \cdot I_p^- \cong \mathbf{A}_0 \cdot \mathbb{F}_p[G]^-.$$

*Furthermore, if* $U = E_p/E^p$ *is the module of* $p$*-primary units which are not global* $p$*-th powers, then*

$$\mathbf{D}_1 = \text{supp}(\mathcal{A}) \cup \text{supp}(U_p). \tag{21}$$

*In particular, if* $\mathfrak{A} \subset \mathbb{Z}[\zeta]$ *is an ideal of order* $p$ *and* $\mathfrak{A}^p = (\alpha)$, *then*

$$\forall \theta \in \mathbf{A}_1 \quad \exists \nu \in \mathbb{Z}[\zeta],\ \varepsilon \in \mathbb{Z}[\zeta]^\times : \quad \alpha^\theta = \varepsilon \cdot \nu^p \quad \textit{and} \tag{22}$$

$$\varepsilon \in E_p \Rightarrow \varepsilon \in E^p. \tag{23}$$

*Proof.* Since for odd $i \in P$, $\varepsilon_i \cdot \mathcal{A} \neq \{1\} \Leftrightarrow p | B_{p-i}$, this is equivalent to the property of $\mathbf{D}_0$ enounced by the lemma. The consequence for the Stickelberger module requires some computations; we omit the proof here and refer the reader to [Wa] or [Jh].

The statement on $\mathbf{D}_1$ requires reflection [Wa, Chapter 10]. Let $i \in P$ be even and such that $\varepsilon_i \in \text{supp}\,\mathcal{A}$. So $\mathcal{A}_i = \varepsilon_i \mathcal{A} \neq \{1\}$ and, by reflection, there is a primary singular number in the $\varepsilon_{p-i}$ component, whose $p$-th root generates a class field for the ideals in $\mathcal{A}_i$. But then in particular $j = p-i$ is odd and $\mathcal{A}_j \neq \{1\}$, thus $B_{p-j} = B_i \equiv 0 \mod p$. Thus $\varepsilon_i \cdot \mathbb{F}_p[G]$ is a term in $\mathbf{D}_1$. If $i \in \text{supp}\,(U_p)$, then the $p$-th roots of units in $\varepsilon_i U_p$ generate unramified Abelian extensions. By reflection, these are class fields for some ideals in $\varepsilon_j \mathcal{A}$, and thus $B_{p-j} = B_i \equiv 0 \mod p$ and $p - i \in \mathcal{B}$. The analysis for the minus part has been completed above, and we have altogether shown that $\text{supp}\,(\mathcal{A}) \cup \text{supp}\,(U_p) \subset \mathbf{D}_1 \cdot \mathbb{F}_p[G]$. Conversely, let $i \in \mathcal{B}$; we have shown that $\mathbf{D}_0 = \text{supp}\,(\mathcal{A}^-)$; furthermore, if $p - i \in \mathcal{B}$, so $i$ is even, then $\mathcal{A}_i$ is not empty and either there is a primary unit in $\varepsilon_{p-i} U_p$ whose $p$-th root generates a class field in which ideals from $\mathcal{A}_i$ capitulate – case in which $\varepsilon_{p-i}\mathbb{F}_p[G] \subset \text{supp}\,(U_p)$ – or the class field in which these ideals capitulate are generated by $p$-th roots of primary singular numbers; in the latter case, $\mathcal{A}_{p-i}$ must be nonempty and $\varepsilon_{p-i}\mathbb{F}_p[G] \subset \text{supp}\,(\mathcal{A})$. This completes the proof of (21).

Relations (22) and (23) are a consequence of (21) in the sense that $\mathbf{A}_1$ annihilates both $U_p$ and $\mathcal{A}$. Since $\mathbf{A}_1\mathbf{D}_1 = 0$, for all $\theta \in \mathbf{A}_1$ and $\mathfrak{A}$ an ideal of order $p$, the ideal $\mathcal{A}^\theta$ is principal; by raising the resulting identity $\mathfrak{A}^\theta = (\nu)$ to the power $p$ we obtain (22). We now show (23) for $\theta = \varepsilon_k$ and $k, p - k \notin \mathcal{B}$. The statement will follow by linearity (multiplicativity). We have shown that $\alpha^\theta = \varepsilon \cdot \nu^p$; since in our choice $\theta$ is an idempotent, we have $\theta^2 \equiv \theta \mod p$. By applying $\theta$ again to the previous identity, we obtain $\alpha^{\theta^2} = \varepsilon^\theta \cdot \nu^{p\theta}$ and after regrouping $p$-th powers, $\alpha^\theta = \varepsilon^\theta \cdot \nu_1^p$. But $\theta = \varepsilon_k$ is orthogonal to the support of $U_p$, and since $\varepsilon \mod E^p \in U_p$, it follows that $\varepsilon^\theta \in E^p$ and, plainly, (23) must hold. □

### 2.5.1 Lifts

We shall want to consider the action of elements of $\theta \in \mathbb{F}_p[G]$ on explicit algebraic numbers $\beta \in \mathbb{K}$. Unless otherwise specified, an element $\theta = \sum_c m_c \sigma \in \mathbb{F}_p[G]$ is lifted to $\sum_c n_c \sigma_c$, where $n_c \in \mathbb{Z}$ are the unique integers with $0 \leq n_c < p$ and $n_c \equiv m_c \mod p$. In particular, lifts are always positive, of bounded weight $w(\theta) \leq (p-1)^2$. Rather than introducing an additional notation for the lift defined herewith, we shall always assume, unless otherwise specified, that $\theta \in \mathbb{F}_p[G]$ acts upon $\beta \in \mathbb{K}$ via this lift. In a certain case, when it is important that $\theta \in \mathbf{R} \subset \mathbb{F}_p[G]$ we shall want to use the additional degree of freedom (the $\mathbf{N}$ is 0 in $\mathbf{R}$) for imposing the condition $\sum_c n_c \equiv 0 \mod p$ to the lift defined above.

## 2.6 Cyclotomic properties of the equations

If $x, y, p, e$ verify (1), we shall write $\alpha = (x - \zeta)/(1-\zeta)^e$ and $\alpha_c = \sigma_c(\alpha)$. Likewise, if $x, y, z$ verify (3), we set $\alpha = (x + \zeta y)$. Note that (1) and (3) become

$$\mathbf{N}(\alpha) = y^p \quad \text{resp.} \quad \mathbf{N}(\alpha) = z^p$$

and there is an ideal $\mathfrak{A} = (\alpha, u) \subset \mathbb{Z}[\zeta]$, for $u = y$ or $u = z$, with $\mathfrak{A}^p = (\alpha)$. This follows from the fact that for distinct $a, b \in P$, the ideals $\sigma_a(\mathfrak{A}), \sigma_b(\mathfrak{A})$ are coprime. Indeed, in the case of the equation (1), the ideal containing the above also contains $\left((1-\zeta^a)^e \alpha_a - (1-\zeta^b)^e \alpha_b\right) = \zeta^b - \zeta^a$ and it is thus at most divisible by the ramified prime $\wp$ lying above $p$. But by the choice of $e$, $(\alpha, \wp) = 1$. Thus

$$(\alpha) \mid \mathfrak{A}^p = \left(\alpha^p, y\alpha^{p-1}, \ldots, y^p = \mathbf{N}(\alpha)\right).$$

Since $(\sigma_a(\alpha), \sigma_b(\alpha)) = 1$, it follows that $\mathbf{N}(\alpha)/\alpha$ is coprime to $\alpha$. By comparing to the list of generators of $\mathfrak{A}^p$, it follows that $\mathfrak{A}^p/(\alpha) = (1)$, as claimed. The case of (3) is similar; we find that $(\sigma_a(\alpha), \sigma_b(\alpha)) \subset \wp(x, y) = \wp$ and since $(\alpha, \wp) = 1$, the claim follows.

We restrict now our attention to the Second Case of equation (1) and let $x, y, p$ be a solution. In the Second Case one has by definition $x \equiv s \mod p$ with $s \in \{-1, 0, 1\}$. It will be useful to introduce the following notation:

$$\alpha' = \begin{cases} 1 + \frac{x-1}{1-\zeta} & \text{if } s = 1, \\ 1 - \frac{x+1}{1+\zeta} & \text{if } s = -1, \\ 1 - x\overline{\zeta} & \text{if } s = 0. \end{cases} \qquad \chi = \begin{cases} 1 & \text{if } s = 1, \\ -(1+\zeta) & \text{if } s = -1, \\ -\zeta & \text{if } s = 0. \end{cases} \tag{24}$$

Then $\alpha = \chi \cdot \alpha'$ and $\alpha' \equiv 1 \mod p^2/\lambda$ in all cases. In particular, $\alpha'$ is primary and

$$\mathfrak{A}^p = (\alpha) = (\alpha'). \tag{25}$$

We show first that the congruence $x \equiv s$ must hold modulo $p^2$. For this we start with a small auxiliary computation.

**Lemma 6.** *Let*

$$\varpi_s = \begin{cases} \sum_{c \geq (p+1)/2} \sigma_c^{-1}\left(\frac{1}{1-\zeta}\right) & \textit{if } s = 1, \\ \sum_{c \geq (p+1)/2} \sigma_c^{-1}\left(\frac{1}{1+\zeta}\right) & \textit{if } s = -1, \\ \sum_{c \geq (p+1)/2} \sigma_c^{-1}(\zeta) & \textit{if } s = 0. \end{cases}$$

*Then* $\varpi_s \not\equiv 0 \mod \lambda$.

*Proof.* From the last identity in (14) it follows that for $A \in \mathbb{K}$ and with the definition $\varpi = \sum_{c \geq (p+1)/2} \sigma_c^{-1}(A)$ we have $\varpi + \overline{\varpi} = \mathbf{Tr}_{\mathbb{K}/\mathbb{Q}}(A)$. The statement thus follows if we show that $\mathbf{Tr}_{\mathbb{K}/\mathbb{Q}}(A) \not\equiv 0 \mod \lambda$ for $A = 1/(1 \pm \zeta)$ and $A = \zeta$. Since $\mathbf{Tr}_{\mathbb{K}/\mathbb{Q}}(\zeta) = -1$, the last case is easily settled.

For $A = 1/(1 - \zeta)$, note that $A$ is a zero of the polynomial

$$f(X) = \prod_{c \in P} \left(X(1 - \zeta^c) - 1\right) = \left(X^p - (X-1)^p\right)/X.$$

It follows then from Vietà's formulae that $\mathbf{Tr}(A) = (p-1)/2$ in this case.

For $A = 1/(1 + \zeta)$, the minimal polynomial is

$$f(X) = \prod_{c \in P} \left(X(1 + \zeta^c) - 1\right) = \frac{X^p + (X-1)^p}{2X - 1}.$$

Vietà's formulae yield $\mathbf{Tr}(A) = (p-1)/2$ in this case too. Note that the result can also be deduced from $1/(1 + \zeta) + 1/(1 - \zeta) = 2/(1 - \zeta^2)$, which implies

$$\mathbf{Tr}\frac{1}{1+\zeta}+\mathbf{Tr}\frac{1}{1-\zeta}=2\cdot\mathbf{Tr}\frac{1}{1-\zeta^2}.$$

This completes the proof. □

**Lemma 7.** *If* (1) *holds and* $x \equiv s \mod p$ *with* $s \in \{-1, 0, 1\}$, *then* $x \equiv s \mod p^2$.

*Proof.* The proof is a variant of the proof of the *double Wieferich criterion* for Catalan's equation ([Mih], [Mih2]). Let $\alpha, \mathfrak{A}$ be defined like above. We choose $\theta = \psi_2 = \sum_{c\geq(p+1)/2}\sigma_c^{-1} \in I$, where $I$ is the Stickelberger ideal, as before. Then there is a $\beta \in \mathbb{Z}[\zeta]$ such that $\mathfrak{A}^\theta = (\beta)$ and by raising to the $p$-th power we find:

$$\left(\alpha^\theta\right) = \left(\alpha'^\theta\right) = \mathfrak{A}^{p\theta} = \left(\beta^p\right).$$

By Lemma 2, there is a root of unity $\chi' \in <-\zeta>$ such that

$$\chi'\cdot\alpha^\theta = \beta^p. \tag{26}$$

It follows from the same Lemma that $\alpha'^\theta = \beta^p$, as Jacobi sums generating the same ideal and in fact, since $\alpha = \chi\cdot\alpha'$, we find that $\chi' = \chi^{-\theta}$. Since $\beta$ is a Jacobi sum, the previous condition together with (13) imply in fact that

$$\alpha'^\theta = \beta^p \equiv 1 \mod p\lambda^2.$$

The statement of the Lemma follows by developing $\alpha'^\theta \mod p\lambda^2$ and applying the Lemma 6. Details are straight forward and are left to the reader. □

## 3 Classical results revisited; proofs of Theorems 1 and 2

According to the Barlow formulae [Ri1], the First Case of Fermat's Last Theorem holds, if there are coprime integers $x, y; r, s$ with $(xys, p) = 1$ and such that

$$\frac{x^p+y^p}{x+y} = r^p \quad \text{and} \tag{27}$$

$$x+y = s^p. \tag{28}$$

One easily notices that (1) is derived from (27) – which restricts from (3) – by means of the substitution $y = -1$. In the year 1964, Eichler published a famous theorem on the First Case of Fermat and the proof of that theorem in fact concentrates only on consequences of (27), under the restriction that $t := y/x \mod p$ is such that $t \not\equiv -1, 0, 1 \mod p$. This condition follows from $(xyz, p) = 1$ and the symmetry of Fermat's equation $x^p + y^p + z^p = 0$.

The condition $t \not\equiv -1, 0, 1 \mod p$ is satisfied in (3); furthermore, under the substitution $y = -1$ we obtain $t \equiv -1/x$ and the previous condition on $t$ is satisfied exactly in the First Case of (1). Thus Eichler's proof applies literally to (3) and to the First Case of (1) and the consequence is Theorem 1. Rather than duplicating the proof with at most a minor change in notation, we refer the reader to Eichler's paper [Ei] and Washington's beautiful version of the proof in [Wa].

The statement of Theorem 2 was proved by Wieferich and Furtwängler, roughly hundred years ago, for the First Case of Fermat. Their proofs can probably be adapted to equation (1). We choose however this time to give an alternative proof, using a local application of properties of the Stickelberger ideal. This proof has not been used so far in the literature; Lenstra and Stevenhagen have suggested in a recent paper in the *Mathematical Intelligencer* that a similar idea should work (H. W. Lenstra, Jr., Leiden University, pers. commun.).

We begin with two lemmata.

**Lemma 8.** *Let $r \neq p$ be a rational prime and $\mathfrak{R} \in \mathbb{Z}[\zeta]$ a prime above it. If the equation $\beta^p \equiv \zeta \mod \mathfrak{R}$ holds for some $\beta \in \mathbb{Z}[\zeta]$, then*

$$r^{p-1} \equiv 1 \mod p^2 \tag{29}$$

*Proof.* Let $\mathbf{k} = \mathbb{Z}[\zeta]/\mathfrak{R}$ and $f = [\mathbf{k} : \mathbb{F}_r]$ be the extension degree over the prime field $\mathbb{F}_r$. Let $\tilde{\zeta} = \zeta \mod \mathfrak{R}$ and $\tilde{\beta} = \beta \mod \mathfrak{R}$. Note that $\tilde{\zeta} \neq 1$, otherwise $\zeta - 1 \equiv 0 \mod \mathfrak{R}$ and a fortiori

$$p = (1-\zeta) \cdot \frac{p}{1-\zeta} \equiv 0 \mod \mathfrak{R},$$

which is absurd, since $r$ and $p$ are distinct primes. It follows that $\tilde{\beta}$ has order $p^2$ in $\mathbf{k}^\times$. The multiplicative group is cyclic with $r^f - 1$ elements and thus $p^2 | (r^f - 1)$. Also, by elementary Galois theory, $f | (p-1)$ and $(r^f - 1) | (r^{p-1} - 1)$. Finally, $p^2 | (r^{p-1} - 1)$, as claimed. □

**Lemma 9.** *Let $p$ be an odd prime, $\mathbb{K} = \mathbb{Q}(\zeta)$ the $p$-th cyclotomic extension and $I$ the Stickelberger ideal. Then there is a $\Theta \in I$ such that $\varphi(\Theta) \neq 0$, with $\varphi$ defined by* (16). *In particular, $\zeta^\Theta \neq 1$.*

*Proof.* Remember that $\varphi(\Theta_n) = \varphi_n$, where $\Theta_n$ are the Fuchsian elements and $\varphi_n$ the $n$-th Fermat quotients: $(n^p - n)/p \mod p$, for $n = 1, 2, \ldots, p-1$. If the statement of the lemma is false, then all the Fermat quotients vanish. Consider the polynomial

$$f(X) = \frac{(X+1)^p - (X^p + 1)}{p} = \sum_{k=1}^{p-1} \binom{p}{k} / p \cdot X^k \equiv - \sum_{k=1}^{p-1} (-X)^k / k \mod p,$$

as a polynomial in $\mathbb{F}_p[X]$. Note also that

$$\begin{aligned} f(X) &= \frac{(X+1)^p - (X^p+1)}{p} \\ &= \frac{(X+1)^p - (X+1)}{p} - \frac{X^p - X}{p} \equiv \varphi_{X+1} - \varphi_X \mod p. \end{aligned}$$

If all Fermat quotients would vanish, then $f$ has precisely the $p-1$ zeroes $X = 0, 1, 2, \ldots, p-2$ over $\mathbb{F}_p$. However, by comparing the coefficient $-1/2$ of $X^{p-2}$ in $f(X)$, the sum of the zeroes should be, by Vietà,

$$1/2 \equiv \sum_{k=1}^{p-1} X_i \equiv -(p-1) \equiv 1 \mod p.$$

This is absurd, and thus all Fermat quotients cannot vanish simultaneously, which completes the proof. □

The Theorem 2 will follow from the following.

**Proposition 1.** *Let $x, y, z, p$ be a solution of* (3) *and $r$ be a rational prime with $r|xy(x^2-y^2)$. Then* (29) *holds for $r$.*

*Proof.* We first remark that the product $xy(x^2-y^2) = y(x-y)x(x+y)$ being equal to $y$ times by three consecutive integers in an arithmetic progression with ratio $y$, it is certainly divisible by the primes $r = 2$ and 3. This shows that Proposition 1 implies (4) and thus Theorem 2.

Suppose that (3) has a solution and let $\alpha = x + \zeta y$ and $\mathfrak{A} = (\alpha, z)$. Let $\Theta \in I$ be such that $\zeta^{\Theta} \neq 1$ – the existence being guaranteed by Lemma 9. Then by Lemma 2 there is a Jacobi sum $\beta \in \mathbb{Z}[\zeta]$ such that the identity $\mathfrak{A}^{\Theta} = (\beta)$ holds between principal ideals. By raising to the $p$-th power and using $\mathfrak{A}^p = (\alpha)$, we find:

$$\mathfrak{A}^{p\Theta} = (\alpha^{\Theta}) = (\beta^p).$$

Since $x \not\equiv sy \mod p$ when $s \in \{-1, 0, 1\}$, there is a $c \in P$, $c \equiv y/(x+y) \mod p$ and $c \neq p-1$ and $c \neq (p-1)/2 \mod p$ ($c = (p-1)/2$ is equivalent to $x \equiv y \mod p$, while $c = p-1$ is equivalent to $x \equiv 0 \mod p$; both cases are excluded in (3)). With this, we expand $\alpha$ in powers of $\lambda$, under use of (7)

$$\begin{aligned}\alpha^{\Theta} = (x+\zeta y)^{\Theta} &= \left((x+y)\cdot\left(1 - y\frac{1-\zeta}{x+y}\right)\right)^{\Theta}\\ &= \left((x+y)\cdot\zeta^c\left(\zeta^{-y/(x+y)}\cdot\left(1 - y\frac{1-\zeta}{x+y}\right)\right)\right)^{\Theta}\\ &= (x+y)^{\Theta}\cdot\zeta^{\varphi_c}\cdot(1+O(\lambda^2))^{\Theta} \equiv (x+y)^{\Theta}\cdot\zeta^{\varphi_c} \mod \lambda^2.\end{aligned}$$

But $(x+y)^{\Theta} = (x+y)^{w(\Theta)}$ and we may assume that the weight of $\Theta$ is a multiple of $p-1$ (eventually multiply $\Theta$ by 2), and thus $(x+y)^{\Theta} \equiv 1 \mod p$. It follows that

$$\left(\zeta^c\cdot\alpha\right)^{\Theta} \equiv 1 \mod \lambda^2. \tag{30}$$

If follows then also from Lemma 2 that

$$\left(\zeta^c\cdot\alpha\right)^{\Theta} = \beta^p, \tag{31}$$

as algebraic numbers – and in fact, as Jacobi sums.

Let now $r|x$ be a prime and $\mathfrak{R} \subset \mathbb{Z}[\zeta]$ be a prime above $r$, so $\mathfrak{R}|(x)$. Then

$$\left(\zeta^c\cdot\alpha\right)^{\Theta} = \left(\zeta^c\cdot(x+\zeta y)\right)^{\Theta} \equiv \zeta^{(1+c)\Theta} \mod \mathfrak{R},$$

and (31) implies

$$\beta^p \equiv \zeta^{(1+c)\Theta} \mod \mathfrak{R}.$$

Finally, by the choice of $\Theta$ and $c \neq p-1$, the root of unity $\zeta^{(1+c)\Theta}$ is primitive and we may apply Lemma 8. This settles the case of divisors of $r|x$.

The remaining cases require one more simple trick. Let for instance $r|(x+y)$ and $\mathfrak{R}|(r)|(x+y)$ be a prime above $r$. Then, from (31) and using complex conjugates, we find successively:

$$\begin{aligned}\beta^p \equiv \left(\zeta^c\cdot\alpha\right)^{\Theta} &\equiv \left(-\zeta^c\cdot y\lambda\right)^{\Theta} \mod \mathfrak{R} \text{ and}\\ \left(\beta/\overline{\beta}\right)^p &\equiv (-1)^{\Theta}\cdot\zeta^{(1+2c)\Theta} \mod \mathfrak{R}.\end{aligned}$$

Since $c \neq (p-1)/2$ the root of unity $\zeta^{(1+2c)\Theta}$ is primitive too and the case is settled by the same Lemma 8. For $\mathfrak{R}|(r)|(x-y)$, we have

$$\beta^p \equiv \left(\zeta^c \cdot \alpha\right)^{\Theta} \equiv \left(\zeta^c \cdot y(1+\zeta)\right)^{\Theta} \mod \mathfrak{R} \quad \text{and}$$
$$\left(\beta/\overline{\beta}\right)^p \equiv \zeta^{(1+2c)\Theta} \mod \mathfrak{R},$$

and the argument is the same. Finally, if $\mathfrak{R}|(r)|y$, we have

$$\beta^p \equiv \left(\zeta^c \cdot \alpha\right)^{\Theta} \equiv \left(\zeta^c \cdot x\right)^{\Theta} \mod \mathfrak{R} \quad \text{and}$$
$$\left(\beta/\overline{\beta}\right)^p \equiv \zeta^{(2c)\Theta} \mod \mathfrak{R},$$

and since $c \not\equiv 0 \mod p$, this settles the case and completes the proof of Proposition 1 and of the Theorem 2. □

# 4 General upper bounds

We restrict from now on our attention to the Nagell–Ljunggren equation and assume that the prime $p > 17$ leads to a solution $x, y$ of (1). In what follows we shall use various kinds of *annihilators* $\Theta \in \mathbb{Z}/(p \cdot \mathbb{Z})[G]$ for obtaining pure $p$-th power equations of the type

$$\alpha^{\Theta} = \beta^p, \quad \text{with} \quad \beta \in \mathbb{Z}[\zeta],$$

and $\alpha$ as defined in the previous section. Starting from this equation, we can use methods of Diophantine approximation, in particular the development in binomial series of the previous equation, thus gaining lower and upper bounds for $|x|$. More precisely, for the upper bounds we shall use the (binomial) power series development

$$f[\Theta](x) = (1 - \zeta/x)^{\Theta/p}.$$

This series converges uniformly in the complex plane, when $|x| > 1$. The global convergence of the series will yield some tight *upper* bounds for the solution $x$. We shall start in the next subsection with some general considerations on formal power series and their evaluations, which give more detail about the previous statement. We conclude the description of the general method by mentioning that *lower* bounds shall be gained from local power series expansions at the ramified prime $p$. Unfortunately, it is only for the Second Case and under the assumption that Vandiver's conjecture holds for $p$, that we are able to derive a contradiction between the lower and upper bounds.

## 4.1 Binomial power series

For $r \in \mathbb{R}$, the series of Abel (or generalized binomial series) is

$$f(z) = \sum_{k \geq 0} \binom{r}{k} z^k;$$

it converges uniformly for $|z| < 1$ and the sum verifies $f(z) = (1+z)^r$. We want to investigate series expansions of algebraic numbers $\mu \in \mathbb{Z}[\zeta]$ which verify $\mu^p = (1-\zeta/x)^{\Theta}$, where $\Theta = \sum_c n_c \sigma_c \in \mathbb{F}_p[G]$.

We first consider the formal power series for $(1 - \zeta/x)^{\Theta}$ and start with some definitions.

**Definition 1.** Let $\mathbf{R} = \mathbb{Q}(\zeta)[[T]]$ be the ring of formal power series over the $p$-th cyclotomic field. The *elementary p-th root* series is defined as:

$$f(T) = \sum_{k\geq 0} \binom{1/p}{k} \cdot (-\zeta \cdot T)^k = \sum_{k\geq 0} a_k \cdot T^k \in \mathbf{R},$$

where $a_k = \binom{1/p}{k} \cdot (-\zeta)^k$. Let $0 < n < p$ and $\sigma \in G$; then $\sigma$ acts upon $\zeta$ and *not* upon the formal variable $T$; thus, by definition:

$$f[n\sigma](T) = \sigma\,(f(T))^n = f^{n\sigma}(T) = \sum_{k\geq 0} \binom{n/p}{k} \cdot (\zeta^{\sigma} \cdot T)^k = \sum_{k\geq 0} a_k(n\sigma) \cdot T^k \in \mathbf{R},$$

and $a_k(n\sigma) = \binom{n/p}{k}(-\zeta)^{k\sigma}$.

For $\Theta = \sum_{c\in P} n_c \cdot \sigma_c \in \mathbb{F}_p[G]$ we define the additive map $\rho : \mathbb{F}_p[G] \to \mathbb{Z}[\zeta]$ by

$$\rho \, : \, \Theta \mapsto -\sum_{c\in P} n_c \cdot \zeta^c.$$

We note two important properties of the coefficients of these elementary series:

**Lemma 10.** *For $k \geq 0$, let $E(k) = k + v_q(k!)$. Then $E(k)$ is strictly monotonous and verifies*

$$E(k) < k \cdot \frac{p}{p-1}. \tag{32}$$

*Furthermore*

$$p^{E(k)} \cdot \binom{1/p}{k} \in \mathbb{Z} \quad \textit{and} \quad E(k) = -v_p\left(\binom{1/p}{k}\right). \tag{33}$$

*Writing $b_k(n\sigma) = (p^k \cdot k!) \cdot a_k(n\sigma)$, the following congruence holds:*

$$b_k(n\sigma) \equiv (-n\zeta^{\sigma})^k = \rho(n\sigma)^k \mod p \cdot \mathbb{Z}[\zeta]. \tag{34}$$

*Proof.* Since $k + 1 > k$ and $v_p((k+1)!) \geq v_p(\,k!\,)$, the function $E(k)$ is strictly monotonous. Note that $v_p(\,k!\,) \leq \sum_{i>0} \lfloor k/p^i \rfloor < k/(p-1)$; this leads to the upper bound for $E(k)$. For $n \not\equiv 0 \mod p$ we have

$$v_p(b_k(n)) = v_p\left(p^k \cdot k! \cdot \binom{n/p}{k}\right) = v_p\,(n \cdot (n-p) \ldots (n-(k-1)p)) = 0.$$

This shows that $v_p(a_k(n\sigma)) = -E(k)$.

We now estimate $v_\ell\,(b_k(n\sigma))$ for a prime $\ell \neq p$. The pigeon hole principle shows that the number of multiples of $\ell^i$ in the above product is $\lfloor k/\ell^i \rfloor$. Since $\ell$ may divide $n$ to a high power, by adding up we find:

$$v_\ell\,(n \cdot (n-p) \ldots (n-(k-1)p)) \geq \sum_{i>0} \lfloor k/\ell^i \rfloor = v_\ell(k!).$$

Together with the estimate of $v_p(b_k(n))$, this yields (33).

The $p$-adic development of $b_k(n\sigma)$ starts with:

$$b_k(n\sigma) = (n \cdot (n-p) \ldots (n-(k-1)p)) \cdot (-\zeta)^{k\sigma} \equiv (-n\zeta^\sigma)^k \mod p\mathbb{Z}[\zeta].$$

This is (34), which completes the proof. □

Let $\Theta = \sum_{c \in P} n_c \sigma_c \in \mathbb{F}_p[G]$. The definition of $f[n\sigma]$ is naturally generalized to

$$f[\Theta](T) = \prod_{c \in P} f[n_c \sigma_c](T) = \prod_{c \in P} \left( \sum_{k \geq 0} \binom{n_c/p}{k} (\zeta^c \cdot T)^k \right). \tag{35}$$

Since $f[\Theta] \in \mathbf{R}$, it also has a development as a simple power series. In consistency with the definition of $a_k$, $b_k$ in Lemma 10, we shall write this development as

$$f[\Theta](T) = \sum_{k \geq 0} a_k(\Theta) \cdot T^k \quad \text{and let} \tag{36}$$

$$b_k(\Theta) = a_k(\Theta) \cdot \left( p^k \cdot k! \right).$$

The arithmetic properties of the coefficients $a_k$, $b_k$ are given by the following:

**Lemma 11.** *The coefficients $a_k(\Theta)$, $b_k(\Theta)$ of the series $f[\Theta] \in \mathbf{R}$ have the properties:*

1. *Both $a_k$, $b_k$ commute with the Galois action, i.e., $a_k(\sigma\Theta) = (a_k(\Theta))^\sigma$ and the same for $b_k$.*
2. *$a_k(\Theta) \in \mathbb{Z}[\zeta, 1/p]$. More precisely, $p^{E(k)} \cdot a_k(\Theta) \in \mathbb{Z}[\zeta]$.*
3. *$b_k(\Theta) \in \mathbb{Z}[\zeta]$ and*

$$b_k(\Theta) \equiv \rho(\Theta)^k \mod p \cdot \mathbb{Z}[\zeta]. \tag{37}$$

4. *If $\Theta = \jmath\Theta$, then $a_k(\Theta) \in \mathbb{R}$.*

*Proof.* Property 1 follows from (35) and the fact that $\sigma$ acts on $\zeta$ but not on the formal parameter $T$. Suppose that $\Theta = \Theta_1 + \Theta_2$; then the coefficients of $f[\Theta]$ are derived from the ones of $f[\Theta_i]$, $i = 1, 2$, as follows:

$$\begin{aligned} a_k(\Theta) &= a_k(\Theta_1 + \Theta_2) = \sum_{l=0}^{k} a_l(\Theta_1) \cdot a_{(k-l)}(\Theta_2) \\ b_k(\Theta) &= b_k(\Theta_1 + \Theta_2) = \sum_{l=0}^{k} \binom{k}{l} \cdot \left( b_l(\Theta_1) \cdot b_{(k-l)}(\Theta_2) \right). \end{aligned}$$

We have

$$v_p\left( a_l(\Theta_1) \cdot a_{(k-l)}(\Theta_2) \right) = -\left( E(l) + E(k-l) \right) = -E(k) + v_p\left( \binom{k}{l} \right) \geq -E(k);$$

thus $v_p(a_k(\Theta)) \geq -E(k)$, by induction on the canonical weights $w(\Theta_1)$, $w(\Theta_2)$. This proves 2. By (34), we have $b_k(\Theta) \equiv \rho(\Theta)^k \mod p\mathbb{Z}[\zeta]$ for $\Theta = n\sigma$, $\forall \sigma \in G$.

Assume that (37) holds for $\Theta_i$, $i = 1, 2$. Then by the previous identity for $b_k(\Theta_1+\Theta_2)$, we have

$$b_k(\Theta_1+\Theta_2) = \sum_{l=0}^{k} \binom{k}{l} \cdot \left(b_l(\Theta_1) \cdot b_{(k-l)}(\Theta_2)\right)$$
$$\equiv \sum_{l=0}^{k} \binom{k}{l} \rho(\Theta_1)^l \cdot \rho(\Theta_2)^{k-l} = \rho(\Theta_1+\Theta_2)^k \mod p\mathbb{Z}[\zeta].$$

The relation (37) follows from this by induction on the weights of $\Theta_1, \Theta_2$.

Finally, if $\Theta = (1+\jmath)\theta$ is real, a case which is useful if $p|x$ (thus $s = 0$ in the Second Case), note that

$$a_k(1+\jmath) = \sum_{l=0}^{k} a_l \cdot \overline{a}_{(k-l)} = \overline{a_k(1+\jmath)} \in \mathbb{R}.$$

It follows from $f[\Theta] = f[(1+\jmath)\theta]$ that $a_k(\Theta) \in \mathbb{R}$. □

We now give an estimate of the error term in the evaluation of the general series $f[\Theta](1/x)$. The convergence radius of $f(T)$ being one, it follows by multiplicativity, that the series $f[\Theta](T)$ also have the same domain of convergence, for all $\Theta \in \mathbb{F}_p[G]$. Let

$$S_m(\Theta; T) = \sum_{k=0}^{m} a_k(\Theta) \cdot T^m$$

be the $m$-th partial sum of $f[\Theta]$ and $R_m(\Theta; T) = f[\Theta] - S_m(\Theta; T)$ the remainder term. We estimate this remainder, when $T$ is replaced by a complex number inside the domain of convergence.

**Lemma 12.** *Let $\Theta = \sum_{c\in P} n_c\sigma_c \in \mathbf{R}$ have weight $w(\Theta) = H$. If $z \in \mathbb{C}$, $|z| < 1$, then*

$$|f[\Theta](z) - S_m(\Theta; z)| \leq \binom{-H/p}{m+1} \cdot \frac{|z|^{m+1}}{(1-|z|)^{m+1+h}}. \quad (38)$$

*Proof.* A power series $f(T) = \sum_{k=0}^{\infty} a_k T^k$ with complex coefficients is *dominated* by the series $g(T) = \sum_{k=0}^{\infty} A_k T^k$ with nonnegative real coefficients if $|a_k| \leq A_k$ for $k = 0, 1, \ldots$; if this is the case, we write $f << g$. The relation of dominance is preserved by addition and multiplication of power series.

Let $r$ be a real number, and $s$ a complex number satisfying $|s| \leq 1$. Then for the binomial series we have

$$(1+sT)^r = \sum_{k=0}^{\infty} \binom{r}{k} s^k T^k (1-T)^{-|r|} = \sum_{k=0}^{\infty} (-1)^k \binom{-|r|}{k} T^k.$$

Indeed, the coefficients of the latter series are positive and $\left|\binom{r}{k}\right| \leq \left|\binom{-|r|}{k}\right|$.

It follows that $f[\Theta](T) << (1-T)^{-H/p}$ and $g[\Theta](T) << (1-T)^{-H/p}$. From common remainder estimates for Taylor series, we obtain the following:

$$
\begin{aligned}
|f[\Theta](z) - S_m(\Theta; z)| &\le \left|(1-|z|)^{-H/p} - S_m(|z|)\right| \\
&\le \sup_{0\le|\xi|\le|z|} \left|\left(\frac{d^{m+1}(1-T)^{-H}}{dT^{m+1}}\bigg|_{T=\xi}\right)\right| \frac{|z|^{m+1}}{(m+1)!} \\
&= \binom{-H/p}{m+1} \cdot \frac{|z|^{m+1}}{(1-|z|)^{H/p+m+1}},
\end{aligned}
$$

as claimed. □

## 4.2 General bounding strategy

We consider the general equation (1) and let $p$ be an odd prime and $(x, y)$ be a nontrivial solution, for which we define $\alpha \in \mathbb{Z}[\zeta]$, $\mathfrak{A} \subset \mathbb{Z}[\zeta]$ like above, where $\zeta$ is a primitive $p$-th root of unity. Let $\mathbf{T} \subset \mathbf{R} = \mathbb{Z}[G]$ be the set such that

$$\forall \Theta \in \mathbf{T}, \quad \exists \beta \in \mathbb{Z}[\zeta] : \quad (x-\zeta)^{\Theta} = \beta^p.$$

One easily verifies that $\mathbf{T}$ is a submodule of $\mathbb{Z}[G]$. It is in fact one of the annihilators of $\mathfrak{A}$ in the class group: this is immediate if $x \not\equiv 1 \mod p$ and requires some additional comments otherwise. We shall see below that $\mathbf{T} \setminus \left(p, \mathbf{N}_{\mathbb{Q}(\zeta),\mathbb{Q}}\right) \neq \emptyset$. First we sketch the general approach for gaining upper bounds $B \in \mathbb{R}$ such that $|x| < B$.

If $w = w(\Theta)$, then the definition of $\beta$ yields $\beta^p = x^w \cdot (1-\zeta/x)^{\Theta}$. Let $f[\Theta](1/x) = (1-\zeta/x)^{\Theta/p}$ be the binomial series and

$$\tilde{f}[\Theta](x) = \left(\frac{x}{p^e}\right)^{w/p} \cdot f[\Theta](1/x).$$

The root $x^{1/p}$ is intended as the unique real $p$-th root of $x$; since $p$ is odd, such a root always exists. Furthermore, the denominator is 1 except when $s = 1$ in the Second Case; in this case the root $p^{1/p}$ is the positive real value of this $p$-th root. One can consider $\beta[\Theta]$, $\tilde{f}[\Theta]$ as maps $\mathbf{T} \to \mathbb{C}$, where $\Theta$ acts on elements in $\mathbb{K}$ by its lift, as defined in Sect. 2.5; the maps are linear morphisms from the additive group of $\mathbf{T}$ to the multiplicative group of $\mathbb{C}$ verifying $\beta^p(\Theta) = g^p(\Theta)$. They thus differ by a $p$-th root of unity, which we design by $\zeta^{\kappa(\Theta)}$; more precisely, we set $\tilde{f}(\Theta) = \zeta^{\kappa(\Theta)} \cdot \beta$. This induces a further map $\kappa : \mathbf{T} \to \mathbb{Z}/(p \cdot \mathbb{Z})$ which is additive and verifies $\kappa(\jmath\Theta) + \kappa(\Theta) = 0$, due to the continuity of complex conjugation. In particular, if $\Theta \in (1+\jmath)\mathbf{T}$, then $\kappa(\Theta) = 0$. We denote the map $\kappa$ by *Galois exponent map*; the *signature* of $\Theta$ will be the vector of Galois exponents for all conjugates of $\Theta$, i.e., $K(\Theta) = (\kappa(\sigma\Theta))_{\sigma \in G}$.

Let us suppose there is a $\Theta \in \mathbf{T}$ for which the signature vanishes. Since the power series $f[\Theta](1/x)$ has coefficients which depend only on $p$ and $\Theta$ but not on $x$, one can build a nonvanishing linear combination $\gamma$ of the conjugates of $\tilde{f}(\Theta)$ in which only negative powers of $x$ occur, multiplied by some coefficients which do not depend on $x$. For sufficiently large $x$, one has then $|\sigma(\gamma)| < 1$ for all $\sigma \in G$, in contradiction with $\gamma$ being a nonvanishing algebraic integer. Hence, one deduces an upper bound on $x$. Unfortunately, the assumption of a completely vanishing signature is more than one can achieve in fact. We are in the following situation: on the one hand, the tuples $\{\beta(\sigma\Theta) : \sigma \in G\}$ are built of Galois conjugates, but one has little direct information about them; on the other hand, $\{\tilde{f}(\sigma\Theta) : \sigma \in G\}$ are explicitly known, but they need not be mutually conjugate. The deviation from Galois conjugation in the second tuple is given exactly by the signature. Since this is not vanishing, the general strategy we shall adopt is to build larger linear combinations of numbers of the type $\tilde{f}(\sigma\Theta) : \sigma \in G, \Theta \in \mathbf{T}$, in order to cancel both the Galois exponents and sufficiently many leading terms of the series expansion. This leads to upper bounds for $|x|$ in a similar way as described above; we shall have to eventually consider tuples $\{\tilde{f}(\sigma\Theta)\}$ for more than one $\Theta \in \mathbf{T}$.

Having exposed the general idea of the machinery, we now put it at work and shall distinguish three cases, as claimed in the Theorem 3. It turns also out that in two of the three cases a slight modification of the generic definition of $\tilde{f}(\Theta)$ leads to better bounds.

### 4.3 Proof of Theorem 3

We shall give in this section the proof of the Theorem, following the sub-case distinction given in (5). We start with the case $x \equiv 0 \mod p$, which has a different approach from the remaining cases.

**Lemma 13.** *Suppose that* (1) *has a solution with* $x \equiv 0 \mod p$ *and* $p$ *an odd prime. Let* $\zeta$ *be a primitive* $p$*-th root of unity and* $\alpha' = 1 - \overline{\zeta}$. *Then for each* $\Theta \in \mathbf{A}_1$ *there is a* $\nu \in \mathbb{Z}[\zeta]$ *such that* $\alpha'^{\Theta} = \nu^p$.

*Proof.* The ideal $\mathfrak{A} = (\alpha, y) \in \mathcal{A}$, since it has order dividing $p$ by (25) and by Lemma 5, the identity (22) must hold for some unit. But $x \equiv 0 \mod p^2$, as proved in Lemma 7, it follows that $\alpha'$ is $p$-primary and by consequence of (22), so must be $\varepsilon$. But then (23) implies that $\varepsilon \in E^p$ which implies the claim of this lemma. Note in particular that $\mathbf{A}_1$ is nontrivial, as shown prior to Lemma 5. □

The proof for $x \equiv 0 \mod p$ is now very close to the pattern of proof for Catalan's conjecture [Mih2].

Let $\Theta = (1 + \jmath)\Theta_0 \in \mathbb{Z}[G]^+$ with $\Theta_0 \mod p \in \mathbf{A}_1$. As shown in Sect. 2.5, when lifting an element from $\mathbf{A}_1$ to $\Theta_0 \in \mathbb{Z}[G]$ we may impose the following conditions:

(1) $\Theta_0 = \sum_{c \in P} n_c \sigma_c$ with $0 \le n_c < p$.
(2) $w(\Theta_0) = \sum_{c \in P} n_c \equiv 0 \mod p$.

The same holds then for $\Theta$; furthermore, since both $\Theta$ and $p\mathbf{N} - \Theta$ have their images modulo $p$ in $\mathbf{A}_1$ and $w(\Theta) + w(p\mathbf{N} - \Theta) = p(p-1)$, we may assume, by choosing the element of smaller weight among the two, that $w(\Theta) \le \frac{p(p-1)}{2}$. By condition (2) above, it follows that there is a $h \in \mathbb{Z}$ with $0 < h \le \frac{p-1}{2}$ such that $w(\Theta) = h \cdot p$.

For $\Theta$ chosen as above, we may apply Lemma 12 with $H = p \cdot h$. First note that

$$\nu = \tilde{f}[\Theta](x) = x^h \cdot f[\Theta](1/x),$$

where $f[\Theta]$ is the series $(1 - \zeta T)^{\Theta/p}$. The above identity holds since the right and left hand side have the same $p$-th power and are real at the same time. They thus differ by a real $p$-th root of unity, which can only be $\zeta = 1$. Next, we let $\beta = p^h \cdot \nu \in \mathbb{Z}[\zeta]^+$ and consider the $m$-th partial sum and remainder of $\beta$ for $m = h$. Thus

$$S'_m = p^h \cdot x^h \cdot S_m(1/x) = \sum_{i=0}^{h} x^{h-i} \cdot p^h \cdot a_i(\Theta), \tag{39}$$

and $R'_m = \beta - S'_m = R_m(1/x) \cdot (x \cdot p)^h$. By point 2 of Lemma 11, the coefficient $p^h \cdot a_i(\Theta) \in \mathbb{Z}[\zeta]$ for $i \leq h$ and thus $S'_m \in \mathbb{Z}[\zeta]$. Consequently we shall also have $R'_m \in \mathbb{Z}[\zeta]$. The estimate of Lemma 12 yields for this algebraic integer:

$$|R'_m| = (p \cdot x)^h \cdot |R_m(1/x)| \leq \binom{-h}{h+1} \cdot \left| \frac{p^h}{x \cdot (1 - 1/|x|)^{2h+1}} \right|.$$

We claim that the last term is dominated by $\frac{(4p)^h}{|x|}$. Indeed, the binomial coefficient is $\binom{-h}{h+1} = \binom{2h}{h+1} < 4^h/2$. Since $1/\left(1 - 1/|x|\right)^{2h+1} < 2$ whenever $|x| > 2$, the claim follows.

We thus have $|R'_m(\Theta)| < (4p)^h/|x|$, and this inequality is Galois-invariant; we may indeed replace $\Theta$ by $\sigma\Theta$ in the above deduction and the upper bound does not change, due to the fact that the formal series $f[\Theta]$ is Galois covariant. Assuming that $|x| > (4p)^h$, it follows that $R'_m$ is an algebraic integer with the property that all of its conjugates have sub-unitary absolute value. This implies plainly $R'_m = 0$ and thus $\beta = S'_m$. By definition of $\beta = p^h\nu$ we also have $S'_m/p = \beta/p \in \mathbb{Z}[\zeta]$. We may compare the expression for $S'_m/p$ with the expansion (39); it shows that all the terms in the expansion of $S'_m/p \in \mathbb{Z}[\zeta]$ except for the free term $p^{h-1} \cdot a_h(\Theta)$ are algebraic integers. But then the same must hold for this free term; however, by definition of $b_k(\Theta)$ we have

$$p^{h-1} a_h(\Theta) = \frac{b_h(\Theta)}{p \cdot h!}.$$

Now $h! \not\equiv 0 \mod p$, while by point 3 of Lemma 11, $b_h(\Theta) \equiv \rho^h \mod p\mathbb{Z}[\zeta]$ and this expression also vanishes only if $\Theta \equiv 0 \mod p$. Since we showed that $\mathbf{A}_1$ is not trivial, we may choose $\Theta$ so that $\rho^h = \rho^h(\Theta) \not\equiv 0 \mod p\mathbb{Z}[\zeta]$ and in this case the free term of $S'_m$ is of the shape $c/p$ with $c = p^h a_h(\Theta) \equiv \frac{1}{h!} \cdot \rho(\Theta)^h \not\equiv 0 \mod p$; hence, it cannot be an algebraic integer. This is a contradiction to the assumption $|x| > (4p)^h$ and it confirms the estimate of Theorem 3 for the case $x \equiv 0 \mod p$.

In the remaining cases we need to adopt an other strategy. Since no useful real expansion is possible, we shall use Stickelberger annihilators. In this case, the Galois exponents do not vanish and we shall have to cancel them by linear combination. The next lemma studies more explicitly the annihilation by Stickelberger elements.

**Lemma 14.** *Let $x, y$ be a solution of* (1) *and $\Theta \in I_0$, such that $\varsigma(\Theta)$ is even in the Second Case if $s = \pm 1$. Then there is a $\beta = \beta[\Theta] \in \mathbb{Z}[\zeta]$ such that:*

$$(x - \zeta)^{\Theta} = \tilde{f}[\Theta](x)^p = p^{e\varsigma(\Theta)} \cdot \beta^p. \tag{40}$$

*Proof.* In the First Case, let $\alpha = (x - \zeta)$; then there is a $p$-th root of unity $\zeta^a$ such that

$$\zeta^a \cdot \alpha \equiv 1 \mod (1 - \zeta)^2.$$

By definition of the Fermat module, $\zeta^{a\Theta} = 1$ for $\Theta \in I_0$ and thus $\alpha^{\Theta} \equiv 1 \mod (1 - \zeta)^2$. It follows that the left hand side is a Jacobi Sum. On the other hand there is some

Jacobi Sum $\beta = \beta[\Theta] \in \mathbb{Z}[\zeta]$ such that $\mathfrak{A}^{p\Theta} = (\beta^p) = (\alpha^\Theta)$; since both Jacobi Sums $\beta^p$ and $\alpha^\Theta$ generate the same principal ideal, they are equal and this confirms the existence of $\beta$ in (40). In the second case we use the definition (24) of the $p$-primary numbers $\alpha'$ which generate the ideal $\mathfrak{A} = (\alpha', y)$. If $s = 0$, then $\alpha' = 1 - \overline{\zeta} x$ and the proof of (40) follows like above. If $s = \pm 1$ an additional computation is required. Suppose first that $s = -1$ and $\alpha' = (x - \zeta)/(1 + \zeta)$ and let $u = (1 + \zeta)^\Theta$. Then $u \cdot \overline{u} = (\mathbf{N}(1 + \zeta))^{\varsigma(\Theta)} = 1$ and $u/\overline{u} = (1 + \zeta)/(1 + \overline{\zeta})^\Theta = \zeta^\Theta = 1$, by definition of the Fermat module. Consequently $u = \pm 1$ and since $u \equiv 2^\Theta \equiv 2^{w(\Theta)} = 1 \mod p$ – recall that $\varsigma(\Theta)$ is even in this case – we must have $u = 1$. It follows that $\alpha'^\Theta = (x - \zeta)^\Theta$ and (40) follows by the previous arguments. If $s = 1$, one proves in the same way that $(1 - \zeta)^\Theta = p^{\varsigma(\Theta)}$ and, if $\beta$ is the Jacobi sum generating $\mathfrak{A}^p$, one has $(x - \zeta)^\Theta = p^{\varsigma(\Theta)} \cdot \beta^p$. Together with $e = 1$ when $s = 1$, this implies the claim of (40) in this case too and completes the proof of Lemma 15. □

The Wieferich-type congruence $a^{p-1} \equiv 1 \mod p^2$ implies $\varphi(a) = 0$ and due to (4), in the First Case, the Fermat module is known to contain the elements $\Theta_2, \Theta_3 - \Theta_2$ of weight $\frac{p-1}{2}$. In the Second Case, by Lemma 3, we have, for $p > 7$ at least two elements $\Theta_1, \Theta_2 \in I_0$ which are not Galois conjugate to each other (i.e., $\Theta_2 \neq \sigma\Theta_1$ for all $\sigma \in G$) and have weight $p - 1$. For $p = 7$ we have one element of weight $(p-1)/2$ and may adopt the proof of the First Case.

We shall follow the case distinction above and start with the First Case, in which we have $\Theta = \Theta_2 \in I_0$, an element with relative weight $\varsigma = 1/2$. Let $K = \{\kappa_c = \kappa(\sigma_c\Theta) : c \in P\}$ be the signature of $\Theta$, so $\tilde{f}[\sigma_c\Theta](x) = \zeta^{\kappa_c}\beta[\sigma_c\Theta]$. Recall that by complex conjugation we have $\kappa_{p-c} = -\kappa_c$ and in this case the development of $\tilde{f}$ starts by:

$$\tilde{f}[\sigma_c\Theta](1/x) = x^{(p-1)/2p} \cdot \left(1 - \frac{\eta(\sigma\Theta)}{px} + O(1/x^2)\right),$$

where $\eta(\sum_c n_c\sigma_c) = \sum_c n_c\sigma_c(\zeta)$. The purpose of elimination will be in this case to find $\lambda_c \in \mathbb{Z}[\zeta]$, $c = 1, 2, \ldots, \frac{p-1}{2}$ such that if $\beta = \beta[\Theta]$ and

$$\delta = \sum_{c=1}^{(p-1)/2} \lambda_c \cdot \sigma_c(\beta) + \overline{\lambda_c \cdot \sigma_c(\beta)}, \quad \text{then} \quad \sigma(\delta) = O(\sqrt{y}/x), \quad \forall \sigma \in G,$$

in the appropriate sense of uniform cancellation of the first term in the expansion of $g(x)$. Explicitly, if we write $\beta_c = \sigma_c(\beta)$ and $f_c(x) = \tilde{f}[\sigma_c\Theta](x)$, the definition of $\delta$ states

$$\delta = \sum_{c=1}^{(p-1)/2} \lambda_c \cdot \beta_c + \overline{\lambda_c \cdot \beta_c} = \sum_{c=1}^{(p-1)/2} \lambda_c \cdot \zeta^{-\kappa_c} f_c(x) + \overline{\lambda_c \cdot \zeta^{-\kappa_c} f_c(x)},$$

$$\begin{aligned}\sigma_a(\delta) &= \sum_{c=1}^{(p-1)/2} \sigma_a(\lambda_c) \cdot \beta_{ac} + \overline{\sigma_a(\lambda_c) \cdot \beta_{ac}} \\ &= \sum_{c=1}^{(p-1)/2} \sigma_a(\lambda_c) \cdot \zeta^{-\kappa_{ac}} f_{ac}(x) + \overline{\sigma_a(\lambda_c) \cdot \zeta^{-\kappa_{ac}} \cdot f_{ac}(x)},\end{aligned}$$

and the $O(\sqrt{y}/x)$ condition can be reformulated to

$$\sum_{c=1}^{(p-1)/2} \lambda_c \zeta^{-\kappa_c} = 0 \quad \text{and}$$

$$\sum_{c=1}^{(p-1)/2} \sigma_a(\lambda_c)\zeta^{-\kappa_{ac}} = 0, \quad a \in P.$$

Recall that $\lambda_c$ are undeterminates, thus $\lambda_c \neq \sigma_c(\lambda)$, but we may restrict to $(p-1)/2$ equations, since the remaining conditions will hold due to complex conjugation. We can finally apply $\sigma_a^{-1}$ to the second equation above, thus obtaining a linear system in the $\lambda$'s:

$$\sum_{c=1}^{(p-1)/2} \lambda_c \cdot \zeta^{-\kappa_{ac}/a} = 0, \quad a = 1, 2, \ldots, (p-1)/2. \tag{41}$$

Let $\mathbf{A} = \left(\zeta^{-\kappa_{ac}}\right)_{a,c=1}^{(p-1)/2}$ be the matrix of the above linear system and $\vec{\lambda} = (\lambda_c)_{c=1}^{(p-1)/2}$. If $\mathbf{A}$ is regular, then (41) has only the trivial solution, so we drop our condition by replacing the vanishing constant vector on the right hand side in (41) by $\vec{d} = (\delta_{a,(p-1)/2})_{a=1}^{(p-1)/2}$ (here $\delta_{a,b}$ is the Kronecker $\delta$): all conjugates except one vanish in the first order. The new system is $\mathbf{A} \cdot \vec{\lambda} = \vec{d}$, and it has a (unique) solution $\vec{\lambda}$, when $\mathbf{A}$ is regular. By Cramer's rule, $\lambda_c = A_c/A$, where $A = \det(\mathbf{A})$ and $A_c = \det(\mathbf{A}_c)$ are the determinants of some minors of $\mathbf{A}$ obtained by replacing the $c$-th column by the vector $\vec{d}$.

The determinant of $\mathbf{A}_c$ can be estimated by Hadamard's inequality [Bl] (22.8): if $\mathbf{a}_i$ are the column vectors of a matrix $\mathbf{A}$, then $|\det(\mathbf{A})| \leq \prod_i |\mathbf{a}_i|$, where the absolute value on the right hand side is the Euclidean metric of the vectors. It follows in the concrete case under investigation that

$$|A_c| = |\det(\mathbf{A}_c)| \leq \prod_{a=1}^{(p-3)/2} \left| \sqrt{\sum_{b=1}^{(p-3)/2} \zeta^{-2\kappa_{ab}/a}} \right| \leq \left(\frac{p-3}{2}\right)^{p-3/4} = D_1.$$

The determinant $A$ is estimated in the same way, and we obtain

$$|A| \leq \left(\frac{p-1}{2}\right)^{p-1/4} = D.$$

Then

$$A\delta = \sum_{c=1}^{(p-1)/2} A\lambda_c\zeta^{-\kappa_c} f_c(x) + \overline{A\lambda_c\zeta^{-\kappa_c} f_c(x)} \in \mathbb{Z}[\zeta];$$

this and the choice of $\lambda_c$ leads to vanishing of the first terms in the expansions of $f_c$. By writing $R_{c,0}(x)$ for the first order remainder of those expansions, we find

$$|A| \cdot |\delta| \leq |A| x^{(p-1)/2p} \cdot \sum_{c=1}^{(p-1)/2} |A_c| |R_{c,0}(x)|.$$

The remainder $R_{c,0}(x)$ can be estimated using Lemma 12. The binomial coefficient $|\binom{-H/p}{1}| = |\binom{-(p-1)/p}{1}| = 1 - 1/p$ and in fact $\binom{-H/p}{1}/(1-1/|x|)^{(p+1)/2p} < 1$. The estimate thus simply yields: $|R_{c,0}(x)| < \frac{1}{|x|}$ uniformly for all $c$. Since $|A\lambda_c| = |A_c| < D_1$ we finally have

$$|A\delta| < (p-1)D_1 \cdot \frac{|x|^{(p-1)/2p}}{|x|}; \tag{42}$$

by choice of $\vec{d}$, the estimate holds for all conjugates $\sigma_a(\delta)$ except for $a = (p-1)/2$. In this particular case, we use $\beta_c \cdot \overline{\beta}_c = y$ and thus

$$|\beta_c| < \sqrt{y} < \left(|x|^{p-1}(1-1/|x|)\right)^{1/2p} < 2|x|^{(p-1)/2p},$$

yielding

$$|A\delta| < D_1 \cdot (p-1) \cdot \sqrt{y} < 2(p-1)D_1|x|^{(p-1)/2p}.$$

By assembling the last estimates, we obtain the following upper bound for the norm of $A\delta \in \mathbb{Z}[\zeta]^+$:

$$N = |\mathbf{N}(A\delta)| < 2\left((p-1)D_1\right)^{(p-1)/2} \cdot |x|^{(p-1)/2p-((p+1)(p-1))/4p}.$$

Assuming that $\delta \neq 0$, we must have $N \geq 1$ and thus:

$$|x|^{(p-1)/2p} \leq 2^{2/(p-1)} \cdot (p-1)D_1 < 2\left(\frac{p-3}{2}\right)^{(p+1)/4},$$

so $|x| < 4 \cdot \left(\frac{p-3}{2}\right)^{(p+2)/2}$. We have used in the last inequalities the fact that $p \geq 17$ and $|x| < p^p$, which can easily be obtained from the same inequalities, in order to control terms in the range of $|x|^{O(1)/p^2}$. Suppose that $\delta = 0$. Then we have in particular $\sigma_a(\delta) = 0$, and by the choice of $\vec{\lambda}$ and the estimates of $|R_{0,c}|$ it follows that $0 = A\delta = A \cdot |x|^{(p-1)/2} - \sum_c A_c R_{0,c}$, and thus

$$|x| \leq \sum_c |A_c|/|A| < (p-1)D_1 < 3\left(\frac{p-3}{2}\right)^{(p-3)/4},$$

an upper bound which is stronger then the general one.

Assume now that $\mathbf{A}$ is singular. There is then a maximal regular submatrix $\mathbf{A}'$ of $\mathbf{A}$ with rank $k < (p-1)/2$ and determinant $D' = \det(\mathbf{A}')$. Let us assume without loss of generality that $\mathbf{A}'$ is built up from the first $k$ rows and columns of $\mathbf{A}$, so the variables $\lambda_i$, $i > k$ are dependent on the first $k$ ones. We fix all the dependent unknowns to $\lambda_i = 0$ except for one, say, $\lambda_{k+1} = D'$ and proceed like in the regular case. If $\vec{\lambda}'$ is the vector of independent unknowns and $\vec{d} = -\lambda_{k+1}\zeta^{-\kappa_{a(k+1)}/a} + \delta_{0,k}$, where $1 \leq a \leq k$, is the constant vector, we have the linear system $\mathbf{A}' \times \vec{\lambda}' = \vec{d}$. The inequalities of Hadamard yield in this case smaller values for $D_1$, $D$, while the details of the previous proof can be adapted straight forward. The details are left to the reader.

Let us consider the Second Case, namely the solutions for which $x \equiv s \mod p$ with $s = \pm 1$. For $p > 7$, we may let $\Theta_1, \Theta_2 \in I_0$ be Galois independent with weight $p-1$, according to Lemma 3. The strategy of the proof in the present case is essentially identical to the one in the First Case. However, since the minimal weight of an element $\Theta \in I_0$ is not known to be less than $p-1$, for such $\Theta$ we only have the bound $|\beta[\Theta]| < y$; therefore, cancellation in the power series development is necessary up to the second order. This leads to the requirement for twice as many equations and to the use of $\Theta_1, \Theta_2$ defined above. First one seeks

$$\delta' = \sum_{c=1}^{(p-1)/2} \lambda_c \beta[\sigma_c \Theta_1] + \lambda'_c \beta[\sigma_c \Theta_2] = O(y/x),$$

and $\sigma_a(\delta') = O(y/x^2)$ for $a = 2, \ldots, (p-1)/2$. The condition for $\delta'$ is weaker than for $\sigma_a(\delta')$, $a > 1$ and this leads to an inhomogeneous system granted to have a solution in the regular case. Once $\delta'$ is found, we use the techniques above for deriving bounds for $|x|$.

Here are the details. We shall write again for simplicity $\beta_c = \beta[\sigma_c \Theta_1]$, $\beta'_c = \beta[\sigma_c \Theta_2]$ and $\kappa_c = \kappa(\sigma_c \Theta_1)$, $\kappa'_c = \kappa(\sigma_c \Theta_2)$, etc. In our series developments up to the second order one also encounters $\eta_c = \eta[\sigma_c \Theta_1]$ and $\eta'_c = \eta[\sigma_c \Theta_2]$. Recall the definition for $\Theta = \sum_c n_c \sigma_c^{-1}$ of $\eta[\Theta] = \sum_c n_c \sigma_c^{-1}(\zeta)$. If $w(\Theta) = p-1$, then $n_c \in \{0, 1, 2\}$ and $n_c + n_{p-c} = 2$. This leads to the simple upper bound $|\eta_c|, |\eta'_c| < p-1$. We now give the linear system for $\lambda_c, \lambda'_c$, omitting the deduction steps which are similar to the previous case:

$$\sum_{c=1}^{(p-1)/2} \lambda_c \zeta^{-\kappa_{ac}/a} + \lambda'_c \zeta^{-\kappa'_{ac}/a} = 0 \tag{43}$$

$$\sum_{c=1}^{(p-1)/2} \lambda_c \eta_c \zeta^{-\kappa_{ac}/a} + \lambda'_c \eta'_c \zeta^{-\kappa'_{ac}/a} = \delta_{(p-1)/2,a}.$$

Let $\mathbf{A}$ be the matrix of the above system and $\vec{d}$ its constants vector. Assuming that it is regular, we fix $A = \det(\mathbf{A})$ and $A_c = \det(\mathbf{A}_c)$, $A'_c = \det(\mathbf{A}'_c)$ the determinants of the minor corresponding to $\lambda_c, \lambda'_c$ by Cramer's rule. The Hadamard inequality yields in this case, by using the upper bound for $|\eta_c|$,

$$A_c \le (p-2)^{(p-2)/2} \cdot (p-1)^{(p-3)/2} < (p-2)^{p-2} = D_1,$$
$$A \le (p-1)^{(p-1)/2} \cdot (p-1)^{(p-1)/2} = (p-1)^{p-1} = D.$$

We have to estimate the second-order remainder instead of the first-order one. Let thus $R_{c,1}(x)$ be the second-order remainder for $f_c(x)$; by a straightforward application of Lemma 12 one finds that $|R_{c,1}| < R = |x|^{-(p+1)/p}$ uniformly for all $c \in P$.

$$|\sigma_a(A\delta)| \le (p-1)R \cdot \max(|A_c|) \le (p-1)R \cdot D_1,$$

for all $a < (p-1)/2$. For $a = (p-1)/2$, using the bound $|R_{0,c}| < 1/|x|^{1/p} = |x|R$, we have $|\sigma_{(p-1)/2}(A\delta)| \le (p-1)|x|R \cdot D_1$. By assuming first that $\delta \neq 0$, the norm is bounded by

$$1 \le N = \left|\mathbf{N}_{\mathbb{K}^+/\mathbb{Q}}(A\delta)\right| \le |x| \cdot ((p-1)RD_1)^{(p-1)/2}, \quad \text{so}$$
$$|x|^{((p^2-1)/2p)-1} \le ((p-1)D_1)^{(p-1)/2}.$$

By using again $p \geq 17$ we find after some simple steps, that $|x| < 4(p-2)^p$. If $\delta = 0$, one considers like previously

$$0 = \sigma_{(p-1)/2}(\delta) = x^{-1/p} + \sum_c \lambda_c R_{1,c},$$

eventually finding a tighter bound for $|x|$.

Finally, if $\mathbf{A}$ is singular, one uses the same estimates for some regular submatrix and the bounds obtained improve the ones in the regular case. This completes the proof of Theorem 3.

## 5 Lower bounds and proof of Theorem 4

In this section we shall use local methods for providing lower bounds in the Second Case. We shall assume that $x, y$ is a solution of (1) in the Second Case for an odd prime $p$ for which Vandiver's conjecture holds, thus $p \nmid h_p^+$. As a consequence, if $\alpha'$ is defined by (24), then the ideal $\mathfrak{A} = (\alpha', y)$ is principal. Indeed, if $\mathfrak{A} = (\alpha', x)$ is not principal, since $\mathfrak{A}^p = (\alpha')$, then $\alpha'$ is $p$-singular primary. Thus $\mathbb{Q}(\zeta)\left(\alpha'^{1/p}\right)$ is an Abelian unramified extension in which, by reflection, an ideal from $\mathcal{A}^+$ must capitulate. This contradicts $p \nmid h_p^+$, so $\mathfrak{A}$ must be principal. Let $\mathfrak{A} = (\beta)$, so $\mathfrak{A}^p = (\alpha') = (\beta^p)$. Then $\alpha'/\overline{\alpha}' = \varepsilon \cdot \left(\beta/\overline{\beta}\right)^p$ for a unit $\varepsilon$ which must in fact be a root of unity, due to Dedekind's unit theorem. However both the left hand side and the cofactor of $\varepsilon$ are $p$-adic $p$-th powers; thus $\varepsilon$ is a global $p$-th power by (23), so we may assume that $\varepsilon = 1$ and we have plainly

$$\frac{\alpha'}{\overline{\alpha}'} = \left(\frac{\beta}{\overline{\beta}}\right)^p. \tag{44}$$

Note that $\gamma = \beta/\overline{\beta}$ is only determined up to a $p$-th root of unity and we shall fix its value by the assumption $\gamma \equiv 1 \mod (1-\zeta)^2$. We define $\delta = y \cdot \beta/\overline{\beta}$; since $\beta \mid y$, we have $\delta \in \mathbb{Z}[\zeta]$ and $|\delta| = y$. Furthermore, $\beta/\overline{\beta}$ has a local power series expansion at $p$ derived from the fact that $\alpha'$ is locally a prime power. We shall find a linear combination $\Delta$ of the conjugates of $\delta$ which vanishes up to a high order – this leads to a lower bound for $\Delta$ which, combined with $|\delta| = y$, finally yields sufficient lower bounds for $y$.

Suppose that $x \equiv s \mod p$ and let us write $\alpha' = 1 - (x-s)\mu$, where

$$\mu = \begin{cases} \overline{\zeta} & \text{if } s = 0, \\ -\frac{1}{1-\zeta} & \text{if } s = 1, \\ \frac{1}{1+\zeta} & \text{if } s = -1. \end{cases}$$

The equation (44) leads to the Abel series expansion

$$f(x) = (1-(x-s)\mu)^{1/p} \cdot (1-(x-s)\overline{\mu})^{-1/p},$$

which verifies

$$f(x)^p = \left(\frac{\beta}{\overline{\beta}}\right)^p, \tag{45}$$

in any topology in which it converges – and in particular in $\mathbb{Z}_p[\zeta]$. Since $x \equiv s \mod p$, by Lemma 7 we also have $x \equiv s \mod p^2$, which ensures $p$ - adic convergence of the Abel series. Furthermore, the series must converge to $\beta$; indeed, by (45) we must have $f(x) \equiv \zeta^a \cdot (\beta/\overline{\beta}) \mod p \cdot \mathbb{Z}[\zeta]$. The previous expansion implies in particular that $f(x) \equiv 1 \mod p$; we also assumed $\beta/\overline{\beta} \equiv 1 \mod (1-\zeta)^2$, so it follows that $\zeta^a = 1$ as claimed.

We now compute $f(x)$ explicitly for the various values of $s$. Let first $s = 0$; then

$$f(x) = \left(\sum_{n\geq 0}\binom{1/p}{n}\cdot(-\overline{\zeta})^n x^n\right)\times\left(\sum_{n\geq 0}\binom{-1/p}{n}\cdot(-\zeta)^n x^n\right)$$
$$= \sum_{n\geq 0} b_n(\zeta)\cdot x^n, \quad \text{with}$$
$$b_n(\zeta) = (-1)^n\cdot\sum_{m=0}^{n}\binom{-1/p}{m}\cdot\binom{1/p}{n-m}\cdot\zeta^{n-2m}.$$

Next we note that for $d = (p-1)/2$,

$$\mathbf{Tr}_{\mathbb{Q}}^{\mathbb{Q}(\zeta)}\left((\zeta^{d+1}-\zeta^d)\cdot\zeta^a\right) = \begin{cases} p & \text{for } a = d, \\ -p & \text{for } a = d+1, \\ 0 & \text{otherwise.}\end{cases} \tag{46}$$

In particular $\mathbf{Tr}_{\mathbb{Q}}^{\mathbb{Q}(\zeta)}\left((\zeta^{d+1}-\zeta^d)b_n(\zeta)\right) = 0$ for $n < d$. This suggests the definition $\Delta = \mathbf{Tr}_{\mathbb{Q}}^{\mathbb{Q}(\zeta)}\left((\zeta^{d+1}-\zeta^d)\gamma\right)$. Note that $v_p\left(x^d\cdot\binom{1/p}{m}\cdot\binom{-1/p}{d-m}\right) = 2d$ for $m \leq d$ and thus $v_p\left(x^d\cdot b_d(\zeta)\right) \geq 2d = p-1$. We have proved that $\Delta \equiv 0 \mod p^{p-1}$. If $\Delta \neq 0$, by using $|\gamma| = y$ we have the double inequality:

$$p^{p-1} \leq \Delta \leq \sum_{c=1}^{p-1}|\sigma_c(\zeta^{d+1}-\zeta^d)|\cdot y < 2(p-1)y.$$

It follows that $y > p^{p-2}/2$, in contradiction with the upper bound found in Theorem 3. This proves the statement of Theorem 4 for this case. We still have to show that $\Delta \neq 0$. For this, note that

$$(-p)^d \cdot \mathbf{Tr}\left((\zeta^{d+1}-\zeta^d)b_d(\zeta)\right) = p\left(\binom{1/p}{d}-\binom{-1/p}{d}\right)$$
$$\equiv p\cdot\left((1-(-1)^d)-(1-(-1)^{d-1})\cdot\frac{p(p-1)}{2}\right)\not\equiv 0 \mod p^2.$$

Furthermore, by (46), we also have $b_{d+1} \equiv 0 \mod p$; together with the above we find

$$\Delta \equiv (x/p)^d\cdot p\cdot\left((1-(-1)^d)-(1-(-1)^{d-1})\cdot\frac{p(p-1)}{2}\right)\not\equiv 0 \mod p\cdot(x/p)^{d+1}$$

and a fortiori $\Delta \neq 0$.

We now consider the cases $s = \pm 1$ simultaneously. In both cases we have: $\mu = \frac{-s}{1-s\zeta}$ and $\mu/\overline{\mu} = -s\zeta$. One proves like before that $f(x) = \gamma$ and finds the expansion

$$f(x) = \left( \sum_{n\geq 0} \binom{1/p}{n} \cdot \mu^n (x-s)^n \right) \times \left( \sum_{n\geq 0} \binom{-1/p}{n} \cdot \overline{\mu}^n (x-s)^n \right)$$
$$= \sum_{n\geq 0} b_n(\zeta) \cdot \mu^n \cdot (x-s)^n, \quad \text{with}$$
$$b_n(\zeta) = \sum_{m=0}^{n} \binom{-1/p}{m} \cdot \binom{1/p}{n-m} \cdot (-s\zeta)^{n-m}.$$

The fact that the exponent of $\zeta$ in the expansion of $b_n(\zeta)$ only runs in the range $0 \leq a \leq n$ in this case (while in the previous case, the range was $-n \leq a \leq n$) is of great help and will allow cancellation to a higher power of $p$. Like before, we start by noting that for $\nu = (\zeta^2 - \zeta) \cdot (1 - s\zeta)^{p-3}$ we have $\nu \cdot \mu^a = (-s)^a (\zeta^2 - \zeta) \cdot (1 - s\zeta)^{p-3-a}$ and thus

$$\mathbf{Tr}_{\mathbb{Q}}^{\mathbb{Q}(\zeta)} \left( \nu \cdot \mu^a \right) = 0 \quad \text{for} \quad 0 \leq p-3.$$

We consequently define $\Delta = \mathbf{Tr}_{\mathbb{Q}}^{\mathbb{Q}(\zeta)} (\nu \cdot \gamma)$ and find that $\Delta \equiv 0 \mod p^{2(p-2)}$, while $\Delta \neq 0$. The proofs are analog to the ones above. The double bound for $|\Delta|$ is now

$$p^{2(p-2)} \leq \Delta \leq \sum_{c=1}^{p-1} |\sigma_c(\nu)| \cdot y < 2^{p-2}(p-1)y.$$

It follows that $y > p^{2p-5}/2^{p-2}$, and since $y < |x|$ and for $p > 7$ we have $p^{2p-5}/2^{p-2} > (2(p-1))^p$. We have reached a contradiction to the upper bound for $|x|$ in Theorem 3, which completes the proof of Theorem 4.

## 6 Conclusion

Heuristic arguments exposed by Washington [La1] suggest that primes $p$ which verify the conditions of either Theorem 1 or Theorem 4 do not exist or are *very scarce*. No proof in support of these arguments is known and the computer verification is hard. The most intensive computations up to date have been pursued by J. Buhler et al. [BCEMS]: they prove that there are no primes $p < 12\ 000\ 000$ which verify either condition. This improves the previously known lower bound $p = 17$ for possible solutions of (1).

The condition of Theorem 2 is *easier*. If one assumes that the Fermat quotients $\varphi_n(p) \equiv (n^p - n)/p \mod p$ are uniformly random distributed when $p < N$, for large $N$, then the probability that $\varphi_2(p) = \varphi_3(p) = 0$ for some $p$, as required in Theorem 2, is $P(N) \sim \frac{\log(N)}{N} \cdot \sum_{p<N} 1/p^2$. Since the right hand sum is convergent, the heuristic suggests that there may be at most finitely many primes verifying (4). The condition is also easy to verify on a computer, so the lower bound for solutions in the First Case can be pushed, in principle, further than the one for the Second Case.

Certainly, the relatively tight upper bounds leave hope that someone might find an adequate idea for proving sufficient lower bounds in the general case.

**Acknowledgments.** I thank Yann Bugeaud, who directed my attention to this interesting equation, for motivating and enlightening discussions. Part of the research for this paper was accomplished while I was visiting professor at the University of Göttingen during the spring term of 2003. I am grateful to Axel Munk and the Institute for Stochastics of the University of Göttingen, who provided the stimulating environment in which an important part of this paper was conceived and matured.

## References

[BCEMS] Buhler, J., Crandall, R., Ernvall, R., Metsänkylä, T., Shokrollahi, A.: Irregular primes and cyclotomic invariants to 12 million. *J. Symb. Comput.* **31**, 89–96 (2001)

[Bl] Blatter, C.: *Analysis III*. Heidelberger Taschenbücher, vol. 153. Springer, Heidelberg (1974)

[BH] Bugeaud, Y., Hanrot, G.: Un nouveau critère pour l'équation de Catalan. *Matematika* **47**, 63–73 (2000)

[BHM] Bugeaud, Y., Hanrot, G., Mignotte, M.: Sur l'équation diophantienne $\frac{x^n-1}{x-1} = y^q$, III. *Proc. Lond. Math. Soc.* **84**, 59–78 (2002)

[Ca] Cassels, J.W.S.: On the equation $a^x - b^y = 1$, II. *Proc. Camb. Philos. Soc.* **56**, 97–103 (1960)

[Ei] Eichler, M.: Eine Bemerkung zur Fermatschen Vermutung. *Acta Arith.* **11**, 129–131 (1965) err. 261

[Go] Gouvêa, F.Q.: *p-adic Numbers. An Introduction*, 2nd edn. Universitext. Springer, Heidelberg (1991)

[Hy] Hyyrö, S.: Über das Catalan'sche Problem. *Ann. Univ. Turku Ser. AI* **79**, 3–10 (1964)

[IR] Ireland, K., Rosen, M.: *A Classical Introduction to Modern Number Theory*, 2nd edn. Graduate Texts in Mathematics, vol. 84. Springer, Heidelberg (1990)

[Iw] Iwasawa, K.: A note on Jacobi Sums. In: *Informatica Teorica, Strutture in Corpi Algebrici. Istituto Nazionale di Alta Mathematica.*, Symp. Math., vol. 15, pp. 447–459. Academic Press, London (1975)

[Jh] Jha, V.: *The Stickelberger Ideal in the Spirit of Kummer with Applications to the First Case of Fermat's Last Theorem*. Queen's Papers in Pure and Applied Mathematics, vol. 93. Queen's University, Kingston, Ont. (1993)

[La] Lang, S.: *Algebraic Number Theory*, 2nd edn. Graduate Texts in Mathematics, vol. 110. Springer, Heidelberg (1986)

[La1] Lang, S.: *Cyclotomic Fields*, I and II, combined 2nd edn. Graduate Texts in Mathematics, vol. 121. Springer, Heidelberg (1990)

[Mi] Mignotte, M.: Catalan's equation just before 2000. In: Jutila, M., Metsänkylä, T. (eds.) *Number Theory: Proceedings of the Turku Symposium on Number Theory in Memory of Kustaa Inkeri, May 31–June 4, 1999*, pp. 247–254. de Gruyter, Berlin (2001)

[Mih] Mihăilescu, P.: A class number free criterion for Catalan's conjecture. *J. Number Theory* **99**, 225–231 (2003)

[Mih1] Mihăilescu, P.: On the class groups of cyclotomic extensions in presence of a solution to Catalan's equation. *J. Number Theory* **118**, 123–144 (2006)

[Mih2] Mihăilescu, P.: Primary cyclotomic units and a proof of Catalan's conjecture. *J. Reine Angew. Math.* **572**, 167–195 (2004)

[Ri] Ribenboim, P.: *Catalan's Conjecture*. Academic Press, London (1994)

[Ri1] Ribenboim, P.: *13 Lectures on Fermat's Last Theorem*. Springer, Heidelberg (1979)

[Wa] Washington, L.: *Introduction to Cyclotomic Fields*, 2nd edn. Graduate Texts in Mathematics, vol. 83. Springer, Heidelberg (1996)

*Diophantine Approximation*
H. P. Schlickewei et al., Editors


# CONSTRUCTION OF APPROXIMATIONS TO ZETA-VALUES

Yuri V. Nesterenko
*Faculty of Mechanics and Mathematics, Moscow Lomonosov State University, Vorobiovy Gory, GSP-2, 119992 Moscow, Russia*
nester@orc.ru

To Professor Wolfgang M. Schmidt on the occasion of his 70th birthday

## 1 Introduction

Polylogarithmic functions are defined by series

$$L_k(z) = \sum_{\nu=1}^{\infty} \frac{z^\nu}{\nu^k}, \qquad k \geq 1.$$

Due to equalities $L_k(1) = \zeta(k)$, $k \geq 2$, they play an important role in study of arithmetic properties of Riemann zeta-function $\zeta(s)$ at integer points.

More generally for any rational function $R(s)$ that can be presented as a sum of simple fractions

$$R(s) = \sum_{\ell \in \mathcal{P}} \sum_{k=1}^{d(\ell)} \frac{B_{\ell,k}}{(s+\ell)^k}, \qquad B_{\ell,k} \in \mathbb{Q}, \tag{1}$$

where $\mathcal{P}$ is a set of distinct positive integers and $d(\ell) \geq 0$, one can find the following equalities:

$$F(z) = \sum_{\nu=0}^{\infty} R(\nu) z^\nu = \sum_{\ell \in \mathcal{P}} \sum_{k=1}^{d(\ell)} B_{\ell,k} \sum_{\nu=0}^{\infty} \frac{z^\nu}{(\nu+\ell)^k} =$$
$$\sum_{\ell \in \mathcal{P}} \sum_{k=1}^{d(\ell)} B_{\ell,k} z^{-\ell} \sum_{\nu=\ell}^{\infty} \frac{z^\nu}{\nu^k} = \sum_{\ell \in \mathcal{P}} \sum_{k=1}^{d(\ell)} B_{\ell,k} z^{-\ell} \left( L_k(z) - \sum_{\nu=1}^{\ell-1} \frac{z^\nu}{\nu^k} \right).$$

This confirms that the function $F(z)$ is a linear form in 1 and polylogarithms with coefficients in $\mathbb{Q}[1/z]$. It is clear that analogous result can be proved if we put as coefficients of the series $F(z)$ any derivative of $R(s)$ and shift the lower limit of summation on any admissible integer number. The following proposition defines the general construction.

---

*Keywords.* Irrationality, polylogarithms, zeta-function.

*2000 Mathematics subject classification.* 11J72.

**Proposition 1.** *For any complex z from the convergence domain of the series*

$$G_r(z) = \frac{(-1)^{r-1}}{(r-1)!} \sum_{\nu=1}^{\infty} R^{(r-1)}(\nu - a) z^{\nu} \tag{2}$$

*the following identity holds*

$$G_r(z) = A_0(z^{-1}) + \sum_{k=1}^{q} A_k(z^{-1}) L_{k+r-1}(z).$$

*Here* $q = \max_{\ell \in \mathcal{P}} d(\ell)$ *and*

$$A_k(x) = \binom{k+r-2}{r-1} \sum_{\substack{\ell \in \mathcal{P} \\ d(\ell) \geq k}} B_{\ell,k} x^{\ell - a}, \qquad k = 1, \ldots, q, \tag{3}$$

$$A_0(x) = -\sum_{\ell \in \mathcal{P}} \sum_{k=1}^{d(\ell)} \sum_{\nu=1}^{\ell - a} \binom{k+r-2}{r-1} B_{\ell,k} \nu^{1-k-r} x^{\ell - a - \nu}. \tag{4}$$

*Proof.* See [9, Proposition 1]. □

For arithmetic applications of this construction one has to choose the rational function $R(s)$ in such a way that the number $G_r(1)$, a linear form in zeta-values, be rather small, and the coefficients $A_k(1) \in \mathbb{Q}$ have common denominator and magnitude that are not very large. In most cases the choice of the function $R(s)$ may be described as follows.

Let $a_j \geq 1, b_j \geq 1, j = 1, \ldots, m$, be integers. Define

$$R(s) = \gamma \prod_{j=1}^{m} \frac{\Gamma(s + a_j)}{\Gamma(s + b_j)} \quad \in \mathbb{Q}(s), \tag{5}$$

where $\gamma$ is a rational number that will be defined later.

With the choice (5) there exist integral representations for $G_r(z)$, $r \geq 1$ that are useful in applications for the computation of the asymptotic of the constructed linear forms.

Let $r \geq 1$ be integer and $u$ be a complex number. Write

$$I_r(u) = \frac{1}{2\pi i} \int_L R(s) \left( \frac{\pi}{\sin \pi s} \right)^r e^{\pi i u s} \mathrm{d}s, \tag{6}$$

where the path of integration $L$ goes from $-i\infty$ to $i\infty$ and separates the poles of $\Gamma(s + a_j)$, $1 \leq j \leq m$, from points $0, 1, 2, \ldots$.

It is easy to check, that the integral (6) converges for $|\Re u| < r$. In the following proposition we express the function $G_r(z)$ in terms of integrals (6). For simplicity we assume that $a = 0$.

**Proposition 2.** *Let r be integer,* $r \geq 1$, *and the rational function* $R(s)$ *is defined by* (1).

1. *If* $\operatorname{ord}_{s=\infty} R(s) \geq 2$, *then* $G_1(1) = -I_1(1)$.
2. *For* $r \geq 2$ *there exist constants* $c_\lambda$ *depending only on* $r$ *and* $\lambda$, *such that the following equalities hold*

$$G_r(1) = \sum_{\substack{|\lambda|<r \\ \lambda \equiv r \pmod 2}} c_\lambda I_r(\lambda).$$

*All the integrals* $I_r(\lambda)$ *are evaluated on straight line* $\Re s = -1/2$ *from* $-i\infty$ *to* $+i\infty$.

*Proof.* See [9, Theorem 1]. □

The main results of this article describe arithmetical properties of coefficients $B_{\ell,k}$ in (1) and coefficients of polynomials $A_k(z^{-1})$ depending on parameters $a_j, b_j$.

Denote

$$S_1 = \{j \mid \ a_j > b_j\}, \qquad S_2 = \{j \mid \ a_j < b_j\},$$
$$R_j(s) = \frac{(b_j + s)(b_j + 1 + s)\cdots(a_j - 1 + s)}{(a_j - b_j)!}, \quad \text{if } j \in S_1,$$

and

$$R_j(s) = \frac{(b_j - a_j - 1)!}{(a_j + s)(a_j + 1 + s)\cdots(b_j - 1 + s)}, \quad \text{if } j \in S_2.$$

Then the function (5) can be presented in the form

$$R(s) = \gamma \prod_{j=1}^{m} \frac{s(s+1)\cdots(s+a_j-1)}{s(s+1)\cdots(s+b_j-1)} = \prod_{j=1}^{m} R_j(s). \tag{7}$$

The last equality defines the constant $\gamma = \gamma(\overline{a}, \overline{b}) > 0$.

In the sequel the notation $\Delta_j$ will be used for special segments of the real line. Define

$$\Delta_j = [b_j, a_j - 1], \quad \text{if } j \in S_1, \qquad \text{and} \qquad \Delta_j = [a_j, b_j - 1], \quad \text{if } j \in S_2.$$

For the length of $\Delta_j$ we will use notation $|\Delta_j|$.

For any integer $\ell$ define

$$\mathcal{M}(\ell) = \{j \in S_2 \mid \ \ell \in \Delta_j\}$$

and denote

$$d(\ell) = \operatorname{Card} \mathcal{M}(\ell),$$

where $\mathcal{P}$ is the union of sets $\Delta_j$, $j \in S_2$. Note that the equality $d(\ell) = 0$ is possible. For coefficients $B_{\ell,k}$ we have the expression

$$B_{\ell,k} = \frac{1}{(d-k)!} \left(\frac{d}{ds}\right)^{d-k} \left(R(s)(s+\ell)^d\right)_{s=-\ell} \in \mathbb{Q}, \tag{8}$$

where $d = d(\ell)$. Further denote $q = \max_{\ell \in \mathcal{P}} d(\ell)$.

## 2 Common denominator for coefficients of $A_k(z)$

We will denote the least common multiple of all integers $1, 2, \ldots, H$ by the symbol $D_H$.

Let $\mathfrak{M}$ be a finite set of integers that are not necessarily distinct. We write the elements of the set $\mathfrak{M}$ in decreasing order

$$\mathfrak{M} = \{m(1) \geq m(2) \geq \cdots \geq m(v)\},$$

where $v = \operatorname{Card} \mathfrak{M}$. The following terminology will be useful in the sequel. We will say that the number $m(1)$ is the first consecutive maximum of the set $\mathfrak{M}$, the number $m(2)$ is the second consecutive maximum of the set $\mathfrak{M}$ and so on.

Define two sets of integers

$$\mathcal{N}_1 = \{a_i - b_i - \sum_{j \in J} |\Delta_j|, \quad \text{where} \quad i \in S_1, \ J \subset S_2\},$$
$$\mathcal{N}_2 = \{b_i - a_j - 1, \quad \text{where} \quad i, j \in S_2, \quad j \neq i\}.$$

In this definition the subset $J \subset S_2$ can be empty.

**Theorem 1.** *Let the rational function $R(s)$ be defined by* (5)*, parameters $a_i, b_j$ satisfy*

$$b_1 + \cdots + b_m - a_1 - \cdots - a_m \geq 1,$$

*and* (1) *be the representation of $R(s)$ as the sum of simple fractions. Then for any $\ell \in \mathcal{P}$ and $k \in \mathbb{Z}$, $1 \leq k \leq d(\ell)$ the following inclusion holds*

$$D_{m(1)} D_{m(2)} \cdots D_{m(\operatorname{Card} S_2 - k)} B_{\ell,k} \in \mathbb{Z}, \tag{9}$$

*where $m(1), m(2), \ldots$ are consecutive maxima of the set $\mathcal{N}_1 \cup \mathcal{N}_2$ and $D_{m(0)} = 1$.*

Now one can prove a bound for the common denominator for coefficients of polynomials $A_j(x)$ from Proposition 1. For that we should define some sets

$$\mathcal{N}_3 = \{b_i - a - 1 \mid i \in S_2\},$$

and $\mathcal{N} = \mathcal{N}_1 \cup \mathcal{N}_2 \cup \mathcal{N}_3$.

**Corollary 1.** *In notations and conditions of Theorem* 1 *the following inclusions hold*

$$D_{m(1)} \cdots D_{m(\operatorname{Card} S_2 - k)} \cdot A_k(x) \in \mathbb{Z}[x], \qquad 1 \leq k \leq q,$$
$$D_{n(0)}^{\mu - 1} \cdot D_{n(1)} \cdots D_{n(\operatorname{Card} S_2)} \cdot A_0(x) \in \mathbb{Z}[x],$$

*where $n(0)$ is a maximal number of the set $\mathcal{N}_3$ and $n(j)$, $j \geq 1$, are consecutive maxima of the set $\mathcal{N}$.*

Results of this shape were first proved in articles of G. Rhin and C. Viola [11,12], devoted to the measure of irrationality of numbers $\zeta(2)$, $\zeta(3)$.

*Proof.* Equality (3) and (9) imply this assertion for polynomials $A_1(x), \ldots, A_q(x)$. Let $\ell, k, \nu$ be a triplet corresponding to a term in the representation (4) for the polynomial $A_0(x)$.

For any $i \in \mathcal{M}(\ell)$ the following inequality holds: $\nu \le \ell - a \le b_i - a - 1$. Therefore, the set $\mathcal{N}_3$ contains $d(\ell) \ge k$ numbers greater or equal $\nu$. For this $\nu$ we have also $D_{n(0)} \cdot \nu^{-1} \in \mathbb{Z}$. Take into account (9) we derive the assertion for $A_0(x)$. □

For the proof of Theorem 1, one needs several lemmas. For any integer $v \ge 1$ denote

$$H_v(z) = \frac{z(z+1)\cdots(z+v-1)}{v!}, \quad \text{and} \quad H_0(z) = 1. \tag{10}$$

**Lemma 1.** *Let* $v \ge 1$ *be integer and*

$$H(z) = H_v(z) = \frac{z(z+1)\cdots(z+v-1)}{v!}.$$

*Then for any integer k and integer* $t \ge 0$ *the following inclusion holds*

$$D_v^t \cdot \frac{1}{t!} H^{(t)}(k) \in \mathbb{Z}.$$

*Proof.* If $t > v$, the statement is true since $H^{(t)}(z) \equiv 0$. We assume later that $0 \le t \le v$.

Every polynomial $H_n(z)$ is an integral valued polynomial. It is well known that any integral valued polynomial can be represented as a sum of $H_n(z)$, $n \ge 0$, with integer coefficients. Hence the set of polynomials $H_n(z-k+1)$, $n \ge 0$, has the same property. This reduces our statement to the case $k = 1$.

Due to the identity

$$H^{(t)}(z) = t! \cdot H(z) \cdot \sum (z+j_1)^{-1} \cdots (z+j_t)^{-1},$$

where summation is taken over all sets $\{j_1, \dots, j_t\}$ of integers satisfying $0 \le j_1 < j_2 < \dots < j_t < v$, we derive

$$\frac{1}{t!} H^{(t)}(1) = \sum (1+j_1)^{-1} \cdots (1+j_t)^{-1}.$$

Now the statement is evident. □

**Lemma 2.** *Let* $u, v$ *be integers,* $1 \le u \le v$. *Then*

$$D_v \cdot \frac{(u-1)!(v-u)!}{v!} \in \mathbb{Z}.$$

*Proof.* For the proof it is enough to substitute $x = v$ into the identity

$$\frac{(u-1)!}{x(x-1)\cdots(x-u+1)} = \sum_{i=0}^{u-1} \frac{c_i}{x-i},$$

where

$$c_i = (-1)^{u-1-i} \frac{(u-1)!}{i!(u-1-i)!} \in \mathbb{Z}.$$

□

**Lemma 3.** *Let be*

$$G(z) = \frac{(b-a-1)!}{(z+a)\cdots(z+b-1)},$$

*where $b > a$ are integers; let $\ell$ be an integer number such that $\ell \notin [a, b-1]$, and $M = \max(\ell - a, b - \ell - 1)$. Then for any integer $t \geq 0$ we have*

$$D_M^{t+1} \cdot \frac{1}{t!} G^{(t)}(-\ell) \in \mathbb{Z}.$$

*Proof.* In the case $t = 0$ we get

$$G(-\ell) = \frac{(b-a-1)!(a-\ell-1)!}{(b-\ell-1)!} \qquad \text{for } \ell < a,$$

$$G(-\ell) = \frac{(b-a-1)!(\ell-b)!}{(\ell-a)!} \cdot (-1)^{b-a} \quad \text{for } \ell \geq b.$$

Due to Lemma 2 we derive the assertion.

Now we consider $t > 0$. According to the Leibnitz rule one can write

$$\begin{aligned} \frac{1}{t!} G^{(t)}(z) &= (-1)^t (b-a-1)! \sum_{\overline{k}} \prod_{j=a}^{b-1} \frac{1}{(z+j)^{k_j+1}} \\ &= (-1)^t G(z) \sum_{\overline{k}} \prod_{j=a}^{b-1} \frac{1}{(z+j)^{k_j}}, \end{aligned} \tag{11}$$

where the summation is taken over all sets of nonnegative integers $\overline{k} = (k_a, \ldots, k_{b-1})$ whose sum is equal $t$. For any $j$, $a \leq j \leq b-1$ we have $|\ell - j| \leq M$. According to Lemma 2 the representation (11) proves the statement. □

**Lemma 4.** *Let $G(z)$ be the same function as in Lemma 3. Let $\ell$ be integer, $\ell \in [a, b-1]$, and $M = \max(\ell - a, b - \ell - 1)$. Then for any nonnegative integer $t$ we have*

$$D_M^t \cdot \frac{1}{t!} \left(\frac{d}{dz}\right)^t (G(z)(z+\ell))_{z=-\ell} \in \mathbb{Z}.$$

*Proof.* For $t = 0$, the statement is evident due to

$$G(z)(z+\ell)|_{z=-\ell} = (-1)^{\ell-a} \frac{(b-a-1)!}{(\ell-a)!(b-1-\ell)!} \in \mathbb{Z}.$$

For the proof in the case $t > 0$ let us represent $G(z)$ as

$$G(z) = \sum_{i=a}^{b-1} \frac{c_i}{z+i}$$

with coefficients

$$c_i = G(z)(z+i)|_{z=-i} = (-1)^{i-a} \frac{(b-a-1)!}{(i-a)!(b-i-1)!} \in \mathbb{Z}.$$

This representation for $G(z)$ implies

$$G(z)(z+\ell) = \sum_{i=a}^{b-1} c_i \frac{z+\ell}{z+i} = \sum_{i=a}^{b-1} c_i + \sum_{\substack{i=a \\ i \neq \ell}}^{b-1} c_i \cdot \frac{\ell - i}{z+i}$$

and

$$\frac{1}{t!}\left(\frac{d}{dz}\right)^t (G(z)(z+\ell))_{z=-\ell} = -\sum_{\substack{i=a \\ i\neq\ell}}^{b-1} c_i \frac{1}{(\ell-i)^t}.$$

Due to the inequality $|\ell - i| \leq M$ this gives the assertion. □

The following lemma uses notations introduced before the formulation of Theorem 1.

**Lemma 5.** *For any $\ell \in \mathcal{P}$ we have*

$$D_{p(1)}D_{p(2)}\cdots D_{p(\mathrm{Card}\, S_2-d(\ell))}B_{\ell,d(\ell)} \in \mathbb{Z},$$

*where $p(1), p(2), \ldots$ are consecutive maxima of the set*

$$\{\max(b_i - a_j - 1, b_j - a_i - 1), \quad \textit{where} \quad i \in \mathcal{M}(\ell),\ j \in S_2\backslash\mathcal{M}(\ell)\}. \qquad (12)$$

*Proof.* Without loss of generality one can assume that $1 \in \mathcal{M}(\ell)$. For $j \in S_1$ the function $R_j(s)$ is an integral valued polynomial. Therefore

$$R_j(-\ell) \in \mathbb{Z}, \quad j \in S_1.$$

Further, for $j \in \mathcal{M}(\ell)$ we have

$$R_j(s)(s+\ell)|_{s=-\ell} = (-1)^{\ell-a_j}\frac{(b_j-a_j-1)!}{(\ell-a_j)!(b_j-\ell-1)!} \in \mathbb{Z}.$$

In the case $j \in S_2\backslash\mathcal{M}(\ell)$ we have $\ell \notin \Delta_j$, and according to Lemma 3 for $t = 0$ the inclusion holds

$$D_M R_j(-\ell) \in \mathbb{Z},$$

where $M = \max(\ell - a_j, b_j - 1 - \ell)$. Taking into account that $\ell \in \Delta_1$, we derive inequalities

$$\ell - a_j \leq b_1 - a_j - 1, \qquad b_j - \ell - 1 \leq b_j - a_1 - 1.$$

Now define $m(i_j) = \max(b_j - a_1 - 1, b_1 - a_j - 1)$.

It is proved that there exist $u = \mathrm{Card}\, S_2 - \mathrm{Card}\,\mathcal{M}(\ell) = \mathrm{Card}\, S_2 - d(\ell)$ numbers $m(i_j),\ j \in S_2\backslash\mathcal{M}(\ell)$ placed at different positions in the set (12) with the property

$$\prod_{j\in S_2\backslash\mathcal{M}(\ell)} D_{m(i_j)} \cdot R(s)(s+\ell)^{d(\ell)}|_{s=-\ell} \in \mathbb{Z}.$$

This finishes the proof of Lemma 5. □

**Lemma 6.** *Suppose that the parameters of the rational function (5) satisfy conditions*

$$|\Delta_i| < |\Delta_j|, \qquad i \in S_1, \quad j \in S_2.$$

*Then for any $\ell \in \mathcal{P}$ and $k \in \mathbb{Z}$, $1 \leq k \leq d = d(\ell)$ the inclusion*

$$D_{q(1)}D_{q(2)}\cdots D_{q(\mathrm{Card}\, S_2-k)}B_{\ell,k} \in \mathbb{Z}.$$

*is true. Here $q(1), q(2), \ldots$ are consecutive maxima of the set $\mathcal{N}_2$. The assertion is correct in the case* Card $S_1 = 0$ *as well.*

*Proof.* For any $j, 1 \le j \le m$ denote

$$Q_j(s) = \begin{cases} R_j(s), & \text{if } j \notin \mathcal{M}(\ell), \\ R_j(s)(s+\ell), & \text{if } j \in \mathcal{M}(\ell). \end{cases}$$

Then according to (8) the following equality holds

$$B_{\ell,k} = \sum_{t_1+\cdots+t_m=d-k} \frac{1}{t_1!} Q_1(s)^{(t_1)} \ldots \frac{1}{t_m!} Q_m(s)^{(t_m)} |_{s=-\ell} \,.$$

Let us fix some vector $(t_1, \ldots, t_m)$ with the condition $t_1+\cdots+t_m = d-k$. The lemmas proved above give denominators for the multipliers: Lemma 1 gives the denominator in the case $j \in S_1$, whereas for $j \in S_2$ we apply Lemmas 3 and 4. Note that

$$\sum_{j\in S_1} t_j + \sum_{j\in\mathcal{M}(\ell)} t_j + \sum_{j\in S_2\setminus\mathcal{M}(\ell)} (t_j+1) = d-k+(\text{Card } S_2 - d) = \text{Card } S_2 - k.$$

Let $i \in S_2$, $j \in \mathcal{M}(\ell)$, $i \ne j$. Since $\ell \in \Delta_j$, we have

$$\ell - a_i \le b_j - a_i - 1, \quad b_i - 1 - \ell \le b_i - a_j - 1,$$

and

$$\max(\ell - a_i, b_i - 1 - \ell) \le \max(b_j - a_i - 1, b_i - a_j - 1).$$

Hence for any $i \in \mathcal{M}(\ell)$ the set $\mathcal{N}_2$ contains $d-1$ elements corresponding to $j \in \mathcal{M}(\ell)$, $j \ne i$, that can be denominators for $\frac{1}{t_i!} Q_i(s)^{(t_i)}|_{s=-\ell}$ according to Lemma 4. But in the case $i \in S_2\setminus\mathcal{M}(\ell)$ the set $\mathcal{N}_2$ contains $d$ elements $j \in \mathcal{M}(\ell)$ which in view of Lemma 3 can serve as denominators for $\frac{1}{t_i!} Q_i(s)^{(t_i)}|_{s=-\ell}$.

In the case Card $S_1 = 0$, the proof of Lemma 6 is finished.

Let Card $S_1 \ge 1$ and $h$ be a proper index from the set $S_1$. Suppose $i, j$ are any distinct indices from $S_2$. Without loss of generality one can assume that $b_i \ge b_j$. By the hypotheses of Lemma 6 one can write

$$b_i - a_j - 1 \ge b_j - a_j - 1 = |\Delta_j| \ge |\Delta_h| + 1 = a_h - b_h.$$

According to Lemma 1 every element $\max(b_i - a_j - 1, b_j - a_i - 1)$, $i, j \in S_2$, $i < j$ can be denominator for $\frac{1}{t_h!} Q_h(s)^{(t_h)}|_{s=-\ell}$, $h \in S_1$. Taking in account that the number of pairs $i \in S_2$, $j \in S_2$, $i < j$ equals

$$\frac{\text{Card } S_2(\text{Card } S_2 - 1)}{2} \ge \text{Card } S_2 - 1,$$

we conclude that these denominators can be chosen as standing at different places in the set $\mathcal{N}_2$ and also can be picked in such a way that they are distinct from the denominators chosen above for $j \in S_2$. This finishes the proof of the lemma. □

*Proof of Theorem 1.* For the proof we will use induction on the lexicographic ordering of pairs of numbers

$$(\text{Card } S_2, \ \sum_{i\in S_1}(a_i - b_i)). \tag{13}$$

In the sequel we will compare rational functions that are similar to (5). For these functions we use the ordering induced by the ordering of the corresponding pairs (13).

For brevity the notation $U = \sum_{i\in S_1}(a_i - b_i)$ will be used. Note that in the hypotheses we have Card $S_2 \ge 1$, $U \ge$ Card $S_1$.

Suppose that the rational function $R(s)$ satisfies all conditions of Theorem 1 and that the assertion of this theorem is valid for any rational function that precedes $R(s)$. Let us prove the assertion for the function $R(s)$. Choose and fix a pair of integers $\ell,\ k$ satisfying

$$\ell \in \mathcal{P}, \qquad 1 \le k \le d(\ell).$$

We will prove the inclusion (9) for this pair.

In the case $k = d(\ell)$, the inclusion (9) is valid due to Lemma 5. Therefore, in the sequel we will assume that $k < d(\ell)$.

Let Card $S_1 = 0$ or for any $i \in S_1,\ j \in S_2$, we have the inequality $|\Delta_i| < |\Delta_j|$. In this case the assertion is satisfied for the function $R(s)$ due to Lemma 6. Therefore, we can assume that there exist two indices $i \in S_1,\ j \in S_2$ such that $|\Delta_i| \ge |\Delta_j|$. Without loss of generality one can assume that $|\Delta_1| \ge |\Delta_2|$ and $1 \in S_1,\ 2 \in S_2$. The last inequality can be written in the form $a_1 + a_2 \ge b_1 + b_2$.

Every integral valued polynomial is a linear combination with integer coefficients of polynomials $H_n(s),\ n \ge 0$, see (10). The same property is valid for the set of polynomials $H_n(s + a_2),\ n \ge 0$. Hence there exist integers $c_n$ such that

$$R_1(s) = \sum_{n=0}^{a_1-b_1} c_n H_n(s + a_2),$$

see (7). This implies that the rational function $R(s)$ can be represented as a sum with integer coefficients of functions $R(n, s)$ that have the same form as (5) but with parameters $a_2 + n, a_2$ instead of $a_1, b_1$. The rest of the parameters does not change.

In the case $n < a_1 - b_1$ function $R(n, s)$ has the same numbers Card $S_2$, $d(\ell)$ as $R(s)$, but the sum $U$ decreases. Indeed, it depends upon $n$ instead of $a_1 - b_1$ and the numbers $a_1 - b_1 - \sum_{j\in J}|\Delta_j|$ are replaced by smaller values. Therefore one can apply the inductive assumption to these functions $R(n, s)$. Additionally this implies that the integer $D_{m(1)} \cdots D_{m(u)}$ attributed to $R(s)$ is divisible by the corresponding numbers defined for the functions $R(n, s),\ 0 \le n < a_1 - b_1$. For $n < a_1 - b_1$ the coefficient $B_{\ell,k}(n)$ in the representation of $R(n, s)$ as the sum of reduced fractions corresponding to $B_{\ell,k}$ has the property

$$D_{m(1)} \cdots D_{m(u)} \cdot B_{\ell,k}(n) \in \mathbb{Z}, \qquad 0 \le n < a_1 - b_1. \tag{14}$$

In the case $n = a_1 - b_1$, we get

$$R(a_1 - b_1, s) = \frac{(b_2 - a_2 - 1)!(a_1 + a_2 - b_1 - b_2)!}{(a_1 - b_1)!} \cdot \Phi(s),$$

where

$$\Phi(s) = \frac{(s + b_2) \cdots (s + a_2 + a_1 - b_1 - 1)}{(a_1 + a_2 - b_1 - b_2)!} \cdot \prod_{j=3}^{m} R_j(s).$$

The function $\Phi(s)$ has the form as in (5). Here the number corresponding to Card $S_1$ does not increase, but the numbers corresponding to Card $S_2$ and $U$ decrease. Hence one can apply the inductive assumption to the function $\Phi(s)$. Note that under this change from $R(s)$ to $\Phi(s)$ the number $d(\ell)$ decreases by at most 1. Hence for the function $\Phi(s)$ and the pair of numbers $\ell, k$ we have the inclusion corresponding to (9). Also the number $u = \operatorname{Card} S_2 - k$ will be changed into $u-1$. What happens with the sets $\mathcal{N}_1$ and $\mathcal{N}_2$ when we replace $R(s)$ by $\Phi(s)$? In the set $\mathcal{N}_1$ some numbers $a_1 - b_1 - \sum_{j\in J} |\Delta_j|$ disappear (this is true in particular for $a_1 - b_1$). In the set $\mathcal{N}_2$ all numbers $b_i - a_j - 1$ with $i = 2$ or $j = 2$ disappear. Taking into account the inclusion

$$D_{a_1-b_1} \cdot \frac{(b_2-a_2-1)!(a_1+a_2-b_1-b_2)!}{(a_1-b_1)!} \in \mathbb{Z},$$

which is valid according to Lemma 2, we derive

$$D_{a_1-b_1} \cdot D_{m'(1)} \cdots D_{m'(u-1)} \cdot B_{\ell,k}(a_1-b_1) \in \mathbb{Z},$$

where $m'(j)$ are consecutive maxima of the set corresponding to $\mathcal{N}_1 \cup \mathcal{N}_2$, defined for the function $R(a_1-b_1, s)$. Therefore

$$D_{m(1)} \cdots D_{m(u)} \cdot B_{\ell,k}(a_1-b_1) \in \mathbb{Z}. \tag{15}$$

Together with (14) this inclusion finishes the proof of (9) and consequently the proof of Theorem 1. □

## 3 Upper bounds for the coefficients of $A_k(x)$

Assume that the parameters $a_i, b_i$ have the form

$$a_i = \alpha_i n + \xi_i, \quad b_i = \beta_i n + \eta_i, \quad i = 1, \dots, m,$$

where $\alpha_i, \beta_i, \xi_i, \eta_i, n$ are nonnegative integers. In this section we study the asymptotic behavior of the coefficients of the polynomials $A_k(x)$ when $\alpha_i, \beta_i, \xi_i, \eta_i$ are fixed and $n$ tends to infinity.

The function $R(s)$ does not change after any permutation of the pairs $(a_i, b_i)$. Therefore the functions $G_r(z)$ and the polynomials $A_k(x)$ do not change either. But if we will permute the parameters $a_i, b_i$ and mix the pairs, then it follows from (5) that only the number $\gamma$ changes. Simultaneously the rational function $R(s)$ is multiplied by a number that does not depend on $s$. The same change will happen to the functions $G_r(z)$ and the polynomials $A_k(x)$. The change of $\gamma$ can be easily controlled. Therefore, computing bounds for the coefficients of the polynomials $A_k(x)$ one can assume

$$1 \le a_1 \le \cdots \le a_m, \quad 1 \le b_1 \le \cdots \le b_m. \tag{16}$$

**Lemma 7.** *Under the hypothesis (16) the segments $\Delta_j, j \in S_1$ and $\Delta_k, k \in S_2$ have no common points and the representation (7) is irreducible.*

*Proof.* Indeed, in the case $j < k$ we have

$$b_j \le a_j - 1 < a_k \le b_k - 1.$$

In the opposite case $j > k$ we get

$$a_k \le b_k - 1 < b_j \le a_j - 1.$$

□

Introduce the real valued function

$$f(x) = \sum_{j=1}^{m} (\alpha_j - x) \log|\alpha_j - x| - (\beta_j - x) \log|\beta_j - x| + (\beta_j - \alpha_j) \log|\beta_j - \alpha_j|.$$

The following lemma explains the appearance of this function in our context.

**Lemma 8.** *For any $\ell \in \mathcal{P}$ we have*

$$\log|B_{\ell,d}| = n \cdot f(\eta) + O(\log n), \tag{17}$$

*where $\eta = \ell/n$, $d = d(\ell)$ and the constant in $O(.)$ depends only on $\alpha_i, \beta_i, \xi_i, \eta_i$.*

*Proof.* Define the functions $Q_j(s)$, $1 \le j \le m$, by:

$$Q_j(s) = \begin{cases} R_j(s), & \text{if } j \notin \mathcal{M}(\ell), \\ R_j(s)(s+\ell), & \text{if } j \in \mathcal{M}(\ell). \end{cases}$$

According to (8) we derive

$$B_{\ell,d} = \prod_{j=1}^{m} Q_j(-\ell). \tag{18}$$

For $j \in S_1$, we have

$$Q_j(-\ell) = \frac{(-\ell + b_j) \cdots (-\ell + a_j - 1)}{(a_j - b_j)!},$$

and

$$Q_j(-\ell) = (-1)^{a_j - b_j} \frac{(\ell - b_j)!}{(\ell - a_j)!(a_j - b_j)!}, \quad \text{for } \ell > a_j,$$

$$Q_j(-\ell) = \frac{(a_j - 1 - \ell)!}{(b_j - 1 - \ell)!(a_j - b_j)!}, \quad \text{for } \ell < b_j - 1.$$

We infer from Lemma 7 that $\ell \notin \Delta_j$. Apply to any factorial the inequality

$$|\log\Gamma(x) - (x - 1/2)\log x + x - 1/2\log 2\pi| < 1/x, \quad x > 0,$$

(see [16, 12.31]). Then we derive in both cases

$$\begin{aligned} \log|Q_j(-\ell)| = n[&(\alpha_j - \xi)\log|\alpha_j - \xi| - (\beta_j - \xi)\log|\beta_j - \xi| \\ &- (\alpha_j - \beta_j)\log|\alpha_j - \beta_j|] + O(\log n). \end{aligned} \tag{19}$$

If $j \in S_2 \backslash \mathcal{M}(\ell)$, then

$$Q_j(-\ell) = (-1)^{b_j - a_j} \frac{(b_j - a_j - 1)!(\ell - b_j)!}{(\ell - a_j)!}, \quad \text{for } \ell > b_j - 1,$$

$$Q_j(-\ell) = \frac{(b_j - a_j - 1)!(a_j - 1 - \ell)!}{(b_j - 1 - \ell)!}, \quad \text{for } \ell < a_j.$$

Again in both cases one can write the asymptotic formula (19).

In the last case $j \in \mathcal{M}(\ell)$ we have

$$Q_j(-\ell) = (-1)^{b_j - a_j} \frac{(b_j - a_j - 1)!}{(\ell - a_j)!(b_j - 1 - \ell)!}.$$

This representation leads to the formula (19) too.

Summing up relations (19) for $j = 1, \dots, m$, and using (18) one can derive the equality (17). □

Define segments $\Omega_i = [\beta_i, \alpha_i]$, $i \in S_1$, $\Omega_i = [\alpha_i, \beta_i]$, $i \in S_2$ and set $\overline{\mathcal{P}} = \bigcup_{i \in S_2} \Omega_i$.

**Theorem 2.** *Let $M = \sup_{\overline{\mathcal{P}}} f(x)$. Then under the above assumptions about the parameters $a_j, b_j$ we have*

$$\limsup_{n \to \infty} \frac{1}{n} \log \max_{0 \le k \le q} H(A_k(x)) \le M, \tag{20}$$

*where $A_k(x)$ are polynomials defined in Proposition 1 and $H(A_k(x))$ denotes the maximum of the coefficients of $A_k(x)$.*

*Proof.* Lemma 8 gives an upper bound for the coefficients $B_{\ell,k}$ of the polynomial $A_k(x)$, $k \ge 1$, with $d(\ell) = k$. For the other coefficients proceed as follows.

The function $R(s)$ can be written in the form

$$R(s) = \gamma \prod_j (s + j)^{u_j}, \quad u_j \in \mathbb{Z},$$

where $j$ runs through all points $j \in \bigcup_{i=1}^m \Delta_i$. Write $N = \operatorname{Card} \bigcup_{i=1}^m \Delta_i$ and $t = d(\ell) - k$. According to (8) we have

$$B_{\ell,k} = \gamma \sum_{t_1 + \dots + t_N = t} \prod_{j \ne \ell} \frac{u_j(u_j - 1) \cdots (u_j - t_j - 1)}{t_j!} (j - \ell)^{u_j - t_j}.$$

Here the summation is over all sets of nonnegative integers $\{t_1, \dots, t_{\ell-1}, t_{\ell+1}, \dots, t_N\}$ with sum $t$. In the product all numbers $1 \le j \le N$, $j \ne \ell$ are present. This representation can be written in the form

$$B_{\ell,k} = B_{\ell,d} \sum_{t_1 + \dots + t_N = t} \prod_{j \ne \ell} \frac{u_j(u_j - 1) \cdots (u_j - t_j - 1)}{t_j!} (j - \ell)^{-t_j - 1},$$

with $d = d(\ell)$. Taking in account the inequality

$$\left| \frac{u_j(u_j - 1) \cdots (u_j - t_j - 1)}{t_j!} \right| \le (2 + |u_j|)^{t_j} \le (q + 2)^{t_j},$$

we derive

$$|B_{\ell,k}| \le N^t (q + 2)^t |B_{\ell,d}|, \qquad d = d(\ell),$$

so $\log |B_{\ell,k}| \le nM + O(\log n)$ and consequently

$$\log |H(A_k(x))| \le nM + O(\log n), \quad 1 \le k \le q.$$

Due to (4) this inequality is valid for $A_0(x)$ too. Thus inequality (20) is proved. □

In some cases the inequality proved in Theorem 2 becomes an equality. We describe below such a situation.

**Theorem 3.** *If the maximum value of the function $f(x)$ on the set $\overline{\mathcal{P}}$ is taken in only one point, then for $\eta = 1$ or for $\eta = -1$ we have*

$$\lim_{n\to\infty} \frac{1}{n} \log \max_{0\le k\le q} |A_k(\eta)| = M. \tag{21}$$

*Moreover, in this case (20) becomes an equality.*

*Proof.* Denote by $\rho$ the unique point in $\overline{\mathcal{P}}$ satisfying $f(\rho) = M$. At points that are distinct from $\alpha_j, \beta_j$ we have

$$f'(x) = \sum_{j=1}^{m} (-\log |\alpha_j - x| + \log |\beta_j - x|) = -\log |g(x)|,$$

where

$$g(x) = \prod_{j=1}^{m} \frac{x - \alpha_j}{x - \beta_j}.$$

Hence the function $f(x)$ increases in some neighborhood of every point $\alpha_j$ and it decreases in some neighborhood of every point $\beta_j$. Due to the fact that this function is continuous at the points $\alpha_j, \beta_j$ we derive that $\rho$ is an inner point of any segments $\Omega_j$ containing $\rho$. Moreover, on the set $\overline{\mathcal{P}}$ the function $|g(x)|$ takes on as a value any positive number and the point $\rho$ is a solution of the equation $|g(x)| = 1$.

Choose an open interval $\mathfrak{U}$ with center in $\rho$ such a way that

1. the length of $\mathfrak{U}$ is less than the distance from $\rho$ to any point $\alpha_j, \beta_j$,
2. the function $f(x)$ is strictly increasing on the set $\{x \in \mathfrak{U}, x \le \rho\}$ and is strictly decreasing on the set $\{x \in \mathfrak{U}, x \ge \rho\}$,
3. $M_1 = \sup_{\overline{\mathcal{P}}\setminus\mathfrak{U}} |f(x)| < M$.

For brevity, denote by the letter $d$ the number of segments $\Omega_i$, $i \in S_2$, containing the point $\rho$. Put $\eta = (-1)^d$. Then

$$A_d(\eta) = \binom{d+r-2}{r-1} \sum_{\substack{\ell\in\mathcal{P}\\ d(\ell)\ge d}} B_{\ell,d}\eta^{\ell} = \sigma_1 + \sigma_2,$$

where

$$\sigma_1 = \binom{d+r-2}{r-1} \sum_{\substack{\ell/n\in\mathfrak{U}\\ d(\ell)\ge d}} B_{\ell,d}\eta^{\ell}, \qquad \sigma_2 = \binom{d+r-2}{r-1} \sum_{\substack{\ell/n\in\overline{\mathcal{P}}\setminus\mathfrak{U}\\ d(\ell)\ge d}} B_{\ell,d}\eta^{\ell}.$$

According to Lemma 8 we have a bound

$$|\sigma_2| \le \operatorname{Card}\mathcal{P} \cdot \exp(nM_1 + O(\log n)) = \exp(nM_1 + O(\log n)). \tag{22}$$

Let $2\varepsilon$ be the length of the interval $\mathfrak{U}$. If $\ell/n \in \mathfrak{U}$, then we have the inequalities

$$(\rho - \varepsilon)n < \ell < (\rho + \varepsilon)n.$$

For sufficiently large $n$ the number $\ell$ is contained exactly in $d$ segments $\Delta_i$, $i \in S_2$. Therefore

$$\sigma_1 = \binom{d+r-2}{r-1} \sum_{\ell/n \in \mathfrak{U}} B_{\ell,d}\eta^{\ell}.$$

To estimate $|\sigma_1|$ from below write $\lambda = [\rho n]$. Then $|\lambda/n - \rho| < 1/n$, and for sufficiently large $n$ we have the inclusion $\lambda/n \in \mathfrak{U}$.

If $(\ell - 1)/n, \ell/n \in \mathfrak{U}$, then $\ell \neq a_j, b_j$ and according to (8) we have the inequalities

$$B_{\ell-1,q} = R(s)(s+\ell+1)^q|_{s=1-\ell} = R(s+1)(s+\ell)^q|_{s=-\ell} = B_{\ell,q} \prod_{j=1}^{m} \frac{a_j - \ell}{b_j - \ell}.$$

Note that the fraction $(a_j - \ell)/(b_j - \ell)$ is positive for $\ell \notin \Delta_j$, and that it is negative for $\ell \in \Delta_j$. Therefore for even $d$ all coefficients $B_{\ell,d}$ in the sum $\sigma_1$ have the same sign. In this case $\eta = 1$, and all summands in $\sigma_1$ have the same sign. In the case of odd $d$ the sequence $B_{\ell,d}$ in $\sigma_1$ has alternating sign. But in this case $\eta = -1$ and again we have summands in $\sigma_1$ of the same sign. In any case we have the inequality

$$|\sigma_1| \geq |B_{\lambda,d}| = \exp(nf(\lambda/n) + O(\log n)).$$

The derivative of the function $f(x)$ is bounded on the set $\mathfrak{U}$. Hence one can conclude that $f(\lambda/n) = f(\xi) + O(n^{-1})$ and

$$|\sigma_1| \geq \exp(nM + O(\log n)).$$

The last inequality and (22) imply $|A_d(\eta)| \geq \exp(nM + O(\log n))$.

The number of terms in the polynomial $A_d(x)$ is $O(n)$. Therefore (20) proves (21).

Since the number of terms in $A_d(x)$ is $O(n)$ the equality (21) implies that under the hypotheses of Theorem 3 the upper limit in (20) cannot be less than $M$. This finishes the proof of Theorem 3. □

We mention some facts that can be derived from the proof of Theorem 3.

At first the point $\rho \in \mathcal{P}$ with $f(\rho) = M$ satisfies $g(\rho) = (-1)^d$ and therefore it is a root of the polynomial

$$\prod_{j=1}^{m}(x - \alpha_j) - (-1)^d \prod_{j=1}^{m}(x - \beta_j) = 0.$$

Secondly, since $|g(\rho)| = 1$, the following equality holds

$$M = \sum_{j=1}^{m} \left(\alpha_j \log|\rho - \alpha_j| - \beta_j \log|\rho - \beta_j| + (\beta_j - \alpha_j)\log|\beta_j - \alpha_j|\right).$$

## 4 Some examples

In this section a short survey of results proved with this construction of linear forms in values of polylogarithmic functions will be given. We avoid all computations and stress only on details of the constructions.

### 4.1 E. Nikishin [10]

Let us take

$$R(s) = \gamma \cdot \frac{s(s-1)\cdots(s-N+1)}{(s+1)_{n_1}\cdots(s+1)_{n_m}}, \tag{23}$$

where the $n_j$'s are integers satisfying

$$n_1 \geq n_2 \geq \cdots \geq n_m \geq n_1 - 1,$$

and

$$N+1 = \sum_{k=1}^{m} n_k, \quad (s+1)_r = (s+1)(s+2)\cdots(s+r).$$

It is clear that

$$R(s) = \gamma \cdot \frac{\Gamma(s+1)^{m+1}}{\Gamma(s-N+1)\Gamma(s+n_1+1)\cdots\Gamma(s+n_m+1)}.$$

After a shift of the variable $s \to s+N$ this function gets the form (5), (7) and we define the rational number $\gamma$ as in (7).

According to Proposition 1 we have the representation

$$G_1(z) = \sum_{\nu=1}^{\infty} R(\nu-1)z^{\nu} = A_0(z^{-1}) + A_1(z^{-1})Li_1(z) + \cdots + A_m(z^{-1})Li_m(z), \tag{24}$$

with polynomials $A_k(x) \in \mathbb{Q}[x]$ satisfying

$$\deg A_0(x) \leq n_1 - 2, \quad \deg A_k(x) \leq n_k - 1, \quad 1 \leq k \leq m,$$

and

$$\operatorname{ord}_{z=0} G_1(z) = N+1.$$

To estimate the common denominators for the coefficients of the polynomials $A_j(x)$ Nikishin was the first to introduce in [10] some important ideas.

Nikishin used $1/z$ instead of $z$ in our setting, and the identity (24) then can be considered as the Padé approximation of the first kind to the set of functions $1, Li_k(z^{-1})$ at the point $\infty$. He started from the problem to find such an approximation and found it in the form of the integral

$$G_1(z) = \frac{1}{2\pi i} \int_{\Re s=-1/2} R(s)\frac{\pi}{\sin \pi s}(-z)^s ds.$$

The theorem proved in [10] claims: let $a$, $b$ be integers satisfying

$$b > |a|^{m+1} \cdot e^{(m-1)(m \ln m + 2m \ln 2)} > 0.$$

Then the numbers

$$1, Li_1(a/b), \dots, Li_m(a/b) \tag{25}$$

are linearly independent over $\mathbb{Q}$.

In 1999 T. Hessami-Pilehrood [4] used the rational function (23) without the assumption $n_m \geq n_1 - 1$ to prove a lower bound for the absolute value of the linear form $x_0 + x_1 Li_1(a/b) + \cdots + x_m Li_m(a/b)$, $x_j \in \mathbb{Z}$, expressed in terms of $X = \overline{x}_1 \cdots \overline{x}_m$ for any set of integers $\overline{x}_k$ satisfying the conditions

$$\overline{x}_k \geq \max(1, |x_k|), \quad \overline{x}_1 \geq \overline{x}_2 \geq \cdots \geq \overline{x}_m.$$

In 2001 T. Rivoal [13] proved with

$$R(s) = \gamma \frac{s(s-1)\cdots(s-rn+1)}{(s+1)_n^m}, \quad r = \left[\frac{m}{\ln^2 m}\right]$$

that for any rational $\alpha$, $0 < |\alpha| < 1$ the dimension of the linear space generated over $\mathbb{Q}$ by the numbers $1, Li_1(\alpha), \dots, Li_m(\alpha)$ can be bounded from below by $\frac{\ln m}{1+\ln 2}(1+o(1))$ when $m \to \infty$.

## 4.2 L. Gutnik [2]

Let

$$R(s) = \gamma \cdot \frac{(s-1)^m \cdots (s-n)^m}{s(s+1)\cdots(s+mn+1)} = \gamma \cdot \frac{\Gamma(s)^{m+1}}{\Gamma(s-n)^m \Gamma(s+mn+2)},$$

where $\gamma$ is defined as in (7). To apply Proposition 1 we make the shift of variable $s \to s+n+1$ and take $a = n+1$. According to this proposition we derive

$$G_r(z) = \frac{(-1)^{r-1}}{(r-1)!} \sum_{\nu=1}^{\infty} R^{(r-1)}(\nu) z^\nu = B_r(z^{-1}) - A_1(z^{-1}) Li_r(z), \quad 1 \leq r \leq m,$$

where

$$B_r(x), A_1(x) \in \mathbb{Q}[x], \quad \deg A_1(x) \leq mn+1, \quad \operatorname{ord} G_r(z) = n+1.$$

Gutnik constructed functions $G_r(z)$ (Padé approximation of the second kind to polylogarithmic functions at the point $\infty$) with complex integrals and improved Nikishin's result. The linear independence of numbers $1, Li_1(a/b), \dots, Li_m(a/b)$ was proved for $|b| > |a|^{m+1} m^{-1} e^{m^2} > 0$.

## 4.3 L. Gutnik [1]

For the rational function

$$R(s) = \left[\frac{(s+1)(s+2)\cdots(s+n)}{(s+n+1)(s+n+2)\cdots(s+2n+1)}\right]^2$$

we have

$$G_1(z) = \sum_{\nu=1}^{\infty} R(\nu - n - 1)z^{\nu} = C_1(z^{-1}) + B(z^{-1})Li_1(z) + A(z^{-1})Li_2(z), \tag{26}$$

$$G_2(z) = \sum_{\nu=1}^{\infty} R'(\nu - n - 1)z^{\nu} = C_2(z^{-1}) + B(z^{-1})Li_2(z) + 2A(z^{-1})Li_3(z). \tag{27}$$

The coefficients in these linear forms are polynomials $A(x), B(x), C_j(x) \in \mathbb{Q}[x]$ and satisfy

$$\deg A \le n, \quad \deg B \le n, \quad \deg C_j \le n - 1, \quad \operatorname{ord}_{z=0} G_j(z) = n + 1, \quad j = 1, 2.$$

Due to the equalities $Li_1(-1) = -\ln 2$ and $Li_k(-1) = (2^{1-k} - 1)\zeta(k)$ for $k \ge 2$ we have

$$\begin{pmatrix} -2G_1(-1) \\ -2G_2(-1) \end{pmatrix} = \begin{pmatrix} -2C_1(-1) \\ -2C_2(-1) \end{pmatrix} + B(-1)\begin{pmatrix} 2\ln 2 \\ \zeta(2) \end{pmatrix} + A(-1)\begin{pmatrix} \zeta(2) \\ 3\zeta(3) \end{pmatrix}.$$

Gutnik used in [1] these equalities and proved that for any rational number $q$ at least one of the numbers

$$\zeta(2) + 2q\ln 2, \quad 3\zeta(3) + q\zeta(2)$$

is irrational. He defined $G_j(z)$ as complex integrals (Meijer G-functions) and applied differential equations for $G$-functions to prove recurrence relations in $n$ for $G_j(z)$. Then he used Poincaré's theorem to estimate these integrals at $z = -1$.

This result was generalized to vectors with polylogarithmic coordinates of any dimension $p \ge 2$ in the articles of Gutnik [3] and Hessami-Pilehrood [5,6]. In the latter one a rational function of the type

$$R(s) = \gamma\left(\frac{((s+1)\cdots(s+n)}{(s+n+1)\cdots(s+2n+1)}\right)^{p}$$

was used.

Since $G_1(z)$, $Li_2(z)$ converge at $z = 1$ and $Li_1(z) = -\ln(1 - z)$ diverges at this point we derive from (26) that $B(1) = 0$ and from (27) that

$$-2G_2(1) = 3A(1)\zeta(3) - 2C_2(1).$$

The numbers $A(1)$ and $C_2(1)$ are rationals and the ratio $2C_2(1)/(3A(1))$ gives a good approximation to $\zeta(3)$. This allows it to prove the irrationality of $\zeta(3)$ (Apery's theorem, 1978). The magnitude of $A(1), C_2(1)$ as well as the magnitude of their denominators can be estimated in terms of $n$ due to Theorems 1 and 3. The asymptotic behavior of $G_2(1)$ can be computed with the saddle point method applied to the integral $G_2(1) = I_2(0)$, see Proposition 2. More details can be found in [8].

### 4.4 T. Rivoal [14]

Let us assume that the parameters $a_j, b_j$ in (5) satisfy the equalities

$$a_j + b_j = c, \qquad j = 1, \ldots, m,$$

and $\delta = \sum_{j=1}^{m}(b_j - a_j)$. It is easy to check that with this assumption the rational function (5) satisfies the identity $R(-c-s) = (-1)^{\delta}R(s)$. This leads to equalities $A_k(1) = (-1)^{\delta+k}A_k(1)$ for the coefficients of the linear forms (3). We obtain

$$A_k(1) = 0, \qquad k \equiv \delta + 1 \pmod 2.$$

As an application the function

$$R(s) = \left(s + \frac{n}{2}\right)\frac{(s-n)\cdots(s-1)\cdot(s+n+1)\cdots(s+2n)}{s^4(s+1)^4\cdots(s+n)^4}$$

was proposed by K. Ball to prove the irrationality of $\zeta(3)$. Up to the shift of the variable $s$ and for even $n$ it looks like (5) and we derive

$$G_1(z) = \sum_{\nu=1}^{\infty} R(\nu)z^{\nu} = A_0(z^{-1}) + A_1(z^{-1})Li_1(z) + \cdots + A_4(z^{-1})Li_4(z).$$

In this case $\delta \equiv 1 \pmod 2$ and consequently $A_2(1) = A_4(1) = 0$. Due to the convergence of the series $G_1(1)$ additionally the equality $A_1(1) = 0$ holds. Finally we have $G_1(1) = A_0(1) + A_3(1)\zeta(3)$ and the fractions $-A_0(1)/A_3(1)$ give rational approximations to $\zeta(3)$. Rivoal had proved that these approximations coincide with the approximations of Apery.

Following Rivoal let us take

$$R(s) = \gamma\frac{(s-rn)\cdots(s-1)(s+n+1)\cdots(s+(r+1)n)}{(s(s+1)\cdots(s+n))^m},$$

where $m = 2d + 1 \geq 3$ is an odd number and where $r$ is an integer parameter depending on $m$. In this case $\delta \equiv n + 1 \pmod 2$ and with even $n$ we derive

$$G_1(1) = \sum_{\nu=1}^{\infty} R(\nu) = A_0(1) + A_3(1)\zeta(3) + A_5(1)\zeta(5) + \cdots + A_{2d+1}(1)\zeta(2d+1)$$

a linear form in values of the zeta-function at odd points with rational coefficients. Taking for example $r = [m \ln^{-2} m]$ Rivoal has proved with this construction that there are infinitely many numbers $\zeta(2k+1)$ which together with 1 are linearly independent over $\mathbb{Q}$. So in particular there are infinitely many irrationals among $\zeta(2k+1)$.

## 4.5 W. Zudilin [17]

The rational function

$$T(x) = x\frac{[(x-u)(x-u+1)\cdots(x-v)(x+v)\cdots(x+u-1)(x+u)]^3}{\prod_{j=1}^{10}(x-w_j)(x-w_j+1)\cdots(x+w_j)},$$

$$u > v > w_1 > \ldots > w_{10} > 0, \qquad u, v, w_j \in \mathbb{Z}, \quad \mathrm{ord}_{\infty} T(x) \geq 2,$$

is odd: $T(-x) = -T(x)$. Denote $R(s) = T(s+u+1)$ and $G_3 = \frac{1}{2}\sum_{\nu=0}^{\infty} R''(\nu)$. Due to Proposition 1 the number $G_3(1)$ is a linear form in $1, \zeta(3), \zeta(4), \ldots, \zeta(12)$ with rational coefficients. By the symmetry $R(-2u-2-s) = -R(s)$ we have $A_4(1) = A_6(1) = A_8(1) = A_{10}(1) = A_{12}(1) = 0$. Moreover the inequality $\mathrm{ord}_{s=\infty} R(s) \geq 2$ implies $A_3(1) = 0$. Hence we have

$$G_3(1) = A_0(1) + A_5(1)\zeta(5) + A_7(1)\zeta(7) + A_9(1)\zeta(9) + A_{11}(1)\zeta(11).$$

Choosing

$$u = 91n, \quad v = 37n, \quad w_k = (35 - 2k)n, \quad 1 \leq k \leq 10,$$

Zudilin proved with this construction that at least one of the four numbers $\zeta(5)$, $\zeta(7)$, $\zeta(9)$, $\zeta(11)$ is irrational.

**Acknowledgment.** I express my deep gratitude to H.P. Schlickewei for his assistance at the final stage of my work with this article. This research was partially supported by Alexander von Humboldt Foundation and Russian Foundation of Basic Research, grant nr. 03-01-00359.

## References

1. Gutnik, L.A.: On the irrationality of certain quantities involving $\zeta(3)$. *Usp. Mat. Nauk* **34**, 190 (1979) (in Russian); *Acta Arith.* **42**, 255–264 (1983)
2. Gutnik, L.A.: On linear independence over $\mathbb{Q}$ of dilogaritms at rational points. *Usp. Mat. Nauk* **37**, 179–180 (1982) (in Russian); *Russ. Math. Surv.* **37**, 176–177 (1982)
3. Gutnik, L.A.: On the rank over $\mathbb{Q}$ of some real matrices. VINITI, 1984, No 5736-84, pp. 1–29 (in Russian)
4. Hessami-Pilehrood, T.G.: Lower bound of some linear form. *Mat. Zametki* **66**, 617–623 (1999) (in Russian)
5. Hessami Pilehrood, T.G.: Arithmetic properties of values of hypergeometric functions. Ph. D. thesis, Moscow University, Moscow, Russia (1999)
6. Hessami Pilehrood, T.G.: Linear independence of vectors with polylogarithmic coordinates. *Vestn. Mosk. Univ., Ser. I*, **9**(6), 54–56 (1999); *Mosc. Univ. Math. Bull.* **54**(6), 54–56 (1999)
7. Luke, Yu.L.: *Mathematical Functions and Their Approximations.* Academic Press, New York (1975)
8. Nesterenko, Yu.: A few remarks on $\zeta(3)$. *Mat. Zametki [Math. Notes]* **59**, 865–880 (1996)
9. Nesterenko, Yu.: Integral identities and constructions of approximations to zeta-values. *J. Theor. Nombres Bordx.* **15**, 535–550 (2003)
10. Nikishin, E.M.: On irrationality of values of functions $F(x, s)$. *Mat. Sb.* **109**, 410–417 (1979); *Math. USSR Sb.* **37**, 381–388 (1980)
11. Rhin, G., Viola, C.: On a permutation group related to $\zeta(2)$. *Acta Arith.* **77**, 23–56 (1966)
12. Rhin, G., Viola, C.: The group structure for $\zeta(3)$. *Acta Arith.* **97**, 269–293 (2001)
13. Rivoal, T.: Propriétés diophantiennes des valeours de la fonction zêta de Riemann aux entiers impairs. Thèse de doctorat, l'Université de Caen, Caen, France (2001)
14. Rivoal, T.: La fonction Zêta de Riemann prend une infinité de valeurs irrationnelles aux entiers impairs. *C. R. Acad. Sci. Paris, Sér. I Math.* **331**, 267–270 (2000)
15. Slater, L.J.: *Generalized Hypergeometric Functions.* Cambridge University Press, Cambridge (1966)
16. Whittaker, E.F., Watson, G.N.: *A Course of Modern Analysis.* Cambridge University Press, Cambridge (1927)
17. Zudilin, V.V.: One of the numbers $\zeta(5)$, $\zeta(7)$, $\zeta(9)$, $\zeta(11)$ is irrational. *Usp. Mat. Nauk* **56**, 149–150 (2001); *Russ. Math. Surv.* **56**, 774–776 (2001)

*Diophantine Approximation*
H. P. Schlickewei et al., Editors


# QUELQUES ASPECTS DIOPHANTIENS DES VARIÉTÉS TORIQUES PROJECTIVES

Patrice Philippon[1] et Martín Sombra[2]
[1] *Projet Géométrie et Dynamique, Institut de Mathématiques de Jussieu (U.M.R. 7586), Case 7012, 2 place Jussieu, 75251 Paris Cedex 05, France*
pph@math.jussieu.fr
[2] *Departament d'Àlgebra i Geometria, Universitat de Barcelona, Gran Via 585, 08007 Barcelona, Espagne*
sombra@ub.edu

An Wolfgang Schmidt
Der oft und schick
'nen Edelsatz
Geschmiedet hat

## 1 Introduction et résultats

Les variétés toriques jouent un rôle important au carrefour de l'algèbre, la géométrie et la combinatoire. Elles constituent une classe de variétés suffisamment rigide pour que beaucoup des invariants s'explicitent en termes combinatoires, et en même temps suffisamment riche pour permettre de tester et illustrer diverses conjectures et théories abstraites. Elle trouve application dans de nombreuses branches des mathématiques : géométrie algébrique bien sûr, algèbre commutative, combinatoire, calcul formel, géométries symplectique et kählerienne, topologie et physique mathématique, *voir* par exemple [Ful93], [GKZ94], [Stu96], [Cox01], [Aud91], [Don02].

Par définition, les variétés toriques projectives sont les compactifications équivariantes des translatés de sous-tores des tores multiplicatifs $\mathbf{G}_m^N$. Du point de vue de la géométrie diophantienne, ces variétés se trouvent à la croisée des problèmes de Lehmer et de Bogomolov généralisés sur les tores. En effet, on sait que lorsqu'elles ne sont pas de torsion, les minorations pour la hauteur normalisée de ces variétés sont de nature fondamentalement arithmétique, dépendant essentiellement du corps de définition de la variété. Au contraire, pour les sous-variétés de $\mathbf{G}_m^N$ qui ne sont pas des translatées de sous-tores on dispose de minorations ne dépendant que de leur géométrie, *voir* [AD03], [AD04]. D'un autre côté, R. Ferretti [Fer03] a exploité les variétés toriques afin de trouver des exemples concrets de l'extension du théorème du sous-espace aux variétés projectives, qu'il a obtenue avec J.-H. Evertse [EF02].

*Mots clefs.* Variété torique, hauteur normalisée, multihauteurs, fonction de Hilbert arithmétique, poids de Chow, volume mixte, indice d'obstruction, minimums successifs.

*2000 Classification mathématique par sujets.* Primaire 11G50; Secondaire 14M25, 14G40.

Dans [PS04] (*voir* aussi [PS05]) nous avons étudié l'un des invariants arithmétiques les plus significatifs des variétés dans le cas torique, à savoir leur hauteur normalisée. Cet invariant est l'analogue arithmétique du degré, il mesure la complexité binaire d'une représentation de la variété et contrôle aussi la distribution des points algébriques de petite hauteur sur la variété. Dans [PS04], on a donné une expression explicite pour la hauteur normalisée d'une variété torique et plus généralement pour la multihauteur d'un tore par rapport à plusieurs plongements monomiaux.

Ces résultats sont en parfait parallèle avec la théorie géométrique. En fait, on construit un objet adélique $\Theta_X$ associé à une variété torique $X$ (constitué par une famille finie de fonctions concaves et affines par morceaux) qui est le pendant arithmétique du polytope classiquement associé à l'action du tore et dont l'intégrale donne la hauteur. Grâce à cette approche, il est possible de calculer explicitement cette quantité pour n'importe qu'elle variété torique particulière et de tester utilement des conjectures et résultats.

Le présent texte a le double propos d'introduire le lecteur à l'étude des variétés toriques débutée dans [PS04] (§ 5, § 6), et de présenter de nouvelles applications des variétés toriques à des problèmes diophantiens ou d'origine diophantienne (§ 3, § 4, § 7, § 8).

Dans le § 3 on s'intéresse aux indices d'obstruction successifs des variétés toriques définies sur un corps algébriquement clos $\mathbf{K}$. Il s'agit des plus petits degrés de formes d'une suite sécante (soit globalement, soit dans un ouvert) découpant un ensemble algébrique ayant la variété comme composante. Différentes variantes de ces indices jouent un rôle important dans les généralisations des problèmes de Lehmer et de Bogomolov par exemple, *voir* [AD03].

Les sous-variétés toriques de $\mathbf{P}^N(\mathbf{K})$ correspondent à des idéaux binomiaux premiers et homogènes de l'anneau $\mathbf{K}[x_0, \ldots, x_N]$. De plus, les binômes engendrant l'idéal d'une variété torique $X$ s'explicitent en termes d'un certain $\mathbf{Z}$-module $\Gamma_X \subset \mathbf{Z}^{N+1}$ naturellement associé à $X$, *voir* [ES96] ou § 3.

On montre que le premier indice d'obstruction d'une variété torique $X$ est égal au premier minimum de $\Gamma_X$ par rapport à une métrique convenable; et plus généralement, que les indices d'obstruction successifs $\omega_i(X; (\mathbf{P}^N)^\circ)$ de cette variété relatifs à l'ouvert $(\mathbf{P}^N)^\circ$ coïncident avec les minimums successifs de $\Gamma_X$ et qu'ils se réalisent par des équations binomiales (Proposition 3.4). *Via* ce résultat, le deuxième théorème de Minkowski se traduit en des estimations pour le produit des indices d'obstruction successifs, qui précisent dans le cas torique les estimations de M. Chardin [Cha89] et de Chardin et P. Philippon [CP99]:

**Proposition 1.1.** *Soit $X \subset \mathbf{P}^N$ une variété torique de dimension $n$, alors*

$$\deg(X) \leq \omega_1(X; (\mathbf{P}^N)^\circ) \cdots \omega_{N-n}(X; (\mathbf{P}^N)^\circ) \leq (N+1)^{N-n} \deg(X).$$

En outre, le réseau $\Gamma_X$ s'identifie au réseau des périodes de l'application exponentielle restreinte à l'espace tangent en l'origine de $X^\circ$. On retrouve ainsi au § 4 certains résultats de [BP88] reliant degré et multi-degrés d'un sous-groupe algébrique d'un tore multiplicatif au volume de son réseau des périodes et à la hauteur de son espace tangent.

Dans l'article [PS06] on poursuit une étude approfondie des indices d'obstruction des variétés toriques et de son application à la minoration de la hauteur des points dans ces variétés.

Soit maintenant $X \subset \mathbf{P}^N$ une variété quelconque de dimension $n$ et $\tau = (\tau_0, \ldots, \tau_N) \in \mathbf{R}^{N+1}$ un vecteur *poids*. Soient $n+1$ groupes $U_0, \ldots, U_n$ de $N+1$ variables chacun et considérons la *forme de Chow de X*

$$Ch_X = \sum_{a \in \mathbf{N}^{(n+1)(N+1)}} c_a U_0^{a_0} \cdots U_n^{a_n} \in \mathbf{K}[U_0, \ldots, U_n]$$

Le *poids de Chow relatif à* $\tau$ (ou $\tau$*-poids de Chow*) de $X$ est défini comme le poids de sa forme de Chow par rapport au vecteur $(\tau, \ldots, \tau) \in \mathbf{R}^{(n+1)(N+1)}$, c'est-à-dire

$$e_\tau(X) := \max\{\langle a_0, \tau\rangle + \cdots + \langle a_n, \tau\rangle : a \in \mathbf{N}^{(n+1)(N+1)} \text{ tel que } c_a \neq 0\},$$

où $\langle \cdot, \cdot \rangle$ désigne le produit scalaire ordinaire.

Le poids de Chow a été introduit par D. Mumford [Mum77] en relation avec la stabilité des variétés projectives. On le retrouve (et en particulier l'énoncé 5.3 du présent texte) dans un travail de S.K. Donaldson [Don02] montrant la relation entre stabilité des variétés toriques et existence de métrique kählerienne à courbure constante. Il apparaît également en géométrie diophantienne au travers des travaux de Ferretti [Fer03], Evertse et Ferretti [EF02] et dans notre formule pour la hauteur d'une variété torique [PS04].

Pour $\tau \in \mathbf{Z}^{N+1}$, considérons l'action du sous-groupe à un paramètre

$$*_\tau : \mathbf{G}_m \times \mathbf{P}^N \to \mathbf{P}^N, \qquad (t, (x_0 : \cdots : x_N)) \mapsto (t^{\tau_0} x_0 : \cdots : t^{\tau_N} x_N)$$

et la *déformation torique* $X_\tau$ de $X$ associée, définie comme l'adhérence de Zariski de l'ensemble

$$\{((1 : t), t *_\tau x) \ : \ t \in \mathbf{G}_m, \ x \in X\} \subset \mathbf{P}^1 \times \mathbf{P}^N.$$

La *variété initiale de X relative au poids* $\tau \in \mathbf{Z}^{N+1}$ est par définition

$$\mathrm{init}_\tau(X) := \iota^*(X_\tau \cdot (\{(0 : 1)\} \times \mathbf{P}^N)) \in Z_n(\mathbf{P}^N),$$

où $\iota : \mathbf{P}^N \to \mathbf{P}^1 \times \mathbf{P}^N$ désigne l'inclusion $(x_0 : \cdots : x_N) \mapsto ((0 : 1), (x_0 : \cdots : x_N))$. Autrement-dit, $\mathrm{init}_\tau(X)$ est le *cycle limite* $\lim_{t \to \infty} t *_\tau X$ de $X$ sous l'action $*_\tau$. C'est un cycle de même dimension et degré que $X$.

On montre au § 5 que lorsque $\tau \in \mathbf{N}^{N+1}$, le poids de Chow relatif à $\tau$ s'interprète comme un bi-degré d'une variante de la déformation torique ci-dessus et se comporte donc comme une hauteur. Comme conséquence de cette interprétation, on démontre un théorème de Bézout pour le poids de Chow du cycle intersection $X \cdot H$, qui précise la majoration obtenue par Ferretti [Fer03, Prop. 4.3].

**Théorème 1.2.** *Soit* $X \subset \mathbf{P}^N$ *une variété projective et* $H \in \mathrm{Div}(\mathbf{P}^N)$ *un diviseur ne contenant pas X, alors pour* $\tau \in \mathbf{Z}^{N+1}$ *on a*

$$e_\tau(X \cdot H) = e_\tau(X) \deg(H) + e_\tau(H) \deg(X) - (\tau_0 + \cdots + \tau_N) \deg(H) \deg(X) - \sum_{Y \in \mathrm{Irr}(\mathrm{init}_\tau(X))} m(X_\tau \cdot H_\tau; \iota(Y)) \deg(Y)$$

*où la somme porte sur les composantes irréductibles de* $\mathrm{init}_\tau(X)$.

*En particulier, si H est effectif on a pour tout* $\tau \in \mathbf{R}^{N+1}$

$$e_\tau(X \cdot H) \leq e_\tau(X) \deg(H) + e_\tau(H) \deg(X) - (\tau_0 + \cdots + \tau_N) \deg(H) \deg(X),$$

*avec égalité si et seulement si les variétés initiales de X et H s'intersectent proprement.*

*Remarque 1.3.* Lorsque $H$ est effectif, le terme correctif $\sum_Y m(X_\tau \cdot H_\tau; \iota(Y)) \deg(Y)$ est le degré de la partie du cycle intersection $X_\tau \cdot H_\tau$, contenue dans la variété initiale de $X$.

Suivant l'attitude générale adoptée dans ce texte nous écrivons également au § 8 ce théorème en termes combinatoires pour l'intersection d'une variété torique avec un diviseur monomial. Comme autre application de la formule de la hauteur d'une variété torique, on en déduit dans ce cas un théorème de Bézout *exact* pour la hauteur normalisée du cycle intersection (*voir* Corollaire 8.6).

L'étude des points algébriques de petite hauteur (ou *petits points*) a reçu une attention considérable au cours des dernières années. La taille des petits points dans une variété est quantifiée par ses minimums algébriques successifs.

Soit $X \subset \mathbf{P}^N(\overline{\mathbf{Q}})$ une variété quasi-projective quelconque, de dimension $n$. Pour $\theta \geq 0$ on pose $X(\theta)$ pour l'ensemble des points de $X$ de hauteur normalisée $\widehat{h}$ majorée par $\theta$. Pour $i = 1, \ldots, n+1$, le *$i$-ème minimum algébrique de $X$ par rapport à la hauteur normalisée* est

$$\widehat{\mu}_i(X) := \inf\left\{\theta : \dim(\overline{X(\theta)}) \geq n - i + 1\right\},$$

où $\overline{X(\theta)}$ désigne l'adhérence de Zariski de $X(\theta)$. On écrit $\widehat{\mu}^{\mathrm{ess}}(X) := \widehat{\mu}_1(X)$ et $\widehat{\mu}^{\mathrm{abs}}(X) := \widehat{\mu}_{n+1}(X)$ pour les minimums *essentiel* et *absolu* respectivement; on a $\widehat{\mu}_1(X) \geq \cdots \geq \widehat{\mu}_{n+1}(X) \geq 0$.

La répartition de la hauteur des points algébriques d'une variété projective *fermée* $X$ est en relation avec sa hauteur, le lien est donné par le *théorème des minimums successifs* [Zha95, Thm. 5.2 et Lem. 6.5]:

$$\widehat{\mu}_1(X) + \cdots + \widehat{\mu}_{n+1}(X) \leq \frac{\widehat{h}(X)}{\deg(X)} \leq (n+1)\,\widehat{\mu}_1(X). \tag{1.1}$$

Comme application de la formule pour la hauteur d'une variété torique de [PS04], on construit au § 7 des exemples montrant que toute configuration possible des minimums successifs est arbitrairement approchable et que le quotient $\widehat{h}(X)/\deg(X)$ peut atteindre n'importe quelle valeur dans l'intervalle autorisé par l'encadrement ci-dessus:

**Théorème 1.4.** *Soient $n, N \in \mathbf{N}$ tels que $N \geq 3n+1$ et $\mu_1, \ldots, \mu_{n+1}, \nu \in \mathbf{R}$ tels que*

$$\mu_1 \geq \cdots \geq \mu_{n+1} \geq 0 \quad et \quad \mu_1 + \cdots + \mu_{n+1} \leq \nu < (n+1)\,\mu_1.$$

*Alors pour $0 < \varepsilon_1 \leq (n+1)\mu_1 - \nu$, $\varepsilon_2 > 0$ arbitraires, il existe une variété torique $X \subset \mathbf{P}^N$ de dimension $n$ telle que*

$$0 < \mu_i - \widehat{\mu}_i(X) \leq \varepsilon_1 \quad pour\ i = 1, \ldots, n+1 \ et \quad \left|\frac{\widehat{h}(X)}{\deg(X)} - \nu\right| < \varepsilon_2 \mu_1.$$

*De plus, la variété $X$ peut être choisie de degré $\leq (4n^2\,\varepsilon_2^{-1})^n$ et définie sur une extension kummerienne $K = \mathbf{Q}(2^{1/\ell})$ de degré $\leq \lfloor \log(2)\,\varepsilon_1^{-1} \rfloor + 1$.*

Les exemples construits présentent une codimension minimale $N - n = 2n + 1$ de l'ordre de la dimension de la variété produite. La question se pose donc de savoir ce qu'il en est pour les variétés de petite codimension (*cf.* Proposition 7.3).

Pour faciliter la lecture, nous avons tressé dans le texte plusieurs paragraphes d'introduction aux variétés toriques (§ 2), aux poids de Chow (§ 5) et à la théorie de l'intersection multi-projective (§ 2). Les paragraphes § 3 à § 7 doivent pouvoir être lus de manière essentiellement indépendante. Dans les paragraphes 2 à 5 nous prenons un corps de base **K** algébriquement clos quelconque ou **C** pour le § 4. L'arithmétique apparaît à partir de la fin du § 5, où nous posons nos conventions sur les places et valeurs absolues des corps de nombres (Convention 5.4).

## 2 Généralités sur les variétés toriques projectives

On note $\mathbf{G}_m^n := (\mathbf{K}^\times)^n$ le tore algébrique et $\mathbf{P}^N$ l'espace projectif sur **K**, de dimension $n$ et $N$ respectivement. Une variété est toujours supposée réduite et irréductible. Pour une famille de polynômes homogènes $f_1, \ldots, f_s \in \mathbf{K}[x_0, \ldots, x_N]$ on pose $Z(f_1, \ldots, f_s) \subset \mathbf{P}^N$ l'ensemble de ses zéros communs. Réciproquement, pour un ensemble algébrique $Z \subset \mathbf{P}^N$ on pose $I(Z)$ son idéal de définition dans $\mathbf{K}[x_0, \ldots, x_N]$.

On note $\mathbf{R}_+$ et $\mathbf{R}_+^\times$ les ensembles des nombres réels non-négatifs et positifs, respectivement. On note encore **N** et $\mathbf{N}^\times$ les entiers naturels avec et sans 0, respectivement. Pour $N, D \in \mathbf{N}$ on pose $\mathbf{N}_D^{N+1} := \{a \in \mathbf{N}^{N+1} \ : \ a_0 + \cdots + a_N = D\}$.

Dans ce paragraphe on donne un bref aperçu des propriétés géométriques des variétés toriques. Pour plus de détails, on renvoie le lecteur a [GKZ94], [Ful93], [Ewa96], [Cox01].

Soit $\mathcal{A} = (a_0, \ldots, a_N) \in (\mathbf{Z}^n)^{N+1}$ une suite de $N+1$ vecteurs de $\mathbf{Z}^n$, on considère l'action diagonale de $\mathbf{G}_m^n$ sur $\mathbf{P}^N$

$$*_\mathcal{A} : \mathbf{G}_m^n \times \mathbf{P}^N \to \mathbf{P}^N, \qquad (s, x) \mapsto (s^{a_0} x_0 : \cdots : s^{a_N} x_N).$$

On s'intéressera à l'adhérence de Zariski des orbites de cette action; pour un point $\alpha = (\alpha_0 : \cdots : \alpha_N) \in \mathbf{P}^N$ on pose

$$X_{\mathcal{A},\alpha} := \overline{\mathbf{G}_m^n *_\mathcal{A} \alpha} \ \subset \mathbf{P}^N$$

la *variété torique projective* associée au couple $(\mathcal{A}, \alpha)$. Autrement-dit, $X_{\mathcal{A},\alpha}$ est l'adhérence de Zariski de l'image de l'application monomiale

$$\varphi_{\mathcal{A},\alpha} := *_\mathcal{A}|_\alpha : \mathbf{G}_m^n \to \mathbf{P}^N, \qquad s \mapsto (\alpha_0 s^{a_0} : \cdots : \alpha_N s^{a_N}).$$

C'est une variété torique projective au sens de [GKZ94], c'est-à-dire une sous-variété de $\mathbf{P}^N$ stable sous l'action d'un tore $\mathbf{G}_m^n$, avec une orbite dense $X_{\mathcal{A},\alpha}^\circ := \mathbf{G}_m^n *_\mathcal{A} \alpha$.

Lorsque le point $\alpha$ est contenu dans un des sous-espaces standard de $\mathbf{P}^N$, la sous-variété $X_{\mathcal{A},\alpha}$ toute entière reste dans cet espace, puisque l'action $*_\mathcal{A}$ est diagonale. Quitte à se restreindre au sous-espace standard minimal contenant $\alpha$, on peut supposer sans perte de généralité $\alpha \in (\mathbf{P}^N)^\circ := \mathbf{P}^N \setminus \{x_0 \cdots x_N = 0\}$ et on fixe désormais pour $\alpha$ des coordonnées $(\alpha_0, \ldots, \alpha_N) \in (\mathbf{K}^\times)^{N+1}$.

On notera $X_\mathcal{A}$ la variété torique associée à $\mathcal{A} \in (\mathbf{Z}^n)^{N+1}$ et $(1, \ldots, 1) \in (\mathbf{K}^\times)^{N+1}$. Dans ce cas, l'orbite principale $X_\mathcal{A}^\circ$ est un sous-tore du tore $\mathbf{G}_m^N \cong (\mathbf{P}^N)^\circ$ et en fait tous les sous-groupes connexes de $(\mathbf{P}^N)^\circ$ sont de cette forme. Dans le cas général $X_{\mathcal{A},\alpha}^\circ = \alpha \cdot X_\mathcal{A}^\circ$ où $\cdot$ désigne la multiplication dans $(\mathbf{P}^N)^\circ$, c'est-à-dire que l'orbite principale de $X_{\mathcal{A},\alpha}$ est le translaté d'un sous-tore.

Le couple $(\mathcal{A}, \alpha)$ peut s'interpréter comme une suite finie de monômes $\alpha_0 s^{a_0}, \ldots, \alpha_N s^{a_N}$ de l'anneau des polynômes de Laurent $\mathbf{K}[s_1^{\pm 1}, \ldots, s_n^{\pm 1}]$, on dit que les $a_i$ sont les *exposants* et les $\alpha_i$ les *coefficients*. Le *support*

$$\mathrm{Supp}(\mathcal{A}) := \{a_{i_0}, \ldots, a_{i_M}\} \subset \mathbf{Z}^n$$

est l'ensemble des exposants distincts dans $\mathcal{A}$. Les variétés $X_{\mathcal{A},\alpha}$ et $X_{\mathrm{Supp}(\mathcal{A})}$ sont linéairement isomorphes par l'application

$$\mathbf{P}^N \to \mathbf{P}^M, \qquad (x_0 : \cdots : x_N) \mapsto (\alpha_{i_0}^{-1} x_{i_0} : \cdots : \alpha_{i_M}^{-1} x_{i_M}),$$

et donc leurs propriétés *géométriques* sont les mêmes. Pour ces propriétés, on peut donc se ramener au cas habituel où les $a_i$ sont tous distincts et $\alpha_i = 1$ pour tout $i$. Soit $L_{\mathcal{A}} \subset \mathbf{Z}^n$ le sous-module engendré par les différences des vecteurs $a_0, \ldots, a_N$, on a par exemple

$$\dim(X_{\mathcal{A},\alpha}) = \mathrm{rang}_{\mathbf{Z}}(L_{\mathrm{Supp}(\mathcal{A})}) = \mathrm{rang}_{\mathbf{Z}}(L_{\mathcal{A}}).$$

On introduit encore le polytope $Q_{\mathcal{A}} \subset \mathbf{R}^n$ enveloppe convexe des vecteurs $a_0, \ldots, a_N$.

**Lemme 2.1.** *Soient $\mathcal{A} = \{a_0, \ldots, a_N\}$ et $\mathcal{B} = \{b_0, \ldots, b_M\}$ des sous ensembles de $\mathbf{Z}^n$ de cardinal $N+1$ et $M+1$ respectivement, avec $N \leq M$. Posons $\pi : \mathbf{P}^M \to \mathbf{P}^N$ la projection linéaire standard $(y_0 : \cdots : y_M) \mapsto (y_0 : \cdots : y_N)$, on a*

(a) *si $\mathcal{A} \subset \mathcal{B}$ alors $X_{\mathcal{A}} = \overline{\pi(X_{\mathcal{B}})}$;*
(b) *si $Q_{\mathcal{A}} = Q_{\mathcal{B}}$ alors $\pi$ est un morphisme régulier et fini (au sens des fibres), de degré $\deg(\pi) = [L_{\mathcal{B}} : L_{\mathcal{A}}]$. Si de plus $L_{\mathcal{A}} = L_{\mathcal{B}}$ alors $\pi$ est un isomorphisme entre $X_{\mathcal{B}}^\circ$ et $X_{\mathcal{A}}^\circ$;*
(c) *si $Q_{\mathcal{A}} = Q_{\mathcal{B}} = Q$ et si pour toute face $F$ de $Q$ on a $L_{\mathcal{A}\cap F} = L_{\mathcal{B}\cap F}$, alors $\pi : X_{\mathcal{B}} \to X_{\mathcal{A}}$ est un isomorphisme.*

*Démonstration.* La partie (a) est conséquence directe des définitions. Pour les parties (b) et (c), l'hypothèse $Q_{\mathcal{A}} = Q_{\mathcal{B}} = Q$ entraîne que la projection $\pi$ est compatible avec la décomposition en orbites (2.3) ci-dessous. Pour chaque face $F$ de $Q$, $\pi : X_{\mathcal{B},F}^\circ \to X_{\mathcal{A},F}^\circ$ est une application monomiale de degré $[L_{\mathcal{B}\cap F} : \pi^*(L_{\mathcal{A}\cap F})] = [L_{\mathcal{B}\cap F} : L_{\mathcal{A}\cap F}]$, en particulier on voit que $\pi$ est régulière à fibres finies (et de degré 1 entre $X_{\mathcal{A}}^\circ$ et $X_{\mathcal{B}}^\circ$ si $L_{\mathcal{A}} = L_{\mathcal{B}}$).

Dans (c), l'hypothèse $L_{\mathcal{B}\cap F} = L_{\mathcal{A}\cap F}$ entraîne déjà que $\pi$ est une bijection. Pour chaque exposant $b_i \in \mathcal{B}$ considérons la face $F$ de $Q$ de dimension minimale contenant $b_i$. On écrit alors $b_i = \sum_{j : a_j \in \mathcal{A}\cap F} \lambda_{i,j}\, a_j$ avec $\lambda_{i,j} \in \mathbf{Z}$, $\sum_j \lambda_{i,j} = 0$. L'inverse est alors $X_{\mathcal{A}} \to X_{\mathcal{B}}$, $x \mapsto (x^{\lambda_0} : \cdots : x^{\lambda_M})$. □

Suivant la philosophie générale, les propriétés géométriques des variétés toriques se traduisent en des énoncés combinatoires sur les vecteurs $a_0, \ldots, a_N \in \mathbf{Z}^n$ définissant l'action. En ce qui concerne la théorie de l'intersection géométrique de ces variétés, le résultat le plus fondamental est que le degré s'identifie au volume du polytope $Q_{\mathcal{A}}$.

Le sous-module $L_{\mathcal{A}}$ est un réseau de l'espace linéaire $L_{\mathcal{A}} \otimes_{\mathbf{Z}} \mathbf{R} \subset \mathbf{R}^n$; on considère la forme volume $\mu_{\mathcal{A}}$ sur cet espace linéaire, invariante par translations et normalisée de sorte que

$$\mu_{\mathcal{A}}(L_{\mathcal{A}} \otimes_{\mathbf{Z}} \mathbf{R}/L_{\mathcal{A}}) = 1;$$

autrement-dit, de sorte que le volume d'un domaine fondamental soit 1. Soit $r :=$ $\mathrm{rang}_{\mathbf{Z}}(L_{\mathcal{A}})$, le degré de $X_{\mathcal{A},\alpha}$ s'explicite comme $r!$ fois le volume normalisé du polytope associé [GKZ94, § 2.6, Thm. 2.3], [Ful93, p. 111]

$$\deg(X_{\mathcal{A},\alpha}) = r!\mu_{\mathcal{A}}(Q_{\mathcal{A}}).$$

Soit maintenant $\eta : \mathbf{Z}^r \hookrightarrow \mathbf{Z}^n$ une application linéaire injective telle que $\eta(\mathbf{Z}^r) = L_{\mathcal{A}}$ et posons $b_i := \eta^{-1}(a_i) \in \mathbf{Z}^r$ puis $\mathcal{B} := (b_0, \dots, b_N) \in (\mathbf{Z}^r)^{N+1}$, alors $X_{\mathcal{B},\alpha} =$ $X_{\mathcal{A},\alpha}$ [GKZ94, Ch. 5, Prop. 1.2], et ainsi on peut toujours supposer sans perte de généralité $L_{\mathcal{A}} = \mathbf{Z}^n$. Dans cette situation, la forme volume $\mu_{\mathcal{A}}$ coïncide avec la forme volume euclidienne $\mathrm{Vol}_n$ sur $\mathbf{R}^n$ et donc

$$\dim(X_{\mathcal{A},\alpha}) = n, \qquad \deg(X_{\mathcal{A},\alpha}) = n!\,\mathrm{Vol}_n(Q_{\mathcal{A}}). \tag{2.1}$$

Plus généralement, les multidegrés du tore $\mathbf{G}_m^n$ plongé dans un produit d'espaces projectifs *via* plusieurs applications monomiales s'expriment comme volumes mixtes des polytopes associés à ces applications. Nous explicitons cela maintenant.

Soit $Z \subset \mathbf{P}^{N_0} \times \cdots \times \mathbf{P}^{N_m}$ une sous-variété de dimension $n$ et $c = (c_0, \dots, c_m) \in$ $\mathbf{N}_n^{m+1}$ avec $0 \le c_i \le N_i$, le *multidegré de* $Z$ *d'indice* $c$ est défini par

$$\deg_c(Z) := \mathrm{Card}\left(X \cap \pi_0^{-1}(E_0) \cap \cdots \cap \pi_m^{-1}(E_m)\right)$$

où $\pi_i$ désigne la projection $\mathbf{P}^{N_0} \times \cdots \times \mathbf{P}^{N_m} \to \mathbf{P}^{N_i}$ et $E_i$ est un sous-espace linéaire générique de $\mathbf{P}^{N_i}$ de codimension $c_i$.

Pour $i = 0, \dots, m$ on fixe un groupe $x_i = \{x_{i,0}, \dots, x_{i,N_i}\}$ de $N_i + 1$ variables chacun et on considère l'anneau $\mathbf{K}[x_0, \dots, x_m]$, multigradué par $\deg(x_{i,j}) := e_i$ où $e_0, \dots, e_m \in \mathbf{Z}^{m+1}$ désignent les vecteurs de la base standard. Ainsi $I(Z) \subset$ $\mathbf{K}[x_0, \dots, x_m]$ est un idéal multi-homogène de rang $N_0 + \cdots + N_m - n$. Soient $d_0, \dots, d_n \in \mathbf{N}^{m+1}$; pour chaque $d_i \in \mathbf{N}^{m+1}$ on introduit un groupe de variables $U_i = \{U_{i,0}, \dots, U_{i,M_i}\}$, $M_i + 1 = \prod_{j=0}^m \binom{d_j+N_j}{N_j}$, et on considère la *forme résultante d'indice* $d_0, \dots, d_n$ de $I(Z)$:

$$\mathrm{rés}_{d_0,\dots,d_n}(I(Z)) \in \mathbf{K}[U_0, \dots, U_n].$$

On renvoie le lecteur à [Rem01a, § 3] ou encore a [PS04, § 2.2] pour la définition et propriétés de base des formes résultantes. Maintenant, à un vecteur $c = (c_0, \dots, c_m) \in$ $\mathbf{N}_n^{m+1}$ comme ci-dessus on associe l'indice partiel

$$d(c) := (\underbrace{e_0, \dots, e_0}_{c_0 \text{ fois}}, \dots, \underbrace{e_m, \dots, e_m}_{c_m \text{ fois}}) \in (\mathbf{N}^{m+1})^n,$$

où $e_0, \dots, e_m$ désignent les vecteurs de la base standard de $\mathbf{R}^{m+1}$. Pour tout choix de $d_0 \in \mathbf{N}^{m+1} \setminus \{\mathbf{0}\}$ on a [Rem01a, Prop. 3.4 et 2.11]

$$\deg_c(Z) = \deg_{U_0}(\mathrm{rés}_{d_0,d(c)}(I(Z))).$$

Soit encore $D = (D_0, \dots, D_m) \in (\mathbf{N}^\times)^{m+1}$ et considérons

$$\Psi_D : \mathbf{P}^{N_0} \times \cdots \times \mathbf{P}^{N_m} \longrightarrow \mathbf{P}^{\binom{D_0+N_0}{N_0}\cdots\binom{D_m+N_m}{N_m}-1}$$

$$(x_0, \dots, x_m) \longmapsto \left(x_0^{b_0} \cdots x_m^{b_m} \, : \, b_0 \in \mathbf{N}_{D_0}^{N_0+1}, \dots, \, b_m \in \mathbf{N}_{D_m}^{N_m+1}\right)$$

le *plongement mixte* associé (composition des plongements de Veronese et Segre), on sait que [Rem01b, § 2.3]

$$\deg(\Psi_D(Z)) = \sum_{c\in \mathbf{N}_n^{m+1}} \binom{n}{c} \deg_c(Z) D^c. \tag{2.2}$$

Considérons maintenant des ensembles convexes $Q_1, \ldots, Q_n \subset \mathbf{R}^n$, leur *volume mixte* (ou *multi-volume*) est défini par la formule de type inclusion-exclusion

$$\mathrm{MV}(Q_1, \ldots, Q_n) := \sum_{j=1}^{n} (-1)^{n-j} \sum_{1\le i_1<\cdots<i_j\le n} \mathrm{Vol}_n(Q_{i_1} + \cdots + Q_{i_j}).$$

Cette notion généralise le volume d'un ensemble convexe car $\mathrm{MV}(Q, \ldots, Q) = n!\,\mathrm{Vol}_n(Q)$. Le volume mixte est positif ou nul, symétrique et linéaire en chaque variable $Q_i$ par rapport à la somme de Minkowski. On renvoie à [CLO98, § 7.4] ou [Ewa96] pour ses propriétés de base.

Interprétons ces notions dans le cas torique: soient $\mathcal{A}_0 \in (\mathbf{Z}^n)^{N_0+1}, \ldots, \mathcal{A}_m \in (\mathbf{Z}^n)^{N_m+1}$ tels que $L_{\mathcal{A}_0} + \cdots + L_{\mathcal{A}_m} = \mathbf{Z}^n$. Posons $\underline{\mathcal{A}} := (\mathcal{A}_0, \ldots, \mathcal{A}_m)$ et considérons l'action associée $*_{\underline{\mathcal{A}}}$ de $\mathbf{G}_m^n$ sur le produit d'espaces projectifs $\mathbf{P}^{N_0} \times \cdots \times \mathbf{P}^{N_m}$:

$$\begin{aligned} *_{\underline{\mathcal{A}}} : \mathbf{G}_m^n \times \mathbf{P}^{N_0} \times \cdots \times \mathbf{P}^{N_m} &\longrightarrow \mathbf{P}^{N_0} \times \cdots \times \mathbf{P}^{N_m} \\ (s, x_0, \ldots, x_m) &\longmapsto (s *_{\mathcal{A}_0} x_0, \ldots, s *_{\mathcal{A}_m} x_m) \end{aligned}$$

On considère la variété torique multi-projective $X_{\underline{\mathcal{A}}} \subset \mathbf{P}^{N_0} \times \cdots \times \mathbf{P}^{N_m}$, adhérence de Zariski de l'orbite du point $((1 : \cdots : 1), \ldots, (1 : \cdots : 1))$, et plus généralement on pose $X_{\underline{\mathcal{A}},\underline{\alpha}}$ pour l'adhérence de l'orbite de $\underline{\alpha} = (\alpha_0, \ldots, \alpha_m) \in \mathbf{P}^{N_0} \times \cdots \times \mathbf{P}^{N_m}$.

**Proposition 2.2.** *Soit $c \in \mathbf{N}_n^{m+1}$, dans la situation ci-dessus on a*

$$\deg_c(X_{\underline{\mathcal{A}}}) = \mathrm{MV}_n(\underbrace{Q_{\mathcal{A}_0}, \ldots, Q_{\mathcal{A}_0}}_{c_0 \text{ fois}}, \ldots, \underbrace{Q_{\mathcal{A}_m}, \ldots, Q_{\mathcal{A}_m}}_{c_m \text{ fois}}).$$

*Démonstration.* Pour $D = (D_0, \ldots, D_m) \in (\mathbf{N}^\times)^{m+1}$ on pose $D \cdot \underline{\mathcal{A}} := D_0\mathcal{A}_0 + \cdots + D_m\mathcal{A}_m$ l'ensemble des sommes de $D_i$ éléments pris dans $\mathcal{A}_i$ pour $i = 0, \ldots, m$ et $Q_{D\cdot\underline{\mathcal{A}}} = D_0 Q_{\mathcal{A}_0} + \cdots + D_m Q_{\mathcal{A}_m}$ l'enveloppe convexe de $D \cdot \underline{\mathcal{A}}$. On vérifie $\Psi_D(X_{\underline{\mathcal{A}}}) = X_{D\cdot\underline{\mathcal{A}}}$, et par la multilinéarité du volume mixte

$$n!\mathrm{Vol}_n(Q_{D\cdot\underline{\mathcal{A}}}) = \sum_{c\in\mathbf{N}_n^{m+1}} \binom{n}{c} \mathrm{MV}_n(\underbrace{Q_{\mathcal{A}_0}, \ldots, Q_{\mathcal{A}_0}}_{c_0 \text{ fois}}, \ldots, \underbrace{Q_{\mathcal{A}_m}, \ldots, Q_{\mathcal{A}_m}}_{c_m \text{ fois}}) D^c.$$

On a $\deg(\Psi_D(X_{\underline{\mathcal{A}}})) = n!\mathrm{Vol}_n(Q_{D\cdot\underline{\mathcal{A}}})$ pour tout $D$ et la comparaison de l'identité ci-dessus avec (2.2) permet de conclure. □

Les orbites de l'action $*_{\mathcal{A}}$ sur $X_{\mathcal{A},\alpha}$ sont en correspondance avec l'ensemble $\mathrm{F}(Q_{\mathcal{A}})$ des faces du polytope $Q_{\mathcal{A}}$. Pour chaque face $P$ on associe un point $\alpha_P := (\alpha_{P,0} : \cdots : \alpha_{P,N}) \in \mathbf{P}^N$ défini par $\alpha_{P,j} := \alpha_j$ si $a_j \in P$ et $\alpha_{P,j} := 0$ sinon; la bijection est donnée par [GKZ94, Ch. 5, Prop. 1.9], [Ful93, § 3.1]

$$P \mapsto X^\circ_{\mathcal{A},\alpha,P} := \mathbf{G}_m^n *_{\mathcal{A}} \alpha_P \subset \mathbf{P}^N.$$

On a la décomposition

$$X_{\mathcal{A},\alpha} = \bigsqcup_{P \in \mathrm{F}(Q_{\mathcal{A}})} X^{\circ}_{\mathcal{A},\alpha,P}. \tag{2.3}$$

Posons $N(P) := \mathrm{Card}\{i \,:\, a_i \in P\} - 1$ et

$$\mathcal{A}(P) = (a_i \,:\, a_i \in P) \in (\mathbf{Z}^n)^{N(P)+1}, \qquad \alpha(P) := (\alpha_i \,:\, a_i \in P) \in (\mathbf{K}^{\times})^{N(P)+1},$$

on vérifie que $X^{\circ}_{\mathcal{A},\alpha,P} \subset \mathbf{P}^N$ est l'orbite principale d'une variété torique contenue dans un sous-espace standard de dimension $N(P)$. Restreinte à ce sous-espace, elle s'identifie à la sous-variété $X^{\circ}_{\mathcal{A}(P),\alpha(P)} \subset \mathbf{P}^{N(P)}$, de dimension égale à la dimension (réelle) de la face $P$.

Plus généralement, soit $X_{\underline{\mathcal{A}}} \subset \mathbf{P}^{N_0} \times \cdots \times \mathbf{P}^{N_m}$ une variété torique multi-projective et posons $Q_{\underline{\mathcal{A}}} := (Q_{\mathcal{A}_0}, \ldots, Q_{\mathcal{A}_m})$ la famille de $m+1$ polytopes associée.

Pour un polytope $Q \subset \mathbf{R}^n$ et un vecteur $b \in \mathbf{R}^n$ on note $Q(b)$ la face de $Q$ définie comme l'ensemble des points $u \in Q$ maximisant la fonctionnelle linéaire $u \mapsto \langle b, u\rangle$. Considérons les combinaisons des faces des $Q_{\mathcal{A}_i}$ obtenues par cette méthode:

$$\mathrm{F}(Q_{\underline{\mathcal{A}}}) := \{(Q_{\mathcal{A}_0}(b), \ldots, Q_{\mathcal{A}_m}(b)) \,:\, b \in \mathbf{R}^n\} \subset \mathrm{F}(Q_{\mathcal{A}_0}) \times \cdots \times \mathrm{F}(Q_{\mathcal{A}_m}).$$

On peut vérifier que l'ensemble des sommes de Minkowski $Q_{\mathcal{A}_0}(b) + \cdots + Q_{\mathcal{A}_m}(b)$ pour $b \in \mathbf{R}^n$, coïncide avec l'ensemble des faces de la somme de Minkowski $Q_{\mathcal{A}_0} + \cdots + Q_{\mathcal{A}_m}$.

Soit $\underline{\alpha} = (\alpha_0, \ldots, \alpha_m) \in \mathbf{P}^{N_0} \times \cdots \times \mathbf{P}^{N_m}$, les orbites de l'action $*_{\underline{\mathcal{A}}}$ sur $X_{\underline{\mathcal{A}},\underline{\alpha}}$ sont en correspondance avec l'ensemble $\mathrm{F}(Q_{\underline{\mathcal{A}}})$. Pour chaque $\underline{P} = (P_0, \ldots, P_m) \in \mathrm{F}(Q_{\underline{\mathcal{A}}})$ on considère le point $\alpha_{\underline{P}} := (\alpha_{P_0}, \ldots, \alpha_{P_m}) \in \mathbf{P}^{N_0} \times \cdots \times \mathbf{P}^{N_m}$; la bijection est donnée par

$$\underline{P} \mapsto X^{\circ}_{\underline{\mathcal{A}},\underline{\alpha},\underline{P}} := \mathbf{G}_m^n *_{\underline{\mathcal{A}}} \alpha_{\underline{P}} \subset \mathbf{P}^{N_0} \times \cdots \times \mathbf{P}^{N_m},$$

et cette correspondance préserve la dimension, car $\dim(X_{\underline{\mathcal{A}},\underline{\alpha},\underline{P}}) = \dim(P_0 + \cdots + P_m)$. La démonstration est une légère variante de [GKZ94, Ch. 5, Prop. 1.9].

Classiquement les variétés toriques sont construites à partir d'*éventails*, *voir* par exemple [Ful93]. Rappelons qu'un éventail est un ensemble de cônes simpliciaux de sommet l'origine engendrés par un nombre fini de points de $\mathbf{Z}^n$, tels que toute face et toute intersection de cônes de l'éventail appartienne encore à l'éventail. Le lien avec notre présentation des variétés toriques *via* des polytopes résulte de ce qu'un polytope $Q$ à sommets dans $\mathbf{Z}^n$ détermine un éventail et donc aussi une variété torique au *sens classique*, qui de plus est munie d'un fibré en droites ample. Dans les alinéa suivants (tirés essentiellement de [Cox01] et [Ful93]) on explicite cette construction.

Supposons sans perte de généralité $\dim(Q) = n$ et posons $\mathrm{F}_i(Q)$ l'ensemble des faces de $Q$ de dimension $i$, de sorte que $\mathrm{F}(Q) := \bigsqcup_{i=0}^n \mathrm{F}_i(Q)$. Pour chaque *hyperface* (c'est-à-dire face de codimension 1) $F$ de $Q$ on prends le vecteur normal intérieur primitif $v_F \in \mathbf{Z}^n$ et on note

$$m_F := -\min\{\langle x, v_F\rangle : x \in Q\} \in \mathbf{Z},$$

où $\langle \cdot, \cdot \rangle$ désigne le produit scalaire ordinaire sur $\mathbf{R}^n$. On peut donc décrire $Q$ comme intersection de demi-espaces

$$Q = \bigcap_{F \in \mathrm{F}_{n-1}(Q)} \{x \in \mathbf{R}^n \,:\, \langle x, v_F\rangle \geq -m_F\}. \tag{2.4}$$

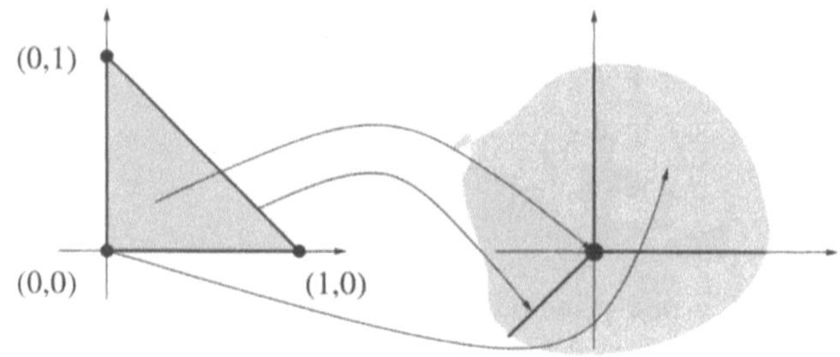


Fig. 1

À chaque face $P$ de $Q$ on associe le cône $\sigma_P$ engendré par les vecteurs $v_F$ correspondant aux hyperfaces $F \supset P$; on a $\dim(P) + \dim(\sigma_P) = n$. L'ensemble de ces cônes forme un éventail *complet* $\Sigma$ de $\mathbf{R}^n$, c'est-à-dire un éventail recouvrant $\mathbf{R}^n$.

La figure 1 montre la correspondance $P \mapsto \sigma_P$ entre faces de $Q$ et cônes de $\Sigma$, pour $Q$ le simplexe standard de $\mathbf{R}^2$.

Notons $\mathcal{X}_\Sigma$ la variété torique (non plongée dans un espace projectif) définie par cet éventail. Pour chaque hyperface $F$ de $Q$ on note $D_F$ le diviseur correspondant à la sous-variété torique de $\mathcal{X}_\Sigma$ associée à l'arête $\sigma_F$ de $\Sigma$ et on introduit un diviseur de Weil sur $\mathcal{X}_\Sigma$ invariant sous l'action de $\mathbf{G}_m^n$

$$D_Q := \sum_{F \in \mathrm{F}_{n-1}(Q)} m_F \, D_F,$$

qui est en fait un diviseur de Cartier ample [Ful93, § 3.4]. Pour $a \in \mathbf{Z}^n$, l'application monomiale $\chi^a : \mathbf{G}_m^n \to \mathbf{G}_m, s \mapsto s^a$ peut se voir comme une fonction rationnelle sur $\mathcal{X}_\Sigma$. Cette fonction rationnelle s'étend en une section globale de $\mathcal{O}(D_Q)$ si et seulement si $a \in Q$, et en fait

$$\Gamma(\mathcal{X}_\Sigma, \mathcal{O}(D_Q)) = \bigoplus_{a \in Q \cap \mathbf{Z}^n} \mathbf{K} \cdot \chi^a.$$

En posant finalement $\mathcal{A}_Q := Q \cap \mathbf{Z}^n = \{a_0, \ldots, a_N\}$, on vérifie que l'application $\mathbf{G}_m^n \to \mathbf{P}^N, s \mapsto (s^{a_0}, \ldots, s^{a_N})$ s'étend en un morphisme régulier $\mathcal{X}_\Sigma \to \mathbf{P}^N$, dont l'image est la variété torique projective $X_{\mathcal{A}_Q}$. De fait, ceci est le morphisme de normalisation de $X_{\mathcal{A}_Q}$ [Stu96, Cor. 13.6].

Cependant, notons que la notion d'éventail permet de définir des variétés toriques qui, même complètes, n'admettent pas nécessairement de diviseur $\mathbf{G}_m^n$-invariant ample. Toutefois, lorsqu'une variété torique abstraite possède un tel diviseur, elle admet une application monomiale régulière vers un espace projectif et son image est une variété torique projective du type que nous étudions ici. On notera encore que les variétés toriques abstraites sont des variétés normales, *voir* [Ful93, § 2.1, p. 29], ce qui n'est pas toujours le cas pour les variétés toriques projectives.

## 3 Équations et indices d'obstruction successifs

Soit $X \subset \mathbf{P}^N$ une variété projective de dimension $n \geq 0$ et $I(X) \subset \mathbf{K}[x_0, \ldots, x_N]$ son idéal de définition, l'*indice d'obstruction* $\omega(X)$ est défini comme le plus petit degré d'une équation homogène $f \in I(X) \setminus \{0\}$. Plus généralement, pour $i = 1, \dots, N-n$ on définit le *i-ème indice d'obstruction de X* par

$$\omega_i(X) := \min\{D \in \mathbf{N} : \dim(Z(f : f \in I(X)_D)) \le N - i\},$$

où $I(X)_D$ désigne la partie de degré $D$ de l'idéal homogène $I(X)$. Alternativement, $\omega_i(X)$ est le plus petit entier $D \geq 0$ tel qu'il existe des polynômes homogènes $f_1, \dots, f_i \in I(X)$ de degré borné par $D$ formant une intersection complète. Évidemment $\omega(X) = \omega_1(X)$ et

$$1 \le \omega_1(X) \le \cdots \le \omega_{N-n}(X).$$

L'invariant $\omega(X)$ joue un rôle important dans les problèmes de Lehmer généralisé et de Bogomolov sur les tores [AD04], [Rat04]. D'un autre côté, les lemmes de zéros consistent à minorer le premier indice $\omega(X)$ pour une variété de dimension 0, éventuellement munie de multiplicités; ces résultats sont des outils fondamentaux en théorie des nombres transcendants, *voir* [Ber87]. Signalons que pour ces applications, il est souvent important de considérer des indices d'obstruction *sur des sous-corps de* $\overline{\mathbf{Q}}$ (*i.e.* dont les équations sont à coefficients dans un sous-corps fixé) et pour des *schémas projectifs* quelconques.

Plus généralement encore, on peut considérer des *indices d'obstruction successifs relatifs à un ouvert* donné $U \subset \mathbf{P}^N$:

$$\omega_i(X; U) := \min\{D \in \mathbf{N} : \dim(Z(f : f \in I(X)_D) \cap U) \le N - i\}.$$

Les idéaux des variétés toriques sont binomiaux, c'est-à-dire engendrés par des familles de polynômes de la forme $\alpha\, x^a - \beta\, x^b$ avec $a, b \in \mathbf{N}^n$ et $\alpha, \beta \in \mathbf{K}^\times$. Dans la suite on explicite la relation entre variétés toriques et idéaux binomiaux, qui est la clé de notre étude des indices d'obstruction successifs de ces variétés.

Soit $\mathcal{A} = (a_0, \dots, a_N) \in (\mathbf{Z}^n)^{N+1}$ tel que $L_\mathcal{A} = \mathbf{Z}^n$ et $\alpha = (\alpha_0, \dots, \alpha_N) \in (\mathbf{K}^\times)^{N+1}$. Considérons l'application linéaire

$$\eta_\mathcal{A} : \mathbf{Z}^{N+1} \to \mathbf{Z} \times \mathbf{Z}^n, \qquad \lambda \mapsto (\lambda_0 + \cdots + \lambda_N, \lambda_0\, a_0 + \cdots + \lambda_N\, a_N),$$

et posons $\Gamma_\mathcal{A} := \ker(\eta_\mathcal{A})$ son noyau. C'est un sous-module *saturé* de $\mathbf{Z}^{N+1}$ (c'est-à-dire que le quotient $\mathbf{Z}^{N+1}/\Gamma_\mathcal{A}$ est sans torsion) de rang $N - n$. Notons $\Delta \subset \mathbf{R}^{N+1}$ l'hyperplan d'équation $\lambda_0 + \cdots + \lambda_N = 0$ et posons $\Delta^\mathbf{Z} := \Delta \cap \mathbf{Z}^{N+1}$; évidemment on a $\Gamma_\mathcal{A} \subset \Delta^\mathbf{Z}$.

Pour $b \in \mathbf{R}^{N+1}$ on écrit de façon unique $b = b_+ - b_-$ avec $b_+, b_- \in \mathbf{R}_+^{N+1}$ à supports disjoints, c'est-à-dire $(b_+)_i = b_i$ si $b_i > 0$ et 0 sinon, et $(b_-)_i = -b_i$ si $b_i < 0$ et 0 sinon. Le résultat suivant est une reformulation de [Stu96, Cor. 4.3]:

**Proposition 3.1.** $I(X_{\mathcal{A},\alpha}) = \big(x^{b_+} - \alpha^b\, x^{b_-} : b \in \Gamma_\mathcal{A}\big)$.

*Démonstration.* Le cône de $X_{\mathcal{A},\alpha}$ coïncide avec l'adhérence de Zariski de l'image de

$$\mathbf{G}_m^n \times \mathbf{G}_m \to \mathbf{A}^{N+1}, \qquad (s, t) = (s_1, \dots, s_n, t) \mapsto (\alpha_0\, t\, s^{a_0}, \dots, \alpha_N\, t\, s^{a_N}),$$

donc $I(X_{\mathcal{A},\alpha})$ coïncide avec le noyau de l'homomorphisme

$$\mathbf{K}[x_0, \dots, x_N] \to \mathbf{K}[s_1^{\pm 1}, \dots, s_n^{\pm 1}, t^{\pm 1}], \qquad x_i \mapsto \alpha_i\, t\, s^{a_i};$$

le résultat devient une conséquence directe de [Stu96, Cor. 4.3]. □

Ainsi, l'idéal d'une variété torique projective est binomial, il en résulte automatiquement que c'est un idéal premier et homogène (puisque $X_{\mathcal{A},\alpha}$ est une variété projective) ne contenant aucune des variables $x_i$, à cause de l'hypothèse $\alpha \in (\mathbf{P}^N)^\circ$.

Soit maintenant $\Gamma \subset \mathbf{Z}^{N+1}$ un sous-module quelconque et $\rho$ un *caractère partiel*, c'est-à-dire un homomorphisme $\rho : \Gamma \to \mathbf{K}^\times$. Ces données définissent un idéal binomial

$$I(\Gamma, \rho) := \left(x^{b+} - \rho(b) x^{b-} \ : \ b \in \Gamma\right) \subset \mathbf{K}[x_0, \dots, x_N].$$

**Proposition 3.2** [ES96, Cor. 2.6]. *La correspondence* $(\Gamma, \rho) \mapsto I(\Gamma, \rho)$ *est une bijection entre les sous-modules saturés* $\Gamma \subset \mathbf{Z}^{N+1}$ *munis d'un caractère partiel* $\rho$, *et les idéaux de* $\mathbf{K}[x_0, \dots, x_N]$ *binomiaux, premiers et ne contenant aucune des variables* $x_i$. *De plus* $\mathrm{rang}_{\mathbf{Z}}(\Gamma) = \mathrm{rang}\,(I(\Gamma, \rho))$.

On vérifie sans peine que $I(\Gamma, \rho)$ est homogène si et seulement si $\Gamma \subset \Delta^{\mathbf{Z}}$. Ainsi, la donnée d'un couple $(\mathcal{A}, \alpha)$ définit un sous-module saturé $\Gamma_{\mathcal{A}} \subset \Delta^{\mathbf{Z}}$ et un caractère partiel $\rho_{\mathcal{A},\alpha} : b \mapsto \alpha^b$; la proposition 3.1 peut être reformulée sous la forme $I(X_{\mathcal{A},\alpha}) = I(\Gamma_{\mathcal{A}}, \rho_{\mathcal{A},\alpha})$.

Réciproquement, à partir d'un idéal binomial de $\mathbf{K}[x_0, \dots, x_N]$ premier, homogène et ne contenant aucune des variables $x_i$, on peut construire $(\mathcal{A}, \alpha)$ tel que $I = I(X_{\mathcal{A},\alpha})$: soient $\Gamma \subset \Delta^{\mathbf{Z}}$ le sous-module saturé et $\rho$ le caractère partiel associés à $I$, $n := N - \mathrm{rang}_{\mathbf{Z}}(\Gamma)$ et prenons $v_0, \dots, v_n \in \mathbf{Z}^{N+1}$ une base de l'orthogonal $\Gamma^\perp$ de $\Gamma$ dans $\mathbf{Z}^{N+1}$ relativement au produit scalaire usuel. Puisque $\Gamma \subset \Delta^{\mathbf{Z}}$ on peut supposer sans perte de généralité $v_0 = (1, \dots, 1)$; on pose alors

$$a_i := (v_{1,i}, \dots, v_{n,i}) \in \mathbf{Z}^n \qquad \text{pour } i = 0, \dots, N,$$

et $\mathcal{A} := (a_0, \dots, a_N) \in (\mathbf{Z}^n)^{N+1}$. En outre, $\rho$ peut s'étendre (de manière pas forcément unique) en un caractère total $\rho : \mathbf{Z}^{N+1} \to \mathbf{K}^\times$ et on prend $\alpha := (\rho(e_0), \dots, \rho(e_N)) \in (\mathbf{K}^\times)^{N+1}$ où les $e_i$ désignent les vecteurs de la base standard de $\mathbf{Z}^{N+1}$.

Par construction $\Gamma_{\mathcal{A}} = \Gamma$ et $L_{\mathcal{A}} = \mathbf{Z}^n$, parce que $v_0, \dots, v_n$ est une base de $\Gamma^\perp$ et $\Gamma$ est saturé. La proposition 3.1 entraîne $I(X_{\mathcal{A},\alpha}) = I$. En particulier, on en déduit que la correspondance $X \mapsto I(X)$ est une bijection entre l'ensemble des variétés toriques de $\mathbf{P}^N$ et l'ensemble des idéaux de $\mathbf{K}[x_0, \dots, x_N]$ binomiaux, premiers, homogènes, ne contenant aucune des variables de $x_i$ pour $i = 0, \dots, N$. Dans la suite, on notera $\Gamma_X$ et $\rho_X$ le $\mathbf{Z}$-module et le caractère partiel associés à une variété torique donnée $X$.

*Exemple 3.3.* Soit $\mathbf{G}_m^2 \to \mathbf{P}^3$, $(s,t) \mapsto (1 : s : 3s^2t : st^2)$ et posons $S \subset \mathbf{P}^3$ la surface torique, adhérence de Zariski de l'image de cette application. Soit

$$[\eta_{\mathcal{A}}] = \begin{bmatrix} 1 & 1 & 1 & 1 \\ 0 & 1 & 2 & 1 \\ 0 & 0 & 1 & 2 \end{bmatrix}$$

la matrice de l'application linéaire $\eta_{\mathcal{A}} : \mathbf{Z}^4 \to \mathbf{Z}^3$ dans les bases canoniques, alors $\Gamma_{\mathcal{A}} = \ker(\eta_{\mathcal{A}})$ est engendré par le seul vecteur

$$\gamma = (M_0, -M_1, M_2. -M_3) = (2, -3, 2, -1) \in \mathbf{Z}^4$$

où $M_i$ désigne le $i$-ème mineur de la matrice $[\eta_{\mathcal{A}}]$. Et donc une équation de $S$ est

$$x^{\gamma_+} - \alpha^\gamma x^{\gamma_-} = x_0^2 x_2^2 - 9 x_1^3 x_3 \in \mathbf{K}[x_0, x_1, x_2, x_3].$$

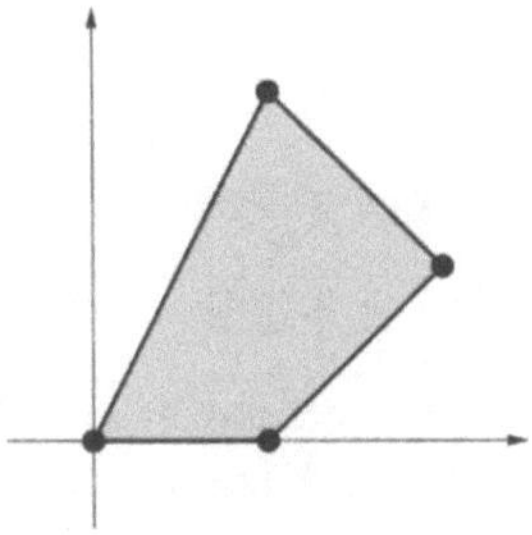

Fig. 2

Réciproquement, partons de l'équation binomiale $f := x_0^2 x_2^2 - 9 x_1^3 x_3 \in \mathbf{K}[x_0, x_1, x_2, x_3]$. Le $\mathbf{Z}$-module saturé $\Gamma \subset \Delta^{\mathbf{Z}}$ correspondant est engendré par $\gamma = (2, -3, 2, -1)$ et le caractère partiel est $\rho : \Gamma \to \mathbf{K}^\times, m \cdot \gamma \mapsto 9^m$.

On vérifie que $(1, 1, 1, 1), (0, 1, 2, 1), (0, 0, 1, 2) \in \mathbf{Z}^4$ forment bien une base de $\Gamma^\perp$ et que $\mathbf{Z}^4 \to \mathbf{K}^\times, b \mapsto (1, 1, 3, 1)^b = 3^{b_2}$, est une extension possible de $\rho$, donc $Z(f) = X_{\mathcal{A},\alpha}$ avec $\mathcal{A} = ((0, 0), (1, 0), (2, 1), (1, 2)) \in (\mathbf{Z}^2)^4$ et $\alpha = (1, 1, 3, 1) \in (\mathbf{K}^\times)^4$.

La figure 2 montre les exposants et le polytope associés à cette surface:

Considérons la norme $\ell^1$

$$||v||_1 = \sum_{i=0}^{N} |v_i| \qquad \text{pour } v \in \mathbf{R}^{N+1}$$

et pour un $\mathbf{Z}$-module quelconque $\Gamma \subset \mathbf{R}^{N+1}$ notons $\mu_i := \mu_i(\Gamma; ||\cdot||_1)$ le $i$-ème minimum successif de $\Gamma$ relativement à cette norme. Rappelons que $\mu_i$ est défini comme le plus petit $\nu \in \mathbf{R}_+$ tel qu'il existe $i$ vecteurs indépendants $v_1, \ldots, v_i \in \Gamma$ de norme bornée par $\nu$. Alternativement

$$\mu_i(\Gamma; ||\cdot||_1) := \min\{\nu \in \mathbf{R}_+ \ : \ \mathrm{rang}_{\mathbf{Z}}(\nu \cdot B_{||\cdot||_1} \cap \Gamma) \geq i\},$$

où $B_{||\cdot||_1}$ désigne la boule unité pour la norme $\ell^1$. Le résultat suivant montre que les indices d'obstruction d'une variété torique relatifs à l'ouvert $(\mathbf{P}^N)^\circ$ coïncident avec les minimums successifs du $\mathbf{Z}$-module associé:

**Proposition 3.4.** *Soit $X \subset \mathbf{P}^N$ une variété torique de dimension $n$, alors*

$$\omega_i(X; (\mathbf{P}^N)^\circ) = \tfrac{1}{2}\, \mu_i(\Gamma_X; ||\cdot||_1) \quad \textit{pour } i = 1, \ldots, N-n.$$

*Démonstration.* On peut se restreindre sans perte de généralité au cas $\alpha = (1, \ldots, 1)$. Soient $v_1, \ldots, v_{N-n}$ des vecteurs formant une base de $\Gamma_X$. D'après [ES96, Thm. 2.1(b)], $X^\circ = X \cap (\mathbf{P}^N)^\circ$ est une intersection complète decoupé par les binômes

$$x^{(v_1)_+} - x^{(v_1)_-}, \ldots, \ x^{(v_{N-n})_+} - x^{(v_{N-n})_-} \in \mathbf{K}[x_0, \ldots, x_N]$$

et on a $\deg(x^{(v_j)_+} - x^{(v_j)_-}) = \frac{1}{2}\, ||v_j||_1$ (car $\sum_{k=0}^{N} v_{j,k} = 0$). En prenant $v_1, \ldots, v_{N-n}$ réalisant les minimums successifs du module $\Gamma_X$ on obtient

$$\omega_i(X; (\mathbf{P}^N)^\circ) \leq \max\{\deg(x^{(v_j)_+} - x^{(v_j)_-}) \ : \ 1 \leq j \leq i\} = \tfrac{1}{2}\, \mu_i(\Gamma_X; ||\cdot||_1).$$

Dans l'autre direction, soit $f_1, \ldots, f_{N-n} \in I(X)$ une suite de polynômes homogènes réalisant les indices d'obstruction successifs de $X$ sur $(\mathbf{P}^N)^\circ$. Par la proposition 3.1

$$f_j(x) = \sum_{b \in \Gamma_X} g_{j,b}(x)\,(x^{b_+} - x^{b_-})$$

pour certains $g_{j,b} \in \mathbf{K}[x_0, \ldots, x_N]$ et, pour tout $b \in \Gamma_{\mathcal{A}}$ tel que $g_{j,b} \neq 0$, on a

$$\deg(f_j) = \deg(g_{j,b}) + \deg(x^{b_+} - x^{b_-}) = \deg(g_{j,b}) + \tfrac{1}{2}\,||b||_1 \geq \tfrac{1}{2}\,||b||_1. \quad (3.5)$$

Soit $1 \leq i \leq N-n$, posons $L_i$ le $\mathbf{Z}$-module engendré par les $b \in \Gamma_X$ tels que $g_{j,b} \neq 0$ pour un certain $1 \leq j \leq i$. Alors, $f_j \in J_i := (x^{b_+} - x^{b_-} : b \in L_i) \subset \mathbf{K}[x_0, \ldots, x_N]$ pour $j = 1, \ldots, i$ et donc

$$i = \mathrm{rang}(f_1, \ldots, f_i) \leq \mathrm{rang}(J_i) = \mathrm{rang}_{\mathbf{Z}}(L_i)$$

par la proposition 3.2 ci-dessus. Ainsi $L_i$ est un sous-module de $\Gamma_X$ de rang au moins $i$ et on peut trouver $i$ vecteurs linéairement indépendants parmi les $b \in \Gamma_{\mathcal{A}}$ tels que $g_{j,b} \neq 0$ pour un certain $1 \leq j \leq i$. L'un au moins de ces vecteurs est de norme $\|b\|_1 \geq \mu_i(\Gamma_X; \|\cdot\|_1)$ et on en déduit avec (3.5)

$$\omega_i(X; (\mathbf{P}^N)^\circ) = \max\{\deg(f_j) : j = 1, \ldots, i\} \geq \tfrac{1}{2}\,\mu_i(\Gamma_X, ||\cdot||_1).$$

□

Soit $X \subset \mathbf{P}^N$ une variété de dimension $n$, comme conséquence directe de sa majoration de la fonction de Hilbert géométrique, Chardin a montré qu'on a toujours

$$\omega_1(X) \leq N\ \deg(X)^{1/(N-n)},$$

et aussi $\omega_2(X)^{1/2} \leq N\ \deg(X)^{1/(N-n)}$ [Cha89]. Ultérieurement, Chardin et Philippon ont considéré le problème de l'*interpolation algébrique*, qui consiste à estimer le degré minimal de sous-variétés

$$X = Y_{N-n} \subset \cdots \subset Y_1 \subset \mathbf{P}^N$$

telles que $\mathrm{codim}(Y_j) = j$. Ils ont montré qu'on peut choisir $Y_1, \ldots, Y_{N-n}$ comme ci-dessus, satisfaisant [CP99]

$$\deg(Y_j)^{1/j} \leq N\,4^{N-1}\ \deg(X)^{1/(N-n)}. \quad (3.6)$$

L'interpolation algébrique est très proche des problèmes d'estimation des indices d'obstruction successifs. En fait, une reformulation de leur démonstration montre qu'il existe des équations $f_1, \ldots, f_{N-n} \in I(X)$ formant une intersection complète au voisinage du point générique de $X$ et telles que

$$\deg(X) \leq \deg(f_1) \cdots \deg(f_{N-n}) \leq c(N)\ \deg(X)$$

pour une constante $c(N) > 0$ explicite. Ceci équivaut aux inégalités

$$\deg(X) \leq \omega_1(X; U) \cdots \omega_{N-n}(X; U) \leq c(N)\ \deg(X) \quad (3.7)$$

pour un ouvert $U \subset \mathbf{P}^N$ tel que $X \cap U \neq \emptyset$. L'interprétation des indices d'obstruction des variétés toriques comme minimums successifs d'un module permet de traduire le deuxième théorème de Minkowski en des estimations pour le produit de ces indices relatifs à l'ouvert $(\mathbf{P}^N)^\circ$:

**Corollaire 3.5.** *Soit $X \subset \mathbf{P}^N$ une variété torique de dimension n, alors*

$$\deg(X) \leq \omega_1(X; (\mathbf{P}^N)^\circ) \cdots \omega_{N-n}(X; (\mathbf{P}^N)^\circ) \leq c(N,n)\deg(X)$$

*avec*

$$c(N,n) := \binom{N+1}{n+1}^{1/2} \left(\frac{(N+1)}{\pi}\right)^{N-n/2} \Gamma\left(1+\frac{N-n}{2}\right) \leq \left(\frac{N+1}{\sqrt{\pi}}\right)^{N-n}.$$

Ceci est la proposition 1.1 de l'introduction, avec une constante plus précise.

*Démonstration.* La première inégalité est conséquence directe du théorème de Bézout; passons à la deuxième. Posons $\Gamma_X^{\mathbf{R}} := \Gamma_X \otimes \mathbf{R} \cong \mathbf{R}^{N-n}$, de sorte que $\Gamma_X$ est un réseau de $\Gamma_X^{\mathbf{R}}$. Posons aussi $B := B_{\|\cdot\|_1} \cap \Gamma_X^{\mathbf{R}}$ la boule unité de $\Gamma_X^{\mathbf{R}}$ par rapport à la restriction de la norme $\ell^1$. Ceci étant un ensemble convexe et symétrique par rapport à l'origine, le deuxième théorème de Minkowski sur les minimums successifs d'un réseau entraîne

$$\frac{2^{N-n}}{(N-n)!} \leq \frac{\mathrm{Vol}_{N-n}(B)}{\mathrm{Vol}_{N-n}(\Gamma_X^{\mathbf{R}}/\Gamma_X)} \prod_{i=1}^{N-n} \mu_i(\Gamma_X; \|\cdot\|_1) \leq 2^{N-n}$$

et donc

$$\prod_{i=1}^{N-n} \omega_i(X; (\mathbf{P}^N)^\circ) \leq \frac{\mathrm{Vol}_{N-n}(\Gamma_X^{\mathbf{R}}/\Gamma_X)}{\mathrm{Vol}_{N-n}(B)} \tag{3.8}$$

par la proposition 3.4.

Pour conclure, il suffit de majorer le quotient $\mathrm{Vol}_{N-n}(\Gamma_X^{\mathbf{R}}/\Gamma_X)/\mathrm{Vol}_{N-n}(B)$. Soient $\mathcal{A} \in (\mathbf{Z}^n)^{N+1}$ et $\alpha \in (\mathbf{K}^\times)^{N+1}$ tels que $X = X_{\mathcal{A},\alpha}$; en particulier $\Gamma_{\mathcal{A}} = \Gamma_X$. Soient

$$v_i := (a_{0,i}, \ldots, a_{N,i}) \in \mathbf{Z}^{N+1} \quad \text{pour } i = 0, \ldots, n$$

les lignes de la matrice (dans les bases standard) de l'application $\eta_{\mathcal{A}} : \mathbf{Z}^{N+1} \to \mathbf{Z}^{n+1}$. Le module $\Gamma_X$ est l'orthogonal du sous-module $V \subset \mathbf{Z}^{N+1}$ engendré par $v_0, \ldots, v_n$. C'est aussi un sous-module saturé à cause de l'hypothèse $L_{\mathcal{A}} = \mathbf{Z}^n$; la formule de Brill–Gordan (*voir* par exemple [PS04, formule (II.2)]) entraîne alors

$$\mathrm{Vol}_{N-n}(\Gamma_X^{\mathbf{R}}/\Gamma_X) = \mathrm{Vol}_{n+1}(V^{\mathbf{R}}/V)$$

et par la formule de Cauchy–Binet

$$\mathrm{Vol}_{N-n}(\Gamma_X^{\mathbf{R}}/\Gamma_X) = \mathrm{Vol}_{n+1}(V^{\mathbf{R}}/V) = \|v_0 \wedge \cdots \wedge v_n\|_2$$
$$= \left(\sum_{J:\mathrm{Card}(J)=n+1} \det([\eta_{\mathcal{A}}]_J)^2\right)^{1/2},$$

où $[\eta_{\mathcal{A}}]_J$ désigne le mineur $(n+1) \times (n+1)$ de la matrice de $\eta_{\mathcal{A}}$ dont les colonnes sont indexées par $J$. Chaque terme $\det([\eta_{\mathcal{A}}]_J)$ est $\pm n!$ fois le volume d'un simplexe contenu dans le polytope $Q_{\mathcal{A}}$; ainsi on peut majorer cette quantité par

$$\mathrm{Vol}_{N-n}(\Gamma_X^{\mathbf{R}}/\Gamma_X) \leq \binom{N+1}{n+1}^{1/2} n!\,\mathrm{Vol}_n(Q_{\mathcal{A}}) = \binom{N+1}{n+1}^{1/2} \deg(X).$$

En outre, notons que $B$ contient la boule euclidienne de $\Gamma_X^{\mathbf{R}}$ de rayon $(N+1)^{-1/2}$ centrée en l'origine, ainsi

$$\mathrm{Vol}_{N-n}(B) \geq \mathrm{Vol}_{N-n}((N+1)^{-1/2} B_{\|\cdot\|_2} \cap \Gamma_X^{\mathbf{R}}) = \left(\frac{\pi}{N+1}\right)^{(N-n)/2} \cdot \frac{1}{\Gamma\left(1+\frac{N-n}{2}\right)}.$$

On en déduit

$$\begin{aligned}\prod_{i=1}^{N-n} \omega_i(X; (\mathbf{P}^N)^\circ) &\leq \frac{\mathrm{Vol}_{N-n}(\Gamma_X^{\mathbf{R}}/\Gamma_X)}{\mathrm{Vol}_{N-n}(B)} \\ &\leq \left(\frac{N+1}{n+1}\right)^{1/2} \cdot \left(\frac{N+1}{\pi}\right)^{(N-n)/2} \cdot \Gamma\left(1+\frac{N-n}{2}\right) \cdot \deg(X) \\ &\leq \frac{(N+1)^{(N-n)/2}}{(N-n)!^{1/2}} \cdot \left(\frac{N+1}{\pi}\right)^{(N-n)/2} \cdot \prod_{k=0}^{[(N-n)/2]-1} \\ &\quad \times \left(\frac{N-n}{2} - k\right) \cdot \deg(X) \\ &\leq \left(\frac{N+1}{\sqrt{\pi}}\right)^{N-n} \cdot \deg(X).\end{aligned}$$

□

Ce résultat améliore la constante $c(N)$ dans (3.7) pour le cas torique et précise l'ouvert $U$ par rapport auquel les indices peuvent être considérés.

En revenant au cas général d'une variété $X \subset \mathbf{P}^N$ quelconque de dimension $n$, il est naturel de se demander si l'estimation (3.7) reste valide pour $U = \mathbf{P}^N$. Autrement-dit, s'il existe toujours des polynômes homogènes $f_1, \ldots, f_{N-n} \in I(X)$ formant une intersection complète *globale* et tels que

$$\deg(X) \leq \deg(f_1) \cdots \deg(f_{N-n}) \leq c(N)\, \deg(X).$$

Il serait intéressant de décider cette question déjà sur la classe des variétés toriques. Pour ce faire, on voudrait expliciter les indices $\omega_i(X; \mathbf{P}^N)$ de manière analogue à la proposition 3.4 pour $\omega_i(X; (\mathbf{P}^N)^\circ)$.

## 4 Volumes, hauteurs d'espaces tangents et degrés

Dans la définition du § 2, les sous-groupes algébriques connexes de $\mathbf{G}_m^N$ correspondent aux variétés toriques $X_{\mathcal{A}}^\circ$ *via* l'identification $\iota : \mathbf{G}_m^N \hookrightarrow (\mathbf{P}^N)^\circ$. D'un autre côté, l'ensemble des points complexes d'un sous-groupe algébrique connexe $G$ de $\mathbf{G}_m^N$, de dimension $n$, est décrit *via* l'application exponentielle par son espace tangent $TG(\mathbf{C}) \subset \mathbf{C}^N$ à l'origine et son réseau de périodes $\Lambda := TG(\mathbf{C}) \cap (2i\pi\mathbf{Z})^N$, de rang $n$ sur $\mathbf{Z}$.

Munissons l'espace tangent à $\mathbf{G}_m^N$ en l'origine $T\mathbf{G}_m^N(\mathbf{C}) \simeq \mathbf{C}^N$ de la structure euclidienne pour laquelle les $2N$ éléments $e_i$ et $2i\pi e_i$ ($i = 1, \ldots, N$) forment une base orthonormée.

Soit $\mathcal{A} := (a_0, \ldots, a_N) \in (\mathbf{Z}^n)^{N+1}$ tel que $L_{\mathcal{A}} = \mathbf{Z}^n$ et $G := \iota^{-1}(X_{\mathcal{A}}^\circ)$. Des générateurs de $\Lambda$ sont donnés par les vecteurs $2i\pi\lambda_1, \ldots, 2i\pi\lambda_n \in (2i\pi\mathbf{Z})^N$ où

$\lambda_{i,j} = a_{j,i} - a_{0,i}$ pour $i = 1, \ldots, n$ et $j = 1, \ldots, N$. En fait, le $\mathbf{Z}$-module $\frac{1}{2i\pi}\Lambda$ est l'orthogonal de la projection par $\pi : \mathbf{R}^{N+1} \to \mathbf{R}^N$, $(x_0, x_1, \ldots, x_N) \mapsto (x_1, \ldots, x_N)$, du $\mathbf{Z}$-module $\Gamma_{\mathcal{A}} \subset \Delta^{\mathbf{Z}}$, introduit au § 3 précédent.

Considérons encore l'adhérence de Zariski $\overline{G}$ de $G$ dans la compactification $(\mathbf{P}^1)^N$ de $\mathbf{G}_m^N$, ses multidegrés sont indexés par des multi-indices $c = (c_1, \ldots, c_N) \in \mathbf{N}_n^N$ où $c_j \in \{0, 1\}$. À toute suite croissante $1 \leq j_1 < \cdots < j_n \leq N$ on associe un multi-indice $c(j_1, \ldots, j_n)$ défini par $c(j_1, \ldots, j_n)_j = 1$ si $j \in \{j_1, \ldots, j_n\}$ et 0 sinon. Indiquons comment on retrouve par ce biais la proposition 4 de [BP88].

**Proposition 4.1** *Avec les notation ci-dessus, le volume euclidien de la projection de* $\Lambda$ *sur le produit des n facteurs de* $\mathbf{R}^N$ *indexés par* $j_1, \ldots, j_n$ *est égal à* $\mathrm{MV}_n(\overline{a_0\, a_{j_1}}, \ldots, \overline{a_0\, a_{j_n}}) = \deg_{c(j_1,\ldots,j_n)}(\overline{G})$, *où* $\overline{a_0\, a_j}$ *désigne le segment de droite joignant* $a_0$ *à* $a_j$ *dans* $\mathbf{R}^n$.

*Démonstration* En effet, la projection de $\Lambda$ considérée est un réseau de $(2i\pi\mathbf{R})^n$ engendré par les vecteurs $(2i\pi\lambda_{i,j_1}, \ldots, 2i\pi\lambda_{i,j_n})$ pour $i = 1, \ldots, n$. Son volume pour la structure euclidienne fixée est donc égal à $|\Lambda_{j_1,\ldots,j_n}| = \mathrm{MV}_n(\overline{a_0\, a_{j_1}}, \ldots, \overline{a_0\, a_{j_n}})$ où

$$\Lambda_{j_1,\ldots,j_n} = \det \begin{bmatrix} \lambda_{1,j_1} & \ldots & \lambda_{1,j_n} \\ \vdots & & \vdots \\ \lambda_{n,j_1} & \ldots & \lambda_{n,j_n} \end{bmatrix} \in \mathbf{Z}.$$

Finalement, par la formule (2.2) ce multivolume est égal au multidegré $\deg_{c(j_1,\ldots,j_n)}(\overline{G})$ correspondant. □

On notera que les $\Lambda_{j_1,\ldots,j_n}$ introduits dans la démonstration précédente sont des coordonnées grassmanniennes de l'espace tangent $TG(\mathbf{C})$ en l'origine de $G$ dans $\mathbf{C}^N$. Comme $\Lambda$ est un réseau primitif de $TG(\mathbf{C})$ on a

$$\mathrm{ppcm}(\Lambda_{j_1,\ldots,j_n} \;:\; 1 \leq j_1 < \cdots < j_n \leq N) = 1.$$

L'image de $G$ par le plongement de Segre $s : (\mathbf{P}^1)^N \to \mathbf{P}^{2^N-1}$ est décrite par la somme de Minkowski

$$\overline{a_0\, a_1} + \cdots + \overline{a_0\, a_N} \subset \mathbf{R}^n,$$

dont le volume est égal à la somme des volumes mixtes pour toutes les projections du type envisagé dans la proposition 4.1. En particulier

$$\begin{aligned} \deg_{\mathbf{P}_{2^N-1}}(\overline{s(G)}) &= n!\mathrm{Vol}_n(\overline{a_0\, a_1} + \cdots + \overline{a_0\, a_N}) \\ &= n! \textstyle\sum_{1\leq j_1<\cdots<j_n\leq N} \mathrm{MV}_n(\overline{a_0\, a_{j_1}}, \ldots, \overline{a_0\, a_{j_n}}) \\ &= n! \textstyle\sum_{1\leq j_1<\cdots<j_n\leq N} |\Lambda_{j_1,\ldots,j_n}|. \end{aligned}$$

Ainsi, le degré de $\overline{s(G)}$ est égal à $n!$ fois la hauteur associée à la norme $\ell^1$ des coordonnées grassmanniennes de son espace tangent dans $\mathbf{C}^N$.

Si l'on considère la hauteur de Schmidt $h_S(T)$ d'un sous-espace $T \subset \overline{\mathbf{Q}}^N$ définie comme la hauteur projective du point qui le représente dans la variété grassmannienne [Sch91, page 28], on a la formule

$$\begin{aligned} h_S(TG) &= \left( \sum_{1 \le j_1 < \cdots < j_n \le N} |\Lambda_{j_1,\dots,j_n}|^2 \right)^{1/2} = \mathrm{Vol}_n(TG(\mathbf{C}) \cap (2i\pi\mathbf{R})^N/\Lambda) \\ &= \mathrm{Vol}_{N-n}(\Gamma_{\mathcal{A}}^{\mathbf{R}}/\Gamma_{\mathcal{A}}), \end{aligned}$$

car $TG(\mathbf{C}) \subset \mathbf{C}^N$ est défini sur $\mathbf{Q}$ et ses coordonnées grassmanniennes $\Lambda_{j_1,\dots,j_n}$ dans $\mathbf{Z}$ sont premières entre elles dans leur ensemble. La seconde égalité n'est autre que la formule de Cauchy–Binet déjà utilisée au § 2. La troisième égalité s'obtient en identifiant l'hyperplan $\Delta = \{\lambda_0 + \cdots + \lambda_N = 0\} \subset \mathbf{R}^{N+1}$ avec $\mathbf{R}^N$ par la projection sur les $N$ dernières coordonnées, de sorte que l'image de $\Delta^{\mathbf{Z}}$ est $\mathbf{Z}^N$.

En remarquant que $Q_{\mathcal{A}} = \mathrm{Conv}(a_0, \dots, a_N)$ contient tous les simplexes de sommets pris parmi les $a_i$ et est contenu dans l'union de ces mêmes simplexes ayant $a_0$ comme sommet fixe, on vérifie facilement (le volume d'un simplexe $\mathrm{Conv}(a_0, a_{i_1}, \dots, a_{i_n})$ est égal à $\frac{1}{n!}|\Lambda_{i_1,\dots,i_n}|$)

$$\frac{1}{n!} \max_{1 \le j_1,\dots,j_n \le N} |\Lambda_{j_1,\dots,j_n}| \le \mathrm{Vol}_n(Q_{\mathcal{A}}) \le \frac{1}{n!} \sum_{1 \le j_1 < \cdots < j_n \le N} |\Lambda_{j_1,\dots,j_n}|,$$

puis

$$\binom{N}{n}^{-1/2} \mathrm{Vol}_{N-n}(\Gamma_{\mathcal{A}}^{\mathbf{R}}/\Gamma_{\mathcal{A}}) \le n!\mathrm{Vol}_n(Q_{\mathcal{A}}) \le \binom{N}{n}^{1/2} \mathrm{Vol}_{N-n}(\Gamma_{\mathcal{A}}^{\mathbf{R}}/\Gamma_{\mathcal{A}}).$$

D'après la formule (2.1), on sait que $\deg_{\mathbf{P}^N}(\overline{G}) = n!\mathrm{Vol}_n(Q_{\mathcal{A}})$ et en particulier

**Proposition 4.2** *Pour tout sous-groupe algébrique $G \subset \mathbf{G}_m^N$ de dimension n on a*

$$n! \deg_{\mathbf{P}^N}(\overline{G}) \le \deg_{\mathbf{P}^{2^N-1}}(\overline{s(G)}) \le n! \binom{N}{n} \deg_{\mathbf{P}^N}(\overline{G}).$$

On pourra comparer avec le résultat analogue pour la hauteur normalisée dans [DP99, Prop. 2.2].

Si l'on introduit la fonction $f_{\mathcal{A}}$ sur l'enveloppe convexe $Q_{\mathcal{A}}$ des points $\mathcal{A} = (a_0, \dots, a_N)$ dans $\mathbf{R}^n$, à valeur dans $\mathbf{N}$, définie pour $u \in Q_{\mathcal{A}}$ par :

$$f_{\mathcal{A}}(u) := \mathrm{Card}\left\{(j_1, \dots, j_n);\ 1 \le j_1 < \dots < j_n \le N \ \text{ et } \ u \in \mathrm{Conv}(a_0, a_{j_1}, \dots, a_{j_n})\right\},$$

on vérifie facilement l'égalité

$$\frac{\deg_{\mathbf{P}^{2^N-1}}(\overline{s(G)})}{n! \deg_{\mathbf{P}^N}(\overline{G})} = \frac{1}{\mathrm{Vol}_n(Q_{\mathcal{A}})} \cdot \int_{Q_{\mathcal{A}}} f_{\mathcal{A}}(u)\, du.$$

*Exemple 4.3* En considérant des points $a_0, \dots, a_N$ dans $\mathbf{Z}^n$ tels que $a_{n+1}, \dots, a_N$ soient très proches de $a_0$ comparativement à $a_1, \dots, a_n$ (*voir* Fig. 3 pour $n = 2$), on a

$$\sum_{1 \le j_1 < \cdots < j_n \le N} |\Lambda_{j_1,\dots,j_n}| \approx \max_{1 \le j_1,\dots,j_n \le N} |\Lambda_{j_1,\dots,j_n}| = |\Lambda_{1,\dots,n}|.$$

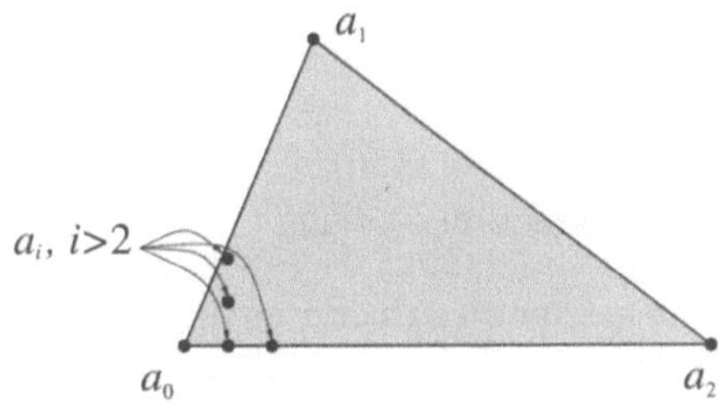


Fig. 3

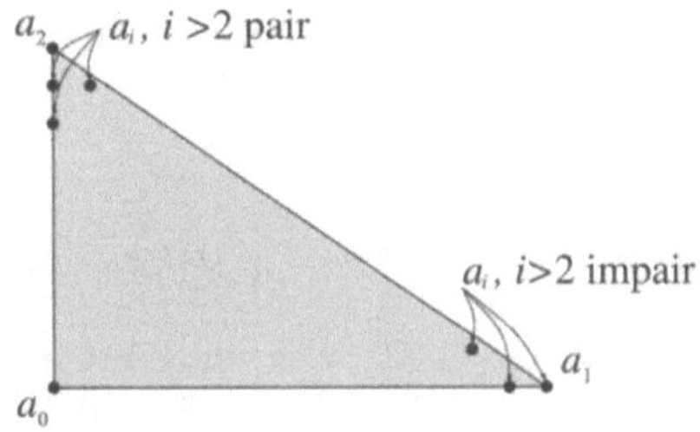


Fig. 4

On obtient ainsi une variété torique pour laquelle $n!\deg_{\mathbf{P}_N}(\overline{G}) \approx \deg_{\mathbf{P}_{2N-1}}(\overline{s(G)})$, montrant que l'inégalité de gauche dans la proposition 4.2 est optimale.

*Exemple 4.4* Considérons maintenant $N = mn$ points $a_1, \ldots, a_N$ de $\mathbf{Z}^n$ également concentrés autour de $a_1, \ldots, a_n$, par exemple $a_{kn+i} \approx a_i$ à l'intérieur de $\text{Conv}(a_0, a_1, \ldots, a_n)$, pour $k = 1, \ldots, m-1$ et $i = 1, \ldots, n$ (*voir* Fig. 4 pour $n = 2$). On a alors

$$\sum_{1 \le j_1 < \cdots < j_n \le N} |\Lambda_{j_1,\ldots,j_n}| \gtrsim \left(\frac{N}{n}\right)^n n!\text{Vol}_n(a_0, a_1, \ldots, a_n).$$

On obtient ainsi une variété torique pour laquelle $\deg_{\mathbf{P}_{2N-1}}(\overline{s(G)}) \gtrsim n!\left(\frac{N}{n}\right)^n \deg_{\mathbf{P}_N}(\overline{G})$, laissant supposer que l'inégalité de droite dans la proposition 4.2 pourrait être améliorée d'un facteur $e^n$. De fait, lorsque $n = 2$ on vérifie pour toute configuration $\mathcal{A}$ de points et tout $u \in Q_{\mathcal{A}}$ la majoration $f_{\mathcal{A}}(u) \le \left(\frac{N}{2}\right)^2$, d'où $\deg_{\mathbf{P}_{2N-1}}(\overline{s(G)}) \le 2\left(\frac{N}{2}\right)^2 \deg_{\mathbf{P}_N}(\overline{G})$, qui est optimal.

## 5 Un théorème de Bézout pour les poids de Chow

Soit $X \subset \mathbf{P}^N$ et $\tau = (\tau_0, \ldots, \tau_N) \in \mathbf{R}^{N+1}$, rappelons que $e_\tau(X) \in \mathbf{R}$ désigne le $\tau$-poids de Chow de $X$ défini dans l'introduction.

Pour $\lambda \in \mathbf{R}^+$ on a $e_{\lambda\tau}(X) = \lambda\, e_\tau(X)$. De même, on vérifie facilement que pour $\tau' \in \mathbf{R}$ on a $e_{\tau+(\tau',\ldots,\tau')}(X) = e_\tau(X) + \tau'(n+1)\deg(X)$. On vérifie encore que si $\sigma_\delta : \mathbf{P}^N \to \mathbf{P}^M$, $M+1 = \binom{N+\delta}{N}$, désigne le plongement de Veronese de degré $\delta$ et $\tau^{(\delta)}$ le vecteur des $\tau$-poids des monômes de degrés $\delta$, on a $e_{\tau^{(\delta)}}(\sigma_\delta(X)) = \delta^{n+1} \cdot e_\tau(X)$. Plus profonde est l'expression de $e_\tau(X)$ comme coefficient dominant d'une *fonction poids de Hilbert* démontrée par Mumford [Mum77, Prop. 2.11] :

$$s_\tau(X; D) := \max_J \sum_{\lambda \in J} (\tau_0\lambda_0 + \cdots + \tau_N\lambda_N) = \frac{e_\tau(X)}{(n+1)!} D^{n+1} + O(D^n),$$

où le maximum porte sur tous les ensembles $J$ d'éléments de $\mathbf{N}_D^{N+1}$ tels que les monômes associés induisent une base de la partie graduée de degré $D$ de l'anneau de la variété $X$.

*Exemple 4.1* Si $H$ est une hypersurface de $\mathbf{P}^N$ d'équation $f \in \mathbf{K}[x_0, \ldots, x_N]$ et $\tau \in \mathbf{R}^{N+1}$, on a

$$e_\tau(H) = (\tau_0 + \cdots + \tau_N)\deg(H) - w_t(\lambda_\tau^*(f))$$

où $\lambda_\tau^*(f) := f(t^{\tau_0}x_0, \ldots, t^{\tau_N}x_N)$ et $w_t$ désigne la valuation $t$-adique.

Lorsque $\tau \in \mathbf{Z}^{N+1}$ le poids de Chow peut s'interpréter en termes d'un bidegré de la déformation torique $X_\tau \subset \mathbf{P}^1 \times \mathbf{P}^N$ de $X$ relative à $\tau$. Une forme résultante d'indice $(1, n+1)$ de $X_\tau$ s'écrit

$$v_1^{e_\tau(X)} v_0^{e_{-\tau}(X)} Ch_X(\ldots, (-v_0/v_1)^{\tau_j} u_{i,j}, \ldots),$$

son degré en $(v_0, v_1)$ est $e_\tau(X) + e_{-\tau}(X)$ d'après la définition du poids de Chow et c'est par ailleurs le bidegré $\deg_{(0,n+1)}(X_\tau)$, d'où

$$\deg_{(0,n+1)}(X_\tau) = e_\tau(X) + e_{-\tau}(X).$$

Rappelons que le degré $\deg_{(0,n+1)}$ est obtenu en intersectant par $n+1$ formes linéaires relevées du second facteur $\mathbf{P}^{2N+1}$, *voir* § 2 pour plus de détails.

Lorsque $\tau \in \mathbf{N}^{N+1}$ on peut aussi interpréter le poids de Chow $e_\tau(X)$ isolément comme un bidegré en considérant une variation de la déformation torique précédente. Précisément, il s'agit maintenant de l'adhérence de Zariski $\widetilde{X}_\tau \subset \mathbf{P}^1 \times \mathbf{P}^{2N+1}$ de

$$\{(1:t) \times (t^{\tau_0}x_0 : \cdots : t^{\tau_N}x_N : x_0 : \cdots : x_N);\, t \in \mathbf{G}_m, x \in X\}.$$

En appliquant ce qui précède on a

$$\deg_{(0,n+1)}(\widetilde{X}_\tau) = e_{\widetilde{\tau}}(\widetilde{X}) + e_{-\widetilde{\tau}}(\widetilde{X}),$$

où $\widetilde{X}$ désigne le *plongement diagonal* de $X$ dans $\mathbf{P}^{2N+1}$ et $\widetilde{\tau} = (\tau_0, \ldots, \tau_N, 0, \ldots, 0) \in \mathbf{N}^{2N+2}$. On vérifie sans difficulté $e_{\widetilde{\tau}}(\widetilde{X}) = e_\tau(X)$ et $e_{-\widetilde{\tau}}(\widetilde{X}) = 0$, d'où $\deg_{(0,n+1)}(\widetilde{X}_\tau) = e_\tau(X)$.

Donnons maintenant la démonstration du théorème 1.2, en commençant par une variante de ce théorème de Bézout pour les poids de Chow. Notons

$$\pi : \mathbf{P}^1 \times \mathbf{P}^{2N+1} \to \mathbf{P}^N, \quad ((t_0 : t_1), (x_0 : \cdots : x_N : y_0 : \cdots : y_N)) \mapsto (y_0 : \cdots : y_N)$$

la projection de $\mathbf{P}^1 \times \mathbf{P}^{2N+1}$ sur $\mathbf{P}^N$ donnée par les $N+1$ dernières coordonnées de $\mathbf{P}^{2N+1}$ et

$$\tilde{\iota} : \mathbf{P}^N \to \mathbf{P}^1 \times \mathbf{P}^{2N+1}, \quad (x_0 : \cdots : x_N) \mapsto ((0:1), (x_0 : \cdots : x_N : 0 : \cdots : 0)).$$

**Théorème 5.2** *Soit $X \subset \mathbf{P}^N$ une variété projective et $H$ un diviseur de $\mathbf{P}^N$ ne contenant pas $X$, alors pour tout $\tau \in \mathbf{Z}^{N+1}$ on a*

$$e_\tau(X \cdot H) = e_\tau(X)\deg(H) - \sum_{Y \in \mathrm{Irr}(\mathrm{init}_\tau(X))} m(\tilde{X}_\tau \cdot \pi^*(H); \tilde{\iota}(Y))\deg(Y).$$

*De plus, si $H$ est une hypersurface et $f \in \mathbf{K}[x_0, \ldots, x_N]$ est une équation de $H$ on a, avec les notations du théorème 1.2,*

$$m(\tilde{X}_\tau \cdot \pi^*(H); \tilde{\iota}(Y)) = m(X_\tau \cdot \mathrm{div}(\lambda_\tau^*(f)); \iota(Y)),$$

*où $\lambda_\tau^*(f) := f(t_0^{\tau_0}x_0, \ldots, t_0^{\tau_N}x_N) \in \mathbf{K}[t_0, t_1][x_0, \ldots, x_N]$.*

*Démonstration* Il suffit d'établir le résultat pour $\tau \in \mathbf{N}^{N+1}$, on se ramène à ce cas en ajoutant à $\tau$ un vecteur $(\tau', \ldots, \tau')$ où $\tau' \in \mathbf{N}$ est suffisamment grand et on remarque qu'avec les propriétés d'homogénéité du poids de Chow l'égalité désirée reste invariante.

Dans ce cas, comme $H$ ne contient pas $X$, le cycle intersection $\widetilde{X}_\tau \cdot \pi^*(H)$ s'écrit comme la somme du cycle $\widetilde{(X \cdot H)_\tau}$ et d'un cycle supporté par $\tilde{\iota}(\mathrm{init}_\tau(X))$, le second vivant donc dans $\{(0 : 1)\} \times \mathbf{P}^{2N+1}$. Comme $\deg_{(0,n+1)}(\widetilde{X}_\tau) = e_\tau(X)$ et $\deg_{(0,n)}(\widetilde{(X \cdot H)_\tau}) = e_\tau(X \cdot H)$ on a, d'après le théorème de Bézout multi-projectif [Rem01b, Thm. 3.4],

$$\begin{aligned}\deg(H)e_\tau(X) &= \deg(H)\deg_{(0,n+1)}(\widetilde{X}_\tau)\\ &= \deg_{(0,n)}(\widetilde{X}_\tau \cdot \pi^*(H))\\ &= \deg_{(0,n)}(\widetilde{(X \cdot H)_\tau}) + \sum_{Y \in \mathrm{Irr}(\mathrm{init}_\tau(X))} m(\widetilde{X}_\tau \cdot \pi^*(H); \tilde{\iota}(Y)) \deg(Y)\\ &= e_\tau(X \cdot H) + \sum_{Y \in \mathrm{Irr}(\mathrm{init}_\tau(X))} m(\widetilde{X}_\tau \cdot \pi^*(H); \tilde{\iota}(Y)) \deg(Y).\end{aligned}$$

Notons finalement que le morphisme

$$\begin{array}{cccc}\pi' : & \mathbf{P}^1 \times \mathbf{P}^{2N+1} & \to & \mathbf{P}^1 \times \mathbf{P}^N\\ & ((t_0 : t_1), (x_0 : \cdots : x_N : y_0 : \cdots : y_N)) & \mapsto & ((t_0 : t_1), (x_0 : \cdots : x_N))\end{array}$$

est un isomorphisme de $\widetilde{X}_\tau$ sur $X_\tau$ dans l'ouvert $t_1 \neq 0$, tel que $\pi' \circ \tilde{\iota} = \iota$ et $\pi'_*(\pi^*(H)) = \mathrm{div}(\lambda_\tau^*(f))$. Cela entraîne, pour toute composante irréductible $Y$ de $\mathrm{init}_\tau(X)$,

$$m(\widetilde{X}_\tau \cdot \pi^*(H); \tilde{\iota}(Y)) = m(X_\tau \cdot \mathrm{div}(\lambda_\tau^*(f)); \iota(Y)).$$

□

*Démonstration du théorème 1.2* On reprend la fin de la démonstration du théorème 5.2 en se placant dans une carte affine contenant une composante $Y$ de $\mathrm{init}_\tau(X)$ et sur laquelle $H$ est décrit par une équation $f$, on a

$$\begin{aligned}m(\widetilde{X}_\tau \cdot \pi^*(H); \tilde{\iota}(Y)) &= \mathrm{long}(\mathbf{K}[\widetilde{X}_\tau]/(\pi^*(f)))_{\tilde{\iota}(Y)}\\ &= \mathrm{long}(\mathbf{K}[X_\tau]/(\lambda_\tau^*(f)))_{\iota(Y)}\\ &= m(X_\tau \cdot \mathrm{div}(\lambda_\tau^*(f)); \iota(Y))\\ &= m(X_\tau \cdot H_\tau; \iota(Y)) + w_{t_0}(\lambda_\tau^*(f))\, m(\mathrm{init}_\tau(X); Y)\\ &= m(X_\tau \cdot H_\tau; \iota(Y)) + ((\tau_0 + \cdots + \tau_N)\deg(H)\\ &\quad -e_\tau(H))\, m(\mathrm{init}_\tau(X); Y)\end{aligned}$$

d'après le calcul de l'exemple 4.1. Enfin, en sommant sur toutes les composantes $Y$ de $\mathrm{init}_\tau(X)$ on a $\sum_Y m(\mathrm{init}_\tau(X); Y)\deg(Y) = \deg(\mathrm{init}_\tau(X)) = \deg(X)$. □

Dans les notations du théorème 5.2, soit $f \in \mathbf{K}[x_0, \ldots, x_N]$ une équation de $H$ et $\tau \in \mathbf{R}^{N+1}$, posons

$$w_{X,\tau}(f) := \frac{1}{\deg(X)} \cdot (e_\tau(X \cdot H) - e_\tau(X)\deg(H))\,, \tag{5.9}$$

c'est une fonction continue de $\tau$. Lorsque $\tau \in \mathbf{N}^{N+1}$ on a donc par le théorème 5.2

$$w_{X,\tau}(f) = - \sum_{Y \in \mathrm{Irr}(\mathrm{init}_\tau(X))} m(\tilde{X}_\tau \cdot \pi^*(H); \tilde{\iota}(Y)) \cdot \frac{\deg(Y)}{\deg(X)}.$$

Toujours dans ce cas, on vérifie $w_{X,k\tau}(f) = k w_{X,\tau}(f)$ pour tout $k \in \mathbf{N}^\times$ et $w_{X,\tau+\tau'(1,\ldots,1)}(f) = w_{X,\tau}(f) - \tau' \deg(H)$ pour $\tau' \in \mathbf{N}$. On peut donc écrire en général (c'est-à-dire pour $\tau \in \mathbf{R}^{N+1}$) l'égalité

$$e_\tau(X \cdot H) = e_\tau(X) \deg(H) + w_{X,\tau}(f) \deg(X)$$

en posant

$$w_{X,\tau}(f) := \lim_{k \to \infty} \frac{1}{k} w_{X,[k(\tau - \tau'(1,\ldots,1))]}(f) + \tau' \deg(H)$$

où $\tau' := \min(\tau_0, \ldots, \tau_N)$ et $[\cdot]$ désigne le vecteur des parties entières dans $\mathbf{N}^{N+1}$.

Dans le cas d'une variété torique $X_{\mathcal{A},\alpha}$ le poids de Chow $e_\tau(X_{\mathcal{A},\alpha}) = e_\tau(X_\mathcal{A})$ s'exprime naturellement en termes d'un polytope construit à l'aide du vecteur $\tau$, au-dessus du polytope $Q_\mathcal{A}$ déjà introduit au paragraphe 2.

**Proposition 5.3** [PS04, Prop. III.1]. *Avec les notations introduites on suppose* $\mathbf{Z}a_0 + \cdots + \mathbf{Z}a_N = \mathbf{Z}^n$. *Soit* $Q_{\mathcal{A},\tau}$ *l'enveloppe convexe des points* $(a_0, \tau_0), \ldots, (a_N, \tau_N)$ *dans* $\mathbf{R}^{n+1}$ *et* $\vartheta_{\mathcal{A},\tau} : Q_\mathcal{A} \to \mathbf{R}$ *la paramétrisation de la toiture de* $Q_{\mathcal{A},\tau}$ *au-dessus de* $Q_\mathcal{A}$, *alors*

$$e_\tau(X_\mathcal{A}) = (n+1)! \int_{Q_\mathcal{A}} \vartheta_{\mathcal{A},\tau}(u)\, \mathrm{d}u_1 \ldots \mathrm{d}u_n.$$

La démonstration qu'on donne de cette proposition dans [PS04, § III] est basée sur un calcul explicite des poids de Hilbert des variétés toriques; nous y montrons également qu'on peut l'obtenir comme conséquence des résultats de I. M. Gelfand, M. M. Kapranov et A. V. Zelevinski sur le polytope de Newton du $\mathcal{A}$-résultant [GKZ94, Ch. 7 et 8]. Elle est également implicite dans [Don02, § 4.2]. L'interprétation ci-dessus du poids de Chow comme bi-degré permet d'en donner encore une autre démonstration simple et directe :

*Démonstration.* Tout d'abord on remarque qu'il suffit de démontrer l'énoncé pour un choix générique (au sens de Zariski) du vecteur $\tau$ dans $\mathbf{R}^{N+1}$, puisque les termes considérés sont continus par rapport à $\tau$. L'identité étant invariante par homothéties et translations, on peut se ramener à supposer $\tau \in \mathbf{N}^{N+1}$ à coordonnées premières entre elles dans leur ensemble.

On a $L_{\mathcal{A},\tau} = \mathbf{Z}^{n+1}$; la proposition 2.2 entraîne alors

$$\begin{aligned} e_\tau(X) &= \deg_{(0,n+1)}(\widetilde{X}_\tau) = \mathrm{MV}_{n+1}(Q_{\mathcal{A},\widetilde{\tau}}, \ldots, Q_{\mathcal{A},\widetilde{\tau}}) \\ &= (n+1)! \int_{Q_\mathcal{A}} \vartheta_{\mathcal{A},\tau}(u) \mathrm{d}u_1 \cdots \mathrm{d}u_n, \end{aligned}$$

où $\widetilde{\tau} = (\tau_0, \ldots, \tau_N, 0, \ldots, 0) \in \mathbf{N}^{2N+2}$. □

La proposition 5.3 permet de reformuler de façon amusante le théorème du sous-espace de Schmidt, dans la version projective qu'en ont donné Evertse et Ferretti [EF02] mais dans le cas très particulier des variétés toriques.

*Convention 5.4.* Si $K$ est un corps de nombres et $v$ une place de $K$ on notera $|\cdot|_v$ la valeur absolue de $K$ dans $v$ étendant la valeur absolue usuelle de $\mathbf{Q}$ si $v$ est archimédienne et la valeur absolue $p$-adique standard de $\mathbf{Q}$ (*i.e.* $|p|_v = p^{-1}$) si $v$ est ultramétrique. On notera encore $M_K$ l'ensemble de ces valeurs absolues représentant les places de $K$.

Soit $K$ un corps de nombres et $S$ un ensemble fini de places de $K$. On considère une variété torique $X_{\mathcal{A},\alpha} \subset \mathbf{P}_N$ définie sur $K$, pour toute place $v \in S$ des réels $\tau_{v,0}, \ldots, \tau_{v,N} \geq 0$ et le système d'inéquations en $x \in X_{\mathcal{A},\alpha}(\overline{\mathbf{Q}})$ :

$$\log\left(\frac{|x_i|_w}{\|x\|_w}\right) \leq -\tau_{v,i}h(x) \quad \text{pour } i = 0, \ldots, N \text{ et } w \mid v \text{ place de } K(x), \tag{5.10}$$

où $\|x\|_w := \max(|x_i|_w;\ i=0,\ldots,N)$ si $w$ est ultramétrique, $\|x\|_w = \left(\sum_{i=0}^{N}|x_i|_w^2\right)^{1/2}$ si $w$ est archimédienne, et $h(x)$ désigne la hauteur projective.

Dans cette situation notons $\vartheta_{\mathcal{A},\tau}(u) := \sum_{v\in S}\frac{[K_v:\mathbf{Q}_v]}{[K:\mathbf{Q}]}\cdot\vartheta_{\mathcal{A},\tau_v}(u)$ pour $u \in Q_{\mathcal{A}}$. Le théorème 3.2 de [EF02] explicite pour tout $\varepsilon > 0$ des réels

$$c_1(N, n, \deg(X_{\mathcal{A},\alpha}), \varepsilon),\ \ c_2(N, n, \deg(X_{\mathcal{A},\alpha}), \varepsilon) \geq 1$$

tels que l'énoncé suivant soit vrai ($\tau_v := (\tau_{v,0}, \ldots, \tau_{v,N})$, $v \in S$) :

**Théorème 5.5.** [EF02, Thm. 3.2] *Si la valeur moyenne de $\vartheta_{\mathcal{A},\tau}$ sur $Q_{\mathcal{A}}$ est $\geq 1+\varepsilon$, alors les points $x \in X_{\mathcal{A},\alpha}(\overline{\mathbf{Q}})$ solutions du système d'inéquations (5.10) et satisfaisant*

$$h(x) \geq c_1(N, n, \deg(X_{\mathcal{A},\alpha}), \varepsilon)(1 + h(X_{\mathcal{A},\alpha}))$$

*appartiennent à un sous-ensemble algébrique propre de $X_{\mathcal{A},\alpha}$, défini sur $K$ et de degré $\leq c_2(N, n, \deg(X_{\mathcal{A},\alpha}), \varepsilon)$.*

*Démonstration.* Notons $X = X_{\mathcal{A},\alpha}$, la condition donnée dans le théorème 3.2 de [EF02] s'écrit

$$\frac{1}{(n+1)\deg(X)}\cdot\sum_{v\in S}\frac{[K_v:\mathbf{Q}_v]}{[K:\mathbf{Q}]}\cdot e_{\tau_v}(X) \geq 1+\varepsilon.$$

Or $\deg(X) = n!\mathrm{Vol}_n(Q_{\mathcal{A}})$ d'après la formule (2.1) et, vue la proposition 5.3, on a

$$\sum_{v\in S}\frac{[K_v:\mathbf{Q}_v]}{[K:\mathbf{Q}]}.\frac{e_{\tau_v}(X)}{(n+1)\deg(X)} = \frac{1}{\mathrm{Vol}_n(Q_{\mathcal{A}})}.\int_{Q_{\mathcal{A}}}\vartheta_{\mathcal{A},\tau}(u)\,du_1\ldots du_n$$

qui est bien la valeur moyenne de $\vartheta_{\mathcal{A},\tau}$ au-dessus $Q_{\mathcal{A}}$. □

Le résultat trivial, provenant de la formule du produit, permet d'affirmer que le système (5.10) n'a pas de solution dans $\mathbf{P}_N^\circ(\overline{\mathbf{Q}})$ dès qu'il existe $i \in \{0, \ldots, N\}$ tel que

$$\sum_{v\in S}\frac{[K_v:\mathbf{Q}_v]}{[K:\mathbf{Q}]}\cdot\tau_{v,i} > 1.$$

On donne un exemple où ce résultat trivial s'applique bien que la condition du théorème 5.5 ne soit pas remplie et un autre exemple où le théorème 5.5 donne un résultat non trivial.

*Exemple 5.6.* Soient $0 < \varepsilon < n+1$, $\mathcal{A} \subset \mathbf{Z}^n$ et $\tau_v$ tels que $\tau_{v,i} = 0$ pour tout $v \in S$ et $i = 1, \ldots, N$ et $\sum_{v \in S} \frac{[K_v:\mathbf{Q}_v]}{[K:\mathbf{Q}]} \tau_{v,0} = 1 + \varepsilon > 1$. On calcule à l'aide de la proposition 5.3 $e_{\tau_v}(X_{\mathcal{A}}) \le \deg(X_{\mathcal{A}}) \cdot \tau_{v,0}$, d'où $\sum_{v \in S} \frac{[K_v:\mathbf{Q}_v]}{[K:\mathbf{Q}]} \cdot \frac{e_{\tau_v}(X_{\mathcal{A}})}{(n+1)\deg(X_{\mathcal{A}})} \le \frac{1+\varepsilon}{n+1} < 1$. On sait donc que dans ce cas le système d'inéquations (5.10) n'a pas de solution telle que $x_0 \neq 0$, bien que l'hypothèse principale du théorème 5.5 ne soit pas satisfaite.

*Exemple 5.7.* Soit maintenant $N' > 3$ un entier, $N := 2N'$ et $D > N'(N'-1)$ un autre entier. On considère $\mathcal{A} := (0, \ldots, N'-1, D-N'+1, \ldots, D) \in \mathbf{Z}$ et $S = \{v_1, \ldots, v_{N'}\}$ un ensemble de $N'$ places de $\mathbf{Q}$. À chaque place $v_i$ de $S$ on associe le vecteur $\tau_{v_i} \in \mathbf{R}^{N+1}$ dont toutes les coordonnées sont nulles sauf celles d'indices $i$ et $D - N' + i$ qui valent $1/(N'-1)$. On a ainsi $e_{\tau_{v_i}}(X_{\mathcal{A}}) \ge 2\frac{D-N'+1}{N'-1}$ et $\sum_{v \in S} \frac{e_{\tau_v}(X_{\mathcal{A}})}{2\deg(X_{\mathcal{A}})} \ge \frac{D-N'+1}{D} \cdot \frac{N'}{N'-1} > 1$, tandis que $\sum_{v \in S} \tau_{v,i} = 1/(N'-1) < 1$ pour tout $i = 0, \ldots, N$. Le théorème 5.5 s'applique donc et donne un résultat *a priori* non trivial dans ce cas.

# 6 Hauteur normalisée

L'espace projectif peut être vu comme une compactification équivariante du groupe multiplicatif $\mathbf{G}_m^N \simeq (\mathbf{P}^N)^\circ$, cette structure de groupe permettant de définir une notion de hauteur pour les sous-variétés de $\mathbf{P}^N$ plus canonique que les autres, appelée *hauteur normalisée*. Cette notion joue un rôle central dans l'approximation diophantienne sur les tores, et tout particulièrement dans les problèmes de Lehmer généralisé et de Bogomolov sur les tores, *voir* [DP99], [AD03] et leurs références.

Suivant [DP99], la hauteur normalisée peut se définir par un procédé « à la Tate ». De façon précise, pour $k \in \mathbf{N}$ on pose $[k] : \mathbf{P}^N \to \mathbf{P}^N$, $(x_0 : \cdots : x_N) \mapsto (x_0^k : \cdots : x_N^k)$ l'application puissance $k$-ième; restreinte au tore $(\mathbf{P}^N)^\circ$ c'est l'application de multiplication par $k$. La hauteur normalisée d'une variété projective $X$ est par définition

$$\widehat{h}(X) := \deg(X) \cdot \lim_{k \to \infty} \frac{h([k]\,X)}{k \deg([k]\,X)} \in \mathbf{R}_+,$$

où deg et $h$ désignent le degré et la hauteur projective, *voir* [DP99, § 2] ou [PS04, § I.2]. Lorsque $X$ est de dimension 0, il s'agit de la hauteur de Gauss–Weil des points.

On sait que le comportement de cette hauteur normalisée est lié à la nature géométrique de la variété $X$. En particulier, $\widehat{h}(X) = 0$ si et seulement si $X$ est une variété de torsion, c'est-à-dire un translaté de sous-tore un point de torsion. Dans [PS04] on a établi un analogue arithmétique de l'expression du degré d'une variété torique comme volume du polytope associé, que nous rappelons maintenant.

Soit $M_K$ l'ensemble des places du corps $K$ et rappelons la convention 5.4. Pour chaque $v \in M_K$ on considère le vecteur $\tau_{\alpha v} := (\log|\alpha_0|_v, \ldots, \log|\alpha_N|_v) \in \mathbf{R}^{N+1}$ et le polytope

$$Q_{\mathcal{A},\tau_{\alpha v}} := \mathrm{Conv}\left((a_0, \log|\alpha_0|_v), \ldots, (a_N, \log|\alpha_N|_v)\right) \subset \mathbf{R}^{n+1},$$

dont la *toiture* au-dessus de $Q_{\mathcal{A}}$ (c'est-à-dire l'enveloppe supérieure) s'envoie bijectivement sur $Q_{\mathcal{A}}$ par la projection standard $\mathbf{R}^{n+1} \to \mathbf{R}^n$. On pose alors

$$\vartheta_{\mathcal{A},\tau_{\alpha v}} : Q_{\mathcal{A}} \to \mathbf{R}, \qquad x \mapsto \max\left\{y \in \mathbf{R} \,:\, (x, y) \in Q_{\mathcal{A},\tau_{\alpha v}}\right\}$$

la paramétrisation de cette toiture; c'est une fonction *concave* et *affine par morceaux*.

**Théorème 6.1** [PS04, Thm. 0.1]. *Soit* $\mathcal{A} \in (\mathbf{Z}^n)^{N+1}$ *tel que* $L_{\mathcal{A}} = \mathbf{Z}^n$ *et* $\alpha \in (K^\times)^{N+1}$, *alors*

$$\begin{aligned}\widehat{h}(X_{\mathcal{A},\alpha}) &= \sum_{v\in M_K} \frac{[K_v : \mathbf{Q}_v]}{[K : \mathbf{Q}]}\, e_{\tau_{\alpha\, v}}(X_{\mathcal{A}})\\ &= (n+1)! \sum_{v\in M_K} \frac{[K_v : \mathbf{Q}_v]}{[K : \mathbf{Q}]} \int_{Q_{\mathcal{A}}} \vartheta_{\mathcal{A},\tau_{\alpha v}}(u)\; du_1 \ldots du_n.\end{aligned}$$

Notons que $\tau_{\alpha v} = 0$ pour presque tout $v$, donc cette somme ne contient qu'un nombre fini de termes non nuls. Dans les cas des points ($n = 0$) la formule se réduit à la définition usuelle de la hauteur normalisée (hauteur de Gauss–Weil) d'un point de $\mathbf{P}^N$.

Comme on a $\widehat{h}(X_{\mathcal{A}}) = 0$, on peut s'interroger si cette formule n'est pas la manifestation d'une propriété générale de la hauteur normalisée par translation. Dans ce sens, B. Sturmfels nous a demandé si pour toute variété projective $X$ on a

$$\widehat{h}(\alpha X) = \widehat{h}(X) + \sum_{v\in M_K} \frac{[K_v : \mathbf{Q}_v]}{[K : \mathbf{Q}]} e_{\tau_{\alpha\, v}}(X)?$$

Soit $\prod_{i=0}^n \prod_{j=0}^N U_{i,j}^{b_{i,j}}$ un monôme apparaissant dans la forme de Chow de $X$, on vérifie facilement

$$e_{\tau_{\alpha\, v}}(X) \geq \left(\sum_{i=0}^n b_{i,0}\right)\tau_{\alpha\, v,0} + \cdots + \left(\sum_{i=0}^n b_{i,N}\right)\tau_{\alpha\, v,N}.$$

Comme $\sum_{v\in M_K} \frac{[K_v:\mathbf{Q}_v]}{[K:\mathbf{Q}]}\tau_{\alpha\, v,i} = 0$ pour tout $i = 0, \ldots, N$ par la formule du produit, on en déduit

$$\sum_{v\in M_K} \frac{[K_v : \mathbf{Q}_v]}{[K : \mathbf{Q}]} e_{\tau_{\alpha\, v}}(X) \geq 0 \quad \text{pour tout } \alpha \in (\overline{\mathbf{Q}}^\times)^{N+1}.$$

Une formule du type précédent ne peut donc être valable en général.

Illustrons cette formule par un exemple en dimension 1. Soient $\mathcal{A}_N := (0, 1, 2, \ldots, N) \in \mathbf{Z}^{N+1}$ et $\alpha_N := (1, 2, 3, \ldots, N+1) \in (\overline{\mathbf{Q}}^\times)^{N+1}$; ainsi

$$\varphi_{\mathcal{A}_N,\alpha_N} : \mathbf{G}_m \to \mathbf{P}^N, \quad s \mapsto (1 : 2s : \cdots : (N+1)s^N).$$

La figure 5 montre pour $N = 3$ les polytopes associés et leurs toitures, pour chaque place $v \in M_{\mathbf{Q}}$:

Ainsi $\vartheta_v \equiv 0$ pour $v \neq \infty, 2$, d'où

$$\widehat{h}(X_{\mathcal{A}_3,\alpha_3}) = 2!\left(\int_0^3 \vartheta_\infty(u)\, du + \int_0^3 \vartheta_2(u)\, du\right) = 2\log(2) + 2\log(3) = \log(36).$$

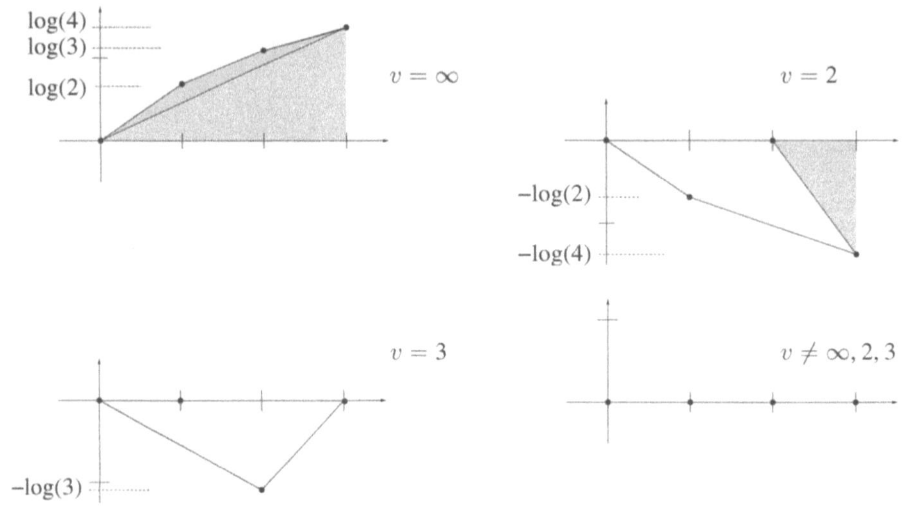


Fig. 5

En général

$$\begin{aligned}\widehat{h}(X_{\mathcal{A}_N,\alpha_N}) &= 2!\left(\int_0^N \vartheta_\infty(u)\,\mathrm{d}u + \sum_{p|N+1}\int_0^N \vartheta_p(u)\,\mathrm{d}u\right)\\ &= 2\left(\log(2)+\cdots+\log(N) + \frac{1}{2}\log(N+1) + \sum_{p|N+1}\frac{1}{2}\log|N+1|_p\right)\\ &= 2\,\log(N!).\end{aligned}$$

À l'instar des multidegrés, les multihauteurs du tore $\mathbf{G}_m^n$ plongé dans un produit d'espaces projectifs *via* plusieurs applications monomiales peuvent aussi s'expliciter à l'aide *d'intégrales mixtes* des fonctions concaves apparaissant dans le théorème 6.1.

Soit $Z \subset \mathbf{P}^{N_0} \times \cdots \times \mathbf{P}^{N_m}$ de dimension $n$ et $c = (c_0, \ldots, c_m) \in \mathbf{N}_{n+1}^{m+1}$ avec $0 \le c_i \le N_i$. Soit $\mathrm{rés}_{d(c)}(I(Z))$ la forme résultante associée, dans les notations du § 2. La *multihauteur projective de* $Z$ *d'indice* $c$ est définie par

$$h_c(Z) := h(\mathrm{rés}_{d(c)}(I(Z))),$$

où $h$ désigne la hauteur des polynômes multihomogènes définie à l'aide de la $S_{N_0+1} \times \cdots \times S_{N_n+1}$*-mesure* pour les places archimediennes; on renvoie à [Rem01a] ou encore [PS04, § I.2] pour les détails. Notons $s : \mathbf{P}^{N_0} \times\cdots\times \mathbf{P}^{N_m} \to \mathbf{P}^{(N_0+1)\cdots(N_m+1)-1}$ le plongement de Segre, la suite

$$k \mapsto \deg(s(Z)) \cdot \frac{h_c([k]\,Z)}{k\,\deg([k]\,s(Z))}$$

converge lorsque $k$ tend vers l'infini [PS04, Prop. I.2]). Sa limite est par définition la *multihauteur normalisée de* $Z$ *d'indice* $c$, notée $\widehat{h}_c(Z)$.

Pour $D=(D_0,\ldots,D_m)\in(\mathbf{N}^\times)^{m+1}$ soit $\Psi_D:\mathbf{P}^{N_0}\times\cdots\times\mathbf{P}^{N_m}\to\mathbf{P}^{\binom{D_0+N_0}{N_0}\cdots\binom{D_m+N_m}{N_m}-1}$ le plongement mixte introduit au § 2, alors on a la formule [PS04, (I.4)]

$$\widehat{h}(\Psi_D(Z))=\sum_{c\in\mathbf{N}^{m+1}_{n+1}}\binom{n+1}{c}\widehat{h}_c(Z)D^c. \tag{6.11}$$

Les multihauteurs des variétés toriques s'explicitent en termes de ce qu'on appelle des *intégrales mixtes* d'une famille de fonctions concaves [PS04, § IV.3], notion analogue au volume mixte. Soient $f:Q\to\mathbf{R}$ et $g:R\to\mathbf{R}$ des fonctions concaves définies sur des ensembles convexes $Q,R\subset\mathbf{R}^n$ respectivement. On pose

$$f\boxplus g:Q+R\to\mathbf{R},\qquad x\mapsto\max\{f(y)+g(z)\ :\ y\in Q,\ z\in R,\ y+z=x\},$$

qui est une fonction concave définie sur la somme de Minkowski $Q+R$; on obtient ainsi une structure de semi-groupe commutatif sur l'ensemble des fonctions concaves (définies sur des ensembles convexes). Pour une famille de fonctions concaves $f_0: Q_0\to\mathbf{R},\ldots,f_n:Q_n\to\mathbf{R}$ l'intégrale mixte (ou *multi-intégrale*) est définie *via* la formule

$$\mathrm{MI}(f_0,\ldots,f_n):=\sum_{j=0}^{n}(-1)^{n-j}\sum_{0\le i_0<\cdots<i_j\le n}\int_{Q_{i_0}+\cdots+Q_{i_j}}(f_{i_0}\boxplus\cdots\boxplus f_{i_j})(u)\,du_1\cdots du_n.$$

Comme pour le volume mixte, l'intégrale mixte est une fonctionnelle positive, symétrique et linéaire en chaque variable $f_i$, *voir* [PS04, § IV.3].

Soit $\mathcal{A}_0\in(\mathbf{Z}^n)^{N_0+1},\ldots,\mathcal{A}_m\in(\mathbf{Z}^n)^{N_m+1}$ tels que $L_{\mathcal{A}_0}+\cdots+L_{\mathcal{A}_m}=\mathbf{Z}^n$ et $\underline{\mathcal{A}}:=(\mathcal{A}_0,\ldots,\mathcal{A}_m)$. Soit $\alpha_0\in(K^\times)^{N_0+1},\ldots,\alpha_m\in(K^\times)^{N_m+1}$ et posons $\underline{\alpha}:=(\alpha_0,\ldots,\alpha_m)$. Considérons alors l'action monomiale $*_{\underline{\mathcal{A}}}$ de $\mathbf{G}_m^n$ sur le produit d'espaces projectifs $\mathbf{P}^{N_0}\times\cdots\times\mathbf{P}^{N_m}$ associée; on note $X_{\underline{\mathcal{A}},\underline{\alpha}}$ l'adhérence de Zariski de l'orbite du point $\underline{\alpha}\in\mathbf{P}^{N_0}\times\cdots\times\mathbf{P}^{N_m}$.

Pour chaque $v\in M_K$ on note aussi $\vartheta_{\mathcal{A}_i,\tau_{\alpha_i\,v}}:Q_{\mathcal{A}_i}\to\mathbf{R}$ la fonction paramétrant la toiture du polytope $Q_{\mathcal{A}_i,\tau_{\alpha_i\,v}}\subset\mathbf{R}^{n+1}$ associé au vecteur $\mathcal{A}_i$ et au poids $\tau_{\alpha_i\,v}$.

**Théorème 6.2.** [PS04, Thm. 0.3 et Rem. IV.7]. *Soit $c\in\mathbf{N}^{m+1}_{n+1}$, dans la situation ci-dessus on a*

$$\widehat{h}_c(X_{\underline{\mathcal{A}},\underline{\alpha}})=\sum_{v\in M_K}\frac{[K_v:\mathbf{Q}_v]}{[K:\mathbf{Q}]}\mathrm{MI}_c(\vartheta_{\underline{\mathcal{A}},\tau_{\underline{\alpha}\,v}})$$

*avec*

$$\mathrm{MI}_c(\vartheta_{\underline{\mathcal{A}},\tau_{\underline{\alpha}\,v}}):=\mathrm{MI}\Bigg(\underbrace{\vartheta_{\mathcal{A}_0,\tau_{\alpha_0\,v}},\ldots,\vartheta_{\mathcal{A}_0,\tau_{\alpha_0\,v}}}_{c_0\text{ fois}},\ldots,\underbrace{\vartheta_{\mathcal{A}_m,\tau_{\alpha_m\,v}},\ldots,\vartheta_{\mathcal{A}_m,\tau_{\alpha_m\,v}}}_{c_m\text{ fois}}\Bigg).$$

*Exemple 6.3.* Soient $\xi_1,\ldots,\xi_N\in K^\times$ et considérons l'application monomiale

$$\begin{array}{ccccc}\mathbf{G}_m & \xrightarrow{\varphi} & (\mathbf{P}^1)^N & \xrightarrow{\text{Segre}} & \mathbf{P}^{2^N-1}\\ s & \longmapsto & ((1:\xi_1 s),\ldots,(1:\xi_N s)) & \longmapsto & \Big((\prod_{j\in J}\xi_j)\cdot s^{\mathrm{Card}(J)}\ :\ J\subset\{1,\ldots,N\}\Big).\end{array}$$

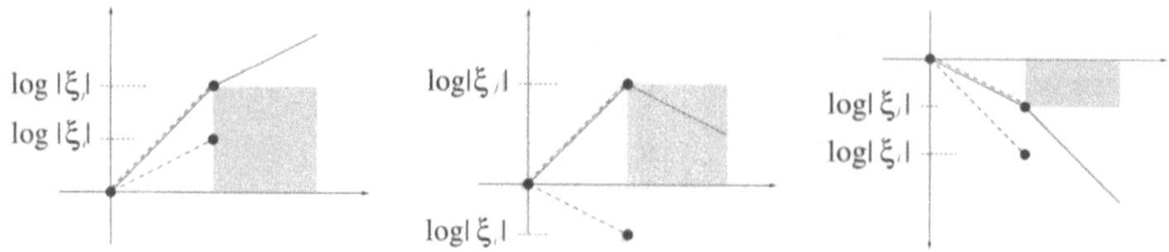

Fig. 6

Soit $X$ l'image de $\varphi$ dans $(\mathbf{P}^1)^N$ et pour $1 \leq i, j \leq N$ notons $c(i, j) \in \mathbf{N}^N$ le vecteur dont les coordonnées d'indices $i$ et $j$ valent 1 et les autres 0, on vérifie à l'aide du théorème 6.2 et de la définition des multi-intégrales

$$\begin{aligned}\widehat{h}_{c(i,j)}(X) &= \sum_{v \in M_K} \frac{[K_v : \mathbf{Q}_v]}{[K : \mathbf{Q}]} \mathrm{MI}_c(\vartheta_{\underline{\mathcal{A}}, \tau_{\underline{\alpha}\, v}}) \\ &= \sum_{v \in M_K} \frac{[K_v : \mathbf{Q}_v]}{[K : \mathbf{Q}]} \max(\log |\xi_i|_v ; \log |\xi_j|_v) \\ &= h(\xi_i : \xi_j).\end{aligned}$$

Dans la figure 6 les parties grisées montrent pour $N = 2$ et en supposant $\log |\xi_i| \leq \log |\xi_j|$, pour chaque cas le calcul de la multi-intégrale $\mathrm{MI}_c(\vartheta_{\underline{\mathcal{A}}, \tau_{\underline{\alpha}\, v}})$.

Par la formule (6.11), on en déduit que la hauteur de l'image de Segre $\circ\, \varphi$ dans $\mathbf{P}^{2^N - 1}$ est égale à

$$\widehat{h}(\mathrm{Segre}(X)) = 2 \sum_{1 \leq i < j \leq N} \widehat{h}_{c(i,j)}(X) = 2 \sum_{1 \leq i < j \leq N} h((\xi_i : \xi_j)) = \sum_{1 \leq i, j \leq N} h((\xi_i : \xi_j)).$$

Une conséquence du théorème 6.1 est que $\widehat{h}(X_{\mathcal{A},\alpha})$ est le logarithme d'un nombre algébrique. En effet, $e_{\tau_{\alpha\, v}}(X_{\mathcal{A}}) = 0$ pour toutes les places $v$ sauf un nombre fini et, comme le poids de Chow est combinaison à coefficients entiers des composantes du vecteur poids $\tau_{\alpha\, v}$, c'est à dire des $\log |\alpha|_v$, notre assertion est claire. Un tel logarithme de nombre algébrique étant nul ou transcendant, on peut énoncer:

**Proposition 6.4.** *Soit $X$ une variété torique qui n'est pas de torsion, alors $\widehat{h}(X) \notin \overline{\mathbf{Q}}$.*

Les hauteurs normalisées des variétés toriques qui ne sont pas de torsion, sont donc des nombres transcendants parmi les plus simples que l'on connaisse, à savoir les logarithmes de nombres algébriques. On sait qu'il devrait résulter des conjectures les plus générales sur les variétés et les motifs que les hauteurs des variétés projectives s'expriment rationnellement en termes de valeurs de fonctions $L$ et de leurs dérivées, ou encore en termes de polylogarithmes.

Dès qu'on considère des variétés projectives qui ne sont plus toriques on doit s'attendre à ce que leur hauteur normalisée fasse effectivement intervenir des polylogarithmes supérieurs. De même, si l'on s'intéresse à la hauteur *projective* d'une variété torique, une quantité supplémentaire s'ajoute aux intégrales des fonctions $\vartheta_{\mathcal{A}, \tau_{\alpha\, v}}$, qui s'écrit bien sur des exemples à l'aide de polylogarithmes mais est en général difficile à évaluer exactement (travail en cours).

# 7 Optimalité du théorème des minimums algébriques successifs

Dans ce paragraphe on étudie les minimums algébriques successifs des variétés toriques, définis dans l'introduction. Comme résultat principal, on démontre le théorème 1.4 énoncé dans l'introduction qui entraîne l'optimalité des estimations dans le théorème des minimums algébriques successifs: on construit des exemples montrant qu'à des $\varepsilon$-près, toute configuration possible des minimums successifs se réalise et que le quotient $\widehat{h}(X)/\deg(X)$ peut atteindre n'importe quel valeur dans l'intervalle autorisé par les inégalités (1.1).

Notons que l'analogue abélien du théorème 1.4 est faux: soit $A$ une variété abélienne définie sur $\overline{\mathbf{Q}}$ munie d'un fibré en droites $L$ ample et symétrique, permettant de définir une notion de hauteur normalisé $\widehat{h} = \widehat{h}_L$ pour les sous-variétés de $A$ [Phi91]. En particulier, on a une notion de minimums successifs et le théorème de Zhang y est toujours valable, *voir* [Zha95, Thm. 5.2].

Soit $\alpha + B \subset A$ le translaté d'une sous-variété abélienne $B$ par un point $\alpha$ et $\mathrm{Tors}(B)$ le sous-groupe des points de torsion de $B$. Soit $\beta$ un point quelconque dans $\alpha + B$, alors $\beta + \mathrm{Tors}(B)$ est un sous-ensemble de points de hauteur $\widehat{h}(\beta)$ dense dans $\alpha + B$ (puisque $\mathrm{Tors}(B)$ est Zariski dense dans $B$), donc $\widehat{\mu}^{\mathrm{ess}}(\alpha + B) \leq \widehat{h}(\beta)$. On en déduit $\widehat{\mu}^{\mathrm{ess}}(\alpha + B) = \widehat{\mu}^{\mathrm{abs}}(\alpha + B)$ et donc

$$\widehat{\mu}_i(\alpha + B) = \widehat{\mu}^{\mathrm{ess}}(\alpha + B) \quad \text{pour } i = 1, \dots, n + 1.$$

Ainsi, dans cette situation l'intervalle du théorème des minimums successifs se réduit à un point, et on a les égalités

$$\frac{\widehat{h}(\alpha + B)}{\deg(\alpha + B)} = (n + 1)\, \widehat{\mu}^{\mathrm{ess}}(\alpha + B) = \widehat{\mu}_1(\alpha + B) + \cdots + \widehat{\mu}_{n+1}(\alpha + B).$$

La situation est plus riche dans le cas torique, la différence tenant au fait que les translatés de sous-tores de $(\mathbf{P}^N)^\circ$ ne sont pas des ensembles fermés.

Soit $\mathcal{A} = (a_0, \dots, a_N) \in (\mathbf{Z}^n)^{N+1}$ et $\alpha = (\alpha_0, \dots, \alpha_N) \in (\overline{\mathbf{Q}}^\times)^{N+1}$. L'ouvert principal $X^\circ_{\mathcal{A},\alpha}$ est le translaté de sous-tore $\alpha \cdot X_{\mathcal{A}}$ et avec le même raisonnement que pour le cas abélien

$$\widehat{\mu}^{\mathrm{ess}}(X_{\mathcal{A},\alpha}) = \widehat{\mu}^{\mathrm{ess}}(X^\circ_{\mathcal{A},\alpha}) = \widehat{\mu}_i(X^\circ_{\mathcal{A},\alpha}) \quad \text{pour } i = 1, \dots, n + 1. \tag{7.12}$$

Cependant, les autres minimums successifs de $X_{\mathcal{A},\alpha}$ dépendent des orbites de dimension inférieure, et peuvent donc différer du minimum essentiel.

**Lemme 7.1.** *Avec les notations ci-dessus, pour* $i = 1, \dots, n + 1$

$$\widehat{\mu}_i(X_{\mathcal{A},\alpha}) = \min \left\{ \widehat{\mu}^{\mathrm{abs}}(X^\circ_{\mathcal{A},\alpha,P}) \ : \ P \in \mathrm{F}(Q_{\mathcal{A}}),\ \dim(P) = n - i + 1 \right\},$$

*où $P$ parcours l'ensemble des faces* $\mathrm{F}(Q_{\mathcal{A}})$ *du polytope $Q_{\mathcal{A}}$ de dimension* $n - i + 1$.

*Démonstration.* Considérons la décomposition en orbites: $X_{\mathcal{A},\alpha} = \bigcup_{P \in \mathrm{F}(Q_{\mathcal{A}})} X^\circ_{\mathcal{A},\alpha,P}$ (formule (2.3)). C'est un recouvrement de $X_{\mathcal{A},\alpha}$ et donc par [Som05, Lem. 2.2] on a

$$\begin{aligned} \widehat{\mu}_i(X_{\mathcal{A}}) &= \min \left\{ \widehat{\mu}_{\dim(P)-n+i} \left( X^\circ_{\mathcal{A},\alpha,P} \right) \ : \ \dim(P) \geq n - i + 1 \right\} \\ &= \min \left\{ \widehat{\mu}^{\mathrm{abs}} \left( X^\circ_{\mathcal{A},\alpha,P} \right) \ : \ \dim(P) = n - i + 1 \right\} \end{aligned}$$

car ce minimum est atteint sur les faces de dimension $n - i + 1$. □

Ainsi, le calcul des minimums successifs de $X_{\mathcal{A},\alpha}$ se réduit à celui du minimum essentiel (ou absolu) d'un sous-tore. Le lemme suivant donne le minimum essentiel pour certaines variétés toriques particulières.

**Lemme 7.2.** *Soit $\mathcal{A} = (a_0, \dots, a_N) \in (\mathbf{Z}^n)^{N+1}$ et $\alpha = (\alpha_0, \dots, \alpha_N) \in (K^\times)^{N+1}$, et supposons qu'il existe $a \in \mathrm{Supp}(\mathcal{A})$ tel que pour toute place $v \in M_K$ le maximum des $|\alpha_i|_v$ pour $i = 0, \dots, N$ est atteint au-dessus de $a$, autrement-dit*

$$\max\{|\alpha_i|_v \,:\, 0 \le i \le N\} = \max\{|\alpha_\ell|_v \,:\, 0 \le \ell \le N, a_\ell = a\}.$$

*Alors $\widehat{\mu}^{\mathrm{ess}}(X_{\mathcal{A},\alpha}) = \widehat{\mu}^{\mathrm{abs}}(X^\circ_{\mathcal{A},\alpha}) = \widehat{h}(\alpha)$.*

*Démonstration.* Notons $0 \le \ell_0, \dots, \ell_M \le N$ les indices pour lesquels $a_\ell = a$ et soit $\varpi : \mathbf{P}^N \to \mathbf{P}^M$ la projection $x \mapsto (x_{\ell_0} : \cdots : x_{\ell_M})$. Alors pour un point quelconque $\xi = (\alpha_0\, s^{a_0} : \cdots : \alpha_N\, s^{a_N}) \in X^\circ_{\mathcal{A},\alpha}$ on a $\varpi(\xi) = (\alpha_{\ell_0}\, s^a : \cdots : \alpha_{\ell_M}\, s^a) = (\alpha_{\ell_0} : \cdots : \alpha_{\ell_M})$ d'où

$$\widehat{h}(\xi) \ge \widehat{h}(\varpi(\xi)) = \widehat{h}(\alpha_{\ell_0} : \cdots : \alpha_{\ell_M})$$

et $\widehat{h}(\alpha_{\ell_0} : \cdots : \alpha_{\ell_M}) = \widehat{h}(\alpha)$ grâce à l'hypothèse du lemme, donc $\widehat{\mu}^{\mathrm{abs}}(X^\circ_{\mathcal{A},\alpha}) \ge \widehat{h}(\alpha)$. En outre $\widehat{h}(\alpha) \ge \widehat{\mu}^{\mathrm{ess}}(X_{\mathcal{A},\alpha})$ d'où la conclusion. □

*Démonstration du théorème 1.4.* On peut supposer sans perte de généralité $N = 3n + 1$, puisque le cas général se déduit de celui-ci par immersion de $\mathbf{P}^{3n+1}$ comme un sous-espace standard de $\mathbf{P}^N$.

Soient $d$ un nombre premier, $1 \le k \le n$, $1 \le f \le d - 1$ des paramètres entiers à fixer ultérieurement et encore $q_0 \ge \cdots \ge q_n \ge 0$ des paramètres rationnels. Rappelons que $e_1, \dots, e_n$ désigne la base standard de $\mathbf{Z}^n$ et $S$ le simplexe standard $\mathrm{Conv}(0, e_1, \dots, e_n) \subset \mathbf{R}^n$, on pose

$$a_i := d\, e_i \qquad \text{pour } i = 1, \dots, n$$

$$b_i := \begin{cases} (d-1)\, e_i & \text{pour } 1 \le i \le k-1 \\ f\, e_i & \text{pour } i = k \\ e_i & \text{pour } k+1 \le i \le n \end{cases}$$

puis

$$\mathcal{A} := (0, a_1 \dots, a_n, 0, a_1, \dots, a_n, b_1, \dots, b_n) \in (\mathbf{Z}^n)^{3n+2},$$

$$\alpha := (\underbrace{1, \dots\dots\dots\dots, 1}_{n+1 \text{ fois}}, 2^{q_0}, 2^{q_1}, \dots\dots, 2^{q_n}, \underbrace{2^{q_0}, \dots, 2^{q_0}}_{k \text{ fois}}, \underbrace{1, \dots, 1}_{n-k \text{ fois}}) \in (\overline{\mathbf{Q}}^\times)^{3n+2}$$

et on considère $X \subset \mathbf{P}^{3n+1}$ la variété torique associée au couple $(\mathcal{A}, \alpha)$ ainsi défini. On a $L_{\mathcal{A}} = \mathbf{Z}^n$ et $Q_{\mathcal{A}} = d\, S$, donc la dimension et le degré de cette variété sont égaux à $n$ et $n!\,\mathrm{Vol}_n(Q_{\mathcal{A}}) = d^n$ respectivement, et si $\ell$ est un dénominateur commun de $q_0, \dots, q_n$, cette variété est définie sur l'extension kummerienne $K := \mathbf{Q}(2^{1/\ell})$. De plus, il résulte du lemme 7.1 et du lemme 7.2 appliqué à toutes les faces de $Q_{\mathcal{A}}$

$$\widehat{\mu}_i(X) = q_{i-1} \log(2), \quad i = 1, \dots, n+1, \tag{7.13}$$

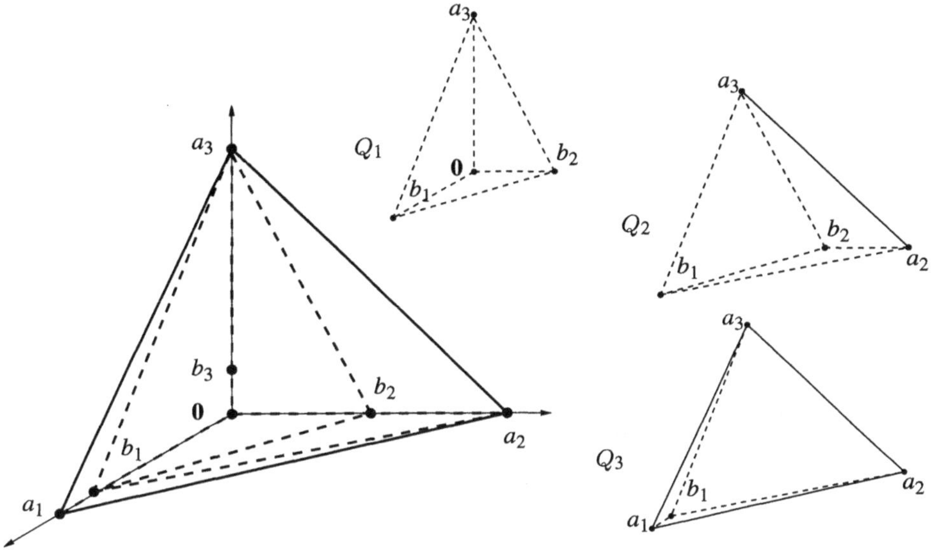


Fig. 7

le $i$-ème minimum se réalisant sur la face $\mathrm{Conv}(a_{i-1}, \ldots, a_n)$. En outre, on estime la hauteur en utilisant la formule dans le théorème 6.1. On décompose le polytope de base en

$$Q_{\mathcal{A}} = Q_1 \cup Q_2 \cup Q_3$$

avec $Q_1 := \mathrm{Conv}(0, b_1, \ldots, b_k, a_{k+1}, \ldots, a_n)$, $Q_2 := \mathrm{Conv}(b_1, \ldots, b_k, a_k, \ldots, a_n)$ et $Q_3 := Q_{\mathcal{A}} \setminus (Q_1 \cup Q_2)$. Pour $n = 3$ on a une figure du type montré dans la figure 7.

Pour toute place finie $v$ de $K$ on vérifie $\vartheta_{\mathcal{A},\tau_{\alpha v}} \equiv 0$; la formule pour $\widehat{h}(X)$ se réduit alors aux contributions des places archimédiennes. Pour $v \in M_K^\infty$, la décomposition du domaine considérée sépare l'intégrale en trois morceaux $I_i := \int_{Q_i} \vartheta_{\mathcal{A},\tau_{\alpha v}}(u)du$ ($i = 1, 2, 3$). On vérifie d'abord que la fonction $\vartheta_{\mathcal{A},\tau_{\alpha v}}$ est *linéaire* sur chacun des simplexes $Q_1$ et $Q_2$. L'intégrale d'une fonction linéaire sur un simplexe étant égale au volume du simplexe multiplié par la valeur moyenne de la fonction, on trouve

$$\begin{aligned} I_1 &= \mathrm{Vol}_n(Q_1)\,\frac{((k+1)\,q_0 + q_{k+1} + \cdots + q_n)\,\log(2)}{n+1} \\ &= \frac{f\,(d-1)^{k-1}\,d^{n-k}}{(n+1)!}\,((k+1)\,q_0 + q_{k+1} + \cdots + q_n)\,\log(2), \\ I_2 &= \mathrm{Vol}_n(Q_2)\,\frac{(k\,q_0 + q_k + \cdots + q_n)\,\log(2)}{n+1} \\ &= \frac{(d-f)\,(d-1)^{k-1}\,d^{n-k}}{(n+1)!}\,(k\,q_0 + q_k + \cdots + q_n)\,\log(2). \end{aligned}$$

En outre on estime brutalement la troisième intégrale

$$0 \le I_3 \le \mathrm{Vol}_n(Q_3) \cdot \max(\vartheta_{\mathcal{A},\tau_{\alpha v}}) = \frac{d^n - (d-1)^{k-1}\,d^{n-k+1}}{n!}\,q_0\,\log(2),$$

puisque $\mathrm{Vol}_n(Q_3) = \mathrm{Vol}_n(Q_{\mathcal{A}}) - \mathrm{Vol}_n(Q_1 \cup Q_2) = d^n - (d-1)^{k-1} d^{n-k+1}$. Il s'ensuit

$$(n+1)! \int_{Q_{\mathcal{A}}} \vartheta_{\mathcal{A},\tau_{\alpha v}}(u)\mathrm{d}u = (n+1)!\,(I_1 + I_2 + I_3) = d^n\,(\theta + E(\mathcal{A},\alpha,v))$$

avec

$$\begin{aligned}\theta &= \left(1-\frac{1}{d}\right)^{k-1} \\ &\quad \times \left(\frac{d-f}{d}\,(k\,q_0 + q_k + \cdots + q_n) + \frac{f}{d}\,((k+1)\,q_0 + q_{k+1} + \cdots + q_n)\right)\log(2) \\ &= \left(1-\frac{1}{d}\right)^{k-1}\left((kq_0 + q_k + \cdots + q_n) + \frac{f}{d}(q_0 - q_k)\right)\log(2)\end{aligned}$$

et

$$\begin{aligned}0 \le E(\mathcal{A},\alpha,v) &= \frac{(n+1)!}{d^n}\,I_3 \le (n+1)\left(1-\left(1-\frac{1}{d}\right)^{k-1}\right) q_0 \log(2) \\ &\le \frac{(n+1)\,(k-1)}{d}\,q_0 \log(2).\end{aligned}$$

On déduit alors du théorème 6.1

$$0 \le \theta - \frac{\widehat{h}(X)}{\deg(X)} \le \frac{(n+1)\,(k-1)}{d}\,q_0 \log(2). \tag{7.14}$$

Maintenant on fixe les paramètres : d'abord on prend $\ell := \lceil \log(2)\,\varepsilon_1^{-1} \rceil + 1$, et pour $0 \le i \le n$ on pose $q_i := \frac{1}{\ell}\left\lfloor \frac{\ell \cdot \mu_{i+1}}{\log(2)} \right\rfloor$ de sorte que $\widehat{\mu}_{i+1}(X) = q_i \log(2)$ satisfait

$$0 \le \mu_{i+1} - q_i \log(2) < \varepsilon_1 \tag{7.15}$$

comme voulu. On vérifie

$$(q_0 + \cdots + q_n)\log(2) \le \mu_1 + \cdots + \mu_{n+1} \le \nu \le (n+1)\mu_1 - \varepsilon_1 \le (n+1)q_0 \log(2)$$

et on fixe $1 \le k \le n$ tel que

$$(k\,q_0 + q_k + \cdots + q_n)\log(2) \le \nu \le ((k+1)\,q_0 + q_{k+1} + \cdots + q_n)\log(2).$$

Soit $\lambda \in [0,1]$ tel que

$$\nu = (1-\lambda)\,(k\,q_0 + q_k + \cdots + q_n)\log(2) + \lambda\,((k+1)\,q_0 + q_{k+1} + \cdots + q_n)\log(2).$$

On prend alors $1 \le f \le d-1$ tel que

$$|\lambda - f/d| \le 1/d \tag{7.16}$$

et on considère le $\theta = \theta(d)$ associé à ces paramètres. On vérifie facilement

$$\begin{aligned}|\nu - \theta| &\leq \left|\lambda - \frac{f}{d}\right| (q_0 - q_k)\log(2) \\ &\quad + \left(1 - \left(1 - \frac{1}{d}\right)^{k-1}\right) \cdot ((k+1)q_0 + q_{k+1} + \cdots + q_n)\log(2) \\ &\leq \frac{\mu_1}{d} + \frac{(k-1)(n+1)\mu_1}{d} = \frac{n^2\mu_1}{d}.\end{aligned}$$

Finalement on obtient le résultat cherché en sommant avec l'inégalité (7.14)

$$\left|\frac{\widehat{h}(X)}{\deg(X)} - \nu\right| \leq \frac{(n+1)(k-1)}{d}\,\mu_1 + \frac{(n+1)(k-1)+1}{d}\,\mu_1 \leq \frac{2n^2}{d}\,\mu_1.$$

En prenant, grâce au postulat de Bertrand, $d$ un premier entre $2n^2\,\mu_1\,\varepsilon_2^{-1}$ et $4n^2\mu_1\varepsilon_2^{-1}$ on arrive au résultat annoncé. □

Ce résultat montre que déjà dans le cadre torique, toute configuration possible des minimums $\widehat{\mu}_1(X), \ldots, \widehat{\mu}_{n+1}(X)$ se réalise, et l'encadrement (1.1) est optimal en toute dimension, lorsque le degré de $X$ et celui du corps de définition sont assez grands. Toutefois, notre exemple présente une codimension minimale $N - n = 2n + 1$ de l'ordre de la dimension de la variété produite. La question se pose donc de savoir ce qu'il en est pour les variétés de petite codimension. Dans le cas de codimension 1 on a le résultat suivant qui laisse ouverte la possibilité de raffinements de (1.1) en petit codimension.

**Proposition 7.3.** *Soit $X \subset \mathbf{P}^N$ une hypersurface* torique *d'équation homogène minimale $f_X = x^b - \lambda\, x^c \in \overline{\mathbf{Q}}[x_0, \ldots, x_N]$, alors*

$$\widehat{\mu}^{\mathrm{ess}}(X) = \frac{h(1:\lambda)}{\deg(f_X)} = \frac{\widehat{h}(X)}{\deg(X)}, \quad \widehat{\mu}_2(X) = \cdots = \widehat{\mu}_N(X) = 0.$$

*Démonstration.* Les contributions des places archimédiennes à la hauteur normalisée de $X$ sont égales aux mesures de Mahler des conjuguées de $f_X$ et donc $\widehat{h}(X) = h(1 : \lambda)$, *voir* [PS04], suite de l'exemple 3.7. Par la proposition 3.1 on a $\lambda = \alpha^{b-c}$ pour tout point $\alpha \in X^\circ$. En choisissant $\alpha$ de hauteur normalisée aussi proche du minimum essentiel $\widehat{\mu}^{\mathrm{ess}}(X)$ que l'on veut, on a

$$\frac{\widehat{h}(X)}{\deg(X)} = \frac{h(1:\alpha^{b-c})}{\deg(f_X)} = \frac{h(\alpha^b:\alpha^c)}{\deg(f_X)} \leq \widehat{h}(\alpha) \leq \widehat{\mu}^{\mathrm{ess}}(X) + \varepsilon$$

pour tout $\varepsilon > 0$. Avec la formule (1.1) on obtient $\widehat{\mu}_1(X) + \cdots + \widehat{\mu}_{n+1}(X) \leq \widehat{h}(X)/\deg(X) \leq \widehat{\mu}_1(X)$, ce qui entraîne l'énoncé. □

Il serait très intéressant d'expliciter les minimums successifs d'une variété torique quelconque, en généralisant à la fois le lemme 7.2 et la proposition 7.3. Grâce au lemme 7.1 on sait qu'il suffit de le faire pour le minimum essentiel.

Une question liée est celle de construire explicitement des points de $X^\circ_{\mathcal{A},\alpha}$ de hauteur comparable au minimum essentiel. De plus, on peut se demander s'il existe un point dans la variété réalisant ce minimum essentiel, c'est-à-dire de savoir s'il existe un point de hauteur minimale. Dans le même ordre d'idée, existe-t-il un point de $X_{\mathcal{A},\alpha}$ dont la hauteur normalisée soit égale au quotient $\widehat{h}(X_{\mathcal{A},\alpha})/\deg(X_{\mathcal{A},\alpha})$?

Pour conclure ce paragraphe, explicitons l'encadrement pour le quotient hauteur-sur-degré qui découle de la formule pour la hauteur d'une variété torique:

**Proposition 7.4.** *Soient* $\mathcal{A} \in (\mathbf{Z}^n)^{N+1}$ *et* $\alpha \in (\overline{\mathbf{Q}}^\times)^{N+1}$, *alors*

$$\widehat{h}(\alpha) - n\,\widehat{h}\left(\alpha_j^{-1} \,:\, a_j \in \mathrm{F}_0(Q_{\mathcal{A}})\right) \le \frac{\widehat{h}(X_{\mathcal{A},\alpha})}{\deg(X_{\mathcal{A},\alpha})} \le (n+1)\,\widehat{h}(\alpha),$$

*où* $a_j$ *parcours l'ensemble* $\mathrm{F}_0(Q_{\mathcal{A}})$ *des sommets de* $Q_{\mathcal{A}}$.

*Démonstration.* Soit $K$ le corps de définition de $\alpha$ et $v \in M_K$. On a $\max(\vartheta_{\mathcal{A},\tau_{\alpha\,v}}) = \log\max\{|\alpha_0|_v, \ldots, |\alpha_N|_v\} =: \log(||\alpha||_v)$, ce qui entraîne

$$\int_{Q_{\mathcal{A}}} \vartheta_{\mathcal{A},\tau_{\alpha\,v}}(u)\,\mathrm{d}u \le \mathrm{Vol}_n(Q_{\mathcal{A}})\ \log(\|\alpha\|_v) = \frac{1}{n!}\ \deg(X_{\mathcal{A},\alpha})\ \log(\|\alpha\|_v)$$

et donc $\widehat{h}(X_{\mathcal{A},\alpha}) \le (n+1)\deg(X_{\mathcal{A},\alpha})\,\widehat{h}(\alpha)$, ce qui établit la majoration. Pour la minoration, soit $v \in M_K$ et posons

$$m_v := \min_{u\in Q_{\mathcal{A}}} (\vartheta_{\mathcal{A},\tau_{\alpha\,v}}(u)) = \min\{\max\{\log|\alpha_i|_v \,:\, 0 \le i \le N, a_i = a\} \,:\, a \in \mathrm{F}_0(Q_{\mathcal{A}})\}$$

et considérons le polytope $Q_v := \mathrm{Conv}\,((a_i, \log|\alpha_i|_v), (a_i, m_v) \,:\, i = 0, \ldots, N) \subset \mathbf{R}^{N+1}$. Soit $0 \le \ell \le N$ tel que $\log|\alpha_\ell|_v$ soit maximal, alors $Q_v \supset \mathrm{Conv}\,((a_\ell, \log|\alpha_\ell|_v), Q_{\mathcal{A}} \times \{m_v\})$ et donc $\mathrm{Vol}_{n+1}(Q_v) \ge \frac{1}{n+1}\,(\log\|\alpha\|_v - m_v)\ \mathrm{Vol}_n(Q_{\mathcal{A}})$. On en déduit

$$\begin{aligned}\int_{Q_{\mathcal{A}}} \vartheta_{\mathcal{A},\tau_{\alpha\,v}}(u)\,\mathrm{d}u &= \mathrm{Vol}_{n+1}(Q_v) + m_v\,\mathrm{Vol}_n(Q_{\mathcal{A}})\\ &\ge \left(\frac{1}{n+1}\ \log(\|\alpha\|_v) + \frac{n}{n+1}\,m_v\right)\ \mathrm{Vol}_n(Q_{\mathcal{A}}),\end{aligned}$$

d'où $\widehat{h}(X_{\mathcal{A},\alpha}) \ge \deg(X_{\mathcal{A},\alpha})\ \left(\widehat{h}(\alpha) - n\,\widehat{h}(\alpha_j^{-1} \,:\, a_j \in \mathrm{F}_0(Q_{\mathcal{A}}))\right)$ car $\sum_{v\in M_K} \frac{[K_v:\mathbf{Q}_v]}{[K:\mathbf{Q}]} \cdot (-m_v) \le \widehat{h}(\alpha_j^{-1} \,:\, a_j \in \mathrm{F}_0(Q_{\mathcal{A}}))$. □

Notons qu'en remplaçant le point $\alpha$ par un point de $X^\circ_{\mathcal{A},\alpha}$ de hauteur aussi proche que l'on veut du minimum essentiel, on retrouve simplement la majoration de (1.1); par contre on obtient une minoration différente.

# 8 Poids de Chow et hauteur des diviseurs monomiaux

Dans ce paragraphe on considère l'intersection d'une variété torique avec un diviseur monomial de $\mathbf{P}^N$. On montrera comment dans cette situation, le théorème de Bézout pour les poids de Chow (théorèmes 1.2 et 5.2) s'explicite comme la décomposition polyédrale d'un certain volume, et dans cette situation peut se démontrer de façon indépendante. En combinant ceci avec la formule pour la hauteur d'une variété torique (théorème 6.1), on obtient un théorème de Bézout arithmétique pour la hauteur normalisée du cycle intersection d'une variété torique avec un diviseur monomial.

Soient $\mathcal{A} \in (\mathbf{Z}^n)^{N+1}$ tel que $L_{\mathcal{A}} = \mathbf{Z}^n$ et $b \in \mathbf{Z}^{N+1}$, notons $X_{\mathcal{A}} \subset \mathbf{P}^N$ et $\mathrm{div}(x^b) \in \mathrm{Div}(\mathbf{P}^N)$ la variété torique et le diviseur monomial associés. On s'intéressera au cycle intersection découpé sur $X_{\mathcal{A}}$ par le monôme $x^b$; le lemme ci-dessous explicite ce cycle.

Pour chaque hyperface $F \in F_{n-1}(Q_{\mathcal{A}})$ on considère la variété $X_{\mathcal{A},F} \subset \mathbf{P}^N$, adhérence de Zariski de l'orbite associée $X^\circ_{\mathcal{A},F}$. On considère aussi l'hyperplan d'appui $H_F \subset \mathbf{R}^n$ de cette face et $H_F^{\mathbf{Z}} := H_F \cap \mathbf{Z}^n$; fixons également un point quelconque $a_F \in F$. Notons $L_{\mathcal{A},F}$ le $\mathbf{Z}$-module engendré par les différences des éléments de $\mathcal{A} \cap F$, qui est un sous-réseau de $H_F^{\mathbf{Z}} - a_F$ d'indice

$$i(\mathcal{A}; F) := [H_F^{\mathbf{Z}} - a_F : L_{\mathcal{A},F}].$$

Rappelons que $v_F \in \mathbf{Z}^n$ désigne le plus petit vecteur entier, orthogonal à $H_F$ et dirigé vers l'intérieur de $Q_{\mathcal{A}}$. On pose $M_{\mathcal{A}}(b) = b_0 a_0 + \cdots + b_N a_N \in \mathbf{Z}^n$ et $D := \deg(x^b) = \sum_{j=0}^N b_j$.

**Lemme 8.1.** *Avec les notations ci-dessus, on a*

$$X_{\mathcal{A}} \cdot \mathrm{div}(x^b) = \sum_{F \in \mathrm{F}_{n-1}(Q_{\mathcal{A}})} \langle M_{\mathcal{A}}(b) - D\, a_F, v_F \rangle\, i(\mathcal{A}; F)\, [X_{\mathcal{A},F}] \in Z_{n-1}(\mathbf{P}^N)$$

*où* $\langle \cdot, \cdot \rangle$ *désigne le produit scalaire ordinaire de* $\mathbf{R}^n$ *et* $[X_{\mathcal{A},F}]$ *le cycle défini par* $X_{\mathcal{A},F}$.

En particulier, un cycle $[X_{\mathcal{A},F}]$ intervient comme composante de $X_{\mathcal{A}} \cdot \mathrm{div}(x^b)$ si et seulement si $M_{\mathcal{A}}(b)$ n'appartient pas à l'hyperplan $H_F + (D-1)a_F$.

*Démonstration.* Grâce à la décomposition en orbites (2.3) on vérifie que le cycle considéré est supporté par la réunion des orbites de codimension 1, celles-ci correspondant aux hyperfaces de $Q_{\mathcal{A}}$, soit

$$\bigcup_{F \in \mathrm{F}_{n-1}(Q_{\mathcal{A}})} X_{\mathcal{A},F}.$$

Explicitons les multiplicités correspondantes. Par linéarité on peut se ramener sans perte de généralité au cas où $b = e_i$ est un des vecteurs de la base standard de $\mathbf{R}^{N+1}$ pour un certain $0 \le i \le N$, c'est-à-dire $x^b = x_i$. Fixons une hyperface $F$ et soit $a_j \in F$ un sommet quelconque. Considérons le cône dual de l'angle de $Q_{\mathcal{A}}$ en $a_j$:

$$\sigma := \left\{ u \in \mathbf{R}^n \,:\, \langle u, a_k - a_j \rangle \ge 0 \text{ pour } k = 0, \ldots, N \right\} \subset \mathbf{R}^n,$$

et soit $U_\sigma$ la variété torique affine correspondante, définie par

$$U_\sigma := \mathrm{Spec}(\overline{\mathbf{Q}}[S_\sigma])$$

où $S_\sigma := \sigma^\vee \cap \mathbf{Z}^n$ est le semi-groupe des points entiers dans l'angle de $Q_{\mathcal{A}}$ en $a_j$ [Ful93, § 1.3]. Considérons encore l'application naturelle $N : U_\sigma \to (X_{\mathcal{A}})_{x_j} \subset (\mathbf{P}^N)_{x_j}$ donnée par l'inclusion d'algèbres

$$\overline{\mathbf{Q}}[(X_{\mathcal{A}})_{x_j}] = \overline{\mathbf{Q}}[s^{a_0 - a_j}, \ldots, s^{a_N - a_j}] \hookrightarrow \overline{\mathbf{Q}}[S_\sigma] = \overline{\mathbf{Q}}[U_\sigma].$$

La variété $U_\sigma$ est *normale* car $S_\sigma$ est un semi-groupe saturé, et de ce fait $N$ est le morphisme de normalisation de la carte affine $(X_{\mathcal{A}})_{x_j}$ [Stu96, Cor. 13.6].

Soit $\rho$ l'arête du cône $\sigma$ duale de la face $F$, notons $V(\rho)$ la clôture dans $U_\sigma$ de l'orbite correspondante [Ful93, § 3.1]. On a un diagramme commutatif

$$\begin{array}{ccc} \overline{\mathbf{Q}}[(X_{\mathcal{A}})_{x_j}] & \overset{N_*}{\hookrightarrow} & \overline{\mathbf{Q}}[U_\sigma] \\ \downarrow & & \downarrow \\ \overline{\mathbf{Q}}[(X_{\mathcal{A},F})_{x_j}] & \hookrightarrow & \overline{\mathbf{Q}}[V(\rho)] \end{array}$$

qui implique $N^{-1}(X_{\mathcal{A},F}) = V(\rho)$ et $\deg(N|_{V(\rho)}) = [H_F^{\mathbf{Z}} : L_{\mathcal{A},F}] = i(\mathcal{A}; F)$. Le monôme $\chi := N^*(x_i) = s^{a_i - a_j} \in \overline{\mathbf{Q}}[s_1^{\pm 1}, \dots, s_n^{\pm 1}]$ définit une fonction rationnelle $U_\sigma \dashrightarrow \overline{\mathbf{Q}}$. On a $\mathrm{ord}_{X_{\mathcal{A},F}}(x_i) = \deg(N|_{V(\rho)}) \, \mathrm{ord}_{V(\rho)}(\chi)$ puisque $U_\sigma$ est normal [Ful84, Exerc. 1.2.3.]. On en déduit

$$\begin{aligned} \mathrm{long}_{\overline{\mathbf{Q}}[(X_{\mathcal{A},F})_{x_j}]} \left(\overline{\mathbf{Q}}[(X_{\mathcal{A}})_{x_j}]/(x_i)\right) &= \mathrm{ord}_{X_{\mathcal{A},F}}(x_i) \\ &= \deg(N|_{V(\rho)}) \, \mathrm{ord}_{V(\rho)}(\chi) = i(\mathcal{A}; F) \, \mathrm{ord}_{V(\rho)}(\chi). \end{aligned}$$

Finalement, le lemme de [Ful93, § 3.3, p. 61] entraîne $\mathrm{ord}_{V(\rho)}(\chi) = \langle a_i - a_j, v_F \rangle$ car $v_F$ est le générateur du semi-groupe $\rho \cap \mathbf{Z}^n \cong \mathbf{N}$, d'où

$$m(X_{\mathcal{A}} \cdot \mathrm{div}(x_i); X_{\mathcal{A},F}) = \mathrm{long}_{\overline{\mathbf{Q}}[(X_{\mathcal{A},F})_{x_j}]} \left(\overline{\mathbf{Q}}[(X_{\mathcal{A}})_{x_j}]/(x_i)\right) = \langle a_i - a_j, v_F \rangle \, i(\mathcal{A}; F).$$

□

Le théorème de Bézout géométrique

$$\deg\left(X_{\mathcal{A}} \cdot \mathrm{div}(x^b)\right) = D \deg(X_{\mathcal{A}}) \tag{8.17}$$

s'interprête en termes de décomposition du volume du polytope $Q_{\mathcal{A}}$ : on a $\deg(X_{\mathcal{A}}) = n! \, \mathrm{Vol}_n(Q_{\mathcal{A}})$ et

$$\begin{aligned} \deg(X_{\mathcal{A},F}) &= (n-1)! \mu_{\mathcal{A}(F)}(F) \\ &= \frac{(n-1)!}{\mathrm{Vol}_{n-1}((H_F - a_F)/L_{\mathcal{A},F})} \mathrm{Vol}_{n-1}(F) = \frac{(n-1)!}{i(\mathcal{A}; F) \, ||v_F||_2} \mathrm{Vol}_{n-1}(F), \end{aligned}$$

à cause de la normalisation de la forme volume $\mu_{\mathcal{A}(F)}$ sur $H_F$ et du fait que

$$\mathrm{Vol}_{n-1}(H_F/L_{\mathcal{A},F}) = i(\mathcal{A}; F) \cdot \mathrm{Vol}_{n-1}(H_F/H_F^{\mathbf{Z}}) = i(\mathcal{A}; F) \cdot ||v_F||_2,$$

conséquence de la formule de Brill et Gordan. On a encore

$$\langle M_{\mathcal{A}}(b) - D a_F, v_F \rangle = \varepsilon(b, F) \, \|v_F\|_2 \, \mathrm{dist}(M_{\mathcal{A}}(b), H_F + (D-1) a_F),$$

où $\varepsilon(b, F) = +1$ si $M_{\mathcal{A}}(b)$ et $v_F$ sont d'un même côté de l'hyperplan $H_F + (D-1)a_F$ et $\varepsilon(b, F) = -1$ sinon. Combiné avec le lemme 8.1 cela donne

$$\begin{aligned} \deg\left(X_{\mathcal{A}} \cdot \mathrm{div}(x^b)\right) &= \sum_{F \in \mathrm{F}_{n-1}(Q_{\mathcal{A}})} \langle M_{\mathcal{A}}(b) - D a_F, v_F \rangle \, i(\mathcal{A}, F) \deg(X_{\mathcal{A},F}) \\ &= \sum_{F \in \mathrm{F}_{n-1}(Q_{\mathcal{A}})} \varepsilon(b, F) \, (n-1)! \, \mathrm{Vol}_{n-1}(F) \\ &\qquad \times \mathrm{dist}(M_{\mathcal{A}}(b), H_F + (D-1) a_F) \\ &= \sum_{F \in \mathrm{F}_{n-1}(Q_{\mathcal{A}})} n! \, \varepsilon(b, F) \, \mathrm{Vol}_n(\mathrm{Conv}(F + (D-1) a_F, M_{\mathcal{A}}(b))). \end{aligned}$$

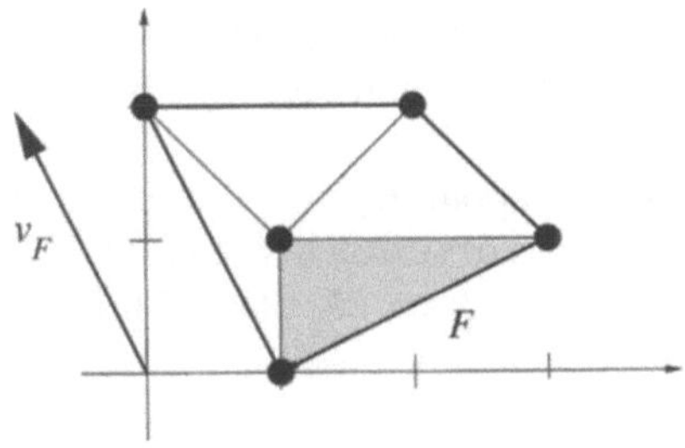


Fig. 8

L'identité (8.17) ci-dessus (multipliée par $n!^{-1}$) se traduit ainsi en la décomposition de volume

$$\sum_{F\in \mathrm{F}_{n-1}(Q_{\mathcal{A}})} \varepsilon(b,F)\,\mathrm{Vol}_n(\mathrm{Conv}(F+(D-1)a_F, M_{\mathcal{A}}(b))) = D\,\mathrm{Vol}_n(Q_{\mathcal{A}}). \quad (8.18)$$

La figure 8 illustre cette décomposition, pour $\mathcal{A} = ((1,1),(1,0),(3,1),(2,2),(0,2)) \in (\mathbf{Z}^2)^6$ et $b = (1,0,0,0,0)$ (donc $M_{\mathcal{A}}(b) = (1,1) = a_0$ et $D = 1$) :

Soit $\tau = (\tau_0,\ldots,\tau_N) \in \mathbf{Z}^{N+1}$. Pour le cas $\tau \in (\mathbf{N}^\times)^{N+1}$, le théorème de Bézout pour les poids de Chow (Théorème 5.2) s'écrit dans les notations du § 5

$$e_\tau(X_{\mathcal{A}}\cdot \mathrm{div}(x^b)) = D\,e_\tau(X_{\mathcal{A}}) - \sum_{Y\in \mathrm{Irr}(\mathrm{init}_\tau(X_{\mathcal{A}}))} m((X_{\mathcal{A}})_\tau\cdot \mathrm{div}(\lambda_\tau^*(x^b));\iota(Y))\cdot \deg(Y). \quad (8.19)$$

On va expliciter les termes intervenant dans cet enoncé. Soit

$$Q_{\mathcal{A},\tau} = \mathrm{Conv}((a_0,\tau_0),\ldots,(a_N,\tau_N)) \subset \mathbf{R}^{n+1}$$

le polytope associé au couple $(\mathcal{A},\tau)$, dont la *toiture* $E_{\mathcal{A},\tau}$ (c'est-à-dire l'enveloppe supérieure) s'envoie bijectivement sur $Q_{\mathcal{A}}$ par la projection $\mathbf{R}^n \times \mathbf{R} \to \mathbf{R}^n$. On appelle *pan* de la toiture toute face de dimension $n$ de $E_{\mathcal{A},\tau}$. De même, on appelle *mur* tout polytope de la forme $\mathrm{Conv}(Q_{\mathcal{A}(F)}, F\times\{0\})$ pour une hyperface $F$ de $Q_{\mathcal{A}}$; dans le cas $\tau \geq 0$ ceci est un des murs de la *maison* $\mathcal{M}_{\mathcal{A},\tau} := \mathrm{Conv}(Q_{\mathcal{A},\tau} \cup (Q_{\mathcal{A}}\times\{0\}))$, c'est-à-dire une face de dimension $n$ se projetant sur une face de dimension $n-1$ de $Q_{\mathcal{A}}$.

La déformation torique $(X_{\mathcal{A}})_\tau$ est l'adhérence de Zariski de l'application

$$\mathbf{G}_m \times \mathbf{G}_m^n \to \mathbf{P}^1 \times \mathbf{P}^N, \quad (t,s) \to \big((1:t),(t^{\tau_0}s^{a_0}:\cdots:t^{\tau_N}s^{a_N})\big).$$

C'est donc la variété torique bi-projective associée aux vecteurs $((0,0),(0,1)) \in (\mathbf{Z}\times\mathbf{Z})^2$ et $(\tau,\mathcal{A}) = ((\tau_0,a_0),\ldots,(\tau_N,a_N)) \in (\mathbf{Z}\times\mathbf{Z}^n)^{N+1}$, *voir* § 2. Le couple de polytopes associé est donc $\mathbf{0}\times[0,1]$ et $Q_{\mathcal{A},\tau}$; comme conséquence de la décomposition en orbites décrite au § 2, on vérifie que les points de $(X_{\mathcal{A}})_\tau$ contenus dans l'hyperplan $\{(0:1)\}\times\mathbf{P}^N$ correspondent nécessairement aux couples de faces de la forme $(\{(\mathbf{0},1)\},F)$ avec $F \in \mathrm{F}_n(E_{\mathcal{A},\tau})$. On voit ainsi qu'il y a bijection entre les pans de $E_{\mathcal{A},\tau}$ et les orbites de $X_{\mathcal{A},\tau}$ contenues dans l'hyperplan $\{(0:1)\}\times\mathbf{P}^N$, en particulier le support de $X_{\mathcal{A},\tau}\cdot(\{(0:1)\}\times\mathbf{P}^N)$ est contenu dans

$$\bigcup_{P\in \mathrm{F}_n(E_{\mathcal{A},\tau})} X_{\mathcal{A},\tau,P}.$$

L'identification $\iota : \mathbf{P}^N \to \{(0:1)\} \times \mathbf{P}^N \subset \mathbf{P}^1 \times \mathbf{P}^N$ met en correspondance le cycle $X_{\mathcal{A},\tau} \cdot (\{(0:1)\} \times \mathbf{P}^N)$ et la variété initiale $\mathrm{init}_\tau(X_\mathcal{A})$, on en déduit qu'il y a bijection entre les composantes de $\mathrm{init}_\tau(X_\mathcal{A})$ et les pans de $E_{\mathcal{A},\tau}$.

Pour chaque pan $P$ on considère son hyperplan d'appui $H_P \subset \mathbf{R}^{n+1}$, posons $L_{\mathcal{A},\tau,P} \subset \mathbf{Z}^{n+1}$ le $\mathbf{Z}$-module engendré par les différences des éléments de $(\tau, \mathcal{A}) \cap P$. Modulo une translation, ce dernier est un sous-réseau de $H_P^{\mathbf{Z}} := H_P \cap \mathbf{Z}^{n+1}$ d'indice

$$i(\mathcal{A}, \tau; P) := [H_P^{\mathbf{Z}} : L_{\mathcal{A},\tau,P}].$$

Avec ces notations, d'après [Stu94, formule (27), page 222] (*voir* aussi [KSZ92, Thm. 5.3.]), on a

$$X_{\mathcal{A},\tau} \cdot (\{(0:1)\} \times \mathbf{P}^N) = \iota(\mathrm{init}_\tau(X_\mathcal{A})) = \sum_{P \in \mathrm{F}_n(E_{\mathcal{A},\tau})} i(\mathcal{A}, \tau; P)\,[X_{\mathcal{A},\tau,P}].$$

Pour chaque pan $P$ on note $(v_P, w_P) \in \mathbf{Z}^n \times \mathbf{Z}$ le plus petit vecteur entier orthogonal au plan d'appui $H_P \subset \mathbf{R}^{n+1}$ tel que $w_P < 0$. On note aussi $(a_P, \tau_P)$ un point quelconque de $P$.

Le vecteur $\tau$ induit une *décomposition polyédrale cohérente* $\mathrm{DPC}_\tau(Q_\mathcal{A})$ du polytope de base $Q_\mathcal{A}$. Les faces $S$ de dimension $n$ de cette décomposition sont en correspondance avec les pans de la toiture; pour $S \in \mathrm{DPC}_\tau(Q_\mathcal{A})$ on écrit $\mathrm{Pan}(S) \in \mathrm{F}_n(E_{\mathcal{A},\tau})$ pour le pan correspondant.

**Lemme 8.2.** *Soient $\tau \in \mathbf{Z}^{N+1}$ et $P \in \mathrm{F}_n(E_{\mathcal{A},\tau})$ un pan de la toiture de $Q_{\mathcal{A},\tau}$ alors,*

$$m(X_{\mathcal{A},\tau} \cdot \mathrm{div}(\lambda_\tau^*(x^b)); X_{\mathcal{A},\tau,P}) = (\langle M_\mathcal{A}(b) - D\,a_P, v_P\rangle - D\,\tau_P\,w_P)\,i(\mathcal{A}, \tau; P),$$

*avec* $\lambda_\tau^*(x^b) = (t^{\tau_0}x_0, \dots, t^{\tau_N}x_N)^b = t_0^{b_0\tau_0 + \cdots + b_N\tau_N} x^b$.

*Démonstration.* Cette démonstration étant tout à fait analogue à celle du lemme 8.1, on n'indiquera que les pas principaux.

Par linéarité on peut se ramener au cas où $b = e_i$ est un des vecteurs de la base standard de $\mathbf{R}^{N+1}$, donc $x^b = x_i$. Soit $(a_j, \tau_j)$ un sommet quelconque du pan $P$, on se place dans la carte affine $\mathbf{A}^1 \times \mathbf{A}^N \subset \mathbf{P}^1 \times \mathbf{P}^N$ correspondant à $t_1 \neq 0$ et $x_j \neq 0$ (puisqu'on veut calculer une multiplicité le long d'une sous-variété de $Z(t_0)$). Dans cette carte, l'application monomiale s'écrit

$$\mathbf{G}_m \times \mathbf{G}_m^n \to \mathbf{A}^1 \times \mathbf{A}^N, \qquad (t, s) \mapsto \big(t; t^{\tau_j - \tau_0} s^{a_0 - a_j}, \dots, t^{\tau_j - \tau_N} s^{a_N - a_j}\big)$$

et donc l'algèbre de cette carte affine de $X_{\mathcal{A},\tau}$ est

$$\overline{\mathbf{Q}}[(X_{\mathcal{A},\tau})_{t_1,x_j}] = \overline{\mathbf{Q}}\big[t, t^{\tau_j - \tau_0} s^{a_0 - a_j}, \dots, t^{\tau_j - \tau_N} s^{a_N - a_j}\big].$$

La normalisation de cette algèbre correspond au cône

$$\sigma := \{(u, v) \in \mathbf{R}^n \times \mathbf{R} : v \geq 0, \langle u, a_k - a_j\rangle + v\,(\tau_j - \tau_k) \geq 0,\ k = 0, \dots, N\}.$$

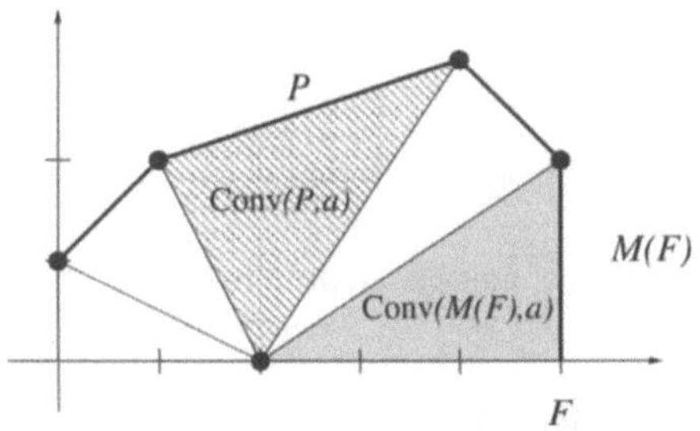


Fig. 9

Le reste de la démonstration suit les lignes de celle du lemme 8.1, en considérant la normalisation de $(X_{\mathcal{A},\tau})_{t_1,x_j}$ donnée par le semi-groupe des points entiers $S_\sigma := \sigma^\vee \cap \mathbf{Z}^n$.

L'hyperplan $H_P \subset \mathbf{R}^{n+1}$ est un des plans d'appui du cône $\sigma^\vee$, car $P$ est un pan. Il définit donc une arête $\rho$ du cône dual $\sigma$, dont le semi-groupe $\rho \cap \mathbf{Z}^n$ est engendré par $(v_P, -w_P)$. Comme $\lambda_\tau^*(x_i) = t^{\tau_i} x_i = t^{\tau_j} s^{a_i - a_j}$ sur la carte considérée, on en conclut

$$\begin{aligned} m(X_{\mathcal{A},\tau} \cdot \operatorname{div}(\lambda_\tau^*(x_i)); X_{\mathcal{A},\tau,P}) &= \langle (a_i, 0) - (a_j, -\tau_j), (v_P, -w_P) \rangle \, i(\mathcal{A}, \tau; P) \\ &= (\langle a_i - a_j, v_P \rangle - \tau_j w_P) \, i(\mathcal{A}, \tau; P). \end{aligned}$$

□

**Proposition 8.3.** *Soit $\mathcal{A} \subset (\mathbf{Z}^n)^{N+1}$ tel que $L_\mathcal{A} = \mathbf{Z}^n$, $b \in \mathbf{Z}^{N+1}$ et $\tau \in \mathbf{N}^{N+1}$, alors l'égalité (8.19) correspond terme à terme à la suivante, multipliée par $(n+1)!$,*

$$\begin{aligned} &D \operatorname{Vol}_{n+1}(\mathcal{M}_{\mathcal{A},\tau}) \\ &= \sum_{F \in \mathrm{F}_{n-1}(\mathcal{Q}_\mathcal{A})} \varepsilon(b, F) \operatorname{Vol}_{n+1}(\operatorname{Conv}(M(F) + (D-1)(a_F, 0), (M_\mathcal{A}(b), 0))) \\ &\quad + \sum_{P \in \mathrm{F}_n(E_{\mathcal{A},\tau})} \varepsilon(b, P) \operatorname{Vol}_{n+1}(\operatorname{Conv}(P + (D-1)(a_P, \tau_P), (M_\mathcal{A}(b), 0))) \end{aligned}$$

*où $\varepsilon(b, F) = +1$ (resp. $\varepsilon(b, P) = +1$) si $M_\mathcal{A}(b)$ et $v_F$ (resp. $(M_\mathcal{A}(b), 0)$ et $(v_P, w_P)$) sont d'un même côté de l'hyperplan $H_F + (D-1)a_F$ (resp. $H_P + (D-1)(a_P, \tau_P)$) et $\varepsilon(b, F) = -1$ (resp. $\varepsilon(b, P) = -1$) sinon.*

La figure 9 (pour $b = e_i$, $D = 1$ et $M_\mathcal{A}(b) = a_i$) illustre cette décomposition.

*Démonstration.* Montrons comment (8.19) se traduit en la décomposition de l'intégrale de la fonction $\vartheta_{\mathcal{A},\tau} : \mathcal{Q}_\mathcal{A} \to \mathbf{R}$ paramétrant la toiture $E_{\mathcal{A},\tau}$ de la proposition 8.3. On a d'abord

$$e_\tau(X_\mathcal{A}) = (n+1)! \int_{\mathcal{Q}_\mathcal{A}} \vartheta_{\mathcal{A},\tau}(u) \, du_1 \cdots du_n = (n+1)! \operatorname{Vol}_{n+1}(\mathcal{M}_{\mathcal{A},\tau})$$

puis, par le lemme 8.1 il vient

$$e_\tau(X_\mathcal{A} \cdot \operatorname{div}(x^b)) = \sum_{F \in \mathrm{F}_{n-1}(\mathcal{Q}_\mathcal{A})} \langle M_\mathcal{A}(b) - D \, a_F, v_F \rangle \, i(\mathcal{A}; F) \, e_\tau(X_{\mathcal{A},F}),$$

et pour chaque face $F$ de $Q_{\mathcal{A}}$

$$e_\tau(X_{\mathcal{A},F}) = \frac{n!}{i(\mathcal{A};F)\,||v_F||_2} \int_F \vartheta_{\mathcal{A},\tau}\, d\mu_{n-1} = \frac{n!}{i(\mathcal{A};F)\,||v_F||_2} \mathrm{Vol}_n(M(F))$$

où $M(F) \subset \mathbf{R}^{n+1}$ désigne le mur de la maison $\mathcal{M}_{\mathcal{A},\tau}$ au-dessus de $F$. Ainsi

$$\begin{aligned}
&\langle M_{\mathcal{A}}(b) - D\, a_F, v_F\rangle\, i(\mathcal{A};F)\, e_\tau(X_{\mathcal{A},F}) \\
&= n!\, \varepsilon(b,F)\mathrm{dist}(M_{\mathcal{A}}(b), F + (D-1)a_F)\,\mathrm{Vol}_n(M(F)) \\
&= (n+1)!\, \varepsilon(b,F)\mathrm{Vol}_{n+1}(\mathrm{Conv}(M(F) + (D-1)(a_F,0), (M_{\mathcal{A}}(b),0))).
\end{aligned}$$

Finalement, soit $Y$ une composante irréductible de $\mathrm{init}_\tau(X_{\mathcal{A}})$ puis $S \in \mathrm{DPC}_\tau(Q_{\mathcal{A}})$ et $P := \mathrm{Pan}(S) \in \mathrm{F}_n(E_{\mathcal{A},\tau})$ les faces correspondantes dans la subdivision et dans la toiture respectivement, on a

$$\deg(Y) = \frac{n!}{i(\mathcal{A};S)} \mathrm{Vol}_n(S) = \frac{n!}{i(\mathcal{A},\tau;P)\,||(v_P,w_P)||_2} \mathrm{Vol}_n(P).$$

Le lemme 8.2 entraîne alors

$$\begin{aligned}
&m(X_{\mathcal{A},\tau} \cdot \mathrm{div}(\lambda_\tau^*(x^b)); \iota(Y))\, \deg(Y) \\
&= \frac{n!}{||(v_P,w_P)||_2} \langle (M_{\mathcal{A}}(b),0) - D\,(a_P,\tau_P), (v_P,w_P)\rangle\, \mathrm{Vol}_n(P) \\
&= (n+1)!\, \varepsilon(b,P)\mathrm{Vol}_{n+1}(\mathrm{Conv}(P + (D-1)(a_P,\tau_P), (M_{\mathcal{A}}(b),0))).
\end{aligned}$$

En regroupant ces calculs on voit que l'identité (8.19) (multipliée par $\frac{1}{(n+1)!}$) se traduit dans la décomposition cherchée. □

Pour $S \in \mathrm{DPC}_\tau(Q_{\mathcal{A}})$ on définit $\theta_{\tau,S}(b) \in \mathbf{R}$ l'unique réel tel que $\big(M_{\mathcal{A}}(b), \theta_{\tau,S}(b)\big) \in H_{\mathrm{Pan}(S)} + (D-1)(a_{\mathrm{Pan}(S)}, \tau_{\mathrm{Pan}(S)})$. Le lemme suivant explicite cette quantité.

**Lemme 8.4.** *Soit $S \in \mathrm{DPC}_\tau(Q_{\mathcal{A}})$ et $a_{j_0}, \ldots, a_{j_n} \in S$ des vecteurs affinement indépendants. Soit $b \in \mathbf{Z}^{N+1}$ et $D := \sum_{j=0}^N b_j$, alors*

$$\theta_{\tau,S}(b) \cdot \det \begin{bmatrix} 1 & \ldots & 1 \\ a_{j_0,1} & \cdots & a_{j_n,1} \\ \vdots & \ddots & \vdots \\ a_{j_0,n} & \cdots & a_{j_n,n} \end{bmatrix} = -\det \begin{bmatrix} 1 & \ldots & 1 & D \\ a_{j_0,1} & \ldots & a_{j_n,1} & M_{\mathcal{A}}(b)_1 \\ \vdots & \ddots & \vdots & \vdots \\ a_{j_0,n} & \ldots & a_{j_n,n} & M_{\mathcal{A}}(b)_n \\ \tau_{j_0} & \ldots & \tau_{j_n} & 0 \end{bmatrix}.$$

En triangulant $S$ par des simplexes et en utilisant la relation entre déterminant et volume, on peut réécrire ceci sous la forme (on pose $P = \mathrm{Pan}(S)$) :

$$\begin{aligned}
\theta_{\tau,S}(b)\,\mathrm{Vol}_n(S) &= (n+1)\,\varepsilon(b,P)\mathrm{Vol}_{n+1}\,(\mathrm{Conv}(P+(D-1)(a_P,\tau_P), (M_{\mathcal{A}}(b),0))) \\
&= m(X_{\mathcal{A},\tau} \cdot \mathrm{div}(\lambda_\tau^*(x^b)); \iota(Y))\, \deg(Y) \qquad (8.20)
\end{aligned}$$

qui s'interprète comme l'égalité des volumes montrés dans la figure 10 (lorsque $D = 1$ et en posant $a = M_{\mathcal{A}}(b)$).

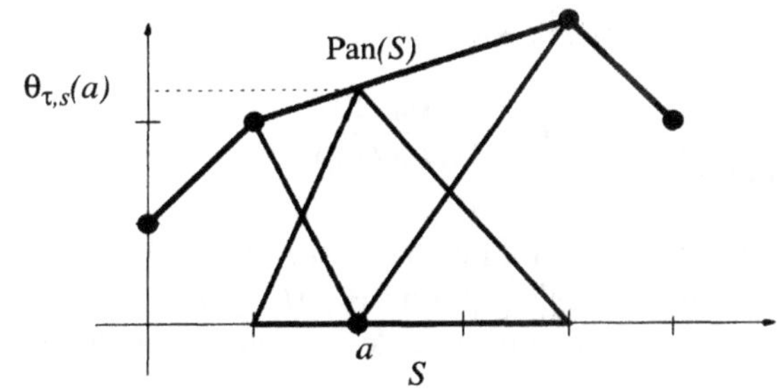


Fig. 10

*Démonstration.* C'est un calcul direct de l'intersection des espaces linéaires $H_{\mathrm{Pan}(S)} + (D-1)(a_{\mathrm{Pan}(S)}, \tau_{\mathrm{Pan}(S)})$ et $\{M_{\mathcal{A}}(b)\} \times \mathbf{R}$ : on a $\theta_{\tau,S}(b) = \sum_{i=0}^{n} v_i\, \tau_{j_i}$ où $v = (v_0, \ldots, v_n) \in \mathbf{R}^{n+1}$ est l'unique solution du système linéaire

$$\sum_{i=0}^{n} v_i = D, \qquad \sum_{i=0}^{n} v_i\, a_{j_i} = M_{\mathcal{A}}(b).$$

On résout ce système par les formules de Cramer et on trouve ainsi

$$\theta_{\tau,S}(b) \cdot \det \begin{bmatrix} 1 & \ldots & 1 \\ a_{j_0,1} & \cdots & a_{j_n,1} \\ \vdots & \ddots & \vdots \\ a_{j_0,n} & \cdots & a_{j_n,n} \end{bmatrix}$$
$$= \sum_{i=0}^{n} \tau_{j_i} \det \begin{bmatrix} 1 & \ldots & 1 & D & 1 & \ldots & 1 \\ a_{j_0,1} & \ldots & a_{j_{i-1},1} & M_{\mathcal{A}}(b)_1 & a_{j_{i+1},1} & \ldots & a_{j_n,1} \\ \vdots & \ddots & \vdots & \vdots & \vdots & \ddots & \vdots \\ a_{j_0,n} & \ldots & a_{j_{i-1},n} & M_{\mathcal{A}}(b)_n & a_{j_{i+1},n} & \ldots & a_{j_n,n} \end{bmatrix}$$

qui est le développement du déterminant dans le membre droite de l'énoncé, par rapport à la dernière ligne. □

Ceci permet d'écrire l'identité (8.19) de la façon suivante:

**Proposition 8.5.** *Soit* $\mathcal{A} \in (\mathbf{Z}^n)^{N+1}$, $b \in \mathbf{Z}^{N+1}$ *et* $\tau = (\tau_0, \ldots, \tau_N) \in \mathbf{R}^{N+1}$, *posons* $D := \sum_{j=0}^{N} b_j$, *alors*

$$e_\tau\left(X_{\mathcal{A}} \cdot \mathrm{div}(x^b)\right) = D\, e_\tau(X_{\mathcal{A}}) - n! \sum_{S \in \mathrm{DPC}_\tau(Q_{\mathcal{A}})} \theta_{\tau,S}(b)\, \mathrm{Vol}_n(S)$$

*où la seconde somme porte sur les faces* $S$ *de dimension* $n$ *de la décomposition polyédrale* $\mathrm{DPC}_\tau(Q_{\mathcal{A}})$.

*Démonstration.* Le cas $\tau \in \mathbf{N}^{N+1}$ résulte de la proposition 8.3 et de (8.20). Ceci s'étend successivement à $\tau \in \mathbf{Z}^{N+1}$ à cause de l'invariance de la formule par remplacement de $\tau$ par $\tau + c \cdot (1, \ldots, 1)$ avec $c \in \mathbf{N}$. Puis, la formule s'étend à $\tau \in \mathbf{Q}^{N+1}$ par homogénéité et finalement à $\tau \in \mathbf{R}^{N+1}$ par continuité. □

Avec la notation (5.9) on peut encore écrire cet énoncé sous la forme :

$$w_{X_{\mathcal{A},\tau}}(x^b) = -\sum_{S\in \mathrm{DPC}_\tau(Q_\mathcal{A})} \theta_{\tau,S}(b)\cdot \frac{\mathrm{Vol}_n(S)}{\mathrm{Vol}_n(Q_\mathcal{A})},$$

dont on vérifie, par continuité et homogénéité, la validité pour tout $\tau \in \mathbf{R}^{N+1}$. Dans le cas où $\mathrm{div}(x^b)$ est effectif, c'est-à-dire quand $b \in \mathbf{N}^{N+1}$, on a $\theta_{\tau,S}(b) \geq \tau_0\, b_0 + \cdots + \tau_N\, b_N$ pour tout $S \in \mathrm{DPC}_\tau(Q_\mathcal{A})$ à cause de la concavité de la toiture du polytope $Q_{\mathcal{A},\tau}$, et donc

$$(\tau_0\, b_0 + \cdots + \tau_N\, b_N)\deg(X_\mathcal{A}) \leq n! \sum_{S\in \mathrm{DPC}_\tau(Q_\mathcal{A})} \theta_{\tau,S}(b)\,\mathrm{Vol}_n(S),$$

ainsi

$$e_\tau\left(X_\mathcal{A}\cdot \mathrm{div}(x^b)\right) \leq D\, e_\tau(X_\mathcal{A}) - (\tau_0\, b_0 + \cdots + \tau_N\, b_N)\deg(X_\mathcal{A}). \qquad (8.21)$$

Alternativement, on peut démontrer cette inégalité par application directe du théorème 1.2 et de l'exemple 4.1.

On en déduit un théorème de Bézout arithmétique *exact* pour la hauteur normalisée de l'intersection d'une variété torique avec un diviseur monomial:

**Corollaire 8.6.** *Soit $K$ un corps de nombres, $\mathcal{A} \in (\mathbf{Z}^n)^{N+1}$ tel que $L_\mathcal{A} = \mathbf{Z}^n$, $\alpha \in (K^\times)^{N+1}$ et $b \in \mathbf{Z}^{N+1}$. Posons $\tau_{\alpha\, v} = (\log|\alpha_0|_v, \ldots, \log|\alpha_N|_v)$ pour toute place $v \in M_K$ et $D := \sum_{j=1}^N b_j$, alors*

$$\begin{aligned}\widehat{h}(X_{\mathcal{A},\alpha}\cdot \mathrm{div}(x^b)) &= D\widehat{h}(X_{\mathcal{A},\alpha}) - n! \sum_{v\in M_K} \frac{[K_v:\mathbf{Q}_v]}{[K:\mathbf{Q}]} \sum_{S\in \mathrm{DPC}_{\tau_{\alpha\, v}}(Q_\mathcal{A})} \theta_{\tau_{\alpha\, v},S}(b)\mathrm{Vol}_n(S)\\ &= D\widehat{h}(X_{\mathcal{A},\alpha}) + \left(\sum_{v\in M_K} \frac{[K_v:\mathbf{Q}_v]}{[K:\mathbf{Q}]} w_{X_{\mathcal{A},\tau_{\alpha\, v}}}(x^b)\right)\deg(X_{\mathcal{A},\alpha}).\end{aligned}$$

*En particulier, si $\mathrm{div}(x^b)$ est effectif (c'est-à-dire $b \in \mathbf{N}^{N+1}$) on a $\widehat{h}(X_{\mathcal{A},\alpha}\cdot \mathrm{div}(x^b)) \leq D\,\widehat{h}(X_{\mathcal{A},\alpha})$.*

*Démonstration.* Pour l'identité on remarque que $X_{\mathcal{A},\alpha}\cdot \mathrm{div}(x^b) = \alpha(X_\mathcal{A}\cdot \mathrm{div}(x^b))$. En sommant sur $v \in M_K$ l'égalité de la proposition 8.5 avec les coefficients $\frac{[K_v:\mathbf{Q}_v]}{[K:\mathbf{Q}]}$, on conclut grâce au théorème 6.1.

Pour établir l'inégalité, on a par (8.21)

$$\begin{aligned}&\widehat{h}(X_{\mathcal{A},\alpha}\cdot \mathrm{div}(x^b))\\ &\leq D\widehat{h}(X_{\mathcal{A},\alpha}) - \sum_{v\in M_K} \frac{[K_v:\mathbf{Q}_v]}{[K:\mathbf{Q}]}(\tau_{\alpha\, v,0}\, b_0 + \cdots + \tau_{\alpha\, v,N}\, b_N)\cdot \deg(X_\mathcal{A})\\ &\leq D\widehat{h}(X_{\mathcal{A},\alpha})\end{aligned}$$

car $\sum_{v\in M_K} \frac{[K_v:\mathbf{Q}_v]}{[K:\mathbf{Q}]}\tau_{\alpha\, v,i} = 0$ pour $i = 0, \ldots, N$, grâce à la formule du produit. □

Si l'on définit la hauteur de $x^b$ relative à la variété $X_{\mathcal{A},\alpha}$ par la formule

$$\widehat{h}_{X_{\mathcal{A},\alpha}}(x^b) := \sum_{v \in M_K} \frac{[K_v : \mathbf{Q}_v]}{[K : \mathbf{Q}]} w_{X_{\mathcal{A},\tau_\alpha v}}(x^b)$$

le résultat précédent se réécrit

$$\widehat{h}(X_{\mathcal{A},\alpha} \cdot \mathrm{div}(x^b)) = D\widehat{h}(X_{\mathcal{A},\alpha}) + \widehat{h}_{X_{\mathcal{A},\alpha}}(x^b) \cdot \deg(X_{\mathcal{A},\alpha}).$$

**Remerciements.** P. Philippon a été partiellement financé par une allocation de recherche de la Fondation Alexander von Humboldt pendant la réalisation de ce travail. M. Sombra a été financé par le Programme Ramón y Cajal du Ministerio de Educación y Ciencia, Espagne.

**Abstract Some diophantine aspects of projective toric varieties.** We present several facets of projective toric varieties, of interest from the point of view of Diophantine geometry. We make explicit the theory in a number of meaningful examples and we also prove a Bézout type theorem for Chow weight of projective varieties.

## Références

[AD03] Amoroso, F., David, S.: Minoration de la hauteur normalisée dans un tore. *J. Inst. Math. Jussieu* **2**, 335–381 (2003)

[AD04] Amoroso, F., David, S.: Distribution des points de petite hauteur dans les groupes multiplicatifs. *Ann. Sc. Norm. Super. Pisa Cl. Sci. V. Ser.* **3**, 325–348 (2004)

[Aud91] Audin, M.: *The Topology of Torus Actions on Symplectic Manifolds*. Progress in Mathematics, vol. 93. Birkhäuser, Basel (1991)

[Ber87] Bertrand, D.: Lemmes de zéros et nombres transcendants. *Sémin. Bourbaki 1985/86, Astérisque* **145–146**, 21–44 (1987)

[BP88] Bertrand, D., Philippon, P.: Sous-groupes algébriques de groupes algébriques commutatifs. Ill. *J. Math.* **32**, 263–280 (1988)

[Cha89] Chardin, M.: Une majoration de la fonction de Hilbert et ses conséquences pour l'interpolation algébrique. *Bull. Soc. Math. Fr.* **117**, 305–318 (1989)

[CP99] Chardin, M., Philippon, P.: Régularité et interpolation. *J. Algebr. Geom.* **8**, 471–481 (1999)

[Cox01] Cox, D.: Minicourse on toric varieties, notes d'un cours donné à l'université de Buenos Aires en Juillet 2001. Téléchargeable à http://www.amherst.edu/~dacox/

[CLO98] Cox, D., Little, J., O'Shea, D.: *Using Algebraic Geometry*. Graduate Texts in Mathematics, vol. 185. Springer, Heidelberg (1998)

[DP99] David, S., Philippon, P.: Minorations des hauteurs normalisées des sous-variétés des tores. *Ann. Sc. Norm. Super. Pisa Cl. Sci IV Ser.* **28**, 489–543 (1999)

[Don02] Donaldson, S.K.: Scalar curvature and stability of toric varieties. *J. Differ. Geom.* **62**, 289–349 (2002)

[ES96] Eisenbud, D., Sturmfels, B.: Binomial ideals. *Duke Math. J.* **84**, 1–45 (1996)

[EF02] Evertse, J.-H., Ferretti, R.G.: Diophantine inequalities on projective varieties. *Int. Math. Res. Not.* **25**, 1295–1330 (2002)

[Ewa96] Ewald, G.: *Combinatorial Convexity and Algebraic Geometry*. Graduate Texts in Mathematics, vol. 168. Springer, Heidelberg (1996)

[Fer03] Ferretti, R.G.: Diophantine approximation and toric deformations. *Duke Math. J.* **118**, 493–522 (2003)

[Ful84] Fulton, W.: *Intersection Theory*. Ergebnisse der Mathematic und ihrer Grenzgebiete, 3e Serie, vol. 2. Springer, Heidelberg (1984)

[Ful93] Fulton, W.: *Introduction to Toric Varieties*. Annals of Mathematical Studies, vol. 131. Princeton University Press, Princeton, N.J. (1993)

[GKZ94] Gelfand, I.M., Kapranov, M.M., Zelevinsky, A.V.: *Discriminants, Resultants and Multidimensional Determinants*. Birkhäuser, Basel (1994)

[KSZ92] Kapranov, M.M., Sturmfels, B., Zelevinsky, A.V.: Chow polytopes and general resultants. *Duke Math. J.* **67**, 189–218 (1992)

[Mum77] Mumford, D.: Stability of projective varieties. *Enseign. Math. II. Sér* **23**, 39–110 (1977)
[Phi91] Philippon, P.: Sur des hauteurs alternatives, I. *Math. Ann.* **289**, 255–283 (1991)
[PS04] Philippon, P., Sombra, M.: Hauteur normalisée des variétés toriques projectives, *J. Inst. Math. Jussieu.* **7**, 327–378 (2008)
[PS05] Philippon, P., Sombra, M.: Géométrie diophantienne et variétés toriques. *C. R. Math. Acad. Sci. Paris* **340**, 507–512 (2005)
[PS06] Philippon, P., Sombra, M.: Minimum essentiel et degrés d'obstruction des translatés de sous-tores. *Acta Arith.* (à paraître)
[Rat04] Ratazzi, N.: Minoration de la hauteur de Néron-Tate pour les points et les sous-variétés: variations sur le problème de Lehmer. Thèse de doctorat, Université de Paris VI, Paris (2004)
[Rem01a] Rémond, G.: Élimination multihomogène. In: Nesterenko, Yu. V., Philippon, P. (eds.) *Introduction to Algebraic Independence Theory*. Lect. Notes Math., vol. 1752, pp. 53–81. Springer, Heidelberg (2001)
[Rem01b] Rémond, G.: Géométrie diophantienne multiprojective. In: Nesterenko, Yu. V., Philippon, P. (eds.) *Introduction to Algebraic Independence Theory*. Lect. Notes Math., vol. 1752, pp. 95–131. Springer, Heidelberg (2001)
[Sch91] Schmidt, W.M.: *Diophantine Approximation and Diophantine Equations*. Lect. Notes Math., vol. 1467. Springer, Heidelberg (1991)
[Som05] Sombra, M.: Minimums successifs des variétés toriques projectives. *J. Reine Angew. Math.* **586**, 207–233 (2005)
[Stu94] Sturmfels, B.: On the Newton polytope of the resultant. *J. Algebr. Comb.* **3**, 207–236 (1994)
[Stu96] Sturmfels, B.: Gröbner bases and convex polytopes. *Am. Math. Soc.* (1996)
[Zha95] Zhang, S.-W.: Positive line bundles on arithmetic varieties. *J. Am. Math. Soc.* **8**, 187–221 (1995)

*Diophantine Approximation*
H. P. Schlickewei et al., Editors


# UNE INÉGALITÉ DE ŁOJASIEWICZ ARITHMÉTIQUE

Gaël Rémond
*Institut Fourier, UMR 5582, Université Grenoble 1, BP 74, 38402 Saint-Martin-d'Hères Cedex, France*
Gael.Remond@ujf-grenoble.fr

Für Wolfgang Schmidt zu seinem 70. Geburtstag

## 1 Résultat

Une inégalité de Łojasiewicz minore la valeur $|f(x)|$ d'une fonction analytique $f\colon \mathbb{R}^n \to \mathbb{R}$ par une puissance de la distance de $x$ à l'ensemble des zéros de $f$. Nous nous intéressons ici au cas arithmétique où $f$ est un polynôme à coefficients entiers.

Nous munissons $\mathbb{R}^n$ de la norme du supremum par

$$|x| = \max(|x_1|, \ldots, |x_n|).$$

Si $Z$ est une partie de $\mathbb{R}^n$, nous écrivons

$$\mathrm{dist}(x, Z) = \inf_{z \in Z} |x - z|.$$

Nous notons encore $|x|^* = \max(1, |x|)$ et $\mathrm{dist}^*(x, Z) = \min(1, \mathrm{dist}(x, Z))$. Si $f$ est une application $\mathbb{R}^n \to \mathbb{R}$, nous désignons par $Z_{\mathbb{R}}(f)$ l'ensemble de ses zéros $f^{-1}(0)$. En particulier, si $P$ est un polynôme à coefficients réels, $Z_{\mathbb{R}}(P)$ désigne l'ensemble de ses zéros réels.

Avec ces conventions, notre résultat s'énonce comme suit.

**Théorème 1.1.** *Soient $n$, $D$ et $H$ des entiers naturels non nuls. Si $P$ est un polynôme de $\mathbb{Z}[X_1, \ldots, X_n]$ de degré au plus $D$ et dont les coefficients sont de valeurs absolues au plus $H$, alors, pour tout $x \in \mathbb{R}^n$, on a*

$$|P(x)| \geq (e^{2nD}H)^{-(n+1)D^n} \left( \frac{\mathrm{dist}^*(x, Z_{\mathbb{R}}(P))}{|x|^*} \right)^{nD^n}.$$

Cette inégalité se rapproche de résultats dus à W. D. Brownawell dans le cas complexe. Pour ces inégalités sur $\mathbb{C}^n$ ainsi que leurs liens avec le Nullstellensatz, nous renvoyons à l'article [B] et à sa bibliographie. Elles ne semblent pas contenir notre minoration puisqu'un point réel peut se trouver plus loin des zéros réels que des zéros complexes.

*Mots clefs.* Polynômes, inégalité de Łojasiewicz, hauteurs.

*2000 Classification mathématique par sujets.* 11C08, 11J25, 11G50.

Signalons encore que, à la différence du cas complexe, nous pouvons facilement donner une minoration de $\max |P_i(x)|$ pour une famille de polynômes $P_1, \ldots, P_m$ à partir du cas d'un seul polynôme $P$ : il suffit de poser $P = P_1^2 + \cdots + P_m^2$.

L'énoncé du théorème 1.1 ne prétend pas être optimal. Les exemples suivants montrent toutefois que la formule serait fausse si l'on remplaçait en exposant de l'une des quantités $H$, $|x|^*$ ou $\mathrm{dist}^*(x, Z_{\mathbb{R}}(P))$ le facteur $D^n$ par $D^{n-\varepsilon}$ pour $\varepsilon > 0$.

*Example.* Dans les trois cas nous utilisons une modification d'un exemple dû à D. Masser et P. Philippon (voir [B, p. 242]). Nous choisissons deux entiers $B$ et $d$ au moins égaux à 2 et nous écrivons $D = 2d$.

1. Nous posons $P = X_1^{2d} + (1 - X_{n-1}X_n^{d-1})^2 + \sum_{i=1}^{n-2}(X_iX_n^{d-1} - X_{i+1}^d)^2$ puis $x_i = B^{1-d^{n-i}}$ pour $1 \leq i \leq n-1$ et $x_n = B$. De la sorte, nous avons $H = 2$, $\mathrm{dist}^*(x, Z_{\mathbb{R}}(P)) = 1$ et $|x|^* = B$ avec
$$P(x) = (|x|^*)^{-2(D/2)^n + D}.$$
2. Nous posons $P = X_1^{2d} + \sum_{i=1}^{n-1}(X_i - X_{i+1}^d)^2$ puis $x_i = B^{-d^{n-i}}$ pour $1 \leq i \leq n$. De la sorte, nous avons $H = 2$, $|x|^* = 1$ et $\mathrm{dist}^*(x, Z_{\mathbb{R}}(P)) = B^{-1}$ avec
$$P(x) = \mathrm{dist}^*(x, Z_{\mathbb{R}}(P))^{2(D/2)^n}.$$
3. Nous posons $P = X_1^{2d} + (1 - BX_n)^2 + \sum_{i=1}^{n-1}(BX_i - X_{i+1}^d)^2$ puis $x_i = B^{-(d^{n-i+1}-1)/(d-1)}$ pour $1 \leq i \leq n$. De la sorte, nous avons $H = B^2$, $\mathrm{dist}^*(x, Z_{\mathbb{R}}(P)) = 1$ et $|x|^* = 1$ avec
$$P(x) = H^{-d(d^n-1)/(d-1)} \leq H^{-(D/2)^n}.$$

La démonstration du théorème 1.1 repose sur l'idée suivante : si $x$ n'est pas un zéro de $P$, l'on peut trouver un pavé $A$ de $\mathbb{R}^n$ le contenant et disjoint de $Z_{\mathbb{R}}(P)$. Ce pavé est contrôlé en termes de $|x|^*$ et $\mathrm{dist}^*(x, Z_{\mathbb{R}}(P))$. On cherche ensuite le minimum de $|P|$ sur $A$ : s'il est atteint au bord de $A$, le problème se ramène par spécialisation au cas de $n-1$ variables ; sinon c'est un minimum local. Maintenant, si $|P|$ atteint un minimum local isolé en $y$, le point $y$ est solution d'un système d'équations algébriques donné par les annulations des dérivées de $P$. Il est alors possible de contrôler la hauteur du point algébrique $y$ et de minorer $|P(y)|$ par l'inégalité de Liouville. Ceci vaut également pour un minimum non isolé quitte à changer de point où le minimum est atteint.

Dans la partie suivante, nous introduisons les outils qui permettent de contrôler un point algébrique défini par une intersection de polynômes comme ci-dessus. Ensuite, nous en déduisons la minoration de la valeur d'un minimum local non nul de $P$ puis une borne inférieure pour le minimum de $|P|$ sur un pavé où $P$ ne s'annule pas et nous concluons en associant à chaque point $x$ un tel pavé le contenant.

## 2 Hauteurs

La preuve du théorème utilise différentes notions de hauteurs dont nous rappelons ici les définitions et les propriétés que nous emploierons. Pour une partie finie $F$ de $\bar{\mathbb{Q}}$ contenant un élément non nul, nous notons
$$h(F) = \sum_v \frac{[K_v : \mathbb{Q}_v]}{[K : \mathbb{Q}]} \log \max_{x \in F} |x|_v$$

où la somme porte sur toutes les places du corps de nombres $K = \mathbb{Q}(F)$, les valeurs absolues étant normalisées par $|2|_v = 2$ si $v$ est infinie et $|p|_v = p^{-1}$ si $v$ est au-dessus de la place $p$ de $\mathbb{Q}$. La hauteur ainsi définie est projective en vertu de la formule du produit (c'est-à-dire $h(aF) = h(F)$ pour tout $a \in \bar{\mathbb{Q}}^\times$) et nous permet d'introduire la hauteur $h(y)$ d'un point $y$ de $\mathbb{P}^n(\bar{\mathbb{Q}})$ comme étant celle de l'un quelconque de ses systèmes de coordonnées. De la même façon, si $P$ est un polynôme non nul à coefficients dans $\bar{\mathbb{Q}}$ (en un nombre quelconque de variables), $h(P)$ désignera la hauteur de la famille de ses coefficients. Dans les parties suivantes, nous manipulons des polynômes à coefficients dans $\mathbb{Z}$, pour lesquels nous noterons $|P|$ le maximum des valeurs absolues des coefficients de $P$ de sorte que l'on a facilement $h(P) \leq \log |P|$.

Lorsque $a$ est un élément de $\bar{\mathbb{Q}}^\times$, nous ferons usage de l'inégalité de Liouville qui assure $\log |a|_v \geq -[\mathbb{Q}(a) : \mathbb{Q}]h(1, a)$ pour toute place $v$ de $\mathbb{Q}(a)$ (cette minoration découle facilement de la définition en écrivant $h(1, a) = h(1, a^{-1})$).

Enfin, nous aurons besoin de la notion de hauteur $h(X)$ d'un sous-schéma fermé de $\mathbb{P}^n_{\bar{\mathbb{Q}}}$. Elle se définit soit à l'aide d'une forme de Chow (en suivant Philippon) soit en termes d'intersection arithmétique dans le cadre de la théorie d'Arakelov (en suivant Faltings, Bost, Gillet et Soulé). Ici nous utilisons [R1] comme référence et nous rappelons dans ce paragraphe toutes les propriétés utilisées plus bas. Considérons d'abord les deux cas extrêmes où $\dim X$ vaut $n$ ou 0. Pour la hauteur de l'espace projectif, nous avons

$$h(\mathbb{P}^n_{\bar{\mathbb{Q}}}) = \sum_{j=1}^{n} \sum_{\ell=1}^{j} \frac{1}{2\ell} = \sum_{\ell=2}^{n+1} \frac{n+1}{2\ell} \leq n \log(n+1).$$

Si maintenant $y$ est un point fermé de $\mathbb{P}^n_{\bar{\mathbb{Q}}}$, la hauteur $h(\{y\})$ du sous-schéma fermé ponctuel $\{y\}$ diffère de $h(y)$ définie plus haut seulement en ce qu'elle fait intervenir aux places infinies la norme euclidienne $(|y_0|_v^2 + \cdots + |y_n|_v^2)^{1/2}$ au lieu de $\max |y_i|_v$.

Les hauteurs de sous-schémas apparaîtront essentiellement à travers le résultat d'intersection (ou de Bézout arithmétique) suivant, énoncé pour simplifier uniquement avec des schémas sur $\mathbb{Q}$ (dont la hauteur est par définition celle de leur extension à $\bar{\mathbb{Q}}$). On y note $\mathcal{V}(\mathcal{P})$ pour une partie $\mathcal{P} \subset \mathbb{Q}[X_0, \ldots, X_n]$ le fermé de $\mathbb{P}^n_{\mathbb{Q}}$ défini par l'idéal homogène engendré par $\mathcal{P}$.

**Lemme 2.1.** *Soient $V$ un sous-schéma fermé intègre de $\mathbb{P}^n_{\mathbb{Q}}$ et $\mathcal{P}$ une famille de polynômes homogènes de $\mathbb{Q}[X_0, \ldots, X_n]$. On note $\delta$ un entier et $H$ un réel tels que $\deg P \leq \delta$ et $h(P) \leq H$ pour tout $P \in \mathcal{P}$. Alors pour toute composante irréductible $X$ du fermé $V \cap \mathcal{V}(\mathcal{P})$ on a*

$$\deg X \leq \deg V \delta^{\dim V - \dim X} \qquad \text{et}$$

$$h(X) \leq h(V)\delta^{\dim V - \dim X} + (\dim V - \dim X) \deg V \delta^{\dim V - \dim X - 1}(H + \sqrt{n}).$$

*Démonstration.* Imaginons d'abord que la famille $\mathcal{P}$ soit réduite à un polynôme $P$. Dans ce cas, ou bien $V \cap \mathcal{V}(P) = V$ et les formules sont claires ou bien $V \cap \mathcal{V}(\mathcal{P})$ est équidimensionnel de dimension $\dim V - 1$. Dans cette dernière situation, le théorème 3.4 de [R1, p. 112] combiné avec le corollaire 3.6 qui le suit [R1, p. 116] montre

$$\deg X \leq \delta \deg X \qquad \text{et} \qquad h(X) \leq \delta h(V) + (\deg V) h_m(P)$$

avec la hauteur modifiée introduite dans [R1, p. 111]. Maintenant, grâce au lemme 5.2 de [R2, p. 300], nous avons $h_m(P) \leq h(P) + \sqrt{n} \leq H + \sqrt{n}$ et cela donne bien les formules de l'énoncé.

Dans le cas général, nous procédons par récurrence sur $d = \dim V - \dim X$. A nouveau, il n'y a rien à faire si $d = 0$. Si $d \geq 1$, on peut trouver une famille $\mathcal{P}' \subset \mathcal{P}$, un élément $P \in \mathcal{P} \setminus \mathcal{P}'$ et une composante $Y$ de $V \cap \mathcal{V}(\mathcal{P}')$ de sorte que $\dim Y = \dim X + 1$ et $X$ est une composante de $Y \cap \mathcal{V}(P)$. D'après ce qui précède

$$\deg X \leq \delta \deg Y \qquad \text{et} \qquad h(X) \leq \delta h(Y) + (\deg Y)(H + \sqrt{n})$$

tandis que par hypothèse de récurrence

$$\deg Y \leq \deg V \delta^{d-1} \qquad \text{et}$$

$$h(X) \leq h(V)\delta^{d-1} + (d-1) \deg V \delta^{d-2}(H + \sqrt{n}).$$

En combinant, nous obtenons exactement le résultat. □

Pour notre dernier énoncé préliminaire, nous utilisons le fait élémentaire suivant : si $f$ est un polynôme non nul (sur un corps de caractéristique 0) de degré au plus $d$ en chacune de ses variables, il existe un point $x$ à coordonnées dans $\{0, 1, \ldots, d\}$ tel que $f(x) \neq 0$.

**Lemme 2.2.** *Soient $V$ un sous-schéma fermé intègre de $\mathbb{P}^n_{\mathbb{Q}}$ et $Z$ un hyperplan de $\mathbb{P}^n_{\mathbb{Q}}$ ne contenant pas $V$. Alors il existe un sous-schéma fermé irréductible $X$ de dimension 0 de $V$ ne rencontrant pas $Z$ tel que $\deg X \leq \deg V$ et $h(X) \leq h(V) + (\dim V)(\deg V)(\log \deg V + \sqrt{n})$.*

*Démonstration.* Remarquons que la condition $X \cap Z = \emptyset$ entraîne $\dim X \leq 0$. Notons $f$ une forme éliminante de $V$, $L_0$ une forme linéaire telle que $Z = V(L_0)$ et $L_1, \ldots, L_{\dim V}$ des formes linéaires telles que

$$f(L_0, L_1, \ldots, L_{\dim V}) \neq 0.$$

Il en existe car $V \not\subset Z$ et nous pouvons même les choisir à coefficients dans $\{0, \ldots, \deg V\}$. Alors n'importe quelle composante irréductible $X$ de l'intersection $V \cap \mathcal{V}(L_1, \ldots, L_{\dim V})$ répond à la question car $\dim V \cap \mathcal{V}(L_1, \ldots, L_{\dim V}) \geq 0$ (dimension d'une intersection dans $\mathbb{P}^n$) et $X \cap Z = \emptyset$ par le théorème de l'élimination. Ainsi $\dim X = 0$ et les formules pour $\deg X$ et $h(X)$ découlent du lemme précédent avec $\delta = 1$ et $H = \log \deg V$. □

## 3 Minimum local

Nous établissons ici l'énoncé suivant.

**Lemme 3.1.** *Soient $n$ et $D$ deux entiers naturels. Si un polynôme $P$ de degré au plus $D$ de $\mathbb{Z}[X_1, \ldots, X_n]$ admet un minimum local en $x \in \mathbb{R}^n$ tel que $P(x) > 0$ alors*

$$\log P(x) \geq -((n+1)D^n - 1) \log |P| - 2n^2 D^{n+1}.$$

*Démonstration.* Le cas d'un polynôme constant est clair et un polynôme de degré exactement 1 n'a pas de minimum local donc nous pouvons supposer $D \geq 2$ et $n \geq 1$.

Par hypothèse, nous avons clairement $(\partial P/\partial X_i)(x) = 0$ pour $1 \le i \le n$. Nous considérons alors le sous-schéma fermé $Y$ de $\mathbb{P}^n_{\mathbb{Q}}$ défini par les homogénéisés $Q_i$ des $\partial P/\partial X_i$ puis une composante irréductible (sur $\mathbb{Q}$) $Y_1$ de $Y$ telle que $Y_1 \times \operatorname{Spec}\mathbb{R}$ contienne le point fermé $(1 : x)$. D'après les formules d'intersection du lemme 2.1 (puisque $\deg Q_i \le D - 1$ et $|Q_i| \le D|P|$) nous trouvons

$$\deg Y_1 \le (D-1)^{\operatorname{codim} Y_1} \qquad \text{et}$$
$$h(Y_1) \le (D-1)^{\operatorname{codim} Y_1} n \log(n+1) + (\operatorname{codim} Y_1)(D-1)^{\operatorname{codim} Y_1 - 1}(\log D|P| + \sqrt{n})$$

(rappelons $h(\mathbb{P}^n_{\mathbb{Q}}) \le n \log(n+1)$). Par conséquent, par le lemme 2.2, il existe un sous-schéma fermé $X$ de $Y_1$ irréductible sur $\mathbb{Q}$ de dimension 0, ne rencontrant pas l'hyperplan à l'infini et tel que

$$\deg X \le (D-1)^n \qquad \text{et}$$
$$\begin{aligned} h(X) \le{} & (D-1)^n n \log(n+1) + (\operatorname{codim} Y_1)(D-1)^{n-1}(\log D|P| + \sqrt{n}) \\ & + (\dim Y_1)(D-1)^{n-1}((n-1)\log(D-1) + \sqrt{n}) \end{aligned}$$

(nous avons majoré $\operatorname{codim} Y_1$ par $n$ dans les exposants sauf dans le dernier terme car si $\dim Y_1 \ne 0$ on a $\operatorname{codim} Y_1 \le n-1$). Pour conclure, nous allons vérifier qu'il existe un point $y \in X(\mathbb{C})$ tel que, si $y = (1 : y_1 : \cdots : y_n)$, alors $P(y_1, \ldots, y_n) = P(x)$. Il suffit pour cela de choisir $y$ dans la même composante irréductible $Z$ de $Y_1 \times \operatorname{Spec}\mathbb{C}$ que $(1 : x)$. Ceci est possible car $X$ est défini sur $\mathbb{Q}$ et $Y_1$ irréductible sur $\mathbb{Q}$. Maintenant $P$ est constant sur $Z \cap \mathbb{C}^n$ puisque toutes ses dérivées y sont nulles.

Ainsi il reste à minorer $P(y_1, \ldots, y_n)$. Bien entendu, comme $\dim X = 0$, ce nombre est algébrique de degré au plus $\deg X$. Par suite l'inégalité de Liouville donne $\log P(y_1, \ldots, y_n) \ge -(\deg X) h(1, P(y_1, \ldots, y_n))$ et, d'autre part,

$$h(1, P(y_1, \ldots, y_n)) \le \log|P| + (D/2)\log(n+1) + Dh(\{y\}).$$

Ainsi

$$\log P(x) \ge -(\deg X)\log|P| - (D/2)(\deg X)\log(n+1) - Dh(X)$$

car $(\deg X) h(\{y\}) = h(X)$ puisque $X$ n'est autre que la réunion des conjugués de $y$. Si l'on substitue dans cette expression les majorations pour $h(X)$ et $\deg X$ on constate que le coefficient de $\log|P|$ est

$$-(D-1)^n - (\operatorname{codim} Y_1) D(D-1)^{n-1} \ge -(n+1)D^n + 1$$

comme prévu. Pour le reste, il suffit de montrer

$$\begin{aligned} & (D/2)(D-1)^n \log(n+1) + D(D-1)^n n \log(n+1) \\ & + (\operatorname{codim} Y_1) D(D-1)^{n-1}(\log D + \sqrt{n}) \\ & + (\dim Y_1) D(D-1)^{n-1}((n-1)\log(D-1) + \sqrt{n}) \le 2n^2 D^{n+1}. \end{aligned}$$

Si $n = 1$, on remarque que $\dim Y_1 = 0$ et donc ceci équivaut à $(3/2)D(D-1)\log 2 + D(\log D + 1) \le 2D^2$ facilement vrai. Sinon on majore aussi bien $\log D + \sqrt{n}$ que $(n-1)\log(D-1) + \sqrt{n}$ par $(n - 1/2)D$ de sorte que le membre de gauche ci-dessus est au plus

$$(n + 1/2) D^{n+1} \log(n+1) + (\operatorname{codim} Y_1 + \dim Y_1) D^n (n - 1/2) D.$$

Enfin $\operatorname{codim} Y_1 + \dim Y_1 = n$ et $\log(n+1) \le n$ donc, en substituant, notre majorant devient exactement $2n^2 D^{n+1}$. □

## 4 Minimum sur un pavé

**Lemme 4.1.** *Soient $n$ et $D$ deux entiers naturels puis $r_1, \ldots, r_n$ et $s_1, \ldots, s_n$ des éléments de $\mathbb{Q}$. Notons $A$ le pavé de $\mathbb{R}^n$ donné par*

$$A = \prod_{i=1}^{n} [r_i, s_i].$$

*Si un polynôme $P$ de degré au plus $D$ de $\mathbb{Z}[X_1, \ldots, X_n]$ ne s'annule pas sur $A$, alors*

$$\min_{x \in A} \log |P(x)| \ge -(n+1) D^n \log |P| - n D^n \eta - 2n^2 D^{n+1}$$

*où $\eta = h(1, r_1, \ldots, r_n, s_1, \ldots, s_n)$.*

*Démonstration.* Nous notons $a$ le plus petit dénominateur commun des $r_i, s_i$ de sorte que $\eta = \log \max(a, |ar_1|, \ldots, |ar_n|, |as_1|, \ldots, |as_n|)$. Considérons maintenant un point $x \in A$ où le minimum est atteint. Quitte à changer $P$ en $-P$, nous supposons $P(x) > 0$. Dans ces conditions, si $x$ est un point intérieur de $A$, un minimum local de $P$ est atteint en $x$ et le résultat découle immédiatement du lemme 3.1. Sinon $x$ est au bord de $A$. Quitte à permuter les coordonnées, nous supposons que pour un entier $m$ avec $0 \le m \le n-1$ nous avons

$$r_i < x_i < s_i \quad \text{pour} \quad 1 \le i \le m$$
$$x_i = r_i \quad \text{ou} \quad x_i = s_i \quad \text{pour} \quad m < i \le n.$$

Considérons alors le polynôme $R \in \mathbb{Z}[X_1, \ldots, X_m]$ défini par

$$R(X_1, \ldots, X_m) = a^D P(X_1, \ldots, X_m, x_{m+1}, \ldots, x_n).$$

Par construction $R$ admet un minimum local au point $(x_1, \ldots, x_m)$. Donc d'après le lemme 3.1 nous avons

$$\log P(x) = \log(R(x_1, \ldots, x_m) a^{-D}) \ge -((m+1)D^m - 1) \log |R| - 2m^2 D^{m+1} - D \log a.$$

A présent, chaque coefficient de $R$ s'obtient en évaluant un polynôme de degré au plus $D$ de $\mathbb{Z}[X_{m+1}, \ldots, X_n]$ en $(x_{m+1}, \ldots, x_n)$. Comme un tel polynôme a au plus $(D+1)^{n-m}$ coefficients, il vient

$$\log |R| \le \log |P| + D\eta + (n-m) \log(D+1).$$

Cela entraîne alors (avec $\log a \le \eta$)

$$\log P(x) \ge -((m+1)D^m - 1) \log |P| - (m+1) D^{m+1} \eta - 2m^2 D^{m+1} - (n-m)(m+1) D^m \log(D+1).$$

Finalement $m + 1 \le n$ et $2m^2 + (n-m)(m+1) \le 2n^2$ donc

$$\log P(x) \ge -n D^{n-1} \log |P| - n D^n \eta - 2n^2 D^n$$

ce qui termine la démonstration. □

# 5 Conclusion

Nous nous plaçons sous les hypothèses du théorème 1.1 et nous posons $\rho = \operatorname{dist}^*(x, Z_{\mathbb{R}}(P))$. Si $\rho = 0$, nous avons $P(x) = 0$ et le résultat est évident. Nous supposons donc $\rho > 0$ et définissons $a$ comme l'unique entier tel que

$$\frac{1}{\rho} < a \leq \frac{1}{\rho} + 1.$$

Ensuite, pour $1 \leq i \leq n$, nous introduisons $b_i$ comme l'unique entier tel que

$$b_i \leq a x_i < b_i + 1.$$

De cette façon, nous avons

$$x_i - \rho < \frac{b_i}{a} \leq x_i \leq \frac{b_i + 1}{a} < x_i + \rho$$

et donc $x$ est élément du pavé

$$A = \prod_{i=1}^{n} \left[ \frac{b_i}{a}, \frac{b_i + 1}{a} \right]$$

dont tout point $y$ vérifie $|x - y| < \rho$. Par suite, $P$ ne s'annule pas sur $A$ et nous pouvons appliquer le lemme 4.1. Il reste à évaluer

$$\eta = \max(a, |b_1|, |b_1 + 1|, \ldots, |b_n|, |b_n + 1|).$$

Or, par définition, $a \leq (1/\rho) + 1 \leq 2/\rho$ et $\max(|b_i|, |b_i + 1|) \leq a|x_i| + 1 \leq (2/\rho)|x|^* + 1 \leq 3|x|^*/\rho$. Par conséquent, nous avons $\eta \leq 3|x|^*/\rho$ et finalement

$$|P(x)| \geq \min_{y \in A} |P(y)| \geq |P|^{-(n+1)D^n} \eta^{-nD^n} e^{-2n^2 D^{n+1}} \geq (e^{2nD} H)^{-(n+1)D^n} \left( \frac{\rho}{|x|^*} \right)^{nD^n}.$$

**Abstract An arithmetic Łojasiewicz inequality.** An explicit lower bound is given for the value of a polynomial $P$ in $n$ variables with integer coefficients at any point $x$ with real coordinates in terms of the distance of $x$ to the real zeroes of $P$, the norm of $x$ and the size of the coefficients of $P$.

# Références

[B] Brownawell, W.D.: The Hilbert Nullstellensatz, inequalities for polynomials, and algebraic independence. In: Nesterenko, Yu., Philippon, P. (eds.) *Introduction to Algebraic Independence Theory.* Lect. Notes Math., vol. 1752, pp. 239–248. Springer, Heidelberg (2001)

[R1] Rémond, G.: Géométrie diophantienne multiprojective. In: Nesterenko, Yu., Philippon, P. (eds.) *Introduction to Algebraic Independence Theory.* Lect. Notes Math., vol. 1752, pp. 95–131. Springer, Heidelberg (2001)

[R2] Rémond, G.: Sur le théorème du produit. *J. Théor. Nombres Bordx.* **13**, 287–302 (2001)

*Diophantine Approximation*
H. P. Schlickewei et al., Editors


# ON THE CONTINUED FRACTION EXPANSION OF A CLASS OF NUMBERS

Damien Roy
*Département de Mathématiques, Université d'Ottawa, 585 King Edward Ottawa, Ontario K1N 6N5, Canada*
droy@uottawa.ca

Au Professeur Wolfgang Schmidt, avec mes meilleurs vœux et toute mon admiration

## 1 Introduction

A classical result of Dirichlet asserts that, for each real number $\xi$ and each real $X \geq 1$, there exists a pair of integers $(x_0, x_1)$ satisfying

$$1 \leq x_0 \leq X \quad \text{and} \quad |x_0\xi - x_1| \leq X^{-1}$$

(a general reference is Chapter I of [10]). If $\xi$ is irrational, then, by letting $X$ tend to infinity, this provides infinitely many rational numbers $x_1/x_0$ with $|\xi - x_1/x_0| \leq x_0^{-2}$. By contrast, an irrational real number $\xi$ is said to be *badly approximable* if there exists a constant $c_1 > 0$ such that $|\xi - p/q| > c_1 q^{-2}$ for each $p/q \in \mathbb{Q}$ or, equivalently, if $\xi$ has bounded partial quotients in its continued fraction expansion. Thanks to H. Davenport and W. M. Schmidt, the badly approximable real numbers can also be described as those $\xi \in \mathbb{R} \setminus \mathbb{Q}$ for which the result of Dirichlet can be improved in the sense that there exists a constant $c_2 < 1$ such that the inequalities $1 \leq x_0 \leq X$ and $|x_0\xi - x_1| \leq c_2 X^{-1}$ admit a solution $(x_0, x_1) \in \mathbb{Z}^2$ for each sufficiently large $X$ (see Theorem 1 of [2]).

If $\xi$ is rational or quadratic real, then, upon writing $\xi^2 = (q\xi + r)/p$ for integers $p$, $q$ and $r$ with $p \neq 0$ and putting $c_3 = |p| \max\{|p|, |q|\}$, one deduces from the result of Dirichlet that, for each $X \geq 1$, there exists a point $(x_0, x_1, x_2) \in \mathbb{Z}^3$ satisfying

$$1 \leq x_0 \leq X, \quad |x_0\xi - x_1| \leq c_3 X^{-1} \quad \text{and} \quad |x_0\xi^2 - x_2| \leq c_3 X^{-1}.$$

Conversely, Davenport and Schmidt proved that for each real number $\xi$ which is neither rational nor quadratic over $\mathbb{Q}$, there is a constant $c_4 > 0$ such that, upon writing $\gamma = (1 + \sqrt{5})/2$, the system of inequations

$$|x_0| \leq X, \quad |x_0\xi - x_1| \leq c_4 X^{-1/\gamma}, \quad |x_0\xi^2 - x_2| \leq c_4 X^{-1/\gamma}, \tag{1}$$

admits no nonzero integer solution $(x_0, x_1, x_2) \in \mathbb{Z}^3$ for arbitrarily large values of $X$ (Theorem 1a of [3]). Since $1/\gamma \simeq 0.618 < 1$, this establishes a clear gap between the

*Keywords.* Badly approximable numbers, continued fractions, extremal real numbers, Fibonacci sequences, words.

*1991 Mathematics subject classification.* Primary 11J70; Secondary 11J04, 11J13.

set of rational or quadratic real numbers and the remaining real numbers. Moreover, this result of Davenport and Schmidt is best possible in the following sense. There exist real numbers $\xi$ which are neither rational nor quadratic and for which there is a constant $c_5 > 0$ such that the system (1), with $c_4$ replaced by $c_5$, admits a nonzero integer solution for each $X \geq 1$ (Theorem 1.1 of [7]). These real numbers, which we call *extremal*, present from this point of view a closest behavior to quadratic real numbers. An application of Schmidt's subspace theorem proves them to be transcendental over $\mathbb{Q}$ (see Theorem 1B in Chapter VI of [10]). Still they possess several properties that make them resemble to quadratic real numbers. In the present paper, we are interested in their approximation by rational numbers.

It is well known that each quadratic real number has an ultimately periodic continued fraction expansion and so is badly approximable. Since there exist extremal real numbers which are badly approximable [6], this raises the question as to whether or not each extremal real number is such. At present, we simply know that an extremal real number $\xi$ satisfies a measure of approximation by rational numbers $p/q$ of the form

$$\left|\xi - \frac{p}{q}\right| \geq c_6 q^{-2}(1 + \log|q|)^{-t},$$

with constants $c_6 > 0$ and $t \geq 0$ depending only on $\xi$ (Theorem 1.3 of [7]). In this paper, we establish a sufficient condition for an extremal real number to have bounded partial quotients and construct new examples of such numbers.

## 2 Notation and statements of the main results

A *Fibonacci sequence* in a monoid is a sequence $(w_i)_{i\geq 1}$ of elements of this monoid which satisfies the recurrence relation $w_{i+2} = w_{i+1}w_i$ for each $i \geq 1$. Here, we shall work with two types of monoids.

One is the monoid of words $E^*$ on an alphabet $E$, with the product given by concatenation of words. A Fibonacci sequence $(w_i)_{i\geq 1}$ in $E^*$ has the property that $w_i$ is a prefix (left-factor) of $w_{i+1}$ for each $i \geq 2$ and so it admits a limit $w_\infty = \lim w_i$ in the completion of $E^*$ for pointwise convergence. This limit is an infinite word unless $w_1$ and $w_2$ are empty. For example, if $E = \{a, b\}$ consists of two distinct elements $a$ and $b$, then the Fibonacci sequence of words starting with $w_1 = b$ and $w_2 = a$ converges to the infinite word $f_{a,b} = abaababa\ldots$. In general, the limit of any Fibonacci sequence of words $(w_i)_{i\geq 1}$ derives from this generic infinite word $f_{a,b}$ by substituting into it the words $w_1$ and $w_2$ for the letters $b$ and $a$ respectively. It will come out indirectly of our analysis that such a limit is an infinite nonultimately periodic word if and only if $w_1$ and $w_2$ do not commute (see the remark after Theorem 2.2). A direct proof of this fact has been recently provided by B. Lucier [5].

The other monoid is in fact a group. It is constructed as follows. Define the *content* $c(A)$ of a nonzero matrix $A$ in $\mathrm{Mat}_{2\times 2}(\mathbb{Z})$ to be the greatest positive common divisor of its coefficients and say that such a matrix $A$ is *primitive* if $c(A) = 1$. For each nonzero $A \in \mathrm{Mat}_{2\times 2}(\mathbb{Z})$, denote by $A^{\mathrm{red}}$ the unique primitive integer matrix such that $A = c(A)A^{\mathrm{red}}$. Then, the set $\mathcal{P}$ of all primitive matrices with nonzero determinant in $\mathrm{Mat}_{2\times 2}(\mathbb{Z})$ is a group for the operation $*$ given by $A * B = (AB)^{\mathrm{red}}$. Its quotient $\mathcal{P}/\{\pm I\}$ is isomorphic to $\mathrm{PGL}_2(\mathbb{Q})$.

**Definition 1.** We say that a Fibonacci sequence $(W_i)_{i\geq 1}$ in $\mathcal{P}$ is *admissible* if there exists a nonsymmetric and non-skew-symmetric matrix $N \in \mathcal{P}$ such that, upon putting $N_i = {}^tN$ for $i$ odd and $N_i = N$ for $i$ even, the product $W_iN_i$ is a symmetric matrix for each $i \geq 1$.

This definition differs slightly from that in §3 of [9]. However, the same argument as in the proof of Proposition 3.1 of [9] shows that most Fibonacci sequences in $\mathcal{P}$ are admissible in the sense that there exists a nonempty Zariski open subset $\mathcal{U}$ of $\mathrm{GL}_2(\mathbb{C})^2$ such that any pair $(W_1, W_2) \in \mathcal{U} \cap \mathcal{P}$ generates an admissible Fibonacci sequence in $\mathcal{P}$.

We define also the *norm* $\|A\|$ of a matrix $A$ with real coefficients to be the largest absolute value of its coefficients. With this notation, we will prove in Sect. 3 the following characterization of extremal real numbers which translates in the present setting several results from [7] and [8].

**Theorem 2.1.** *For each extremal real number $\xi$ there exists an unbounded admissible Fibonacci sequence $(W_i)_{i\geq 1}$ in $\mathcal{P}$ which satisfies*

$$\|W_{i+1}\| \gg\ll \|W_i\|^{\gamma}, \quad \|(\xi, -1)W_i\| \gg\ll \|W_i\|^{-1} \quad \text{and} \quad |\det W_i| \gg\ll 1, \tag{2}$$

*with implied constants that are independent of $i$. Such a sequence is uniquely determined by $\xi$ up to its first terms, and up to term-by-term multiplication by a Fibonacci sequence in $\{\pm 1\}$. Conversely, any unbounded admissible Fibonacci sequence $(W_i)_{i\geq 1}$ in $\mathcal{P}$ which satisfies*

$$\|W_{i+2}\| \gg \|W_{i+1}\|\|W_i\| \quad \text{and} \quad |\det W_i| \ll 1, \tag{3}$$

*also satisfies the conditions* (2) *for some extremal real number $\xi$.*

Thus, any unbounded admissible Fibonacci sequence in $\mathcal{P}$ satisfying (3) is *associated* to some extremal real number $\xi$ in the sense that it satisfies (2). Note also that, since $\gamma^2 = \gamma + 1$, the first condition in (2) is stronger than the first condition in (3).

It is shown in §2 of [6] and in §6 of [7] that, for any choice of distinct positive integers $a$ and $b$, the real number $\xi_{a,b}$ whose continued fraction expansion $\xi_{a,b} = [0, a, b, a, a, \dots]$ is given by 0 followed by the elements of $f_{a,b}$ is an extremal real number. More generally, we will prove the following result (see Sect. 4).

**Theorem 2.2.** *A real number $\xi$ is extremal with an associated Fibonacci sequence in $\mathrm{GL}_2(\mathbb{Z})$ if and only if the sequence of its partial quotients in its continued fraction expansion coincides, up to its first terms, with the limit of a Fibonacci sequence of words $(w_i)_{i\geq 1}$ in $(\mathbb{N} \setminus \{0\})^*$ starting with two noncommuting words $w_1$ and $w_2$.*

Let $w_\infty = a_1a_2a_3\dots$ be the limit of a Fibonacci sequence of words $(w_i)_{i\geq 1}$ in $(\mathbb{N} \setminus \{0\})^*$ starting with nonempty words $w_1$ and $w_2$. If $w_1$ and $w_2$ commute, then $w_\infty = \lim_{i\to\infty}(w_1)^i$ is a periodic word and so $\xi = [0, a_1, a_2, \dots]$ is a quadratic real number. Conversely, if $w_1$ and $w_2$ do not commute, the above theorem shows that this real number $\xi$ is extremal. Since an extremal real number is not quadratic, the infinite word $w_\infty$ cannot in this case be ultimately periodic.

The next result provides a sufficient condition for an extremal real number to be badly approximable.

**Theorem 2.3.** *Let $E = \{a, b\}$ be an alphabet of two letters, let $(w_k)_{k\geq 1}$ be the Fibonacci sequence in $E^*$ generated by $w_1 = b$ and $w_2 = a$, and let $f_{a,b} = \lim_{k\to\infty} w_k$. Let $\xi$ be an extremal real number and let $(W_k)_{k\geq 1}$ be a Fibonacci sequence in $\mathcal{P}$ which is associated to $\xi$. Consider the morphism of monoids $\Phi\colon E^* \to \mathcal{P}$ mapping $w_k$ to $W_k$ for each $k \geq 1$. For each $i \geq 1$, denote by $u_i$ the prefix of $f_{a,b}$ with length $i$, and put $U_i = \Phi(u_i)$. Then, we have*

$$\|(\xi, -1)U_i\| \gg\ll \frac{|\det U_i|}{\|U_i\|} \tag{4}$$

*with implied constants that do not depend on $i$. Moreover, if the sequence* $(\det U_i)_{i\geq 1}$ *is bounded, then $\xi$ is badly approximable.*

It would be interesting to know if, conversely, the sequence $(\det U_i)_{i\geq 1}$ is bounded when $\xi$ is badly approximable. The proof of the above result is given in Sect. 5.

Going back to the definitions, we note that, if $\xi$ is badly approximable (resp. extremal) and if $a, b \in \mathbb{Q}$ with $a \neq 0$, then $a\xi + b$ and $1/\xi$ are as well badly approximable (resp. extremal). This implies that the set of badly approximable real numbers is stable under the action of $\mathrm{GL}_2(\mathbb{Q})$ on $\mathbb{R} \setminus \mathbb{Q}$ by linear fractional transformations. Our last main result, proved in Sect. 6, is that there exist orbits which do not contain any of the numbers produced by Theorem 2.2.

**Theorem 2.4.** *There exist badly approximable extremal real numbers which are not conjugate under the action of* $\mathrm{GL}_2(\mathbb{Q})$ *to any extremal real number having an associated Fibonacci sequence in* $\mathrm{GL}_2(\mathbb{Z})$.

## 3 Proof of Theorem 2.1

The following lemma gathers essentially all facts that we will need from [7] and [8].

**Lemma 3.1.** *Let $\xi$ be an extremal real number. Then, there exists an unbounded sequence of symmetric matrices $(\mathbf{y}_i)_{i\geq 1}$ in $\mathcal{P}$ such that, for each $i \geq 1$, we have*

$$\|\mathbf{y}_{i+1}\| \gg\ll \|\mathbf{y}_i\|^{\gamma}, \quad \|(\xi, -1)\mathbf{y}_i\| \gg\ll \|\mathbf{y}_i\|^{-1} \quad \text{and} \quad |\det \mathbf{y}_i| \gg\ll 1, \tag{5}$$

*with implied constants that are independent of $i$. Such a sequence $(\mathbf{y}_i)_{i\geq 1}$ is uniquely determined by $\xi$ up to its first terms and up to multiplication of each of its terms by $\pm 1$. Moreover, for any such sequence, there exists a nonsymmetric and non-skew-symmetric matrix $M \in \mathcal{P}$ such that*

$$\mathbf{y}_{i+2} = \pm \begin{cases} \mathbf{y}_{i+1} * M * \mathbf{y}_i & \text{if } i \text{ is odd,} \\ \mathbf{y}_{i+1} * {}^t M * \mathbf{y}_i & \text{if } i \text{ is even,} \end{cases} \tag{6}$$

*for any sufficiently large index $i$. Conversely, if $(\mathbf{y}_i)_{i\geq 1}$ is an unbounded sequence of symmetric matrices in $\mathcal{P}$ which satisfies a recurrence relation of the type* (6) *for some nonsymmetric matrix $M \in \mathcal{P}$, and if*

$$\|\mathbf{y}_{i+2}\| \gg \|\mathbf{y}_{i+1}\|\|\mathbf{y}_i\| \quad \text{and} \quad |\det \mathbf{y}_i| \ll 1, \tag{7}$$

*then $(\mathbf{y}_i)_{i\geq 1}$ also satisfies the estimates* (5) *for some extremal real number $\xi$.*

*Proof.* The first assertion in this proposition comes from Theorem 5.1 of [7] upon noting that for an arbitrary symmetric matrix $\mathbf{y} = \begin{pmatrix} y_0 & y_1 \\ y_1 & y_2 \end{pmatrix}$, we have

$$\|(\xi, -1)\mathbf{y}\| = \max\{|y_0\xi - y_1|, |y_1\xi - y_2|\} \gg\ll \max\{|y_0\xi - y_1|, |y_0\xi^2 - y_2|\},$$

with implied constants depending only on $\xi$. The second assertion follows from Proposition 4.1 of [8], the third one from Corollary 4.3 of [8], and the last one from Proposition 5.1 of [8]. □

In the proof of Theorem 2.1, we use repeatedly the following observation (the proof of which is omitted).

**Lemma 3.2.** *Let $(W_i)_{i\ge1}$, $(\mathbf{y}_i)_{i\ge1}$ and $(N_i)_{i\ge1}$ be sequences in $\mathcal{P}$, and let $\xi \in \mathbb{R}$. Assume that the sequence $(N_i)_{i\ge1}$ is bounded and that $\mathbf{y}_i = W_i * N_i$ for each $i \ge 1$. Then, we have*

$$\|W_i\| \gg\ll \|\mathbf{y}_i\|, \quad \|(\xi, -1)W_i\| \gg\ll \|(\xi, -1)\mathbf{y}_i\| \quad \text{and} \quad |\det(W_i)| \gg\ll |\det(\mathbf{y}_i)|,$$

*with implied constants that do not depend on $i$.*

*Proof of Theorem 2.1.* Let $\xi \in \mathbb{R}$ be extremal. Then, Lemma 3.1 provides an unbounded sequence of symmetric matrices $(\mathbf{y}_i)_{i\ge1}$ in $\mathcal{P}$ and a nonsymmetric and non-skew-symmetric matrix $M \in \mathcal{P}$ satisfying both the estimates (5) and the recurrence relation (6) for each sufficiently large $i$. Omitting if necessary a finite even number of initial terms in the sequence $(\mathbf{y}_i)_{i\ge1}$, we may assume, without loss of generality, that (6) holds for each $i \ge 1$. Then, for a suitable choice of signs, the formula $W_i = \pm(\mathbf{y}_i * M_i)$ with $M_i = {}^tM$ if $i$ is odd and $M_i = M$ if $i$ is even, defines an admissible Fibonacci sequence in $\mathcal{P}$ (the corresponding matrix $N$ is the inverse of $M$ in the group $\mathcal{P}$). Moreover, the estimates (5) together with Lemma 3.2 show that this sequence satisfies the conditions (2) of Theorem 2.1. This proves the first assertion of the theorem.

Now, let $(W'_i)_{i\ge1}$ be any unbounded admissible Fibonacci sequence satisfying, like $(W_i)_{i\ge1}$, the conditions (2), and let $N' \in \mathcal{P}$ such that, upon putting $N'_i = {}^tN'$ for $i$ odd and $N'_i = N'$ for $i$ even, the matrix $\mathbf{y}'_i = W'_i * N'_i$ is symmetric for each $i \ge 1$. Then, Lemma 3.2 shows that $(\mathbf{y}'_i)_{i\ge1}$ satisfies, like $(\mathbf{y}_i)_{i\ge1}$, the estimates (5). Consequently, by Lemma 3.1, there exist integers $k, \ell \ge 0$ such that $\mathbf{y}'_{i+k} = \pm\mathbf{y}_{i+\ell}$ for each $i \ge 1$. Since we have $\mathbf{y}_{i+2} = \pm(W_{i+1} * \mathbf{y}_i)$ and $\mathbf{y}'_{i+2} = \pm(W'_{i+1} * \mathbf{y}'_i)$ for $i \ge 1$, this implies that $W'_{i+k} = \pm W_{i+\ell}$ for each $i \ge 2$. Moreover, the signs $\pm$ in the last formula must come from a Fibonacci sequence in $\{\pm1\}$. This proves the second assertion of the theorem.

Finally, let $(W_i)_{i\ge1}$ be any unbounded admissible Fibonacci sequence satisfying the conditions (3) in Theorem 2.1, without reference to a given extremal real number $\xi$, and let $N \in \mathcal{P}$ such that, upon putting $N_i = {}^tN$ for $i$ odd and $N_i = N$ for $i$ even, the matrix $\mathbf{y}_i = W_i * N_i$ is symmetric for each $i \ge 1$. Then, Lemma 3.2 shows that $(\mathbf{y}_i)_{i\ge1}$ satisfies the conditions (7) in Lemma 3.1. Consequently it satisfies the stronger conditions (5) for some extremal real number $\xi$, and thus, by Lemma 3.2, satisfies the estimates (2) of Theorem 2.1 for the same $\xi$. □

## 4 Proof of Theorem 2.2

Serret's theorem asserts that two real numbers have continued fraction expansions which coincide up to their first terms if and only if these numbers belong to the same orbit under the action of $\mathrm{GL}_2(\mathbb{Z})$ by linear fractional transformations (Theorem 6B of [10]). Our proof for Theorem 2.2 is inspired from the proof of this result given by Cassels in §3, Chap. I of [1]. We break it into two propositions. To establish the first one, we need the following auxiliary result which provides the link with continued fractions.

**Lemma 4.1.** *Let $A = \left(\begin{smallmatrix} a & b \\ c & d \end{smallmatrix}\right) \in \mathrm{GL}_2(\mathbb{Z})$ with $d \geq 1$. Then, there is one and only one choice of integers $s \geq 1$ and $a_0, a_1, \dots, a_s$ with $a_1, \dots, a_{s-1} \geq 1$ such that*

$$A = \begin{pmatrix} a_0 & 1 \\ 1 & 0 \end{pmatrix} \begin{pmatrix} a_1 & 1 \\ 1 & 0 \end{pmatrix} \cdots \begin{pmatrix} a_s & 1 \\ 1 & 0 \end{pmatrix}. \tag{8}$$

*These integers are also characterized by the properties*

$$b/d = [a_0, \dots, a_{s-1}], \quad c/d = [a_s, \dots, a_1], \quad \det(A) = (-1)^{s+1}. \tag{9}$$

*Proof.* Induction on $s$ shows that, if $A$ can be written in the form (8) for a choice of integers $s \geq 1$ and $a_0, a_1, \dots, a_s$ with $a_1, \dots, a_{s-1} \geq 1$, then we have $b/d = [a_0, \dots, a_{s-1}]$. Taking the transpose of both sides of (8), this observation also provides $c/d = [a_s, \dots, a_1]$. Moreover the last equality in (9) follows from the multiplicativity of the determinant. Since each rational number has exactly two continued fraction expansions with lengths differing by one, this proves the uniqueness of the factorization (8), when it exists.

Now, without making assumptions on $A$, define an integer $s \geq 1$ and a sequence of integers $a_1, \dots, a_s$ with $a_1, \dots, a_{s-1} \geq 1$ by the conditions $c/d = [a_s, \dots, a_1]$ and $\det(A) = (-1)^{s+1}$. Define also $a_0$ to be the integer for which the distance between $b/d$ and $[a_0, \dots, a_{s-1}]$ is at most $1/2$. Then, by the above observations, the right hand side of (8) is a matrix $A' = \left(\begin{smallmatrix} a' & b' \\ c' & d' \end{smallmatrix}\right)$ with the same determinant as $A$, satisfying $d' \geq 1$, $c'/d' = c/d$ and $|b'/d' - b/d| \leq 1/2$. Since $(c, d)$ and $(c', d')$ are rows of matrices in $\mathrm{GL}_2(\mathbb{Z})$, they are primitive points of $\mathbb{Z}^2$ and the relation $c'/d' = c/d$ implies $(c', d') = \pm(c, d)$. Since $d$ and $d'$ are positive, we deduce that $A$ and $A'$ have the same second row $(c', d') = (c, d)$. Since these matrices also have the same determinant, this forces $(a', b') = (a, b) + k(c, d)$ for some integer $k$. Then, we find $|b'/d' - b/d| = |k|$, thus $k = 0$ and therefore $A' = A$. □

**Corollary 4.2.** *Let $\mathcal{S}_1$ denote the set of matrices $\left(\begin{smallmatrix} a & b \\ c & d \end{smallmatrix}\right) \in \mathrm{Mat}_{2\times 2}(\mathbb{R})$ with $a \geq \max\{b, c\}$ and $\min\{b, c\} \geq d \geq 0$, and define $\mathcal{S} = \mathcal{S}_1 \cap \mathrm{GL}_2(\mathbb{Z})$. Then $\mathcal{S}$ and $\mathcal{S}_1$ are closed under multiplication and transposition. Moreover, the map from $\mathbb{N} \setminus \{0\}$ to $\mathcal{S}$ sending $a$ to $\left(\begin{smallmatrix} a & 1 \\ 1 & 0 \end{smallmatrix}\right)$ for each $a \in \mathbb{N} \setminus \{0\}$ extends to an isomorphism of monoids $\sigma\colon (\mathbb{N} \setminus \{0\})^* \to \mathcal{S} \cup \{I\}$.*

*Proof.* The only delicate point here is the surjectivity of the map $\sigma$. Clearly, any matrix $A = \left(\begin{smallmatrix} a & b \\ c & d \end{smallmatrix}\right) \in \mathcal{S}$ is in the image of $\sigma$ if $d = 0$ because then we have $a \geq b = c = 1$. If $d \geq 1$, we note that the integers $a_0$ and $a_s$ given by Lemma 4.1 are respectively the integral parts of $b/d$ and $c/d$ except in the case where $d = 1$ and $\det A = -1$. In the latter case, we have $s = 2$, $a_0 = b - 1$, $a_1 = 1$, $a_2 = c - 1$, and also $a = bc - 1$. Then, the condition $a \geq \max\{b, c\} \geq 1$ implies $b, c \geq 2$ and so $a_0, a_s \geq 1$. Otherwise, the condition $\min\{b, c\} \geq d \geq 1$ ensures that $a_0, a_s \geq 1$. So, in both cases, the integers $a_0, \ldots, a_s$ are positive, and $A$ is the image of $(a_0, \ldots, a_s)$ under $\sigma$. □

The following proposition presents a first step towards the proof of Theorem 2.2.

**Proposition 4.3.** *The set of extremal real numbers with an associated (admissible) Fibonacci sequence in* $\mathrm{GL}_2(\mathbb{Z})$ *is stable under the action of* $\mathrm{GL}_2(\mathbb{Z})$ *by linear fractional transformations. Any orbit contains an extremal real number with an associated Fibonacci sequence in* $\mathcal{S}$.

*Proof.* Let $\xi$ be an extremal real number with an associated Fibonacci sequence $(W_i)_{i\geq 1}$ in $\mathrm{GL}_2(\mathbb{Z})$. This sequence being admissible, there exists $N \in \mathcal{P}$ with $N \neq \pm\,{}^tN$ such that, upon putting $N_i = N$ if $i$ is even and $N_i = {}^tN$ if $i$ is odd, the product $\mathbf{y}_i = W_i N_i$ is symmetric for each $i \geq 1$.

For each $U \in \mathrm{GL}_2(\mathbb{Z})$, the sequence $(W_i')_{i\geq 1} = (U^{-1} W_i U)_{i\geq 1}$ is a Fibonacci sequence in $\mathrm{GL}_2(\mathbb{Z})$. It is admissible with corresponding matrix $N' = U^{-1} N\, {}^tU^{-1}$, and it satisfies the conditions (2) of Theorem 2.1 with $W_i$ replaced by $W_i'$ and $\xi$ replaced by the real number $\eta$ such that $(\eta, -1)$ is proportional to $(\xi, -1)U$. By varying $U$, we get in this way all real numbers $\eta$ which are conjugate to $\xi$ under $\mathrm{GL}_2(\mathbb{Z})$. So, these numbers are extremal with an associated Fibonacci sequence in $\mathrm{GL}_2(\mathbb{Z})$. This proves the first assertion of the lemma.

For the second assertion, let $1/\xi = [a_0, a_1, a_2, \ldots]$ be the continued fraction expansion of $1/\xi$. Put $d = \det(N)$ and $M = dN^{-1}$, so that we have $M \in \mathcal{P}$ and $W_i = d^{-1}\mathbf{y}_i M_i$, where $M_i = M$ if $i$ is even and $M_i = {}^tM$ if $i$ is odd. Since $M$ is not skew-symmetric and since $[\mathbb{Q}(\xi) : \mathbb{Q}] > 2$, the product

$$\theta = \begin{pmatrix} 1/\xi & 1 \end{pmatrix} M \begin{pmatrix} 1/\xi \\ 1 \end{pmatrix}$$

is nonzero. Replacing $N$ by $-N$ if necessary, so that $M$ is replaced by $-M$, we may assume without loss of generality that this number $\theta$ is positive. For each $k \geq 1$, define

$$U_k = \begin{pmatrix} a_0 & 1 \\ 1 & 0 \end{pmatrix} \begin{pmatrix} a_1 & 1 \\ 1 & 0 \end{pmatrix} \cdots \begin{pmatrix} a_k & 1 \\ 1 & 0 \end{pmatrix}.$$

Then, the standard recurrence relations in the theory of continued fractions show that we have

$$U_k = \begin{pmatrix} p_k & p_{k-1} \\ q_k & q_{k-1} \end{pmatrix},$$

where $p_k/q_k = [a_0, \ldots, a_k]$ denotes the $k$-th convergent of $1/\xi$ written in reduced form. Since $|q_k(1/\xi) - p_k| \leq 1/q_{k+1}$ for each $k \geq 0$, this gives

$$U_k = \begin{pmatrix} 1/\xi \\ 1 \end{pmatrix} \begin{pmatrix} q_k & q_{k-1} \end{pmatrix} + \mathcal{O}(1/q_k)$$

and thus

$${}^tU_k M U_k = \theta \begin{pmatrix} q_k^2 & q_{k-1}q_k \\ q_{k-1}q_k & q_{k-1}^2 \end{pmatrix} + \mathcal{O}(1).$$

The latter matrix belongs to $\mathcal{S}_1$ if $k$ sufficiently large, because we have $q_k > q_{k-1}$ for each $k \geq 2$, and $q_{k-1}$ tends to infinity with $k$. Fix such a value of $k$. Since $\mathcal{S}_1$ is closed under transposition, we get ${}^tU_k M_i U_k \in \mathcal{S}_1$ for each $i \geq 1$. We claim that $\epsilon_i U_k^{-1} W_i U_k$ also belongs to $\mathcal{S}_1$ for an appropriate choice of $\epsilon_i \in \{-1, 1\}$ and each sufficiently large $i$. To prove this, we note that the product $U_k^{-1} \binom{1/\xi}{1}$ is proportional to $\binom{r}{1}$ where $r = [a_{k+1}, a_{k+2}, \dots]$ is a real number with $r > 1$. Since, for each $i \geq 1$, we have

$$\mathbf{y}_i = y_{i,2} \begin{pmatrix} \xi^{-2} & \xi^{-1} \\ \xi^{-1} & 1 \end{pmatrix} + \mathcal{O}(\|W_i\|^{-1})$$

with $y_{i,2} \in \mathbb{Z}$, we find

$$U_k^{-1} \mathbf{y}_i \, {}^tU_k^{-1} = c_i \begin{pmatrix} r^2 & r \\ r & 1 \end{pmatrix} + \mathcal{O}(\|W_i\|^{-1}),$$

for some $c_i \in \mathbb{R}$. Thus, if $i$ is sufficiently large, say, $i \geq i_0$, the matrix $\pm U_k^{-1} \mathbf{y}_i \, {}^tU_k^{-1}$ belongs to $\mathcal{S}_1$ for an appropriate choice of sign $\pm$. Multiplying this matrix on the right by ${}^tU_k M_i U_k$ which also belongs to $\mathcal{S}_1$, we deduce that $\pm U_k^{-1} \mathbf{y}_i M_i U_k \in \mathcal{S}_1$ for the same choice of sign and thus that $\epsilon_i U_k^{-1} W_i U_k \in \mathcal{S}_1$ for some $\epsilon_i \in \{-1, 1\}$. Since $U_k \in \mathrm{GL}_2(\mathbb{Z})$ and since $W_i \in \mathrm{GL}_2(\mathbb{Z})$ for each $i \geq 1$, we conclude that $(\epsilon_i U_k^{-1} W_i U_k)_{i \geq i_0}$ is an admissible Fibonacci sequence in $\mathcal{S}$. By the first part of the proof, it is associated to an extremal real number $\eta$ in the same $\mathrm{GL}_2(\mathbb{Z})$-orbit as $\xi$. □

We also need the following technical result.

**Lemma 4.4.** *Let $(W_i)_{i \geq 1}$ be a Fibonacci sequence in $\mathcal{P}$. If $W_1$ and $W_2$ do not have a common eigenvector in $\mathbb{Q}^2$ and satisfy $W_1 W_2 \neq \pm W_2 W_1$, then $(W_i)_{i \geq 1}$ is an admissible Fibonacci sequence.*

*Proof.* We first note that there exists a nonzero primitive matrix $N \in \mathrm{Mat}_{2\times 2}(\mathbb{Z})$ such that $W_1 \, {}^tN$, $W_2 N$ and $W_3 \, {}^tN$ are symmetric because these three conditions translate into a system of three homogeneous linear equations in the four unknown coefficients of $N$. Fix such a choice of $N$ and define accordingly $N_i = {}^tN$ for $i$ odd and $N_i = N$ for $i$ even. Then, the product $W_i N_i$ is symmetric for $i = 1, 2, 3$ and using the relation of proportionality

$$W_{i+3} N_{i+3} \propto (W_{i+1} N_{i+1}) N_{i+1}^{-1} (W_i N_i) N_i^{-1} (W_{i+1} N_{i+1}),$$

we deduce by induction on $i$ that $W_i N_i$ is symmetric for each $i \geq 1$.

If $\det N = 0$, then we can write $N = A \, {}^tB$ with nonzero column vectors $A$ and $B$ in $\mathbb{Q}^2$. Since $W_1 \, {}^tN = (W_1 B) \, {}^tA$ is symmetric, we deduce that $W_1 B \propto A$. Similarly, since $W_2 N = (W_2 A) \, {}^tB$ and $W_3 \, {}^tN = (W_3 B) \, {}^tA$ are symmetric, we find that $W_2 A \propto B$ and $W_3 B \propto A$. Using the first two relations of proportionality, we also get $W_3 B \propto W_2(W_1 B) \propto W_2 A \propto B$. As $W_3 B \neq 0$, this shows that $A \propto B$, and thus that $B$ is a common eigenvector of $W_1$ and $W_2$, against the hypothesis. Thus we have $N \in \mathcal{P}$. We also note that

$$W_2 W_1 \, {}^tN = {}^t(W_2 W_1 \, {}^tN) = {}^t(W_1 \, {}^tN) \, {}^tW_2 = (W_1 \, {}^tN) \, {}^tW_2 = W_1 {}^t(W_2 N) = W_1 W_2 N.$$

Since $W_1 W_2 \neq \pm W_2 W_1$, this implies that ${}^tN \neq \pm N$. Thus, the sequence $(W_i)_{i \geq 1}$ is admissible. □

The hypotheses of Lemma 4.4 are satisfied for example when the matrices $W_1$, $W_2$, $W_1W_2$ and $W_2W_1$ are linearly independent over $\mathbb{Q}$. The corollary below provides another instance where this lemma applies.

**Corollary 4.5.** *Any Fibonacci sequence $(W_i)_{i\geq 1}$ in $\mathcal{S}$ generated by two noncommuting matrices $W_1, W_2 \in \mathcal{S}$ is admissible.*

*Proof.* Since $\mathcal{S} \subset \mathrm{GL}_2(\mathbb{Z})$, the eigenvalues of a matrix $W \in \mathcal{S}$ are algebraic units. So, if one of them is rational, both of them belong to $\{-1, 1\}$. Since the only matrices of $\mathcal{S}$ with trace at most 2 are $\left(\begin{smallmatrix}1&1\\1&0\end{smallmatrix}\right)$ and $\left(\begin{smallmatrix}2&1\\1&0\end{smallmatrix}\right)$ which have no rational eigenvalue, we deduce that no matrix of $\mathcal{S}$ has a rational eigenvalue. In particular, any $W_1, W_2 \in \mathcal{S}$ do not share a common eigenvector in $\mathbb{Q}^2$. Since such matrices have nonnegative coefficients and nonzero product, they also satisfy $W_1W_2 \neq -W_2W_1$. Thus, if they do not commute, Lemma 4.4 shows that they generate an admissible Fibonacci sequence. □

Serret's theorem combined with Proposition 4.3 reduces the proof of Theorem 2.2 to the following statement.

**Proposition 4.6.** *A real number $\xi$ is extremal with an associated Fibonacci sequence in $\mathcal{S}$ if and only if its continued fraction expansion is of the form $[0, a_1, a_2, \dots]$ where $(a_1, a_2, \dots)$ is the limit of a Fibonacci sequence of words $(w_i)_{i\geq 1}$ in $(\mathbb{N}\setminus\{0\})^*$ starting with two noncommuting words $w_1$ and $w_2$.*

*Proof.* Let $\xi = [0, a_1, a_2, \dots]$ where $(a_1, a_2, \dots)$ is the limit of a sequence of words $(w_i)_{i\geq 1}$ in $(\mathbb{N}\setminus\{0\})^*$ starting with two noncommuting words $w_1$ and $w_2$. Denote by $(W_i)_{i\geq 1}$ the image of the sequence $(w_i)_{i\geq 1}$ under the isomorphism of monoids $\sigma\colon (\mathbb{N}\setminus\{0\})^* \to \mathcal{S}\cup\{I\}$ defined in Corollary 4.2. Since $w_1$ and $w_2$ do not commute, the same is true of $W_1$ and $W_2$ and so, by Corollary 4.5, $(W_i)_{i\geq 1}$ is an admissible Fibonacci sequence in $\mathcal{S}$. We also note that for each pair of matrices $A, B \in \mathcal{S}$, we have $\|AB\| > \|A\|\|B\|$. Then, the relation $W_{i+2} = W_{i+1}W_i$ implies $\|W_{i+2}\| > \|W_{i+1}\|\|W_i\|$ for each $i \geq 1$. In particular, the sequence $(W_i)_{i\geq 1}$ is unbounded. As $|\det W_i| = 1$ for each $i$, it also satisfies the conditions (3) of Theorem 2.1. Thus, the sequence $(W_i)_{i\geq 1}$ is associated to some extremal real number $\eta$. On the other hand, the theory of continued fractions shows that $\|(\xi, -1)W_i\| \ll \|W_i\|^{-1}$ since the ratios of the elements in the columns of $W_i$ are successive convergents of $1/\xi$. Thus, $\xi = \eta$ is extremal.

Conversely, let $\xi$ be an extremal real number with an associated Fibonacci sequence $(W_i)_{i\geq 1}$ in $\mathcal{S}$. The inverse image of this sequence under $\sigma$ is a Fibonacci sequence $(w_i)_{i\geq 1}$ in $(\mathbb{N}\setminus\{0\})^*$ and, as above, we deduce that $\xi = [0, a_1, a_2, \dots]$ where $(a_1, a_2, \dots) = \lim_{i\to\infty} w_i$. Since $\xi$ is neither rational nor quadratic, this sequence is infinite and ultimately not periodic. In particular, $w_1$ and $w_2$ are not both powers of the same word, and so they do not commute (Proposition 1.3.2 of Chapter 1 of [4]). □

*Remark.* Let $w_\infty = (1, 2, 3, 1, 2, 1, 2, 3, \dots)$ be the limit of the Fibonacci sequence $(w_i)_{i\geq 1}$ generated by $w_1 = (3)$ and $w_2 = (1, 2)$. Since $w_1w_2 \neq w_2w_1$, the corresponding real number $\xi = [0, 1, 2, 3, 1, 2, 1, 2, 3, \dots]$ is extremal. However, contrary to the generic Fibonacci word $f_{a,b}$ which contains palindromes of arbitrary length as prefixes, the infinite word $w_\infty$ contains no factor of length greater than 3 which is a palindrome.

## 5 Proof of Theorem 2.3

Throughout this section, the notation is the same as in Theorem 2.3. Namely, we fix an alphabet $E = \{a, b\}$ of two letters and denote by $(w_k)_{k\geq 1}$ the Fibonacci sequence in $E^*$ generated by $w_1 = b$ and $w_2 = a$, with limit $f_{a,b}$. We also fix an extremal real number $\xi$ with an associated Fibonacci sequence $(W_k)_{k\geq 1}$ in $\mathcal{P}$ and denote by $\Phi\colon E^* \to \mathcal{P}$ the morphism of monoids mapping $w_k$ to $W_k$ for each $k \geq 1$. We start with the following observation.

**Lemma 5.1.** *Let $k$ and $\ell$ be integers with $k \geq \ell \geq 2$, and let $w_k = uv$ be a factorization of $w_k$ in $E^*$. Then, there exist a prefix $u_0$ of $w_\ell$ and strictly decreasing sequences of integers $i_1 > i_2 > \cdots > i_s$ and $j_1 > j_2 > \cdots > j_t$ bounded below by $\ell$ such that*

$$u = w_{i_1} w_{i_2} \cdots w_{i_s} u_0 \quad \text{and} \quad u_0 v = w_{j_t} \cdots w_{j_2} w_{j_1}.$$

*If $u$ is not a prefix of $w_\ell$, we can ask that $i_1 \leq k-1$ and $j_1 \leq k-2$.*

*Proof.* If $u$ is a prefix of $w_\ell$, we take $u_0 = u$ so that $u_0 v = w_k$. Otherwise, we have $k > \ell$, thus $k \geq 3$ and the factorization $w_k = w_{k-1} w_{k-2}$ implies that either there is a word $u'$ such that $u = w_{k-1} u'$ and $u'v = w_{k-2}$, or we have $k \geq \ell + 2$ and there is a word $v'$ such that $v = v' w_{k-2}$ and $uv' = w_{k-1}$. The result then follows by induction on $k$. □

Since the sequence $(W_i)_{i\geq 1}$ is admissible, there exists a nonsymmetric and non-skew-symmetric matrix $N$ such that, upon putting $N_i = N$ if $i$ is even and $N_i = {}^tN$ if $i$ is odd, the product $\mathbf{y}_i = W_i N_i$ is symmetric for each $i \geq 1$. This matrix $\mathbf{y}_i$ may not be primitive but, for the next result, it is convenient not to normalize it.

**Lemma 5.2.** *Define $L = \max\{1, |\xi|\}^{-1}(1, \xi)$ and $\theta = LN^{-1}({}^tL)$. Then, there exist an index $\ell \geq 1$ and a constant $c \geq 1$ such that for any sequence of integers $(i_1, \ldots, i_s)$ with entries bounded below by $\ell$ and repeated at most twice, we have*

$$\frac{1}{c} \leq \frac{\|W_{i_1} W_{i_2} \cdots W_{i_s}\|}{|\theta|^s \|\mathbf{y}_{i_1}\| \|\mathbf{y}_{i_2}\| \cdots \|\mathbf{y}_{i_s}\|} \leq c.$$

Note that we have $\theta \neq 0$ since $\xi$ is transcendental and $N$ is not skew-symmetric.

*Proof.* Write

$$\mathbf{y}_i = \begin{pmatrix} y_{i,0} & y_{i,1} \\ y_{i,1} & y_{i,2} \end{pmatrix}$$

for each $i \geq 1$. As $\|(\xi, -1)\mathbf{y}_i\| \ll \|(\xi, -1)W_i\| \ll \|W_i\|^{-1}$, we have

$$\mathbf{y}_i = y_{i,0} \begin{pmatrix} 1 \\ \xi \end{pmatrix} (1 \ \xi) + \mathcal{O}\left(\|W_i\|^{-1}\right),$$

and so $\|\mathbf{y}_i\| = |y_{i,0}| \max\{1, |\xi|\}^2 + \mathcal{O}(\|W_i\|^{-1})$. In particular, this shows that $y_{i,0} \neq 0$ for each sufficiently large $i$, say, for $i \geq \ell$. For those values of $i$, we find

$$\frac{W_i}{\theta \|\mathbf{y}_i\|} = A_i + R_i \tag{10}$$

where $R_i = \mathcal{O}\left(\|W_i\|^{-2}\right)$ and where $A_i = \pm\theta^{-1}({}^tL)LN_i^{-1}$ belongs to the set

$$\mathcal{A} = \left\{\pm I,\ \pm\frac{1}{\theta}\,{}^tLLN^{-1},\ \pm\frac{1}{\theta}\,{}^tLL\,{}^tN^{-1}\right\}.$$

Since $\theta = LN^{-1}({}^tL) = L({}^tN^{-1})({}^tL)$, the set $\mathcal{A}$ is stable under multiplication.

Now, let $(i_1, \ldots, i_s)$ be any sequence of integers bounded below by $\ell$, with no entry repeated more than twice. Using (10), we find

$$\frac{W_{i_1}\cdots W_{i_s}}{\theta^s\|\mathbf{y}_{i_1}\|\cdots\|\mathbf{y}_{i_s}\|} = A + R$$

where $A = A_{i_1}\cdots A_{i_s}$ belongs to $\mathcal{A}$ and where $R$ is a sum, indexed by all nonempty subsequences $(j_1, \ldots, j_t)$ of $(i_1, \ldots, i_s)$, of products of the form $B_1R_{j_1}\cdots B_tR_{j_t}B_{t+1}$ with $B_1, \ldots, B_{t+1} \in \mathcal{A}$. Thus, for an appropriate constant $\kappa > 0$, we have

$$\|R\| \le \left(1+\kappa\|W_{i_1}\|^{-2}\right)\cdots\left(1+\kappa\|W_{i_s}\|^{-2}\right) - 1 \le \exp\left(2\kappa\sum_{i=\ell}^{\infty}\|W_i\|^{-2}\right) - 1.$$

In particular, if $\ell$ is sufficiently large, this gives $\|R\| \le \|A\|/2$. Then we have $\|A + R\| \gg\ll \|A\| \gg\ll 1$ as requested. □

**Lemma 5.3.** *Let $k$ be a positive integer and let $w_k = uv$ be a factorization of $w_k$ in $E^*$. Put $U = \Phi(u)$ and $V = \Phi(v)$. Then, we have $\|UV\| \gg\ll \|U\|\|V\|$ with implied constants that are independent of $k$, $u$ and $v$.*

*Proof.* Let $\ell$ be as in Lemma 5.2. Without loss of generality, we may assume that $k \ge \ell$. Then, according to Lemma 5.1, we can write $u = w_{i_1}\cdots w_{i_s}u_0$ and $u_0v = w_{j_t}\cdots w_{j_1}$, where $u_0$ is a prefix of $w_\ell$ and where $(i_1, \ldots, i_s)$ and $(j_1, \ldots, j_t)$ are strictly decreasing sequences of integers bounded below by $\ell$. Put

$$P = W_{i_1}\cdots W_{i_s},\quad Q = W_{j_t}\cdots W_{j_1}\quad\text{and}\quad U_0 = \Phi(u_0).$$

Then, we have $U = aPU_0$ and $U_0V = bQ$ with nonzero rational numbers $a$ and $b$. Since $U_0$ belongs to a finite set of matrices in $\mathcal{P}$, we deduce that

$$\|U\| \gg\ll |a|\|P\| \quad\text{and}\quad \|V\| \gg\ll |b|\|Q\|.$$

Moreover, since the sequence $(i_1, \ldots, i_s, j_t, \ldots, j_1)$ has its entries repeated at most twice and bounded below by $\ell$, Lemma 5.2 gives

$$\|PQ\| \gg\ll |\theta|^{s+t}\|\mathbf{y}_{i_1}\|\cdots\|\mathbf{y}_{i_s}\|\|\mathbf{y}_{j_t}\|\cdots\|\mathbf{y}_{j_1}\| \gg\ll \|P\|\|Q\|.$$

The conclusion follows because $UV = abPQ$. □

*Proof of Theorem 2.3.* We first note that, for any $(x, y) \in \mathbb{R}^2$ and $U \in \mathrm{GL}_2(\mathbb{R})$, we have

$$\|(x,y)U\| \ge \frac{\|(x,y)UU^{-1}\|}{2\|U^{-1}\|} = \frac{\|(x,y)\|\,|\det U|}{2\|U\|}. \tag{11}$$

Applying this to the point $(x, y) = (\xi, -1)$ and the matrix $U_i = \Phi(u_i)$, where $u_i$ denotes the prefix of $f_{a,b}$ of length $i$, we get

$$\|(\xi,-1)U_i\| \ge \frac{|\det U_i|}{2\|U_i\|}$$

for each $i \geq 1$. To prove an upper bound of the same type for $\|(\xi, -1)U_i\|$, we denote by $k = k(i)$ the smallest positive integer such that $u_i$ is a prefix of $w_k$, and write $w_k = u_i v_i$ with $v_i \in E^*$. Putting $V_i = \Phi(v_i)$, we then have

$$U_i V_i = m_i W_k \tag{12}$$

for some integer $m_i \geq 1$. Applying (11) to the point $(x, y) = (\xi, -1)U_i$ and the matrix $U = V_i$, we find

$$\|(\xi, -1)W_k\| = \frac{1}{m_i}\|(\xi, -1)U_i V_i\| \geq \frac{\|(\xi, -1)U_i\| \, |\det V_i|}{2m_i \|V_i\|}.$$

Since $\|(\xi, -1)W_k\| \ll \|W_k\|^{-1}$, this gives

$$\|(\xi, -1)U_i\| \ll \frac{m_i \|V_i\|}{\|W_k\| \, |\det V_i|}. \tag{13}$$

Applying Lemma 5.3 to the factorization (12) on one hand, and taking determinants of both sides of (12) on the other hand, we also find

$$\|V_i\| \gg\ll \frac{m_i \|W_k\|}{\|U_i\|} \quad \text{and} \quad |\det V_i| = \frac{m_i^2 |\det W_k|}{|\det U_i|} \geq \frac{m_i^2}{|\det U_i|}.$$

These estimates combined with (13) lead to

$$\|(\xi, -1)U_i\| \ll \frac{|\det U_i|}{\|U_i\|}, \tag{14}$$

which completes the proof of (4) in Theorem 2.3.

Now, assume that the integers $\det U_i$ are bounded independently of $i$ and, for each $i \geq 1$, choose a column $\binom{q_i}{p_i}$ of $U_i$ with the largest norm. Then, (14) leads to

$$|q_i \xi - p_i| \ll \|U_i\|^{-1}.$$

Since $U_{i+1}$ is either equal to $U_i * W_1$ or to $U_i * W_2$, we also have $\|U_{i+1}\| \ll \|U_i\|$ and thus $|q_{i+1}| \ll \|U_i\| \ll |q_i|$. Combining these estimates and noting that $\gcd(p_i, q_i)$ is a divisor of $\det U_i$, we deduce the existence of a constant $c \geq 1$ such that

$$|q_{i+1}(q_i \xi - p_i)| \leq c \quad \text{and} \quad |q_{i+1}| \leq c \left| \frac{q_i}{\gcd(p_i, q_i)} \right|, \tag{15}$$

for each $i \geq 1$. Moreover, we have $\limsup_{i \to \infty} |q_i| = \infty$ since $(U_i)_{i \geq 1}$ contains the unbounded sequence $(W_k)_{k \geq 1}$ as a subsequence. These facts imply that $\xi$ is badly approximable. Indeed, if $p/q$ is an arbitrary rational number, then, at the expense of replacing $c$ by a larger constant if necessary, we may assume that there exists an index $i \geq 1$ such that $0 < |q_i| \leq 2c|q| \leq |q_{i+1}|$. Using (15), this gives $2|q| \leq |q_i/\gcd(p_i, q_i)|$, thus $p/q \neq p_i/q_i$ and so we find

$$\left|\xi - \frac{p}{q}\right| \geq \left|\frac{p_i}{q_i} - \frac{p}{q}\right| - \left|\frac{p_i}{q_i} - \xi\right| \geq \frac{1}{|qq_i|} - \frac{c}{|q_i q_{i+1}|} \geq \frac{1}{2|qq_i|} \geq \frac{1}{4cq^2}.$$

□

## 6 Proof of Theorem 2.4

Again, let $E$ be a set of two elements $a$ and $b$, and let $(w_i)_{i\geq 1}$ be the Fibonacci sequence in $E^*$ determined by the conditions $w_1 = b$ and $w_2 = a$, with limit $f_{a,b}$. The following lemma is our main tool for constructing more extremal real numbers.

**Lemma 6.1.** *Let $m$ be a nonzero integer and let $W \in \mathrm{Mat}_{2\times 2}(\mathbb{Z})$ with $W^2 \equiv 0$ mod $m$. Assume that there exist primitive matrices $W_1, W_2 \in \mathrm{Mat}_{2\times 2}(\mathbb{Z})$ of determinant $m$ with $W_1 \equiv W_2 \equiv W$ mod $m$, and consider the morphism of monoids $\Phi\colon E^* \to \mathcal{P}$ mapping $a$ to $W_2$ and $b$ to $W_1$. Then, for each word $u \in E^*$, the determinant of $\Phi(u)$ is 1 if $u$ has even length and it is $m$ if $u$ has odd length.*

*Proof.* We proceed by recurrence on the length $\ell$ of $u$. If $\ell \leq 1$, the result is clear (for the empty word 1, the matrix $\Phi(1)$ is the identity). If $\ell = 2$, we have $\Phi(u) = (W_i W_j)^{\mathrm{red}}$ for some choice of indices $i, j \in \{1, 2\}$. Then, since $W_i W_j \equiv W^2 \equiv 0 \bmod m$ and since $\det(W_i W_j) = m^2$, the matrix $W_i W_j$ has content $|m|$, and so $\Phi(u) = |m|^{-1} W_i W_j$ has determinant 1. Now, assume that $\ell > 2$ and that the result is true for words of smaller length. Write $u = u'u''$, where $u'$ has even length and $u''$ has length 1 or 2. By induction hypothesis, $\Phi(u')$ has determinant 1, while $\Phi(u'')$ is primitive with $\det \Phi(u'') = 1$ if $\ell$ is even and $\det \Phi(u'') = m$ if $\ell$ is odd. Then the product $\Phi(u')\Phi(u'')$ is primitive and so $\Phi(u) = \Phi(u')\Phi(u'')$ has the same determinant as $\Phi(u'')$. □

We also need the following technical result.

**Lemma 6.2.** *Let $r$ be a real number with $0 < r \leq 1$ and let $\mathcal{S}_r$ denote the set of matrices $A \in \mathrm{Mat}_{2\times 2}(\mathbb{R})$ with positive coefficients whose elements of the first row are bounded below by $r$ times those of the second row, and whose elements of the first column are bounded below by $r$ times those of the second column. Then, $\mathcal{S}_r$ is closed under multiplication and, for each $A, A' \in \mathcal{S}_r$, we have $\|AA'\| > r\|A\|\|A'\|$.*

*Proof.* The set $\mathcal{S}_r$ consists of all $2 \times 2$ matrices $A$ with positive coefficients such that the products $(1, -r)A$ and $(1, -r)\,{}^tA$ have nonnegative coefficients. The fact that this set is closed under multiplication then follows from the associativity of the matrix product. To prove the second assertion, take $A, A' \in \mathcal{S}_r$. Let $(a, b)$ and $(a', b')$ denote respectively rows of $A$ and ${}^tA'$ with largest norm. Since $a \geq rb$, we have

$$\|AA'\| \geq aa' + bb' \geq rb(a' + b') > rb\|A'\|.$$

Similarly, since $a' \geq rb'$, we find $\|AA'\| > rb'\|A\|$. If $b = \|A\|$ or $b' = \|A'\|$, this gives $\|AA'\| > r\|A\|\|A'\|$ as requested. Otherwise, we have $a = \|A\|$ and $a' = \|A'\|$ and we get the stronger inequality $\|AA'\| > \|A\|\|A'\|$. □

The next proposition is more specific than Theorem 2.4 and thereby proves it.

**Proposition 6.3.** *Put* $W_1 = \begin{pmatrix} m & m \\ m-1 & m \end{pmatrix}$ *and* $W_2 = \begin{pmatrix} 2m & m \\ 2m-1 & m \end{pmatrix}$ *for a nonzero integer* $m$. *Then the Fibonacci sequence* $(W_i)_{i\geq 1}$ *of* $\mathcal{P}$ *generated by these two matrices is associated to a badly approximable real number* $\xi$, *and it satisfies* $\det W_i = m$ *for each index* $i$ *which is not divisible by* 3. *If* $|m|$ *is not the square of an integer, then* $\xi$ *is not conjugate under the action of* $\mathrm{GL}_2(\mathbb{Q})$ *to an extremal real number having an associated Fibonacci sequence in* $\mathrm{GL}_2(\mathbb{Z})$.

*Proof.* A short computation shows that $W_1$, $W_2$, $W_1W_2$ and $W_2W_1$ are linearly independent over $\mathbb{Q}$. Then, $W_1$ and $W_2$ fulfill the hypotheses of Lemma 4.4 and so the sequence $(W_i)_{i\geq 1}$ is admissible. One can check that a corresponding matrix $N$ is $\begin{pmatrix} m & -m \\ -2m & 2m-1 \end{pmatrix}$. Moreover, for the given choice of $m$, the matrices $W_1$ and $W_2$ satisfy the hypotheses of Lemma 6.1 with $W = \begin{pmatrix} 0 & 0 \\ -1 & 0 \end{pmatrix}$. Thus, defining the map $\Phi\colon E^* \to \mathcal{P}$ as in this lemma, we have $\det \Phi(u) = 1$ for each word $u \in E^*$ of even length and $\det \Phi(u) = m$ for each $u \in E^*$ of odd length. Since the length of $w_i$ is even if and only if $i$ is divisible by 3, we deduce that $W_i = \Phi(w_i)$ has determinant 1 when $i$ is divisible by 3 and determinant $m$ otherwise. In particular, we have $|\det W_i| \leq |m|$ for each $i \geq 1$.

A short computation also gives $W_3 = \pm\begin{pmatrix} 3m-1 & 3m \\ 3m-2 & 3m-1 \end{pmatrix}$ and shows, in the notation of Lemma 6.2, that $\pm W_2$ and $\pm W_3$ both belong to $\mathcal{S}_{1/2}$ for some appropriate choice of signs. Thus, for each $i \geq 2$, one of the matrices $\pm W_i$ belongs to $\mathcal{S}_{1/2}$ and we have

$$\| W_{i+1} W_i \| > \frac{1}{2} \| W_{i+1} \| \| W_i \|.$$

Since the determinant of $W_{i+1}W_i$ is a divisor of $m^2$, the content of this product is a divisor of $m$ and so the matrix $W_{i+2} = (W_{i+1}W_i)^{\mathrm{red}}$ satisfies $\|W_{i+2}\| \geq |m|^{-1} \|W_{i+1}W_i\|$. Combining this inequality with the previous one, we deduce that

$$\| W_{i+2} \| > \frac{1}{2|m|} \| W_{i+1} \| \| W_i \|,$$

for each $i \geq 2$. By induction, this implies $\|W_{i+1}\| > \|W_i\| \geq 2|m|$ for each $i \geq 2$, and so the sequence $(W_i)_{i\geq 1}$ is unbounded. Applying Theorem 2.1, we deduce that the sequence $(W_i)_{i\geq 1}$ is associated to some extremal real number $\xi$. Moreover, since we have $|\det \Phi(u)| \leq |m|$ for each $u \in E^*$, Theorem 2.3 shows that $\xi$ is badly approximable.

Finally, suppose that $\xi$ is $\mathrm{GL}_2(\mathbb{Q})$-conjugate to an extremal real number $\eta$ with an associated Fibonacci sequence $(W_i')_{i\geq 1}$ in $\mathrm{GL}_2(\mathbb{Z})$. Then, there exists a matrix $A \in \mathcal{P}$ such that $(\eta, -1)$ is proportional to $(\xi, -1)A$ and, upon denoting by $B$ the inverse of $A$ in $\mathcal{P}$, we find that $(A * W_i' * B)_{i\geq 1}$ is a Fibonacci sequence in $\mathcal{P}$ which is associated to $\xi$. So, by Theorem 2.1, the sequences $(W_i)_{i\geq 1}$ and $(A * W_i' * B)_{i\geq 1}$ differ only up to their first terms and up to multiplication by a Fibonacci sequence in $\{-1, 1\}$. Comparing determinants, this implies that $|m|$ is the square of an integer. □

*Remark.* The Fibonacci sequence $(W_i)_{i\geq 1}$ in $\mathcal{P}$ starting with

$$W_1 = \begin{pmatrix} 1 & 0 \\ 0 & 2 \end{pmatrix}, \quad W_2 = \begin{pmatrix} 0 & 1 \\ 2 & 0 \end{pmatrix}, \quad W_3 = \begin{pmatrix} 0 & 1 \\ 1 & 0 \end{pmatrix},$$

$$W_4 = \begin{pmatrix} 2 & 0 \\ 0 & 1 \end{pmatrix}, \quad W_5 = \begin{pmatrix} 0 & 2 \\ 1 & 0 \end{pmatrix}, \quad W_6 = \begin{pmatrix} 0 & 1 \\ 1 & 0 \end{pmatrix},$$

is periodic of period 6 as one finds that $W_7 = W_1$ and $W_8 = W_2$. Therefore, $(W_i)_{i\geq 1}$ is a Fibonacci sequence of matrices with bounded determinant. It does not correspond to an extremal real number as the sequence itself is bounded. However, if $\Phi\colon E^* \to \mathcal{P}$ denotes the morphism of monoids sending $w_i$ to $W_i$ for each $i \geq 1$, then, for each $i \geq 1$, the word $v_i = w_{6i+1} \cdots w_7 w_1$ is a prefix of $f_{a,b}$ whose image under $\Phi$ is the matrix $W_1^{i+1}$ which has determinant $2^{i+1}$ tending to infinity with $i$.

**Acknowledgements.** Work partially supported by NSERC and CICMA.

## References

1. Cassels, J.W.S.: *An Introduction to Diophantine Approximation.* Cambridge University Press, Cambridge (1957)
2. Davenport, H., Schmidt, W.M.: Dirichlet's theorem on diophantine approximation. In: *Symposia Mathematica su Teoria dei Numeri, Istituto Nazionale di Alta Matematica, Rome, 1968/69.* Symp. Math., 4, pp. 113–132. Academic Press, London (1970)
3. Davenport, H., Schmidt, W.M.: Approximation to real numbers by algebraic integers. *Acta Arith.* **15**, 393–416 (1969)
4. Lothaire, M.: *Combinatorics on Words.* Encyclopedia of Mathematics and Its Applications, vol. 17. Addison-Wesley, Reading (1983)
5. Lucier, B.: Binary morphisms to ultimately periodic words. arXiv: 0805.1373v1 (2008)
6. Roy, D.: Approximation simultanée d'un nombre et de son carré. *C. R. Acad. Sci. Paris Ser. I* **336**, 1–6 (2003)
7. Roy, D.: Approximation to real numbers by cubic algebraic integers I. *Proc. Lond. Math. Soc.* **88**, 42–62 (2004)
8. Roy, D.: Diophantine approximation in small degree. In: Goren, E.Z., Kisilevsky, H. (eds.) *Number Theory: Proceedings from the 7th Conference of the Canadian Number Theory Association.* CRM Proceedings and Lecture Notes, vol. 36, pp. 269–285. American Mathematical Society, Providence (2004)
9. Roy, D.: On two exponents of approximation related to a real number and its square. *Can. J. Math.* **59**, 211–224 (2007)
10. Schmidt, W.M.: *Diophantine Approximation.* Lect. Notes Math., vol. 785. Springer, Heidelberg (1980)

*Diophantine Approximation*
H. P. Schlickewei et al., Editors


# THE NUMBER OF SOLUTIONS OF A LINEAR HOMOGENEOUS CONGRUENCE

Andrzej Schinzel
*Instytut Matematyczny, Polskiej Akademii Nauk, ulica Śniadeckich 8, 00-956 Warsaw, Poland*
A.Schinzel@impan.gov.pl

Dedicated to Wolfgang Schmidt on the occasion of his 70th birthday

The aim of this paper is to propose and to study the following

*Conjecture.* Let $n \in \mathbb{N}, a_i \in \mathbb{Z}$ and $b_i \in \mathbb{N}$ $(1 \le i \le k)$. The number $N(n; a_1, b_1; \ldots; a_k, b_k)$ of solutions of the congruence

$$\sum_{i=1}^{k} a_i x_i \equiv 0 (\bmod\ n) \text{ with } 0 \le x_i \le b_i \tag{1}$$

satisfies the inequality

$$N(n; a_1, b_1; \ldots; a_k, b_k) \ge 2^{1-n} \prod_{i=1}^{k} (b_i + 1) \,. \tag{2}$$

Let us observe that for $k = n - 1$

$$N(n; \underbrace{1, 1; \ldots; 1, 1}_{n-1} \text{ times}) = 1 = 2^{1-n} \prod_{i=1}^{k} 2,$$

thus, if (2) is true, $2^{1-n}$ is the best possible coefficient independent of $a_i, b_i$. For $b_i = 1$ $(1 \le i \le k)$ the inequality (2) is indeed true, as follows from a more general result of Olson [4] concerning abelian groups. For general $b_i$ and $a_i = 1$ the existence of a positive coefficient, depending only on $n$, for which (2) holds has been proved in an unpublished manuscript of Drmota and Skałba [1]. An analysis of their proof gives for $n$ being a prime the coefficient $\exp\left(-\frac{2+o(1)}{\pi^2} n^2 \log^2 n\right)$. Using a completely different method we shall prove

**Theorem 1.** *Inequality* (2) *holds provided* $(a_i, n) = 1$ *for all* $i \le k$.

*Keywords.* Linear homogeneous congruence, finite abelian group.

*2000 Mathematics subject classification.* 11D79, 25K01.

**Corollary.** *For all $a_i \in \mathbb{Z}$, $b_i \in \mathbb{N}$ $(1 \le i \le k)$ we have*

$$N(n; a_1, b_1; \dots; a_k, b_k) \ge 2^{\tau(n)-\sigma(n)} \prod_{i=1}^{k} (b_i + 1).$$

*where $\sigma(n)$, $\tau(n)$ are the sum and the number of divisors of $n$, respectively.*

Let us note the great difference between the condition $0 \le x_i \le b_i$ and $|x_i| \le b_i$. With the last condition one can easily prove by the box principle that the number of solutions of (1) is at least $\frac{1}{n}\prod_{i=1}^{k}(b_i+1)$; see for a similar result Korobov [2, Lemma 2 p. 103].

The method of proof of Theorem 1 can be modified to cover the case, where for all $i, j \le k$ either $(a_i, n) \mid (a_j, n)$ or $(a_j, n) \mid (a_i, n)$ or $n \mid [a_i, a_j]$. It has been worked out by M. Zakarczemny [7], see [6]. This includes the case, where $n = p^\alpha$ or $n = pq$ $(p, q$ primes), thus (2) holds in this case. I [5] have dealt with the case $n = \prod_{j=1}^{l} q_j^{\alpha_j}$, $\sum_{j=1}^{l} \frac{1}{q_j} \le 1 + \frac{\min\{l, 2l-5\}}{n}$, $q_j$ primes. Theorem 1 implies

**Theorem 2.** *For every finite abelian group $\Gamma$ and every vector $[\alpha_1, \dots, \alpha_k] \in \Gamma^k$ the number of solutions of the equation*

$$\sum_{i=1}^{k} \alpha_i x_i = 0$$

*in nonnegative integers $x_i \le b_i$ is at least*

$$2^{-(\tau(\chi(\Gamma))-1)(|\Gamma|-1)} \prod_{i=1}^{k} (b_i + 1)$$

*where $\chi(\Gamma)$ is the characteristic of $\Gamma$.*

If the initial Conjecture is true, the factor $\tau(\chi(\Gamma)) - 1$ can be omitted in the exponent. Olson's theorem suggests that the right exponent is $1 - D(\Gamma)$, where $D(\Gamma)$ is the Davenport constant of the group $\Gamma$.

The proof of Theorem 1 is based on five lemmas.

**Lemma 1.** *Inequality* (2) *holds for $n = 2$ or 3, $a_i$ prime to $n$ $(1 \le i \le k)$.*

*Proof.* We shall proceed by induction on $k$. For $k = 1$, the condition

$$a_1 x_1 \equiv 0 (\text{mod } n), \quad 0 \le x_1 \le b_1$$

has $[b_1/n] + 1 \ge (b_1 + 1)/n \ge 2^{1-n}(b_1 + 1)$ solutions. Suppose now that the lemma is true for $k - 1$ terms $(k > 1)$. Let $N_r$ be the number of solutions of the condition

$$a_1 r + \sum_{i=2}^{k} a_i x_i \equiv 0 (\text{mod } n), \quad 0 \le x_i \le b_i \ (2 \le i \le k).$$

Now, if $0 \le r < n$, there exist $[(b_1 - r)/n] + 1$ solutions of the condition

$$a_1 x_1 \equiv a_1 r (\text{mod } n), \quad 0 \le x_1 \le b_1.$$

Hence

$$N(n; a_1, b_1; \dots; a_k, b_k) \geq \left(\left[\frac{b_1}{n}\right] + 1\right) N_0 + \left[\frac{b_1+1}{n}\right] \sum_{r=1}^{n-1} N_r$$

$$= \left(\left[\frac{b_1}{n}\right] + 1 - \left[\frac{b_1+1}{n}\right]\right) N_0 + \left[\frac{b_1+1}{n}\right] \sum_{r=0}^{n-1} N_r.$$

However, by the inductive assumption $N_0 \geq 2^{1-n} \prod_{i=2}^{k} (b_i + 1)$. It follows that

$$N(n; a_1, b_1; \dots; a_k, b_k) \geq 2^{1-n} \prod_{i=2}^{k} (b_i + 1) \left(\left[\frac{b_1}{n}\right] + 1 + \left(2^{n-1} - 1\right) \left[\frac{b_1+1}{n}\right]\right).$$

If $n = 2$, we have

$$\left[\frac{b_1}{n}\right] + 1 + \left(2^{n-1} - 1\right) \left[\frac{b_1+1}{n}\right] = \left[\frac{b_1}{2}\right] + 1 + \left[\frac{b_1+1}{2}\right] = b_1 + 1.$$

If $n = 3$, we have

$$\left[\frac{b_1}{n}\right] + 1 + \left(2^{n-1} - 1\right) \left[\frac{b_1+1}{n}\right] = \left[\frac{b_1}{3}\right] + 1 + 3\left[\frac{b_1+1}{3}\right] = b_1 + 1,$$

except for $b_1 = 1$.

Therefore, it remains to consider the case $n = 3$, $b_i = 1$ $(1 \leq i \leq k)$.

Now, $a_1x_1 + a_2x_2$ for $x_i \in \{0, 1\}$ $(i = 1, 2)$ represents all residue classes mod 3, hence for every choice of $x_i \in \{0, 1\}$ $(2 \leq i \leq k)$ there exist $x_1, x_2$ in $\{0, 1\}$ such that $\sum_{i=1}^{k} a_i x_i \equiv 0 (\text{mod } 3)$. It follows that

$$N(n; a_1, b_1; \dots; a_k, b_k) \geq 2^{k-2} = 2^{1-n} \prod_{i=1}^{k} 2.$$

□

**Lemma 2.** *Inequality* (2) *holds for* $n = 4$, $a_1, a_2$ odd, $b_1 = b_2 = 2$.

*Proof.* Every residue class mod 4 is represented at least twice by $a_1x_1 + a_2x_2$, where $0 \leq x_i \leq 2$ $(i = 1, 2)$. Indeed,

$$0 \equiv a_1 \cdot 0 + a_2 \cdot 0 \equiv a_1 \cdot 2 + a_2 \cdot 2 (\text{mod } 4),$$
$$2 \equiv a_1 \cdot 2 + a_2 \cdot 0 \equiv a_1 \cdot 0 + a_2 \cdot 2 (\text{mod } 4)$$

and if $a_2 \equiv a_1$ mod 4, $\varepsilon = \pm 1$, then

$$a_1 \equiv a_1 \cdot 1 + a_2 \cdot 0 \equiv a_1(1 - \varepsilon) + a_2 \cdot 1 (\text{mod } 4),$$
$$-a_1 \equiv a_1 \cdot 1 + a_2 \cdot 2 \equiv a_1(1 + \varepsilon) + a_2 \cdot 1 (\text{mod } 4).$$

Hence, for every choice of nonnegative integers $x_i \leq b_i$ $(2 < i \leq k)$ there exist at least two pairs $[x_1, x_2]$ with $0 \leq x_i \leq b_i$ $(i = 1, 2)$ such that $\sum_{i=1}^{k} a_i x_i \equiv 0 (\text{mod } n)$. It follows that

$$N(n; a_1, b_1; \dots; a_k, b_k) \geq 2\prod_{i=3}^{k}(b_i+1) = \frac{2}{9}\prod_{i=1}^{k}(b_i+1) > 2^{1-n}\prod_{i=1}^{k}(b_i+1).$$

□

**Lemma 3.** *Let* $m \in \mathbb{N}$, $b \in \mathbb{N}$. *For every integer* $a$ *prime to* $m$ *and every set* $B$ *of residues* mod $m$ *the number of distinct residues* mod $m$ *of the form* $ax + y$, *where* $0 \leq x \leq b$, $y \in B$ *is at least* $\min(m, b + |B|)$.

*Proof.* We apply Kneser's theorem (see [3, p. 6]) with $G = \mathbb{Z}_m$, $A = \{ax; 0 \leq x \leq b\}$ and infer the existence of a subgroup $H$ of $\mathbb{Z}_m$ such that

$$A + B + H = A + B \tag{3}$$

and

$$|A + B| \geq |A + H| + |B + H| - |H|. \tag{4}$$

Let $H = \left\{dx : 0 \leq x < \frac{n}{d}\right\}$, where $d \mid m$. If $b < d - 1$, then

$$|A + H| = (b + 1)|H|, \text{ hence by (4)}$$
$$|A + B| \geq (b + 1)|H| + |B| - |H| \geq b + |B|.$$

If $b \geq d - 1$, then $A + H = G$, hence by (3), $A + B = G$. □

**Lemma 4.** *Let* $N_h$ *be the number of distinct residue classes* mod $n$ *of*

$$\sum_{i=1}^{h} a_i x_i \quad (0 \leq x_i \leq b_i).$$

*If* $(n, a_i) = 1$ *for all* $i \leq h$, *then either there exists a* $g \leq h$ *such that* $N_g = n$ *or*

$$N_h \geq 1 + \sum_{i=1}^{h} b_i. \tag{5}$$

*Proof.* We proceed by induction on $h$. For $h = 1$ and $b_1 + 1 \geq n$ the numbers $a_1 x_1$ $(0 \leq x_1 \leq b_1)$ represent all residues mod $n$, hence $N_1 = n$. If $b_1 + 1 < n$, the residues of $a_1 x_1$ $(0 \leq x_1 \leq b_1)$ mod $n$ are all distinct, hence

$$N_1 \geq b_1 + 1.$$

Assume that the alternative above holds for $h - 1$ terms $(h \geq 2)$. If for $g \leq h - 1$, $N_g = n_g$, then the inductive assertion holds. Otherwise, the number of distinct residue classes mod $n$ of $\sum_{i=1}^{h-1} a_i x_i$ $(0 \leq x_i \leq b_i)$ is

$$N_{h-1} \geq 1 + \sum_{i=1}^{h-1} b_i.$$

We apply Lemma 3 with $B$ being the set of these residue classes, $m = n$ and $a = a_h$. It follows that

$$N_h \geq \min\left(n, 1 + \sum_{i=1}^{h} b_i\right)$$

and we obtain either $N_h = n$ or (5). □

**Lemma 5.** *For positive integers $a$ and $x \le a$ we have*

$$\left(\frac{a}{x}+1\right)^{x+1} \le 2^{a+1}, \tag{6}$$

*except for $a = 2$, $x = 1$.*

*Proof.* We consider

$$f(x) = (x+1)\log\left(\frac{a}{x}+1\right)$$

and find that

$$f'(x) = \log\left(\frac{a}{x}+1\right) - \frac{ax+a}{x^2+ax},$$
$$f''(x) = \frac{-a((a-2)x-a)}{\left(x^2+ax\right)^2}.$$

Hence $f''(x) \le 0$ for $a > 2, x \ge a/(a-2)$. It follows that for $a > 2$ and $a/(a-2) \le x \le a$,

$$f'(x) \ge f'(a) = \log 2 - \frac{a+1}{2a} \ge \log 2 - \frac{2}{3} > 0.$$

Hence for $a > 2$ and $a/(a-2) \le x \le a$, $f(x) \le f(a)$, which implies (6). It remains to consider (6) for $a \le 2$, or $x = 1$, or $a = 3, 4$, $x = 2$, which is elementary and reveals the exception $a = 2$, $x = 1$. □

*Proof of Theorem 1.* In view of Lemma 1 we may assume $n \ge 4$. $x_i$ will denote nonnegative integers. Assume that $b_1 \ge \cdots \ge b_k$ and let $j$ be the least integer $g < k$ such that, in the notation of Lemma 3, $N_g = n$, if such $g$ exists, otherwise $j = k$. We assert that

$$N\left(n; a_1, b_1; \ldots; a_k, b_k\right) \ge \begin{cases} [(b_1+1)/n] \prod\limits_{i=2}^{k} (b_i+1), & \text{if } j = 1 \\ \prod\limits_{i=j+1}^{k} (b_i+1), & \text{if } j > 1, \end{cases} \tag{7}$$

where an empty product is 1.

Indeed, if $j = 1$, then for every choice of $x_i \le b_i$ $(i \ge 2)$ with

$$a_1 r + \sum_{i=2}^{k} a_i x_i \equiv 0 (\mathrm{mod}\ n),$$

where $0 \le r < n$, there exist $[(b_1 - r)/n] + 1 \ge [(b_1+1)/n]$ nonnegative integers $x_1 \le b_1$ such that $a_1 x_1 \equiv a_1 r (\mathrm{mod}\ n)$, which implies (7).

If $k > j > 1$, then $N_j = n$, hence all residue classes mod $n$ are represented by $\sum_{i=1}^{j} a_i x_i$ $(0 \le x_i \le b_i)$. It follows that for every choice of $x_i \le b_i$ $(j < i \le k)$ there exist $x_i \le b_i$ $(i \le j)$ such that $\sum_{i=1}^{k} a_i x_i \equiv 0 (\mathrm{mod}\ n)$, which implies (7).

For $j = k$, we have $N(n; a_1, b_1; \ldots; a_k, b_k) \ge 1$, because of the solution $x_1 = \cdots = x_k = 0$.

Formula (7) implies

$$N\left(n; a_1, b_1; \ldots; a_k, b_k\right)^{-1} \prod_{i=1}^{k} (b_i + 1) \begin{cases} \left[\frac{b_1+1}{n}\right]^{-1} (b_i + 1), & \text{if } j = 1 \\ \prod_{i=1}^{j} (b_i + 1), & \text{if } j > 1. \end{cases}$$

For $j = 1$, we obtain, using $n \geq 4$,

$$(b_1 + 1)\left[\frac{b_1 + 1}{n}\right]^{-1} \leq \left(\left[\frac{b_1 + 1}{n}\right] n + n - 1\right)\left[\frac{b_1 + 1}{n}\right]^{-1} \leq 2n - 1 \leq 2^{n-1}.$$

For $j = 2$, we have $b_2 \leq b_1 \leq n - 2$, hence

$$\prod_{i=1}^{j} (b_i + 1) \leq (n - 1)^2$$

and since, by Lemma 5, $(n - 1)^2 \leq 2^{n-1}$ except for $n = 4$, we obtain

$$\prod_{i=1}^{j} (b_i + 1) \leq 2^{n-1}$$

except for $n = 4$, $b_1 = b_2 = 2$. This case is covered by Lemma 2.

If $j \geq 3$, then $N_2 \leq N_{j-1} < n$, hence by Lemma 4, $j - 1 \leq b_1 + \cdots + b_{j-1} \leq n - 2$. It follows that $b_j \leq b_{j-1} \leq (n - 2)/(j - 1)$ and by the inequality for the arithmetic and geometric mean

$$\prod_{i=1}^{j} (b_i + 1) \leq \left(\frac{b_1 + \cdots + b_{j-1}}{j - 1} + 1\right)^{j-1} \left(\frac{n - 2}{j - 1} + 1\right) \leq \left(\frac{n - 2}{j - 1} + 1\right)^{j}.$$

By Lemma 5 the right-hand side does not exceed $2^{n-1}$, since $j - 1 > 1$. □

*Proof of Corollary.* For each $d \mid n$ we consider the system of congruences

$$\sum_{i \in I_d} \frac{a_i}{d} x_i \equiv 0 \left(\bmod \frac{n}{d}\right), \tag{8}$$

where $I_d = \{i \leq k : (n, a_i) = d\}$.

By Theorem 1 the number of solutions of (8) with $0 \leq x_i \leq b_i (i \in I_d)$ is at least

$$2^{1-n/d} \prod_{i \in I_d} (b_i + 1).$$

Now, $\bigcup_{d|n} I_d = \{1, \ldots, k\}$ and $I_d$ are disjoint, hence the number of solutions of the system of congruences

$$\sum_{i \in I_d} \frac{a_i}{d} x_i \equiv 0 \left(\bmod \frac{n}{d}\right), d \mid n \tag{9}$$

in nonnegative integers $x_i \leq b_i$ $(1 \leq i \leq k)$ is at least

$$\prod_{d|n} 2^{1-n/d} \prod_{i \in I_d} (b_i + 1) = 2^{-\sum_{d|n}(n/d-1)} \prod_{i=1}^{k} (b_i + 1).$$

It remains to observe that the system (9) implies (1) and that

$$\sum_{d|n}\left(\frac{n}{d}-1\right)=\sigma(n)-\tau(n).$$

□

*Proof of Theorem 2.* Let $\Gamma=\Gamma_1\oplus\Gamma_2\oplus\cdots\oplus\Gamma_l$, where $\Gamma_i$ is a cyclic group generated by $\gamma_i$ $(1\le i\le l)$, $|\Gamma_i|=g_i$ and $g_i\mid g_{i+1}$. We put

$$\Delta=\bigcup_{j\le l}\Gamma_1\oplus\cdots\oplus\Gamma_{j-1}\oplus\left\{d\gamma_j : d\mid g_j,\ 0<d<g_j\right\}$$

and order elements of $\Delta$ in whatever manner, writting $\delta\prec\varepsilon$ or $\varepsilon\prec\delta$ for any two distinct elements $\delta,\varepsilon$ of $\Delta$. Let further for $\delta\in\Delta$, $o(\delta)$ be the order of $\delta$ in $\Gamma$,

$$I_\delta=\{i\le k:\alpha_i=\delta a \text{ for some } a\in\mathbb{Z}, (a,o(\delta))=1\},$$
$$J_\delta=I_\delta\setminus\bigcup_{\varepsilon\prec\delta}I_\varepsilon.$$

For every $\alpha\in\Gamma$ there exist $a$ and $\delta$ such that

$$a\in\mathbb{Z}, \delta\in\Delta, (a,o(\delta))=1,\ \alpha=\delta a. \tag{10}$$

Indeed, for $\alpha=0$ we take $\delta=0$, $a=1$, otherwise let

$$\alpha=\sum_{\lambda=1}^{l}\gamma_\lambda c_\lambda;\ 0\le c_\lambda<g_\lambda. \tag{11}$$

Let $\mu$ be the greatest $\lambda$ such that $c_\lambda\ne 0$. We choose $t_0$ such that

$$\left(\frac{c_\mu+g_\mu t_0}{(c_\mu,g_\mu)},g_\mu\right)=1, \tag{12}$$

then $t_\lambda\in\mathbb{Z}$ such that

$$\frac{c_\mu+g_\mu t_0}{(c_\mu,g_\mu)}t_\lambda\equiv c_\lambda \pmod{g_\lambda}\quad(1\le\lambda<\mu) \tag{13}$$

and put

$$\delta=\sum_{\lambda<\mu}\gamma_\lambda t_\lambda+\gamma_\mu\left(c_\mu,g_\mu\right),\ a=\frac{c_\mu+g_\mu t_0}{(c_\mu,g_\mu)}.$$

We have (10) by virtue of (11)–(13). It follows that

$$\{1,\ldots,k\}=\bigcup_{\delta\in\Delta}I_\delta=\bigcup_{\delta\in\Delta}J_\delta.$$

Since $J_\delta$ are disjoint, for each $i$ there exists exactly one $\delta\in\Delta$ and exactly one $a\in\mathbb{Z}$, named $a_i$, such that $i\in J_\delta$ and

$$\alpha_i=\delta a;\ 0\le a<o(\delta),\ (a,o(\delta))=1.$$

Consider now the system of congruences

$$\sum_{i \in J_\delta} a_i x_i \equiv 0(\text{mod } o(\delta)), \quad \delta \in \Delta.$$

By Theorem 1 the number of solutions of this system with $0 \le x_i \le b_i$ $(1 \le i \le k)$ is at least

$$\prod_{\delta \in \Delta} 2^{1-o(\delta)} \prod_{i \in J_\delta} (b_i + 1) = 2^{-\sum_{\delta \in \Delta}(o(\delta)-1)} \prod_{i=1}^{k} (b_i + 1)$$

and it remains to estimate the sum in the exponent.

Now, for $\delta \in \Gamma_1 \oplus \cdots \oplus \Gamma_{j-1} \oplus \{d\gamma_j : d \mid g_j,\ d < g_j\}$ we have $o(\delta) \le g_j$ and the number of such $\gamma$ for a fixed $j \le l$ is

$$(\tau(g_j) - 1) \prod_{i<j} g_i.$$

Hence

$$\sum_{\delta \in \Delta}(o(\delta) - 1) \le (\tau(g_l) - 1) \sum_{j=1}^{l} (g_j - 1) \prod_{i<j} g_i = (\tau(g_l) - 1) \left( \prod_{i=1}^{l} g_i - 1 \right)$$
$$= (\tau(\chi(\Gamma)) - 1)(|\Gamma| - 1).$$

## References


1. Drmota, M., Skałba, M.: Equidistribution of divisors in residue classes. Preprint. Technical University of Vienna, Vienna
2. Korobov, N.M.: *Teoretikochislovye Metody v Priblizhennom Analise*. Gos. Izd'vo Fiziko-matematicheskoi Lit'ry, Moscow (1963)
3. Mann, H.B.: *Addition Theorems. The Addition Theorems of Group Theory and Number Theory*. Interscience, New York (1965)
4. Olson, J.: A combinatorial problem in finite abelian groups, II. *J. Number Theory* **1**, 195–199 (1969)
5. Schinzel, A.: The number of solutions of a linear homogeneous congruence II. In: Chen, W., Gowers, T., Halberstam, H., Schmidt, W., Vaughan, R.C. (eds.) *Analytical Number Theory: Essays in Honour of Klaus F. Roth*. Cambridge University Press, Cambridge (to appear)
6. Schinzel, A., Zakarczemny, M.: On a linear homogeneous congruence. *Colloq. Math.* **106**, 283–292 (2006)
7. Zakarczemny, M.: Master dissertation, Department of Mathematics, University of Warsaw, Warsaw, Poland (2004)

*Diophantine Approximation*
H. P. Schlickewei et al., Editors


# A NOTE ON LYAPUNOV THEORY FOR BRUN ALGORITHM

Fritz Schweiger
*Department of Mathematics, University of Salzburg, Hellbrunnerstrasse 34, 5020 Salzburg, Austria*
fritz.schweiger@sbg.ac.at

## 1 Introduction

Regular continued fractions exhibit a number of remarkable properties. We mention three of them.

1. The calculation follows a simple algorithm. For any real number $x$ with $0 < x < 1$ one defines

$$a_1(x) := \left[\frac{1}{x}\right] \quad \text{and}$$
$$Tx := \frac{1}{x} - a_1(x).$$

These definitions immediately lead to the well-known expansion

$$x = \cfrac{1}{a_1 + \cfrac{1}{a_2 + \cfrac{1}{a_3 + \dots}}}.$$

2. The algorithm becomes eventually periodic, i.e., $T^{n+m}x = T^m x$ for some $n \geq 0$, $m \geq 1$, if and only if $x$ is a quadratic irrational number ("Theorem of Lagrange").

3. If one puts

$$\frac{p_n}{q_n} := \cfrac{1}{a_1 + \cfrac{1}{a_2 + \dots + \cfrac{1}{a_n}}},$$

then one obtains "good" approximations to $x$. These fractions satisfy the Dirichlet property

$$|x - \frac{p_n}{q_n}| < \frac{1}{q_n^2}.$$

C. G. J. Jacobi (1804–1851) was the first one who studied multidimensional continued fractions. His main objective was to find a generalization of Lagrange's theorem. He proposed an algorithm for pairs of real numbers $(x_1, x_2)$ in the hope that this algorithm will become eventually periodic if and only if $x_1$ and $x_2$ belong to a cubic

---

*Keywords.* Metric theory of generalized continued fractions.

*2000 Mathematics subject classification.* 11K55, 11J70.

number field. Jacobi found some examples, like $\sqrt[3]{2}$, $\sqrt[3]{4}$ and $\sqrt[3]{3}$, $\sqrt[3]{9}$ but he could not show that Lagrange's theorem is true for this algorithm (see [1]).

A 2-dimensional continued fraction (see [7,8] for an exposition of the background and references) is given by a map $T : B \to B$, $B \subset \mathbb{R}^2$ such that there is a partition $B = \bigcup_{j \in I} B(j)$ and a set of matrices $\alpha(k) \in GL(3, \mathbb{Z}), \alpha(k) = ((A_{rs})), 0 \le r, s \le 2$ such that

$$T(x_1, x_2) = \left( \frac{A_{10} + A_{11}x_1 + A_{12}x_2}{A_{00} + A_{01}x_1 + A_{02}x_2}, \frac{A_{20} + A_{21}x_1 + A_{22}x_2}{A_{00} + A_{01}x_1 + A_{02}x_2} \right)$$

for $x = (x_1, x_2) \in B(k),\ k \in I$.

We put $k(x) := k$ if $x \in B(k)$, and $\beta(k) = ((B_{rs})) := \alpha(k)^{-1}$. Then Oseledets' theorem (see e.g. [4]) can be applied to the stochastic process $\beta(k(x)),\ \beta(k(Tx)),\ \beta(k(T^2x)), \ldots$. If we assume that $B_{rs} \ge 0$ for $0 \le r,\ s \le 2$ and some further technical conditions (see [3,7]), then the three Lyapunov exponents $\lambda_0 > \lambda_1 \ge \lambda_2$ satisfy

$$\lambda_0 > 0$$
$$\lambda_1 \le 0$$
$$\lambda_2 + \lambda_1 + \lambda_0 = 0.$$

We consider the natural extension of the system $(T, B)$, i.e., the space $(\Omega, \tau)$ of all admissible sequences $\omega = (k_i)_{i=-\infty}^{\infty}$, $k_i \in I$, equipped with the shift operator $\tau : \Omega \to \Omega$.

Then there are vector valued functions $e_0(\omega), e_1(\omega), e_2(\omega) \in \mathbb{R}^3$ such that

$$\alpha(x)e_j(\omega) = \varphi_j(\omega)e_j(\tau\omega), \quad 0 \le j \le 2, \quad \varphi_j(\omega) \in \mathbb{R}.$$

Broise-Alamichel and Guivarc'h [2] prove that for the Jacobi–Perron algorithm and the multiplicative Brun algorithm we find $\lambda_2 < \lambda_1 < 0$. In this note an explicit construction for $e_0(\omega), e_1(\omega), e_2(\omega)$ is given for the multiplicative Brun algorithm.

For later purposes we also introduce the Jacobian matrix $J_x$ of the map $T$. Then it is known that the Lyapunov exponents of $J_x$ are given as $\lambda_0 - \lambda_1$ and $\lambda_0 - \lambda_2$.

The link between Lyapunov exponents and Diophantine approximation can be described as follows. First we define

$$\omega(x, g) := \sup \left\{ \delta > 0 : \sum_{i=1}^{2} \left| x_i - \frac{B_{ig}^{(N)}}{B_{0g}^{(N)}} \right| < \frac{1}{(B_{0g}^{(N)})^\delta} \text{ for all } N \ge N(x, g, \delta) \right\}$$
$$\omega(x) := \min_{0 \le g \le n} \omega(x, g).$$

Clearly, $\omega(x) \le \frac{3}{2}$ must be true almost everywhere. For a broad class of multidimensional continued fractions the following result was proved by Lagarias.

**Theorem [3].** *Let $\lambda_0 > \lambda_1$ be the first Lyapunov exponents of the stochastic process $\beta(k_s), s = 1, 2, \ldots$, then for almost all $x$*

$$\omega(x) = 1 - \frac{\lambda_1}{\lambda_0}.$$

## 2 A skew product

As a kind of introductory example we consider first the skew product algorithm

$$B = \{x = (x_1, x_2) : 0 < x_1 \leq 1,\ 0 \leq x_2 \leq 1\}$$

defined by the map

$$T(x_1, x_2) = \left(\frac{1}{x_1} - b_1, \frac{x_2}{x_1} - a_1\right), \quad b_1 = \left[\frac{1}{x_1}\right], a_1 = \left[\frac{x_2}{x_1}\right]$$

with the conditions

(i) $b_1 \geq 1$;
(ii) $b_1 \geq a_1 \geq 0$;
(iii) if $b_1 = a_1$, then $a_2 = 0$.

The associated matrices $\alpha$ and $\beta$ are given as

$$\beta(\omega) = \beta(x) = \begin{pmatrix} b_1 & 1 & 0 \\ 1 & 0 & 0 \\ a_1 & 0 & 1 \end{pmatrix}, \quad \alpha(\omega) = \alpha(x) = \begin{pmatrix} 0 & 1 & 0 \\ 1 & -b_1 & 0 \\ 0 & -a_1 & 1 \end{pmatrix}.$$

Clearly

$$e_0(\omega) = \begin{pmatrix} 1 \\ x_1 \\ x_2 \end{pmatrix}, \quad \alpha(x)e_0(\omega) = x_1 e_0(\tau\omega).$$

Let $\omega = ((b_i, a_i))_{i=-\infty}^{\infty}$ be a sequence in the natural extension. If

$$\gamma^{(s)}(\omega) := \alpha(\tau^{-1}\omega)\ldots\alpha(\tau^{-s-1}\omega) = \begin{pmatrix} C_0^{(s-1)} & C_0^{(s)} & 0 \\ C_1^{(s-1)} & C_1^{(s)} & 0 \\ C_2^{(s-1)} & C_2^{(s)} & 1 \end{pmatrix},$$

then the recursion relations are given as follows. For sake of simplicity we put $b_{-s-1} =: b,\ a_{-s-1} =: a$.

$$\begin{aligned} C_0^{(s+1)} &= -b\,C_0^{(s)} + C_0^{(s-1)} \\ C_1^{(s+1)} &= -b\,C_1^{(s)} + C_1^{(s-1)} \\ C_2^{(s+1)} &= -b\,C_2^{(s)} + C_2^{(s-1)} - a \end{aligned}$$

Then the following properties hold.

(i) $C_0^{(s-1)}C_1^{(s)} - C_0^{(s)}C_1^{(s-1)} = (-1)^s$

(ii) $C_0^{(s)} \geq 1 \quad \text{for } s \equiv 1 \mod 2, \quad s \geq 1$

$C_0^{(s)} \leq -1 \quad \text{for } s \equiv 0 \mod 2, \quad s \geq 2$

(iii) $|C_0^{(s+1)}| = b\,|C_0^{(s)}| + |C_0^{(s-1)}|$

$|C_0^{(s)}| \geq \Theta^{s-1}, \quad \Theta := \frac{1+\sqrt{5}}{2}$

(iv) $|C_0^{(s)}C_2^{(s-1)} - C_2^{(s)}C_0^{(s-1)}| \leq |C_0^{(s)}| + |C_0^{(s-1)}|$

The proofs are by induction. We just indicate the inductive step for (iv).

$$\begin{aligned}
|C_0^{(s)}C_2^{(s+1)} - C_0^{(s+1)}C_2^{(s)}| &= |C_0^{(s)}C_2^{(s-1)} - C_2^{(s)}C_0^{(s-1)} - aC_0^{(s)}| \\
&\le |C_0^{(s)}C_2^{(s-1)} - C_2^{(s)}C_0^{(s-1)}| + a\,|C_0^{(s)}| \\
&\le |C_0^{(s)}| + |C_0^{(s-1)}| + b\,|C_0^{(s)}| = |C_0^{(s)}| + |C_0^{(s+1)}|
\end{aligned}$$

Then

$$\left|\frac{C_1^{(s)}}{C_0^{(s)}} - \frac{C_1^{(s+1)}}{C_0^{(s+1)}}\right| \le \frac{1}{|C_0^{(s)}C_0^{(s+1)}|} \le \Theta^{-2s+1}$$

$$\left|\frac{C_2^{(s)}}{C_0^{(s)}} - \frac{C_2^{(s+1)}}{C_0^{(s+1)}}\right| \le \frac{|C_0^{(s)}| + |C_0^{(s-1)}|}{|C_0^{(s)}C_0^{(s+1)}|} \le 2\,\Theta^{-s}.$$

Then one sees that

$$\lim_{s\to\infty}\left(\frac{C_1^{(s)}}{C_0^{(s)}}, \frac{C_2^{(s)}}{C_0^{(s)}}\right) =: (\xi_1, \xi_2)$$

exists. The vector

$$e_2(\omega) = \begin{pmatrix} 1 \\ \xi_1 \\ \xi_2 \end{pmatrix}$$

corresponds to the Lyapunov exponent $\lambda_2$.

In this special case an alternative approach is possible. Let $T^\#$ denote the dual algorithm defined on the quadrilateral with vertices $(0, 0)$, $(1, -1)$, $(1, 1)$, $(0, 1)$ and the map

$$T^\#(y_1, y_2) = \left(\frac{1}{y_1} - b_0, -\frac{y_2}{y_1} + a_0\right),$$

$$b_0 = \left[\frac{1}{y_1}\right], \quad a_0 = \max\left(\left[\frac{1}{y_1}\right] - \left[\frac{1-y_2}{y_1}\right], 0\right)$$

subject to the conditions

(i) $b_0 \ge 1$;
(ii) $b_0 \ge a_0 \ge 0$;
(iii) if $a_0 \ge 1$, then $b_{-1} > a_{-1}$.

Then the map $Q = \psi^{-1} \circ T^\# \circ \psi$, $\psi(s, t) = (-\frac{1}{s}, \frac{t}{s})$, satisfies

$$Q(\xi_1, \xi_2) = \left(\frac{1}{b_0 + \xi_1}, \frac{a_0 - \xi_2}{b_0 + \xi_1}\right).$$

Therefore the local inverse branches of $Q$ are given by

$$(\eta_1, \eta_2) \mapsto \left(-b_0 + \frac{1}{\eta_1}, -a_0 + \frac{\eta_2}{\eta_1}\right).$$

Hence the vector

$$e_2(\omega) = \begin{pmatrix} 1 \\ \xi_1 \\ \xi_2 \end{pmatrix}$$

satisfies $\alpha(x)e_2(\omega) = \xi_1 e_2(\tau\omega)$.

Since $Q$ is conjugate to $T^\#$, clearly $T$ and $T^\#$ have the same Lyapunov exponents. This shows that $\lambda_0 = -\lambda_2$. Therefore, $\lambda_1 = 0$.

The corresponding vector is given by

$$e_1(\omega) = \begin{pmatrix} 0 \\ 0 \\ 1 \end{pmatrix}.$$

Therefore the Lyapunov vectors of

$$J_x = \frac{1}{x_1^2} \begin{pmatrix} -1 & 0 \\ -x_2 & x_1 \end{pmatrix}$$

are given by

$$u_1(\omega) = \begin{pmatrix} 0 \\ 1 \end{pmatrix} \quad \text{and} \quad u_2(\omega) = \begin{pmatrix} x_1 - \xi_1 \\ x_2 - \xi_2 \end{pmatrix}.$$

The vector $u_1(\omega)$ belongs to $-\lambda_1 + \lambda_0$ and the vector $u_2(\omega)$ corresponds to $-\lambda_2 + \lambda_0$. Since $J_x u_1(\omega) = \frac{1}{x_1} u_1(\tau\omega)$, one sees again that $\lambda_1 = 0$.

If $u = \binom{\cos\theta}{\sin\theta}$, then in Broise-Alamichel and Guivarc'h [2] the cocycle

$$f(\omega, u) = \frac{1}{2} \log \frac{x_1^2}{(x_2\cos\theta - x_1\sin\theta)^2 + \cos^2\theta}$$

is considered. Note that $f(\omega, u) = 0$ along the direction $u = u_1$ as required by the general theory.

## 3 Brun algorithm

The multiplicative version of the Brun algorithm is defined on $\Delta = \{x = (x_1, x_2) : 0 < x_1 \le 1,\ 0 \le x_2 \le x_1\}$ by the map

$$T(x_1, x_2) = \left(\frac{1}{x_1} - N_1, \frac{x_2}{x_1}\right), \quad N_1 = \left[\frac{1}{x_1}\right], \quad N_1 x_1 + x_2 \le 1$$

$$T(x_1, x_2) = \left(\frac{x_2}{x_1}, \frac{1}{x_1} - N_1\right), \quad N_1 = \left[\frac{1}{x_1}\right], \quad 1 \le N_1 x_1 + x_2 .$$

To distinguish these cases we write $(N, \varepsilon)$ for the associated digits, i.e., $N = N_1$ and $\varepsilon = 0$ in the first case, $\varepsilon = 1$ in the second case. Then the associated matrices are given as

$$\alpha(\omega) = \alpha(x) = \begin{pmatrix} 0 & 1 & 0 \\ 1-\varepsilon_1 & -N_1(1-\varepsilon_1) & \varepsilon_1 \\ \varepsilon_1 & -N_1\varepsilon_1 & 1-\varepsilon_1 \end{pmatrix},$$

$$\beta(\omega) = \beta(x) = \begin{pmatrix} N_1 & 1-\varepsilon_1 & \varepsilon_1 \\ 1 & 0 & 0 \\ 0 & \varepsilon_1 & 1-\varepsilon_1 \end{pmatrix}.$$

Again we find

$$e_0(\omega) = \begin{pmatrix} 1 \\ x_1 \\ x_2 \end{pmatrix}, \quad \alpha(x)e_0(\omega) = x_1 e_0(\tau\omega).$$

*Remark.* If $((B_{ij}^{(s)})) := \beta(x)\beta(Tx)\dots\beta(T^{s-1}x)$, then $|x_i B_{00}^{(s)} - B_{i0}^{(s)}|, i = 1, 2; s \geq 1$ is bounded [5,9]. Therefore for periodic expansions the eigenvalues $\sigma_0, \sigma_1, \sigma_2$ of the periodicity matrix properly arranged satisfy the conditions $|\sigma_2| \leq |\sigma_1| \leq 1 < \sigma_0$.

For the construction of $e_2(\omega)$ we proceed as follows. Let

$$\omega = ((N_i, \varepsilon_i))_{i=-\infty}^{\infty}$$

be a sequence in the natural extension. Then we define the matrices

$$\gamma^{(s)}(\omega) = \left(\left(C_{ij}^{(s)}\right)\right) := \alpha(\tau^{-1}\omega)\dots\alpha(\tau^{-s-1}\omega).$$

However, in this case it is not true that $\lim_{s\to\infty}\left(\frac{C_{10}^{(s)}}{C_{00}^{(s)}}, \frac{C_{20}^{(s)}}{C_{00}^{(s)}}\right)$ does exist for all sequences $\omega$. It cannot exist, if $\omega$ is a periodic sequence which corresponds to a cubic number field with complex conjugates. As an example consider the periodic algorithm with $(N_i, \varepsilon_i) = (1, 1)$ for all $i \leq 0$. The eigenvalues of the matrix

$$\alpha = \begin{pmatrix} 0 & 1 & 0 \\ 0 & 0 & 1 \\ 1 & -1 & 0 \end{pmatrix}$$

are the real value $\rho_0$ with $0 < \rho_0 < 1$ and the two complex numbers $\rho_1, \rho_2 = \bar{\rho}_1$. Note that $|\rho_1| = |\rho_2| > 1$. Then a calculation shows that $C_{j0}^{(s)} = a_j\rho_0^s + b_j\rho_1^s + c_j\rho_2^s$, for $0 \leq j \leq 2$.

Since $b_j \neq 0$ and $c_j \neq 0$, one verifies that the required limit cannot exist.

The remark on periodic expansions implies that if $\omega$ is a periodic sequence which corresponds to a totally real cubic number field, then $|\rho_2| > |\rho_1| > 1$ and $0 < \rho_0 < 1$ and therefore

(i) $|C_{00}^{(s)}| > 0$ for all $s \geq s_0$,

(ii) $\lim_{s\to\infty} \frac{C_{j0}^{(s)}}{C_{00}^{(s)}}$ exists.

**Theorem.** *For almost all sequences $\omega$ there is a subsequence such that*

$$\lim_{s\to\infty}\left(\frac{C_{10}^{(s)}}{C_{00}^{(s)}}, \frac{C_{20}^{(s)}}{C_{00}^{(s)}}\right) =: (\xi_1, \xi_2)$$

*exists and the vector*

$$e_2(\omega) = \begin{pmatrix} 1 \\ \xi_1 \\ \xi_2 \end{pmatrix}$$

*satisfies* $\alpha(x)e_2(\omega) = \xi_1 e_2(\tau\omega)$.

*Proof.* Since $\gamma^{(s)}(\tau\omega) = \alpha(\omega)\alpha(\tau^{-1}\omega)\ldots\alpha(\tau^{-s-2}\omega)$ we find

$$e_2(\tau\omega) = \begin{pmatrix} 1 \\ -N_1 + \frac{1}{\xi_1} \\ \frac{\xi_2}{\xi_1} \end{pmatrix} \quad \text{if } \varepsilon_1 = 0,$$

$$e_2(\tau\omega) = \begin{pmatrix} 1 \\ \frac{\xi_2}{\xi_1} \\ -N_1 + \frac{1}{\xi_1} \end{pmatrix} \quad \text{if } \varepsilon_1 = 1.$$

If $\varepsilon_{-s-1} = 0$, then $C_{00}^{(s+1)} = C_{01}^{(s)}, C_{02}^{(s+1)} = C_{02}^{(s)}$, and if $\varepsilon_{-s-1} = 1$, then $C_{00}^{(s+1)} = C_{02}^{(s)}, C_{02}^{(s+1)} = C_{01}^{(s)}$. Therefore, for almost every $\omega$ we find $\limsup_{s\to\infty} \frac{1}{s} \log \left|C_{00}^{(s)}\right| = \limsup_{s\to\infty} \frac{1}{s} \log \left|C_{01}^{(s)}\right| = \limsup_{s\to\infty} \frac{1}{s} \log \left|C_{02}^{(s)}\right| = -\lambda_2 = \lambda_0 + \lambda_1$.

The last equalities come from Oseledets' theorem (note that the rows $\left(C_{i0}^{(s)}, C_{i1}^{(s)}, C_{i2}^{(s)}\right)$, $0 \leq i \leq 2$, have the same growth rate). For given $\varepsilon > 0$ we choose a subsequence such that

$$\log \left|C_{00}^{(s)}\right| \gg s(\lambda_0 + \lambda_1 - \varepsilon)$$

and

$$\log \left|C_{00}^{(s+1)}\right| \gg s(\lambda_0 + \lambda_1 - \varepsilon).$$

If $\delta^{(s)}(\omega) = \left(\left(D_{ij}^{(s)}\right)\right) = \beta(\tau^{-s-1}\omega)\ldots\beta(\tau^{-1}\omega)$ denotes the inverse matrix of $\gamma^{(s)}(\omega)$, then

$$\left|C_{i0}^{(s)}C_{00}^{(s+1)} - C_{i0}^{s+1}C_{00}^{(s)}\right| \ll D_{00}^{(s)} \ll e^{(\lambda_0+\varepsilon)s}$$

is valid almost everywhere. Then

$$\left|\frac{C_{i0}^{(s)}}{C_{00}^{(s)}} - \frac{C_{i0}^{(s+1)}}{C_{00}^{(s+1)}}\right| \leq \frac{\left|C_{i0}^{(s)}C_{00}^{(s+1)} - C_{i0}^{(s+1)}C_{00}^{(s)}\right|}{\left|C_{00}^{(s)}C_{00}^{(s+1)}\right|}, \quad i = 1, 2,$$

and

$$\left|\frac{C_{i0}^{(s)}}{C_{00}^{(s)}} - \frac{C_{i0}^{(s+1)}}{C_{00}^{(s+1)}}\right| \ll e^{-s(\lambda_0+2\lambda_1-3\varepsilon)}.$$

According to Broise-Alamichel and Guivarc'h [2], $\lambda_2 < \lambda_1 < \lambda_0$. Therefore for small $\varepsilon > 0$ we have $\lambda_0 + 2\lambda_1 - 3\varepsilon > 0$ and the subsequence converges.

If $\gamma^{(s+1)}(\tau\omega) = \left(\left(E_{ij}^{(s+1)}\right)\right)$ and $\left(\frac{C_{10}^{(s)}}{C_{00}^{(s)}}, \frac{C_{20}^{(s)}}{C_{00}^{(s)}}\right) \to (\xi_1, \xi_2)$ in a subsequence, then clearly $\left(\frac{E_{10}^{(s)}}{E_{00}^{(s)}}, \frac{E_{20}^{(s)}}{E_{00}^{(s)}}\right) \to (\vartheta_1, \vartheta_2)$ in the same subsequence and

$$\alpha(\omega)\begin{pmatrix} 1 \\ \xi_1 \\ \xi_2 \end{pmatrix} = \xi_1 \begin{pmatrix} 1 \\ \vartheta_1 \\ \vartheta_2 \end{pmatrix}.$$

*Remarks.* 1. An argument due to Perron shows that for periodic expansions $\rho_1 = \rho_2$ holds if and only if the numbers $x_1, x_2$ belong to a cubic number field with complex conjugates.

2. One may conjecture that in fact $\left|C_{00}^{(s)}\right| > 0$ for all $s \geq s(\omega)$ for almost every $\omega \in \Omega$.

3. For Brun algorithm periodicity of a point $(x_1, x_2)$ does not necessarily imply that its coordinates belong to a cubic number field.

The dual algorithm $T^\#$ is defined on $B = \{x = (x_1, x_2) : 0 < x_1 \leq 1, 0 < x_2 \leq 1\}$ by the map

$$T^\#(y_1, y_2) = \left(\frac{1}{y_1} - N_0, \frac{y_2}{y_1}\right), \quad N_0 = \left[\frac{1}{y_1}\right], \quad y_2 \leq y_1,$$
$$T^\#(y_1, y_2) = \left(\frac{1}{y_2} - N_0, \frac{y_1}{y_2}\right), \quad N_0 = \left[\frac{1}{y_2}\right], \quad y_1 \leq y_2.$$

The matrices $\alpha^\#$ and $\beta^\#$ are the transposed matrices of $\alpha$ and $\beta$. In a similar way we can construct vectors

$$e_0^\#(\omega) = \begin{pmatrix} 1 \\ y_1 \\ y_2 \end{pmatrix}, \quad e_2^\#(\omega) = \begin{pmatrix} 1 \\ \eta_1 \\ \eta_2 \end{pmatrix}.$$

Then we verify that $e_1(\omega) := e_0^\#(\omega) \wedge e_2^\#(\omega)$ satisfies $\alpha(x)e_1(\omega) = \rho(\omega)e_1(\tau\omega)$ with a factor $\rho(\omega)$.

Calculation shows that

$$e_1(\omega) = \begin{pmatrix} y_1\eta_2 - y_2\eta_1 \\ y_2 - \eta_2 \\ \eta_1 - y_1 \end{pmatrix}.$$

The dual algorithm $T^\#$ is related to the sequence $\omega$ read backwards. Therefore

$$e_0^\#(\tau\omega) = \begin{pmatrix} 1 \\ \frac{1}{N_1+y_1} \\ \frac{y_2}{N_1+y_1} \end{pmatrix}, \quad e_2^\#(\tau\omega) = \begin{pmatrix} 1 \\ \frac{1}{N_1+\eta_1} \\ \frac{\eta_2}{N_1+\eta_1} \end{pmatrix}$$

or

$$e_0^\#(\tau\omega) = \begin{pmatrix} 1 \\ \frac{y_2}{N_1+y_1} \\ \frac{1}{N_1+y_1} \end{pmatrix}, \quad e_2^\#(\tau\omega) = \begin{pmatrix} 1 \\ \frac{\eta_2}{N_1+\eta_1} \\ \frac{1}{N_1+\eta_1} \end{pmatrix}.$$

This shows that $e_1(\omega)$ is the required vector and the factor is given as $\rho(\omega) = -(N_1 + y_1)(N_1 + \eta_1)$ or $\rho(\omega) = (N_1 + y_1)(N_1 + \eta_1)$. The Lyapunov vectors of

$$J_x = \frac{1}{x_1^2}\begin{pmatrix} -1 & 0 \\ -x_2 & x_1 \end{pmatrix}, \quad \varepsilon_1(x) = 0,$$

$$J_x = \frac{1}{x_1^2}\begin{pmatrix} -x_2 & x_1 \\ -1 & 0 \end{pmatrix}, \quad \varepsilon_1(x) = 1$$

are given by

$$u_1(\omega) = \begin{pmatrix} x_1 - \frac{y_2 - \eta_2}{y_1\eta_2 - y_2\eta_1} \\ x_2 + \frac{y_1 - \eta_1}{y_1\eta_2 - y_2\eta_1} \end{pmatrix}$$

$$u_2(\omega) = \begin{pmatrix} x_1 - \xi_1 \\ x_2 - \xi_2 \end{pmatrix}.$$

Since as before $u_2(\omega)$ obviously belongs to $-\lambda_2 + \lambda_0$, the other vector $u_1(\omega)$ corresponds to $-\lambda_1 + \lambda_0$.

## References

1. Bernstein, L.: *The Jacobi–Perron Algorithm: Its Theory and Application.* Lect. Notes Math. 207. Springer, Heidelberg (1971)
2. Broise-Alamichel, A., Guivarc'h, Y.: Exposants caractéristiques de l'algorithme de Jacobi-Perron et de la transformation associée. *Ann. Inst. Fourier* **51**, 565–686 (2001)
3. Lagarias, J.C.: The quality of the Diophantine approximations found by the Jacobi–Perron algorithm and related algorithms. *Monatsh. Math.* **115**, 299–328 (1993)
4. Mañé, R.: *Ergodic Theory and Differentiable Dynamics.* Springer, Heidelberg (1987)
5. Nakaishi, K.: The exponent of convergence for 2-dimensional Jacobi–Perron type algorithms. *Monatsh. Math.* **132**, 141–152 (2001)
6. Schratzberger, B.: The exponent of convergence for Brun's algorithm in two dimensions. *Sitzungsber. Österr. Akad. Wiss. Math.-naturw. Kl. Abt. II* **207**, 229–238 (1998)
7. Schweiger, F.: *Multidimensional Continued Fractions.* Oxford University Press, Oxford (2000)
8. Schweiger, F.: Diophantine properties of multidimensional continued fractions. In: Dubickas, A., et al. (eds.) *Analytic and Probabilistic Methods in Number Theory*, pp. 242–255. TEV, Vilnius (2002)
9. Toussaint, H.-J.: Der Algorithmus von Viggo Brun und verwandte Kettenbruchentwicklungen. Dissertation, Technische Universität München, Munich, Germany (1986)

*Diophantine Approximation*
H. P. Schlickewei et al., Editors


# ORBIT SUMS AND MODULAR VECTOR INVARIANTS

Serguei A. Stepanov[1,2]
[1] *Department of Mathematics, Bilkent University, 06533 Bilkent, Ankara, Turkey*
[2] *Department of Algebra, V. A. Steklov Mathematical Institute, Russian Academy of Sciences, Ulitsa Gubkina 8, Moscow, GSP-1, 117966 Russia*
sa-stepanov@iitp.ru

To Wolfgang M. Schmidt on the occasion of his 70th birthday

## 1 Introduction

Let $m, n$ be positive integers, $R$ a commutative ring with the unit element 1, and

$$A_{mn} = R[x_{11}, \ldots, x_{m1}; \ldots; x_{1n}, \ldots, x_{mn}]$$

the algebra of polynomials in $mn$ variables $x_{ij}$ over $R$. The symmetric group $S_n$ operates on the algebra $A_{mn}$ as a group of $R$-automorphisms by the rule: $\sigma(x_{ij}) = x_{i,\sigma(j)}$, $\sigma \in G$. Denote by $A_{mn}^{S_n}$ the subalgebra of invariants of the algebra $A_{mn}$ with respect to $S_n$ and define polarized elementary symmetric polynomials $u_{r_1,\ldots,r_m} \in A_{mn}^{S_n}$ in $n$ vector variables $(x_{11}, \ldots, x_{m1}), \ldots, (x_{1n}, \ldots, x_{mn})$ by means of the following formal identity

$$\prod_{j=1}^{n}(1 + x_{1j}z_1 + \cdots + x_{mj}z_m) = 1 + \sum_{1 \le r_1 + \cdots + r_m \le n} u_{r_1,\ldots,r_m} z_1^{r_1} \cdots z_m^{r_m}.$$

The elements of $A_{mn}^{S_n}$ are usually called *vector invariants* of $S_n$. If $R$ is Noetherian, it follows from the Hilbert–Noether finiteness theorem [5,7,8] that $A_{mn}^{S_n}$ is a finitely generated commutative $R$-algebra and $A_{mn}$ is finitely generated as a module over $A_{mn}^{S_n}$; moreover, if every nonzero integer is invertible in $R$, Weyl's theorem [13] states that the invariants $u_{r_1,\ldots,r_m}$ form a complete system of generators of $A_{mn}^{S_n}$ over $R$. In other words, each element $u \in A_{mn}^{S_n}$ can be written as a polynomial in $u_{r_1,\ldots,r_m}$, $1 \le r_1 + \cdots + r_m \le n$, with coefficients in $R$. The last result was recently generalized by D. Richman [10] and S. A. Stepanov [12] as follows: *if $|S_n| = n!$ is invertible in $R$, then $A_{mn}^{S_n}$ is generated as an $R$-algebra by the polarized elementary symmetric polynomials $u_{r_1,\ldots,r_m}$, $1 \le r_1 + \cdots + r_m \le n$, of degree at most $n$.*

Let $A = R[x_1, \ldots, x_N]$ be a finitely generated commutative $R$-algebra, $G$ a finite group of $R$-algebra automorphisms of $A$, and $A^G$ the subalgebra of invariants

*Keywords.* Polynomial invariants, generalized orbit Chern classes, finite groups, Noether degree bound.
*2000 Mathematics subject classification.* 11T55, 13A50, 14L24, 16R30, 20G40.

of $G$. If $z_1, \ldots, z_N$ are commuting variables, then we set

$$P(z_1, \ldots, z_N) = \prod_{\sigma \in G} (1 + \sigma(x_1)z_1 + \sigma(x_2)z_2 + \cdots + \sigma(x_N)z_N).$$

Let $\beta(A^G)$ denote the smallest positive integer $\beta$ such that $A^G$ can be generated as an $R$-algebra by polynomials of degree at most $\beta$. If each nonzero integer is invertible in $R$, the Noether result [7] implies that $A^G$ is generated by the coefficients of $P(z_1, \ldots, z_N)$, so that $\beta(A^G) \leq |G|$. The last inequality is known as the *Noether bound*. The above mentioned result of Richman and Stepanov and the standard arguments based on the use of the Reynolds operator and the Noether map (see [11]) show that Noether's bound holds under the condition that $|G|!$ is invertible in $R$. A recent result of P. Fleischmann [4] demonstrates that Noether's bound remains true under the weaker condition that $|G|$ is invertible in $R$.

Now let $R = F$ be a field, and $G$ a finite group acting linearly on a vector space $V$ over $F$ of finite dimension $n$. If the characteristic of $F$ is positive and divides $|G|$, then we speak of the *modular case*. Otherwise, we have the *nonmodular case*, which includes the case of classical invariants over a field of characteristic zero. Almost everything that is usually used in the nonmodular case is missing in the modular case: the Cohen–Macaulay property fails in general, we have no Reynolds operator (averaging over $G$) and no Molien formula for the Poincaré series. Nevertheless, if $F$ is a field of prime characteristic $p$, and $H$ a $p$-subgroup of $G$, the modular case admits an extensive application of *generalized orbit Chern classes* related to $H$, especially, orbit traces (*orbit sums* of monomials) and top orbit classes (*orbit norms* of monomials). Let $F = F_p$ be a prime finite field, and $H \leq GL(n, F_p)$ a cyclic group of prime order $p$ acting linearly on $V$. We set $A_{mn} = F_p[V^m]$ and denote by $A_{mn}^H$ the algebra of vector invariants of $A_{mn}$ with respect to $H$. It turns out (Theorems 5, 8 and 16) that there exist an $F_p$-linear space $\tilde{V}$ containing $V$ as a subspace, a cyclic group $\tilde{H}$ of order $p$ closely related to $H$ and acting linearly on $\tilde{V}$, such that every invariant $u \in A_{mn}^H$ can be written as a special $F_p$-linear combination of orbit sums $S_{\tilde{H}}(f)$, orbit norms $N_{\tilde{H}}(g)$ related to the group $\tilde{H}$, and also their products $S_{\tilde{H}}(f)N_{\tilde{H}}(g)$, for various monomials $f, g \in F_p[\tilde{(V)}^m]$. This is a new point of view in modular invariant theory that reveals an important role of $p$-subgroups $H$ of $G$ and the associated orbit Chern classes of monomials. It should be pointed out that if $H$ is a cyclic group of prime order $p$, and $F = F_p$ a prime field, then $S_{\tilde{H}}(f)$ and $N_{\tilde{H}}(g)$ can be calculated explicitly. This gives a possibility to determine a system of generating elements of $A_{mn}^H$ in an explicit form. We also point out that the inclusion $A_{mn}^G \subseteq A_{mn}^H$ implies that the structure of the algebra $A_{mn}^G$ inherits many features of the structure of $A_{mn}^H$.

The most significant distinction between nonmodular and modular cases is that in contrast to the nonmodular case, the Noether bound does no longer hold in the modular one. In particular, if $r \leq n$ and $\alpha$ are positive integers, $F$ a field of prime characteristic $p$, and $G = S_n$ the symmetric group of degree $n$, then the following result holds (G. Kemper [6], Stepanov [12]): *if $p^\alpha$ divides $r$, then every system of $R$-algebra generators of $A_{mn}^{S_n}$ contains a generator whose degree is greater than or equal to* $\max\{n, m(p^\alpha - 1)\}$. The last result implies, in particular, that if $R =$ is the ring of integers, then every system of $R$-algebra generators of $A_{mn}^{S_n}$ contains a generator whose degree is greater than or equal to $\max\{n, m(n-1)/2\}$. This shows that feasibility of Noether's bound depends essentially on the arithmetic structure of the ring $R$. It was recently shown by Fleischmann [3] that the lower bound $\max\{n, m(p^\alpha - 1)\}$ is sharp when $m > 1$ and $n = p^\alpha$. In fact, he proved that in this case $\beta(A_{mn}^{S_n}) \leq \max\{n, m(n-1)\}$. The last result can be considered as a refinement of the Campbell–Hughes–Pollack [1] upper bound $\beta(A_{mn}^{S_n}) \leq \max\{n, mn(n-1)/2\}$, which holds over an arbitrary commutative ring $R$.

Let $m, n$ be positive integers, $p$ a prime number, $F = F_p$ a finite field with $p > 2$ elements, and $V = F_p x_1 + \cdots + F_p x_n$ a vector space over $F_p$ of dimension $n \geq 2$. Let $G \leq GL(n, F_p)$ be a group of order divisible by $p$, and $H = \langle\gamma\rangle$ a cyclic subgroup of $G$ of order $p$. Assume that the generating matrix $\gamma$ of $H$ has the Jordan canonical form

$$\gamma = \begin{pmatrix} J_1 & & & \\ & J_2 & & \\ & & \ddots & \\ & & & J_s \end{pmatrix}$$

with basic Jordan blocks $J_1, \ldots, J_s$. If $n_1, \ldots, n_s$ are sizes of these basic blocks, then

$$n_1 + n_2 + \cdots + n_s = n,$$

and we can assume without loss of generality that

$$p \geq n_1 \geq n_2 \geq \cdots \geq n_r \geq 2 \quad \text{and} \quad n_{r+1} = \cdots = n_s = 1.$$

Let $A_{mn}^G$ be the algebra of invariants of the polynomial algebra

$$A_{mn} = F_p[V^m] = F_p[x_{11}, \ldots, x_{1n}; \ldots; x_{m1}, \ldots, x_{mn}]$$

with respect to $G$. In the case when $n_1 = \cdots = n_r = 2$ and $n_{r+1} = \cdots = n_s = 1$, we set

$$N_H^{(0)}(x_{i,2\tau-1}) = \prod_{\alpha \in F_p} \left( x_{i,2\tau-1} + \binom{\alpha}{1} x_{i,2\tau} \right) = x_{i,2\tau-1}^p - x_{i,2\tau-1} x_{i,2\tau}^{p-1},$$

where $1 \leq i \leq m$, $1 \leq \tau \leq r$, and

$$S_H^{(0)}(f) = \sum_{\alpha \in F_p} \prod_{i=1}^m \prod_{\tau=1}^r \left( x_{i,2\tau-1} + \binom{\alpha}{1} x_{i,2\tau} \right)^{s_{i,\tau p-1}},$$

where

$$f = \prod_{i=1}^m \prod_{\tau=1}^r x_{i,2\tau-1}^{s_{i,2\tau-1}}$$

is a monomial in $x_{i,2\tau-1}$, for $1 \leq i \leq m$, $1 \leq \tau \leq r$. Finally, if $(i_1, j_2)$, $(i_2, j_2)$ are two different pairs of positive integers, we write

$$(i_1, j_1) \prec (i_2, j_2) \Leftrightarrow i_1 < i_2 \quad \text{or} \quad i_1 = i_2 \text{ and } j_1 < j_2.$$

The purpose of this paper is to extend the arguments of [12] and to prove the following result.

**Theorem 1.** *Let $H \leq GL(n, F_p)$ be a cyclic group of prime order $p$ generated by $\gamma$. If $n \geq 2$ and the sizes $n_1, \dots, n_s$ of the basis Jordan blocks $J_1, \dots, J_s$ of the matrix $\gamma$ satisfy the conditions*

$$n_1 = \cdots = n_r = 2 \quad \text{and} \quad n_{r+1} = \cdots = n_s = 1,$$

*then*

$$A_{mn}^H = F_p[x_{i,2\tau}, x_{ij}, N_H^{(0)}(x_{i,2\tau-1}), (x_{i_1,2\tau_1-1}x_{i_2,2\tau_2} - x_{i_1,2\tau_1}x_{i_2,2\tau_2-1}), S_H^{(0)}(f')],$$

*where $1 \leq i \leq m$, $1 \leq \tau \leq r$, $2r+1 \leq j \leq s$, $1 \leq i_1, i_2 \leq m$, $1 \leq \tau_1, \tau_2 \leq r$ with $(i_1, \tau_1) \prec (i_2, \tau_2)$, and $f'$ runs through the set of monomials $f$ of the above form such that $0 \leq s_{i,2\tau-1} \leq p-1$.*

As an easy consequence of Theorem 1 we obtain the following.

**Corollary 2.** *If additionally $mr > 2$, then every system of generators of $A_{mn}^H$ contains an element of degree at least $mr(p-1)$.*

In the case when $G$ is an arbitrary finite group containing $H$ as a subgroup, we are able to find a lower degree bound for generating elements of the algebra $A_{mn}^G$.

**Theorem 3.** *Let $G \leq GL(n, F_p)$ be a finite group whose order is divisible by the characteristic $p$ of $F_p$, and $H = \langle\gamma\rangle$ a cyclic subgroup of $G$ of prime order $p$. If $m \geq n \geq 2$, $mr > 2$ and the sizes $n_1, \dots, n_s$ of the basic Jordan blocks $J_1, \dots, J_s$ of the matrix $\gamma$ satisfy the condition*

$$n_1 = \cdots = n_r = 2, \quad \text{and} \quad n_{r+1} = \cdots = n_s = 1,$$

*then every system of $F_p$-algebra generators of $A_{mn}^G = F_p[V^m]^G$ contains a generator whose degree is greater than or equal to $(m-n+2r)(p-1)/r$.*

Theorem 1 provides an explicit construction of generating elements of the algebra $A_{mn}^H$ in terms of orbit sums and orbit norms of monomials. It can be conjectured that the lower degree bound in Corollary 2 is sharp. Theorem 3 improves the lower degree bound

$$\max\left\{2, \frac{m}{n-1}, \frac{m}{|G|-1}, \frac{mp}{n(p-1)}\right\}$$

obtained earlier by Richman [10]. The case when $r = 1$ was studied by Richman [9] (if $p = 2$), and by Campbell and Hughes [2] (if $p > 2$). Our arguments are considerably different from the ones of these authors. In particular, all our constructions are based only on the analysis of orbit Chern classes, without any references to deep results of representation theory or combinatorial analysis. The case when

$$p \geq n_1 \geq \cdots \geq n_\sigma \geq 3,$$

with $1 \leq \sigma \leq s$, requires more complicated calculations and will be considered later. It should be noted that all results of the paper can be easily extended to the case of an arbitrary field $F$ of prime characteristic $p$.

We now explain briefly the main ideas underlying the proofs of Theorem 1 and 3. The use of orbit sums

$$S_G(f) = \sum_{w \in \{\sigma(f) | \sigma \in G} w,$$

where $f \in A_{mn}$ is a monomial, is most efficient in the case when the group $G$ acts on elements of an $R$-algebra by permutation of the vector variables $x_j = (x_{1j}, x_{2j}, \dots, x_{mj})$, $1 \le j \le n$. In that case each invariant $u \in A_{mn}^G$ is an $R$-linear combination of the above orbit sums for various monomials $f$. This important result is a consequence of the following fact: if a monomial $f$ appears in an invariant $u$ with a nonzero coefficient $a$, then for each $\sigma \in G$ the corresponding monomial $\sigma(f)$ also appears in $u$ with the same coefficient $a$. Unfortunately, the above property of orbit sums does not longer hold for finite groups of a more general form, in particular, for cyclic groups $H$ generated by matrices of the following form:

$$\gamma = \begin{pmatrix} J_1 & & & \\ & J_2 & & \\ & & \ddots & \\ & & & J_s \end{pmatrix}$$

with basic Jordan blocks $J_1, J_2, \dots, J_s$ of sizes $n_1, n_2, \dots, n_s$ such that $1 < n_\rho < p$ for some $\rho = 1, 2, \dots, s$. On the other hand, if $n_1 = n_2 = \cdots = n_r = p$ and $n_{r+1} = \cdots = n_s = 1$, then after an appropriate nonsingular linear transformation we can proceed to a new system of vector variables $\tilde{x}_j = (\tilde{x}_{1j}, \tilde{x}_{2j}, \dots, \tilde{x}_{mj})$, $1 \le j \le n$ on which $H$ acts be cyclic permutations.

Let $H$ be the cyclic group of prime order $p > 2$ generated by a nonsingular square matrix $\gamma$ with Jordan blocks $J_1, J_2, \dots, J_s$ of sizes $n_1, n_2, \dots, n_s$, respectively. Assume that $n_1 = n_2 = \cdots = n_r = 2$, $n_{r+1} = \cdots = n_s = 1$, and recall that $H$ acts linearly on the vector space $V^m$ of dimension $m(r+s)$. The proof of Theorem 1 falls into two steps.

(i) At the first step we "blow up" each Jordan block $J_\rho$, $1 \le \rho \le r$, of size $n_\rho = 2$ of the matrix $\gamma$ to a Jordan block of the largest possible size $p$. As a result, the generating matrix $\gamma$ of the group $H$ is transformed into the corresponding square matrix $\tilde{\gamma}$ of size $\nu = (p-1)r + s$, and the group $H$ into the corresponding cyclic group $\tilde{H}$ generated by $\tilde{\gamma}$ and acting on the vector space $\tilde{V}^m$ of dimension $m\nu$. It follows by the above that then one can find new vector variables $\tilde{x}_j = (\tilde{x}_{1j}, \tilde{x}_{2j}, \dots, \tilde{x}_{mj})$, $1 \le j \le \nu$, obtained from the original variables $x_j = (x_{1j}, x_{2j}, \dots, x_{mj})$ by a non-degenerate linear transformation such that $\tilde{H}$ acts by cyclic permutations of the new vector variables. This property of $\tilde{H}$ allows us to show (Theorem 5) that each invariant $v$ of the algebra $A_{m\nu}^{\tilde{H}}$ is an $F_p$-linear combination of the orbit sums $S_{\tilde{H}}(f)$, the orbit norms $N_{\tilde{H}}(g)$ and their products $S_{\tilde{H}}(f)N_{\tilde{H}}(g)$ for various monomials $f, g \in A_{m\nu}$.

(ii) At the second step we demonstrate that the appropriate embedding of the algebra $A_{mn}$ into $A_{m\nu}$ results in a fairly simple test (Theorem 8) distinguishing among the $\tilde{H}$-invariants $v \in A_{m\nu}^{\tilde{H}}$ ones invariants with respect to the action of $H$. The use of this test makes possible an explicit construction of invariants $u \in A_{mn}^H$ as $F_p$-linear combinations of orbit sums $S_{\tilde{H}}(f)$, orbit norms $N_{\tilde{H}}(g)$ and their products $S_{\tilde{H}}(f)N_{\tilde{H}}(g)$ of a special form.

The idea of the proof of Theorem 3 is as follows. Since $A_{mn}^G \subset A_{mn}^H$, the system of generators of the algebra $A_{mn}^H$ indicated in Theorem 1 contains a corresponding system of generators of the algebra $A_{mn}^G$. To prove that the latter contains at least one polynomial of a sufficiently high degree we demonstrate that a certain polynomial $f_0 \in A_{mn}$ of a fairly special form, which is invariant under the action of an arbitrary subgroup of the general linear group $GL(F_p, n)$ (see Section 5), cannot be presented as a polynomial over $F_p$ in elements of moderate degrees of the above mentioned system of generators of the algebra $A^G_{mn}$, not even as a polynomial over $F_p$ in similar elements of the broader system of generators of the algebra $A^H_{mn}$.

## 2 Orbit sums

Let $m, n$ be positive integers, $p \geq 3$ be a prime number, $F_p$ a prime finite field with $p$ elements, $GL(n, F_p)$ the group of invertible $n \times n$ matrices with entries in $F_p$, and

$$A_{mn} = F_p[x_{11}, \dots, x_{m1}; \dots; x_{1n}, \dots, x_{mn}]$$

the algebra of polynomials in commuting variables $x_{11}, \dots, x_{m1}; \dots; x_{1n}, \dots, x_{mn}$. In the sequel we identify $F_p$ with the set $\{0, 1, \dots, p-1\}$. If $g \in A_{mn}$ and

$$\sigma = \begin{pmatrix} \sigma_{11} & \sigma_{12} & \cdots & \sigma_{1n} \\ \sigma_{21} & \sigma_{22} & \cdots & \sigma_{2n} \\ \vdots & \vdots & \vdots & \\ \sigma_{n1} & \sigma_{n2} & \cdots & \sigma_{nn} \end{pmatrix}$$

an element of $GL(n, F_p)$, let $\sigma(g)$ denote the image of $g$ under the $F_p$-algebra endomorphism $\sigma$ which operates on the basis elements $x_{i1}, \dots, x_{in}$ of the vector spaces $V_i = F_p x_{i1} + F_p x_{i2} + \cdots + F_p x_{in}$, $1 \leq i \leq m$, as follows:

$$\begin{pmatrix} \sigma(x_{i1}) \\ \sigma(x_{i2}) \\ \vdots \\ \sigma(x_{in}) \end{pmatrix} = \sigma \begin{pmatrix} x_{i1} \\ x_{12} \\ \vdots \\ x_{in} \end{pmatrix} = \begin{pmatrix} \sigma_{11}x_{i1} + \cdots + \sigma_{1n}x_{in} \\ \sigma_{21}x_{i1} + \cdots + \sigma_{2n}x_{in} \\ \vdots \\ \sigma_{n1}x_{i1} + \cdots + \sigma_{nn}x_{in} \end{pmatrix}.$$

Let $G$ be a subgroup of $GL(n, F_p)$, and $A^G_{mn}$ the set of polynomials $u \in A_{mn}$ such that $\sigma(u) = u$ for every $\sigma \in G$. The set $A^G_{mn}$ forms a subalgebra of $A_{mn}$ which is called the *algebra of vector invariants* of $G$.

Let $p$ be a prime divisor of $|G|$, and $H = \langle \gamma \rangle$ a cyclic subgroup of the group $G$ of order $p$. In an appropriate basis, the matrix $\gamma$ has the following Jordan canonical form

$$\gamma = \begin{pmatrix} J_1 & & & \\ & J_2 & & \\ & & \ddots & \\ & & & J_s \end{pmatrix},$$

where the basic Jordan blocks

$$J_\rho = \begin{pmatrix} 1 & 1 & \cdots & 0 & 0 \\ 0 & 1 & \cdots & 0 & 0 \\ \vdots & \vdots & & \vdots & \vdots \\ 0 & 0 & \cdots & 1 & 1 \\ 0 & 0 & \cdots & 0 & 1 \end{pmatrix}, \quad 1 \leq \rho \leq s,$$

are square matrices of sizes $n_1, n_2, \dots, n_s$, respectively, with $n_1 + n_2 + \cdots + n_s = n$ and $1 \leq n_\rho \leq p$ for all $\rho = 1, 2, \dots, s$. We can assume without loss of generality that

$$n_1 \geq n_2 \geq \cdots \geq n_r \geq 2 \quad \text{and} \quad n_{r+1} = \ldots = n_s = 1.$$

Let $A_{mn}^H$ be the algebra of polynomials $u \in A_{mn}$ which are invariant under the action of the cyclic group $H = \langle \gamma \rangle$. Our aim is to describe explicitly all the elements of $A_{mn}^H$. Set $n' = n_1 + \cdots + n_r$ and consider the polynomial algebra

$$A_{mn'} = F_p[x_{ij} \mid 1 \leq i \leq m, 1 \leq j \leq n'].$$

Let $A_{mn'}^H$ be the algebra of invariants of $A_{mn'}$ with respect to $H$. Since all variables $x_{ij}$, for $1 \leq i \leq m$, $n' + 1 \leq j \leq n$, are invariant under the action of $H$, then every invariant $u \in A_{mn}^H$ is a polynomial of the form

$$u = u_1 f_1 + u_2 f_2 + \cdots + u_l f_l,$$

where $u_k \in A_{mn'}^H$, and $f_k$, for $1 \leq k \leq l$, are monomials in $F_p[x_{ij} \mid 1 \leq i \leq m, n'+1 \leq j \leq n]$. This shows that the problem concerning the structure of invariants $u \in A_{mn}^H$ is reduced to the corresponding problem concerning the structure of invariants $u_k \in A_{m,n'}^H$. As a result, we can assume without loss of generality that $n = n'$, $u \in A_{mn}$ and

$$\gamma = \begin{pmatrix} J_1 & & & \\ & J_2 & & \\ & & \ddots & \\ & & & J_r \end{pmatrix},$$

where $J_1, J_2, \cdots, J_r$ are the basic Jordan blocks of sizes $n_1, n_2, \ldots, n_r$, respectively, with

$$n = n_1 + n_2 + \cdots + n_r \quad \text{and} \quad n_1 \geq n_2 \geq \cdots \geq n_r \geq 2. \tag{1}$$

Set $\nu = rp$ and blow up each of Jordan blocks $J_l, \ldots, J_r$ of the matrix $\gamma$ to the same Jordan block

$$\tilde{J} = \begin{pmatrix} 1 & 1 & \cdots & 0 & 0 \\ 0 & 1 & \cdots & 0 & 0 \\ \vdots & \vdots & & \vdots & \vdots \\ 0 & 0 & \cdots & 1 & 1 \\ 0 & 0 & \cdots & 0 & 1 \end{pmatrix}$$

of size $p$. As a result, the matrix $\gamma$ is blown up to the square $(\nu \times \nu)$-matrix

$$\tilde{\gamma} = \begin{pmatrix} \tilde{J} & & & \\ & \tilde{J} & & \\ & & \ddots & \\ & & & \tilde{J} \end{pmatrix}$$

of size $\nu$ which operates on each vector space

$$\tilde{V}_i = F_p z_{i1} + F_p z_{i2} + \cdots + F_p z_{i\nu}, \quad \text{for} \quad i = 1, 2. \ldots, m,$$

in the same way as a nonsingular linear transformation of $\tilde{V}_i$. Denote by $\tilde{H}$ the cyclic group of order $p$ generated by $\tilde{\gamma}$ and note that the action of $\tilde{H}$ on the spaces $\tilde{V}_i$, for

$i = 1, 2, \ldots, m$, can be considered as an extension of the action of the group $H$ on the corresponding subspaces $V_i$ of $\tilde{V}_i$, for $i = 1, 2, \ldots, m$. If

$$A_{m\nu} = F_p[z_{11}, \ldots, z_{m1}; \ldots; z_{1\nu}, \ldots, z_{m\nu}]$$

is the algebra of all polynomials over $F_p$ in $m\nu$ commuting variables $z_{11}, \ldots, z_{1n}; \ldots; z_{1\nu}, \ldots, z_{m\nu}$, then every element $\tilde{\sigma}$ of the group $\tilde{H}$ gives an $F_p$-algebra endomorphism of $A_{m\nu}$. Let $A_{m\nu}^{\tilde{H}}$ denote the subalgebra of invariants of the algebra $A_{m\nu}$ under the action of $\tilde{H}$. If $f$ is a monomial in $A_{m\nu}$, denote by

$$Orb_{\tilde{H}}(f) = \{\tilde{\sigma}(f) \mid \tilde{\sigma} \in \tilde{H}\}$$

the orbit of $f$ with respect to the group $\tilde{H}$. Set $q = |Orb_{\tilde{H}}(F)|$ and note that $q = 1$ or $q = p$. If $Orb_{\tilde{H}}(f)$ is the orbit of a monomial $f \in A_{m\nu}$ then

$$N_{\tilde{H}}(f) = \prod_{f' \in Orb_{\tilde{H}}(f)} f' \quad \text{and} \quad S_{\tilde{H}}(f) = \sum_{f' \in Orb_{\tilde{H}}(f)} f'$$

are called an *orbit norm* and an *orbit sum* of $f$ with respect to $\tilde{H}$. It is clear that $N_{\tilde{H}}(f)$ and $S_{\tilde{H}}(f)$ are invariant under the action of $\tilde{H}$. Moreover, $N_{\tilde{H}}(f)$ and $S_{\tilde{H}}(f)$ are homogeneous polynomials in $A_{m\nu}$. Now we describe explicitly the elements of $A_{m\nu}$ that are invariant under the action of the cyclic group $\tilde{H}$. At first we prove the following arithmetical result.

**Lemma 4.** *Let $p \geq 3$ be a prime number, and $l_1, l_2, \ldots, l_p$ integers such that*

$$0 \leq l_\rho \leq p - 1, \quad \textit{for all} \quad \rho = 1, 2, \ldots, p,$$

*and*

$$l = \sum_{\rho=1}^{p} l_\rho.$$

*Then*

$$\sum_{\alpha \in F_p} \prod_{\rho=1}^{p} \binom{\alpha}{l_\rho} = \begin{cases} 0 & \text{if } l \leq p-2, \\ (p-1)/l_1! \cdots l_p! & \text{if } l = p-1. \end{cases}$$

*Proof.* As usual, we assume that

$$\binom{\alpha}{l_\rho} = \begin{cases} 1 & \text{if } l_\rho = 0, \\ 0 & \text{if } \alpha < l_\rho. \end{cases}$$

Consider

$$\prod_{\rho=1}^{p} \binom{\alpha}{l_\rho}$$

as a polynomial in $F_p[\alpha]$ of degree $l$, say,

$$\prod_{\rho=1}^{p} \binom{\alpha}{l_\rho} = c_0\alpha^l + c_1\alpha^{l-1} + \cdots + c_l.$$

Now the result is immediate from the relations

$$\sum_{\alpha \in F_p} \alpha^k = \begin{cases} p-1 & \text{if } (p-1) \mid k, \\ 0 & \text{otherwise.} \end{cases}$$

This completes the proof. □

Let

$$f = \prod_{i=1}^{m} \prod_{\tau=1}^{r} \prod_{j=(\tau-1)p+1}^{\tau p-1} z_{ij}^{s_{ij\tau}}$$

be a monomial in the algebra

$$A_{m\nu} = F_p[z_{ij} \mid 1 \leq i \leq m, 1 \leq j \leq \nu].$$

Assume that $f$ is not invariant under action of the group $\tilde{H}$ and consider the corresponding orbit sum

$$S_{\tilde{H}}(f) = \sum_{\alpha \in F_p} \tilde{\gamma}^{\alpha}(f).$$

Since

$$\tilde{\gamma}^{\alpha}(z_{ij}) = \sum_{l=j}^{\tau p} \binom{\alpha}{l-j} z_{il},$$

for $1 \leq i \leq m$, $(\tau - 1)p + 1 \leq j \leq \tau p$, $1 \leq \tau \leq r$, we see that

$$S_{\tilde{H}}(f) = \sum_{\alpha \in F_p} \prod_{i=1}^{m} \prod_{\tau=1}^{r} \prod_{j=(\tau-1)p+1}^{\tau p} \left( \sum_{l=0}^{\tau p-j} \binom{\alpha}{l} z_{i,j+l} \right)^{s_{ij\tau}}. \tag{2}$$

Let $\{\tilde{z}_{ij} \mid 1 \leq i \leq m, 1 \leq j \leq \nu\}$ be new variables defined by the following relations

$$\tilde{z}_{i,(\tau-1)p+\alpha+1} = \sum_{l=0}^{\alpha} \binom{\alpha}{l} z_{i,(\tau-1)p+l+1}, \tag{3}$$

for $1 \leq i \leq m$, $0 \leq \alpha \leq p-1$ and $1 \leq \tau \leq r$. This is a nondegenerated linear transformation, so any $z_{ij}$ is an $F_p$-linear combinations of $\tilde{z}_{ij}$. It follows that every orbit sum $S_{\tilde{H}}(f)$ is an $F_p$-linear combination of the orbit sums

$$S_{\tilde{H}}(\tilde{f}) = \sum_{\tilde{g} \in Orb(\tilde{f})} \tilde{g},$$

where $\tilde{f}$ is a monomial of the form

$$\tilde{f} = \prod_{i=1}^{m} \prod_{\tau=1}^{r} \prod_{j=(\tau-1)p+1}^{\tau p} \tilde{z}_{ij}^{\tilde{s}_{ij\tau}}.$$

Consider also the orbit norm $N_{\tilde{H}}(\tilde{f}) = \prod_{\tilde{g} \in Orb(\tilde{f})} \tilde{g}$ of the monomial $\tilde{f}$ and observe that $N_{\tilde{H}}(\tilde{f})$ is an element of the algebra $A_{m\nu}^{\tilde{H}}$. We now note that the elements $\{\tilde{z}_{ij} \mid 1 \leq$

$j \le \nu\}$ defined by (3) form a basis of the space $\tilde{V}_i = F_p z_{i1} + \dots + F_p z_{i\nu}$, $1 \le i \le m$, and that for each $\tau = 1, 2, \dots r$, the group $\tilde{H}$ operates on the basis elements $\tilde{z}_{ij}$, for $(\tau - 1)p + 1 \le j \le \tau p$, by their cyclic permutations. Let $\tilde{f}$ be a monomial in $F_p[\tilde{z}_{ij} \mid 1 \le i \le m, 1 \le j \le \nu]$ which appears in an invariant $v \in A_{m\nu}^{\tilde{H}}$ with a nonzero coefficient. Since $\tilde{\gamma}^\alpha(v) = v$ for any $\tilde{\gamma}^\alpha \in \tilde{H}$, the coefficient of $\tilde{f}$ in $v$ equals the coefficient of $\tilde{\gamma}^\alpha(\tilde{f})$ in $v$. This shows that if $\tilde{f} = \tilde{f}'\tilde{f}''$ and $\tilde{f}'' = N_{\tilde{H}}(\tilde{g})$ for some monomial $\tilde{g} \in F_p[\tilde{z}_{ij} \mid 1 \le i \le m, 1 \le j \le \nu]$, then $v$ involves the invariant $S_{\tilde{H}}(\tilde{f}')N_{\tilde{H}}(\tilde{g})$. In other words, each invariant $v \in A_{m\nu}^{\tilde{H}}$ is an $F_p$-linear combination of the orbit sums $S_{\tilde{H}}(\tilde{f})$, the orbit norms $N_{\tilde{H}}(\tilde{g})$ and also their products $S_{\tilde{H}}(\tilde{f})N_{\tilde{H}}(\tilde{g})$, for various monomials $\tilde{f}, \tilde{g} \in F_p[\tilde{z}_{ij} \mid 1 \le i \le m, 1 \le j \le \nu]$. On the other hand, every orbit sum $S_{\tilde{H}}(\tilde{f})$ is an $F_p$-linear combination of orbit sums $S_{\tilde{H}}(f)$ of monomials $f \in A_{m\nu}$, which shows that any invariant $v \in A_{m\nu}^{\tilde{H}}$ is an $F_p$-linear combination of the orbit sums $S_{\tilde{H}}(f)$, the orbit norms $N_{\tilde{H}}(g)$ and also their products $S_{\tilde{H}}(f)N_{\tilde{H}}(g)$, for various monomials $f, g \in A_{m\nu}$. This gives the following result.

**Theorem 5.** *Every invariant of the algebra $A_{m\nu}^{\tilde{H}}$ is an $F_p$-linear combination of the orbit sums $S_{\tilde{H}}(f)$, the orbit norms $N_{\tilde{H}}(g)$ and also their products $S_{\tilde{H}}(f)N_{\tilde{H}}(g)$, for various monomials $f, g \in A_{m\nu}$ of the form*

$$f = \prod_{i=1}^{m} \prod_{\tau=1}^{r} \prod_{j=(\tau-1)p+1}^{\tau p} z_{ij}^{s_{ij\tau}}, \quad g = \prod_{i=1}^{m} \prod_{\tau=1}^{r} \prod_{j=(\tau-1)p+1}^{\tau p} z_{ij}^{t_{ij\tau}}.$$

If $0 \le \lambda < p - 1$ is an integer, then among all possible orbit sums $S_{\tilde{H}}(f) \in A_{m\nu}$ we select those ones that involve no variable $z_{i,j}$, for $1 \le i \le m$, $(\tau - 1)p + 1 \le j < \tau p - 1 - \lambda$, $1 \le \tau \le r$. Our next aim is to find an exact form of such orbit sums $S_{\tilde{H}}(f)$, for various monomials $f \in A_{m\nu}$.

Set

$$s_{ij\tau} = \sum_{e=0}^{\eta} s_{ij\tau}^{(e)} p^e, \quad 0 \le s_{ij}^{(e)} \le p - 1,$$

for $0 \le e \le \eta$, $1 \le i \le m$, $(\tau - 1)p + 1 \le j \le \tau p$, $1 \le \tau \le r$, and write $f$ in the following form:

$$f = \prod_{e=0}^{\eta} \prod_{i=1}^{m} \prod_{\tau=1}^{r} \prod_{j=(\tau-1)p+1}^{\tau p} (z_{ij}^{p^e})^{s_{ij\tau}^{(e)}}. \tag{4}$$

Now we define the *weight* of the monomial $f$ and the associated orbit sum $S_{\tilde{H}}(f)$ as follows:

$$w(f) = w(S_{\tilde{H}}(f)) = \sum_{e=0}^{\eta} \sum_{i=1}^{m} \sum_{\tau=1}^{r} \sum_{j=(\tau-1)p+1}^{\tau p} (\tau p - j) s_{ij\tau}^{(e)}.$$

If a monomial $f$ is invariant under the action of $\tilde{H}$, it has the form

$$f = \prod_{e=0}^{\eta} \prod_{i=1}^{m} \prod_{\tau=1}^{r} z_{i,\tau p}^{s_{i,\tau p}^{(e)}}.$$

If $f$ is not invariant under the action of $\tilde{H}$, it follows from (2), (4) and Lemma 4 that the condition $w(f) < p-1$ implies $S_{\tilde{H}}(f) = 0$. On the other hand, if $w(f) > p+\lambda$ and $s_{ij\tau} \geq 1$ at least for one triple $(i, j, \tau)$ with $1 \leq i \leq m$, $(\tau-1)p+1 \leq j < \tau p-1-\lambda$, $1 \leq \tau \leq r$, then the invariant $S_{\tilde{H}}(f)$ involves at least one variable $z_{ij}$, with $1 \leq i \leq m$, $(\tau-1)p+1 \leq j < \tau p-1-\lambda$, $1 \leq \tau \leq r$. This proves the following result.

**Proposition 6.** *Let $0 \leq \lambda < p-1$, $\mu \geq 1$ be integers, and*

$$f = \prod_{e=0}^{\eta}\prod_{i=1}^{m}\prod_{\tau=1}^{r}\prod_{j=(\tau-1)p+1}^{\tau p} (z_{ij}^{p^e})^{s_{ij\tau}^{(e)}}$$

*a monomial in $A_{m\nu}$ that is not invariant under the action of $\tilde{H}$. Every orbit norm $N_{\tilde{H}}(f) \in A_{m\nu}^{\tilde{H}}$ that involves no variable $z_{ij}$, for $1 \leq i \leq m$, $(\tau-1)p+1 \leq j < \tau p-1-\lambda$, $1 \leq \tau \leq r$, has the following form:*

$$N_{\tilde{H}}^{(\lambda)}(f) = \prod_{\alpha\in F_p}\prod_{e=0}^{\eta}\prod_{i=1}^{m}\prod_{\tau=1}^{r}\prod_{j=\tau p-1-\lambda}^{\tau p}\left(\sum_{l=0}^{\tau p-j-1}\binom{\alpha}{l} z_{i,j+l}^{p^e}\right)^{s_{ij\tau}^{(e)}},$$

*and each orbit sum $S_{\tilde{H}}(f) \in A_{m\nu}^{\tilde{H}}$ that involves no variable $z_{ij}$, for $1 \leq i \leq m$, $(\tau-1)p+1 \leq j < \tau p-1-\lambda$, $1 \leq \tau \leq r$, has either the form*

$$S_{\tilde{H}}^{(\lambda)}(f) = \sum_{\alpha\in F_p}\prod_{e=0}^{\eta}\prod_{i=1}^{m}\prod_{\tau=1}^{r}\prod_{j=\tau p-1-\lambda}^{\tau p}\left(\sum_{l=0}^{\tau p-j-1}\binom{\alpha}{l} z_{i,j+l}^{p^e}\right)^{s_{ij\tau}^{(e)}}$$

*or the form*

$$S_{\tilde{H}}^{(\mu,\lambda)}(f) = \sum_{\alpha\in F_p}\prod_{e=0}^{\eta}\prod_{i=1}^{m}\prod_{\tau=1}^{r}\prod_{j=(\tau-1)p+1}^{\tau p}\left(\sum_{l=0}^{\tau p-j}\binom{\alpha}{l} z_{i,j+l}^{p^e}\right)^{s_{ij\tau}^{(e)}},$$

*where $w(f) = p-1+\mu \geq p+\lambda$ and $s_{ij\tau}^{(e)} \geq 1$ at least for one quadruple $(e, i, j, \tau)$ such that $0 \leq e \leq \eta$, $1 \leq i \leq m$, $(\tau-1)p+1 \leq j < \tau p-1-\lambda$, $1 \leq \tau \leq r$.*

We assume in what follows that

$$n_1 = \cdots = n_r = 2. \tag{5}$$

Under this assumption, the following result is an easy consequence of Proposition 6 with $\lambda = 0$.

**Corollary 7.** *Let $\mu$ be a positive integer, and*

$$f = \prod_{e=0}^{\eta}\prod_{i=1}^{m}\prod_{\tau=1}^{r}\prod_{j=(\tau-1)p+1}^{\tau p} (z_{ij}^{p^e})^{s_{ij\tau}^{(e)}}$$

*a monomial in $A_{mn}$ that is not invariant under the action of $\tilde{H}$. Every orbit norm $N_{\tilde{H}}(f)$ that involves no variable $z_{ij}$, for $1 \leq i \leq m$, $(\tau-1)p+1 \leq j < \tau p-1$, $1 \leq \tau \leq r$, has the following form:*

$$N_{\tilde{H}}^{(0)}(f) = \prod_{\alpha\in F_p}\prod_{e=0}^{\eta}\prod_{i=1}^{m}\prod_{\tau=1}^{r}\left(z_{i,\tau p-1}^{p^e} + \binom{\alpha}{1} z_{i,\tau p}^{p^e}\right)^{s_{i,\tau p-1}^{(e)}} \left(z_{i,\tau p}^{p^e}\right)^{s_{i,\tau p}^{(e)}},$$

*and every orbit sum $S_{\tilde{H}}(f)$ that involves no variables $z_{ij}$, for $1 \le i \le m$, $(\tau-1)p+1 \le j < \tau p - 1$, $1 \le \tau \le r$, has either the form*

$$S_{\tilde{H}}^{(0)}(f) = \sum_{\alpha\in F_p}\prod_{e=0}^{\eta}\prod_{i=1}^{m}\prod_{\tau=1}^{r}\left(z_{i,\tau p-1}^{p^e} + \binom{\alpha}{1} z_{i,\tau p}^{p^e}\right)^{s_{i,\tau p-1}^{(e)}} \left(z_{i,\tau p}^{p^e}\right)^{s_{i,\tau p}^{(e)}}$$

*or the form*

$$S_{\tilde{H}}^{(1)}(f) = \sum_{\alpha\in F_p}\prod_{e=0}^{\eta}\prod_{i=1}^{m}\prod_{\tau=1}^{r}\prod_{j=(\tau-1)p+1}^{\tau p}\left(\sum_{l=0}^{\tau p-j}\binom{\alpha}{l} z_{i,j+l}^{p^e}\right)^{s_{ij\tau}^{(e)}},$$

*where $w(f) = p$ and $s_{ij\tau} \ge 1$ at least for one tuple $(e, i, j, \tau)$ such that $0 \le e \le \eta$, $1 \le i \le m$, $(\tau-1)p+1 \le j < \tau p - 1$, $1 \le \tau \le r$.*

*On the other hand, if $S_{\tilde{H}}(f)$ involves at least one variable $z_{i,j}$, for $1 \le i \le m$, $(\tau-1)p+1 \le j < \tau p - 1$, $1 \le \tau \le r$, then $S_{\tilde{H}}(f)$ has the form*

$$S_{\tilde{H}}^{(\mu)}(f) = \sum_{\alpha\in F_p}\prod_{e=0}^{\eta}\prod_{i=1}^{m}\prod_{\tau=1}^{r}\prod_{j=(\tau-1)p+1}^{\tau p}\left(\sum_{l=0}^{\tau p-j}\binom{\alpha}{l} z_{i,j+l}^{p^e}\right)^{s_{ij\tau}^{(e)}},$$

*where $w(f) = p - 1 + \mu \ge p + 1$ and $s_{ij\tau} \ge 1$ at least for one tuple $(e, i, j, \tau)$ such that $0 \le e \le \eta$, $1 \le i \le m$, $(\tau-1)p+1 \le j < \tau p - 1$, $1 \le \tau \le r$.*

For each $i = 1, 2, \dots m$, consider now the embedding

$$\vartheta : V_i \hookrightarrow \tilde{V}_i \tag{6}$$

of the space $V_i$ into the space $\tilde{V}_i$ given by the relations

$$\vartheta(x_{ij}) = \begin{cases} z_{i,\tau p-1} & \text{if } j = 2\tau - 1, \quad 1 \le \tau \le r, \\ z_{i,\tau p} & \text{if } j = 2\tau, \quad 1 \le \tau \le r \end{cases}$$

and define as follows the action of the cyclic group $H$ on $\tilde{V}_i$. If $\gamma$ is a generating element of $H$, its action on elements $(x_{i1}, \dots, x_{in})$ of $V_i$ is given by

$$\gamma(x_{ij}) = \begin{cases} x_{ij} + x_{i,j+1} & \text{if } j = 2\tau - 1, \quad 1 \le \tau \le r, \\ x_{ij} & \text{if } j \ne 2\tau - 1, \quad 1 \le \tau \le r. \end{cases}$$

In that case, the map $\vartheta : V_i \hookrightarrow \tilde{V}_i$ induces the corresponding action of $\gamma$ on the space $\tilde{V}_i$ defined by

$$\gamma(z_{ij}) = \begin{cases} z_{ij} + z_{i,j+1} & \text{if } j = \tau p - 1, \quad 1 \le \tau \le r, \\ z_{ij} & \text{if } j \ne \tau p - 1, \quad 1 \le \tau \le r, \end{cases} \tag{7}$$

This yields a unique extension of the action of $H$ on the space $V_i \subseteq \tilde{V}_i$ to its action on the space $\tilde{V}_i$ and defines the corresponding unique extension of the action of $H$ on elements of $A_{mn}$ to its action on elements of the algebra $A_{m\nu}$.

On the other hand, if $\tilde{\gamma}$ is a generating element of the group $\tilde{H}$, its action on elements $(z_{i1}, \dots, z_{i\nu})$ of the space $\tilde{V}_i$ is given by

$$\tilde{\gamma}(z_{ij}) = \begin{cases} z_{ij} + z_{i,j+1} & \text{if } (\tau-1)p+1 \le j \le \tau p - 1, \quad 1 \le \tau \le r, \\ z_{ij} & \text{if } j = \tau p, \quad 1 \le \tau \le r. \end{cases}$$

Now we consider the invariants $u \in A_{m\nu}^{\tilde{H}}$ that are also invariant under the action of $H$.

**Theorem 8.** *Let $v \in A_{m\nu}^{\tilde{H}}$ be a polynomial of positive degree. Then $v$ is invariant under action of $H$ if and only if the polynomial $v$ involves no variable $z_{ij}$, for $1 \le i \le m$, $(\tau-1)p+1 \le j \le \tau p - 2$, $1 \le \tau \le r$.*

*Proof.* Every invariant $v \in A_{m\nu}^{\tilde{H}}$ of degree $s \ge 1$ is a sum of its homogeneous components $v_k \in A_{m\nu}^{\tilde{H}}$ of degree $k$, for $0 \le k \le s$. This reduces the problem to the case of homogeneous polynomials, so we can assume without loss of generality that $v$ is a homogeneous $\tilde{H}$-invariant.

Suppose that $v \in A_{m\nu}^{\tilde{H}}$ involves at least one variable $z_{ij}$, for $1 \le i \le m$, $(\tau-1)p+1 \le j \le \tau p - 2$, $1 \le \tau \le r$, and write

$$v = \sum_{(s_{ij})} v_{(s_{ij})} \prod_{i=1}^{m} \prod_{\tau=1}^{r} \prod_{(\tau-1)p+1 \le j \le \tau p-2} z_{ij}^{s_{ij}},$$

where $v_{(s_{ij})}$ are homogeneous polynomials in

$$F_p[z_{i,\tau p-1}, z_{i,\tau p} \mid 1 \le i \le m, 1 \le \tau \le r],$$

and the sum is over all tuples

$$(s_{ij}) = (s_{ij} \mid 1 \le i \le m, (\tau-1)p+1 \le j \le \tau p - 2, 1 \le \tau \le r)$$

of nonnegative integers $s_{ij}$ such that

$$\sum_{i=1}^{m} \sum_{\tau=1}^{r} \sum_{(\tau-1)p+1 \le j \le \tau p-2} s_{ij} \le s.$$

Let $j_0 \le rp - 2$, $j_0 \ne \tau p - 1, \tau p$, for $1 \le \tau \le r$, be the largest positive integer such that the polynomial $v$ involves no monomial

$$\prod_{i=1}^{m} \prod_{j=1}^{rp} z_{ij}^{s_{ij}}$$

with $s_{ij} \ge 1$, for all $1 \le i \le m$ and $j > j_0$. In that case,

$$v = v_{(0)} + \sum_{(s_{ij}) \ne (0)} v_{(s_{ij})} \prod_{i=1}^{m} \prod_{\tau=1}^{r} \prod_{\substack{1 \le i \le j_0 \\ (\tau-1)p+1 \le j \le \tau p-2 \\ 1 \le \tau \le r}} z_{ij}^{s_{ij}},$$

and the polynomial $v$ contains at least one nonzero term of the form

$$v_{(s_{ij})} \prod_{i=1}^{m} \prod_{\tau=1}^{r} \prod_{\substack{1\le i\le j_0 \\ (\tau-1)p+1\le j\le \tau p-2 \\ 1\le\tau\le r}} z_{ij}^{s_{ij}},$$

involving $z_{ij_0}$ for some $i = 1, 2, \ldots, m$.

Now we assume that $v$ is invariant under the action of $H$. Since the variables $z_{ij}$, for $1 \le i \le m$, $(\tau - 1)p + 1 \le j \le \tau p - 2$, $1 \le \tau \le r$, are fixed under action of $\gamma \in H$, we have

$$\gamma(v) = \gamma(v_{(0)}) + \sum_{(s_{ij})\neq(0)} \gamma(v_{(s_{ij})}) \prod_{i=1}^{m} \prod_{\tau=1}^{r} \prod_{\substack{1\le i\le j_0 \\ (\tau-1)p+1\le j\le \tau p-2 \\ 1\le\tau\le r}} z_{ij}^{s_{ij}}.$$

This shows, in particular, that the coefficients $v_{(s_{ij})}$ of the polynomial $v$ are invariant under the action of $H$, so

$$\gamma(v) = v_{(0)} + \sum_{(s_{ij})\neq(0)} v_{(s_{ij})} \prod_{i=1}^{m} \prod_{\tau=1}^{r} \prod_{\substack{1\le i\le j_0 \\ (\tau-1)p+1\le j\le \tau p-2 \\ 1\le\tau\le r}} z_{ij}^{s_{ij}}.$$

On the other hand, since $\tilde{\gamma}(z_{ij}) = \gamma(z_{ij})$ for all $1 \le i \le m$, $j = \tau p - 1, \tau p$, $1 \le \tau \le r$, and $\tilde{\gamma}(z_{ij}) = z_{ij} + z_{i,j+1}$, for $1 \le i \le m$, $(\tau - 1)p + 1 \le j \le \tau p - 2$, $1 \le \tau \le r$, then

$$\tilde{\gamma}(v) = v_{(0)} + \sum_{(s_{ij})\neq(0)} v_{(s_{ij})} \prod_{i=1}^{m} \prod_{\substack{1\le i\le j_0 \\ (\tau-1)p+1\le j\le \tau p-2 \\ 1\le\tau\le r}} (z_{ij} + z_{i,j+1})^{s_{ij}},$$

so the polynomial $\tilde{\gamma}(v)$ contains at least one nonzero term

$$v_{(s_{ij})} \prod_{i=1}^{m} \prod_{\tau=1}^{r} \prod_{\substack{1\le i\le j_0 \\ (\tau-1)p+1\le j\le \tau p-2 \\ 1\le\tau\le r}} z_{i,j+1}^{s_{ij}},$$

which involves $z_{i,j_0}$ and which does not appear in $\gamma(v) = v$. This shows that $v$ cannot be invariant under the action of $H$. Conversely, if the polynomial $v \in A_{mv}^{\tilde{H}}$ involves no variable $z_{ij}$, for $1 \le i \le m$, $(\tau - 1)p + 1 \le j \le \tau p - 1$, it is invariant under the action of $H$, and this completes the proof. □

Let $v \in A_{mv}^{\tilde{H}}$ be a polynomial that is invariant under the action of $H$. Theorem 5 shows that $v$ is an $F_p$-linear combinations of $S_{\tilde{H}}^{(0)}(f)$, $S_{\tilde{H}}^{(\mu)}(f)$, $N_{\tilde{H}}^{(0)}(g)$ and also $S_{\tilde{H}}^{(0)}(f)N_{\tilde{H}}^{(0)}(g)$, $S_{\tilde{H}}^{(\mu)}(f)N_{\tilde{H}}^{(0)}(g)$, for various $\mu \ge 1$ and $f, g \in A_{mv}$. Since the polynomials $S_{\tilde{H}}^{(0)}(f)$, $S_{\tilde{H}}^{(1)}(f)$ and $N_{\tilde{H}}^{(0)}(g)$ involve no variable $z_{ij}$, for $1 \le i \le m$, $(\tau-1)p+1 \le j \le \tau p-2$, $1 \le \tau \le r$, it follows from Theorem 8 that the polynomials $S_{\tilde{H}}^{(0)}(f)$, $S_{\tilde{H}}^{(1)}(f)$ and $N_{\tilde{H}}^{(0)}(g)$ are $H$-invariants. Now we prove the following result.

**Proposition 9.** *Let*

$$f = \prod_{e=0}^{\eta}\prod_{i=1}^{m}\prod_{\tau=1}^{r}\prod_{j=(\tau-1)p+1}^{\tau p}(z_{ij}^{p^e})^{s_{ij\tau}^{(e)}}$$

*be a monomial in* $A_{mv}$. *Then* $S_{\tilde{H}}^{(1)}(f)$ *is an* $F_p$*-linear combination of* $H$*-invariants of the form*

$$(z_{i_1,\tau_1 p-1}^{p^{e_1}} z_{i_2,\tau_2 p}^{p^{e_2}} - z_{i_1,\tau_1 p}^{p^{e_1}} z_{i_2,\tau_2 p-1}^{p^{e_2}}) \prod_{e=0}^{\eta}\prod_{i=1}^{m}\prod_{\tau=1}^{r}(z_{i,\tau p}^{p^e})^{\omega_{i,\tau p}^{(e)}},$$

*where* $0 \le e_1, e_2 \le \eta$, $1 \le i_1, i_2 \le m$, $1 \le \tau_1, \tau_2 \le r$, $(i_1, \tau_1) \prec (i_2, \tau_2)$, *and* $H$*-invariants of the form*

$$\prod_{e=0}^{\eta}\prod_{i=1}^{m}\prod_{\tau=1}^{r}(z_{i,\tau p}^{p^e})^{\upsilon_{i,\tau p}^{(e)}}.$$

*Proof.* Let

$$f = \prod_{e=0}^{\eta}\prod_{i=1}^{m}\prod_{\tau=1}^{r}\prod_{j=(\tau-1)p+1}^{\tau p}(z_{ij}^{p^e})^{s_{ij\tau}^{(e)}}$$

and

$$S_{\tilde{H}}^{(1)}(f) = \sum_{\alpha\in F_p}\prod_{e=0}^{\eta}\prod_{i=1}^{m}\prod_{\tau=1}^{r}\prod_{j=(\tau-1)p+1}^{\tau p}\left(\sum_{l=0}^{\tau p-j}\binom{\alpha}{l} z_{i,j+l}^{p^e}\right)^{s_{ij\tau}^{(e)}},$$

where $s_{ij\tau}^{(e)} \ge 1$ at least for one quadruple $(e, i, j, \tau)$ with $0 \le e \le \eta$, $1 \le i \le m$, $(\tau-1)p+1 \le j \le \tau p - 2$, $1 \le \tau \le r$, and $w(f) = p$. Set

$$\upsilon_{i,\tau p}^{(e)} = \sum_{j=(\tau-1)p+1}^{\tau p} s_{ij\tau}^{(e)}.$$

Since $S_{\tilde{H}}^{(1)}(f)$ involves no variable $z_{ij}$, for $1 \le i \le m$, $(\tau-1)p+1 \le j \le \tau p - 2$, $1 \le \tau \le r$, then

$$S_{\tilde{H}}^{(1)}(f) = \sum_{\alpha\in F_p}\prod_{e=0}^{\eta}\prod_{i=1}^{m}\prod_{\tau=1}^{r}\prod_{j=(\tau-1)p+1}^{\tau p}\left(\binom{\alpha}{\tau p - j - 1} z_{i,\tau p-1}^{p^e} + \binom{\alpha}{\tau p - j} z_{i,\tau p}^{p^e}\right)^{s_{ij\tau}^{(e)}}.$$

This shows that

$$S_{\tilde{H}}^{(1)}(f) = \sum_{(e_1,e_2,i_1,i_2,\tau_1,\tau_2)} (a_{i_1 i_2 \tau_1 \tau_2}^{(e_1,e_2)} z_{i_1,\tau_1 p-1}^{p^{e_1}} z_{i_2,\tau_2-p}^{p^{e_2}} + b_{i_1 i_2 \tau_1 \tau_2}^{(e_1,e_2)} z_{i_1,\tau_1}^{p^{e_1}} z_{i_2,\tau_2 p-1}^{p^{e_2}})$$
$$\times \prod_{e=0}^{\eta}\prod_{i=1}^{m}\prod_{\tau=1}^{r}(z_{i,\tau p}^{p^e})^{\omega_{i,\tau p}^{(e)}} + a \prod_{e=0}^{\eta}\prod_{i=1}^{m}\prod_{\tau=1}^{r}(z_{i,\tau p}^{p^e})^{\upsilon_{i,\tau p}^{(e)}},$$

where

$$\omega_{i,\tau p}^{(e)} = \begin{cases} \upsilon_{i,\tau p}^{(e)} - 1 & \text{if } (e, i, \tau) = (e_1, i_1, \tau_1), (e_2, i_2, \tau_2); \\ \upsilon_{i,\tau p}^{(e)} & \text{if } (e, i, \tau) \ne (e_1, i_1, \tau_1), (e_2, i_2, \tau_2). \end{cases}$$

Since $S^{(1)}_{\tilde{H}}$ is invariant under the action of $H$, and

$$\gamma(z_{ij}) = \begin{cases} z_{ij} + z_{i,j+1} & \text{if } j = \tau p - 1, \quad 1 \le \tau \le r; \\ z_{ij} & \text{if } j = \tau p, \quad 1 \le \tau \le r, \end{cases}$$

then $a^{(e_1,e_2)}_{i_1 i_2 \tau_1 \tau_2} + b^{(e_1,e_2)}_{i_1 i_2 \tau_1 \tau_2} = 0$ for all $(e_1, e_2, i_1, i_2, \tau_1, \tau_2)$, and therefore

$$S^{(1)}_{\tilde{H}}(f) = \sum_{(e_1,e_2,i_1,i_2,\tau_1,\tau_2)} a^{(e_1,e_2)}_{i_1 i_2 \tau_1 \tau_2} (z^{p^{e_1}}_{i_1,\tau_1 p-1} z^{p^{e_2}}_{i_2,\tau_2 p} - z^{p^{e_1}}_{i_1,\tau_1 p} z^{p^{e_2}}_{i_2,\tau_2 p-1})$$
$$\times \prod_{e=0}^{\eta} \prod_{i=1}^{m} \prod_{\tau=1}^{r} (z^{p^e}_{i,\tau p})^{\omega^{(e)}_{i,\tau p}} + a \prod_{e=0}^{\eta} \prod_{i=1}^{m} \prod_{\tau=1}^{r} (z^{p^e}_{i,\tau p})^{\upsilon^{(e)}_{i,\tau p}}.$$

This completes the proof. □

Consider the invariants $v \in A^{\tilde{H}}_{m\nu}$ which are $F_p$-linear combinations of orbit sums $S^{(\mu)}_{\tilde{H}}(f)$, for monomials $f$ and $\mu \ge 2$. Our next aim is to describe those $v \in A^{\tilde{H}}_{m\nu}$ that are also invariant under the action of $H$.

**Proposition 10.** *Let $v \in A^{\tilde{H}}_{m\nu}$ be a polynomial that is an $F_p$-linear combination*

$$v = \sum_{k=1}^{K} a_k S^{(\mu_k)}_{\tilde{H}}(f_k), \qquad a_k \ne 0,$$

*of orbit sums $S^{(\mu_k)}_{\tilde{H}}(f_k)$, for various*

$$f_k = \prod_{e=0}^{\eta} \prod_{i=1}^{m} \prod_{\tau=1}^{r} \prod_{j=(\tau-1)p+1}^{\tau p} (z^{p^e}_{ij})^{s^{(e,k)}_{ij\tau}}$$

*and $\mu_1 \ge \mu_2 \cdots \ge \nu_K \ge 2$. If $v$ is invariant under the action of $H$, then $\mu_{k'} = \mu_k$ at least for one pair $(k', k)$ with $k' > k$, and*

$$\sum_{j=(\tau-1)p+1}^{\tau p} s^{(e,k')}_{ij\tau} = \sum_{j=(\tau-1)p+1}^{\tau p} s^{(e,k)}_{ij\tau}.$$

*Proof.* Since $\mu_k \ge 2$, it follows from Corollary 7 that the orbit sum $S^{(\mu_k)}_{\tilde{H}}$ involves at least one variable $z_{ij}$, for $1 \le i \le m$, $(\tau-1)p+1 \le j \le \tau p - 2$, $1 \le \tau \le r$, say, $z_{\iota\kappa}$. On the other hand, since $v$ is invariant under the action of $H$, Theorem 8 implies that $v$ involves no such variable. This shows that $z_{\iota\kappa}$ should be eliminated by means of other orbit sums occurring in $v$ with nonzero coefficients. In that case the linear combination

$$v = \sum_{k=1}^{K} a_k S^{(\mu_k)}_{\tilde{H}}(f_k)$$

must contain, alongside $S^{(\mu_k)}_{\tilde{H}}$, at least one orbit sum $S^{(\mu_{k'})}_{\tilde{H}}(f_{k'})$, for some

$$f_{k'} = \prod_{e=0}^{\eta} \prod_{i=1}^{m} \prod_{\tau=1}^{r} \prod_{j=(\tau-1)p+1}^{\tau p} (z^{p^e}_{ij})^{s^{(e,k')}_{ij\tau}}$$

with $k' > k$ and $\mu_{k'} = \mu_k$.

Consider the orbit sums

$$S_{\tilde{H}}^{(\mu_k)}(f_k) = \sum_{\alpha \in F_p} \prod_{e=0}^{\eta} \prod_{i=1}^{m} \prod_{\tau=1}^{r} \prod_{j=(\tau-1)p+1}^{\tau p} \left( \sum_{l=0}^{\tau p - j} \binom{\alpha}{l} z_{i,j+l}^{p^e} \right)^{s_{ij\tau}^{(e,k)}},$$

$$S_{\tilde{H}}^{(\mu_k)}(f_{k'}) = \sum_{\alpha \in F_p} \prod_{e=0}^{\eta} \prod_{i=1}^{m} \prod_{\tau=1}^{r} \prod_{j=(\tau-1)p+1}^{\tau p} \left( \sum_{l=0}^{\tau p - j} \binom{\alpha}{l} z_{i,j+l}^{p^e} \right)^{s_{ij\tau}^{(e,k')}}$$

and suppose that, on the contrary,

$$\sum_{j=(\tau-1)p+1}^{\tau p} s_{ij\tau}^{(e,k')} \neq \sum_{j+(\tau-1)p+1}^{\tau p} s_{ij\tau}^{(e,k)},$$

for some triple $(e, i, \tau)$. The polynomials $S_{\tilde{H}}^{(\mu_k)}(f_k)$ and $S_{\tilde{H}}^{(\mu_k)}$ are $F_p$-linear combinations of monomials

$$\left( \sum_{\alpha \in F_p} \prod_{e=0}^{\eta} \prod_{i=1}^{m} \prod_{\tau=1}^{r} \prod_{j=(\tau-1)p+1}^{\tau p} \prod_{l=0}^{\tau p - j} \binom{\alpha}{l}^{\sigma_{i,j+l,\tau}^{(e,k)}} \right)$$
$$\times \prod_{e=0}^{\eta} \prod_{i=1}^{m} \prod_{\tau=1}^{r} \prod_{j=(\tau-1)p+1}^{\tau p} \prod_{l=0}^{\tau p - j} (z_{i,j+l}^{p^e})^{\sigma_{i,j+l,\tau}^{(e,k)}}$$

and

$$\left( \sum_{\alpha \in F_p} \prod_{e=0}^{\eta} \prod_{i=1}^{m} \prod_{\tau=1}^{r} \prod_{j=(\tau-1)p+1}^{\tau p} \prod_{l=0}^{\tau p - j} \binom{\alpha}{l}^{\sigma_{i,j+l,\tau}^{(e,k')}} \right) \prod_{e=0}^{\eta} \prod_{i=1}^{m} \prod_{\tau=1}^{r}$$
$$\times \prod_{j=(\tau-1)p+1}^{\tau p} \prod_{l=0}^{\tau p - j} (z_{i,j+l}^{p^e})^{s_{\sigma_{i,j+l,\tau}^{(e,k')}}},$$

respectively, where

$$\sum_{l=0}^{\tau p - j} \sigma_{i,j+l,\tau}^{(e,k)} = s_{ij\tau}^{(e,k)} \quad \text{and} \quad \sum_{l=0}^{\tau p - j} \sigma_{i,j+l,\tau}^{(e,k')} = s_{ij\tau}^{(e,k')}.$$

Under the above supposition, each monomial of $S_{\tilde{H}}^{(\mu_k)}(f_k)$ involving $z_{\iota\kappa}$ differs from every monomial of $S_{\tilde{H}}^{(\mu_{k'})}(f_{k'})$, so that no monomial of $S_{\tilde{H}}^{(\mu_k)}$ involving $z_{\iota\kappa}$ can be eliminated. This yields a contradiction proving Proposition 10.

Set $z_{ij}^{(e)} = z_{ij}^{p^e}$ and consider the product

$$f = \prod_{e=0}^{\eta} \prod_{i=1}^{m} \prod_{\tau=1}^{r} \prod_{j=(\tau-1)p+1}^{\tau p} (z_{ij}^{p^e})^{s_{ij\tau}^{(e)}}$$

as a monomial with respect to $z_{ij}^{(e)}$. Similarly, we consider the associated orbit sum $S_{\tilde{H}}^{(\mu)}(f)$ as a polynomial with respect to $z_{ij}^{(e)}$. We also observe that the weight of every monomial that appears in $S_{\tilde{H}}^{(\mu)}(f)$ with a nonzero coefficient does not exceed $\mu$. A monomial $f$ and the associated orbit sum $S_{\tilde{H}}^{(\mu)}(f)$ are said to be *flat* if $s_{ij\tau}^{(e)} = 0$ or $s_{ij\tau}^{(e)} = 1$, for $0 \le e \le \eta$, $1 \le i \le m$, $(\tau-1)p+1 \le j \le \tau p$, $1 \le \tau \le r$. We note that there is no essential difference between arbitrary orbit sums and flat orbit sums, since each orbit sum $S_{\tilde{H}}^{(\mu)}(f)$ can be obtained from a flat orbit sum by the identification of the corresponding powers $z_{i,(\tau-1)p+l}^{p^e}$, for various $(e, i, \tau)$. This shows that the study of orbit sums $S_{\tilde{H}}^{(\mu)}(f)$, for various monomials $f \in A_{m\nu}$, is reduced to the study of similar orbit sums with flat monomials $f$. The following result is an immediate consequence of Proposition 10. □

**Corollary 11.** *Let $v \in A_{m\nu}^{\tilde{H}}$ be a polynomial of positive degree that is an $F_p$-linear combination*

$$v = \sum_{k=1}^{K} a_k S_{\tilde{H}}^{(\mu_k)}(f_k), \qquad a_k \neq 0,$$

*of flat orbit sums $S_{\tilde{H}}^{(\mu_k)}$, for various*

$$f_k = \prod_{e=0}^{\eta} \prod_{i=1}^{m} \prod_{\tau=1}^{r} \prod_{j=(\tau-1)p+1}^{\tau p} (z_{ij}^{p^e})^{s_{ij\tau}^{(e,k)}},$$

*and let $\mu_1 \ge \mu_2 \ge \cdots \ge \mu_K \ge 2$. If $v$ is invariant under the action of $H$ and if two orbit sums $S_{\tilde{H}}^{(\mu_k)}$ and $S_{\tilde{H}}^{(\mu_{k'})}$ such that $\mu_{k'} = \mu_k$ for $k' > k$ appear in $v$ with nonzero coefficients, then the monomial $f_{k'}$ is obtained from $f_k$ by means of substitutions*

$$z_{ij} p^e \mapsto z_{i,j+l}^{p^e} \quad \text{and} \quad z_{i'j'}^{p^e} \mapsto z_{i',j'-l}^{p^e},$$

*with $(\tau-1)p+1 \le j+l, j'-l \le \tau p$, $0 \le e \le \eta$, $1 \le i, i' \le m$, $1 \le \tau \le r$, that preserve weights and degrees of the monomials $f_k$ and $f_{k'}$.*

It follows from the stated above that a more general $F_p$-linear combination

$$v = \sum_{k=1}^{K} a_k S_{\tilde{H}}^{(\mu_k)}(f_k) + \sum_{l=1}^{L} c_l S_{\tilde{H}}^{(\mu_l)}(f_l) N_{\tilde{H}}(g_l), \quad a_k \neq 0, c_l \neq 0, \deg(g_l) \ge 1,$$

is invariant under the action of $H$ only in the case when it has the following special form

$$v = v_0 + \sum_{\lambda=1}^{\Lambda} v_\lambda N_{\tilde{H}}^{(0)}(g_\lambda),$$

where each $F_p$-linear combination

$$v_\lambda = \sum_{k=1}^{K_\lambda} a_{k\lambda} S_{\tilde{H}}^{(\mu_{k\lambda})}(f_{k\lambda}), \quad \text{for} \quad 0 \le \lambda \le \Lambda,$$

of orbit sums $S_{\tilde{H}}^{(\mu_{k\lambda})}(f_{k\lambda})$ involves no variable $z_{ij}$ with $1 \le i \le m$, $(\tau - 1)p + 1 \le j \le \tau p - 2$, $1 \le \tau \le r$.

Now we are able to give a complete description of the $H$-invariants $v \in A_{mv}^{\tilde{H}}$ that are $F_p$-linear combinations of flat orbit sums $S_{\tilde{H}}^{(\mu)}(f)$, for various $\mu \ge 2$ and $f$. The following result plays a crucial role in constructing a system of generators of the algebra $A_{mn}^{H}$.

**Proposition 12.** *Let $v \in A_{mv}^{\tilde{H}}$ be a polynomial of positive degree that is an $F_p$-linear combination*

$$v = \sum_{k=1}^{K} a_k S_{\tilde{H}}^{(\mu_k)}(f_k), \qquad a_k \neq 0,$$

*of flat orbit sums $S_{\tilde{H}}^{(\mu_k)}(f_k)$, for various $\mu_k \ge 2$ and*

$$f_k = \prod_{e=0}^{\eta}\prod_{i=1}^{m}\prod_{\tau=1}^{r}\prod_{j=(\tau-1)p+1}^{\tau p} (z_{ij}^{p^e})^{s_{ij\tau}^{(e,k)}}.$$

*If $v$ is invariant under the action of $H$, then $v$ is an $F_p$-linear combination of polynomials*

$$\prod_{\kappa=1}^{\sigma}(z_{i_{2\kappa-1},\tau_{2\kappa-1}p-1}^{p^{e_{2\kappa-1}}} z_{i_{2\kappa},\tau_{2\kappa}p}^{p^{e_{2\kappa}}} - z_{i_{2\kappa-1},\tau_{2\kappa-1}p}^{p^{e_{2\kappa-1}}} z_{i_{2\kappa},\tau_{2\kappa}p-1}^{p^{e_{2\kappa}}}) \prod_{e=0}^{\eta}\prod_{i=1}^{m}\prod_{\tau=1}^{r}(z_{i,\tau p}^{p^e})^{\omega_{i,\tau p}^{(e)}},$$

*where $0 \le \sigma \le [s/2]$, $(i_{2\kappa-1}, \tau_{2\kappa-1}) \prec (i_{2\kappa}, \tau_{2\kappa})$, and $0 \le \omega_{i,\tau p}^{(e)} \le 1$.*

*Proof.* Since $v$ is invariant under the action of $H$, it follows from Theorem 8 that $v$ involves only monomials of the form

$$z_{i_1,\tau_1 p-\varepsilon_1}^{p^{e_1}} \cdots z_{i_s,\tau_s p-\varepsilon_s}^{p^{e_s}},$$

with $1 \le s \le \deg v$, $1 \le e_1, \dots, e_s \le \eta$, $1 \le i_1 \le \cdots \le i_s \le m$, $1 \le \tau_1, \dots, \tau_s \le r$, and $\varepsilon_1, \dots, \varepsilon_s \in \{0, 1\}$. Moreover, since $v$ is an $F_p$-linear combination of flat orbit sums, then $(e_k, i_k, \tau_k) \neq (e_l, i_l, \tau_l)$, for all $k$ and $l$ such that $k \neq l$.

Let

$$g = z_{i_1,\tau_1 p-\varepsilon_1}^{p^{e_1}} \cdots z_{i_s,\tau_s p-\varepsilon_s}^{p^{e_s}}$$

be a monomial of maximal possible weight $\sigma \le s$ that appears in $v$ with a nonzero coefficient $a \in F_p$. We can assume without loss of generality that

$$g = z_{i_1,\tau_1 p-1}^{p^{e_1}} \cdots z_{i_{\sigma-1},\tau_{\sigma-1}p-1}^{p^{e_{\sigma-1}}} z_{i_\sigma,\tau_\sigma p-1}^{p^{e_\sigma}} z_{i_{\sigma+1},\tau_{\sigma+1}p}^{p^{e_{\sigma+1}}} z_{i_{\sigma+2},\tau_{\sigma+2}p}^{p^{e_{\sigma+2}}} \cdots z_{i_s,\tau_s p}^{p^{e_s}}.$$

Since $\gamma(v) = v$, then along with $ag$ the invariant $v$ contains also the polynomial

$$\gamma(ag) = ag + \sum_{\kappa=1}^{\sigma} \frac{a}{\kappa!}\left( z_{i_1,\tau_1 p}^{p^{e_1}} \frac{\partial}{\partial z_{i_1,\tau_1 p-1}^{(e_1)}} + \cdots + z_{i_s,\tau_s p}^{p^{e_s}} \frac{\partial}{\partial z_{i_\sigma,\tau_\sigma p-1}^{(e_\sigma)}} \right)^{\kappa} g,$$

involving several associated extra terms

$$a g_{\kappa} = \frac{a}{\kappa!}\left( z_{i_1,\tau_1 p}^{p^{e_1}} \frac{\partial}{\partial z_{i_1,\tau_1 p-1}^{(e_1)}} + \ldots + z_{i_s,\tau_s p}^{p^{e_s}} \frac{\partial}{\partial z_{i_s,\tau_s p-1}^{(e_s)}} \right)^{\kappa} g,$$

each of which is a linear combination of monomials $g_{\iota\kappa}$ of the same weight $\sigma - \kappa$. On the other hand, since $v$ is invariant under the action of $H$ it cannot contain the above extra terms, so the invariant $v$ should involve at least one extra monomial $g'$ of the same weight $\sigma$ that gives a possibility to cancel some of the monomials $g_{\iota\kappa}$. This process of cancellation of extra monomials $g_{\iota\kappa}$ can be described inductively as follows.

Adding and subtracting, if it is necessary, several terms of the form $ag'$ for different monomials $g'$ of the same weight $\sigma$ we can assume without loss of generality that

$$g' = z_{i_1,\tau_1 p-1}^{p^{e_1}} \cdots z_{i_{\sigma-1},\tau_{\sigma-1}p-1}^{p^{e_{\sigma-1}}} z_{i_\sigma,\tau_\sigma p}^{p^{e_\sigma}} z_{i_{\sigma+1},\tau_{\sigma+1}p-1}^{p^{e_{\sigma+1}}} z_{i_{\sigma+2},\tau_{\sigma+2}p}^{p^{e_{\sigma+2}}} \cdots z_{i_s,\tau_s p}^{p^{e_s}}.$$

Since

$$a(g-g') = a(z_{i_\sigma,\tau_\sigma p-1}^{p^{e_\sigma}} z_{i_{\sigma+1},\tau_{\sigma+1}p}^{p^{e_{\sigma+1}}} - z_{i_\sigma,\tau_\sigma p}^{p^{e_\sigma}} z_{i_{\sigma+1},\tau_{\sigma+1}p-1}^{p^{e_{\sigma+1}}})$$

$$\times z_{i_1,\tau_1 p-1}^{p^{e_1}} \cdots z_{i_{\sigma-1},\tau_{\sigma-1}p-1}^{p^{e_{\sigma-1}}} z_{i_{\sigma+2},\tau_{\sigma+2}p}^{p^{e_{\sigma+2}}} \cdots z_{i_s,\tau_s p}^{p^{e_s}},$$

it follows that

$$a(g-g') = a' z_{i_1,\tau_1 p-1}^{p^{e_1}} \cdots z_{i_{\sigma-1},\tau_{\sigma-1}p-1}^{p^{e_{\sigma-1}}} z_{i_{\sigma+1},\tau_{\sigma+1}p}^{p^{e_{\sigma+1}}} \cdots z_{i_s,\tau_s p}^{p^{e_s}},$$

where

$$a' = a(z_{i_\sigma,\tau_\sigma p-1}^{p^{e_\sigma}} z_{i_{\sigma+1},\tau_{\sigma+1}p}^{p^{e_{\sigma+1}}} - z_{i_\sigma,\tau_\sigma p}^{p^{e_\sigma}} z_{i_{\sigma+1},\tau_{\sigma+1}p-1}^{p^{e_{\sigma+1}}})$$

is an $H$-invariant. Repeating this procedure we eliminate after finitely many steps all the extra terms $ag_\kappa$ in the above representation of $\gamma(g)$. As a result we obtain that $\sigma \le [s/2]$ and that $v$ is a $F_p$-linear combination of invariants

$$\prod_{\kappa=1}^{\sigma} (z_{i_\kappa,\tau_\kappa p-1}^{p^{e_\kappa}} z_{i_{\kappa+1},\tau_{\kappa+1}p}^{p^{e_{\kappa+1}}} - z_{i_\kappa,\tau_\kappa p}^{p^{e_\kappa}} z_{i_{\kappa+1},\tau_{\kappa+1}p-1}^{p^{e_{\kappa+1}}}) \prod_{e=0}^{\eta}\prod_{i=1}^{m}\prod_{\tau=1}^{r} z_{i,\tau p}^{\omega_{i,\tau p}^{(e)}}.$$

This finishes the proof. □

Using the above arguments in the case of repeating $z_{i,\tau p-1}^{p^e}$ and $z_{i,\tau p}^{p^e}$ we obtain the following result.

**Corollary 13.** *Let $v \in A_{mv}^{\tilde{H}}$ be a polynomial of positive degree that is an $F_p$-linear combination*

$$v = \sum_{k=1}^{K} a_k S_{\tilde{H}}^{(\mu_k)}(f_k), \qquad a_k \neq 0,$$

*of orbit sums $S_{\tilde{H}}^{(\mu_k)}(f_k)$, for various $\mu_k \ge 2$ and*

$$f_k = \prod_{e=0}^{\eta}\prod_{i=1}^{m}\prod_{\tau=1}^{r}\prod_{j=(\tau-1)+1}^{\tau p} (z_{ij}^{(e)})^{s_{ij\tau}^{(e)}}.$$

*If $v$ is invariant under the action of $H$, then it is an $F_p$-linear combination of invariants*

$$\prod_{\kappa=1}^{\sigma}(z_{i_{2\kappa-1},\tau_{2\kappa-1}p-1}^{p^{e_{2\kappa-1}}} z_{i_{2\kappa},\tau_{2\kappa}p}^{p^{e_{2\kappa}}} - z_{i_{2\kappa-1},\tau_{2\kappa-1}p}^{p^{e_{2\kappa-1}}} z_{i_{2\kappa},\tau_{2\kappa}p-1}^{p^{e_{2\kappa}}})^{s_{i_{2\kappa-1}i_{2\kappa}\tau_{2\kappa-1}\tau_{2\kappa}}^{(e_{2\kappa-1},e_{2\kappa})}}$$
$$\times \prod_{e=0}^{\eta}\prod_{i=1}^{m}\prod_{\tau=1}^{r}(z_{i,\tau p}^{p^e})^{\omega_{i,\tau p}^{(e)}},$$

*with* $0 \le \sigma \le [s/2]$, $(i_{2\kappa-1}, \tau_{2\kappa-1}) \prec (i_{2\kappa}, \tau_{2\kappa})$, *and* $0 \le s_{i_{2\kappa-1}i_{2\kappa}\tau_{2\kappa-1}\tau_{\kappa}}^{(e_{2\kappa-1},e_{2\kappa})}, \omega_{i,\tau p}^{(e)} \le p-1$.

Now we study the structure of invariants $N_{\tilde{H}}^{(0)}(f)$ and $S_{\tilde{H}}^{(0)}(f)$, for various monomials $f \in F_p[z_{i,\tau p-1}, z_{i,\tau p} \mid 1 \le i \le m, 1 \le \tau \le r]$. At first we observe that

$$N_{\tilde{H}}^{(0)}(z_{i,\tau p-1}) = z_{i,\tau p-1}^{p} - z_{i,\tau p-1} z_{i,\tau p}^{p-1} \quad\text{and}\quad N_{\tilde{H}}^{(0)}(z_{i,\tau p}) = z_{i,\tau p}^{p}.$$

Hence if

$$f = \prod_{e=0}^{\eta}\prod_{i=1}^{m}\prod_{\tau=1}^{r}(z_{i,\tau p-1}^{p^e})^{s_{i,\tau p-1}^{(e)}}(z_{i,\tau p}^{p^e})^{s_{i,\tau p}^{(e)}},$$

then

$$N_{\tilde{H}}^{(0)}(f) = \prod_{e=0}^{\eta}\prod_{i=1}^{m}\prod_{\tau=1}^{r}(z_{i,\tau p-1}^{p^{e+1}} - z_{i,\tau p-1}^{p^e} z_{i,\tau p}^{p^e(p-1)})^{s_{i,\tau p-1}}(z_{i,\tau p}^{p^{e+1}})^{s_{i,\tau p}}.$$

We also observe that

$$\begin{aligned}
&z_{i_1,\tau_1 p-1}^{p^{e_1}} z_{i_2,\tau_2 p}^{p^{e_2}} - z_{i_1,\tau_1 p}^{p^{e_1}} z_{i_2,\tau_2 p-1}^{p^{e_2}}\\
&= (z_{i_1,\tau_1 p-1}^{p^{e_1}} - z_{i_1,\tau_1 p-1}^{p^{e_1-1}} z_{i_1,\tau_1 p}^{p^{e_1-1}(p-1)}) z_{i_2,\tau_2 p}^{p^{e_2}}\\
&\quad - (z_{i_1,\tau_1 p}^{p^{e_1-1}} z_{i_2,\tau_2 p}^{p^{e_2}} - z_{i_1,\tau_1 p-1}^{p^{e_1-1}} z_{i_2,\tau_2 p}^{p^{e_2}}) z_{i_1,\tau_1 p}^{p^{e_1-1}(p-1)}
\end{aligned}$$

$$= N_{\tilde{H}}^{(0)}(z_{i_1,\tau_1 p-1})^{p^{e_1-1}} z_{i_2,\tau_2 p}^{p^{e_2}} - (z_{i_1,\tau_1 p}^{p^{e_1-1}} z_{i_2,\tau_2 p-1}^{p^{e_2}} - z_{i_1\tau_1 p-1}^{p^{e_1-1}} z_{i_2,\tau_2 p}^{p^{e_2}}) z_{i_1,\tau_1 p}^{p^{e_1-1}(p-1)}.$$

Iterating the last relation we find that

$$\begin{aligned}
z_{i_1,\tau_1 p-1}^{p^{e_1}} z_{i_2,\tau_2 p}^{p^{e_2}} - z_{i_1,\tau_1 p}^{p^{e_1}} z_{i_2,\tau_2 p-1}^{p^{e_2}} &= (z_{i_1,\tau_1 p-1} z_{i_2,\tau_2 p} - z_{i_1,\tau_1 p} z_{i_2,\tau_2 p-1}) z_{i_1,\tau_1 p}^{p^{e_1}-1} z_{i_2,\tau_2 p}^{p^{e_2}-1}\\
&+ \left(\sum_{\varepsilon_1=1}^{e_1} N_{\tilde{H}}^{(0)}(z_{i_1,\tau_1 p-1})^{p^{e_1-\varepsilon_1}} z_{i_1,\tau_1 p}^{p^{e_1-\varepsilon_1+1}(p^{\varepsilon_1-1}-1)}\right) z_{i_2,\tau_2 p}^{p^{e_2}}\\
&- \left(\sum_{\varepsilon_2=1}^{e_2} N_{\tilde{H}}^{(0)}(z_{i_2,\tau_2 p-1})^{p^{e_2-\varepsilon_2}} z_{i_2,\tau_2 p}^{p^{e_1-\varepsilon_2+1}(p^{\varepsilon_2-1}-1)}\right) z_{i_1,\tau_1 p}^{p^{e_1-1}}.
\end{aligned}$$

Taking into account Proposition 9 and Corollary 13 we arrive at the following result.

**Proposition 14.** *Let $v \in A_{mv}^{\tilde{H}}$ be a polynomial of positive degree that is an $F_p$-linear combination of orbit sums $S_{\tilde{H}}^{(\mu_k)}(f_k)$, for various $\mu_k \geq 1$ and $f_k \in A_{mv}$. If $v$ is invariant under the action of $H$, then $v$ is a polynomial over $F_p$ in $H$-invariants $N_{\tilde{H}}^{(0)}(z_{i,\tau p-1}), z_{i,\tau p}$, for $1 \leq i \leq m$, $1 \leq r \leq r$, and $(z_{i_1,\tau_1 p-1} z_{i_2,\tau_2 p} - z_{i_1,\tau_1 p} z_{i_2,\tau_2 p-1})$, for $1 \leq i_1, i_2 \leq m$, $1 \leq \tau_1, \tau_2 \leq r$.*

Let again

$$f = \prod_{e=0}^{\eta} \prod_{i=1}^{m} \prod_{\tau=1}^{r} (z_{i,\tau p-1}^{p^e})^{s_{i,\tau p-1}^{(e)}} (z_{i,\tau p}^{p^e})^{s_{i,\tau p}^{(e)}}.$$

We set

$$\sum_{e=0}^{\eta} s_{i,\tau p-1}^{(e)} = s_{i,\tau p-1}^{(0)} + t_{i,\tau p-1}^{(0)} p,$$

where $0 \leq s_{i,\tau p-1}^{(0)} \leq p-1$, and write $f$ in the form

$$f = \prod_{i=1}^{m} \prod_{\tau=1}^{r} z_{i,\tau p-1}^{s_{i,\tau p}^{(0)}} (z_{i,\tau p-1}^{p})^{t_{i,\tau p-1}^{(0)}} \prod_{e=0}^{\eta} \prod_{i=1}^{m} \prod_{\tau=1}^{r} z_{i,\tau p}^{s_{i,\tau p}^{(e)}}.$$

Since $z_{i,\tau p-1}^{p} = N_{\tilde{H}}^{(0)}(z_{i,\tau p-1}) + z_{i,\tau p-1} z_{i,\tau p}^{p-1}$, it follows that

$$f = \prod_{i=1}^{m} \prod_{\tau=1}^{r} z_{i,\tau p-1}^{s_{i,\tau p-1}^{(0)}} (N_{\tilde{H}}^{(0)}(z_{i,\tau p-1}) + z_{i,\tau p-1} z_{i,\tau p}^{p-1})^{t_{i,\tau p-1}^{(0)}} \prod_{e=0}^{\eta} \prod_{i=1}^{m} \prod_{\tau=1}^{r} z_{i,\tau p}^{s_{i,\tau p}^{(e)}}.$$

Iterating the last relation we find that $f$ is an $F_p$-linear combination of polynomials

$$\prod_{i=1}^{m} \prod_{\tau=1}^{r} z_{i,\tau p-1}^{s_{i,\tau p-1}} z_{i,\tau p}^{t_{i,\tau p}} N_{\tilde{H}}^{(0)}(z_{i,\tau p-1})^{\omega_{i,\tau p-1}},$$

with $0 \leq s_{i,\tau p-1} \leq p-1$. Hence $S_{\tilde{H}}^{(0)}(f)$ is an $F_p$-linear combination of $H$-invariants

$$S_{\tilde{H}}^{(0)}(f') \prod_{i=1}^{m} \prod_{\tau=1}^{r} z_{i,\tau p}^{t_{i,\tau p}} N_{\tilde{H}}^{(0)}(z_{i,\tau p-1})^{\omega_{i,\tau p-1}},$$

where

$$f' = \prod_{i=1}^{m} \prod_{\tau=1}^{r} z_{i,\tau p-1}^{s_{i,\tau p-1}}$$

and $0 \leq s_{i,\tau p-1} \leq p-1$, for $1 \leq i \leq m$, $1 \leq \tau \leq r$. Thus we obtain the following result.

**Proposition 15.** *Let*

$$f = \prod_{e=0}^{\eta} \prod_{i=1}^{m} \prod_{\tau=1}^{r} (z_{i,\tau p-1}^{p^e})^{s_{i,\tau p-1}^{(e)}} (z_{i,\tau p}^{p^e})^{s_{i,\tau p}^{(e)}}$$

*be a monomial in* $F_p[z_{i,\tau p-1}, z_{i,\tau p} \mid 1 \le i \le m, 1 \le \tau \le r]$, *and let*

$$S_{\tilde{H}}^{(0)}(f) = \sum_{\alpha \in F_p} \prod_{e=0}^{\eta} \prod_{i=1}^{m} \prod_{\tau=1}^{r} \left( z_{i,\tau p-1}^{p^e} + \binom{\alpha}{1} z_{i,\tau p}^{p^e} \right)^{s_{i,\tau p-1}^{(e)}} \left( z_{i,\tau p}^{p^e} \right)^{s_{i,\tau p}^{(e)}}$$

*be the corresponding orbit sum. Then* $S_{\tilde{H}}^{(0)}(f)$ *is a polynomial over* $F_p$ *in* $H$*-invariants* $N_{\tilde{H}}^{(0)}(z_{i,\tau p-1})$, $z_{i,\tau p}$, *for* $1 \le i \le m$, $1 \le \tau \le r$, *and* $S_{\tilde{H}}^{(0)}(f')$, *for various monomials*

$$f' = \prod_{i=1}^{m} \prod_{\tau=1}^{r} z_{i,\tau p-1}^{s_{i,\tau p-1}}$$

*such that* $0 \le s_{i,\tau p-1} \le p-1$, *for* $1 \le i \le m$, $1 \le \tau \le r$.

With the use of Theorem 5, Theorem 8, Proposition 14 and Proposition 15 we now arrive at the following result.

**Theorem 16.** *Let* $v \in A_{mv}^{\tilde{H}}$ *be a polynomial of positive degree that is invariant under the action of* $H$. *Then* $v$ *is a polynomial over* $F_p$ *in the* $H$*-invariants* $N_{\tilde{H}}^{(0)}(z_{i,\tau p-1})$, $z_{i,\tau p}$, *for* $1 \le i \le m$, $1 \le \tau \le r$, *the* $H$*-invariants* $(z_{i_1,\tau_1 p-1} z_{i_2,\tau_2 p} - z_{i_1,\tau_1 p} z_{i_2,\tau_2 p-1})$, *for* $1 \le i_1, i_2 \le m$, $1 \le \tau_1, \tau_2 \le r$, $(i_1, \tau_1) \prec (i_2, \tau_2)$, *and the* $H$*-invariants* $S_{\tilde{H}}^{(0)}(f')$, *for various monomials*

$$f' = \prod_{i=1}^{m} \prod_{\tau=1}^{r} z_{i,\tau p-1}^{s_{i,\tau p-1}}$$

*such that* $0 \le s_{i,\tau p-1} \le p-1$.

Finally, we determine an exact form of the polynomials $S_{\tilde{H}}^{(0)}(f')$ for all monomials $f'$ of the above form.

**Proposition 17.** *Let*

$$f' = \prod_{i=1}^{m} \prod_{\tau=1}^{r} z_{i,\tau p-1}^{s_{i,\tau p-1}}$$

*be a monomial of degree* $s \ge 1$ *such that* $0 \le s_{i,\tau p-1} \le p-1$, *for* $1 \le i \le m$, $1 \le \tau \le r$. *Then*

$$S_{\tilde{H}}^{(0)}(f') = -\sum_{l=1}^{[s/(p-1)]} \sum_{(\sigma_{i,\tau p-1})} \prod_{i=1}^{m} \prod_{\tau=1}^{r} a_{i,\tau p-1} z_{i,\tau p-1}^{s_{i,\tau p-1}-\sigma_{i,\tau p-1}} z_{i,\tau p}^{\sigma_{i,\tau p-1}},$$

*where the inner sum is over all integral tuples* $(\sigma_{i,\tau p-1})$ *such that*

$$\sum_{i=1}^{m} \sum_{\tau=1}^{r} \sigma_{i,\tau p-1} = l(p-1) \le s$$

*and* $0 \le \sigma_{i,\tau p-1} \le s$, *for* $1 \le i \le m$, $1 \le \tau \le r$.

*Proof.* We set

$$s_i = \sum_{\tau=1}^{r} s_{i,\tau p-1} \quad \text{and} \quad s = \sum_{i=1}^{m} s_i .$$

Since

$$\left(z_{i,\tau p-1} + \binom{\alpha}{1} z_{i,\tau p}\right)^{s_{i,\tau p-1}} = \sum_{\alpha \in F_p} a_{\sigma_{i,\tau p-1}} \alpha^{\sigma_{i,\tau p-1}} z_{i,\tau p-1}^{s_{i,\tau p-1}-\sigma_{i,\tau p-1}} z_{i,\tau p}^{\sigma_{i,\tau p}},$$

where

$$a_{\sigma,\tau p-1} = \binom{s_{i,\tau p-1}}{\sigma_{i,\tau p-1}},$$

then

$$\prod_{\tau=1}^{r} \left(z_{i,\tau p-1} + \binom{\alpha}{1} z_{i,\tau p}\right)^{s_{i,\tau p-1}}$$

$$= \sum_{\sigma_i=0}^{s_i} \alpha^{\sigma_i} \sum_{\sigma_{i,p-1}+\cdots+\sigma_{i,rp-1}=\sigma_i} \prod_{\tau=1}^{r} a_{i,\tau p-1} z_{i,\tau p-1}^{s_{i,\tau p-1}-\sigma_{i,\tau p-1}} z_{i,\tau p}^{\sigma_{i,\tau p-1}} .$$

In that case,

$$\prod_{i=1}^{m} \prod_{\tau=1}^{r} \left(z_{i,\tau p-1} + \binom{\alpha}{1} z_{i,\tau p}\right)^{s_{i,\tau p-1}}$$

$$= \sum_{\sigma=0}^{s} \alpha^{\sigma} \sum_{(\sigma_{i,\tau p-1})} \prod_{i=1}^{m} \prod_{\tau=1}^{r} a_{i,\tau p-1} z_{i,\tau p-1}^{s_{i,\tau p-1}-\sigma_{i,\tau p-1}} z_{i,\tau p}^{\sigma_{i,\tau p-1}}$$

and hence

$$S_{\tilde{H}}^{(0)}(f') = \sum_{\alpha \in F_p} \prod_{i=1}^{m} \prod_{\tau=1}^{r} \left(z_{i,\tau p-1} + \binom{\alpha}{1} z_{i,\tau p}\right)^{s_{i,\tau p-1}}$$

$$= \sum_{\sigma=0}^{s} \sum_{\alpha \in F_p} \alpha^{\sigma} \sum_{(\sigma_{i,\tau p-1})} \prod_{i=1}^{m} \prod_{\tau=1}^{r} a_{i,\tau p-1} z_{i,\tau p-1}^{s_{i,\tau p-1}-\sigma_{i,\tau p-1}} z_{i,\tau p}^{\sigma_{i,\tau p-1}},$$

where the inner sum is over all tuples $(\sigma_{i,\tau p-1})$ such that $0 \leq \sigma_{i,\tau p-1} \leq s_{i,\tau p-1}$, for $1 \leq i \leq m$, $1 \leq \tau \leq r$, and

$$\sum_{i=1}^{m} \sum_{\tau=1}^{r} \sigma_{i,\tau p-1} = \sigma .$$

Now the required result follows from the fact that

$$\sum_{\alpha \in F_p} \alpha^{\sigma} = \begin{cases} p-1, & \text{if } \sigma = l(p-1) \\ 0, & \text{otherwise.} \end{cases}$$

□

## 3 Proof of Theorem 1 and Corollary 2

Since $x_{ij}$, for $1 \le i \le m$, $2r+1 \le j \le n$, are invariant under the action of $H$, every invariant $u \in A^H_{mn}$ is a polynomial of the form

$$u = u_1 f_1 + \cdots + u_l f_l ,$$

where $u_\iota \in A^H_{m,2r}$ and $f_k$, for $1 \le k \le l$, are monomials in $F_p[x_{i,j} \mid 1 \le i \le m, 2r+1 \le j \le n]$. This shows that we can assume without loss of generality that $u \in A^H_{m,2r}$. Therefore, it suffices to show that

$$A^H_{m,2r} = F_p[x_{i,2\tau-1}, N^{(0)}_H(x_{i,2\tau-1}), (x_{i_1,2\tau_1-1}x_{i_2,2\tau_2} - x_{i_1,2\tau_1}x_{i_2,2\tau_2-1}), S^{(0)}_H(f')],$$

where $1 \le i \le m$, $1 \le \tau \le r$, $1 \le i_1, i_2 \le m$, $1 \le \tau_1, \tau_2 \le r$, and $(i_1, \tau_1) \prec (i_2, \tau_2)$, and $f'$ runs through the set of monomials

$$f' = \prod_{i=1}^{m}\prod_{\tau=1}^{r} x_{i,2\tau-1}^{s_{i,2\tau-1}} \quad \text{with} \quad 0 \le s_{i,2\tau-1} \le p-1.$$

If $\vartheta : V_i \hookrightarrow \tilde{V}_i$ is the embedding defined by (6), then $\vartheta$ induces the corresponding $F_p$-algebra monomorphism $\vartheta : A_{m,2r} \to A_{m\nu}$. Let $u$ be an element of $A^H_{n,2r}$, and $v = \vartheta(u) \in A_{m\nu}$ the image of $u$. Then $v$ is invariant under the action of $H$ on $A_{m\nu}$ defined by (7) as well as under the action of $\tilde{H}$. It follows from Theorem 16 that $v$ is a polynomial in $H$-invariants $N^{(0)}_{\tilde{H}}(z_{i,\tau p-1})$, $z_{i,\tau p}$, for $1 \le i \le m$, $1 \le \tau \le r$, in $H$-invariants $(z_{i_1,\tau_1 p-1}z_{i_2,\tau_2 p} - z_{i_1,\tau_1 p}z_{i_2,\tau_2 p-1})$, for $1 \le i_2, i_2 \le m$, $1 \le \tau_1, \tau_2 \le r$, $(i_1, \tau_1) \prec (i_2, \tau_2)$, and in $H$-invariants $S^{(0)}_{\tilde{H}}(f')$, for various monomials

$$f' = \prod_{i=1}^{m}\prod_{\tau=1}^{r} z_{i,\tau p-1}^{s_{i,\tau p-1}},$$

such that $0 \le s_{i,\tau p-1} \le p-1$. Identifying now $z_{i,\tau p-1}$, $z_{i,\tau p}$ with $x_{i,2\tau-1}$, $x_{i,2\tau}$ via the isomorphism $\vartheta$ and taking into account that

$$\vartheta(N^{(0)}_H(x_{i,2\tau-1})) = N^{(0)}_{\tilde{H}}(z_{i,2\tau p-1}), \quad \vartheta(S^{(0)}_H(f')) = S^{(0)}_{\tilde{H}}(f')$$

under this identification, we find that

$$A^H_{mn} = F_p[x_{i,2\tau}, x_{ij}, N^{(0)}_H(x_{i,2\tau-1}), (x_{i_1,2\tau_1-1}x_{i_2,2\tau_2} - x_{i_1,2\tau_1}x_{i_2,2\tau_2-1}), S^{(0)}_H(f')].$$

This proves Theorem 1.

To prove Corollary 2 we consider the polynomial

$$S^{(0)}_H(g') = \sum_{\alpha \in F_p}\prod_{i=1}^{m}\prod_{\tau=1}^{r}\left(x_{i,2\tau-1} + \binom{\alpha}{1}x_{i,2\tau}\right)^{p-1}$$

and show that $S^{(0)}_H(g')$ cannot be expressed as a polynomial over $F_p$ in $H$-invariants $x_{i,2\tau}$, $N^{(0)}_H(x_{i,2\tau-1})$, $(x_{i_1,2\tau_1-1}x_{i_2,2\tau_2} - x_{i_1,2\tau_1}x_{i_2,2\tau_2-1})$, and $S^{(0)}_H(f')$, where

$$f' = \prod_{i=1}^{m}\prod_{\tau=1}^{r} x_{i,2\tau-1}^{s_{i,2\tau-1}}$$

is a monomial such that $s_{i,2\tau-1} < p-1$ at least for one pair $(i,\tau)$ with $1 \le i \le m$, $1 \le \tau \le r$. At first we observe that $S_H^{(0)}(g')$ involves the monomial

$$h' = x_{12}^{p-1} \prod_{\substack{i=1\\(i,\tau)\neq(1,1)}}^{m} \prod_{r=1}^{r} x_{i,2\tau-1}^{p-1}$$

of total degree $mr(p-1)$ which occurs in $S_H^{(0)}(g')$ with coefficient $-1$. Moreover, if

$$f' = \prod_{i=1}^{m}\prod_{\tau=1}^{r} x_{i,2\tau-1}^{s_{i,2\tau-1}} \quad \text{and} \quad S_H^{(0)}(f') \neq 0,$$

then we conclude in view of Proposition 17 that the polynomial $S_H^{(0)}(f')$ involves a monomial of the following form:

$$\prod_{i=1}^{m}\prod_{\tau=1}^{r} x_{i,2\tau-1}^{s_{i,2\tau-1}-\sigma_{i,2\tau-1}} x_{i,2\tau}^{\sigma_{i,2\tau-1}},$$

where $0 \le \sigma_{i,2\tau-1} \le s_{i,2\tau-1}$, and

$$\sum_{i=1}^{m}\sum_{\tau=1}^{r} \sigma_{i,2\tau-1} = p-1.$$

Now we show that no monomial $M$ in $x_{i,2\tau}$, $(x_{i_1,2\tau_1-1}x_{i_2,2\tau_2} - x_{i_1,2\tau_1 2}x_{i_2,2\tau_2-1})$, $N_H^{(0)}(x_{i,2\tau-1})$ and $S_H^{(0)}(f')$, where $\deg f' < mr(p-1)$, occurring in $S_H^{(0)}(g')$ when expanded in terms $x_{i,2\tau-1}$, $x_{i,2\tau}$, for $1 \le i \le m$, $1 \le \tau \le r$, contains the monomial $h'$ with a nonzero coefficient. We observe that $S_H^{(0)}(f')$ is a homogeneous polynomial and therefore each monomial $M$ of this kind has the same total degree $mr(p-1)$ in $x_{i,2\tau}$ and $x_{i,2\tau}$.

If $M$ contains as a factor of some power of $N_H^{(0)}(x_{i,2\tau-1}) = x_{i,2\tau-1}^{p} - x_{i,2\tau-1}x_{i,2\tau}^{p-1}$, then it follows from the rather special form of the polynomials $N_H^{(0)}(x_{i,2\tau-1})$, $S_H^{(0)}(f')$ and $(x_{i_1,2\tau_1-1}x_{i_2,2\tau_2} - x_{i_1,2\tau_2}x_{i_2,2\tau_2-1})$, that the expansion of $M$ in powers of $x_{i,2\tau-1}$ and $x_{i,2\tau}$ is an $F_p$-linear combination of monomials each of which either involves $x_{i,2\tau-1}^{p}$ or has a total degree of at least $p$ in the variables $x_{i,2\tau}$.

Assume now that $M$ involves a polynomial $S_H^{(0)}(f')$ with $\deg f' < mr(p-1)$. Then using similar arguments we see that each monomial occurring in $M$ with a nonzero coefficient has a total degree at least $p$ in the $x_{i,2\tau}$.

Finally, since $mr > 2$, each monomial $M$ in the variables $x_{i,2\tau}$, for $1 \le i \le m$, $1 \le \tau \le r$, and the invariants $(x_{i_2,2\tau_1}x_{i_2,2\tau_2} - x_{i_1,2\tau_1}x_{i_2,2\tau_2-1})$, for $1 \le i_1, i_2 \le m$, $1 \le \tau_1, \tau_2 \le r$, $(i_1,\tau_1) \prec (i_2,\tau_2)$, that has the total degree $mr(p-1)$ in the variables $x_{i,2\tau-1}$ and $x_{i,2\tau}$ has a total degree at least $mr(p-1)/2 > p-1$ in the variables $x_{i,2\tau}$.

In each of the above cases the monomial $h'$ cannot occur in the monomial $M$ with a nonzero coefficient, and this completes the proof of Corollary 2.

## 4 A universal invariant

Assume that $m \geq n$ and consider the polynomial

$$f_0 = \sum_{\alpha_1,\dots,\alpha_n \in F_p} (\alpha_1 x_{11} + \cdots + \alpha_n x_{1n})^{p-1} \cdots (\alpha_1 x_{m1} + \cdots + \alpha_n x_{mn})^{p-1}. \quad (8)$$

At first we show that $f_0$ is invariant under the action of the general linear group $GL(n, F_p)$. In that case, the polynomial $f_0$ is also invariant under the action of any subgroup $G$ of $GL(n, F_p)$. It suffices to prove that $f_0$ is invariant under the action of an arbitrary element of the group $G$.

**Proposition 18.** *If $\sigma$ is an arbitrary element of the group $GL(n, F_p)$, then $\sigma(f_0) = f_0$.*

*Proof.* Since

$$\sigma((\alpha_1 x_{11} + \cdots + \alpha_n x_{1n})^{p-1} \cdots (\alpha_1 x_{m1} + \cdots + \alpha_n x_{mn})^{p-1})$$
$$= (\alpha_1 \sigma(x_{11}) + \cdots + \sigma(x_{1n})^{p-1} \cdots (\alpha_1 \sigma(x_{m1}) + \cdots + \alpha_n \sigma(x_{mn}))^{p-1}$$

and action of $\sigma$ permutes the elements of each space $V_i = F_p x_{i1} + \cdots + F_p x_{in}$, for $1 \leq i \leq m$, in the same way, we deduce that

$$\begin{aligned}\sigma(f_0) &= \sum_{\alpha_1,\dots,\alpha_n \in F_p} (\alpha_1 \sigma(x_{11}) + \cdots + \alpha_n \sigma(x_{1n}))^{p-1} \cdots (\alpha_1 \sigma(x_{m1}) \\ &\quad + \cdots + \alpha_n \sigma(x_{mn}))^{p-1} \\ &= \sum_{\alpha'_1,\dots,\alpha'_n \in F_p} (\alpha'_1 x_{11} + \cdots + \alpha'_n x_{1n})^{p-1} \cdots (\alpha'_1 x_{m1} + \cdots + \alpha'_n x_{mn})^{p-1} = f_0.\end{aligned}$$

This proves the proposition. □

If $F$ is a polynomial in $A_{mn}$, let $\pi(F)$ denote its image under the projection

$$\begin{aligned}&\pi(x_{11}, \dots, x_{1n}; \dots; x_{m1}, \dots, x_{mn}) \\ &= (\pi_1(x_{11}, \dots, x_{1n}); \dots; \pi_m(x_{m1}, \dots, x_{mn})),\end{aligned} \quad (9)$$

where

(i) $\pi_i(x_{i1}, \dots, x_{in}) = (0, \dots, 0, x_{ii}, 0, \dots, 0)$, for $i = 1, 2, \dots, n$;
(ii) $\pi_i(x_{i1}, \dots, x_{in}) = (x_{i1}, 0, \dots, 0, 0, \dots, 0, )$, for $i = n+1, \dots, m$.

Clearly the map

$$\begin{aligned}&\pi(f(x_{11}, \dots, x_{1n}; \dots; x_{m1}, \dots, x_{mn})) \\ &= f(\pi_1(x_{11}, \dots, x_{1n}); \dots; \pi_m(x_{m1}, \dots, x_{mn}))\end{aligned}$$

defines an $F_p$-algebra homomorphism.

Denote by $\pi(f_0)$ the image of the polynomial $f_0$ under the projection $\pi$ and find an exact form of the polynomial $\pi(f_0)$.

**Proposition 19.** *If $m \geq n$, then the polynomial $\pi(f_0)$ has the form*

$$\pi(f_0) = (-1)^n \prod_{i=1}^{n} x_{ii}^{p-1} \prod_{i=n+1}^{m} x_{i1}^{p-1}.$$

*Proof.* At first we observe that

$$\pi(f_0) = \sum_{\alpha_1,\ldots,\alpha_n \in F_p} \prod_{i=1}^{n} (\alpha_i x_{ii})^{p-1} \prod_{i=n+1}^{m} (\alpha_1 x_{i1})^{p-1}.$$

In that case,

$$\pi(f_0) = \prod_{i=1}^{n} x_{ii}^{p-1} \prod_{i=n+1}^{m} x_{i1}^{p-1} \left( \sum_{\alpha_1 \ldots,\alpha_n \in F_p} \alpha_1^{(m-n+1)(p-1)} \prod_{j=2}^{n} \alpha_j^{p-1} \right),$$

and since $\sum_{\alpha \in F_p} \alpha^{l(p-1)} = -1$ for every positive integer $l$, then

$$\pi(f_0) = (-1)^n \prod_{i=1}^{n} {x_{ii}}^{p-1} \prod_{i=n+1}^{m} x_{i1}^{p-1}$$

This completes the proof. □

## 5 Proof of Theorem 3

Let $G \leq GL(n, F_p)$ be a group such that its order $|G|$ is divisible by $p$, and $H = \langle \gamma \rangle$ a cyclic subgroup of $G$ of order $p$. Since $H \leq G \leq GL(n, F_p)$, we have

$$A_{mn}^{GL(n,F_p)} \subseteq A_{mn}^{G} \subseteq A_{mn}^{H}.$$

Assume that $m \geq n$ and that the sizes $n_1, \ldots, n_s$ of the basic Jordan blocks of the matrix $\gamma$ satisfy the condition

$$n_1 = \cdots = n_r = 2 \quad \text{and} \quad n_{r+1} = \cdots = n_s = 1,$$

so that $n = 2r + s - r = r + s$.

To prove Theorem 3 it suffices to show that every system of $F_p$-algebra generators of $A_{mn}^{G}$ contains a generator of degree at least $(m - n + 2r)(p - 1)/r$. Set

$$u_{(i_1,\tau_1),(i_2,\tau_2)} = (x_{i_1,2\tau_1-1} x_{i_2,2\tau_2} - x_{i_1,2\tau_1} x_{i_2,2\tau_2-1})$$

and consider the system of generators

$$B_{mn}^{H} = \{x_{i,2\tau}, x_{ij}, N_H^{(0)}(x_{i_1,2\tau_1-1}), u_{(i_1,\tau_1),(i_2,\tau_2)}, S_H^{(0)}(f')\}$$

of the algebra $A_{mn}^{H}$, described by Theorem 1. It follows from Corollary 2 that $B_{mn}^{H}$ consists of $F_p$-algebra generators of the minimal possible degree. Let $B_{mn}^{G}$ be a system of $F_p$-algebra generators of $A_{mn}^{G}$. Since $A_{mn}^{G} \subseteq A_{mn}^{H}$, we can assume without loss of generality that $B_{mn}^{G} \subseteq B_{mn}^{H}$. Denote by $\tilde{B}_{mn}^{H}$ a subset of $B_{mn}^{H}$ which consists of the invariants $u \in B_{mn}^{H}$ satisfying the condition

$$\deg u < (m - n + 2r)(p - 1)/2$$

and set $\tilde{B}^G_{mn} = \tilde{B}^H_{mn} \cap B^G_{mn}$. Let $f_0 \in A_{mn}$ be the polynomial defined by (8). Propositions 18 and 19 imply that $f_0$ is a homogeneous polynomial of degree $m(p-1)$ that is invariant under action of $GL(n, F_p)$. In the case $f_0$ is also an invariant under the action of $G$ as well as $H$. The crucial point is that the invariant $f_0 \in A^G_{mn}$ is *indecomposable* in $A^H_{mn}$ with respect to $\tilde{B}^H_{mn}$, i.e., $f_0$ cannot be written as a polynomial over $F_p$ in vector invariants $u \in \tilde{B}^H_{mn}$. All the more, the invariant $f_0$ is indecomposable in $A^G_{mn}$ with respect to $\tilde{B}^G_{mn}$, i.e., $f_0$ cannot be written as a polynomial in invariants $u \in \tilde{B}^G_{mn}$.

Denote by $\eta$ the cardinality of $\tilde{B}^H_{mn}$ and enumerate the elements $u$ of $\tilde{B}^H_{mn}$ by the numbers $1, 2, \ldots, \eta$. Assume for the contrary that $f_0$ is a polynomial over $F_p$ in elements of $u_1, \ldots, u_\eta \in \tilde{B}^H_{mn}$ and write

$$f_0 = \sum_{\substack{1 \le \delta_1 + \cdots + \delta_\eta \le m(p-1) \\ \delta_1 \deg u_1 + \cdots + \delta_\eta \deg u_\eta = m(p-1)}} a_{\delta_1,\ldots,\delta_\eta} u_1^{\delta_1} \cdots u_\eta^{\delta_\eta}, \tag{10}$$

Comparing degrees of the monomials which appear in both sides of the last identity (with respect to each of the variables $x_{11}, \ldots, x_{m1}; \ldots; x_{1n}, \ldots, x_{mn}$), then taking into account that $f_0$, $u_1, \ldots, u_\eta$ are homogeneous polynomials in $x_{11}, \ldots, x_{m1}; \ldots; x_{1n}, \ldots, x_{mn}$, and $\deg f_0 = m(p-1)$, $\deg_{x_{ij}} f_0 = p-1$, for $1 \le i \le m$, $1 \le j \le n$, we find that $0 \le \delta_\rho \le p-1$, for $1 \le \rho \le \eta$, and $\delta_1 + \cdots + \delta_\eta > 1$; moreover, we see that

$$\deg_{x_{ij}} u_\rho \le p-1,$$

for all $1 \le i \le m$, $1 \le j \le n$ and $1 \le \rho \le \eta$. Now we assume that

$$\{1, 2, \ldots, \eta\} = I_1 \cup I_2 \cup I_3 \cup I_4 \cup I_5,$$

and $u_\epsilon = S^{(0)}_H(f'_\epsilon)$, for $\epsilon \in I_1$; $u_\kappa = N^{(0)}_H(x_{i,2\tau-1})$, for $\kappa \in I_2$, $1 \le i = i(\kappa) \le m$, and $1 \le \tau = \tau(\kappa) \le r$; $u_\lambda = u_{(i_1,\tau_1),(i_2,\tau_2)}$, for $\lambda \in I_3$, $1 \le i_1 = i_1(\lambda)$, $i_2 = i_2(\lambda) \le m$, $1 \le \tau_1 = \tau_1(\lambda)$, $\tau_2 = \tau_2(\lambda) \le r$, and $(i_1, \tau_1) \prec (i_2, \tau_2)$; $u_\mu = x_{i,2\tau}$, for $\mu \in I_4$, $1 \le i = i(\mu) \le m$, and $1 \le \tau = \tau(\mu) \le r$; $u_\nu = x_{ij}$, for $\nu \in I_5$, $1 \le i = i(\nu) \le m$, and $2r+1 \le j = j(\nu) \le n$. The relation (10) takes the following form:

$$\begin{aligned} f_0 = \sum a_{\delta_1,\ldots,\delta_\eta} \prod_{\epsilon \in I_1} (S^{(0)}_H(f'_\epsilon))^{\delta_\epsilon} \prod_{\kappa \in I_2} (N^{(0)}_H(x_{i(\kappa),2\tau(\kappa)-1}))^{\delta_\kappa} \\ \times \prod_{\lambda \in I_3} (u_{(i_1(\lambda),\tau_1(\lambda)),(i_2(\lambda),\tau_2(\lambda))})^{\delta_\lambda} \prod_{\mu \in B_4} x^{\delta_\mu}_{i(\mu),2\tau(\mu)} \prod_{\nu \in I_5} x^{\delta_\nu}_{i(\nu),j(\nu)}, \end{aligned} \tag{11}$$

where the sum on the right-hand side is over all nonnegative integers $\delta_1, \ldots, \delta_\delta$ satisfying the condition

$$\begin{aligned} \sum_{\epsilon \in I_1} \delta_\epsilon \deg S^{(0)}_H(f'_\epsilon) + \sum_{\kappa \in I_2} \delta_\kappa \deg N^{(0)}_H(x_{i(\kappa),2\tau(\kappa)-1}) + \sum_{\lambda \in I_3} 2\delta_\lambda + \sum_{\mu \in I_4} \delta_\mu + \sum_{\nu \in I_5} \delta_\nu \\ = m(p-1). \end{aligned}$$

Now we prove that equality (11) is impossible by showing that the monomial

$$\prod_{i=1}^{n} x_{ii}^{p-1} \prod_{i=n+1}^{m} x_{i1}^{p-1}$$

appears in the left-hand side of (11), but does not appear in its right-hand side. Let $\pi$ be the $F_p$-algebra homomorphism defined by (9). Then Proposition 19 implies

$$\pi(f_0) = (-1)^n \prod_{i=1}^{n} x_{ii}^{p-1} \prod_{i=n+1}^{m} x_{i1}^{p-1}.$$

On the other hand, we have

$$\pi(N_H^{(0)}(x_{i,2\tau-1})) = \begin{cases} x_{2\tau-1,2\tau-1}^{p}, & \text{if } i = 2\tau-1, 1 \le \tau \le r \\ x_{i1}^{p}, & \text{if } n+1 \le i \le m \\ 0, & \text{otherwise,} \end{cases}$$

$$\pi(u_{(i,\sigma),(k,\tau)}) = \begin{cases} x_{2\sigma-1,2\sigma-1}x_{2\tau,2\tau}, & \text{if } i = 2\sigma-1, k = 2\tau, 1 \le \sigma \le \tau \le r \\ -x_{2\sigma,2\sigma}x_{2\tau-1,2\tau-1}, & \text{if } i_1 = 2\sigma, k = 2\tau-1, 1 \le \sigma < \tau \le r \\ -x_{k1}x_{2\sigma,2\sigma}, & \text{if } n+1 \le k \le m, \tau = 1, 1 \le \sigma \le r \\ 0, & \text{otherwise,} \end{cases}$$

and if

$$f' = \prod_{i=1}^{m}\prod_{\tau=1}^{r} x_{i,2\tau-1}^{s_{i,2\tau-1}},$$

then it follows by Proposition 17,

$$\pi(S_H^{(0)}(f')) = -\prod_{\tau=1}^{r} x_{2\tau-1,2\tau-1}^{s_{2\tau-1,2\tau-1}} x_{2\tau,2\tau}^{s_{2\tau,2\tau-1}},$$

where $\sum_{\tau=1}^{r} s_{2\tau,2\tau-1} = l(p-1)$ for some positive integer $l \le m$.

Applying the $F_p$-algebra homomorphism $\pi$ to both sides of the relation (11) we obtain

$$\begin{aligned} \prod_{i=1}^{2r} x_{ii}^{p-1} \prod_{i=n+1}^{m} x_{i1}^{p-1} = {} & \prod_{\epsilon\in I_1}\left(\prod_{\tau=1}^{r} x_{2\tau-1,2\tau-1}^{s_{2\tau-1,2\tau-1}^{(\epsilon)}} x_{2\tau,2\tau}^{s_{2\tau,2\tau-1}^{(\epsilon)}} \prod_{i=n+1}^{m} x_{i1}^{s_{i1}^{(\epsilon)}}\right)^{\delta_\epsilon} \\ & \times \prod_{\kappa\in I_2}\left(\prod_{\tau=1}^{r}(x_{2\tau-1,2\tau-1}^{p})^{s_{2\tau-1,2\tau-1}^{(\kappa)}} \prod_{i=n+1}^{m}(x_{i1}^{p})^{s_{i1}^{(\kappa)}}\right)^{\delta_\kappa} \\ & \times \prod_{\lambda\in J_3}\left(\prod_{\rho=1}^{r}\prod_{\sigma=1}^{r}(x_{2\rho-1,2\rho-1}x_{2\sigma,2\sigma})^{s_{2\rho-1,2\sigma}^{(\lambda)}}\right. \\ & \left. \times \prod_{i=n+1}^{m}\prod_{\tau=1}^{r}(x_{i1}x_{2\tau,2\tau})^{s_{i,2\tau}^{(\lambda)}}\right)^{\delta_\lambda} \prod_{\mu\in I_4}\left(\prod_{\tau=1}^{r} x_{2\tau,2\tau}^{s_{2\tau,2\tau}^{(\mu)}}\right)^{\delta_\mu}, \end{aligned} \tag{12}$$

where

$$\sum_{\tau=1}^{r} s_{2\tau-1,2\tau}^{(\epsilon)} = l(p-1), \quad \sum_{\tau=1}^{r} s_{2\tau-1,2\tau-1}^{(\kappa)} = 1, \quad \sum_{i=n+1}^{m} s_{ii}^{(\kappa)} = 1,$$

$$\sum_{\rho=1}^{r}\sum_{\sigma=1}^{r} s_{2\rho-1,2\sigma}^{(\lambda)} = 1, \quad \sum_{i=n+1}^{m}\sum_{\tau=1}^{r} s_{i,2\tau}^{(\lambda)} = 1,$$

$$\sum_{\tau=1}^{r} s_{2\tau,2\tau}^{(\mu)} = 1, \quad \sum_{i=1}^{m}\sum_{j=2r+1}^{n} s_{ij}^{(\nu)} = 1$$

and $s_{2\tau-1,2\tau-1}^{(\kappa)} + s_{ii}^{(\kappa)} \leq 1$ and $s_{2\rho-1,2\sigma}^{(\lambda)} + s_{i,2\tau}^{(\lambda)} \leq 1$. If $s_{2\tau-1,2\tau-1}^{(\kappa)}\delta_\kappa \geq 1$ or $s_{i1}^{(\kappa)}\delta_\kappa \geq 1$, for some $(\tau, \kappa)$, $(i, \kappa)$, then degree of the right-hand side of (12) with respect to $x_{2\tau,2\tau}$ or $x_{i1}$ is at least $p$, which is impossible. This shows that $\delta_\kappa = 0$ for all $\kappa \in I_2$.

Set

$$L_1 = \sum_{\epsilon\in I_1}\sum_{i=n+1}^{m} \delta_\epsilon s_{i,1}^{(\epsilon)}, \quad L_3 = \sum_{\lambda\in I_3}\sum_{i=n+1}^{m} \delta_\lambda s_{i,2\tau}^{(\lambda)},$$

$$M_1 = \sum_{\epsilon\in I_1}\sum_{\tau=1}^{r} \delta_\epsilon s_{2\tau-1,2\tau-1}^{(\epsilon)}, \quad M_3 = \sum_{\lambda\in I_3}\sum_{\rho=1}^{r}\sum_{\sigma=1}^{r} \delta_\lambda s_{2\rho-1,2\sigma}^{(\lambda)}$$

and

$$N_1 = \sum_{\epsilon\in I_1}\sum_{\tau=1}^{r} \delta_\epsilon s_{2\tau,2\tau-1}^{(\epsilon)}, \quad N_4 = \sum_{\mu\in I_4}\sum_{\tau=1}^{r} \delta_\mu s_{2\tau,2\tau}^{(\mu)}.$$

Comparing total degrees of the left-hand side and the right-hand side of relation (12) with respect to $x_{i1}$'s, $x_{2\tau-1,2\tau-1}$'s and $x_{2\tau,2\tau}$'s, respectively, we obtain

$$L_1 + L_3 = (m-n)(p-1), \quad M_1 + M_3 = r(p-1)$$

and

$$N_1 + L_3 + M_3 + N_4 = r(p-1).$$

We observe also that the condition

$$\sum_{\tau=1}^{r} s_{2\tau,2\tau-1}^{(\epsilon)} \geq l(p-1) \geq p-1$$

implies $N_1 \geq (p-1)\sum_{\epsilon\in I_1}\delta_\epsilon$. Set $\theta = \sum_{\epsilon\in I_1}\delta_\epsilon$ and note that $\theta \leq r$. Now we consider the following possibilities:

*Case* 1. If $\vartheta = r$, then $L_3 = M_3 = N_4 = 0$ and therefore

$$\prod_{i=1}^{2r} x_{ii}^{p-1} \prod_{i=n+1}^{m} x_{i1}^{p-1} = \prod_{\epsilon\in I_1}\left(\prod_{i=n+1}^{m} x_{i1}^{s_{i1}} \prod_{\tau=1}^{r} x_{2\tau-1,2\tau-1}^{s_{2\tau-1,2\tau-1}^{(\epsilon)}} x_{2\tau,2\tau}^{s_{2\tau,2\tau-1}^{(\epsilon)}}\right)^{\tau_\mu}.$$

Comparing total degrees of both sides of the last equality and taking into account that

$$\deg S_H^{(0)}(f_\epsilon') < \frac{(m-n+2r)(p-1)}{r}$$

we obtain $(m-n+2r)(p-1) < (m-n+2r)(p-1)$. This yields a contradiction.

*Case* 2. Let $\vartheta < r$. Since

$$L_1 + L_3 + M_1 + M_3 + N_1 = (m-n+r)(p-1) + N_1,$$
$$L_3 + M_3 \le r(p-1) - N_1, \quad N_1 \ge \theta(p-1)$$

and since by assumption,

$$L_1 + M_1 + N_1 < \theta\frac{(m-n+2r)(p-1)}{r},$$

then

$$(m-n+r)(p-1) + N_1 < \theta\frac{(m-n+2r)(p-1)}{r} + M_1 + M_3.$$

In that case,

$$(m-n+r)(p-1) < \theta\frac{(m-n+r)(p-1)}{r} + (r-\theta)(p-1)$$

and hence $(r-\theta)(m-n+r)(p-1) < (r-\theta)r(p-1)$. Since $m \ge n$, we arrive at a contradiction, which completes the proof of Theorem 3.

## References


1. Campbell, H.E.A., Hughes, I., Pollack, R.D.: Vector invariants of symmetric groups. *Can. Math. Bull.* **33**, 391–397 (1990)
2. Campbell, H.E.A., Hughes, I.P.: Vector invariants of $U_2(F_p)$: a proof of a conjecture of Richman. *Adv. Math.* **126**, 1–20 (1997)
3. Fleischmann, P.: A new degree bound for the vector invariants of symmetric groups. *Trans. Am. Math. Soc.* **350**, 1703–1712 (1998)
4. Fleischmann, P.: The Noether bound in invariant theory of finite groups. *Adv. Math.* **156**, 23–32 (2000)
5. Hilbert, D.: Über die vollen Invariantensysteme. *Math. Ann.* **42**, 313–373 (1893)
6. Kemper, G.: Lower degree bounds for modular invariants and a question of I. Hughes. *Transform. Groups* **3**, 135–144 (1998)
7. Noether, E.: Der Endlichkeitssatz der Invarianten endlicher Gruppen. *Math. Ann.* **77**, 89–92 (1916)
8. Noether, E.: Der Endlichkeitssatz der Invarianten endlicher linearer Gruppen der Characteristik $p$. *Nachr. Ges. Wiss. Göttingen* **1926**, 28–35 (1926)
9. Richman, D.: On vector invariants over finite fields. *Adv. Math.* **81**, 30–65 (1990)
10. Richman, D.: Invariants of finite groups over fields of characteristic $p$. *Adv. Math.* **124**, 25–48 (1996)
11. Smith, L.: *Polynomial Invariants of Finite Groups*. A.K. Peters, Wellesley (1995)
12. Stepanov, S.A.: Vector invariants of symmetric groups in prime characteristic. *Discrete Math. Appl.* **10**, 455–468 (2000)
13. Weyl, H.: *The Classical Groups*, 2nd edn. Princeton University Press, Princeton (1953)

*Diophantine Approximation*
H. P. Schlickewei et al., Editors


# NEW IRRATIONALITY RESULTS FOR DILOGARITHMS OF RATIONAL NUMBERS

Carlo Viola
*Dipartimento di Matematica, Università di Pisa, Largo B. Pontecorvo 5, 56127 Pisa, Italy*
viola@dm.unipi.it

Dedicated to Wolfgang Schmidt on the occasion of his 70th birthday

## 1 Introduction

A natural method to investigate diophantine properties of transcendental (or conjecturally transcendental) constants occurring in various mathematical contexts consists in the search for sequences of good rational approximations, or algebraic approximations with bounded degree, to suitable values of some special transcendental functions, such as the logarithm, or the polylogarithm of order $q \geq 2$, or the hypergeometric functions, etc. Traditionally, one employs for this purpose Padé or Padé-type approximations to the functions involved.

In recent years, however, special attention was directed towards an alternative method, namely to the study of arithmetic properties of simple or multiple definite integrals of Euler's type related to the hypergeometric functions. These integrals proved to be a very flexible tool both for getting sharp asymptotic estimates, obtained by means of classical techniques such as the Laplace method or the saddle point method in complex analysis, and for improving such estimates through the $p$-adic valuation of the gamma factors appearing in the Euler integral representation of the hypergeometric functions. The latter idea, together with a group-theoretic approach, was introduced by Rhin and me in [6] and developed in [7], where the best known irrationality measures of $\zeta(2) = \pi^2/6$ and of $\zeta(3)$ are proved (see also [3] for interesting geometric interpretations of the Rhin–Viola permutation groups).

The group method of Rhin and Viola is related to the actions of suitable birational transformations on the simple or multiple Euler-type integrals involved (see [10] and [11] for a detailed discussion on this aspect of the method), and becomes more and more relevant as the dimension of the integrals increases. Accordingly, the group structure in the diophantine study of $\zeta(3)$ given in [7], where triple integrals are involved, is richer than the one for double integrals in [6], and this is in turn more relevant than the group structure underlying the one-dimensional case for the diophantine study of logarithms of rational numbers [9], or of logarithms of algebraic numbers [1].

*Keywords.* Dilogarithm, hypergeometric function, permutation groups, irrationality measures.

*2000 Mathematics subject classification.* 11J82, 33B30, 33C05, 20B35.

Rhin and I have recently applied in [8] our group method to the diophantine study of double integrals of Euler's type related to the dilogarithm, thus obtaining qualitative and quantitative improvements on all the best previously known irrationality results for dilogarithms of positive rational numbers. The aim of the present paper is to give a brief survey of this matter, together with some hints for further research developments.

## 2 Double integrals and permutation groups related to the dilogarithm

Let $q \geq 1$ be an integer. We recall that the polylogarithm $\mathrm{Li}_q(x)$ of order $q$ is defined, in the disc $|x| < 1$, by the power series

$$\mathrm{Li}_q(x) = \sum_{n=1}^{\infty} \frac{x^n}{n^q}.$$

Thus $\mathrm{Li}_1'(x) = \sum_{m=0}^{\infty} x^m = (1-x)^{-1}$, whence

$$\mathrm{Li}_1(x) = \int_0^x \frac{\mathrm{d}t}{1-t} = -\log(1-x).$$

For any $q \geq 2$ we have $x\mathrm{Li}_q'(x) = \mathrm{Li}_{q-1}(x)$, i.e.,

$$\mathrm{Li}_q(x) = \int_0^x \frac{\mathrm{Li}_{q-1}(t)}{t}\,\mathrm{d}t.$$

In particular, for $q = 2$, the dilogarithm is given by

$$\mathrm{Li}_2(x) = \sum_{n=1}^{\infty} \frac{x^n}{n^2} = -\int_0^x \frac{\log(1-t)}{t}\,\mathrm{d}t. \tag{2.1}$$

By the integral representation (2.1), it is plain that $\mathrm{Li}_2(x)$ is holomorphic in the cut plane $\mathbb{C} \setminus [1, +\infty)$.

The dilogarithm was introduced by Euler, and subsequently studied by many authors. Maier [5, §8, Satz 3] proved that for any $r, s \in \mathbb{Z}$ satisfying $r \neq 0$ and $s > 2^8 e^2 |r|^3$, $\mathrm{Li}_2(r/s)$ is irrational. Chudnovsky [2, Theorem 7.1] found again Maier's result for $r > 0$ and $s > 2^2 e^2 r^3$. He also announced [2, Theorem 7.5] that $\mathrm{Li}_2(1/s)$ is irrational for any integer $s \geq 14$. By applying an ingenious analysis of the $p$-adic valuation of products of binomial coefficients occurring as coefficients of suitable Legendre-type polynomials, Hata [4] proved the irrationality of $\mathrm{Li}_2(1/s)$ for all integers $s \in (-\infty, -5] \cup [7, +\infty)$, and gave irrationality measures of $\mathrm{Li}_2(1/s)$ for the integers $s$ such that $7 \leq s \leq 18$ or $-16 \leq s \leq -5$.

In [8], Rhin and I introduce a variant of the double integral considered in [6] for the diophantine study of $\zeta(2) = \mathrm{Li}_2(1) = \pi^2/6$. For any $z \in \mathbb{R}$, $z > 1$, and any integers $h, j, k, l, m \geq 0$, we define in Section 2 of [8]:

$$I_z^{(0)}(h, j, k, l, m) = z^{-l-m} \int_0^1\!\!\int_0^1 \frac{x^j(1-x)^h y^k (1-y)^l}{(x(1-y)+yz)^{j+k-m+1}}\, \mathrm{d}x\, \mathrm{d}y, \tag{2.2}$$

$$I_z^{(1)}(h, j, k, l, m) = z^{-l-m} \int_0^1 \left( \frac{1}{2\pi i} \oint_{|y-x/(x-z)|=\varrho} \frac{x^j(1-x)^h y^k (1-y)^l}{(x(1-y)+yz)^{j+k-m+1}}\, \mathrm{d}y \right) \mathrm{d}x, \tag{2.3}$$

$$I_z^{(2)}(h, j, k, l, m) = z^{-l-m} \frac{1}{2\pi i} \oint_{|x-z|=\sigma} \left( \frac{1}{2\pi i} \oint_{|y-x/(x-z)|=\varrho} \frac{x^j(1-x)^h y^k (1-y)^l}{(x(1-y)+yz)^{j+k-m+1}}\, \mathrm{d}y \right) \mathrm{d}x \tag{2.4}$$

for any $\varrho, \sigma > 0$, and

$$I_z(h, j, k, l, m) = I_z^{(0)}(h, j, k, l, m) - (\log z)\, I_z^{(1)}(h, j, k, l, m). \tag{2.5}$$

In the arithmetical study of the above integrals a crucial role is played by the transformation $\lambda = \lambda_{x,z} : y \mapsto Y$ defined by

$$\lambda : Y = \frac{1-y}{1 - \dfrac{x-z}{x} y} = \frac{x(1-y)}{x(1-y)+yz} \tag{2.6}$$

for $x \neq 0$, $x \neq z$. It is easy to see that $\lambda$ is an involution and satisfies

$$\frac{\mathrm{d}Y}{x(1-Y)+Yz} = -\frac{\mathrm{d}y}{x(1-y)+yz}.$$

By applying the transformation $\lambda$ to the integrals (2.2), (2.3) and (2.4), i.e., by making the change of variable (2.6), one easily gets

$$I_z^{(\nu)}(h, j, k, l, m) = I_z^{(\nu)}(h, m, l, k, j) \qquad (\nu = 0, 1, 2), \tag{2.7}$$

whence, by (2.5),

$$I_z(h, j, k, l, m) = I_z(h, m, l, k, j). \tag{2.8}$$

Therefore, with the action of the transformation $\lambda$ on (2.2)–(2.5) one associates the permutation $\boldsymbol{\lambda}$ of $h, j, k, l, m$ defined by

$$\boldsymbol{\lambda} = (j\ m)(k\ l) \tag{2.9}$$

and extended by linearity to any linear combination of $h, j, k, l, m$. Thus the values of the integrals (2.2)–(2.5) are invariant under the action of $\boldsymbol{\lambda}$.

Natural quantities associated with (2.2)–(2.5) are the nonnegative integers

$$\begin{aligned} H &= \max\{l+m-j,\ m+h-k,\ h+j-l,\ j+k-m\},\\ K &= \max\{l+m-j,\ \min\{m+h-k,\ h+j-l\},\ j+k-m\},\\ \alpha &= \max\{j+k,\ k+l,\ l+m\},\\ \beta &= \max\{0,\ k+l-h\},\\ \delta &= \max\{h,\ m+h-k,\ h+j-l,\ j+k,\ k+l,\ l+m\}, \end{aligned} \tag{2.10}$$

which are easily seen to be all invariant under the action of the permutation $\boldsymbol{\lambda}$.

For any positive integer $n$, let

$$d_n = \text{l.c.m.}\{1,\dots,n\},$$

and $d_0 = 1$. In [8, Theorem 2.1], we prove that

$$\begin{aligned} d_H d_K z^\alpha (z-1)^\beta I_z(h,j,k,l,m) &= P(z) - Q(z)\,\mathrm{Li}_2(1/z),\\ d_H d_K z^\alpha (z-1)^\beta I_z^{(1)}(h,j,k,l,m) &= R(z) - Q(z)\,\mathrm{Li}_1(1/z),\\ d_H d_K z^\alpha (z-1)^\beta I_z^{(2)}(h,j,k,l,m) &= Q(z), \end{aligned} \tag{2.11}$$

where

$$P(z),\ Q(z),\ R(z) \in \mathbb{Z}[z], \quad \max\{\deg P(z),\ \deg Q(z),\ \deg R(z)\} \le \delta, \tag{2.12}$$

with $H,\ K,\ \alpha,\ \beta,\ \delta$ defined by (2.10).

Following the method of [6] and [7], besides the birational transformation $\lambda$ given by (2.6) we also consider a hypergeometric integral transformation $\varphi$ acting on (2.2)–(2.5). Using the Euler integral representation of the hypergeometric function ${}_2F_1(a,b;c;t)$ and the invariance of this function under the interchange of the parameters $a,\ b$ we obtain, if $m+h-k \ge 0$ and $j+k-m \ge 0$,

$$\begin{aligned} &I_z^{(\nu)}(h,j,k,l,m)\\ &\quad = \frac{h!\,j!}{(m+h-k)!\,(j+k-m)!}\, I_z^{(\nu)}(m+h-k,\ j+k-m,\ m,\ l,\ k) \end{aligned} \tag{2.13}$$

for $\nu = 0, 1, 2$, whence, by (2.5),

$$\begin{aligned} &I_z(h,j,k,l,m)\\ &\quad = \frac{h!\,j!}{(m+h-k)!\,(j+k-m)!}\, I_z(m+h-k,\ j+k-m,\ m,\ l,\ k). \end{aligned} \tag{2.14}$$

We henceforth assume the nonnegative integers $h,\ j,\ k,\ l,\ m$ to be chosen so that $l+m-j,\ m+h-k,\ h+j-l,\ j+k-m$ are also nonnegative. For simplicity of notation, let $J_z(h,j,k,l,m)$ denote any one of the quantities (2.2)–(2.5). Then the transformation formulae (2.7), (2.8), (2.13) and (2.14) can be written as

$$\frac{J_z(h,j,k,l,m)}{h!\,j!\,k!\,l!\,m!} = \frac{J_z(h,m,l,k,j)}{h!\,m!\,l!\,k!\,j!} \tag{2.15}$$

and

$$\frac{J_z(h,j,k,l,m)}{h!\,j!\,k!\,l!\,m!} = \frac{J_z(m+h-k,\ j+k-m,\ m,\ l,\ k)}{(m+h-k)!\,(j+k-m)!\,m!\,l!\,k!}. \tag{2.16}$$

We denote by $\varphi$ the transformation acting on

$$\frac{J_z(h, j, k, l, m)}{h!\, j!\, k!\, l!\, m!} \tag{2.17}$$

as in (2.16), and by $\boldsymbol{\varphi}$ the corresponding permutation, mapping $h, j, k, l, m$ to $m+h-k$, $j+k-m$, $m$, $l$, $k$ respectively, and extended by linearity to any linear combination of $h, j, k, l, m$. Thus we see that both the permutation $\boldsymbol{\lambda}$ in (2.9) and the permutation $\boldsymbol{\varphi}$ act on the set

$$\mathcal{S} = \{h,\ j,\ k,\ l,\ m,\ l+m-j,\ m+h-k,\ h+j-l,\ j+k-m\}, \tag{2.18}$$

and their actions on $\mathcal{S}$ are given by

$$\boldsymbol{\lambda} = (j\ \ m)(k\ \ l)(l+m-j\ \ j+k-m)(m+h-k\ \ h+j-l) \tag{2.19}$$

and

$$\boldsymbol{\varphi} = (h\ \ m+h-k)(j\ \ j+k-m)(k\ \ m). \tag{2.20}$$

Let $\boldsymbol{\Phi} = \langle \boldsymbol{\varphi}, \boldsymbol{\lambda} \rangle$ be the permutation group generated by $\boldsymbol{\lambda}$ and $\boldsymbol{\varphi}$. By (2.15) and (2.16), for any permutation $\boldsymbol{\chi} \in \boldsymbol{\Phi}$ we have

$$\frac{J_z(h, j, k, l, m)}{h!\, j!\, k!\, l!\, m!} = \frac{J_z(\boldsymbol{\chi}(h), \boldsymbol{\chi}(j), \boldsymbol{\chi}(k), \boldsymbol{\chi}(l), \boldsymbol{\chi}(m))}{\boldsymbol{\chi}(h)!\, \boldsymbol{\chi}(j)!\, \boldsymbol{\chi}(k)!\, \boldsymbol{\chi}(l)!\, \boldsymbol{\chi}(m)!}, \tag{2.21}$$

so that the value of (2.17) is invariant under the action of the group $\boldsymbol{\Phi}$.

The algebraic structure of the group $\boldsymbol{\Phi}$ becomes clear by considering the actions of $\boldsymbol{\lambda}$ and $\boldsymbol{\varphi}$ on the auxiliary integers

$$h+j,\ \ j+k,\ \ k+l,\ \ l+m,\ \ m+h,$$

which we denote by $u_1, \ldots, u_5$ respectively. From (2.19) and (2.20) it is plain that the actions of $\boldsymbol{\lambda}$ and $\boldsymbol{\varphi}$ on such integers are

$$\boldsymbol{\lambda} = (u_1\ \ u_5)(u_2\ \ u_4), \quad \boldsymbol{\varphi} = (u_3\ \ u_4),$$

and it is easy to check that $\boldsymbol{\Phi}$ acts faithfully on the set $\{u_1, \ldots, u_5\}$.

Also, $\boldsymbol{\Phi}$ is intransitive over $\{u_1, \ldots, u_5\}$, since each of the two subsets $\{u_1, u_5\}$ and $\{u_2, u_3, u_4\}$ is mapped onto itself by both $\boldsymbol{\lambda}$ and $\boldsymbol{\varphi}$. Moreover, it is easy to see that $(u_1\ \ u_5)$ and $(u_2\ \ u_4)$ belong to $\boldsymbol{\Phi}$. Therefore, $\boldsymbol{\Phi}$ is isomorphic to the product of the symmetric groups of permutations of $\{u_1, u_5\}$ and of $\{u_2, u_3, u_4\}$: $\boldsymbol{\Phi} \cong \mathfrak{S}_2 \times \mathfrak{S}_3$, whence the order of $\boldsymbol{\Phi}$ is $|\boldsymbol{\Phi}| = 2! \cdot 3! = 12$.

For any $\boldsymbol{\chi} \in \boldsymbol{\Phi}$, we consider the quotient

$$\frac{h!\, j!\, k!\, l!\, m!}{\boldsymbol{\chi}(h)!\, \boldsymbol{\chi}(j)!\, \boldsymbol{\chi}(k)!\, \boldsymbol{\chi}(l)!\, \boldsymbol{\chi}(m)!} \tag{2.22}$$

resulting from the transformation formula (2.21) for $J_z(h, j, k, l, m)$. If $\boldsymbol{\chi}, \boldsymbol{\chi}' \in \boldsymbol{\Phi}$ lie in the same left coset of the subgroup $\boldsymbol{\Lambda} = \langle \boldsymbol{\lambda} \rangle$ of order 2 in $\boldsymbol{\Phi}$, the quotient (2.22) equals the analogous quotient for $\boldsymbol{\chi}'$ (for any permutations $\boldsymbol{\chi}_1$ and $\boldsymbol{\chi}_2$ we denote by $\boldsymbol{\chi}_1\boldsymbol{\chi}_2$ the permutation obtained by applying first $\boldsymbol{\chi}_2$ and then $\boldsymbol{\chi}_1$). Thus the six left cosets of $\boldsymbol{\Lambda}$ in $\boldsymbol{\Phi}$ are characterized by the corresponding quotients (2.22). Each of such quotients can be simplified by removing the factorials of the integers appearing both in the numerator and in the denominator (independently of the numerical values assigned to $h, j, k, l, m$). If, after simplifying (2.22), the resulting quotient has $v$ factorials in the numerator and $v$ in the denominator, we say that $\chi$ is a permutation of level $v$, or that the left coset $\chi\boldsymbol{\Lambda}$ has level $v$.

The subgroup $\boldsymbol{\Lambda}$ obviously has level 0. Among the remaining five left cosets of $\boldsymbol{\Lambda}$ in $\boldsymbol{\Phi}$ there are three cosets of level 2, with corresponding quotients

$$\frac{h!\,j!}{(m+h-k)!\,(j+k-m)!}, \qquad \frac{h!\,m!}{(h+j-l)!\,(l+m-j)!}, \\ \frac{k!\,l!}{(l+m-j)!\,(j+k-m)!}, \tag{2.23}$$

and two cosets of level 3, with quotients

$$\frac{h!\,j!\,k!}{(h+j-l)!\,(l+m-j)!\,(j+k-m)!}, \\ \frac{h!\,l!\,m!}{(m+h-k)!\,(j+k-m)!\,(l+m-j)!}. \tag{2.24}$$

## 3 Irrationality results for $\mathrm{Li}_2(r/s)$

The integers $\alpha$, $\beta$ and $\delta$ defined in (2.10) are plainly invariant under the actions of $\boldsymbol{\lambda}$ and $\boldsymbol{\varphi}$ given by (2.19) and (2.20), whereas $H$ and $K$ are invariant under the action of $\boldsymbol{\lambda}$ but not of $\boldsymbol{\varphi}$. However, to get arithmetical consequences for the dilogarithm from the group-theoretic arguments outlined in Section 2 we require in place of $H$ and $K$ suitable integers, say, $M$ and $N$, invariant under the action of the whole permutation group $\boldsymbol{\Phi}$. We define $M = \max \mathcal{S}$, where $\mathcal{S}$ is the set (2.18), and we take $N$ to be the maximum in the sequence of the eight integers obtained by omitting from $\mathcal{S}$ the maximum of $h$, $m+h-k$, $h+j-l$ (with only one omission, so that $N = M$ if $(h,\ m+h-k,\ h+j-l)$ has more than one maximal element).

The integers $H$ and $K$ defined in (2.10) plainly do not exceed $M$ and $N$ respectively, so that (2.11) and (2.12) hold a fortiori with $M$ in place of $H$ and $N$ in place of $K$. Moreover $\boldsymbol{\lambda}$ and $\boldsymbol{\varphi}$ map the subset $\{h,\ m+h-k,\ h+j-l\} \subset \mathcal{S}$ onto itself. Therefore $N$, as well as $M$, $\alpha$, $\beta$ and $\delta$, is invariant under the action of $\boldsymbol{\Phi}$.

We replace in (2.11) $h, j, k, l, m$ by $hn, jn, kn, ln, mn$ respectively, where $h, j, k, l, m$ are fixed and $n = 1, 2, \ldots$, so that $M, N, \alpha, \beta, \delta$ are replaced by $Mn$, $Nn$, $\alpha n$, $\beta n$, $\delta n$. Also, we take the number $z > 1$ to be rational,

$$z = s/r \quad \text{with } r, s \in \mathbb{Z},\ 1 \le r < s.$$

Then (2.11) and (2.12) yield

$$\begin{cases} d_{Mn}\, d_{Nn}\, s^{\alpha n}(s-r)^{\beta n}\, r^{(\delta-\alpha-\beta)n}\, I_{s/r}(hn, jn, kn, ln, mn) \\ \qquad\qquad\qquad\qquad = A_n - B_n \,\mathrm{Li}_2(r/s) \\ d_{Mn}\, d_{Nn}\, s^{\alpha n}(s-r)^{\beta n}\, r^{(\delta-\alpha-\beta)n}\, I^{(2)}_{s/r}(hn, jn, kn, ln, mn) = B_n \end{cases} \tag{3.1}$$

with $A_n, B_n \in \mathbb{Z}$.

In order to prove the irrationality of $\mathrm{Li}_2(r/s)$ and to get a good irrationality measure of this number, one requires two sequences of integers, e.g., the $A_n$ and $B_n$ in (3.1), such that $A_n - B_n \operatorname{Li}_2(r/s) \neq 0$ and, for $n \to \infty$,

$$|A_n - B_n \operatorname{Li}_2(r/s)| \to 0 \tag{3.2}$$

as rapidly as possible, with

$$|B_n| \to \infty \tag{3.3}$$

as slowly as possible. For suitable $h,\ j,\ k,\ l,\ m$, the permutation group method allows one to improve considerably the sequences $A_n$ and $B_n$ in (3.1). One finds a large common divisor $\Delta_n \Delta_n'$ of the above $A_n$ and $B_n$ (see (3.6)), so that, defining

$$a_n = \frac{A_n}{\Delta_n \Delta_n'}, \qquad b_n = \frac{B_n}{\Delta_n \Delta_n'},$$

the integers $a_n$ and $b_n$ yield a linear form $|a_n - b_n \operatorname{Li}_2(r/s)|$ tending to 0 more rapidly than (3.2), with a growth of $|b_n|$ slower than (3.3).

The construction of the common divisor $\Delta_n \Delta_n'$ of $A_n$ and $B_n$ is as follows. Let $[x]$ denote the integral part of $x$, and for any prime number $p$ let $\omega = \{n/p\} = n/p - [n/p]$ be the fractional part of $n/p$. Following the method of [6, Section 4] we prove (see [8, Lemma 4.1]) that any prime $p > \sqrt{Mn}$ satisfying at least one of the following five inequalities:

$$\begin{aligned}
&[(m+h-k)\omega] + [(j+k-m)\omega] < [h\omega] + [j\omega],\\
&[(h+j-l)\omega] + [(l+m-j)\omega] < [h\omega] + [m\omega],\\
&[(l+m-j)\omega] + [(j+k-m)\omega] < [k\omega] + [l\omega],\\
&[(h+j-l)\omega] + [(l+m-j)\omega] + [(j+k-m)\omega] < [h\omega] + [j\omega] + [k\omega],\\
&[(m+h-k)\omega] + [(j+k-m)\omega] + [(l+m-j)\omega] < [h\omega] + [l\omega] + [m\omega]
\end{aligned} \tag{3.4}$$

divides the integers $A_n$ and $B_n$ in (3.1), and any prime $p > \sqrt{Mn}$ satisfying at least one of the following two conditions:

$$\begin{cases} [(h+j-l)\omega] + [(l+m-j)\omega] < [h\omega] + [m\omega] \ \text{and}\\ [m\omega] + [(j+k-m)\omega] < [j\omega] + [k\omega], \end{cases}$$
$$\begin{cases} [(m+h-k)\omega] + [(j+k-m)\omega] < [h\omega] + [j\omega] \ \text{and}\\ [j\omega] + [(l+m-j)\omega] < [l\omega] + [m\omega] \end{cases} \tag{3.5}$$

is such that $p^2$ divides $A_n$ and $B_n$. Note that each of the two conditions (3.5) implies one of the last two inequalities (3.4). Naturally the inequalities (3.4) arise from the five quotients (2.23) and (2.24).

Let $\Omega$ be the set of real numbers $\omega \in [0, 1)$ satisfying at least one of the five inequalities (3.4), and let $\Omega' \subset \Omega$ be the set of real numbers $\omega \in [0, 1)$ satisfying at least one of the two conditions (3.5). Let

$$\Delta_n = \prod_{\substack{p>\sqrt{Mn}\\ \{n/p\}\in\Omega}} p, \qquad \Delta_n' = \prod_{\substack{p>\sqrt{Mn}\\ \{n/p\}\in\Omega'}} p \qquad (n = 1, 2, \dots) \tag{3.6}$$

where $p$ denotes a prime, so that $\Delta_n \Delta'_n \mid A_n$ and $\Delta_n \Delta'_n \mid B_n$, and let

$$D_n = \frac{d_{Mn}\, d_{Nn}}{\Delta_n \Delta'_n}.$$

By well-known arguments one obtains

$$\lim_{n\to\infty} \frac{1}{n} \log D_n = M + N - \left( \int\limits_{\Omega} \mathrm{d}\psi(x) + \int\limits_{\Omega'} \mathrm{d}\psi(x) \right), \tag{3.7}$$

where $\psi(x) = \Gamma'(x)/\Gamma(x)$ is the logarithmic derivative of the Euler gamma-function. Dividing (3.1) by $\Delta_n \Delta'_n$ yields

$$\begin{cases} D_n\, s^{\alpha n}(s-r)^{\beta n}\, r^{(\delta-\alpha-\beta)n} I_{s/r}(hn, jn, kn, ln, mn) = a_n - b_n \operatorname{Li}_2(r/s) \\ D_n\, s^{\alpha n}(s-r)^{\beta n}\, r^{(\delta-\alpha-\beta)n} I^{(2)}_{s/r}(hn, jn, kn, ln, mn) = b_n \end{cases} \tag{3.8}$$

with $a_n,\ b_n \in \mathbb{Z}$.

The advantage of using (3.8) instead of (3.1) is quantified by the arithmetical correction

$$\int\limits_{\Omega} \mathrm{d}\psi(x) + \int\limits_{\Omega'} \mathrm{d}\psi(x) \tag{3.9}$$

in (3.7). We require from (3.8) the asymptotic behaviour of $|a_n - b_n \operatorname{Li}_2(r/s)|$ and an asymptotic upper bound for $|b_n|$ as $n \to \infty$. These are given by (3.7), by an exact asymptotic estimate of $|I_{s/r}(hn, jn, kn, ln, mn)|$ and by an asymptotic upper bound for $\left|I^{(2)}_{s/r}(hn, jn, kn, ln, mn)\right|$.

Under the further assumption that the nine integers belonging to the set $\mathcal{S}$ in (2.18) are all $> 0$, a detailed analysis developed in Section 5 of [8] shows that the function

$$f_z(x, y) := \frac{x^j (1-x)^h y^k (1-y)^l}{(x(1-y) + yz)^{j+k-m}} \qquad (z > 1) \tag{3.10}$$

has exactly three stationary points $(x_0, y_0)$, $(x_1, y_1)$, $(x_2, y_2)$ with respect to $x,\ y$ for which $x(1-x)y(1-y) \neq 0$, and that such points satisfy the inequalities

$$0 < x_0 < \frac{j}{h+j} < x_1 < 1 < z < x_2$$

and

$$y_1 < \frac{x_1}{x_1 - z} < 0 < y_0 < 1 < \frac{x_2}{x_2 - z} < y_2.$$

Moreover, for $\nu = 0, 1, 2$, the required asymptotic upper and lower bounds for $\left|I^{(\nu)}_z(hn, jn, kn, ln, mn)\right|$ as $n \to \infty$ are determined by the corresponding values $|f_z(x_\nu, y_\nu)|$.

In conclusion, for integers $r$ and $s$ satisfying $1 \le r < s$ let

$$c_0 = -\log f_{s/r}(x_0, y_0), \quad c_1 = -\log |f_{s/r}(x_1, y_1)|, \quad c_2 = \log |f_{s/r}(x_2, y_2)|$$

and

$$c_3 = M + N - \Big( \int_{\Omega} \mathrm{d}\psi(x) + \int_{\Omega'} \mathrm{d}\psi(x) \Big) + (\alpha - l - m) \log s + (m + h - k) \log r + \beta \log(s - r), \qquad (3.11)$$

with $\alpha$ and $\beta$ defined by (2.10). In [8, Theorem 5.2] we prove the following statement about the irrationality of $\mathrm{Li}_2(r/s)$ and its least irrationality measure $\mu(\mathrm{Li}_2(r/s))$.

**Theorem.** *If* $c_3 < c_0 < c_1$, *then* $\mathrm{Li}_2(r/s) \notin \mathbb{Q}$ *and*

$$\mu(\mathrm{Li}_2(r/s)) \le \frac{c_0 + c_2}{c_0 - c_3}.$$

We get in [8] several new results from the above theorem, by making suitable choices of $h, j, k, l, m$ and $r, s$. First we extend Hata's result $\mathrm{Li}_2(1/s) \notin \mathbb{Q}$ for $s \ge 7$ (see [4, Theorem 1.1]) by proving that $\mathrm{Li}_2(1/6)$ is irrational, with irrationality measure

$$\mu(\mathrm{Li}_2(1/6)) < 783.2903\ldots.$$

This is obtained with the choices

$$h = 98, \quad j = 49, \quad k = 71, \quad l = 34, \quad m = 91,$$

whence $M = m + h - k = 118$, $N = h + j - l = 113$, $\alpha = \delta = l + m = 125$, $\beta = k + l - h = 7$. Here

$$c_0 = 199.0132\ldots, \quad c_1 = 199.2086\ldots, \quad c_2 = 403.4271\ldots$$

and $c_3 = 198.2441\ldots$ with

$$\int_{\Omega} \mathrm{d}\psi(x) = 42.7123\ldots, \qquad \int_{\Omega'} \mathrm{d}\psi(x) = 1.3096\ldots.$$

Note that without using the arithmetical correction (3.9) in (3.11) one would get in place of $c_3$ the constant

$$198.2441\ldots + 42.7123\ldots + 1.3096\ldots > c_0 = 199.0132\ldots,$$

and therefore one could not prove the irrationality of $\mathrm{Li}_2(1/6)$.

Secondly, we improve all the irrationality measures of $\mathrm{Li}_2(1/s)$ given by Hata for $7 \le s \le 18$ (see [4, p. 386]). For instance, we get $\mu(\mathrm{Li}_2(1/7)) < 69.6887\ldots$ (Hata: $95.0605\ldots$), $\mu(\mathrm{Li}_2(1/8)) < 47.4251\ldots$ (Hata: $51.0848\ldots$), etc.

Also, we show ([8, Corollary 6.1]) that for any fixed positive integer $r$ there exists an effectively computable constant $s_1 = s_1(r) > r$ such that for any integer $s \ge s_1$ we have $\mathrm{Li}_2(r/s) \notin \mathbb{Q}$, with an explicit upper bound for $\mu(\mathrm{Li}_2(r/s))$.

A suitable choice of the integers $h, j, k, l, m$ in Corollary 6.1 of [8] (namely, $h = 2j$, $k = l = j$, $m = j + 1$ for any sufficiently large $j$) easily yields

$$\limsup_{\substack{s \in \mathbb{Z} \\ s \to +\infty}} \mu(\mathrm{Li}_2(r/s)) \le 3 \qquad (3.12)$$

for any fixed integer $r \ge 1$. The bound (3.12) extends to any positive integer $r$ a result announced by Chudnovsky for $r = 1$ (see [2, Corollary 7.2]).

## 4 Concluding remarks

The results on the dilogarithm outlined in the previous sections can be extended in several ways. A natural direction for further research consists in applying the permutation group method of Rhin and Viola to obtain irrationality results for $\mathrm{Li}_2(1/z)$ when $z \in \mathbb{Q}$, $z < 0$, or when $z \in i\mathbb{Q}$ $(i^2 = -1)$, or, more generally, to get $\mathbb{K}$-irrationality results and $\mathbb{K}$-irrationality measures of $\mathrm{Li}_2(1/z)$ for suitable number fields $\mathbb{K}$ and suitable $z \in \mathbb{K}$. The latter generalization requires the use of a conveniently defined height of an algebraic number, e.g., the Weil absolute logarithmic height (see [1]).

A further direction consists in the search for linear independence measures over $\mathbb{Q}$ (or, more generally, over a number field $\mathbb{K}$) of 1, $\mathrm{Li}_1(1/z)$ and $\mathrm{Li}_2(1/z)$ for suitable $z \in \mathbb{Q}$ (or $z \in \mathbb{K}$). For this purpose one can use (2.11)–(2.12), i.e., Theorem 2.1 of [8], since the first two equations (2.11) are linear forms in 1, $\mathrm{Li}_2(1/z)$ and in 1, $\mathrm{Li}_1(1/z)$ with the *same* coefficient $Q(z)$ for $\mathrm{Li}_2(1/z)$ and $\mathrm{Li}_1(1/z)$, also given by the third of (2.11).

The main technical tool to obtain the above generalizations and extensions should be a suitable version of the saddle point method in $\mathbb{C}^2$. This should yield asymptotic estimates of $I_z^{(\nu)}(hn, jn, kn, ln, mn)$ for more general $z \in \mathbb{C}$, thereby proving that, even in a more general setting, the exact asymptotic behaviour, for $\nu = 0, 1, 2$, of $I_z^{(\nu)}(hn, jn, kn, ln, mn)$ as $n \to \infty$ is indeed determined by the value $f_z(x_\nu, y_\nu)$ of the function (3.10) at the corresponding stationary point $(x_\nu, y_\nu)$.

## References


1. Amoroso, F., Viola, C.: Approximation measures for logarithms of algebraic numbers. *Ann. Sc. Norm. Super. Pisa Cl. Sci. IV. Ser* **30**, 225–249 (2001)
2. Chudnovsky, G.V.: Padé approximations to the generalized hypergeometric functions. I. *J. Math. Pures Appl.* **58**, 445–476 (1979)
3. Fischler, S.: Groupes de Rhin-Viola et intégrales multiples. *J. Théor. Nombres Bordx.* **15**, 479–534 (2003)
4. Hata, M.: Rational approximations to the dilogarithm. *Trans. Am. Math. Soc.* **336**, 363–387 (1993)
5. Maier, W.: Potenzreihen irrationalen Grenzwertes. *J. Reine Angew. Math.* **156**, 93–148 (1927)
6. Rhin, G., Viola, C.: On a permutation group related to $\zeta(2)$. *Acta Arith.* **77**, 23–56 (1996)
7. Rhin, G., Viola, C.: The group structure for $\zeta(3)$. *Acta Arith.* **97**, 269–293 (2001)
8. Rhin, G., Viola, C.: The permutation group method for the dilogarithm. *Ann. Sc. Norm. Super. Pisa Cl. Sci. V. Ser* **4**, 389–437 (2005)
9. Viola, C.: Hypergeometric functions and irrationality measures. In: Motohashi, Y. (ed.) *Analytic Number Theory*. London Mathematical Society Lecture Note Series, vol. 247, pp. 353–360. Cambridge University Press, Cambridge (1997)
10. Viola, C.: Birational transformations and values of the Riemann zeta-function. *J. Théor. Nombres Bordx.* **15**, 561–592 (2003)
11. Viola, C.: The arithmetic of Euler's integrals. *Riv. Mat. Univ. Parma (7)* **3***, 119–149 (2004)

MIX
Papier aus verantwortungsvollen Quellen
Paper from responsible sources
FSC® C105338

If you have any concerns about our products,
you can contact us on
**ProductSafety@springernature.com**

In case Publisher is established outside the EU,
the EU authorized representative is:
**Springer Nature Customer Service Center GmbH**
**Europaplatz 3, 69115 Heidelberg, Germany**

Printed by Libri Plureos GmbH
in Hamburg, Germany